A
COURSE OF LECTURES
ON
NATURAL PHILOSOPHY
AND THE
MECHANICAL ARTS

A

COURSE OF LECTURES

ON

NATURAL PHILOSOPHY

AND THE

MECHANICAL ARTS

THOMAS YOUNG

IN TWO VOLS

VOLUME-I

GYAN

Published by

GYAN

Gyan Publishing House
5, Ansari Road
Daryaganj, New Delhi-110002
Phone: 011-47034999, 9811692060
E-mail: books@gyanbooks.com

Distribution Network
gyanbooks.com
India, USA, Canada, UK, Australia

ISBN: 978-93-6780-702-6 (Set)
978-93-6780-101-7 (HB)
First Published, 1807

2nd Impression 2024

Printed at: Gyan Press, Delhi.

A COURSE OF LECTURES ON NATURAL PHILOSOPHY AND THE MECHANICAL ARTS
Author: THOMAS YOUNG

A

COURSE OF LECTURES

ON

NATURAL PHILOSOPHY

AND THE

MECHANICAL ARTS.

BY THOMAS YOUNG, M.D.

FOR. SEC. R.S. F.L.S. MEMBER OF EMMANUEL COLLEGE, CAMBRIDGE, AND LATE PROFESSOR OF NATURAL PHILOSOPHY IN THE ROYAL INSTITUTION OF GREAT BRITAIN

IN TWO VOLUMES.

VOLUME I.

1807.

TO THE RIGHT HONOURABLE THOMAS GRENVILLE

A MAN EQUALLY ESTEEMED FOR HIS PRIVATE VIRTUES

AND RESPECTED FOR HIS DISTINGUISHED TALENTS

WHO LATELY PRESIDED

AS FIRST LORD OF THE ADMIRALTY

OVER THAT DEPARTMENT OF THE PUBLIC SERVICE

TO WHICH THE PRINCIPLES OF MECHANICAL SCIENCE

MAY WITH THE GREATEST NATIONAL BENEFIT

BE PRACTICALLY APPLIED

THIS WORK IS DEDICATED

BY THE AUTHOR.

PREFACE.

HAVING undertaken to prepare a course of lectures on natural philosophy, to be delivered in the theatre of the Royal Institution, I thought that the plan of the Institution required something more than a mere compilation from the elementary works at present existing; and that it was my duty to collect from original authors, to examine with attention, and to digest into one system, every thing relating to the principles of the mechanical sciences, that could tend to the improvement of the arts subservient to the conveniences of life. I found also, in delivering the lectures, that it was most eligible to commit to writing, as nearly as possible, the whole that was required to be said on each subject; and that, even when an experiment was to be performed, it was best to describe that experiment uninterruptedly, and to repeat the explanation during its exhibition. Hence it became necessary that the written lectures should be as clearly and copiously expressed, and in a language as much adapted to the comprehension of a mixed audience, as the nature of the investigations would allow; and that each experiment, which was to be performed, should also be minutely described in them. If therefore there was any novelty either in the matter or the arrangement of the lectures, as they were delivered for two successive years, it is obvious that they must have possessed an equal claim to the attention of a reader, had they been published as a book; and upon resigning the situation of Professor of Natural Philosophy, I immediately began to prepare them for publication.

I had in some measure pledged myself, in the printed syllabus of the lectures, to make a catalogue of the best works already published on the several subjects; with references to such passages as appeared to be most important: it was therefore necessary, as well for this purpose, as in order to procure all possible information that could tend to the improvement of the work, to look over a select library of books entirely with this view, making notes of the principal subjects discussed in them, and examining carefully such parts as appeared to deserve more than ordinary attention. Hence arose a catalogue of references; respecting which it is sufficient to say, that the labour of arranging about twenty thousand articles, in a systematic form, was by no means less considerable than that of collecting them. The transactions of scientific societies, and the best and latest periodical publications, which have so much multiplied the number of the sources of information, constituted no small part of the collection, which was thus to be reduced into one body of science.

With the addition of the materials acquired in making this compilation, and of the results of many original investigations, to which they had given rise, it became almost indispensable to copy the whole of the lectures once more, and to exchange some of them for others, which were wholly new; at the same time all possible pains were taken to discover and to correct every obscurity of expression or of argument. Drawings were also to be made, for representing to the reader the apparatus and experiments exhibited at the time of delivering the lectures, for showing the construction of a variety of machines and instruments connected with the different subjects to be explained, and for illustrating them in many other ways. These figures have been extended to more than forty plates, very closely engraved, and the execution of the engravings has been minutely superintended. But the text of the lectures has been made so independent of the figures, that the reader is never interrupted in the middle of a chain of reasoning, but is referred, at the end of a paragraph, to a plate, which has always a sufficient explanation on the opposite page.

The bulk of this work is not so great, as to require, for its entire perusal, any unreasonable portion of time or of labour. There may, however, be some persons who would be satisfied with attending to those parts in which it differs most from former publications, without having leisure or inclination to study the whole. To such it may be desirable to have those subjects pointed out, which appear to the author to be the most deserving of their notice.

The fundamental doctrines of motion have, in the first place, been more immediately referred to axioms simply mathematical, than has hitherto been usual; and the application of these doctrines to practical purposes has perhaps in some instances been facilitated. The passive strength of materials of all kinds has been very fully investigated, and many new conclusions have been formed respecting it, which are of immediate importance to the architect and to the engineer, and which appear to contradict the results of some very elaborate calculations.

The theory of waves has been much simplified, and somewhat extended, and their motions have been illustrated by experiments of a peculiar nature. A similar method of reasoning has been applied to the circulation of the blood, to the propagation of sound, either in fluids or in solids, and to the vibrations of musical chords; the general principle of a velocity, corresponding to half the height of a certain modulus, being shown to be applicable to all these cases: and a connexion has been established between the sound to be obtained from a given solid, and its strength in resisting a flexure of any kind; or, in the case of ice and water, between the sound in a solid and the compressibility in a fluid state. The doctrine of sound and of sounding bodies in general has also received some new illustrations, and the theory of music and of musical intervals has been particularly discussed.

With respect to the mathematical part of optics, the curvature of the images, formed by lenses and mirrors, has been correctly investigated, and the inaccuracy of some former estimations has been demonstrated.

In the department of physical optics, the phenomena of halos and parhelia have been explained, upon principles not entirely new, but long forgotten : the functions of the eye have been minutely examined, and the mode of its accommodation to the perception of objects at different distances ascertained : the various phenomena of coloured light have been copiously described, and accurately represented by coloured plates ; and some new cases of the production of colours have been pointed out, and have been referred to the general law of double lights, by which a great variety of the experiments of former opticians have also been explained ; and this law has been applied to the establishment of a theory of the nature of light, which satisfactorily removes almost every difficulty that has hitherto attended the subject.

The theory of the tides has been reduced into an extremely simple form, which appears to agree better with all the phenomena, than the more intricate calculations which they have commonly been supposed to require. With respect to the cohesion and capillary action of liquids, I have had the good fortune to anticipate Mr. Laplace in his late researches, and I have endeavoured to show, that my assumptions are more universally applicable to the facts, than those which that justly celebrated mathematician has employed. I have also attempted to throw some new light on the general properties of matter in other forms : and on the doctrine of heat, which is materially concerned in them ; and to deduce some useful conclusions from a comparison of various experiments on the elasticity of steam, on evaporation, and on the indications of hygrometers. I have enumerated, in a compendious and systematical form, the principal facts which have been discovered with respect to galvanic electricity; and I have fortunately been able to profit by Mr. Davy's most important experiments, which have lately been communicated to the Royal Society, and which have already given to this branch of science a much greater perfection, and a far greater extent, than it before possessed. The historical part of the work can scarcely be called new, but several of the circumstances, which are related, have escaped the notice of former writers on the history of the sciences.

Besides these improvements, if I may be allowed to give them that name, there are others, perhaps of less importance, which may still be interesting to those who are particularly engaged in those departments of science, or of mechanical practice, to which they relate. Among these may be ranked, in the division of mechanics, properly so called, a simple demonstration of the law of the force by which a body revolves in an ellipsis; another of the properties of cycloidal pendulums; an examination of the mechanism of animal motions; a comparison of the measures and weights of different countries; and a convenient estimate of the effect of human labour: with respect to architecture, a simple method of drawing the outline of a column: an investigation of the best forms for arches; a determination of the curve which affords the greatest space for turning; considerations on the structure of the joints employed in carpentry, and on the firmness of wedges; and an easy mode of forming a kirb roof: for the purposes of machinery of different kinds, an arrangement of bars for obtaining rectilinear motion; an inquiry into the most eligible proportions of wheels and pinions; remarks on the friction of wheel work, and of balances; a mode of finding the form of a tooth for impelling a pallet without friction; a chronometer for measuring minute portions of time; a clock scapement; a calculation of the effect of temperature on steel springs; an easy determination of the best line of draught for a carriage; an investigation of the resistance to be overcome by a wheel or roller; and an estimation of the ultimate pressure produced by a blow.

In the hydraulic and optical part, may be enumerated an overflowing lamp; a simplification of the rules for finding the velocity of running water; remarks on the application of force to hydraulic machines; a mode of letting out air from water pipes; an analysis of the human voice; and some arrangements for solar microscopes, and for other optical instruments of a similar nature.

In the astronomical and physical division of the work, will be found a general rule for determining the correction on account of aberration; a comparison of observations on the figure of the earth; a table of the order of electrical excitation; a chart of the variation of the compass, and of the trade winds; formulae for finding the heat of summer and winter; remarks on the theory of the winds; and a comparative table of all the mechanical properties of a variety of natural bodies.

A few of these subjects have been more fully discussed in the miscellaneous papers, which have already been published, in the Philosophical Transactions and elsewhere, and which are now reprinted with corrections and additions; others are summarily investigated in the mathematical elements, which form a part of the second volume, or in the remarks, which are inserted, in their proper places, in the catalogue of references.

The arrangement of the whole work is probably different in many respects from any other that has yet been adopted; the extent of the subjects, which have been admitted, rendered it necessary to preserve a very strict attention to a methodical and uniform system; and it is presumed, that this arrangement will be considered as in itself of some value, especially in a work calculated to serve as a key, by means of which, access may be obtained to all the widely scattered treasures of science; and which will enable those, who are desirous of extending their researches in any particular department, to obtain expeditiously all the information that books can afford them.

It will not be thought surprising, that the execution of this plan, allowing for some professional engagements of a different kind, and for a variety of accidental interruptions, should have occupied more than three years, from the resignation of the professorship to the publication of the work. Some part of it is in its nature incapable of permanent perfection, since the catalogue must require to be continually extended by the enumeration of new publications; and it might perhaps be desirable that an appendix should be added to it at least every ten years: but the lectures themselves may be expected to remain tolerably commensurate to the state of the sciences for a much longer period; since, in investigations so intimately connected with mathematical principles, the essential improvements will always bear a very small proportion to the number of innovations. I do not, however, mean to assert, that the catalogue is by any means complete, even with regard to older works, but I believe that the references, which it contains, are at least sufficient to lead those, who may consult the passages quoted, to the works of every author of eminence that has treated of the respective subjects. Nor do I profess to have excluded all references that are of little importance; but I trust that the number, which I have admitted, will be found inconsiderable; and it would have been very difficult to have rejected any of them, without some chance of omitting others of greater value.

Whatever the deficiencies of this work may be, I think it right to observe, that my present pursuits will not allow me to look forwards to any period, at which I shall be able to remove them, or even to attend to the correction of the press, or the revision of the engravings, in case of the necessity of a second edition. I have already begun to collect materials for a work, in a form nearly similar, relating to every department of medical knowledge: this work will not, however, be speedily ready for publication; it will be comparatively more concise than these lectures, in proportion to what has been said and written respecting physic, but, I hope, much more complete, with regard to all that is known with certainty, and can be applied with utility.

Welbeck Street,
30th March, 1807.

CONTENTS.

PART THE FIRST. MECHANICS.

3

PART THE SECOND. HYDRODYNAMICS.

PART THE THIRD. PHYSICS; 485.

ADDITIONS AND CORRECTIONS.

P. 40. L. 5 from the bottom; for "therefore," read, afterwards.

P. 71. L. 5 from the bottom; for "IV," read, V.

P. 79. L. 2, for "IV," read, V.

P. 87. After l. 4, insert,

When an insulated body revolves round an axis in any direction, the state of revolution cannot be permanent, unless the axis be so situated, that the centrifugal forces on each side of it balance each other. It is obvious that this must happen in a homogeneous sphere, whatever may be the situation of the axis; and it has been demonstrated, that when the body is of an irregular form, there are at least three different axes, situated at right angles to each other, round which the body may revolve in an equilibrium either stable or tottering. It may also be shown that if a body, revolving round any axis, receive at the same time an impulse which would cause it to revolve round a second axis in another direction, the two revolutions will be combined, and will form a single revolution round a third axis, in an intermediate position, which will remain at rest until it be displaced by some new force, provided that it be one of the axes of permanent revolution: so that no body can revolve round a moveable axis without a continual disturbing force. And when an irregular body begins to move on an axis incapable of equilibrium, its revolution will be gradually altered, so as to approach continually to a revolution round one of the natural axes; but it will never pass beyond the state of equilibrium, as in many other cases of deviation from such a state; since the momentum, produced by the excess of centrifugal force in one part of the revolution, is destroyed in another. For a similar reason, if a stick be thrown in a horizontal position, with a rotatory motion, it will fall in the same position much more certainly than if it were thrown without any rotation; for any small disturbing force, which might be sufficient to turn it into a vertical position during the course of its path, will only produce, when combined with the rotatory motion, a slight change of the direction of the rotation, which will confine the deviation of the stick from a horizontal position within narrow limits.

P. 138. L. 9, after "concerned," insert, it has indeed been asserted that the specific gravity of elastic gum is even diminished by tension, so that the actual distances of the particles cannot, in this case, be supposed to be materially increased.

P. 146. L. 3, after "124," insert, 125.

L. 8 from the bottom, for "IX," read, X.

P. 169. L. 7, for "XIV, read, XIII.

P. 176. L. 19, for "the circle," read, a second circle.

P. 196. L. 5 . . 2, from the bottom, for "If the friction . . to obviate this," read, Since friction is always increased by an increase of pressure, the effect of any addition to the sustaining force must tend, in some degree, to retard the vibrations, even if the friction be somewhat less increased than the force propelling the balance. In order to obviate this retardation.

P. 238. L. 5 from the bottom, after "arches," insert, since they must have left too small a space for the passage of the water. If, however, we may believe Herodotus, whom Mr. King has quoted, this was in reality a kind of drawbridge. According to this author, it was built by Nitocris, the immediate successor of Semiramis: the stones were united by iron and lead, and beams were laid across them, which were removed at night, in order to prevent the mutual depredations of the inhabitants of different parts of the city.

P. 261. L. 19, for "XX," read, XIX.

P. 267. L. 18, for "heel," read, pitch.

P. 273. L. 3 from the bottom, omit "logarithm of."

L. 2 from the bottom, for "numbers," read, corresponding logarithms.

P. 292. L. 9, for "de," read, du.

P. 420. L. 2, for "more," read, most.

P. 424, after line 5, insert,

Dr. Wollaston has very ingeniously applied the effect of the reflection of two plane surfaces, inclined to each other, to the construction of an instrument for drawing, which he calls a camera lucida. He usually employs the internal reflection of a prism of glass, of which the four surfaces are ground so as to form proper angles with each other. The image formed by the first surface is inverted, and the second reflection restores it to its original position, but places it in a direction which is at right angles with the direction of the object; so that when we look down through the prism on a sheet of paper, we see a perfect picture of the objects immediately before us, while at the same time the aperture, through which we look, is only partly occupied by the edge of the prism, the remaining part being left open, or simply covered with a lens, for the admission of the direct rays of light, by which we may see, at the same time, the paper and the pencil to be employed, for making a drawing or a copy of any object placed before us.

P. 425. L. 16, for "XXVII," read, XXVIII.

P. 464. L. 15, for "other points at," read, at other points.

P. 477. Last line but one, after "telescopes," insert, but with respect to the theory of halos and parhelia, he was less successful than Mariotte had been some years before.

P. 535. L. 7 from the bottom, for "ecipses," read, species.

P. 545 ... Running title. The numbers of all the pages are too great by 20.

P. 587. L. 15 .. 17, for "the attraction .. is produced," read, a current is observed in its most exposed parts.

P. 588. L. 14, 15, for "on account .. moon" read, These currents, as well as the general current of the sea, have been attributed by some astronomers to the immediate attraction of the sun and moon, and of the satellites of Jupiter, which they have supposed to act in the same manner as the attraction of the sun operates in retarding the lunar motions; but the fact is, that, according to Mr. Laplace, the disturbing force of the sun produces this effect on the moon only in proportion as it increases her distance from the earth; consequently no such retardation can possibly be produced by the force of gravitation in the rotation of the sea or of the atmosphere, and the whole effect must be attributed to the operation of meteorological causes, producing first the trade winds, and secondly occasioning, by means of the friction of these winds, a similar motion in the sea. In the case of the atmosphere of Jupiter, the effects of heat can indeed scarcely be supposed to be very perceptible, and the rotation of this planet being extremely rapid, it is not at all impossible that the satellites may exert an action on the atmosphere somewhat analogous to the retardation of the moon's motion by the disturbing force of the sun.

P. 565, L. 12, for "Almamoun, was the son", read, Almamoun, the son.

P. 603. L. 2. for "The observations of the transit of Venus were twice made in the South Seas", read, Observations of the transit of Venus were made with great care in the South Seas.

At the end, insert, For the latest improvement that has been made in astronomy, we are also indebted to the zeal and ingenuity of Dr. Olbers, who, in pursuit of an opinion which he had formed, respecting the origin of the three small planets from the separation of a larger one into fragments, has been in the habit of examining monthly that part of the heavens, in which he supposes the event to have taken place, and through which each of the bodies must necessarily pass. He has had the good fortune to discover, in this manner, a fourth planet, which nearly resembles the three others in its appearance, except that it seems to be considerably larger.

P. 621. L. 22, after "descriptions," insert,

We may form some idea of the effects of this mutual action, by neglecting the force of repulsion, as Clairaut has done, and attending only to that of cohesion.

P. 622. L. 5 .. 15, for "It may also .. densities," read, This mode of reasoning is however by no means sufficient to explain all the phenomena; for it may be inferred from it that when the attractive power of the solid is greater or less than half that of the liquid, the surface of the liquid must, at its origin, be in the same direction with that of the solid, instead of forming an angle with it, as it often does in reality. But the difficulty may be removed by reverting to the general principle of superficial cohesion, and by comparing the common surface of the liquid and solid with the surface of a single liquid, of which the attractive power is equal only to the difference of the respective powers of the substances concerned. In this manner it may be shown, that if the attractive power of the solid be equal to that of the liquid, or still greater, it will be wetted by the liquid, which will rise until its surface acquires the same direction with that of the solid; and that in other cases, the angle of contact will be greater, in proportion as the solid is less attractive. A similar comparison is also equally applicable to the contact of two liquids of different densities.

P. 630. L. 11, for "one," read, we.

P. 678. L. 7. for "when in contact," read, either during their contact, or after separation.

P. 702. L. 9 .. 28. For "Astronomers .. years," read, "nor can any sufficient cause be found, in the attractions of the celestial bodies, either for the general easterly trade winds, or for the current of the sea in a similar direction, which appears to be the immediate effect of their friction on the surface of the water."

P. 770. Fig. 172, add, The strap itself must however be made stronger when in the situation B.

A

COURSE OF LECTURES

ON

NATURAL PHILOSOPHY

AND THE

MECHANICAL ARTS

PART I.

MECHANICS.

A

COURSE OF LECTURES

ON

NATURAL PHILOSOPHY

AND THE

MECHANICAL ARTS.

LECTURE I.

INTRODUCTION.

It is to be presumed, that most of those who honour the theatre of the Royal Institution with their attendance, are already acquainted with the nature of the objects which its founders and promoters have been endeavouring to attain: yet it appears to be by no means superfluous, that I should define with accuracy my own views of the utility that is likely to be derived from it, and of the most effectual means of accomplishing its purposes; in order that we may be able to distinguish, without difficulty, the most eligible track for our common progress through the regions of science; and that those who are desirous of accompanying me in the journey, may know precisely what route we are to follow, and what departments will more particularly arrest our attention.

Societies, which are merely literary and philosophical, have in general principally proposed to themselves, to enlighten the understanding by the discovery of unknown phenomena, and to exercise the reasoning powers, by opening new fields for speculation. Other associations have been more particularly intended for the encouragement of the arts, of manufactures, and of commerce. The primary and peculiar object of the Royal Institution of Great Britain is professedly of an humbler, but not of a less interesting nature. It is, to apply to domestic convenience the improvements which have been made in science, and to introduce into general practice such mechanical inventions as are of decided utility. But while it is chiefly engaged in this pursuit, it extends its views, in some measure, to the promotion of the same ends which belong to the particular provinces of other literary societies; and it is the more impossible that such objects should be wholly excluded, as it is upon the advancement of these that the specific objects of the Institution must ultimately depend. Hence the dissemination of the knowledge of natural philosophy and chemistry becomes a very essential part of the design of the Royal Institution: and this department must, in the natural order of arrangement, be anterior to the application of the sciences to practical uses. To exclude all knowledge but that which has already been applied to immediate utility, would be to reduce our faculties to a state of servitude, and to frustrate the very purposes which we are labouring to accomplish. No discovery, however remote in its nature from the subjects of daily observation, can with reason be declared wholly inapplicable to the benefit of mankind.

It has therefore always appeared to me, to be not only the best beginning, but also an object of high and permanent importance in the plan of the Institution, to direct the public attention to the cultivation of the elementary doctrines of natural philosophy, as well speculative as practical. Those who possess the genuine spirit of scientific investigation, and who have tasted the pure satisfaction arising from an advancement in intellectual acquirements, are contented to proceed in their researches, without inquiring at every step, what they gain by their newly discovered lights, and to what practical purposes they are applicable: they receive a sufficient gratification from the enlargement of their views of the constitution of the universe, and experience, in the immediate pursuit of knowledge, that pleasure which others wish to obtain more circuitously by its means. And it is one of the principal advantages of a liberal education, that it creates a susceptibility of an enjoyment so elegant and so rational.

A considerable portion of my audience, to whose information it will be my particular ambition to accommodate my lectures, consists of that sex, which, by the custom of civilised society, is in some measure exempted from the more laborious duties that occupy the time and attention of the other sex. The many leisure hours, which are at the command of females in the superior orders of society, may surely be appropriated, with greater satisfaction, to the improvement of the mind, and to the acquisition of knowledge, than to such amusements as are only designed for facilitating the insipid consumption of superfluous time. The hours thus spent will unquestionably become, by means of a little habit, as much more agreeable at the moment, as they must be more capable of affording self approbation upon reflection. And besides, like the seasoning which reconciled the Spartans to their uninviting diet, they will even heighten the relish for those pursuits which they interrupt: for mental exercise is as necessary to mental enjoyment, as corporal labour to corporal health and vigour. In this point of view, the Royal Institution may in some degree supply the place of a subordinate university, to those whose sex or situation in life has denied them the advantage of an academical education in the national seminaries of learning.

But notwithstanding the necessity of introducing very copiously speculations of a more general nature, we must not lose sight of the original objects of the Royal Institution; and we must therefore direct our attention more particularly to the theory of practical mechanics, and of manufactures. In these departments we shall find some deficiencies which may without much difficulty be supplied from scientific principles; and by an ample collection and display of models, illustrative of machines, and of inventions of all kinds, we may proceed in the most direct manner to contribute to the dissemination of that kind of knowledge which is most particularly our object. So that we must be more practical than academies of sciences, and more theoretical than societies for the improvement of arts; while we endeavour at the same time to give stability to our proceedings, by an annual recurrence to the elementary knowledge which is subservient to the purposes of both; and, as far as we are able, to apply to practice the newest lights, which may from time to time be thrown on particular branches of mechanical science. It is thus that we may most effectually perform, what the idolized sophists of antiquity but verbally professed, to bring down philosophy from the heavens, and to make her an inhabitant of the earth.

To those who are engaged in the practical cultivation of various arts, subservient to the conveniences of life, these lectures may be of some utility, by furnishing them with well established principles, applicable to a variety of cases, which may occasionally occur to them, where a little deviation from the ordinary routine of their profession may be necessary. Unfortunately, the hands that execute are too often inadequately supported by the head that directs; and much labour is lost for want of a little previous application to the fundamental doctrines of the mechanical sciences. Nor is any exorbitant portion of time or industry necessary for this purpose; for it happens singularly enough, that almost all practical applications of science depend on principles easily learnt; and, except in astronomy only, it has seldom been found that very abstruse investigations have been of great importance to society. Our most refined analytical calculations are by far too imperfect to apply to all possible cases of mechanical actions that can be proposed; and those problems which most frequently occur, may in general be solved by methods sufficiently obvious; although, from a want of proper order and perspicuity in the treatment of first principles, it has often happened that the most elementary propositions have been considered as requiring great study and application.

We may also be able to render an important service to society, and to confer a still more essential benefit on individuals, by repressing the premature zeal of unskilful inventors. We need only read over the monthly accounts of patents, intended for securing the pecuniary advantages of useful discoveries, in order to be convinced what expense of time and fortune is continually lavished on the feeblest attempts to innovate and improve. If we can be successful in convincing such inconsiderate enthusiasts of their real ignorance, or if we can show them, that even their own fairy ground has been preoccupied, we may save them from impending ruin, and may relieve the public from the distraction of having its attention perpetually excited by unworthy objects. The ridicule attendant on the name of a projector has been in general but too well deserved; for few, very few, who have aspired at improvement, have ever had the patience to submit their inventions to such experimental tests as common sense would suggest to an impartial observer. We may venture to affirm that out of every hundred of fancied improvements in arts or in machines, ninety at least, if not ninety nine, are either old or useless; the object of our researches is, to enable ourselves to distinguish and to adopt the hundredth. But while we prune the luxuriant shoots of youthful invention, we must remember to perform our task with leniency, and to show that we wish only to give additional vigour to the healthful branches, and not to extirpate the parent plant.

The Repository of the Royal Institution, as soon as it can be properly furnished, will be considered as a supplementary room for apparatus, in which the most interesting models, exhibited and described in the lectures, will be placed for more frequent inspection, and where a few other articles may perhaps deserve admission, which will not require so particular an explanation. To those who have profited by the lectures, or who are already too far advanced to stand in need of them, our rooms for reading and for literary conversation may be a source of mutual instruction. Our Library in time must contain all those works of importance which are too expensive for the private collections of the generality of individuals; which are necessary to complete the knowledge of particular sciences, and to which references will occasionally be given in the lectures on those sciences. Our Journals, free from commercial shackles, will present the public, from time to time, with concise accounts of the most interesting novelties in science and in the useful arts; and they will furnish a perpetual incitement to their editors to appropriate, as much as possible, to their own improvement, whatever is valuable in the publications of their cotemporaries. When all the advantages, which may reasonably be expected from this Institution, shall be fully understood and impartially considered, it is to be hoped that few persons of liberal minds will be indifferent to its success, or unwilling to contribute to it and to participate in it.

To that regulation, which forbids the introduction of any discussions connected with the learned professions, I shall always most willingly submit, and most punctually attend. It requires the study of a considerable portion of a man's life to qualify him to be of use to mankind in any of them; and nothing can be more pernicious to individuals or to society, than the attempting to proceed practically upon an imperfect conception of a few first principles only.

In physic, the wisest can do but little, and the ignorant can only do worse than nothing: and anxiously as we are disposed to seek whatever relief the learned and experienced may be able to afford us, so cautiously ought we to avoid the mischievous interference of the half studied empiric: in politics and in religion, we need but to look back on the history of kingdoms and republics, in order to be aware of the mischiefs which ensue, when "fools rush in where angels fear to tread."

Deeply impressed with the importance of mathematical investigations, both for the advancement of science and for the improvement of the mind, I thought it in the first place an indispensable duty to present the Royal Institution, in my Syllabus, with a connected system of natural philosophy, on a plan seldom, if ever, before executed in the most copious treatises. Proceeding from the simplest axioms of abstract mathematics, the syllabus contains a strict demonstration of every proposition which I have found it necessary to employ throughout the whole extent of natural philosophy. In the astronomical part only, some observations occur, unsupported by mathematical evidence: here, however, it was as impracticable, as it would have been useless, to attempt to enter into investigations, which in many instances have been extended far beyond the limits even of Newton's researches. But for the sake of those who are not disposed to undertake the labour of following, with mathematical accuracy, all the steps of the demonstrations on which the doctrines of the mechanical sciences are founded, I shall endeavour to avoid, in the whole of this course of lectures, every intricacy which might be perplexing to a beginner, and every argument which is fitter for the closet than for a public theatre. Here I propose to support the same propositions by experimental proofs: not that I consider such proofs as the most conclusive, or as more interesting to a truly philosophic mind than a deduction from general principles; but because there is a satisfaction in discovering the coincidence of theories with visible effects, and because objects of sense are of advantage in assisting the imagination to comprehend, and the memory to retain, what in a more abstracted form might fail to excite sufficient attention.

This combination of experimental with analogical arguments, constitutes the principal merit of modern philosophy. And here let the citizen of the world excuse the partiality of an Englishman, if I pride myself, and congratulate my audience, on the decided superiority of our own country, in the first establishment, and in the subsequent cultivation, of the true philosophy of the operations of nature. I grant that we have at times been culpably negligent of the labours of others; that we have of late suffered our neighbours to excel us in abstract mathematics, and perhaps, in some instances, in patient and persevering observation of naked phenomena. We have not at this moment a Lagrange or a Laplace: what we have, I do not think it necessary to enumerate: but there is a certain combination of theoretical reasoning with experimental inquiry, in which Great Britain, from the time of the reformation of philosophy, has never been inferior to any nation existing. I need only refer to the Transactions of the Royal Society, for abundant instances of the mode of investigation to which I allude; and I will venture to affirm, that their late publications are equal in importance to any that have preceded. It was in England that a Bacon first taught the world the true method of the study of nature, and rescued science from that barbarism in which the followers of Aristotle, by a too servile imitation of their master, had involved it; and with which, even of late, a mad spirit of innovation, under the name of the critical philosophy, has, in a considerable part of Europe, again been threatening it. It was in this country that Newton advanced, with one gigantic stride, from the regions of twilight into the noon day of science. A Boyle and a Hooke, who would otherwise have been deservedly the boast of their century, served but as obscure forerunners of Newton's glories. After these, a crowd of eminent men succeeded, each of great individual merit; but, absorbed in the prosecution of the Newtonian discoveries, they chose rather to be useful by their humble industry, than to wander in search of the brilliancy of novelty. It is difficult to judge of our cotemporaries; but we appear at present to be in possession of more than one philosopher, whose names posterity will be eager to rank in the same class with the few that have been enumerated. But it is not our present business to enter into the history of science; respecting what is supposed to be wholly unknown, we can have little curiosity: a short sketch of the progress of each branch of natural philosophy will be more properly introduced, after we have finished our investigation of the principal doctrines belonging to it.

With regard to the mode of delivering these lectures, I shall in general intreat my audience to pardon the formality of a written discourse, in favour of the advantage of a superior degree of order and perspicuity. It would unquestionably be desirable that every syllable advanced should be rendered perfectly easy and comprehensible even to the most uninformed; that the most inattentive might find sufficient variety and entertainment in what is submitted to them, to excite their curiosity, and that in all cases the pleasing, and sometimes even the surprising, should be united with the instructive and the important. But whenever there appears to be a real impossibility of reconciling these various objects, I shall esteem it better to seek for substantial utility than temporary amusement; for if we fail of being useful for want of being sufficiently popular, we remain at least respectable: but if we are unsuccessful in our attempts to amuse, we immediately appear trifling and contemptible. It shall however at all times be my endeavour to avoid each extreme; and I trust, that I shall then only be condemned, when I am found abstruse from ostentation, or uninteresting from supineness. The most difficult thing for a teacher is, to recollect how much it cost himself to learn, and to accommodate his instruction to the apprehension of the uninformed: by bearing in mind this observation, I hope to be able to render my lectures more and more intelligible and familiar; not by passing over difficulties, but by endeavouring to facilitate the task of overcoming them; and if at any time I appear to have failed in this attempt, I shall think myself honoured by any subsequent inquiries, that my audience may be disposed to make.

We have to extend our views over the whole circle of natural and artificial knowledge, to consider in detail the principles and application of the philosophy of nature and of art. We are to discuss a great number of subjects, to each of which a separate title and rank among the sciences has sometimes been assigned; and it is necessary, in order to obtain a distinct conception of the foundation and relation of each subdivision, to pay particular attention to the order in which the sciences are to be treated, and to the connexion which subsists between them, as well as to the degree of importance, which each of them claims, with regard either to theory or to practice. To insist on the propriety of a distinct and logical order is unnecessary; for however superfluous we may deem the scholastic forms of rhetoric, it is confessedly advantageous to the judgment as well as to the memory, to unite those things which are naturally connected, and to separate those which are essentially distinct. When a traveller is desirous of becoming acquainted with a city or country, before unknown to him, he naturally begins by taking, from some elevated situation, a distant view of the distribution of its parts; and in the same manner, before we enter on the particular consideration of the subjects of our researches, it may be of use to form to ourselves a general idea of the sciences and arts which are to be placed among them.

Upon the advantages of mathematical and philosophical investigation in general, it is unnecessary to enlarge, because no liberal mind can require any arguments to be convinced how much the judgment is strengthened, and the invention assisted, by habits of reasoning with caution and accuracy. The public opinion is rather, on the contrary, in danger, at least in some parts of the world, of being too exclusively biassed in favour of natural philosophy; and has sometimes been inclined to a devotion too much limited to science, without a sufficient attention to such literature as an elegant mind always desires to see united with it. As to the practical importance of philosophical theories of the arts, it may have been overrated by some, but no person is authorised to affirm, that it has been too highly estimated, unless he has made himself master of every thing that theory is capable of doing; such a one, although he may in some cases be obliged to confess the insufficiency of our calculations, will never have reason to complain of their fallacy.

The division of the whole course of lectures into three parts, was originally suggested by the periodical succession in which the appointed hours recur: but it appears to be more convenient than any other for the regular classification of the subjects. The general doctrines of motion, and their application to all purposes variable at pleasure, supply the materials of the first two parts, of which the one treats of the motions of solid bodies, and the other of those of fluids, including the theory of light. The third part relates to the particular history of the phenomena of nature, and of the affections of bodies actually existing in the universe, independently of the art of man; comprehending astronomy, geography, and the doctrine of the properties of matter, and of the most general and powerful agents that influence it.

The synthetical order of proceeding, from simple and general principles, to their more intricate combinations in particular cases, is by far the most compendious for conveying information, with regard to sciences that are at all referable to certain fundamental laws. For these laws being once established, each fact, as soon as it is known, assumes its place in the system, and is retained in the memory by its relation to the rest as a connecting link. In the analytical mode, on the contrary, which is absolutely necessary for the first investigation of truth, we are obliged to begin by collecting a number of insulated circumstances, which lead us back by degrees to the knowledge of original principles, but which, until we arrive at those principles, are merely a burden to the memory. For the phenomena of nature resemble the scattered leaves of the Sibylline prophecies; a word only, or a single syllable, is written on each leaf, which, when separately considered, conveys no instruction to the mind; but when, by the labour of patient investigation, every fragment is replaced in its appropriate connexion, the whole begins at once to speak a perspicuous and a harmonious language.

Proceeding therefore in the synthetical order, we set out from the abstract doctrines of mathematics, relating to quantity, space, and number, which we pass over, as supposed to be previously understood, or as sufficiently explained in the mathematical elements, and go on to their immediate application to mechanics and hydrodynamics, or to such cases of the motions of solids and fluids as are dependent on arbitrary assumptions, that is, where we do not confine our inquiries to any particular cases of existing phenomena. By means of principles which are deducible in a satisfactory manner from mathematical axioms, with the assistance only of the general logic of induction, we shall be able to draw such conclusions, as are capable of giving us very important information respecting the operations of nature and of art, and to lay down such laws, as, to an uninformed person, it would appear to be beyond the powers of reason to determine, without the assistance of experiment. The affections of falling bodies, and of projectiles, the phenomena of bodies revolving round a centre, the motions of pendulums, the properties of the centre of gravity, the equilibrium of forces in machines of different kinds; the laws of preponderance, and the effects of collision; all these are wholly referable to axiomatical evidence, and are frequently applicable to important uses in practice. Upon these foundations, we shall proceed to the general principles of machinery, and the application of forces of different kinds; we shall inquire what are the principle sources of motion that we can subject to our command, and what advantages are peculiar to each of them; and then, according to the purposes for which they are employed, we shall separately examine the principal machines and manufactures in which those forces are applied to the service of mankind.

Such instruments and machines as are more or less immediately subservient to mathematical purposes, will be the first in order, including all the mechanism of literature, the arts of writing, engraving, and printing, in their various branches, and the comparison of measures, with each other and with different standards; the principles of perspective will also form a useful appendage to the description of geometrical instruments. The determination of weights, and of the magnitude of moving forces of various kinds, constituting the science of statics, will be the next subject, and will be followed by the consideration of the retarding force of friction, and of the passive strength of the various materials, that are employed in building and in machinery.

All these subjects are in part preparatory to the immediate examination of the mechanical arts and manufactures, which are so numerous and complicated as not to admit of regular arrangement without some difficulty: they may however be divided into such as are principally employed for resisting, for modifying, or for counteracting, any motion or force; thus architecture and carpentry are chiefly intended to resist the force of gravitation: these comprehend the employments of the mason, the bricklayer, the joiner, the cabinet maker, and the locksmith. In these departments it is often of the utmost importance to the mechanic, to recur, especially in works of magnitude, to philosophical principles; and in many other cases, where there is no need of much calculation, we may still be of service, by collecting such inventions of ingenious artists, as are convenient and elegant, and which, although simple in their principles, are not obvious in their arrangements; and in the same manner we may be able, in taking a general view of other arts and manufactures, to explain their principles, where theory is concerned, and to exhibit practical precedents, where the nature of the subject requires no refined investigation.

The modification of motion and force includes its communication and alteration, by joints of various kinds, by wheelwork, and by cordage, and its equalisation by means of timekeepers. The subject of wheelwork gives considerable scope for mathematical research, and requires the more notice, as it has often been inaccurately treated: the consideration of cordage leads us to that of union by twisting, and by intermixture of fibres; including the important arts of carding, combing, spinning, ropemaking, weaving, fulling, felting, and papermaking; which constitute the employment of many millions of manufacturers, of all ages and sexes, in every part of the world, and by which the animal and vegetable productions of a large portion of the surface of the globe, are made to contribute, as well to the power and riches of the individuals who supply them, as to the health and comfort of the public that consumes them. The admirable art of the watch and clock maker is a peculiarly interesting department of practical mechanics, it affords employment for mathematical investigation, for experimental inquiry, and for ingenious invention; and the perfection, which it has derived from a combination of these means, does honour as well to the nations who have encouraged it, as to the individuals who have been engaged in it.

To counteract the powers of gravitation and of friction, is the object of such machines as are used for raising and removing weights: cranes, friction wheels, and carriages of all kinds, are referable to this head, and some of them have been the subjects of much speculation and experiment. Lastly, to overcome and to modify the corpuscular forces of cohesion and repulsion, and to change the external forms of bodies, is the object of machinery intended for compression, extension, penetration, attrition, trituration, agitation, and demolition. For these purposes we employ presses, forges, rolling, stamping, coining, and milling machines; the processes of digging, ploughing, and many other agricultural arts; boring, mining, grinding, polishing, and turning; mills of various kinds, threshing mills, corn mills, oil mills, and powder mills; besides the chemical agents concerned in blasting rocks, and in the operations of artillery. All these arts are comprehended in the department of mechanics, which constitutes the first division of this course. Not that we shall be able to enter at large into the detail of each; but having formed a general outline, we may fill up its particular parts with more or less minuteness, as we may find more or less matter of importance to insert in each; and those who wish to pursue the subjects further, will every where be able to derive great assistance from the authors whose works will be mentioned.

The doctrines of hydrodynamics relate to the motions and affections of fluids, in which we no longer consider each distinct particle that is capable of separate motion, but where we attend to the effect of an infinite number of particles, constituting a liquid or aeriform aggregate. The general theory of such motions will be premised, under the heads hydrostatics, or the affections of liquids at rest, pneumatostatics, or the properties of elastic fluids at rest; and hydraulics, or the theory of fluids in motion. The practical application of this theory to hydraulic and pneumatic machines is of very considerable importance, and is as interesting to the philosopher as it is necessary to the engineer. The employment of the force of water and wind to the best advantage, the draining of lands and mines, the supply of water for domestic convenience, the manoeuvres of seamanship, the construction of the steam engine, are all dependent upon hydrodynamical principles, and are often considered as comprehended in the science of hydraulics. Harmonics and optics, the remaining parts of this division, are more insulated: the doctrine of sound, the theory of music, and the construction of musical instruments, are as pleasing to the intellect in theory, as they are gratifying to the senses in practice; but the science of optics is not less interesting, and at the same time far more useful; the instruments which it furnishes are of almost indispensable necessity to the navigator, to the naturalist, to the physiologist, and even to the man of business or pleasure. It is perhaps in this science that many persons of the greatest genius have been the most happily employed. The reasons for which it is classed as a division of hydrodynamics will be explained hereafter.

The contemplation of the particular phenomena of nature, as they are displayed in the universe at large, contributes perhaps less to the perfection of any of the arts, which are immediately subservient to profit or convenience, than the study of mechanics and hydrodynamics. But the dignity and magnificence of some of these phenomena, and the beauty and variety of others, render them highly interesting to the philosophical mind, at the same time that some of them are of the utmost importance in their application to the purposes of life. In all these respects the science of astronomy holds the first rank; its uses in assisting navigation, and in regulating chronology, are beyond all calculation. Geography, and hydrography, or the particular histories of the earth and sea, are immediately connected with astronomy. The discussion of the properties of matter in general, and of the alterations of temperature to which all bodies are liable, has not hitherto received a distinct appellation as a science; but both these subjects require a separate consideration, and afford a vast scope for speculation and for observation. Electricity and magnetism are partly referable to the affections of matter, and partly to the agency of substances which appear to agree with common matter in some properties, and to differ from it in others. The phenomena produced by these agents are often such as excite a high degree of curiosity to inquire into their causes, although the inquiry too often terminates only in astonishment; but we have reason to expect considerable advancement in these sciences from the singular discoveries of modern chemists. The utility of the philosophy of electricity is sufficiently exemplified in the general introduction of conductors for securing us against lightning, to say nothing of the occasional employment of electricity in medicine; and since the important discovery of the compass, we have only to lament that the changeable nature of magnetic effects so much limits the utility of that instrument for nautical and geographical purposes. Of meteorology, and of geology, our knowledge is hitherto very imperfect; notwithstanding many diffuse treatises which relate to them, we cannot boast of having reduced them to any determinate laws; and yet there are some meteorological facts which well deserve our attention. Natural history is the last of the sciences that it will be necessary for us to notice; some may think it superfluous to attempt to give so superficial a sketch of this most extensive subject, as our plan will allow; but it is still possible to select some general observations respecting the methods of classification, as well as the philosophy of natural history, which, although very concise, may yet be in some measure instructive. This third division of the course would properly include, together with the general properties of matter, and the particular actions of its particles, the whole science of chemistry, but the variety and importance of chemical researches, demand a separate and minute discussion; and the novelty and beauty of many of the experiments with which the labours of our cotemporaries have presented us, and which will be exhibited in the theatre of the Royal Institution by the Professor of Chemistry, are sufficient to make this department of natural philosophy the most entertaining of all the sciences.

Such is the whole outline of our plan, and such are the practical uses, to which the arts and sciences, comprehended in it, are principally applicable.

Before we proceed to the examination of its several parts, we must pause to consider the mode of reasoning which is the most generally to be adopted. It depends on the axiom which has always been essentially concerned in every improvement of natural philosophy, but which has been more and more employed, ever since the revival of letters, under the name induction, and which has been sufficiently discussed by modern metaphysicians. That like causes produce like effects, or, that in similar circumstances similar consequences ensue, is the most general and most important law of nature; it is the foundation of all analogical reasoning, and is collected from constant experience, by an indispensable and unavoidable propensity of the human mind.

It does not appear that we can have any other accurate conception of causation, or of the connexion of a cause with its effect, than a strong impression of the observation, from uniform experience, that the one has constantly followed the other. We do not know the intimate nature of the connexion by which gravity causes a stone to fall, or how the string of a bow urges the arrow forwards; nor is there any original absurdity in supposing it possible that the stone might have remained suspended in the air, or that the bowstring might have passed through the arrow as light passes through glass. But it is obvious that we cannot help concluding the stone's weight to be the cause of its fall, and that every heavy body will fall unless supported; and the pressure of the string to be the cause of the arrow's motion, and, that if we shoot, the arrow will fly; if we hesitated to make these conclusions, we should often pay dear for our scepticism. This explanation is sufficient to show the identity of the two expressions, that like causes produce like effects, and, that in similar circumstances similar consequences ensue. And such is the ground of argument from experience, the simplest principle of reasoning, after pure mathematical truths; which appear to be so far prior to experience, as their contradiction always implies an absurdity repugnant to the imagination.

In the application of induction, the greatest caution and circumspection are necessary; for it is obvious that, before we can infer with certainty the complete similarity of two events, we must be perfectly well assured that we are acquainted with every circumstance which can have any relation to their causes. The error of some of the ancient schools consisted principally in the want of sufficient precaution in this respect; for although Bacon is, with great justice, considered as the author of the most correct method of induction, yet, according to his own statement, it was chiefly the guarded and gradual application of the mode of argument, that he laboured to introduce. He remarks, that the Aristotelians, from a hasty observation of a few concurring facts, proceeded immediately to deduce universal principles of science, and fundamental laws of nature, and then derived from these, by their syllogisms, all the particular cases, which ought to have been made intermediate steps in the inquiry. Of such an error we may easily find a familiar instance. We observe, that, in general, heavy bodies fall to the ground unless they are supported; it was therefore concluded that all heavy bodies tend downwards: and since flame was most frequently seen to rise upwards, it was inferred that flame was naturally and absolutely light. Had sufficient precaution been employed in observing the effects of fluids on falling and on floating bodies, in examining the relations of flame to the circumambient atmosphere, and in ascertaining the specific gravity of the air at different temperatures, it would readily have been discovered, that the greater weight of the colder air was the cause of the ascent of the flame; flame being less heavy than air, but yet having no positive tendency to ascend. And accordingly the Epicureans, whose arguments, as far as they related to matter and motion, were often more accurate than those of their cotemporaries, had corrected this error; for we find in the second book of Lucretius a very just explanation of the phenomenon.

> "See with what force yon river's crystal stream
> Resists the weight of many a massy beam.
> To sink the wood the more we vainly toil,
> The higher it rebounds, with swift recoil.
> Yet that the beam would of itself ascend
> No man will rashly venture to contend.
> Thus too the flame has weight, though highly rare,
> Nor mounts but when compelled by heavier air."

It may be proper to notice here those axioms which are denominated by Newton rules of philosophizing; although it must be confessed that they render us very little immediate assistance in our investigations. The first is, that "no more causes are to be admitted as existing in nature, than are true, and sufficient for explaining the phenomena to be considered:" the second, "therefore effects of the same kind are to be attributed, as far as is possible, to the same causes:" thirdly, "those qualities of bodies which cannot be increased nor diminished, and which are found in all bodies within the reach of our experiments, are to be considered as general qualities of all bodies existing:" fourthly, "in experimental philosophy, propositions collected by induction from phenomena, are to be esteemed either accurately or very nearly true, notwithstanding any contrary hypotheses, until other phenomena occur, by which they may either be corrected or confuted."

As an illustration of the remark, that these axioms, though strictly true, are of little real utility in assisting our investigations, I shall give an instance from the subject of electricity. Supposing that we wish to determine, whether or no the electric fluid has weight; we are to inquire whether or no gravitation is one of those properties which are described in the third rule, and whether that rule will authorise us to apply it to the electric fluid, as one of those qualities of bodies, which cannot be increased nor diminished, which are found in all bodies within the reach of our experiments, and which are therefore to be considered as general qualities of all bodies existing. Now it appears to be in the first place uncertain whether or no the increase and diminution of gravity, from a change of distance, is strictly compatible with the terms of the definition; and in the second place, we are equally at a loss to decide, whether or no the electric fluid can with propriety be called a body, for it appears in some respects to be wholly different from tangible matter, while it has other qualities in common with it. Such are the difficulties of laying down general laws on so comprehensive a scale, that we shall find it more secure to be contented to proceed gradually by closer inductions in particular cases. We shall however seldom be much embarrassed in the choice of a mode of argumentation. The laws of motion, which will be the first immediate subjects of discussion, have indeed sometimes been referred to experimental evidence; but we shall be able to deduce them all in a satisfactory manner, by means of our general axiom, from reasonings purely mathematical, which, wherever they are applicable, are unquestionably preferable to the imperfect evidence of the senses, employed in experimental investigations.

LECTURE II.

ON MOTION.

THE whole science of mechanics depends on the laws of motion, either actually existing, or suppressed by the opposition of the forces which tend to produce it. The nature of motion requires therefore to be particularly examined at the entrance of the science of mechanical philosophy; and although the subject is so abstract as to demand some effort of the attention, being seldom capable of receiving much immediate illustration from the objects of sense, yet we shall find it indispensable to our progress in the investigation of many particular problems of importance, to obtain, in the first place, a clear conception of the properties and affections of motions of all kinds.

One of the ancient philosophers, on being asked for a definition of motion, is said to have walked across the room, and to have answered, you see it, but what it is, I cannot tell you. It does not, however, appear absolutely necessary to appeal to the senses for the idea of motion: for a definition is the resolution of a complex idea into the more simple elements which compose it; and, in the present instance, these elements are, the existence of two points at a certain distance, and after a certain interval of time, the existence of the same points at a different distance; the difference of the distances being supposed to be ascertained according to that postulate of geometry, which has in general been tacitly understood, but which I have expressly inserted in the geometrical part of my syllabus; requiring that the length of a line be capable of being identified, whether by the effect of any object on the senses, or merely in imagination.

Motion, therefore, is the change of rectilinear distance between two points. Allowing the accuracy of this definition, it appears that two points are necessary to constitute motion; that in all cases when we are inquiring whether or no any body or point is in motion, we must recur to some other point which we can compare with it, and that if a single atom existed alone in the universe, it could neither be said to be in motion nor at rest. This may seem in some measure paradoxical, but it is the necessary consequence of our definition, and the paradox is only owing to the difficulty of imagining the existence of a single atom, unsurrounded by innumerable points of a space which we represent to ourselves as immoveable.

It has been for want of a precise definition of the term motion, that many authors have fallen into confusion with respect to absolute and relative motion. For the definition of motion, as the change of rectilinear distance between two points, appears to be the definition of what is commonly called relative motion; but, on a strict examination, we shall find, that what we usually call absolute motion is merely relative to some space, which we imagine to be without motion, but which is so in imagination only. The space which we call quiescent, is in general the earth's surface; yet we well know, from astronomical considerations, that every point of the earth's surface is perpetually in motion, and that in very various directions: nor are any material objects accessible to our senses, which we can consider as absolutely motionless, or even as motionless with regard to each other; since the continual variation of temperature to which all bodies are liable, and the minute agitations arising from the motions of other bodies with which they are connected, will always tend to produce some imperceptible change of their distances.

When therefore we assert, that a body is absolutely at rest, we only mean to compare it with some large space in which it is contained: for that there exists a body absolutely at rest, in as strict a sense as an absolutely straight line may be conceived to exist, we cannot positively affirm; and if such a quiescent body did exist, we have no criterion by which it could be distinguished. Supposing a ship to move at the rate of three miles in an hour, and a person on board to walk or to be drawn towards the stern at the same rate, he would be relatively in motion, with respect to the ship, yet we might very properly consider him as absolutely at rest: but he would, on a more extended view, be at rest only in relation to the earth's surface; for he would still be revolving round the axis of the earth, and with the earth round the sun; and with the sun and the whole solar system, he would be slowly moving among the starry worlds which surround them. Now with respect to any effects within the ship, all the subsequent relations are of no consequence, and the change of his rectilinear distance from the various parts of the ship, is all that needs to be considered in determining those effects. In the same manner, if the ship appear, by comparison with the water only, to be moving through it with the velocity of three miles an hour, and the water be moving at the same time in a contrary direction at the same rate, in consequence of a tide or current, the ship will be at rest with respect to the shore, but the mutual actions of the ship and the water will be the same as if the water were actually at rest, and the ship in motion.

It is not sufficient to observe the increase or decrease of distance of a moving point from another single point only: we must compare its successive situations with many other points surrounding it; and for this purpose these points must be at rest among themselves, in order to be considered as belonging to a quiescent space or surface; which may be defined as a space or surface, of which all the points remain always at equal distances from each other, without any external influence. In this sense we must call the deck of the ship a quiescent surface, whether the ship be at anchor or under sail: but we must not consider a surface revolving round a centre as a quiescent surface; for it will appear hereafter that no such motion can exist without the influence of a centripetal force, which renders it improper for determining the affections of a moving body.

When a point is in motion with respect to a quiescent space, it is often simply denominated a moving point, and the right line joining any two of its places, immediately contiguous to each other, is called its direction. If it remains continually in one right line drawn in the quiescent space, that line is always the line of its direction; if it describes several right lines, each line is the line of its direction as long as it continues in it; but if its path becomes curved, we can no longer consider it as perfectly coinciding at any time with a right line, and we must recur to the letter of the definition, by supposing a right line to be drawn through two successive points in which it is found, and then if these points be conceived to approach each other without limit, we shall have the line of its direction. Now such a line is called in geometry a tangent: for it meets the curve, but does not cut it, provided that the curvature be continued. (Plate I. Fig. 1 ... 3.)

Having formed an accurate idea of the nature of motion, and of the import of the terms employed in speaking of its properties, we may proceed to consider the mechanical laws to which it is subjected, and which are derivable from the essence of the definitions that have been premised. The first is, that a moving point never quits the line of its direction without a disturbing cause: for a right line being the same with respect to all sides, no reason can be imagined why the point should incline to one side more than another; and the general law of induction requires, that the moving point should preserve the same relations towards the points similarly situated on every side of the line. This argument appears to be sufficiently satisfactory to give us ground for asserting, that the law of motion here laid down may be considered as independent of experimental proof. It was once proposed as a prize question by the academy of sciences at Berlin, to determine whether the laws of motion were necessary or accidental; that is, whether they were to be considered as mathematical or as physical truths. Maupertuis, then president of the academy, wrote an elaborate dissertation, in which he endeavoured to deduce them from a complicated principle of the production of every effect in the manner which requires the least possible action, a principle which he supposes to be most consistent with the wise economy of nature. But this principle has itself been shown to be capable of accommodation to any other imaginable laws of motion, and the intricacy of the theory tends only to envelope the subject in unnecessary obscurity; the laws of motion appear to be easily demonstrable from the simplest mathematical truths, granting only the homogeneity or similarity of matter with respect to motion, and allowing the general axiom, that like causes produce like effects. If, however, any person thinks differently, he is at liberty to call these laws experimental axioms, collected from a comparison of various phenomena; for we cannot easily reduce them to direct experiments, since we can never remove from our experiments the action of all disturbing causes; for either gravitation, or the contact of surrounding bodies, will interfere with all the motions which we can examine.

Having established the rectilinear direction of undisturbed motion, we come to consider its uniformity. Here the idea of time enters into our subject. To define time in general is neither easy nor necessary; but we must have some measure of equal times. Our abstract idea of time depends on the memory of past sensations; but it is obvious that the results of an intellectual measure of the duration of time would be liable to the greatest uncertainty. We may observe, that, on a journey, the perpetual succession of various objects will often make a week appear, upon retrospection, as long as a month spent in a continuation of such employments as are uniform, without being laborious; the multitude of new impressions not only serving to increase the apparent magnitude of the interval, by filling up its vacuities, but tending also to diminish the vivacity of the ideas which they have superseded, and to give them the character of the fainter recollections of an earlier date. We are therefore obliged to estimate the lapse of time by the changes in external objects: of these changes, the simplest and most convenient is the apparent motion of the sun, or rather of the stars, derived from the actual rotation of the earth on its axis, which is not indeed an undisturbed rectilinear motion, but which is equally applicable to every practical purpose. Hence we obtain, by astronomical observations, the well known measures of the duration of time, implied by the terms day, hour, minute, and second.

Now the equality of times being thus estimated from any one motion, all other bodies moving without disturbance, will describe equal successive parts of their lines of direction in equal times. And this is the second law of motion, which, with the former law, constitutes Newton's first axiom or law of motion: "that every body perseveres in its state of rest or uniform rectilinear motion, except so far as it is compelled by some force to change it." It appears that this second law is strictly deducible from the axioms and definitions which have been premised, and principally from the consideration of the relative nature of motion, and the total deficiency of a criterion of absolute motion. For, since the velocity of a body, moving without resistance or disturbance, is only a relation to another body, if the second body has no mechanical connexion with the first, its state with respect to motion can have no effect on the velocity of the first body, however great its comparative magnitude may be: and if a body is at rest, there is nothing to determine it to begin to move either to the right hand or to the left; if it is at rest with respect to any other bodies, it will remain in the same condition, whatever the relative motions of those bodies may be, when compared with the surrounding objects; and these relations can only be preserved by its continuance in uniform rectilinear motion. This law is also confirmed by its perfect agreement with all experimental observations, although it is too simple to admit of an immediate proof. For we can never place any body in such circumstances as to be totally exempt from the operation of all accelerating or retarding causes; and the deductions from such experiments as we can make, would require in general, for the accurate determination of the necessary corrections, a previous knowledge of the law which we wish to demonstrate.

When, indeed, we consider the motion of a projectile, we have only to allow for the disturbing force of gravitation, which so modifies the effect, that the body deviates from a right line, but remains in the same vertical plane; whence we may infer, that, in the absence of the force of gravitation, the body would continue to move in every other plane in which its motion began, as well as in the vertical plane, since in that case all these planes would be indifferent to it; it must therefore remain in their common intersection, which could only be a straight line: so that by thus combining arguments with observation, we may obtain a confirmation of the law of the rectilinear direction of undisturbed motion, partly founded on direct experiment. Its uniformity is however still less subjected to immediate examination; yet, from a consideration of the nature of friction and resistance, combined with the laws of gravitation, we may ultimately show the perfect coincidence of the theory with experiment. The tendency of matter to persevere in this manner in the state of rest or of uniform rectilinear motion, is called its inertia.

In all these cases it is of importance to attend to the composition of motion, or the joint effect of more than one motion existing at the same time. The existence of two or more motions, at the same time, in the same body, is not at first comprehended without some difficulty. It is in fact only a combination or separation of relations that is considered: in the same manner as by combining the relation of son to father, and brother to brother, we obtain the relation of nephew to uncle, so by combining the motion of a man walking in a ship, with the motion of the ship, we determine the relative velocity of the man with respect to the earth's surface. It is, however, necessary, for ascertaining these relations, to consider the affections of a space or surface in motion, and to examine how it may move in the most simple manner with respect to another space.

If any number of points move in parallel lines, describing equal spaces in equal times, they are at rest with respect to each other; for it may easily be demonstrated that the rectilinear distance of each, from each of the rest, remains unchanged: and if all the points of a plane move in this manner on another plane, either plane may be said to be in rectilinear motion with respect to the other. This is easily exemplified by causing one plane to move on another, so that two or more of its points shall always remain in a given right line in the second plane: as when a square is made to slide along the straight edge of a board, the surface of the square is in rectilinear motion with respect to the board. (Plate I. Fig. 4.)

If, besides this general motion of the plane, any point be supposed to have a particular motion in it, the point will have two motions with respect to the other plane, the one in common with its plane, and the other peculiar to itself; and the joint effect of these motions with respect to the second plane is called the result of the two motions. Thus, when a carriage moves on a perfectly level road, all its points describe parallel lines, and it is in rectilinear motion with respect to the road: its wheels partake of this motion, but have also a rotatory motion of their own; and the result of the two motions of each point of the wheels is the cycloid or trochoid that it describes in a quiescent vertical plane. (Plate I. Fig. 5.)

When an arm is made to slide upon a bar, and a thread, fixed to the bar, is made to pass, over a pulley at the end of the arm next the bar, to a slider which is moveable along the arm, the slider moves on the arm with the same velocity as the arm on the bar; but if the thread, instead of being fixed to the slider, be passed again over a pulley attached to it, and then brought back to be fixed to the arm, the motion of the slider will be only half that of the arm; and this will be true in whatever position the arm be fixed. Here we have two motions in the slider, one in common with the arm, and the other peculiar to itself, which may be either equal or unequal to the first; and by tracing a line on a fixed plane, with a point attached to the slider, we may easily examine the joint result of both the motions. (Plate I. Fig. 6.)

The joint result of any two motions is the diagonal of the parallelogram, of which the sides would be described, in the same time, by the separate motions; that is, if we have two lines representing the directions and velocities of the separate motions, and from the remoter extremity of each draw a line parallel to the other, the intersection of these lines will be the place of the moving body at the end of the given time. This is the necessary consequence of the coexistence of two motions in the sense that has been defined; it is also capable of a complete illustration by means of the apparatus that has been described. (Plate I. Fig. 7.)

Any given motion may be considered as the result of any two or more motions capable of composing it in this manner. Thus the line described by the tracing point of our apparatus will be precisely the same, whether it be simply drawn along in the given direction, or made to move on the arm with a velocity equal to that of the arm, or, when the arm is in a different position, with only half that velocity. (Plate I. Fig. 8.)

This principle constitutes the important doctrine of the resolution of motion. There is some difficulty in imagining a slower motion to contain, as it were, within itself, two more rapid motions opposing each other: but in fact we have only to suppose ourselves adding or subtracting mathematical quantities, and we must relinquish the prejudice, derived from our own feelings, which associates the idea of effort with that of motion. When we conceive a state of rest as the result of equal and contrary motions, we use the same mode of representation as when we say that a cipher is the sum of two equal quantities with opposite signs; for instance, plus ten and minus ten make nothing.

The law of motion here established differs but little in its enunciation from the original words of Aristotle, in his mechanical problems. He says, that if a moving body has two motions, bearing a constant proportion to each other, it must necessarily describe the diameter of a parallelogram, of which the sides are in the ratio of the two motions. It is obvious that this proposition includes the consideration not only of uniform motions, but also of motions which are similarly accelerated or retarded: and we should scarcely have expected, that, from the time at which the subject began to be so clearly understood, two thousand years would have elapsed, before this law began to be applied to the determination of the velocity of bodies actuated by deflecting forces, which Newton has so simply and elegantly deduced from it.

In the laws of motion, which are the chief foundation of the Principia, their great author introduces at once the consideration of forces; and the first corollary stands thus; "a body describes the diagonal of a parallelogram by two forces acting conjointly, in the same time in which it would describe its sides by the same forces acting separately." It appears, however, to be more natural and perspicuous to defer the consideration of force, until the simpler doctrine of motion has been separately examined.

We may easily proceed to the composition of any number of different motions, by combining them successively in pairs. Hence any equable motions, represented by the sides of a polygon, that is, of a figure consisting of any number of straight sides, being supposed to take place in the same moveable body, in directions parallel to those sides, and in the order of going round the figure, destroy each other, and the body remains at rest. We may understand the truth of this proposition by imagining each motion to take place in succession for an equal small interval of time; then the point would describe a small polygon similar to the original one, and would be found, at the end of every such interval, in its original situation.

When the motions to be combined are numerous and diversified, it is often convenient to resolve each motion into three parts, reduced to the directions of three given lines perpendicular to each other. It is easy to find in this manner by addition and subtraction only, the general result of any number of motions. We may describe the flight of a bird, ascending in an oblique direction, by estimating its progress northwards or southwards, eastwards or westwards, and at the same time upwards, and we may thus determine its place as accurately, as by ascertaining the immediate bearing and angular elevation of its path, and its velocity in the direction of its motion.

LECTURE III.

ON ACCELERATING FORCES.

WE have hitherto only considered motion as already existing, without any regard to its origin or alteration; we have seen that all undisturbed motions are equable and rectilinear; and that two motions represented by the sides of a parallelogram, cause a body to describe its diagonal by their joint effect. We are now to examine the causes which produce or destroy motion. Any cause of a change of the motion of a body, with respect to a quiescent space, is called a force; that is, any cause which produces motion in a body at rest, or which increases, diminishes, or modifies it in a body which was before in motion. Thus the power of gravitation, which causes a stone to fall to the ground, is called a force; but when the stone, after descending down a hill, rolls along a horizontal plane, it is no longer impelled by any force, and its relative motion continues unaltered, until it is gradually destroyed by the retarding force of friction. Its perseverance in the state of motion or rest in consequence of the inertia of matter, has sometimes been expressed by the term vis inertiae, or force of inertia; but it appears to be somewhat inaccurate to apply the term force to a property, which is never the cause of a change of motion in the body to which it belongs.

It is a necessary condition in the definition of force, that it be the cause of a change of motion with respect to a quiescent space. For if the change were only in the relative motion of two points, it might happen without the operation of any force: thus, if a body be moving without disturbance, its motion with respect to another body, not in the line of its direction, will be perpetually changed; and this change, considered alone, would indicate the existence of a repulsive force: and, on the other hand, two bodies may be subjected to the action of an attractive force, while their distance remains unaltered, in consequence of the centrifugal effect of a rotatory motion. (Plate I. Fig. 9.)

The exertion of an animal, the unbending of a bow, and the communication of motion by impulse, are familiar instances of the actions of forces. We must not imagine that the idea of force is naturally connected with that of labour or difficulty; this association is only derived from habit, since our voluntary actions are in general attended with a certain effort, which leaves an impression almost inseparable from that of the force that it calls into action.

It is natural to inquire in what immediate manner any force acts, so as to produce motion; for instance, by what means the earth causes a stone to gravitate towards it. In some cases, indeed, we are disposed to imagine that we understand better the nature of the action of a force, as, when a body in motion strikes another, we conceive that the impenetrability of matter is a sufficient cause for the communication of motion, since the first body cannot continue its course without displacing the second; and it has been supposed that if we could discover any similar impulse that might be the cause of gravitation, we should have a perfect idea of its operation. But the fact is, that even in cases of apparent impulse, the bodies impelling each other are not actually in contact; and if any analogy between gravitation and impulse be ever established, it will not be by referring them both to the impenetrability of matter, but to the intervention of some common agent, perhaps imponderable. It was observed by Newton, that a considerable force was necessary to bring two pieces of glass into a degree of contact, which still was not quite perfect; and Professor Robison has estimated this force at a thousand pounds for every square inch. These extremely minute intervals have been ascertained by observations on the colours of the thin plate of air included between the glasses; and when an image of these colours is exhibited by means of the solar microscope, it is very easily shown that the glasses are separated from each other, by the operation of this repulsive force, as soon as the pressure of the screws which confine them is diminished; the rings of colours dependent on their distance contracting their dimensions accordingly. Hence it is obvious, that whenever two pieces of glass strike each other, without exerting a pressure equal to a thousand pounds on a square inch, they may affect each other's motion without actually coming into contact. Some persons might perhaps be disposed to attribute this repulsion to the elasticity of particles of air adhering to the glass, but I have found that the experiment succeeds equally well in the vacuum of the air pump. We must therefore be contented to acknowledge our total ignorance of the intimate nature of forces of every kind; and we are first to examine the effect of forces, considering only their magnitude and direction, without any regard to their origin.

It was truly asserted by Descartes, that the state of motion is equally natural with that of rest. When a body is once in motion, it requires no foreign power to sustain its velocity. If therefore a moving body is subjected to the influence of any force, which acts upon it in the line of its direction, its motion will be either accelerated or retarded, accordingly as the direction of the force coincides with that of the motion, or is opposed to it. A stone, for instance, beginning to fall, or projected downwards, by no means retains the same velocity throughout its descent, but acquires more and more motion every instant. We well know, that the greater the height from which a body falls, the more danger there is of its destroying whatever opposes its progress. In the same manner, when a ball is thrown upwards, it gradually loses its motion by the operation of gravitation, which is now a retarding force, and at last begins again to descend.

It may here be proper to inquire what is the precise meaning of the term velocity; we appear indeed to understand sufficiently the common use of the word, but it is not easy to give a correct definition of it. The velocity of a body may be said to be the quantity or degree of its motion, independently of any consideration of its mass or magnitude; and it might always be measured by the space described in a certain portion of time; for instance a second, if there were no other motions than undisturbed or uniform motions: but the velocity may vary very considerably within the second, and we must therefore have some other measure of it than the space actually described in any finite interval of time. If however the times be supposed infinitely short, the elements of space described may be considered as the true measures of velocities. These elements, although smaller than any assignable quantity, may yet be accurately compared with each other; and the reason that they afford a true criterion of the velocity, is this, that the change produced in the velocity, during so short an interval of time, must be absolutely inconsiderable in comparison with the whole velocity, and the element of space becomes a true measure of the temporary velocity, in the same manner as any larger portion of space may be the measure of a uniform velocity.

When the increase or diminution of the velocity of a moving body is uniform, its cause is called a uniform force: the spaces which would be described in any given time, with the actual velocity uniformly continued, being always equally increased or diminished by the action of such a force. For example, if the velocities, at the beginning of any two separate seconds, be such, that the body would describe one foot and ten feet in the respective seconds, if undisturbed, and the spaces actually described become two feet and eleven feet, each being increased one foot, the accelerating force must be denominated uniform.

The power of gravitation, acting at or near the earth's surface, may, without sensible error, be considered as such a force. Thus, if a body begins to fall from a state of rest, it describes about 16 feet, or more correctly $16\frac{1}{12}$, in the first second; if it begins a second with a velocity of 32 feet, it describes 32 and 16, or 48 feet, in this second. The decrease of the force of gravitation in proportion to the squares of the distances from the earth's centre, is barely perceptible, at any heights within our reach, by the nicest tests that we can employ.

The velocity produced by any uniformly accelerating force, is proportional to the magnitude of the force, and the time of its operation conjointly. When the forces are the same, a little consideration will convince us that, since every equal portion of time adds equally to the velocity, the whole velocity produced or destroyed must be in proportion to the whole time; and when the forces differ, the velocities differ in the same ratio; for the forces are only measured by the velocities which they generate. Thus a double force, in a double time, produces a quadruple velocity. That a force producing a double velocity is properly called a double force, may be shown from the laws of the composition of motion; for when the equal sides of a parallelogram representing two separate forces or motions, approach to each other, and at last coincide in direction, the diagonal of the parallelogram, representing their joint effect, becomes equal to the sum of the sides. (Plate I. Fig. 10.)

The machine invented by Mr. Atwood furnishes us with a very convenient mode of making experiments on accelerating forces. The velocity, produced by the undiminished force of gravity, is much too great to be conveniently submitted to experimental examination; but by means of this apparatus, we can diminish it in any degree that is required. Two boxes, which are attached to a thread passing over a pulley, may be filled with different weights, which counterbalance each other, and constitute, together with the pulley, an inert mass, which is put into motion by a small weight added to one of them. The time of descent is measured by a second or half second pendulum, the space described being ascertained by the place of a moveable stage, against which the bottom of the descending box strikes: and when we wish to determine immediately the velocity acquired at any point, by measuring the space uniformly described in a given time, the accelerating force is removed, by means of a ring, which intercepts the preponderating weight, and the box proceeds with a uniform velocity, except so far as the friction of the machine retards it. By changing the proportion of the preponderating weight to the whole weight of the boxes, it is obvious that we may change the velocity of the descent, and thus exhibit the effects of forces of different magnitudes. The most convenient mode of letting the weights go, without danger of disturbance from their vibrations, is to hold the lowest weight only, and to allow it to ascend at the instant of a beat of the pendulum. (Plate I. Fig. 11.)

That the velocity generated is proportional to the time of the action of the force, or that the force of gravitation, thus modified, is properly called a uniform accelerating force, may be shown by placing the moveable ring so as to intercept the same bar successively at two different points: thus the space uniformly described in a second, by the box alone, is twice as great, when the force is withdrawn after a descent of ten half seconds, as it is after a descent of five. And if we chose to vary the weight of the bar, we might show in a similar manner, that the velocity generated in a given time is proportional to the force employed.

We are next to determine the magnitude of the whole space described in a given time with a velocity thus uniformly increasing. The law discovered by Galileo, that the space described is as the square of the time of descent, and that it is also equal to half the space which would be described in the same time with the final velocity, is one of the most useful and interesting propositions in the whole science of mechanics. Its truth is easily shown from mathematical considerations, by comparing the time with the base, and the velocity with the perpendicular of a triangle gradually increasing, of which the area will represent the space described; and we may observe, by means of Atwood's machine, that a quadruple space is always described in a double time, whatever may be the magnitude of the force. Of course, if the forces vary, the spaces are as the forces and as the squares of the times conjointly. (Plate I. Fig. 12.)

It may also be demonstrated, that if a body falls through one foot in a second by means of a certain force, it will require a quadruple force to make it fall through the same space in half a second; and in general, where the spaces are equal, the forces are as the squares of the velocities. Wherever the space and the force remain the same, whether the force be uniform or not, the squares of any two velocities, with which different bodies enter the space, will receive equal additions while they pass through it.

When a force acts in a direction contrary to that of the moving body, we may readily determine the retardation that it produces, by comparing the motion with that of a body accelerated by the same force. For the degrees, by which an ascending body loses its motion, are the same as those by which it is again accelerated at the same points, when it has acquired its greatest height and again descends. We may thus calculate to what height a body will rise when projected upwards with a given velocity, and retarded by the force of gravitation. Since the force of gravitation produces or destroys a velocity of 32 feet in every second, a velocity of 320 feet, for instance, will be destroyed in 10 seconds; and according to what has been premised, a body will fall in 10 seconds through a hundred times 16 feet, or 1600 feet, which is therefore the height, to which a velocity of 320 feet in a second will carry a body, moving without resistance in a vertical direction. We may also obtain the same result by squaring one eighth of the velocity: thus one eighth of 320 is 40, of which the square is 1600, the height corresponding to the given velocity; and this velocity is sometimes called the velocity due to the height.

LECTURE IV.

ON DEFLECTIVE FORCES.

IT has been shown that the velocity, generated by an accelerating force, is proportional to the time of its action, and the space described to the square of the time. We are next to consider the more complicated cases of the action of such forces. When they are directed to a certain point out of the line of the motion which they affect, they become central forces, of which we have an example in the force of gravitation, considered as it governs the planetary motions; and when this point becomes extremely distant in comparison with the length of the body's path, so that the force acts very nearly in parallel lines, the body comes under the denomination of a projectile, as, for instance, a cannon ball projected horizontally or obliquely.

An accelerating force, therefore, tending to a point out of the line of direction of a moving body, deflects it from that line, and is then usually called a centripetal force. And the natural tendency of the body to persevere in its rectilinear motion, unless opposed by such a force, is sometimes called a centrifugal force. How far the term force is properly applicable to the perseverance of a body in its rectilinear motion, may perhaps be liable to dispute. If we allow the propriety of the appellation, we must extend the definition of the term force to any change of the relative motion of two points, and we must also allow the inertia of a body to be justly denominated a force. The fact, however, is certain, that all bodies revolving round a centre, have a tendency to recede from the centre, in the direction of the tangent, and when this force is counterbalanced, an equal centrifugal force must be exerted.

The effects of a centrifugal force may be observed in the familiar instance of a stone placed in a sling, which may be made to revolve in a vertical direction, and even at the upper part of its orbit, may adhere, as it were, notwithstanding its weight, to the sling which is above it, in consequence of the excess of the centrifugal force above the force of gravitation.

It is also a centrifugal force that is the foundation of the amusement of a boy driving a hoop. A hoop at rest, placed on its edge, would very quickly fall to the ground; but when it is moving forwards, a slight inclination towards either side causes the parts to acquire a motion towards that side, those which are uppermost being most affected by it; and this lateral motion, assisted sometimes by the curvature of the surface of the hoop, causes its path to deviate from a rectilinear direction, so that instead of moving straight forwards, it turns to that side, towards which it began to incline; and in this position, its tendency to fall still further is counteracted by the centrifugal force, and it generally makes several complete revolutions before it falls. The motion of a bowl, with its bias, is of a similar nature; the centrifugal force counteracting the tendency to curvilinear motion, so as to diminish it very considerably, until the velocity is so much reduced, as to suffer it to describe a path evidently curved, and becoming more and more so as the motion is slower.

When a body is retained in a circular orbit, by a force directed to its centre, its velocity is every where equal to that which it would acquire, in falling, by means of the same force, if uniform, through half the radius, that is, through one fourth of the diameter. This proposition affords a very convenient method of comparing the effects of central forces with those of simple accelerating forces, and deserves to be retained in memory. We may in some measure demonstrate its truth by means of the whirling table: an apparatus which is arranged on purpose for exhibiting the properties of central forces, although it is more calculated for showing their comparative than their absolute magnitude; for accordingly as we place the string on the pullies, the two horizontal arms may be made to revolve either with equal velocities, or one twice as fast as the other. The sliding stages, which may be placed at different distances from the centres, and which are made to move along the arms with as little friction as possible, are in a certain proportion to the weights, which are to be raised, by means of threads passing over pullies in the centres, as soon as the centrifugal forces of the stages with their weights are sufficiently great; and the experiment is to be so arranged, that when the velocity, having been gradually increased, produces a sufficient centrifugal force, both stages may raise their weights, and fly off at the same instant. But, for the present purpose, one of the stages only is required, and the time of revolution may be measured by a half second pendulum. We may make the force, or the weight to be raised, equal to the weight of the revolving body, and we shall find that this body will fly off when its velocity becomes equal to that which would be acquired by any heavy body in falling through a height equal to half the distance from the centre, and as much greater as is sufficient for overcoming the friction of the machine. (Plate I. Fig. 13.)

From this proposition we may easily calculate the velocity, with which a sling of a given length must revolve, in order to retain a stone in its place in all positions; supposing the motion to be in a vertical plane, it is obvious that the stone will have a tendency to fall when it is at the uppermost point of the orbit, unless the centrifugal force be at least equal to the force of gravity. Thus if the length of the sling be two feet, we must find the velocity acquired by a heavy body in falling through a height of one foot, which will be eight feet in a second, since eight times the square root of 1 is eight; and this must be its velocity at the highest point; with this velocity it would perform each revolution in about a second and a half, but its motion in other parts of its orbit will be greatly accelerated by the gravitation of the stone.

It may also be demonstrated, that when bodies revolve in equal circles, their centrifugal forces are proportional to the squares of their velocities. Thus, in the whirling table, the two stages being equally loaded, one of them, which is made to revolve with twice the velocity of the other, will lift four equal weights at the same instant that the other raises a single one. But when two bodies revolve with equal velocities at different distances, the forces are inversely as the distances; consequently the forces are, in all cases, directly as the squares of the velocities, and inversely as the distances.

If two bodies revolve in equal times at different distances, the forces by which they are retained in their orbits are simply as the distances. If one of the stages of the whirling table be placed at twice the distance of the other, it will raise twice as great a weight, when the revolutions are performed in the same time.

In general, the forces are as the distances directly, and as the squares of the times of revolution inversely. Thus the same weight revolving in a double time, at the same distance, will have its effect reduced to one fourth, but at a double distance the effect will again be increased to half of its original magnitude.

From these principles we may deduce the law which was discovered by Kepler in the motions of the planetary bodies, but which was first demonstrated by Newton from mechanical considerations. Where the forces vary inversely as the squares of the distances, as in the case of gravitation, the squares of the times of revolution are proportional to the cubes of the distances. Thus if the distance of one body be four times as great as that of another, the cube of 4 being 64, which is the square of 8, the time of its revolution will be 8 times as great as that of the first body. It would be easy to show the truth of this proposition experimentally by means of the whirling table, but the proof would be less striking than those of the simpler laws which have already been laid down.

Hitherto we have supposed the orbit of a revolving body to be a perfect circle; but it often happens in nature, as, for instance, in all the planetary motions, that the orbit deviates more or less from a circular form; and in such cases we may apply another very important law which was also discovered by Kepler; that the right line joining a revolving body and its centre of attraction, always describes equal areas in equal times, and the velocity of the body is therefore always inversely as the perpendicular drawn from the centre to the tangent. (Plate I. Fig. 14.)

The demonstration of this law, invented by Newton, was one of the most elegant applications of the geometry of infinites or indivisibles; a branch of mathematics of which Archimedes laid the foundations, which Cavalleri and Wallis greatly advanced, and which Newton brought near to perfection. Its truth may be in some measure shown by an experiment on the revolution of a ball suspended by a long thread, and drawn towards a point immediately under the point of suspension by another thread, which may either be held in the hand, or have a weight attached to it. The ball being made to revolve, its motion becomes evidently more rapid when it is drawn by the horizontal thread nearer to the fixed point, and slower when it is suffered to fly off to a greater distance. (Plate II. Fig. 15.)

It was also discovered by Kepler that each of the planets revolves in an ellipsis, of which the sun occupies one of the foci. It is well known that an ellipsis is an oval figure, which may be described by fixing the ends of a thread to two points, and moving a tracing point so that it may always be at the point of the angle formed by the thread; and that the two fixed points are called its foci. The inference respecting the force by which a body may be made to revolve in an ellipsis, was first made by Newton; that is, that the force directed to its focus must be inversely as the square of the distance. We have no other experimental proof of this theorem than astronomical observations, which are indeed perfectly decisive, but do not require to be here anticipated. (Plate II. Fig. 16.)

There is another general proposition which is sometimes of use in the comparison of rectilinear and curvilinear motions. Two bodies being attracted in the same manner towards a given centre, that is, with equal forces at equal distances, if their velocities be once equal at equal distances, they will remain always equal at equal distances, whatever be their directions. For instance, if one cannon ball be shot obliquely upwards, and another perpendicularly upwards, with the same velocity, the one will describe a curve, and the other a straight line, but their velocities will always remain equal, not at the same instants of time, but at equal distances from the earth's centre, or after having ascended through equal vertical heights, although in different directions. This proposition has usually been made a step in the demonstration of the law of the force by which a body is made to revolve in an ellipsis; but there is a much simpler method of demonstrating that law, by means of some properties of the curvature of the ellipsis.

In treating of the motion of projectiles, the force of gravitation may, without sensible error, be considered as an equable force, acting in parallel lines perpendicular to the horizon. In reality, if we ascend a mile from the earth's surface, the actual weight of a body is diminished about a two thousandth part, or three grains and a half for every pound, and we may discover this inequality by means of the vibrations of pendulums, which become a little slower when they are placed on the summits of very high mountains. On the other hand, a body not specifically heavier than water, gains more in apparent weight on account of the diminished density of the atmosphere at great elevations, than it loses by the increase of its distance from the earth. But both these differences may, in all common calculations, be wholly disregarded. The direction of gravity is always exactly perpendicular to the horizon, that is, to the surface of the earth, which is somewhat curved, on account of the earth's spheroidical figure; but any small portion of this surface may be practically considered as a plane, and the vertical lines perpendicular to it, as parallel to each other.

The oblique motion of a projectile may be the most easily understood by resolving its velocity into two parts, the one in a horizontal, the other in a vertical direction. It appears from the doctrine of the composition of motion, that the horizontal velocity will not be affected by the force of gravitation acting in a direction perpendicular to it, and that it will therefore continue uniform; and that the vertical motion will also be the same as if the body had no horizontal motion. Thus if we let fall from the head of the mast of a ship a weight, which partakes of its progressive motion, the weight will descend by the side of the mast in the same manner, and with the same relative velocity, as if neither the ship nor the weight had any horizontal motion.

We may therefore always determine the greatest height to which a projectile will rise, by finding the height from which a body must fall, in order to gain a velocity equal to its vertical velocity, or its velocity of ascent, that is, by squaring one eighth of the number of feet that it would rise in the first second if it were not retarded. For example, suppose a musket to be so elevated that the muzzle is higher than the but-end by half of the length, that is, at an angle of 30°; and let the ball be discharged with a velocity of 1000 feet in a second; then its vertical velocity will be half as great, or 500 feet in a second: now the square of one eighth of 500 is 3906, consequently the height to which the ball would rise, if unresisted by the air, is 3906 feet, or three quarters of a mile. But in fact, a musket ball, actually shot upwards, with a velocity of 1670 feet in a second, which would rise six or seven miles in a vacuum, is so retarded by the air, that it does not attain the height of a single mile.

We may easily find the time of the body's ascent from its initial velocity; for the time of ascent is directly proportional to the velocity, and may be found in seconds by dividing the vertical velocity in feet by 32; or if we divide by 16 only, we shall have the time of ascent and descent; and then the horizontal range may be found, by calculating the distance described in this time, with the uniform horizontal velocity. Thus, in the example that we have assumed, dividing 500 by 16, we have 31 seconds for the whole time of the range; but the hypotenuse of our triangle being 1000, and the perpendicular 500, the base will be 886 feet; consequently the horizontal range is 31 times 886, that is, nearly 28000 feet, or above 5 miles. But the resistance of the air will reduce this distance also to less than one mile.

It may be demonstrated that the horizontal range of a body, projected with a given velocity, is always proportional to the sine of twice the angle of elevation: that is, to the elevation of the muzzle of the piece in a situation twice as remote from a horizontal position as its actual situation. Hence it follows, that the greatest horizontal range will be when the elevation is half a right angle; supposing the body to move in a vacuum. But the resistance of the air increases with the length of the path, and the same cause also makes the angle of descent much greater than the angle of ascent, as we may observe in the track of a bomb. For both these reasons, the best elevation is somewhat less than 45°, and sometimes, when the velocity is very great, as little as 30°. But it usually happens in the operations of natural causes, that near the point where any quantity is greatest or least, its variation is slower than elsewhere: a small difference, therefore, in the angle of elevation, is of little consequence to the extent of the range, provided that it continue between the limits of 45° and 35°; and for the same reason, the angular adjustment requires less accuracy in this position than in any other, which besides the economy of powder, makes it the best elevation for practice. (Plate II. Fig. 17, 18.)

The path of a projectile, supposed to move without resistance, is always a parabola. This interesting proposition was first discovered by Galileo: it follows very readily from the doctrine of the composition of motion, combined with the laws which that philosopher established concerning the fall of heavy bodies. If from any points of a given right line, as many lines be drawn, parallel to each other, and proportional to the squares of the corresponding segments of the first line, the curve in which all their extremities are found, is a parabola. Now supposing the first line to be placed in the direction of the initial motion of a projectile, and parallel vertical lines to be drawn through any points of it, proportional to the squares of the segments which they cut off, these lines will represent the effect of gravitation, during the times in which the same segments would have been described, by the motion of projection alone; consequently the projectile will always be found at the extremity of the vertical line corresponding to the time elapsed, and will therefore describe a parabola. (Plate II. Fig. 17, 19.)

It is easy to show by experiment, that the path of a projectile is a parabola: if we only let a ball descend from a certain point, along a groove, so as to acquire a known velocity, we may trace on a board the parabola which it will afterwards describe, during its free descent; and by placing rings at different parts of the curve, we may observe that it will pass through them all without striking them. (Plate II. Fig. 19.)

In practical cases, on a large scale, where the velocity of a projectile is considerable, the resistance of the atmosphere, is so great as to render the Galilean propositions of little or no use; and a complete determination of the path, including all the circumstances which may influence it, is attended with difficulties almost insuperable. It appears from Robins's experiments, that the resistance of the air to an iron ball of $4\frac{1}{2}$ inches in diameter, moving at the rate of 800 feet in a second, is equal to four times its weight, and that where the velocity is much greater, the resistance increases far more rapidly. But what must very much diminish the probability of our deriving any great practical advantage from the theory of gunnery, is an observation, made also by Mr. Robins, that a ball sometimes deviates three or four hundred yards laterally, without any apparent reason; so that we cannot be absolutely certain to come within this distance of our mark in any direction. The circumstance is probably owing to an accidental rotatory motion communicated to the ball in its passage through the piece, causing therefore a greater friction from the air on one side than on the other; and it may in some measure be remedied by employing a rifle barrel, which determines the rotation of the ball in such a manner that its axis coincides at first with the path of the ball, so that the same face of the ball is turned in succession every way. For the ordinary purposes of gunnery, an estimation governed by experience is found to be the best guide; at the same time there is no doubt but that some assistance may be obtained from theory and from experiment. Those who are desirous of pursuing the subject, may find much information relating to it, collected by Professor Robison, in the article Projectile of the Encyclopaedia Britannica.

LECTURE V.

ON CONFINED MOTION.

WE have hitherto considered the principal cases of motion, either undisturbed, or simply subjected to the action of an accelerating, retarding, or deflective force. We now proceed to examine the effects of an additional modification, which is introduced, when the motion is limited to a given line or surface of any kind; the body either being supposed to slide on the surface of a solid actually extended, or being confined to an imaginary surface by its attachment to a thread, or still more narrowly restricted, by means of two threads, which allow it to move only in a given line. Suspension is the most convenient mode of making experiments on confined motion; but it is not always easy to cause the body to remain in the surface that is required; and to confine it in this manner to a perfectly plane surface, is impossible. When we suffer a body to slide along any surface, there is a loss of force from friction, from the production of rotatory motion, or from both these causes combined. The effect of friction is obvious and well known; and we may be convinced of the retardation attendant on the production of rotatory motion, by allowing two cylinders, of equal dimensions, to roll down an inclined plane; the one being covered with sheet lead, the other having an equal weight of lead in its axis, and being covered with paper; and both having similar projecting surfaces at the ends, which come into contact with the plane: we may easily observe that in the first cylinder, much more of the force is consumed in producing rotatory motion, than in the second, and that it therefore descends much more slowly. (Plate II. Fig. 20.)

When a body is placed on an inclined plane, the force urging it to descend, in the direction of the plane, is to the whole force of gravity, as the height of the plane is to its length. This is demonstrable from the principles of the composition of motion, and may also be shown experimentally with great accuracy, when we consider the doctrine of the equilibrium of forces. But the interference of friction will only allow us to observe, with respect to the velocities produced, that they nearly approach to those which the calculation indicates. Thus if a plane be inclined one inch in 32, a ball will descend on it in two seconds, instead of 64 feet, somewhat less than two feet.

It may be deduced from the laws of accelerating forces, that when bodies descend on any inclined planes, of equal heights, but of different inclinations, the times of descent are as the lengths of the planes, and the final velocities are equal. Thus a body will acquire a velocity of 32 feet in a second, after having descended 16 feet, either in a vertical line or in an oblique direction; but the time of descent will be as much greater than a second, as the oblique length of the path is greater than 16 feet. This may be shown by experiment, as nearly as the obstacles already mentioned will permit, the times being measured by a pendulum, or by a stop watch. (Plate II. Fig. 21.)

There is an elegant proposition, of a similar nature, which is still more capable of experimental confirmation; that is, that the times of falling through all chords drawn to the lowest point of a circle are equal. If two or more bodies are placed at different points of a circle, and suffered to descend at the same instant along as many planes, which meet in the lowest point of the circle, they will arrive there at the same time. (Plate II. Fig. 22.)

The velocity of a body, descending along a given surface, is the same as that of a body falling freely through an equal height, not only when the surface is a plane, but also when it is a continued curve, in which the body is retained by its attachment to a thread, or is supported by any regular surface, supposed to be free from friction. We may easily show, by an experiment on a suspended ball, that its velocity is the same when it descends from the same height, whatever may be the form of its path, by observing the height to which it rises on the opposite side of the lowest point. We may alter the form of the path in which it descends, by placing pins at different points, so as to interfere with the thread that supports the ball, and to form in succession temporary centres of motion; and we shall find, in all cases, that the body ascends to a height equal to that from which it descended, with a small deduction on account of friction. (Plate II. Fig. 23.)

Hence is derived the idea conveyed by the term living or ascending force; for since the height, to which a body will rise perpendicularly, is as the square of its velocity, it will preserve a tendency to rise to a height which is as the square of its velocity, whatever may be the path into which it is directed, provided that it meet with no abrupt angle, or that it rebound at each angle in a new direction, without losing any velocity. The same idea is somewhat more concisely expressed by the term energy, which indicates the tendency of a body to ascend or to penetrate to a certain distance, in opposition to a retarding force.

The most important cases of the motion of bodies, confined to given surfaces, are those which relate to the properties of pendulums. Of these the simplest is the motion of a body in a cycloidal path. The cycloid is a curve which has many peculiarities; we have already seen that it is described by marking the path of a given point in the circumference of a circle which rolls on a right line. Galileo was the first that considered it with attention, but he failed in his attempts to investigate its properties. It is singular enough, that the principal cause of his want of success was an inaccurate experiment: in order to obtain some previous information respecting the area included by it, he cut a board into a cycloidal form, and weighed it, and he inferred from the experiment, that the area bore some irrational proportion to that of the describing circle, while in fact it is exactly triple. In the same manner it has happened in later times, that Newton, in his closet, determined the figure of the earth more accurately, than Cassini from actual measurement. It was Huygens that first demonstrated the properties of the cycloidal pendulum, which are of still more importance in the solution of various mechanical problems, than for the immediate purposes of timekeepers, to which that eminent philosopher intended to apply them. (Plate I. Fig 5.)

If a body be suspended by a thread playing between two cycloidal cheeks, it will describe another equal cycloid by the evolution of the thread, and the time of vibration will be the same, in whatever part of the curve it may begin to descend. Hence the vibrations of a body moving in a cycloid are denominated isochronous, or of equal duration. This equality may be shown by letting go two pendulous balls at the same instant, at different points of the curve, and observing that they meet at the lowest point. (Plate II. Fig. 24.)

The absolute time of the descent or ascent of a pendulum, in a cycloid, is to the time in which any heavy body would fall through one half of the length of the thread, as half the circumference of a circle to its diameter. It is therefore nearly equal to the time required for the descent of a body through $\frac{5}{4}$ of the length of the thread; and if we suffer the pendulum to descend, at the same moment that a body falls, from a point elevated one fourth of the length of the thread above the point of suspension, this body will meet the pendulum at the lowest point of its vibration. (Plate II. Fig. 24.)

Hence it may readily be inferred, that since the times of falling through any spaces, are as the square roots of those spaces, the times of vibration of different pendulums are as the square roots of their lengths. Thus, the times of vibration of pendulums of 1 foot and 4 foot in length, will be as 1 to 2: the time of vibration of a pendulum 39 $\frac{13}{100}$ inches in length is one second; the length of a pendulum vibrating in two seconds must be four times as great.

The velocity, with which a pendulous body moves, at each point of a cycloidal curve, may be represented, by supposing another pendulum to revolve uniformly in a circle, setting out from the lowest point, at the same time that the first pendulum begins to move, and completing its revolution in the time of two vibrations; then the height, acquired by the pendulum revolving equably, will always be equal to the space described by the pendulum vibrating in the cycloid. (Plate II. Fig. 24.)

It may also be shown, that if the pendulum vibrate through the whole curve, it will everywhere move with the same velocity as the point of the circle which is supposed to have originally described the cycloid, provided that the circle roll onwards with an equable motion.

All these properties depend on this circumstance, that the relative force, urging the body to descend along the curve, is always proportional to the distance from the lowest point; and it happens in many other instances of the action of various forces, that a similar law prevails: in all such cases, the vibrations are isochronous, and the space described corresponds to the versed sine of a circular arc increasing uniformly, that is to the height of any point of a wheel revolving uniformly on its axis, or rolling uniformly on a horizontal plane.

The cycloid is the curve in which a body may descend, in the shortest possible time, from a given point, to another obliquely below it. It may easily be shown that a body descends more rapidly in a cycloid than in the right line joining the two points. This property is of little practical utility; the proposition was formerly considered as somewhat difficult to be demonstrated, but of late, from the invention of new modes of calculation, theorems of a similar nature have been much extended with great facility. The experiment naturally suggests a familiar proverb, which cautions us against being led away too precipitately by an appearance of brevity and facility. (Plate II. Fig. 25.)

It has been found that the inconveniences, resulting from the complicated apparatus necessary to introduce a cycloidal motion, for the pendulums of clocks, are more than equivalent to the advantage of perfect isochronism in theory. For since, in small cycloidal arcs, the curvature is nearly constant, the time of vibration of a simple circular pendulum must be ultimately the same, as that of a cycloidal pendulum of the same length; but in larger arcs, the time must be somewhat greater, because the circular arc falls without the cycloidal, and is less inclined to the horizon at equal distances from the lowest point. This may be shown by a comparison of two equal pendulums, vibrating in arcs of different extent: it may also be observed, by an experiment with two simple pendulums of different lengths, that their times of vibration, like those of cycloidal pendulums, are proportional to the square roots of their lengths; a half second pendulum being only one fourth as long as a pendulum vibrating seconds.

We have been obliged to suppose the weight, as well as the inertia, of a pendulum, to be referred to one point, since we are not at present prepared to examine the effect of the slight difference between the situations, and the velocities of the different parts of the substances, employed in our experiments. The nature of rotatory motion requires to be more fully understood, before we can attend to the determination of the centres of oscillation of bodies of various figures, that is, of the points in which their whole weight may be supposed to be concentrated, with regard to its effect on the times of their vibrations.

It is remarkable that the isochronism of pendulums, which is a property so important in its application, may still be preserved, notwithstanding the interference of a constant retarding force, such as the force of friction is in many cases found to be. It has been shown by Newton, that each complete vibration of a cycloidal pendulum, retarded by a resistance of this nature, will be shorter than the preceding one by a certain constant space, but that it will be performed in the same time.

There is a great analogy between the vibrations of pendulums, and the revolution of balls suspended from a fixed point. If a body, suspended by a thread, revolve freely in a horizontal circle, the time of revolution will be the same, whenever the height of the point of suspension, above the plane of revolution is the same, whatever be the length of the thread. Thus, if a number of balls are fixed to threads, or rather wires, connected to the same point of an axis, which is made to revolve by means of the whirling table, they will so arrange themselves, as to remain very nearly in the same horizontal plane. (Plate II. Fig. 26.)

The time of each revolution of the balls is equal to the time occupied by a double vibration of a pendulum, of which the length is equal to the height of the point of suspension above the plane in which they revolve; consequently all the revolutions will be nearly isochronous, while the threads or wires deviate but little from a vertical situation. In fact, we may imagine such a revolution to be composed of two vibrations of a simple pendulum, existing at the same time, in directions at right angles to each other; for while a pendulum is vibrating from north to south, it is liable to the impression of any force, capable of causing a vibration from east to west; and the joint result of both vibrations will be a uniform revolution in a circle, if the vibrations are equal and properly combined; but if they are unequal, the joint vibration will be ultimately an ellipsis, the joint force being directed to its centre, and always proportional to the distance from that centre. (Plate II. Fig. 27.)

The near approach of these revolutions to isochronism has sometimes been applied to the measurement of time, but more frequently, and more successfully, to the regulation of the motions of machines. Thus in Mr. Watt's steam engines, two balls are fixed at the ends of rods in continual revolution, and as soon as the motion becomes a little too rapid, the balls rise considerably, and turn a cock, which diminishes the quantity of steam admitted. (Plate II. Fig. 28.)

The same laws are applicable to many other cases of rotatory motion; for instance, if we wish to determine the height, at which a ball, revolving with a given velocity, will be retained in a spherical bowl; or the inclination of a circular road, capable of counteracting the centrifugal force of a horse, running round it: for the horse, like the ball of the revolving pendulum, has a centrifugal tendency, which is greater as his velocity is greater: this centrifugal force, combined with the force of gravity, composes a result, which, in the case of the pendulum, is completely counteracted by the force of the thread or wire, and must therefore be in the direction of the thread, and which obliges the horse to place his legs in a similar direction, proceeding from an imaginary point of suspension above; since he would otherwise be liable to fall outwards, if his velocity were sufficiently great. But in order to withstand the pressure of the horse's legs, the road must be in a direction perpendicular to them; otherwise its materials will naturally be forced outwards, until they produce an elevation sufficient to give the road the required form. Thus, if the diameter of the ring were 40 feet, and the horse moved at the rate of 12 miles an hour, he would perform about 500 revolutions in an hour, and half a revolution in 3 seconds and a half. Now the length of a pendulum vibrating in $3\frac{1}{2}$ seconds, must be 39 inches multiplied by the square of $3\frac{1}{2}$, or a little more than 80 feet: the road must therefore be perpendicular to the direction of a line drawn to it from a point 80 feet above the centre of the ring; and its external circumference must be higher than its internal circumference by one fourth of its breadth. It would however be improper to have a road of this form in a manege, since the horse must be taught to perform all his evolutions on a perfect plane.

There is a general principle of curvilinear motion, which is in itself of little importance or practical utility, but which so far deserves to be noticed, as it has been magnified by some philosophers into a fundamental law of nature.

Among all the curves that a body can describe, in moving from one point to another, it always selects that, in which, if its velocity be supposed to be every where multiplied by the distance that it describes, the sum of the infinitely small products will be a minimum, that is, less than in any other path that the body could take. For example, if a body move freely, and therefore with a uniform velocity, in any regular curved surface, it will pass from one part of the surface to another by the shortest possible path. This has been called the principle of the least possible action; it is however merely a mathematical inference from the simpler laws of motion, and if those laws were even different from what they are, the principle would be true in another form, and in another sense of the word action.

LECTURE VI.

ON THE MOTIONS OF SIMPLE MASSES.

HITHERTO we have considered the motions of one or more single points or atoms only, without any regard to the bulk or mass of a moveable body: but it now becomes necessary to attend also to the difference of the masses of bodies in motion. This may however be done, without considering the actual magnitude or extent of the body. We may easily conceive different masses or bulks to be concentrated in a mathematical point; and it is most convenient to define a moveable body, as a moveable point or particle, composed of other elementary particles, differing only in number, and thus constituting the proportionally different mass or bulk of the body.

Although, in our experiments on motion, we are obliged to have recourse to material bodies, and although such bodies differ considerably from this definition of a single moveable body, yet they serve sufficiently well to represent such bodies, especially when they are small, and regularly formed; and we are here considering the doctrine of motion rather in a mathematical than in a physical sense, so that we are able to neglect all such properties of matter as are not immediately necessary to our purpose. Indeed though the general properties of matter are usually placed at the entrance of elementary works on mechanics, it has yet been found necessary to omit the consideration of their effects, in examining the laws and affections of motion. The forces of cohesion and repulsion, for example, act, in general, in a very complicated manner, in almost all cases of the communication of motion; but to consider these operations minutely in treating of collision, would be to involve the subject in very great and very unnecessary difficulties; and the complete investigation of these properties of matter would require the employment of various branches of mechanical and hydrodynamical science. We may therefore take a much simpler course, by deferring entirely all theoretical consideration of actual matter; but in the mean time we must have, for our experimental illustrations, some measure of the mass or bulk as here defined. We might employ spherical bodies, composed only of homogeneous substances, that is, of substances of the same kind, and we might estimate the mass by the comparative magnitude, imagining all the particles of each sphere to be united in its centre. But it is more convenient to anticipate, from the gravitation of matter, a measure of the mass derived from the weight: since it can be proved that the weight of a body is proportional to its absolute quantity of matter, supposing all matter to be alike in its affections relative to motion. So that instead of numbering the particles of each body, the same purpose is answered by determining their comparative weight.

Inertia, or a tendency to persevere in a state of rest, or of uniform rectilinear motion, is a property attached to all matter, and may be considered as proportional to the mass or weight of a body. When the motions of a system of bodies are considered, their inertia may in some respects be referred to a single point, which is called the centre of inertia. The centre of inertia of two bodies is that point, in the right line joining them, which divides it into two such portions, that the one is to the other, as the mass of the remoter body to that of the adjacent body. For instance, if one body weighs a pound, and another two pounds, and their distance is a yard, then the centre of inertia is at the distance of two feet from the smaller body, and one foot from the larger: and the distance of each is to the whole distance, as the weight of the other to the whole weight. Also the products obtained by multiplying each weight by its distance are equal: thus two multiplied by one, is equal to one multiplied by two. (Plate II. Fig. 29.)

This point is most commonly called the centre of gravity; it has also sometimes been denominated the centre of position. Since it has many properties independent of the consideration of gravity, it ought not to derive its name from gravitation, and the term centre of inertia begins now, with great propriety, to be generally adopted.

The centre of inertia of any two bodies initially at rest, remains at rest, notwithstanding any reciprocal action of the bodies; that is, notwithstanding any action which affects the single particles of both equally, in increasing or diminishing their distance. For it may be shown, from the principles of the composition of motion, that any force, acting in this manner, will cause each of the two bodies to describe a space proportional to the magnitude of the other body: thus a body of one pound will move through a space twice as great as a body of two pounds weight, and the remaining parts of the original distance will still be divided in the same proportion, by the original centre of inertia, which therefore still remains the centre of inertia, and is at rest. And it follows also, that if the centre of inertia is at first in motion, its motion will not be affected by any reciprocal action of the bodies.

This important property is very capable of experimental illustration; first observing, that all known forces are reciprocal, and among the rest the action of a spring: we place two unequal bodies so as to be separated when a spring is set at liberty, and we find that they describe, in any given interval of time, distances which are inversely as their weights; and that consequently the place of the centre of inertia remains unaltered. They may either be made to float on water, or may be suspended by long threads; the spring may be detached by burning a thread that confines it, and it may be observed whether or no they strike at the same instant two obstacles, placed at such distances as the theory requires; or if they are suspended as pendulums, the arcs which they describe may be measured, the velocities being always nearly proportional to these arcs, and accurately so to their chords. (Plate II. Fig. 30.)

The same might also be shown of attractive as well as of repulsive forces. For instance, if we placed ourselves in a small boat, and pulled a rope tied to a much larger one, we should draw ourselves towards the large boat with a motion as much more rapid than that of the large boat, as its weight is greater than that of our own boat; and the two boats would meet in their common centre of inertia, supposing the resistance of the water inconsiderable.

Having established this property of the centre of inertia, as a law of motion, we may derive from it the true estimate of the quantity of motion in different bodies, in a much more satisfactory manner, than it has usually been explained. For since the same reciprocal action produces, in a body weighing two pounds, only half the velocity that it produces in a body weighing one pound, the cause being the same, the effects must be considered as equal, and

2

the quantity of motion must always be measured by the joint ratio of mass to mass, and velocity to velocity; that is, by the ratio of the products, obtained by multiplying the weight of each body by the number expressing its velocity; and these products are called the momenta of the bodies. We appear to have deduced this measure of motion from the most unexceptionable arguments, and we shall have occasion to apply the momentum thus estimated as a true measure of force; at the same time that we allow the practical importance of considering, in many cases, the efficacy of forces, according to another criterion, when we multiply the mass by the square of the velocity, in order to determine the energy: yet the true quantity of motion, or momentum, of any body, is always to be understood, as the product of its mass into its velocity. Thus a body weighing one pound, moving with a velocity of a hundred feet in a second, has the same momentum, and the same quantity of motion, as a body of ten pounds, moving at the rate of ten feet in a second.

We may also demonstrate experimentally, by means of Mr. Atwood's machine, that the same momentum is generated, in a given time, by the same preponderating force, whatever may be the quantity of matter moved. Thus if the preponderating weight be one sixteenth of the whole weight of the boxes, it will fall one foot in a second, instead of 16, and a velocity of two feet will be acquired by the whole mass, instead of a velocity of 32 feet, which the preponderating weight alone would have acquired. And when we compare the centrifugal forces of bodies revolving in the same time, at different distances from the centre of motion, we find that a greater quantity of matter compensates for a smaller force; so that two balls connected by a wire, with liberty to slide either way, will retain each other in their respective situations, when their common centre of inertia coincides with the centre of motion; the centrifugal force of each particle of the one being as much greater than that of an equal particle of the other, as its weight, or the number of the particles, is smaller.

But it is not enough to determine the centre of inertia of two bodies only, considered as single points; since in general a much greater number of points is concerned: we must therefore define the sense in which the term is in this case to be applied. We proceed by considering the first and second of three or more bodies, as a single body, equal to both of them, and placed in their common centre of inertia; determining the centre of inertia of this imaginary body and the third body, and continuing a similar process for all the bodies of the system. And it matters not with which of the bodies we begin the operation, for it may be demonstrated, that the point thus found will be the same by whatever steps it be determined. When we come to consider the properties of the same point as the centre of gravity, we shall be able to produce an experimental proof of this assertion, since it will be found that there is only one point in any system of bodies which possesses these properties. (Plate III. Fig. 31.)

We may always represent the motion of the centre of inertia of a system of moving bodies, by supposing their masses to be united into one body, and this body to receive at once a momentum equal to that of each body of the system, in a direction parallel to its motion. This may often be the most conveniently done, by referring all the motions of this imaginary body to three given directions, and collecting all the results into three sums, which will represent the motion of the centre of inertia of the system.

We have already presupposed this proposition, when we have employed material bodies of finite magnitude, that is, systems of material atoms, to represent imaginary bodies of the same weight, condensed into their centres; and it now appears, that the velocity and direction of the motions of such bodies as we have employed, agree precisely with those of our imaginary material points. We cannot attempt to confirm this law by experiment, because the deductions from the sensible consequences of an experiment would require nearly the same processes as the mathematical demonstration.

It is obvious that the result of any number of uniform and rectilinear motions, thus collected, must also be a uniform and rectilinear motion. The centre of inertia of a system of bodies moving without disturbance, is, therefore, either at rest, or moving equably in a right line.

The mass, or weight, of each of any number of bodies, being multiplied by its distance from a given plane, the products, collected into one sum, will be equal to the whole weight of the system, multiplied by the distance of the common centre of inertia from the same plane. And the proposition will be equally true, if instead of the shortest distances, we substitute the distances from the same plane, measured obliquely, in any directions always parallel to each other. This property is peculiarly applicable to the consideration of the centre of gravity, and affords also the readiest means of determining its place in bodies of complicated forms. (Plate III. Fig. 32.)

We have already seen that the place of the centre of inertia of two bodies is not affected by any reciprocal action between them; and the same is true of the actions of a system of three or more bodies. We might easily apply our experiment on the reciprocal action of two bodies to a greater number, but we should throw no further light on the subject, and the mode of obtaining the conclusion would be somewhat complicated.

All the forces in nature, with which we are acquainted, act reciprocally between different masses of matter, so that any two bodies repelling or attracting each other, are made to recede or approach with equal momenta. This circumstance is generally expressed by the third law of motion, that action and reaction are equal. There would be something peculiar, and almost inconceivable, in a force which could affect unequally the similar particles of matter; or in the particles themselves, if they could be possessed of such different degrees of mobility, as to be equally moveable with respect to one force, and unequally with respect to another. For instance, a magnet and a piece of iron, each weighing a pound, will remain in equilibrium when their weights are opposed to each other by means of a balance; they will be separated with equal velocities, if impelled by the unbending of a spring placed between them, and it is difficult to conceive that they should approach each other with unequal velocities in consequence of magnetic attraction, or of any other natural force. The reciprocality of force is therefore a necessary law in the mathematical consideration of mechanics, and it is also perfectly warranted by experience. The contrary supposition is so highly improbable, that the principle may almost as justly be termed a necessary axiom, as a phenomenon collected from observation.

Sir Isaac Newton observes, in his third law of motion, that "reaction is always contrary and equal to action, or, that the mutual actions of two bodies are always equal, and directed contrary ways." He proceeds, "if any body draws or presses another, it is itself as much drawn or pressed. If any one presses a stone with his finger, his finger is also pressed by the stone. If a horse is drawing a weight tied to a rope, the horse is also equally drawn backwards towards the weight: for the rope, being distended throughout, will in the same endeavour to contract, urge the horse towards the weight, and the weight towards the horse, and will impede the progress of the one as much as it promotes the advance of the other." Now although Newton has always applied this law in the most unexceptionable manner, yet it must be confessed that the illustrations here quoted are clothed in such language as to have too much the appearance of paradox. When we say that a thing presses another, we commonly mean, that the thing pressing has a tendency to move forwards, into the place of the thing pressed, but the stone would not sensibly advance into the place of the finger, if it were removed; and in the same manner we understand, that a thing pulling another has a tendency to recede further from the thing pulled, and to draw this after it; but it is obvious that the weight which the horse is drawing would not return towards its first situation, with the horse in its train, although the exertion of the horse should intirely cease; in these senses, therefore, we cannot say, that the stone presses, or, that the weight pulls, and we have no reason to offend the just prejudices of a beginner, by introducing paradoxical expressions without necessity. Yet it is true in both cases, that if all friction, and all connexion with the surrounding bodies, could be instantaneously destroyed, the point of the finger and the stone would recede from each other, and the horse and the weight would approach each other, with equal quantities of motion. And this is what we mean by the reciprocality of forces, or the equality of action and reaction.

The quantity of action of two attractive or repulsive bodies on each other is partly dependent on their magnitude. When the bodies are of the same kind, their mutual action is in the compound ratio of their bulks; that is, in the ratio of the products of the numbers expressing their bulks. For instance, if two bodies, each containing a cubic inch of matter, attract or repel each other with a force of a grain, and there be two other bodies, the one containing two inches, the other ten, of the same matter, then the mutual attraction or repulsion of these will be expressed by 20 grains; for each of the 10 inches is attracted by each of the two with a force of a grain. And the mutual action of 3 and 10 will be 30, of 4 and 10, 40; so that when one of the bodies remains the same, the attraction will be simply as the bulk of the other. Hence the quantity of matter, in every body surrounding us, is considered as proportional to its weight; for it is inferred from experiment, that all material bodies are equally subject to the power of gravitation towards the earth, and are, in respect to this force, of the same kind. For the apparent difference in the velocity, with which different substances fall through the atmosphere, is only owing to the resistance of the air, as is sometimes shown by an experiment on a feather and a piece of gold, falling in the vacuum of an air pump; but the true cause was known long before the invention of this machine, and it is distinctly explained in the second book of Lucretius:

> "In water or in air when weights descend,
> The heavier weights more swiftly downwards tend.
> The limpid waves, the gales that gently play,
> Yield to the weightier mass a readier way,
> But if the weights in empty space should fall,
> One common swiftness we should find in all."

We are therefore to suppose, that the different weights of equal bulks of different substances, depend merely on the greater or less number of particles contained in a given space, independently of any other characters that may constitute the specific differences of those substances.

In some cases it is necessary to consider the sum of the masses of two bodies, in order to estimate their mutual action, that is, when we wish to know the whole relative motion of two bodies with respect to each other; for here we must add together their single motions with respect to the centre of inertia, which are inversely in the same ratio. This consideration is sometimes of use in determining the action of the sun on the several planets.

If two bodies act on each other with forces proportional to any power of their distance, for instance to the square or the cube of the distance, the forces will also be proportional to the same power of either of their distances from their common centre of inertia. Thus, in the planetary motions, when one body performs a revolution by means of the attractive force of another, this other cannot remain absolutely at rest; but because it is more convenient to determine the effect of the attraction as directed to a fixed point, we consider the force as residing in the common centre of inertia of the two bodies, which remains at rest, as far as the mutual actions of those bodies only are concerned, and it may be shown, that the force diminishes, as the square of the distance of the bodies, either from this point or from each other, increases. The reciprocal forces of two bodies may therefore be considered as tending to or from their common centre of inertia, as a fixed point; but it often happens that, the difference of magnitude being very great, the motion of one of the bodies may be disregarded. Thus we usually neglect the motion of the sun, in treating of the planetary motions produced by his attraction, although, by means of very nice observations, this motion becomes sensible. But it is utterly beyond the power of our senses to discover the reciprocal motion of the earth produced by any terrestrial cause, even by the most copious eruption of a volcano, although, speaking mathematically, we cannot deny that, whenever a cannon ball is fired upwards, the whole globe must suffer a minute depression in its course. The boast of Archimedes was therefore accompanied by an unnecessary condition: "give me," said he, "but a firm support, and I will move the earth;" but granting him his support, he could only have displaced the earth insensibly by the properties of his machines; and without any such support, when he threw rocks upon the ships of Marcellus, he actually caused the walls of Syracuse and the island of Sicily to move northwards, with as much momentum, as carried his projectiles southwards against the Roman armaments.

LECTURE VII.

ON PRESSURE AND EQUILIBRIUM.

WE have now examined the principal cases in which a simple force is employed in the production of motion; it is of equal consequence to attend to the opposition of forces, where they prevent each other's action. A force counteracted by another force, so that no motion is produced, becomes a pressure: thus we continually exert a pressure, by means of our weight, upon the ground on which we stand, the seat on which we sit, and the bed on which we sleep; but at the instant when we are falling or leaping, we neither exert nor experience a pressure on any part.

It was very truly asserted by the antients, that pressure and motion are absolutely incommensurable as effects; for according to the definition of pressure, the force appears to be what is called in logic a potential cause, which is not in a state of activity: and since an interval of time must elapse after the removal of the opposite force, before the first force can have caused any actual motion, this effect of a finite time cannot with justice be conceived to bear any proportion to the pressure, which is as it were a nascent effect only. It is true that a large weight, pressing on a spring, may keep it bent, in exactly the same place, into which a smaller weight, falling on it with a certain velocity, would inflect it: but, to retain a spring in a certain position, and to bend it into that position, are effects absolutely incommensurable; the one being a measure of the constant repulsive force of the spring, bent to a certain point, the other of the sum of the effects of the same spring, in various degrees of flexure, for a certain time. Hence the smallest possible momentum is said to be more than equivalent to the greatest possible pressure: a very light weight, falling from a very minute distance, will force back a very strong spring, although often through an imperceptible space only. But the impulse of a stream of infinitely small particles, like those of which a fluid is supposed to consist, striking an obstacle in a constant succession, may be counteracted by a certain pressure, without producing any finite motion.

Nothing however forbids us to compare two pressures, by considering the initial motions which they would produce, if the opposition were removed; nor is there any difficulty in extending the laws of the composition of motion to the composition of pressure. For since we measure forces by the motions which they produce, it is obvious that the composition of forces is included in the doctrine of the composition of motions; and when we combine three forces according to the laws of motion, there can be no question but that the resulting motion is truly determined in all cases, whatever may be its magnitude; nor can any reason be given why it should be otherwise, when this motion is evanescent, and the force becomes a pressure. The case is similar to that of a fraction, which may still retain a real value, when both its numerator and denominator become less than any assignable quantity. Some authors on mechanics, and indeed the most eminent, Bernoulli, Dalembert, and Laplace, have deduced the laws of pressure, more immediately, from the principle of the equality of the effects of equal causes; and the demonstration may be found, in an improved form, in the article Dynamics of the Supplement of the Encyclopaedia Britannica; but its steps are still tedious and intricate.

We are therefore to consider the momentum, or quantity of motion, which would be produced by any force in action, as the measure of the pressure occasioned by it, when opposed; and to understand by equal or proportionate pressures, such as are produced by forces which would generate equal or proportionate momenta in a given time. And it may be inferred, that two contrary pressures will balance each other, when the momenta, which the forces would separately produce, in contrary directions, are equal; and that any one pressure will counterbalance two others, when it would produce a momentum, equal and contrary to the momentum which would be derived from the joint result of the other forces. For, supposing each of two forces opposed to each other to act for an instant, and to remain inactive for the next equal instant, while the other force is exerted, it is obvious that these effects will neutralise each other, so that the body, on which they are supposed to operate, will retain its situation; but such an action is precisely half of the continued action of each force; consequently, since the halves completely counteract each other, the wholes will do the same. And a similar mode of reasoning may be extended to any number of forces opposed to each other.

It follows from the laws of the composition of motion, that the result of two pressures, expressed by the sides of a parallelogram, will be represented by its diagonal, and that, if a body remain at rest by means of three pressures, they must be related to each other in magnitude as the sides of a triangle parallel to their directions. This may be very completely shown by experiment. We attach three weights to as many threads, united in one point, and passing over three pullies; then by drawing any triangle, of which the sides are in the directions of the threads, or in parallel directions, we may always express the magnitude of each weight, by the length of the side of the triangle corresponding to its thread. (Plate III. Fig. 33.)

The most important of the problems relating to equilibrium are such as concern the machines which are usually called mechanical powers. We are not, however, to enter at present into all the properties and uses of these machines; we have at first only to examine them in a state of rest, since the determination of their motion requires additional considerations, and their application to practice belongs to another subdivision of our subject.

There is a general law of mechanical equilibrium, which includes the principal properties of most of these machines. If two or more bodies, connected together, be suspended from a given point, they will be at rest when their centre of inertia is in the vertical line passing through the point of suspension. The truth of this proposition may easily be illustrated, by the actual suspension of any body, or system of bodies, from or upon a fixed point; the whole remaining in equilibrium, when the centre of inertia is either vertically below the point of suspension, or above the point of support, or when the fixed point coincides with the centre of inertia. And whatever may be the form of a compound body, it may be considered as a system of bodies connected together, the situation of the common centre of the inertia determining the quiescent position of the body. (Plate III. Fig. 34..38.)

Hence the centre of inertia is called the centre of gravity; and it may be practically found, by determining the intersection of two lines which become vertical in any two positions in which the body is at rest. Thus, if we suspend a board of an irregular form from any two points successively, and mark the situation of the vertical line in each position, we may find by the intersection the place of the centre of gravity: and it will appear that this intersection will be the same, whatever positions we employ. (Plate III. Fig. 39.)

The consideration of the degree of stability of equilibrium is of material importance in many mechanical operations. Like other variable quantities, the stability may be positive, negative, or evanescent. The equilibrium is positively more or less stable, when the centre of gravity would be obliged to ascend more or less rapidly, if it quitted the vertical line: the equilibrium is tottering, and the stability is negative, when the centre of gravity would descend if it were displaced; but when the centre of gravity coincides with the centre of motion, or when its path would be a horizontal right line, the equilibrium has been called insensible, but may more properly be termed neutral, and the body will rest in any position, without tending either to fall, or to return to its original situation. It is obvious that the centre of gravity cannot move, without descending, when it is vertically over the fixed point, nor without ascending, when it is immediately below it; so that in the one case the equilibrium is tottering, and in the other stable. Hence we may understand the reason of fixing the moveable handles of a vessel of any kind at its upper part, in order that the centre of suspension may be always above the centre of gravity. If they be fixed too low, the vessel will be liable to overset, unless there be sufficient friction to retain it in its proper situation. (Plate III. Fig. 40.)

An oval surface, placed on a horizontal plane, is capable of a stable equilibrium, when it rests on its side, or on the extremity of its lesser axis, and of a tottering equilibrium, when it stands on the extremity of its greater axis. But the equilibrium of a circle or a sphere is always neutral, for, when disturbed, it neither recovers its first position, nor deviates further from it. A flat body, resting on a sphere, will have its equilibrium tottering or stable, accordingly as its centre of gravity is more or less than the semidiameter of the sphere above the point of contact. (Plate III. Fig. 41, 42.)

The stability of a body supported on a flat basis, of a given extent, is of a different kind, and is independent of equilibrium. For here, if the centre of gravity move either way, it must begin its motion in an inclined direction, instead of describing a curve which is initially horizontal. The stability of such a body becomes less and less, as it is more and more inclined, till, when the centre of gravity is vertically over the margin of the basis, there is a tottering equilibrium; and if the inclination be still further continued, the body will fall. (Plate III. Fig. 43.)

The broader the basis, and the lower the centre of gravity, the steeper must the path of that centre be, and consequently the greater the stability. Thus the disposition of the weight in a carriage may considerably affect its stability, by altering the place of the centre of gravity. A waggon loaded with iron is much less easily overturned, than when it is loaded with an equal weight of hay; supposing the inequality of the road, or any accidental obstacle, to elevate one side of the waggon, it will always recover its position, provided that the centre of gravity remain within the vertical line, passing through the point of contact of the lower wheel and the ground; and it is obvious that the higher the centre of gravity is situated, the sooner it passes this line. If the velocity of the motion were very great, the wheel which is elevated might be lifted off the ground by the momentum, and the centre of gravity might thus be carried beyond the vertical line, by means of an obstacle which would not have overset the waggon, if it had been moving slowly. (Plate III. Fig. 44.)

If a person be sitting or standing in a carriage, the part of the carriage on which he sits or stands may be considered as representing the place of his weight, provided that his situation be always perpendicular; but if the motion be rapid, he will not be able to remain constantly in a posture perfectly erect, and the centre of gravity of the carriage, with its passengers, will be somewhat more elevated, than it would be on this supposition.

The direction of the initial motion of the centre of gravity readily explains the suspension of a weight, or a bucket of water, on a rod resting on the end of a table, when another rod is employed, to keep the bucket at such a distance from the end of the first, that the centre of gravity may be under the table;

for although the bucket seems suspended by its handle, yet if the handle began to descend, the centre of gravity would be obliged to rise; consequently the whole will retain its position, and remain at rest. (Plate III. Fig. 45.)

The apparent ascent of a loaded cylinder on an inclined plane, and the motion of a roller composed of two united cones, with a common axis, resting on the edge of a triangle which is inclined to the horizon, may be easily understood from the same consideration. (Plate III. Fig. 46.)

We may also observe, in the equilibrium of animals, many circumstances illustrative of the properties of the centre of gravity. When a person stands on one foot, and leans forwards, in the attitude which is usually exhibited in the statues of Mercury, the other foot is elevated behind, in order to bring back the centre of gravity, so as to be vertically over some part of the foot on which he stands. But on account of the convex and irregular form of the foot, the basis that it affords is really very narrow; hence when we attempt to stand on one foot, we find it often necessary to use a muscular exertion, in order to bring the point of support to that side towards which we are beginning to fall; and when the basis is still more contracted, the body never remains at rest, but, by a succession of actions of this kind, sometimes too minute to be visible, it is kept in a state of perpetual vibration, without ever attaining such a position as would give it any degree of positive stability; and thus it may be conceived to be supported even on a single point, recovering its position, from time to time, by means of a slight degree of rotatory motion, which is produced by its flexure, and by the changes of the position of the extremities: hence, by habit, the arts of ropedancers and balancers are acquired. Sometimes, however, the position of the balancer is not so difficult to be preserved as it appears, the curvature of the wire in contact with the foot tending materially to assist him.

When we attempt to rise from a seat, we generally draw our feet inwards, in order to bring the point of support into, or near, the vertical line passing through the centre of gravity, and to create a tottering equilibrium, which is favourable for the beginning of motion. And before we rise, we bend the upper part of the body forwards, in order to procure a momentum, capable of carrying the centre of gravity beyond the vertical line, passing through the point of support.

When a horse is walking, the centre of gravity is sometimes supported only by two feet of the same side, yet for a time so short, that its declension towards the other side is easily recovered, after the legs on that side have resumed their activity. Some authors have thought it impossible that a quadruped should stand for an instant with both feet of the same side raised from the earth; but when a horse is walking fast, it may very often be observed, that the print of the hind foot is considerably more advanced than that of the fore foot, which has been raised to make way for it.

From the general law of the equilibrium of the centre of gravity, we may deduce the properties of levers of all kinds. It follows from the definition of this point, that if two bodies be attached to a straight rod of inconsiderable weight, they may be sustained in equilibrium, by a fixed point, or fulcrum, which divides their distance into portions which are inversely as their weights. And it is obvious that if any other equivalent forces be substituted for weights, acting at the same distance from the fulcrum, and with the same inclination to the rod or lever, the conditions of equilibrium will be precisely the same. Also if either of the forces be transferred to an equal distance on the other side of the fulcrum, and act there in a contrary direction, the equilibrium will still remain. Hence we have two principal kinds of levers; the first, in which the fixed point, or fulcrum, is between the points at which the forces or weights are applied; the second, where the forces are applied, in contrary directions, on the same side of the fulcrum. (Plate III. Fig. 47.)

The demonstrations of the fundamental property of the lever have been very various. Archimedes himself has given us two. Huygens, Newton, Maclaurin, Dr. Hamilton, and Mr. Vince, have elucidated the same subject by different methods of considering it. The demonstration of Archimedes, as improved by Mr. Vince, is ingenious and elegant, but it is neither so general and natural as one of Dr. Hamilton's, nor so simple and convincing as Maclaurin's, which it may be worth our while to notice. Supposing two equal weights, of an ounce each, to be fixed at the ends of the equal arms of a lever of the first kind; in this case it is obvious that there will be an equilibrium, since there is no reason why either weight should preponderate. It is also evident that the fulcrum supports the whole weight of two ounces, neglecting that of the lever; consequently we may substitute for the fulcrum a force equivalent to two ounces, drawing the lever upwards; and instead of one of the weights, we may place the end of the lever under a firm obstacle, and the equilibrium will still remain, the lever being now of the second kind. Here therefore, the weight remaining at the other end of the lever counterbalances a force of two ounces, acting at half the distance from the new fulcrum; and we may substitute for this force a weight of two ounces, acting at an equal distance on the other side of that fulcrum, supposing the lever to be sufficiently lengthened, and there will still be an equilibrium. In this case the fulcrum will sustain a weight of three ounces; and we may substitute for it a force of three ounces acting upwards, and proceed as before. In a similar manner the demonstration may be extended to any commensurable proportion of the arms, that is, any proportion that can be expressed by numbers; and it is easy to show that the same law must be true of all ratios whatever, even if they happen to be incommensurable, such as the side of a square, compared to its diagonal, which cannot be accurately expressed by any numbers whatever; the forces remaining always in equilibrium, when they are to each other inversely as the distances at which they are applied.

It is sometimes more convenient to have a series of levers acting on each other, with a moderate increase of power in each, than to have a single lever equivalent in its effect. We may also bend either arm of a lever in any manner that we please, without altering its power, provided that the direction of the force be perpendicular to the line drawn to the fulcrum; or if the force be applied obliquely, it may always be imagined to act at the end of a lever equal in length to the perpendicular let fall from the fulcrum on the direction of the force. Thus, if two levers are connected by a rope or bar, when the direction of one of them nearly coincides with that of the rope, a force applied transversely to the lever acts with a great mechanical advantage against the rope; but as the inclination increases, the advantage gradually diminishes, and changes, at last, to an equal advantage on the side of the rope and the other lever to which it is attached. When therefore a great force is required in the beginning of the motion, and afterwards a much smaller force with a greater velocity, this apparatus may be extremely convenient: thus, in opening a steam valve, the pressure of the steam is at first to be overcome, and after this, little or no additional force is required; and Mr. Watt has very ingeniously applied this arrangement of levers to the purpose in his steam engines. In the same manner, it is necessary that the platten of a printing press, or the part which presses the paper on the types, should descend from a considerable height, but it is only at the instant of taking off the impression that a great force is required; and both these ends are obtained by similar means in a press lately invented by Lord Stanhope. (Plate III. Fig. 48, 49.)

The wheel and axis bear a very strong resemblance to the lever. If two threads, or perfectly flexible and inextensible lines, be wound in contrary directions round two cylinders, drums, or rollers, moveable together on the same axis, there will be an equilibrium, when the weights attached to the threads, or the forces operating on them, are inversely as the radii of the cylinders, or as the diameters of which they are the halves. It may easily be understood, that the weights have the same power in turning round the cylinders, as if they were immediately attached to the arms of a lever, equal in length to their semidiameter, and that the conditions of equilibrium will be the same. The demonstration may also be more immediately deduced from the position of the centre of gravity, immediately below the axis of the cylinders, which requires the weights to be inversely as the radii. With respect to stability, the equilibrium is neutral, and the cylinders will remain at rest in any situation. A single cylinder is also often combined with a lever or winch, and in this case the radius of the cylinder is to be compared with the length of the lever or winch. (Plate III. Fig. 50.)

Systems of wheels and pinions, of various kinds, resemble, in their mechanical properties, either a series of levers, or the combination of cylinders, which constitutes the wheel and axis; but the form of the teeth may produce a difference in their action, which will be mentioned when the practical construction of wheelwork is discussed.

Sometimes the axis connected with a winch is composed of two cylinders, one end of the rope being uncoiled from the smaller, while the other end winds round the larger; the weight being supported by a pulley running in its angle. Here the conditions of equilibrium are easily determined from the place of the centre of gravity, and the effect of the machine is the same, as if the weight were attached to a rope coiled round a simple cylinder, of a diameter equal to half the difference of the diameters of the double axis. The machine is, however, much stronger than such a cylinder would be, and does not require so great a curvature in the ropes employed. (Plate IV. Fig. 51.)

The laws of the equilibrium of pullies have been referred, by some writers on mechanics, to those of the lever; but the comparison is both unnecessary and imperfect; in the simple case of two equal weights attached to a thread passing over a single pulley, which is the only one that allows us to recur to the properties of the lever, the conditions of equilibrium are axiomatically evident, without any further reasoning; and in more complicated cases, the calculations proceed on perfectly different grounds. We are, therefore, to consider a pulley as a cylinder, moving on an axis, merely in order to change the direction of a thread, without friction; for whatever is demonstrable of pullies or their combinations, would be equally true of as many perfectly smooth grooves, which do not bear the most distant analogy to the lever.

Now when the direction of a thread is altered, by passing over any perfectly smooth surface, it communicates the whole force acting on it; for the resistance of a surface, without friction, can only be in a direction perpendicular to itself and to the thread, and the operation of any force remains undisturbed by a resistance which is always in a direction perpendicular to it.

A fixed pulley, therefore, has no effect in gaining a mechanical advantage; but by means of a moveable pulley, it is obvious that a weight may be supported by two forces, each equivalent to half the weight, applied in a vertical direction to the extremities of the thread; and these forces may be derived from two weights, if the thread be made to pass over two fixed pullies in a proper position; and if one of the ends be attached to a fixed point, and the other remain connected to its weight, the equilibrium will continue unimpaired, each portion of the thread still supporting one half of the original weight; so that, by means of a single moveable pulley, one body may retain in equilibrium another of double its weight. (Plate IV. Fig 52, 53.)

The modes of arranging pullies are very various, but the advantage which they procure may always be estimated, from the consideration that every part of the same thread must be equally stretched; and where there is only one thread, the weight will be divided equally among all the portions which help to support the moveable block, each of them bearing a weight equivalent to the force which is applied at the end of the thread. In the common ship's blocks, the pullies or shieves are equal in magnitude, and placed side by side; here their number cannot conveniently exceed two or three, without causing an obliquity in the block, when the force is applied to the rope. Mr. Smeaton, for this reason, invented a system of pullies, arranged in two rows in each block, one larger, and the other smaller: the force being applied in the middle, the rope passes on the larger pullies, till it arrives at the last, then returns through the whole of the smaller series, to the opposite side, and comes back again on the larger, to be finally attached in the middle. (Plate IV. Fig. 54 ... 56.)

If the diameters of all the pullies, in both blocks, be taken in the ratio of the number of portions of the thread intervening between them and the fixed extremity, their angular velocity will be equal, each of them turning on its axis in the same time. They may therefore be fixed to a single axis in each block; and in this case the axis being longer, there will be less accidental friction from its want of steadiness, and even the necessary friction may, perhaps, be somewhat diminished. (Plate IV. Fig. 57.)

If one end of a thread, supporting a moveable pulley, be fixed, and the other attached to another moveable pulley, and the threads of this pulley be similarly arranged, the weight will be counterpoised by a power, which is found by halving it as many times as there are moveable pullies; for it is obvious that each of these pullies doubles the effect of the power. (Plate IV. Fig. 58.)

There are also other arrangements, by which the effect of pullies may be increased or diversified: for instance, where one end of each rope is attached to the weight to be moved; or where two of the pullies are connected by a rope passing over a third; but these methods are of little practical utility. (Plate IV. Fig. 59, 60.)

We have hitherto supposed the ropes, passing over the pullies, to be either perfectly or very nearly parallel to each other; but when their directions are oblique, the forces applied to them require to be modified accordingly. Thus, if two threads be attached to a weight, and passed over two pullies, fixed at a distance from each other, so that two equal weights may be attached to their extremities, the depression of the first weight below either pulley, will be to its distance from the pulley, in the same proportion as half of the weight to either of the other weights; and if, instead of having a weight attached to it, one end of the thread be fixed to a firm obstacle, the effect will be precisely the same. A machine of this kind is sometimes called a swig, perhaps by corruption from swing. (Plate IV. Fig. 61.)

If all the weights are unequal, we must draw a triangle, of which the three sides are in the same proportions as the weights; and we may determine the directions of the threads, by placing such a triangle, with the side, representing the middle weight, in a vertical position.

A force may also be applied obliquely to a wheel and axis. Supposing a rope to be coiled obliquely round the axis, it will require, in order to preserve the equilibrium, a force as much greater than would be sufficient, if it were simply applied in the direction of the motion, as the length of any part of the rope uncoiled is greater than the perpendicular distance of its extremity from the axis. So that when the rope becomes very oblique, a great force is required in order to counteract a much smaller one acting perpendicularly. This remark may be in some measure illustrated by considering the method used by joiners and stonecutters for keeping a saw straight: two ropes or braces are twisted together by means of a pin or lever passing between them, and serve each other in place of an axis, round which they are coiled obliquely, so that they act with great force, when they are sufficiently tight, and not too much twisted. (Plate IV. Fig. 62.)

It appears from the laws which have already been laid down, respecting the motions of bodies on inclined surfaces, that a weight, acting vertically, will hold in equilibrium another weight, resting on an inclined plane, without friction, when the first is to the second as the height of the plane to its oblique length. The pressure on the plane is in this case to the weight resting on it, as the horizontal length of the plane is to its oblique length. This pressure may be measured experimentally, by substituting for the resistance of the plane, that of a thread perpendicular to it. (Plate IV. Fig. 63.)

The same principles are applicable to the equilibrium of the wedge. A wedge is a solid which has three plane faces inclined to each other, and two triangular ends; and we suppose the faces perfectly polished, so as to be free from friction, and that no force can act on them otherwise than in a perpendicular direction. Now in order that three forces, acting on the faces or sides of a wedge, may hold each other in equilibrium, each of them must be in proportion to the length of the side on which it acts: they must also be applied at such parts that their directions may meet in one point; for otherwise they will not be completely opposed to each other, and a rotatory motion will be produced. (Plate IV. Fig. 64.)

If each face of the wedge were conceived to be capable of receiving a pressure, not only in a perpendicular direction, but in any other direction at pleasure, as some authors have supposed, the instrument would lose its essential character as a wedge; but in such cases, the proportion of the forces required for the state of equilibrium, may always be determined by drawing a triangle with its sides parallel to their directions.

It happens, however, not uncommonly, that the force actually operating on the wedge is derived from another force, acting in a direction more or less oblique, as when a heavy body rests on one of the faces of the wedge which is inclined to the horizon, the body being retained in its situation, by an obstacle or a thread which confines it to a vertical line, and the sliding away of the wedge being prevented by a horizontal force. A wedge so situated, and supposed to be capable of sliding without friction on a horizontal surface, is sometimes called a moveable inclined plane, and it will support the weight resting on it, if the horizontal force be to the weight, as the height of the plane is to its horizontal length. If the thread, or the obstacle helping to support the weight, be placed in any other direction, the magnitude of the forces must be determined from the general law of the composition of three pressures. (Plate IV. Fig 65.)

If a prop or bar, leaning against a smooth vertical surface or wall, be employed to support or to raise a weight, by means of a force which draws its base along a smooth horizontal surface, the horizontal force must be to the weight as the distance of the bottom of the prop from the wall to its perpendicular height. And from similar principles, the conditions of the equilibrium of arches, domes, and roofs may be determined. (Plate IV. Fig. 66, 67.)

The action of a screw depends on the same principles as that of an inclined plane; for by rolling a thin and flexible wedge, for instance, a triangular piece of card, round a cylinder, we form a screw. We may consider the force tending to turn the screw round its axis, as applied horizontally to the base of the wedge, and the weight which is to be raised as acting vertically on its inclined surface: the circumference of the cylinder will represent the horizontal length of the wedge, and the distance between the threads, measured in the direction of the axis, will be its height, provided that the threads be single; consequently, the forces required for the equilibrium are to each other, as the height of one spire to the circumference of the screw. But besides these forces, it is necessary that some obstacle be present, which may prevent the body, on which the screw acts, from following it in its motion round its axis, otherwise there can be no equilibrium. (Plate V. Fig. 68.)

The cylinder, which is the foundation of a screw, may be either convex or concave, making a cylindrical or a tubular screw, and these, when fitted together, are sometimes called a screw and a nut. The nut acts on the screw with the same mechanical power as a single point would do, since it only divides the pressure among the different parts of the spire. In general the screw is applied in combination with a lever, in order to procure an advantage in overcoming the friction, which is always considerable in the simple screw and nut, and which would resist a force applied immediately at the circumference, without any diminution of its power. Sometimes the spires of a screw are made to act on the teeth of a wheel, when a very slow motion of the wheel, or a very rapid motion of the screw, is required for the purposes of the machine. (Plate V. Fig. 69, 70.)

The power of screws may be increased, in a great proportion, by means of an arrangement invented by Mr. Hunter; which is somewhat similar, in its operation, to the double axis already described. A cylindrical screw is bored, and made at the same time a tubular screw, with a little difference in the distances of the threads, so that when it is turned within a fixed nut, it rises or sinks a little more or less than the internal screw which perforates it would rise or sink by the action of its own threads, and a weight attached to this internal screw ascends, in each revolution, only through a space equal to the difference of the height of the two coils. Here the machine is analogous to a very thin wedge, of which the thickness is only equal to the difference of the distances of the threads, and which of course acts with a great mechanical advantage. It might in some cases be more convenient to make two cylindrical screws, of different kinds, at different parts of the same axis, rather than to perforate it. The friction of such machines is, however, a great impediment to their operation. (Plate V. Fig. 71.)

In all the kinds of equilibrium that we have considered, and in all other cases that can be imagined, it will be found that the forces, or rather weights, opposed to each other, are so arranged, that if they were put in motion, their momenta in the direction of gravity would, in the first instance, be equal and contrary, the velocity being as much greater as the magnitude of the weight is smaller. Thus, if an ounce weight, placed on a lever, at the distance of four feet from the fulcrum, counterpoise a weight of four ounces at the distance of one foot, the velocity with which the ounce would descend, if the lever were moved, would be four times as great as that with which the weight of four ounces would descend. A single moveable pulley ascends with half the velocity of the end of the rope which is drawn upwards, and acts with a force twice as great; a block of three shieves enables a weight to sustain another six times as great; but the velocity, with which this weight ascends, is only one sixth of that with which the smaller weight must descend. When a weight rests on an inclined plane, of which the height is one half of the length, it may be retained by the action of a weight of half its magnitude, drawing it up the plane by means of a thread passing over a pulley; here if the weight ascended or descended along the oblique surface, its velocity, reduced to a vertical direction, would be half as great as that of the smaller weight which balances it.

Some authors have considered this law as affording a fundamental demonstration of the conditions of equilibrium in all possible cases. For since, wherever two weights are in equilibrium, if one of them descended, the other must ascend with an equal quantity of motion, it appears absurd to suppose that the force of gravitation could produce these two equal and contrary effects at the same time. But it is more satisfactory to trace, in every case, the steps by which the immediate actions of the different weights are enabled to oppose each other; and the general law may then be inferred, by induction, from the agreement of the particular results, in confirmation of the general reasoning which tends to establish its truth.

LECTURE VIII.

ON COLLISION.

HAVING inquired into the laws and properties of the motions and rest of single bodies, under the operation of one or more forces, and into the equilibrium of these forces, in different circumstances, we are next to examine some simple cases of the motions of various moveable bodies acting reciprocally on each other. In all problems of this kind, it is of importance to recollect the general principle already laid down, respecting the centre of inertia, that its place is not affected by any reciprocal or mutual action of the bodies constituting the system.

Whenever two bodies act on each other, so as to change the direction of their relative motions, by means of any forces which preserve their activity undiminished at equal distances on every side, the relative velocities with which the bodies approach to, or recede from each other, will always be equal at equal distances. For example, the velocity of a comet, when it passes near the earth in its descent towards the sun, is the same as its velocity of ascent in its return, although, at different distances, its velocity has undergone considerable changes. In this case, the force acts continually, and attracts the bodies towards each other; but the force concerned in collision, when a body strikes or impels another, acts only during the time of more or less intimate contact, and tends to separate the bodies from each other. When this force exerts itself as powerfully in causing the bodies to separate, as in destroying the velocity with which they meet each other, the bodies are called perfectly elastic: when the bodies meet each other without a reaction of this kind, they are called more or less inelastic. Ivory, metals, and elastic gum, are highly, and almost perfectly elastic: clay, wax, mixed with a little oil, and other soft bodies, are almost inelastic: and the effects of inelastic bodies may be imitated by elastic ones, if we cause them to unite or adhere after an impulse, so as to destroy the effect of the repulsive force which tends to separate them.

When two bodies approach to each other, their form is in some degree changed, and the more as the velocity is greater. In general, the repulsive force exerted is exactly proportional to the degree in which a body is compressed; and when a body strikes another, this force continues to be increased until the relative motion has been destroyed, and the bodies are for an instant at rest with respect to each other; the repulsive action then proceeds with an intensity which is gradually diminished, and if the bodies are perfectly elastic, they reassume their primitive form, and separate with a velocity equal to that with which they before approached each other. Strictly speaking, the repulsion commences a little before the moment of actual contact, but only at a distance which in common cases is imperceptible. The change of form of an elastic substance, during collision, is easily shown by throwing a ball of ivory on a slab of marble, or a piece of smooth iron, coloured with black lead, or printing ink; or by suffering it to fall from various heights: the degree of compression will then be indicated by the magnitude of the black spot which appears on the ball. It may be shown, from the laws of pendulums, that, on the supposition that the force is proportional to the degree of compression, its greatest exertion is to the weight of a striking body, as the height from which the body must have fallen, in order to acquire its velocity, to half the depth of the impression.

For making experiments on the phenomena of collision, it is most convenient to suspend the bodies employed, by threads, in the manner of pendulums; their velocities may then be easily measured, by observing the chords of the arcs through which they descend or ascend, since the velocities acquired in descending through circular arcs are always proportional to their chords; and for this purpose, the apparatus is provided with a graduated arc, which is commonly divided into equal parts, although it would be a little more correct to place the divisions at the ends of arcs, of which the chords are expressed by the corresponding numbers. (Plate V. Fig. 72.)

The simplest case of the collision of elastic bodies is when two equal balls descend through equal arcs, so as to meet each other with equal velocities. They recede from each other after collision with the same velocities, and rise to the points from which they before descended, with a small deduction for the resistance of the surrounding bodies.

When a ball at rest is struck by another equal ball, it receives a velocity equal to that of the ball which strikes it, and this ball remains at rest. And if two equal balls meet or overtake each other with any unequal velocities, their motions will be exchanged, each rising to a height equal to that from which the other descended.

The effect of collision takes place so rapidly, that if several equal balls be disposed in a right line, in apparent contact with each other, and another ball strike the first of them, they will all receive in succession the whole velocity of the moving ball before they begin to act on the succeeding ones; they will then transmit the whole velocity to the succeeding balls, and remain entirely at rest, so that the last ball only will fly off.

In the same manner, if two or more equal balls, in apparent contact, be in motion, and strike against any number of others placed in a line, the first of the moving balls will first drive off the most remote, and then the second the last but one, of the row of balls which were at rest: so that the same number of balls will fly off together on one side, as descended to strike the row of balls on the other side; the others remaining at rest.

If the line of balls, instead of being loosely in contact, had been firmly united, they would have been impelled with a smaller velocity, and the ball striking them would have been reflected. For when a smaller elastic body strikes a larger, it rebounds with a velocity less than its first velocity, and the larger body proceeds also with a less velocity than that of the body striking it. But if a larger body strikes a smaller, it still proceeds with a smaller velocity, and the smaller body advances with a greater.

The momentum communicated by a smaller elastic body to a larger one is greater than its own, and when the first body is of a magnitude comparatively inconsiderable, it rebounds with a velocity nearly as great as the velocity of its impulse, and the second body acquires a momentum nearly twice as great as that of the first. When a larger body strikes a smaller one, it communicates to it only as much momentum as it loses.

In the communication of motion between inelastic bodies, the want of a repulsive force, capable of separating them with an equal relative velocity, is probably owing to a permanent change of form; such bodies receiving and retaining a depression at the point of contact. When the velocity is too small to produce this change of form, the bodies, however inelastic, may usually be observed to rebound a little.

Bodies, which are perfectly inelastic, remain in contact after collision; they must therefore proceed with the same velocity as the centre of inertia had before collision. Thus, if two equal balls meet, with equal velocities, they remain at rest; if one is at rest, and the other strikes it, they proceed with half the velocity of the ball which was first in motion. If they are of unequal dimensions, the joint velocity is as much smaller than that of the striking ball, as the weight of this ball is smaller than the sum of the weights of both balls. And in a similar manner, the effects of any given velocities in either ball may be determined.

It follows immediately from the properties of the centre of inertia, that in all cases of collision, whether of elastic or inelastic bodies, the sum of the momenta of all the bodies of the system, that is, of their masses or weights multiplied by the numbers expressing their velocities, is the same, when reduced to the same direction, after their mutual collision, as it was before their collision. When the bodies are perfectly elastic, it may also be shown that the sum of their energies or ascending forces, in their respective directions, remains also unaltered.

The term energy may be applied, with great propriety, to the product of the mass or weight of a body, into the square of the number expressing its velocity. Thus, if a weight of one ounce moves with a velocity of a foot in a second, we may call its energy 1; if a second body of two ounces have a velocity of three feet in a second, its energy will be twice the square of three, or 18. This product has been denominated the living or ascending force, since the height of the body's vertical ascent is in proportion to it; and some have considered it as the true measure of the quantity of motion; but although this opinion has been very universally rejected, yet the force thus estimated well deserves a distinct denomination. After the considerations and demonstrations which have been premised on the subject of forces, there can be no reasonable doubt with respect to the true measure of motion; nor can there be much hesitation in allowing at once that since the same force, continued for a double time, is known to produce a double velocity, a double force must also produce a double velocity in the same time. Notwithstanding the simplicity of this view of the subject, Leibnitz, Smeaton, and many others, have chosen to estimate the force of a moving body, by the product of its mass into the square of its velocity; and though we cannot admit that this estimation of force is just, yet it may be allowed that many of the sensible effects of motion, and even the advantage of any mechanical power, however it may be employed, are usually proportional to this product, or to the weight of the moving body, multiplied by the height from which it must have fallen, in order to acquire the given velocity. Thus a bullet, moving with a double velocity, will penetrate to a quadruple depth in clay or tallow: a ball of equal size, but of one fourth of the weight, moving with a double velocity, will penetrate to an equal depth: and, with a smaller quantity of motion, will make an equal excavation in a shorter time. This appears at first sight somewhat paradoxical: but, on the other hand, we are to consider the resistance of the clay or tallow as a uniformly retarding force, and it will be obvious, that the motion, which it can destroy in a short time, must be less than that which requires a longer time for its destruction. Thus also when the resistance, opposed by any body to a force tending to break it, is to be overcome, the space through which it may be bent, before it breaks, being given, as well as the force exerted at every point of that space, the power of any body to break it is proportional to the energy of its motion, or to its weight multiplied by the square of its velocity.

In almost all cases of the forces employed in practical mechanics, the labour expended in producing any motion, is proportional, not to the momentum, but to the energy which is obtained; since these forces are seldom to be considered as uniformly accelerating forces, but generally act at some disadvantage, when the velocity is already considerable. For instance, if it be necessary to obtain a certain velocity, by means of the descent of a heavy body from a height, to which we carry it by a flight of steps, we must ascend, if we wish to double the velocity, a quadruple number of steps, and this will cost us nearly four times as much labour. In the same manner, if we press with a given force on the shorter end of a lever, in order to move a weight at a greater distance on the other side of the fulcrum, a certain portion of the force is expended in the pressure which is supported by the fulcrum, and we by no means produce the same momentum, as would have been obtained, by the immediate action of an equal force, on the body to be moved.

An elastic ball, of 2 ounces weight, moving with a velocity of 3 feet in a second, possesses an energy, as we have already seen, which may be expressed by 18. If it strike a ball of 1 ounce which is at rest, its velocity will be reduced to 1 foot in a second, and the smaller ball will receive a velocity of 4 feet: the energy of the first ball will then be expressed by 2, and that of the second by 16, making together 18, as before. The momentum of the larger ball after collision is 2, that of the smaller 4, and the sum of these is equal to the original momentum of the first ball.

Supposing the magnitude of an elastic body, which is at rest, to be infinite, it will receive twice the momentum of a small body that strikes it; but its velocity, and consequently its energy, will be inconsiderable, since the energy is expressed by the product of the momentum into the velocity. And if the larger body be of a finite magnitude, but still much greater than the smaller, its energy will be very small; that of the smaller, which rebounds with a velocity not much less than its original velocity, being but little diminished. It is for this reason, that a man, having a heavy anvil placed on his chest, can bear, without much inconvenience, the blow of a large hammer striking on the anvil, while a much slighter blow of the hammer, acting immediately on his body, would have fractured his ribs, and destroyed his life. The anvil receives a momentum nearly twice as great as that of the hammer; but its tendency to overcome the strength of the bones, and to crush the man, is only proportional to its energy, which is nearly as much less than that of the hammer, as four times the weight of the hammer is less than the weight of the anvil. Thus, if the weight of the hammer were 5 pounds, and that of the anvil 100, the energy of the anvil would be less than one fifth as great as that of the hammer, besides some further diminution, on account of the want of perfect elasticity, and from the effect of the larger surface of the anvil, in dividing the pressure occasioned by the blow, so as to enable a greater portion of the chest to cooperate in resisting it.

When a body strikes another, in a direction which does not pass through its centre of gravity, the effect produced involves the consideration of rotatory motion, since in this case the body is made to revolve on an axis. But this can never happen when the body is spherical, and its surface perfectly polished; since every impulse must then be perpendicular to the surface, and must consequently be directed to the centre of the body. If the motion of a ball, which strikes another, is not directed to its centre, the surface of contact must be oblique with respect to its motion, and the second ball will only receive an impulse in a direction perpendicular to this surface, while the first receives, from its reaction, an equal impulse in a contrary direction, which is combined with its primitive motion. The magnitude of this impulse may be determined by resolving the motion of the first ball into two parts, the one parallel to the surface of contact, and the other perpendicular; the first part remaining always unaltered, the second being modified by the collision. If, for example, the balls were equal, this second part of the motion would be destroyed, and the remaining motion would be in the direction of the surface of contact, and perpendicular to that of the ball impelled.

Hence it follows, that if we wish to impel a billiard ball in a given direction, by the stroke of another ball, we have only to imagine a third ball to be placed in contact with the first, immediately behind it in the line of the required motion, and to aim at the centre of this imaginary ball: the first ball will then be impelled in the required direction, and the second will also continue to move in a direction perpendicular to it.

By a similar resolution of the motion of an elastic ball, we may determine its path, when it is reflected from a fixed obstacle. That part of the motion, which is in a direction parallel to the surface of the obstacle, remains undiminished: the motion perpendicular to it is changed for an equal motion in a contrary direction, and the joint result of these constitutes a motion, in a direction, which is equally inclined to the surface, with the first motion, but on the opposite side of the perpendicular. Of this we have also a familiar instance in the motions of billiard balls; for we may observe, that a ball rebounds from the cushion, in an angle equal to that in which it arrives at it; and if we wish that our ball, after reflection, should strike another, placed in a given situation, we may suppose a third ball to be situated at an equal distance, on the other side of the cushion, and aim at this imaginary ball: our ball will then strike the second ball, after reflection, with a direct impulse. We here suppose the reflection to take place when the centre of the ball arrives at the cushion, while in fact the surface only comes into contact with it; if we wish to be more accurate, we may place the imaginary ball, at an equal distance beyond the centre of a ball, lying in contact with the nearest part of the cushion, instead of measuring the distance from the cushion itself. (Plate V. Fig. 73.)

When the number of bodies, which meet each other, is greater, and their magnitudes and motions are diversified, the calculation of the effects of collision becomes very intricate, and the problem is scarcely applicable to any practical purpose. Those who are desirous of pursuing the investigation as a mathematical amusement, will find all the assistance that they require in the profound and elegant works of Maclaurin.

LECTURE IX.

ON THE MOTIONS OF CONNECTED BODIES.

THE motions of single bodies, acting in any manner on each other, which we have been considering, as far as they belong to the effects of collision, are of less importance to practical mechanics, than the affections of such bodies as are united, so as either to revolve round a common centre, or to participate in each other's motions, by any kind of machinery.

It is only within half a century, that the phenomena and effects of rotatory motion have been sufficiently investigated. Newton committed a mistake, which is now universally acknowledged, in his computation of the precession of the equinoxes, for want of attending sufficiently to the subject; and it is of importance in the calculation of many of the effects of mechanical arrangements, that it should be treated in an accurate manner.

The effect of a moving body, in producing motion in any other bodies, so connected as to be capable of turning freely round a given centre, is jointly proportional to its distance from that centre, and to its momentum in the direction of the motion to be produced. Thus a body, of one pound weight, moving with a velocity of one foot in a second, will have three times as great an effect on a system of bodies, to which its whole force is communicated, at the distance of one yard from the centre of their motion, as if it acted only at the distance of a foot, on the same system of bodies: a double weight, or a double velocity, would also produce a double effect. For, supposing two unequal bodies to be connected by an inflexible line, and to move with equal velocities, in a direction perpendicular to that of the line, it is demonstrable, from the principles of the composition of motion, that they may be wholly stopped by an obstacle applied to the centre of gravity, consequently their effects, in turning the line round this point, are equal; here the momenta are proportional to the weights, but the products obtained by multiplying them by the distances from the centre, at which they act, are equal: these products therefore represent the rotatory power of the respective bodies. Hence in a connected system of bodies, revolving round a given point, with equal angular velocities, the effect produced by the rotatory motion of each body, as well as the force which is employed in producing it, is expressed by the product of the mass multiplied by the square of the velocity, since the velocity is in this case proportional to the distance from the centre; and this product is the same that I have denominated the energy of a moving body.

These propositions are of great use in all inquiries respecting the operations of machines; and it is of importance to bear in mind, that although the equilibrium of a system of bodies is determined by the equality of the products of their weights, into their effective distances on each side of the centre, yet that the estimation of the mechanical power of each body, when once in motion, requires the mass to be multiplied by the square of the distance, or of the velocity. For this reason, together with some others, which have been already mentioned, some have considered the square of the velocity as affording the true measure of force; but the properties of motion, concerned in the determination of rotatory power, are in reality no more than necessary consequences of the simpler laws, on which the whole theory of mechanics is founded.

The effects of rotatory motion may be very conveniently examined, by means of an apparatus, similar to that which was employed for the same purpose by Mr. Smeaton. A vertical axis is turned by a thread passing over a pulley, and supporting a scale with weights; the thread may be applied at different parts of the axis, having different diameters, and the axis supports two arms, on which two leaden weights are fixed, at distances which may be varied at pleasure. The same force will then produce, in the same time, but half the velocity, in the same situation of the weights, when the thread is applied to a part of the axis of half the diameter: and if the weights are removed to a double distance from the axis, a quadruple force will be required, in order to produce an equal angular velocity in a given time. (Plate V. Fig. 74.)

When a number of connected bodies, or a single body of considerable magnitude, is made to revolve round a centre, it is sometimes necessary to inquire, into what point their masses might be supposed to be concentrated, so as to preserve the same rotatory power, with the same angular velocity. This point is called the centre of gyration. In a circle, or any portion of a circle, turning round its centre, the square of the distance of this point, from the centre, is half the square of the semidiameter; and the whole effect of the momentum of the circle, upon an obstacle at its circumference, is exactly half as great as that of an equal quantity of matter, striking the obstacle with the velocity of the circumference.

There is another point, of which the determination is of considerable utility in many mechanical problems: this is the centre of percussion; or the point at which an obstacle must be applied, in order to receive the whole effect of a stroke of a body, which is revolving round a given centre, without producing any pressure, or strain, on the centre, or axis of motion. In a straight line, or a slender rod, fixed at one extremity, the distance of this point, from the centre of motion, is two thirds of the whole length.

The same point is also the centre of oscillation, the distance of which determines the time of oscillation, or vibration, of the body, suspended as a pendulum, upon the given centre of motion. It may easily be shown, that a rod a yard long, and of equable thickness, suspended at one extremity, vibrates in the same time as a ball suspended by a thread, of which the length is two feet. But if the rod were suspended on a centre, at some point within its extremities, the time of its vibration would be prolonged, so as to become equal to that of a simple pendulum of much greater length. This may be illustrated by two balls, fixed at the end of a rod, with a centre of suspension moveable to any part of the rod, for as the centre approaches the middle of the rod, the vibrations are rendered extremely slow. (Plate V. Fig 75.)

The rotatory motion of bodies, not fixed on an axis, might be considered in this place, but the subject involves in its whole extent some intricacy of calculation, and, except in astronomy, the investigation is scarcely applicable to any problems which occur in practice. We may, however, examine a few of the simplest cases. If two bodies be supposed to be connected by an inflexible line, and to be moving with equal velocities in parallel directions; if an immoveable obstacle be applied, so as to form a fulcrum, at the common centre of gravity, they will, as we have already seen, be wholly stopped: but if the fulcrum be applied to any other part of the line, one of the bodies will move forwards, and the other backwards, with a velocity which may easily be determined by calculating their rotatory power with respect to the fulcrum. If the fulcrum be applied at a point of the line continued beyond the bodies, the one will lose and the other gain velocity, since the quantity of rotatory power will always remain unaltered: that point only which is denominated the centre of oscillation retaining its original velocity. Now the same inequality in the motion of the bodies, and consequently the same angular velocity of rotation will be produced, if the connected bodies be initially at rest, and the fulcrum be applied to them with the same relative velocity. For example, if a straight rod or wire receive an impulse at one end in a transverse direction, the centre of oscillation, which is at the distance of two thirds of the length from the end struck, will at the first instant remain at rest, consequently the centre will move with one fourth of the velocity of the impulse, and this must be the velocity of the progressive motion of the rod, since the centre of gravity of any body, which is at liberty, moves always with an equable velocity in a right line, while the whole rod will also revolve equably round its centre, except such retardations as may arise from foreign causes. In a similar manner the computation may be extended to bodies of a more complicated form. Thus it has been calculated at what point of each planet an impulse must have operated, in order to communicate to it at one blow its rotation and its progressive motion in its orbit.

Those who have asserted that the motion of the centre of gravity of a body can only be produced by an impulse, which is either wholly or partly directed towards it, have obviously been mistaken. The centre of oscillation is the only point which remains at rest with regard to the first effect of the stroke, and the centre of gravity, which never coincides with the centre of oscillation, moves in the direction of the impulse, while the parts beyond the centre of oscillation begin to move in a contrary direction. Hence it is, that a thin stick may be broken, by a blow on the middle, without injuring the glasses on which it is supported: for the ends of the stick, instead of being depressed by the stroke, would rise with half the velocity of the body which strikes them, if the two portions were separated without the loss of any force. But unless some art has been previously employed in producing a partial separation, it will frequently be found, that the stick has strength enough to break the glasses before it gives way.

The subject of preponderance, or of the action of weights or forces counteracted by other forces, and incumbered with foreign matter to be put in motion, requires for its discussion a previous knowledge of the simple operation of forces, of the conditions of equilibrium, and of the estimation of rotatory power. The consideration of the effects of preponderance enables us to determine, in some circumstances, the best possible proportions of the powers of machines, for producing the required effects in the most advantageous manner. For, in order that motion may be produced, it is not sufficient that there be an equilibrium, in procuring which a part only of the power is expended, but there must be an excess of force above that which would be necessary for the equilibrium; and it is often of consequence to know what portion of the power must be employed in each way, in order that the greatest effect may be produced in a given time. We are sometimes told, that what we gain in power, we lose in time. In one sense indeed the remark is true; thus one man can do no more by a powerful machine in ten hours, than ten men can do by a weaker machine in one hour; but in other senses the assertion is often erroneous; for by increasing the mechanical advantage to a given degree, we may in some cases considerably increase the performance of a machine, without adding to the force.

According to the nature of the force employed, and to the construction of the machine, a different calculation may be required for finding the best proportions of the forces to be employed; but a few simple instances will serve to show the nature of the determination. Thus, in order that a smaller weight may raise a greater to a given vertical height, in the shortest time possible, by means of an inclined plane, the length of the plane must be to its height, as twice the greater weight to the smaller, so that the acting force may be twice as great as that which is simply required for the equilibrium. This may be shown experimentally, by causing three equal weights, supported on wheels, to ascend at the same time as many inclined planes of the same height, but of different lengths, by means of the descent of three other equal weights, connected with the former three, by threads passing over pullies. The length of one of the planes is twice its height, that of another considerably more, and that of a third less: if the weights begin to rise at the same time, the first will arrive at the top, before either of the others. (Plate V. Fig. 76.)

If a given weight, or any equivalent force, be employed to raise another equal weight, by means of levers, wheels, pullies, or any similar powers, the greatest effect will be produced, if the acting weight be capable of sustaining, in equilibrium, a weight about twice and a half as great as itself. This proposition may be very satisfactorily illustrated by an experiment. Three double pullies being placed, independently of each other, on an axis, round which they move freely, the diameters of the two cylindrical portions, which compose the first, being in the ratio of 3 to 2, those of the second as 5 to 2, and those of the third as 4 to 1, six equal weights are attached to them in pairs, so that three may be raised by the descent of the other three, on the principle of the wheel and axis. If then we hold the lower weights, by means of threads, or otherwise, and let them go, so that they may begin to rise at the same instant, it will appear evidently that the middle pulley raises its weight the fastest; and consequently, that in this case, the ratio of 5 to 2 is more advantageous, than either a much less, or a much greater ratio. If the weight to be raised were very great in proportion to the descending weight, the arrangement ought to be such, that this weight might retain, in equilibrium, a weight about twice as great as that which is actually to be raised. If the descending weight were a hundred times as great as the ascending weight, the greatest velocity would be obtained in this case, by making the descending weight capable of holding in equilibrium a weight one ninth as great as itself. (Plate VI. Fig. 77.)

The proportion required for the greatest effect is somewhat different, when the heights, through which both the weights are to move, are limited, as they usually must be in practical cases. Here, if we suppose the operation to be continually repeated, the effect will be greatest in a given time, when the ascending weight is between two thirds and one half, of the exact counterpoise to the descending weight. If, however, the force were accumulated during the action of the machine, there would be no limit to the advantage of a slow motion. Thus, if we have a stream of water, filling a single reservoir, which is to raise a weight by means of its descent, the proportion here assigned will be the best for performing the most work in a given time; but if we chose to double our machine, so that one reservoir should be filled during the descent of another, it would be proper to proportion the weights in such a manner, that the whole time required for filling one of the reservoirs should be occupied in the descent and the reascent of the other.

In all these cases, if great accuracy were required, it would be necessary in the calculation to add to the mass to be moved, the quantity of moveable matter in the machine, reduced to a mean distance from the fulcrum or centre, according to its rotatory power, in the same manner as the centre of gyration is determined. But there is seldom occasion for such a degree of precision. The magnitude of the pressure which is exerted on the fulcrum, during the motion of the connected bodies, may always be determined, by comparing the actual velocity of the centre of gravity with that of a body descending without resistance.

These propositions and experiments must be allowed to require an attentive consideration from those who are engaged in practical mechanics; and it is natural to suppose that the proportions laid down may be adopted with safety, and employed with success, and that we may sometimes derive important advantages from their application. But on more mature consideration, we shall find some practical reasons for caution in admitting them without material alterations.

If a machine were constructed for raising a solid weight, and so arranged as to perform its office in the shortest possible time with a given expense of power, the weight would still possess, when it arrived at the place of its destination, a considerable and still increasing velocity: in order that it might retain its situation, it would be necessary that this velocity should be destroyed; if it were suddenly destroyed, the machinery would undergo a strain which might be very injurious to it: and if the velocity were gradually diminished, the time would no longer be the same as is supposed in the calculation. In the second place, the forces generally employed are by no means uniformly accelerating forces, like that of gravitation, to which the propositions which we have been considering are adapted: they are not only less active when a certain velocity has once been attained, but they are often capable of a temporary increase or diminution of intensity at pleasure. We have seen the inconvenience of producing a great final velocity, on account of its endangering the structure of the machine: if therefore our permanent force be calculated according to the common rule, so as to be able to maintain the equilibrium, and overcome the friction, the momentum or inertia of the weights, when once set in motion, will be able to sustain that motion equably; and it will not be difficult to give them a sufficient momentum, by a greater exertion of the moving force, for a short space of time, at the beginning: and this is in fact the true mode of operation of many machines where animal strength is employed. Other forces, for instance those of wind and water, regulate themselves in some measure, at least with respect to the relative velocity of the sails and the wind, or the floatboards and the water; for we may easily increase the resistance, until the most advantageous effect is produced. Many authors, considering the pressure of a stream of water as analogous to the impulse of a number of unconnected particles, striking the floatboards, and then ceasing to produce any further effect, have inferred, that the force obtained by such an impulse must be as the square of the relative velocity, and that the effect of an undershot wheel must be the most advantageous, when its velocity is one third of that of the stream: but it will hereafter appear, that this estimation of hydraulic force is by no means accurate. If we compare the greatest velocity with which a man or a horse can run or walk without fatigue, to the velocity of the stream, and the actual velocity of that part of the machine to which the force is applied, to the velocity of the floatboards of a water wheel, the strength which can be exerted may be represented, according to the experiments of some authors, by the impulse of the stream, as supposed to be proportional to the square of the relative velocity; consequently the same velocity would be most advantageous in both cases, and the man or horse ought, according to these experiments, to move, when his force is applied to a machine, with one third of the velocity with which he could walk or run when at liberty. This, for a man, would be about a mile and a half an hour; for a horse, two or three miles: but in general both men and horses appear to work most advantageously with a velocity somewhat greater than this.

Where a uniformly accelerating force, like that of gravitation, is employed in machines, it might often be of advantage to regulate its operation, so that it might act nearly in the same manner as the forces that we have been considering; at first with greater intensity, and afterwards with sufficient power to sustain the equilibrium, and overcome the friction only. This might be done, by means of a spiral barrel, like the fusee of a watch; and a similar modification has sometimes been applied, by causing the ascending weight, when it arrives near the place of its destination, to act on a counterpoise, which resists it with a force continually increasing, by the operation of a barrel of the same kind, so as to prevent the effect of the shock which too rapid a motion would occasion.

On the whole, we may conclude, that on account of the limited velocity which is usually admissible in the operation of machines, a very small portion of the moving force is expended in producing momentum; the velocity of 3 miles an hour, would be generated in a heavy body, descending by its own weight, in one seventh of a second, and a very short time is generally sufficient for obtaining as rapid a motion as the machine or the nature of the force will allow; and when this has been effected, the whole force is employed in maintaining the equilibrium, and overcoming the resistance: so that the common opinion, which has probably been formed without entering minutely into the consideration of the subject, and which appears, when first we examine its foundation with accuracy, to lead to material errors, is in great measure justified by a more profound investigation.

To seek for a source of motion in the construction of a machine, betrays a gross ignorance of the principles on which all machines operate. The only interest that we can take in the projects which have been tried for procuring a perpetual motion, must arise from the opportunity that they afford us to observe the weakness of human reason; to see a man spending whole years in the pursuit of an object, which a week's application to sober philosophy might have convinced him was unattainable. The most satisfactory confutation of the notion of the possibility of a perpetual motion, is derived from the consideration of the properties of the centre of gravity: we have only to examine whether it will begin to descend or to ascend, when the machine moves, or whether it will remain at rest. If it be so placed, that it must either remain at rest or ascend, it is clear, from the laws of equilibrium, that no motion derived from gravitation can take place: if it may descend, it must either continue to descend for ever, with a finite velocity, which is impossible, or it must first descend and then ascend, with a vibratory motion, and then the case will be reducible to that of a pendulum, where it is obvious that no new motion is generated, and that the friction and resistance of the air must soon destroy the original motion. One of the most common fallacies, by which the superficial projectors of machines for obtaining a perpetual motion have been deluded, has arisen from imagining, that any number of weights ascending by a certain path, on one side of the centre of motion, and descending in the other, at a greater distance, must cause a constant preponderance on the side of the descent: for this purpose, the weights have either been fixed on hinges which allow them to fall over at a certain point, so as to become more distant from the centre, or made to slide or roll along grooves or planes, which lead them to a more remote part of the wheel, from whence they return as they ascend: but it will appear on the inspection of such a machine, that although some of the weights are more distant from the centre than others, yet there is always a proportionally smaller number of them on that side on which they have the greatest power; so that these circumstances precisely counterbalance each other. (Plate VI. Fig. 78.)

LECTURE X.

ON DRAWING, WRITING, AND MEASURING.

HAVING investigated all the general principles and laws of motion, and of mechanical power, we may now proceed to the consideration of particular departments of practical mechanics. But before we can satisfactorily compare the various forces, which we are to employ or to oppose, we must have some mode of determining their magnitude; and we must begin by examining the spaces which are measures of their action: a knowledge of the instruments employed for delineation, and of the rules of perspective projection, is also necessarily required, as a previous step in the study of practical mechanics. We have therefore to consider, as preliminary subjects, first the arts which may be expressed by the terms instrumental geometry, or the geometry of mechanics; secondly, statics, or the mode of ascertaining the magnitude of weights, and of other active forces; and thirdly, the examination of the passive strength of materials of various kinds, and of the negative force of friction.

The art of drawing can scarcely be distinguished by any correct definition from painting. In its simplest state, when we merely imitate an original laid before us, it is called copying; and in writing, we only copy the letters of the alphabet. If we proceed in a mathematical manner in the operation of drawing, we require a number of geometrical instruments, which are still more necessary for the first construction of diagrams or figures. In modelling and sculpture, a solid is simply imitated; but when a solid is represented on a plane, the principles of perspective are employed in determining the position of the lines which are to form the picture. The productions of the arts of drawing and writing are multiplied and perpetuated by means of engraving and printing; inventions which have been the sources of inestimable advantage in the instruction and civilisation of mankind.

In drawing, we may employ the pen, the pencil, chalks, crayons, inks, water colours, or body colours; we may paint in miniature, in distemper, in fresco, in oils, in varnish, in wax, or in enamel; and we may imitate the effects of painting, by mosaic work, or by tapestry.

The first step in copying a drawing, or in painting, is to procure a correct outline: a master of the art can do this with sufficient accuracy, by such an estimate of the proportions of the figures, as the eye alone enables him to form; especially if he be assisted by lines, which divide the original into a number of squares, and enable him to transfer their contents to the corresponding squares of the copy, which may in this manner be reduced, or enlarged, when it is required. But a copy may sometimes be more expeditiously made, by tracing immediately from the original, when the materials employed are sufficiently transparent to admit the outlines to be seen through them; or, where the original is of no value, by pricking a number of points through it, so as to mark the copy, either at once, or by means of charcoal powder rubbed through the holes, which is called stenciling: and for this purpose, an intermediate copy may be formed on semitransparent paper. Another method is to put a thin paper, rubbed with the powder of black lead, or of red chalk, between the original and the paper intended for the copy, and to pass a blunt point over all the lines to be traced, which produces correspondent lines on the paper; this is called calking. Where the work is large, it may be covered with a thin gauze, and its outlines traced on the gauze with chalk, which is then to be placed on the blank surface, and the chalk shaken off it, in the way that a carpenter marks a board with his line.

The pen was formerly much used for making rough sketches, and it is still sometimes employed for the same purpose, as well as for assisting the effect of the pencil. The appearances of uniform lights and shades must necessarily be imitated in drawings with the pen, as well as engravings, by a mixture of the whiteness of the paper, with the blackness or colour of the ink, the eye being too remote to distinguish minutely the separate lines, by which the effect is produced, although they do not entirely escape its observation. In this respect, drawings in pencils and chalks have an advantage over engravings; these substances, after being laid on in lines, are spread, by means of rubbers, or stumps, of paper, leather, or linen, so as to produce a greater uniformity of tint. Some, indeed, are of opinion, that engravings derive a great brilliancy from the hatches that are employed in shading them, and that minute inequalities of colour make every tint more pleasing. In drawings with chalk, however, the advantage of rubbers is unquestionable. The lines of a drawing may be made to have an appearance of greater freedom than those of an engraving; they should be parallel, and when they are crossed, moderately oblique to each other; their direction should be governed by that of the outline. Engravings in mezzotinto exhibit no lines: but they are deficient in spirit and precision: the effect of aqua tinta approaches much nearer to that of drawing, and it has a similar advantage in the mode of producing its lights and shades. (Plate VI. Fig. 79.)

It is well known, that the best pencils are made of English black lead, or plumbago. Of black chalks, the Italian is harder and more generally useful than the French: red chalk has the disadvantage of not being easily removed, either by bread or by Indian rubber, without leaving a brownish mark. All these chalks are of the nature of a soft schistus or slate: they may be made to adhere firmly to the paper, by dipping the drawings in milk freed from cream, or even in water only, which dissolves the size or gum of the paper. Sometimes a grey paper is used, which serves for a middle tint, and lessens the labour, the lights and shades only being added in white and black chalks.

Crayons consist of colours mixed up with gum water, or other adhesive substances, and usually also with some chalk, plaster, or pipe clay, so as to be of a proper consistence for working in the manner of chalks. The principal inconvenience attending them is their want of adhesion to the paper: the paper must therefore not be too smooth.

For drawings washed in light and shade only, the materials employed are Indian ink, the black liquor of the cuttle fish, or bistre, which is extracted from soot: both these last produce a browner and richer tint than the Indian ink. In using these washes, as well as water colours, there is a great diversity in the methods of different artists: some work with a dry pencil, others with a full one: some begin all their coloured drawings in black only, others use colours from the beginning. When a full pencil is used, care must be taken that no part of the same tint dry sooner or later than the rest. When body colours are employed, there is less difficulty in producing a uniformity of tint than with water colours, each coat of the colour being laid on in sufficient quantity to cover all that is below it without mixing: hence it becomes easier to make any alterations that may be required. For water colours of all descriptions, a certain quantity of gum is used, and sometimes a size made of isinglass, with a little sugarcandy. Body colours contain less gum than other water colours. Besides paper, wood, silk, and cotton velvet, are sometimes used for drawings in water colours.

In miniatures, the most delicate tints are laid on in points, with simple water colours; but for the draperies, body colours are sometimes used. They are commonly executed on ivory.

For painting in distemper, the colours are mixed with a size made by boiling shreds of untanned leather, or of parchment, for several hours: this method is chiefly employed for colouring walls or paper, but sometimes for painting on cloth. For delicate purposes, the size may be made with isinglass.

When a wall or cieling is painted in fresco, the rough coat of the plaster is covered with a coat of fine sand and lime, as far as it can be painted before it is dry, the colours being partly imbibed by this coat, and thus becoming durable. When they have been once laid on, no alteration can be made, without taking off the last coat of plaster, and each part must be completed at once; it is therefore always necessary to have a finished drawing for a copy; this is usually executed on paper, and is called a cartoon. The colours can be only of earths or metallic oxids; they are prepared as for painting in distemper. The only paintings of the antients, which have been preserved, were executed in fresco.

The art of painting in oil was first discovered by Van Eyck of Bruges, towards the end of the 14th century: it has now become almost the only manner in which paintings of magnitude are performed. The colours are mixed with linseed or nut oil, and sometimes with oil of poppy seed, together with a small portion of oil of turpentine, to assist in drying them; and with the occasional addition of other oily and resinous substances. The work may be executed on wood, cloth, silk, paper, marble, or metals: these substances being first washed with size, and then primed with an oil colour, which is usually white, but sometimes dark. Some painters have, however, preferred a ground of distemper. The glare of the oil colours, or of the varnish, which is added in order to give them brilliancy, is considered as an inconvenience attending oil paintings; and some of the colours are too liable to fade or to blacken by the effect of time.

The encaustic paintings of the ancients were imperfect approximations to the art of painting in oil. Wax or resins were employed for retaining the colours in their places; and they were applied by means of a moderate heat. An effect nearly similar is produced by dissolving the resins in spirits of wine, as is done in painting in varnish. A much greater degree of heat is required for paintings in enamel: for this purpose the colours are mixed with a glass of easy fusion, and, when finely powdered, they are usually applied with oil of turpentine, or sometimes oil of lavender, to a ground of metal or porcelain; they are afterwards fixed and vitrified by exposure to the heat of a furnace.

Mosaic work is performed by putting together small pieces of stone, or baked clay, of various colours, so as to imitate the effects of painting: in tapestry, and in embroidery, the same is done by weaving, or working in, threads of different kinds.

The art of writing is of great antiquity, but it is probably in all countries, and certainly in some, of a later date than that of drawing representations of nature. The Mexicans, at the first arrival of the Spaniards in South America, are said to have employed drawings as a mode of conveying intelligence; some of them simply resembling the objects to which they related, others intended as hieroglyphics; that is, like the antient Egyptian characters, of a nature intermediate between drawing and writing. The Chinese have always used arbitrary marks to represent whole words, or the names of external objects, not resembling the objects to which they relate, nor composed of letters appropriated to constituent parts of the sound, although they are said to be combined from a few hundred radical characters expressive of the most simple ideas. The art of writing with alphabetical letters must have been sufficiently understood, in the age of Moses, to serve the purpose of the promulgation of laws and of religion: it is generally supposed to have been invented by the Phenicians. Among the Greeks it was in a very imperfect state until the time of the siege of Troy, or about 3000 years ago. The Chinese write from above downwards, beginning on the right side; the other eastern nations have always written from right to left. The most ancient Greek inscriptions are turned alternately backwards and forwards, the letters being reversed in the lines which begin on the right side; but the Greeks soon confined themselves to that mode, which has been since adopted by all European nations, and which appears to be in itself the most natural, at least for writing with a pen, and with the right hand.

The earliest methods of writing were probably such as rather deserve the name of engraving; the letters being cut in stone, in wood, on sheets of lead, on bark, or on leaves. For temporary purposes, they were formed on tablets of wax, with a point called a stile, and this practice was long continued for epistolary correspondence, and was not wholly out of use in the fourteenth century. The stile was made of metal or of bone; its upper extremity was flattened, for the purpose of erasing what had been written. The Egyptian papyrus is said by Varro to have been first used for writing, at the time of the foundation of Alexandria; the leaves of palms, the inner bark of trees, or sometimes linen cloth, having been before employed. The exportation of the papyrus was forbidden by Ptolemy, and in consequence of this prohibition, skins of parchment, or of vellum, were first applied to the purpose of writing at Pergamus, for the library of king Eumenes, whence they were called membrana pergamena. To make the best paper, the widest and finest leaves of the papyrus were matted together, united by a vegetable glue, and pressed till they became sufficiently smooth; the coarser kinds were not used for writing, but for commercial purposes. In China, paper is sometimes made of a thin and almost transparent membrane taken from the bark of a tree. Paper of cotton was introduced into Europe from the east in the middle ages; it has been since superseded by that which is made of linen rags, and which is also an eastern invention; but for coarse and strong paper, old ropes of hemp are also used; and sometimes many other vegetable substances have been employed. The strength and consistence of paper is owing to the lateral adhesion derived from the intermixture of the fibres, assisted by the glutinous size, which is also of use in obviating the bibulous quality of the paper, by filling up its pores.

Ivory, and prepared ass's skin, are sometimes employed for writing with a black lead pencil; for slates, a pencil of a softer kind of slate is used. The ancient mathematicians usually constructed their diagrams on sand for the instruction of their pupils.

Pens of goose quills, swan's quills, or crow quills, were known as early as the seventh century: in Europe they have generally superseded the reeds, which were employed for writing by the ancients: but in India, reeds, canes, and bamboos, are still in use. In China a hair pencil is used instead of a pen.

The inks of the ancients are said to have been made of a carbonaceous substance, and the modern Indian ink owes its blackness to similar materials. Common writing ink consists of a gallate of iron, suspended by means of a little gum; the sulfuric acid, which remains mixed with it, is probably of no consequence to its blackness. It has been observed, that an abundance of the gallic acid produces a much blacker colour, than is obtained where this acid is used in a smaller proportion. Mr. Ribaucourt's method of making ink, is to boil eight ounces of galls, and four of logwood, in twelve pounds of water, until the quantity is reduced to one half; and, having strained the decoction, to add to it four ounces of sulfate of iron, one of sulfate of copper, three of gum arabic, and one of sugar candy. But for ordinary purposes, it is sufficient to infuse three ounces of galls for a day or two in a pint of water, and to add to it an ounce of gum arabic, half an ounce of green sulfate of iron, or copperas, and a drachm of sulfate of copper, or blue vitriol, or even a much smaller quantity of gum and of copperas, if a very fluid ink is required. The sulfate of copper produces a durable stain, but it does not immediately add to the blackness of the ink: its principal use is to counteract the tendency of the ink to become mouldy. Sometimes a mercurial salt is employed for the same purpose, and a little cotton, if the inkstand is too open, is also useful in preserving the ink; but the addition of spirits is often insufficient, and is liable to make the ink run.

It has been proposed to use inks of different colours for indicating different numbers; so that by ten kinds of ink applied in different ways, any numbers at pleasure might be expressed. Thus, in making an index of the words of an author, each page might be readily covered with lines of different colours drawn in different directions, so that each word, when cut out, might indicate the page to which it belongs.

An ingenious instrument has lately been constructed, by means of which copies may be multiplied with great facility; it is called the polygraph, and consists of two or more pens, so connected by frames and springs, as to move always in parallel directions, each having an inkstand and a sheet of paper for itself. In this manner five copies may be made at once with tolerable facility, and themethod may perhaps hereafter be extended to a much greater number.

A mode of writing, perfectly different from any of those which have been mentioned, is performed by means of the telegraph, which is justly considered as the invention of the ingenious Dr. Hooke. The ancients had attempted something similar, by the exhibition of torches on elevated situations; but Dr. Hooke observes, that the addition of the telescope is absolutely necessary for the practical success of the process; and the directions which he gives for its performance differ very little from the plan which has since been generally adopted, first in France, and afterwards, with some variations, in this country. Dr. Hooke proposed the employment of alphabetical and other arbitrary characters; at present it is usual to have six boards, each turning on its axis so as to appear or disappear at pleasure: these admit of sixty four combinations, which are sufficient, besides indicating the letters of the alphabet, for every other purpose that can be required. (Plate VI. Fig. 80, 81.)

Pens for drawing lines and figures differ sometimes from those which are used for writing; they are made of two plates of steel inclined to each other and adjusted by a screw; or sometimes of a plate of tin folded up, so as to include a receptacle for the ink; or of a glass tube drawn to a very fine point, and still remaining perforated. In all these pens, as well as in common pens, the ink is retained by its cohesion, and by the capillary attraction of the pen; and it attaches itself to the paper by the operation of similar powers.

It is by no means easy to comply strictly with that postulate of geometry, which requires us to draw a straight line from one point to another. The edge of a ruler is made straight by the instrument called a plane, which is worked with a considerable velocity, and therefore naturally tends to move in a right line, besides that it is guided by the flatness of its lower surface. We judge of the straightness of a line, by means of the well known property of light, which moves only in right lines, so that if we look along the edge of a ruler, we easily discover its irregularities; and this may be done with still greater accuracy, if we look through a small hole made with a pin in a card. Rulers of silver, brass, or ivory, have a material advantage over those of wood, as they are not liable to be spoilt by warping. A pen filled with ink cannot be applied close to the edge of a ruler without inconvenience; it is therefore best, for diagrams which require great accuracy, to draw the lines first with a steel point, or a very hard black lead pencil, and to finish them with ink if necessary. The paper should also be fixed on a drawing board; and plates of lead or copper may be employed, instead of paper, for very delicate purposes. The carpenter's chalk line is a useful instrument for supplying the place of a very long ruler; it becomes straight when it is stretched, because a right line is the shortest distance between any two points.

For drawing a circle of a given radius, we use compasses, with one point generally of metal, the other of various descriptions. Compasses are sometimes made with a spring, instead of a joint, and opened or shut by a screw: sometimes a graduated arc is fixed in one leg, and passes through the other; and when great accuracy is required, hair compasses may be employed, having a joint with a spring in one of the legs, which is bent a little by means of a fine screw. Beam compasses are useful for drawing circles of larger radii: they have also the advantage of being steadier than the common compasses, and of admitting readily the application of a graduated scale, so as to indicate the measure of the radius of the circle which is described. Sometimes, for drawing portions of very large circles, two wheels, differing a little in diameter, are fixed on a common axis, and thus made to revolve round a point, which is more or less distant, accordingly as the wheels are set at a greater or less distance on the axis, the surface of the wheels tracing the circles on the paper; or two rulers joined together, so as to form an angle, are made to slide against two points, or edges, projecting from a third ruler, so that the angular point remains always in the arc of a circle. The same effect may be produced, somewhat more commodiously, by means of a thin piece of elastic wood, which is made to assume any required curvature, by the action of screws, applied to different parts of its concavity: it would, however, be more simple and accurate to employ only one screw, in the middle of the arc, and to make the flexible ruler, or bow, every where of such a thickness, as to assume a circular form in its utmost state of flexure: it would then retain the circular form, without a sensible error, in every other position. (Plate VI. Fig. 82 . . 85.)

For drawing a line perpendicular to another, we often employ a square; and if we use a rectangular drawing board, there is an additional convenience in making the square to slide on its margin. Rulers also, of various descriptions, are commonly made rectangular, in order to answer occasionally the same purpose.

Triangular compasses are sometimes used, for laying down a triangle equal to a given triangle; and by repeating the operation, any figure, which can be divided into triangles, may be copied without the intersection of arcs: but the same end is more commonly obtained, by pricking off the figure with a steel point. (Plate VI. Fig. 86.)

Various properties of parallel lines are employed in constructing parallel rulers: a parallelogram with jointed angles is the most commonly used; two equal rulers being united by equal cross bars placed in an oblique position, and turning on pins fixed in the rulers: the instrument is much improved by adding a third ruler, similarly united to the second, for then the obliquity of one of the two motions may be made to correct that of the other. A simple cylinder, or a round ruler, answers the purpose in a rough manner, and two small rollers, fixed on the same axis, are also sometimes attached to a flat ruler, and cause it to move so as to be always in parallel positions. A very useful instrument for drawing parallel lines, at any given distances, is now generally known by the name of Marquois's scales, although it is by no means of late invention; by sliding a triangle along a graduated ruler, we read off the divisions on an amplified scale with great accuracy; but where the distances of the lines are great, the obliquity of this motion is a considerable inconvenience. The ruler or square of the drawing board affords us lines parallel to each other, in a certain position; and if it be made with a joint, or as the workmen call it, bevilled, it may be employed for the same purpose, in all other directions. The systems of lines, on which music is written, are drawn at one stroke by a pen with five orifices, usually made of brass. It was long since proposed to rule a whole page at once, with a more complicated pen of the same kind, and the greatest part of the paper, on which music is written in this country, is actually ruled by such a machine, for which a patent has been taken out. (Plate VI. Fig. 87, 88.)

The pantograph is used for copying figures, and at the same time reducing or enlarging them; it consists of four rulers, two of them united by a joint at the extremities, and receiving at the middle the other two, which are but half as long, and are also united together, so as to form with the others a jointed parallelogram, of which two of the sides are produced beyond the angles; if holes be made in these, and in one of the shorter rulers, so situated as to be in the same right line in any position of the instrument, they will remain in a right line in any other position, and they will always divide this line in the same proportion: so that if one of the holes be placed on a fixed axis or pin, a tracing point inserted in another, and a pencil in the third, any figure delineated by the pencil will be similar to that which is described by the tracing point. And instead of holes in the rulers, they may be furnished with sliding sockets, to receive the axis, the point, and the pencil. (Plate VI. Fig. 89.)

Proportional compasses are also of great use, in reducing lines and figures to a different scale. This instrument consists of two legs, pointed at each end, and turning on a centre, which slides in a groove common to both legs, and is furnished with an index. The divisions of the scale are so laid down, that the centre may divide the length of the legs from point to point in a given proportion; hence, by the properties of similar triangles, when the legs are opened to any extent, the intervals between each pair of points must be to each other in the same ratio as the portions of the legs. Sometimes a screw is added, for the sake of adjusting the centre with greater accuracy; and it is usual to lay down scales for dividing the circumference of a circle into a given number of parts, and for some other purposes; but the instrument might be much improved by inserting, in the common scale, fractional or decimal divisions, between the whole numbers, so that the legs might be divided, for example, in the ratio of 2 to 3, 3 to 4, or 4 to 5, or of 10 to 11, 12 or 13, at pleasure. (Plate VI. Fig. 90.)

The use of the sector depends also on the properties of similar triangles. The scale of equal parts, which is laid down on each leg, beginning from the centre, serves to determine the length of the legs of two equilateral triangles, in any required proportion to each other, according to the division which we mark, and the transverse distances from the corresponding points are necessarily in the same proportion. Thus, if we have any line in a figure which we wish to call three feet, or three inches, we may take the interval with a pair of common compasses, and open the sector to such an angle, that it may extend from the third division of one leg to that of the other; then all the other divisions of the scale will furnish us with the lengths corresponding to any distances that we may wish to lay down. The other scales usually engraved on the sector are principally intended for trigonometrical calculations on similar principles. (Plate VII. Fig. 91.)

The magnitude of angles admits an easy determination and description, by the comparison of the respective arcs with a circle, or with a right angle. We may divide an angle geometrically, by continual bisection, into parts as small as may be required, and by numbering these parts, we may define any angle, with an error smaller than any assignable quantity. Bisections of this kind are sometimes actually employed in the construction of instruments; for instance, in one of the arcs of the mural quadrant of the observatory at Greenwich, the right angle is divided into 96 parts, by the continual bisection of one sixth of the circle. There are also some practical methods of dividing angles into three or more equal parts, which are sufficiently accurate for many purposes, although it is well known that in theory the perfect trisection of an angle is beyond the reach of plain geometry. This trisection is necessary in the common division of the circle into 360 degrees, a number which was probably chosen because it admits a great variety of divisors, and because it nearly represents the diurnal and annual motion of the sun among the stars. The circle being divided into 6 parts, the chord of each of which is equal to the radius, these parts are divided into 60 degrees, each degree into 60 minutes, and each minute into 60 seconds: further than this we cannot easily carry the accuracy of our determination, although, in calculations, we sometimes descend as far as tenths or even hundredths of a second. The decimal division of a right angle, which has been lately adopted in France, appears to have very little advantage for the purposes of calculation, beyond the common method, and its execution in practice must be much more difficult.

Whole circles, or theodolites, divided into degrees and their parts, quadrants and sextants, are usually employed in measuring angles; and protractors, semicircles, and lines of chords, in laying them off. The most convenient of quadrants for general use is Hadley's reflecting instrument, which is in fact an octant or a sextant, but in which, for reasons depending on optical principles, each degree of the arc is reckoned for two.

For the graduation of all instruments of this kind, of moderate dimensions, Mr. Ramsden's dividing engine is of great utility; the instrument being fixed on the revolving plate of the engine, its arc is made to advance under the cutting tool by very minute steps, regulated by the turns of a screw, of which each revolution is divided into a considerable number of equal parts. The largest and finest instruments are, however, still usually divided by hand, that is, by means of compasses. Some artists have first divided a straight plate, and then made a hoop of it, which has served as a standard for further processes. An arc of 7° 10′, of which the chord is one eighth of the radius, may be employed as a test of the accuracy of the work. A micrometer screw is often used in large instruments as a substitute for the minutest divisions; a moveable part of the index being brought to coincide with the nearest point marked in the arc, by turning the screw through a part of its revolution, which is measured by means of a graduated circle. But a simpler method of reading off divisions with accuracy in common instruments, is the application of a vernier, an apparatus so called from its inventor. The space occupied by eleven divisions of the scale being divided into ten parts on the index, the coincidence of any of the divisions of the index with those of the scale, shows, by its distance from the end, the number of tenths that are to be added to that of the intire divisions. (Plate VII. Fig. 92.)

There are several ways of measuring the angular elevation of an object above the horizon; at sea, the apparent horizon, formed by the surface of the water, affords the most convenient determination; but since the spectator is somewhat elevated above the convex surface of the sea, the apparent horizon is necessarily lower than the true horizon, and a correction is therefore required according to the height. In the open sea this correction may be determined by measuring the whole angle above and below the apparent horizon, and taking one fourth of the difference for the dip or depression. On shore, a plumb line is the simplest instrument for determining the situation of the horizon, and its accidental vibrations may be prevented by suspending the weight in water or in oil. For small instruments, a spirit level, of which the operation depends on hydrostatical principles, is capable of greater delicacy than a plumb line. It readily indicates, when well made, an error of a single second, but it requires some attention to avoid inequalities of temperature, which would tend to disturb its figure. Well rectified ether is found, on account of its perfect fluidity, to be the best liquid for a spirit level. An artificial horizon is a reflecting surface, employed for obtaining an image, as much below the horizon, as the object is above it, and for measuring the angular distance of this image from the object: sometimes a plane speculum of glass or metal is used for this purpose, being previously adjusted by a spirit level; and sometimes the surface of mercury, treacle, or tar, protected from the wind by a vessel with holes in it, or by a glass cover, either detached, or simply floating on the mercury, when this liquid is employed.

It is in many cases simpler and more convenient to estimate angles, not by the arcs subtending them, but by their sines, or the perpendiculars falling from one leg on the other. Thus, it is usual among miners, to say that the ground rises or falls one foot, or one yard, in ten, when the sine of the angle of its inclination to the horizon is one tenth of the radius. Angles of different magnitudes are indeed proportional to the arcs, and not to the sines, so that in this sense the sine is not a true measure of the comparative magnitude of the angle; but in making calculations, we are more frequently obliged to employ the sine or cosine of an angle than the angle or arc itself. It is, however, easy to pass from one of these elements to the others by means either of trigonometrical tables, or of the scales engraved on the sector.

The sines, tangents, and secants laid down on the sector, may be employed according to the properties of similar triangles, in the computation of proportions. The same purpose is answered by Gunter's scale, by the sliding rule, and by the logarithmic circles of Clairaut and of Nicholson, which are employed mechanically in the same manner as a table of logarithms is used arithmetically, the proportion of any two numbers to each other being determined by the distance of the corresponding divisions on the scale; so that if we wish to double or to halve a number, we have only to find the distance from 1 to 2, and to lay it off from the given number either way. (Plate VII. Fig. 93, 94.)

The measurement of angles is at once applied to the estimation of distances in the dendrometer or engymeter; a part of the instrument forms a base of known dimensions, and the angle at each extremity of this base being measured with great accuracy, the distance of the object may be inferred from an easy calculation, or from a table. The most complete instruments of this kind have two speculums for measuring the difference of the angles at once, in the manner of Hadley's quadrant. Telescopic scales or micrometers are also sometimes used for measuring angles subtended by distant objects, of which the magnitude is known or may be estimated, for example, by the height of a rank of soldiers, and inferring at once the distance at which they stand.

Arithmetical and even algebraical machines, of a much more complicated nature, have been invented and constructed with great labour and ingenuity; but they are rather to be considered as mathematical toys, than as instruments capable of any useful application.

An angle, when once measured, can be verbally and numerically described, by reference to the whole circle as a unit: but for the identification of the measure of a right line, we have no natural unit of this kind, and it is therefore necessary to establish some arbitrary standard with which any given lengths and surfaces may be compared. It might be of advantage in the communication between different countries to fix one single standard to be employed throughout the world, but this does not appear to be practically possible, even if it were determined what the standard ought to be. " The observation of the isochronism of the small vibrations of a pendulum, and the ease and certainty with which the length of a pendulum vibrating seconds may be ascertained, have suggested," says Mr. Laplace, in his account of the system of the world, "the idea of employing this length as a universal measure. We cannot reflect on the prodigious number of measures in use, not only among different nations, but even in the same country, their capricious and inconvenient divisions, the difficulty of determining and comparing them, the embarrassment and the frauds which they occasion in commerce, without regarding, as one of the greatest benefits, that the improvements of the sciences and the ordinances of civil governments can render to humanity, the adoption of a system of measures, of which the divisions, being uniform, may be easily employed in calculations, and which may be derived, in a manner the least arbitrary, from a fundamental magnitude indicated by nature itself. A nation that would introduce such a system of measures, would unite to the advantage of reaping the first fruits of the improvement, the pleasure of seeing its example followed by other countries, of which it would thus become the benefactor: for the slow but irresistible empire of reason must at length prevail over national jealousies, and over all other obstacles that are opposed to a measure, of which the convenience is universally felt. Such were the motives that determined the constituent assembly to intrust the Academy of Sciences with this important charge. The new system of weights and measures is the result of the labours of the Committee, seconded by the zeal and information of several members of the national representation.

"The identity of the calculation of decimal fractions and of whole numbers, leaves no doubt with respect to the advantage of the division of measures of all kinds into decimal parts: it is sufficient, in order to be convinced of this, to compare the difficulty of compound multiplication and division, with the facility of the same operations where whole numbers only are concerned, a facility that becomes still greater by means of logarithms, of which the use may also be rendered extremely popular by simple and cheap instruments. The decimal division was therefore adopted without hesitation; and in order to preserve the uniformity of the whole system, it was resolved to deduce every thing from the same linear measure, and its decimal divisions. The question was then reduced to the choice of this universal measure, to which the name of metre was to be given.

"The length of the pendulum, and that of a meridian of the earth, are the two principal standards that nature affords us, for fixing the unit of linear measures. Both of these being independent of moral revolutions, they cannot experience a sensible alteration without very great changes in the physical constitution of the earth. The first method, which is of easy execution, has the inconvenience of making the measure of length depend on two elements, heterogeneous with respect to itself and to each other, gravitation, and time; besides that the division of time into small portions is wholly arbitrary. It was resolved, therefore, to employ the second method, which," says Mr. Laplace, "appears to be of very high antiquity; it is so natural to man to refer measures of distance to the dimensions of the globe which he inhabits, in order that, in transporting himself from place to place, he may know, by the denomination of the space passed through alone, the relation of this space to the entire circumference of the earth. This method has also the advantage of making nautical measures correspond at once with celestial ones. The navigator has often occasion to compare with each other the distance that he has passed over, and the arc of the heavens corresponding to that distance; it is therefore of consequence that these measures should be readily obtained from each other, by altering only the place of the units. But, for this purpose, the fundamental unit of linear measures must be an aliquot part of the terrestrial meridian, which must correspond to one of the divisions of the circumference of a circle. Thus the choice of the metre was reduced to that of the unit of angular measure, and the right angle, as constituting the limit of the inclination of two lines to each other, was considered as entitled to the preference.

"The arc, which was measured in 1740, from Dunkirk to the Pyrenees, might have served for finding the magnitude of the quadrant of the meridian; but a new and more accurate measurement of a larger arc was more likely to excite an interest in favour of the new measures. Delambre and Méchain were therefore intrusted with the direction of the operations for measuring an arc from Dunkirk to Barcelona, and after making a proper correction for the ellipticity of the earth, according to the measurement of the arc in Peru; the quadrant was determined to be equal to 5130740 of the iron toise used at the equator, its temperature being 61 $\frac{1}{4}$° of Fahrenheit: the ten millionth part of this quadrant was taken for the unit or metre. A standard was deposited in the custody of the legislative body, adjusted at the temperature of melting ice. In order to be able always to identify this length, without recurring to an actual measurement of the arc, it was of importance to compare it very accurately with that of the pendulum vibrating seconds, and this has been done with great care by Borda, at the observatory of Paris. The unit of measures of land is the are, or 100 square metres: a cubic metre of wood is called a stere, and a cubic decimetre, or a cube of which the side is one tenth of a metre, is a litre, or measure of fluids.

"Uniformity appeared to require that the day should be divided into ten hours, the hour into a hundred minutes, and the minute into a hundred seconds. This division, useful as it will be to astronomers, is of less advantage in civil life, where arithmetical operations are seldom performed on the parts of time; and the difficulty of adapting it to clocks and watches, together with our commercial relations with foreign countries, have suspended its introduction for the present. We may, however, expect that it will ultimately be brought into general use."

Such is Mr. Laplace's account of the new system of measures, the result of the joint labours of many of the ablest mathematicians on the continent. There is not at present any great probability that it will ever be employed in this country. It is of little consequence from what the original unit has been derived, unless we can with ease and accuracy recur to its origin: and whether a standard has been first adjusted according to the circumference of the globe, or to the foot of an individual hero, the facility of comparing other measures with it is the same. It is confessed that the pendulum affords the readiest method of recovering the standard when lost; and if it was necessary for the Committee of the French Academy to determine a unit absolutely new, it would perhaps have been more eligible to fix on one which was independent of any ulterior comparison, than to seek for an ideal perfection in attempting to copy from a more magnificent original: to say nothing of the uncertainty with regard to the ellipticity of the earth, and the probable irregularity of its form in various respects. On the other hand, it must be allowed, that the correct determination of the length of the pendulum has sometimes been found more difficult than Mr. Laplace's statement would lead us to suppose it, and we cannot depend on any measurement of it as totally exempt from an error of the ten thousandth part of the whole.

The metre, as definitively established by the government of France, is equal to $39\frac{371}{1000}$ English inches, measured, as it has been usual in this country, on a standard scale of brass, at the temperature of 62° of Fahrenheit; while the French, on the contrary, reduce the length of their measures to that which they would acquire at the freezing point. Hence ten thousand inches are nearly 254 metres, a thousand feet 305 metres. The length of the pendulum vibrating seconds in London, was found by George Graham, from a mean of several experiments, all agreeing very nearly together, to be $39\frac{13}{100}$ inches. This is also nearly a mean between the length which may be deduced, with proper corrections, from Borda's experiments at Paris, and Mr. Whitehurst's experiments made in London, with the apparatus invented by Mr. Hatton, where the length ascertained is the difference between the lengths of two pendulums vibrating in different times. Mr. Whitehurst's measures, however, require some corrections, which Mr. Nicholson has pointed out. The fall of a heavy body in the first second appears, from this determination of the length of the pendulum, to be sixteen feet one inch and a tenth.

Of the old French measure, 15 inches made nearly 16 English, and 76, very exactly, 81; the toise was $76\frac{736}{1000}$ inches. In Germany the Rhinland foot is generally used; 100 of these feet make 103 English.

A wine gallon contains 231 cubic inches; an ale gallon is the content of 10 yards of a cylindrical inch pipe.

A variety of instruments are used for the immediate comparison of the standard measure, or its parts, with other lengths or distances. Such are scales, simple and diagonal, verniers, micrometer screws, beam compasses, rods, lines, chains, and measuring wheels. The greatest accuracy has generally been supposed to be obtained, in large distances, by means of rods, made of glass or of platina, in order to be less susceptible of such changes as are produced by variations of temperature; General Roy, however, found that a steel chain was as little liable to error, as any mode that he could employ; and those who have continued the extensive survey which he began, even prefer it to every other. For the comparison of standards, and for determining small distances with great precision, beam compasses, or scales with sliding indices, furnished with microscopes and cross wires, have been constructed by the artists of this country: in France, a lever has sometimes been introduced, its longer arm having an ample range of motion, corresponding to a very minute difference in the length of the substance which acts on the shorter arm. But for common purposes the diagonal scale is sufficiently accurate, and may be applied without the error of the thousandth of an inch: in cases where a very delicate vernier, or a micrometer screw is applied, a magnifier is usually required. Mr. Coventry has, however, succeeded in making simple scales, which are accurate enough to measure the ten thousandth of an inch. He draws parallel lines on glass, at this distance, which are in some parts sufficiently regular, although they can only be seen by the help of a powerful microscope: but those which are at the distance of the five thousandth of an inch are much more correct and distinct. For dividing rectilinear scales of all kinds, Mr. Ramsden constructed a machine which acts by the turns of a screw: others have employed an apparatus resembling Marquois's parallel rulers. (Plate VII. Fig. 95 . . 97.)

The motion of a ship at sea is measured by a log line, or a rope divided by knots into equal parts, and attached to a log, which is retained nearly at rest by the resistance of the water. Attempts have also been made to cause a little waterwheel to turn by the motion of the ship, and to measure both the rate and the distance run; and an instrument has been invented for doing the same upon hydraulical principles; raising the water of a gage to different heights, by means of the pressure occasioned by the relative motion of the ship and the water, and discharging at the same time a small stream into a reservoir, with a velocity proportional to that of the ship.

LECTURE XI.

ON MODELLING, PERSPECTIVE, ENGRAVING, AND PRINTING.

WE have examined the principal instruments and materials employed for drawing and for measuring; we are now to consider, first, the methods of copying solids, and of projecting their images on a plane surface; and secondly, the arts of perpetuating the works of the pen and of the pencil by engraving and printing.

When it is required to make a copy of a solid of an irregular form, as, for example, of a statue, we must determine the situation of a sufficient number of points to guide us in our work with accuracy, by means of an instrument capable of being fixed in any required situation, so that the extremity of a sliding bar, or pin, may be in contact with each point in the original, and then removed to a similar part of another frame, on which the copy is placed, a perforation being made, by degrees, in the block, so as to suffer the pin to arrive at its proper place, at which it stops. (Plate VII. Fig. 98.)

The model of a statue is generally first made of clay, and a cast of this taken immediately in plaster of Paris, since the clay would crack and change its form in drying. This mode of copying, by means of plaster, is exceedingly useful in various departments of the mechanical arts: the original is well oiled and placed in a proper vessel; a mixture of prepared plaster and water, of the consistence of cream, is then poured on it; this in a short time hardens, and is divided into several parts, in such a manner as not to injure the original figure in its removal. These pieces, being again united, form a mould for the ultimate cast. Sometimes a small figure is first modelled in a mixture of wax, turpentine, and oil; and a mould being formed on this, the ultimate cast is made either of plaster, or of a composition of wax with white lead and a little oil, which serves as an imitation of marble.

We have, however, much less frequent occasion to make an exact copy of a solid of any kind, than to represent its appearance by means of perspective delineation. Supposing ourselves provided with proper materials for drawing, we may easily imitate, with the assistance of a correct eye, and a hand well exercised, the figures and relative positions of objects actually before us, by delineating them in the same form as they would appear to be projected on a transparent surface placed before the eye. Considering the simplicity of this process, it is almost surprising that the doctrine of perspective should have been supposed to require a very serious study, and that material errors should have been committed with respect to it, by men whose general merits in other departments of painting is by no means contemptible. But it must be confessed, that when, instead of imitating objects immediately before us, the pencil is employed in embodying imaginary forms, calculated either for beauty or for utility, a great degree of care and attention may be necessary, in order to produce a true representation of objects, which are either absent, or have no existence: and here memory and fancy only will scarcely ever be sufficient, without a recurrence to mathematical principles. To architects therefore, and to mechanics in general, a knowledge of perspective is almost indispensable, whenever they wish to convey, by a drawing, an accurate idea of their projected works.

If any assistance be required for the delineation of an object actually before us, it may easily be obtained in a mechanical manner, by means of a frame with cross threads or wires, interposed between the eye and the object. The eye is applied to an aperture, which must be fixed, in order to preserve the proportions of the picture; and which must be small, in order that the threads and the more distant objects may be viewed at the same time, with sufficient distinctness. The paper being furnished with corresponding lines, we may observe in what division of the frame any conspicuous point of the object appears, and may then represent its image by a point similarly situated among the lines drawn on our paper; and having obtained, in this manner, a sufficient number of points, we may complete the figures by the addition of proper outlines. Sometimes, for the delineation of large objects requiring close inspection, it has been found useful to employ two similar frames, the one a little smaller than the other, and placed at a certain distance from it, so that every part of the object, when seen through the corresponding divisions of both frames, appears in the same manner as if the eye were situated at a very remote point. It was in this manner that the elegant anatomical figures of Albinus were executed. (Plate VII. Fig. 99.)

But if it be required to lay down, in the plane of a picture, the projection of an *object*, of which the actual dimensions and situation are given, we may obtain the requisite measures from the properties of similar triangles, and the consideration of the rectilinear motion of light. We may consider our picture as a reduced copy of a projection formed on an imaginary plane, which, as well as the picture, is generally supposed to be in a vertical situation, and which stands on the horizontal plane, at the point where the objects to be represented begin. In order to find the position of the image of a given right line, we must determine the point in which a line parallel to it, passing through the place of the eye, cuts the plane of the picture; this is called the vanishing point of the given line, and of all other lines parallel to it, since the image of any such line, continued without limit, will be a right line directed to this point, but never passing it. When the lines to be represented are parallel to the picture, the distance of their vanishing point becomes infinite, and their images are also parallel to the lines and to each other. The centre of the picture, or that point which is nearest to the eye, is the vanishing point of all lines perpendicular to the picture; through this point it is usual to draw a horizontal and a vertical line: we may then lay off downwards on the vertical line the distance of the eye from the picture, in order to find the point of distance, which serves to determine the position of any oblique lines on a horizontal plane: for if we draw a ground plan of any object, considering the picture as a horizontal surface, we may find the vanishing point of each of its lines, by drawing a line parallel to it through the point of distance, until it meets the horizontal vanishing line. (Plate VII. Fig. 100, 101.)

In order to find the position of the image of a given point of a line, we must divide the whole image in such a manner, that its parts may be to each other, in the same proportion as the distance of the given point, and of the eye, from the plane of projection. This may be readily done, when a ground plan has been first made, by drawing a line from any point in the plan, to the point of distance, which will cut the whole image of the line in the point required. (Plate VII. Fig. 102.)

When it is required to determine a point in a line parallel to the picture, we may suppose a line to be drawn through it perpendicular to the picture, and, by finding the image of this line, we may intersect the former image in the point required. It is thus that the height of any number of columns, or figures, at different distances, may be readily determined. (Plate VIII. Fig. 103.)

The projection of curvilinear figures is most conveniently effected, by drawing across them parallel lines, which form small squares or rectangles, throwing these divisions into perspective, and tracing a curve through the corresponding points. There are also methods of determining mathematically, or of drawing mechanically the ellipsis, which results from the projection of a circle, in a given position, but they are considerably intricate, and a steady hand is seldom in want of them. (Plate VIII. Fig. 104.)

This system of perspective must necessarily be employed when we wish to represent objects, which appear to us under angles of considerable magnitude, and to give them as much as possible the appearance of an imitation of nature. But for almost all purposes of science, and of mechanical practice, the most convenient representation is the orthographical projection, where the distance of the eye, from the plane, is supposed to be increased without limit, and the rays of light passing to the eye to be parallel to each other. In order to represent any object in this manner, we must assume one line for the direction of the centre of the picture, to which the images of all lines perpendicular to the plane of projection must be parallel, and another for that of the point of distance, by means of which we may measure the first lines, as if that point were actually within reach; and in this manner we may determine the place of any number of points of the object to be delineated. (Plate VIII. Fig. 105.)

If we wish to apply the mechanical method of drawing by the assistance of a frame to this mode of representation, instead of a fixed aperture for a sight, or a second frame of smaller dimensions, we must employ a second frame of the same magnitude with the first, in the manner which has already been described. Professor Camper has censured Albinus for not adopting this method in his figures: but subjects so large as those which he has represented would have had less of the appearance of nature, if they had been projected orthographically, nor would such projections have been materially more instructive.

It frequently happens, that in geographical and astronomical drawings, we have occasion to represent, on a plane, the whole, or a part of a spherical surface. Here, if we employ the orthographical projection, the distortion will be such, that the parts near the apparent circumference will be so much contracted, as to render it impossible to exhibit them with distinctness. It is, therefore, more convenient, in this case, to employ the stereographical projection, where the eye is supposed to be at a moderate distance from the object. The place of the eye may be assumed either within or without the sphere, at pleasure, and according to the magnitude of the portion which we wish to represent, the point, from which the sphere may be viewed with the least distortion, may be determined by calculation. But in these cases all circles obliquely situated on the sphere must be represented by ellipses: there is, however, one point in which the eye may be placed, which has the peculiar and important advantage, that the image of every circle, greater or lesser, still remains a circle. This point is in the surface itself, at the extremity of the diameter perpendicular to the plane of projection; and this is the point usually employed in the stereographical projection of the sphere, which serves for the geometrical construction of problems in spherical trigonometry. The projection of the whole surface of the sphere would occupy an infinite space, but within the limits of the hemisphere, the utmost distortion of the linear measure is only in the proportion of 2 to 1, each degree at the circumference of the figure occupying a space twice as great as at the centre. The angles, which the circles form in crossing each other, are also correctly represented. (Plate VIII. Fig. 106.)

For projecting figures on curved or irregular surfaces, the readiest method is to trace cross lines on them, with the assistance of such a frame as has been described for drawing in perspective, representing the appearance of uniform squares or rectangles, and to delineate in each of these the corresponding parts of the object, or of the drawing which serves as a copy.

The arts of writing and drawing, in all their varieties, are extended in their performance, and perpetuated in their duration, by means of engraving and printing. If there is any one circumstance to which we can peculiarly attribute the more rapid progress of general civilisation in modern than in ancient times, it is the facility of multiplying copies of literary productions of all kinds, by the assistance of these arts. The distinguishing character of printing consists in the employment of moveable types: the art of engraving is more simple, and in some of its forms, more ancient. The Romans were in the habit of using seals and stamps, for marking letters and words on wax and on pottery; it was usual in the middle ages to employ perforated plates of metal as patterns for guiding a brush, by means of which the capital letters were inserted in some manuscripts, and the Chinese are said to have been long in possession of the art of printing books from wooden blocks. It was in this form that printing was first introduced into Europe, in the beginning of the fifteenth century. There seems to have been formerly a method of engraving on wood with greater ease and accuracy than is now practised; the hatches may be observed in old wooden cuts to cross each other more frequently, and with greater freedom, than in modern works, although some have conjectured, with considerable appearance of probability, that these old engravings were in reality etched in relief on metal. The art of engraving on wood is, however, at present in a high degree of perfection in this country, and blocks are still frequently used for mathematical diagrams and other simple figures; for although they are somewhat more expensive than copper plates, they wear much longer, and they have the advantage of being printed off at the same time with the letter press, and of being included in the same page with the text to which they belong, since the ink is applied to the projecting parts only, both of these cuts and of the common printing types.

The method of engraving on plates of pewter or of copper, and of taking impressions, by means of the portion of ink retained in the furrows cut by the graver, was also introduced in the fifteenth century. For dry engraving, the drawing, if it is not executed in black lead, is generally prepared by passing a pencil over its principal features, and the outline is transferred to the plate, which has a thin coat of white wax laid on it, by placing the drawing on it, and rubbing it with a burnisher; sometimes a drawing in Indian ink, especially if freed from a part of its gum, may be transferred in this manner without the application of a pencil. When written characters are to be engraved, the plate is laid on a cushion, so as to be readily turned under the graver, which is a great convenience in forming curved lines.

In laying on equable shades of considerable extent, much labour is saved by the use of a ruling machine, which enables us to draw lines, at any required distance, very accurately parallel, and either straight, or following each other's gentle undulations, in order to avoid the appearance of stiffness. This machine, like the dividing engine, is sometimes adjusted by the revolutions of a screw, and sometimes by the oblique motion of a triangular slider. Besides the cutting graver, which is of a prismatic form, terminated by an oblique surface, other instruments are occasionally employed; the dry needle makes a very fine line, and leaves the metal that it has displaced, to be rubbed off by another tool. Sometimes a number of detached excavations are formed by a pointed instrument, and the projections are afterwards removed; this is called stippling. A burnisher and some charcoal are required for erasing the strokes of the graver, when it is necessary, and for polishing the surface. It is seldom, however, that a plate is begun and completed by dry engraving only.

For engraving in mezzotinto, the plate is roughened, by scraping it in every direction with a tool made for the purpose, so that an impression from it, in this state, would be wholly dark; the lights are then inserted, by removing the inequalities of the surface, in particular parts, by means of a smooth scraper, and a burnisher. As the plate wears in printing, some of these parts are liable to have the grain a little raised again, so that the lights are less clear in the later impressions than in the proofs. It is well known, that in common engravings the proofs are usually the darkest throughout.

The most expeditious and most generally useful mode of working on copper, is the process of etching. The plate, being covered with a proper varnish, is usually blackened with smoke, and the drawing is placed on it, with the interposition of a paper rubbed over with red chalk, which, when the drawing is traced with a wooden point, adheres to the varnish, in the form of the outline: or if it is required that the ultimate impression be turned the same way as the drawing, an intermediate outline must be procured in the same manner on a separate paper, and then transferred to the plate. All the outlines thus marked are traced with needles, which make as many furrows in the varnish, and leave the copper bare: the shades are inserted with the assistance of the ruling machine, wherever parallel lines can be employed. The plate thus prepared, and furnished with an elevated border of a proper consistence, is subjected to the action of the diluted nitric acid, until all the parts are sufficiently corroded, care being taken in the mean time to sweep off the air bubbles as they collect, and to stop out, or cover with a new varnish, the lighter parts, which are soonest completed. When the varnish is removed, the finishing touches are added with the graver: and if the plate requires further corrosion, the varnish may sometimes be replaced, without filling up the lines, by applying it on a ball or cushion, taking care to avoid any oblique motion. It is said that the acid sometimes operates so as to undermine the metal on each side, and to render the furrows wider as they become deeper, and that for this reason in etchings, as well as in mezzotintos, the later impressions are sometimes darker than the proofs; but this is by no means universally true. It is well known to chemists, that glass may be corroded in a similar manner by means of the fluoric acid.

An etching may also be expeditiously executed by using a varnish mixed with mutton fat, and drawing upon a paper laid on the plate; the varnish then adheres to the back of the paper, under the lines which are drawn, and is immediately removed when the paper is taken off, without the use of needles. Sometimes the outlines only are etched, and the plate is finished in mezzotinto.

In the mode of engraving called aqua tinta, the outline having been first etched, the shades are also produced by corrosion, the parts being prepared by various methods, so as to be partially protected from the action of the acid. Sometimes a little resin, very finely powdered, is sifted on the plate, which is then sufficiently warmed to make the particles adhere to it; sometimes it is varnished with a spirituous solution of resin, which cracks throughout in drying; and if a strong line be any where required, it may be traced with a mixture of whiting with some adhesive substance, before the varnish is laid on; this will cause it to break up at that part; or the varnish may be partially removed, by rubbing it with spirits, or with an essential oil. The lighter parts may be covered, during the corrosion, with a second varnish, which defends them from the acid. This mode of engraving succeeds very well in imitating the effect of drawings, but the plates are soon worn out. In order to judge of the state of the work, an impression of any part of the plate may be taken off, by pouring on it a little plaster of Paris mixed with water.

Musical characters are usually stamped with punches; in this country, on plates of pewter, but in France generally on copper. Mr. Rochon has invented a machine for stamping letters on copper, instead of printing, but the method does not appear to have been practically employed.

In whatever way the plate may have been engraved, when an impression is to be taken from it, it is covered with printing ink of the finest kind, by means of stuffed balls, and then wiped, chiefly with the hand, so that the ink is wholly removed from the polished surface; it is then placed, with the moistened paper, on a board, between flannels, and strongly pressed in passing between two wooden rollers. By frequent use the plate loses its sharpness, and sometimes requires to be retouched; hence arises the greater value of first impressions; but by proper precautions in cleaning the plate, its delicacy may be preserved for a long time.

An impression, while it is moist, may be reversed, by passing it through the press with another paper. And by writing with a peculiar ink, even common letters may be thus copied on thin paper, and the impression will be legible on the opposite side. Mr. Montbret proposes to put some sugar candy into the ink, and to take a copy on unsized paper by means of a hot iron.

A simple and elegant method of multiplying drawings has been lately introduced by Mr. André. The drawings are made with an unctuous composition, in the form of a crayon or of an ink, on a soft stone of a calcarious nature, somewhat like a stone marle. When the drawing is finished, the stone is moistened, and imbibes so much water, that the printing ink will not adhere to it, except at the parts where the crayon or the ink has been applied; and in this manner an impression is procured, which has much of the freedom and spirit of an original drawing. When the ink is used, a little acid is afterwards applied to the stone, in order to corrode its intermediate parts; and the bold stile of the impression much resembles that of the old wooden cuts.

The art of printing with separate types was invented soon after the introduction of wooden blocks into Europe. The improvement was great and important. The year 1443, or 1444, is considered as the date of the oldest printed book; but the precise time and place of the invention remain somewhat doubtful: the art, however, advanced towards perfection by very rapid steps. The letters are first cut, in a reversed form, on steel punches; with these a matrix of copper is stamped, and the matrix forms the lower part of the mould in which the types are cast; the metal is a composition of lead and antimony, which is easily fusible. Thus the printed sheet is the fourth form of the letter, reckoning from the original engraving on the punch: in the stereotype printing, lately invented, or rather improved and revived, it is the sixth. In this method, when a form for the side of a sheet has been composed, made up, corrected, and locked up by wedges in the chase or iron frame, which confines it, a mould of the whole is formed in fine plaster, and as many repetitions of it may be cast very thin, in type metal, as will serve to print for the use of a century, without the expense of keeping a large quantity of types made up, or of providing paper for a numerous impression at once.

The modes of arranging the types in boxes or cases, of composing the separate lines on the stick, and making them up by degrees into pages and forms, of correcting the press, of applying the ink, and taking off the impression, are entirely calculated for the simplicity and convenience of the manual operations concerned, and require little or no detailed explanation.

LECTURE XII.

ON STATICS.

THE examination of the magnitude of the various forces, employed in practical mechanics, constitutes the doctrine of statics. The term statics, in a strict sense, implies the determination of weights only; but it may without impropriety be extended to the estimation of forces of all kinds, especially active forces, that can be compared with weights, in the same manner as the term hydrostatics comprehends every thing that relates to the equilibrium of fluids. The measurement of the passive strength of the materials employed, the changes produced in them by the forces which they resist, and the laws of the negative force of friction, are also subjects immediately introductory to the particular constructions and uses of machinery, and nearly connected with the department of statics.

The art of weighing is peculiarly important, as it furnishes us with the only practical mode of determining the quantity of matter in a given body. We might indeed cause two bodies to meet each other with known velocities, and from the effects of their collision, we might determine their comparative momenta, and the proportion of their masses; but it is obvious that this process would be exceedingly troublesome, and incapable of great accuracy; we therefore recur to the well known law of gravitation, that the weight of every body is proportional to the quantity of matter that it contains, and we judge of its mass from its weight. If all bodies were of equal density, we might determine their masses from their external dimensions; but we seldom find even a single body which is of uniform density throughout; and even if we had such a body, it would in general be much easier to weigh it correctly than to measure it.

The weight of a body is commonly ascertained, by comparing it immediately with other weights of known dimensions: but sometimes the flexure of a spring is employed for the comparison. Standard weights have generally been deduced from a certain measure of a known substance, and in particular of water. According to the most accurate experiments, when the barometer is at 30 inches, and Fahrenheit's thermometer at 62°, 12 wine gallons of distilled water weigh exactly 100 pounds avoirdupois, each containing 7000 grains troy; and a cubic inch weighs $252\frac{1}{2}$ grains. A hogshead of water, wine measure, weighs, therefore, 525 pounds, and a tun 2100 pounds, which is nearly equal to a ton weight. Mr. Barlow supposes that the tun measure of water contained originally 32 cubic feet, and weighed 2000 pounds, which was also called a ton weight, the gallon being somewhat smaller than it is at present, and the cubic foot weighing exactly 1000 ounces, or $62\frac{1}{2}$ pounds. A quarter of wheat weighed about a quarter of a ton, and a bushel as much as a cubic foot of water. A chaldron of coals was also considered as equivalent to a ton, although it now weighs nearly half as much more. But at the mean temperature of this climate, or 52°, a cubic foot of distilled water weighs only 998 ounces. The avoirdupois ounce appears to agree very nearly with the ancient Roman ounce. Of the old French weight, 100 pounds made 108 English pounds avoirdupois. The gramme of the new weights is a cubic centimetre of pure water at its greatest density, that is, about the temperature of 39° of Fahrenheit; it is equal to $15\frac{4}{9}$ English grains: hence the chiliogramme is $2\frac{1}{5}$ pounds, and five myriogrammes are nearly a hundred weight. Five grammes of silver, including one tenth of alloy, make a franc, which is one eightieth better than the old franc or livre, and is intrinsically worth nearly ninepence three farthings English.

The instruments usually employed for the comparison of weights are either balances, or steelyards. In the common balance, the weights of the substances compared are equal; in a compound weighing machine, we use weights which are smaller, in a certain proportion, than those which they represent: in the steelyard, a single weight acquires different values at different parts of the arm, and in the bent lever balance, the position of the arms determines the magnitude of the counterpoise. The spring steelyard measures the weight, by the degree of flexure that it produces in a spring.

The beam of a common balance must have its arms precisely equal. The scales, being freely suspended from fixed points in the beam, act on them always in the direction of gravity; and the effect is the same as if the whole weight were concentrated in those points. The beam supports the scales, and is itself supported, by means of fine edges of hard steel, working on steel, agate, or garnet, in order that the motion may be free, and the distances of the points precisely defined. The best beams are made of two hollow cones of brass, united at their bases; they are lifted off their supports when the balance is not used, in order to avoid accidental injuries; the scales also are supported, so as not to hang from the beam, until they have received their weights. According to the position of the fulcrum, with respect to the points of suspension of the scales, the equilibrium of the balance may be either stable, neutral, or tottering; or if the beam be too flexible, it may pass from one of these states to the other by the effect of the weights. The stable equilibrium is the most usual and the best, because it gives us an opportunity of determining the degree of inequality of the weights, by the position in which the centre of gravity rests, or by the middle point of the vibrations of the beam, which are sometimes measured by an index, pointing to a graduated arc. If, however, the fulcrum be too much elevated above the centre of gravity, the equilibrium may be too stable, and may require too great an inequality, in order to produce a sensible preponderance. If, on the contrary, by the elevation of the points of suspension of the scales, the equilibrium be rendered tottering, the lower scale will not rise, even if it be somewhat less loaded than the upper; and steelyards of this construction have sometimes been employed, in order to impose on the purchaser by the appearance of an ample weight. It is necessary, where great accuracy is desired, to bring the equilibrium very near the state of neutrality, and to make the vibrations of the beam slow and extensive, whether the scales have weights in them or not: for this purpose a small weight is sometimes inclosed within the beam, which is raised or depressed at pleasure, by a screw, so as to bring the centre of gravity of the whole moveable apparatus, as near to the fulcrum as may be required for the occasion. Mr. Ramsden's balance, made for the Royal Society, is capable of weighing ten pounds, and turns with one ten millionth part of the weight. (Plate VIII. Fig. 107 . . 109.)

The arms of a balance have sometimes been made unequal for fraudulent purposes, the weight being placed nearer to the fulcrum than the substance to be weighed. It is obvious that the fraud may be detected, by changing the places of the contents of the two scales. In such a case, if a counterpoise to the same weight be determined in each situation, the sum of both will be greater than twice the weight; and the purchaser would be sure of having even more than his due, by requesting the seller to weigh half in the one scale and half in the other. For example, if one arm of the beam were only three fourths as long as the other, the counterpoise, to a weight of twelve ounces, would be nine ounces in one scale, and sixteen in the other, making together twenty five instead of twenty four ounces. (Plate VIII. Fig. 110.)

Supposing the beams of a balance to be accidentally unequal, either in length or in weight, we may still weigh in it with accuracy, by making a perfect counterpoise of any kind to a weight, and then removing the weight and putting in its place as much of the substance to be weighed, as is sufficient to restore the equilibrium.

The weights may also be reduced, or increased, in proportion to the length of the arms, if they differ from each other, care being taken to put the weights always into the same scale. This is actually performed in weighing machines, where a composition of levers is employed, in order to enable us to determine the weight of large masses by means of weights of moderate dimensions. (Plate IX. Fig. 111.)

When the effective lengths of one or both arms of the beam are capable of being varied, by changing the points of suspension according to the divisions of a scale, the instrument is called a steelyard. Where one weight only is used, it is not necessary that the two arms should exactly balance each other, since the divisions may be so placed as to make the necessary adjustment; but it is sometimes convenient to have two or three weights, of different magnitudes, and for this purpose the instrument should be in equilibrium without any weight. In such cases, great accuracy may be obtained by applying a small weight at the end, in the form of a micrometer screw. (Plate IX. Fig. 112.)

The arms of a balance, though constant in length, may vary in effect without limit, if they can sufficiently alter their inclination to the horizon; for no weight, however great, acting on the arm of a bent lever, can make it perfectly vertical, since, in this position, the weight may be overpowered by the minutest counterpoise acting on the other arm. The centre of gravity being, in the common balance, very nearly in a right line between the weights, in order that it may be immediately below the fulcrum, the arm must have a very considerable angular motion for a slight inequality of the weights; but in the bent lever balance, the centre of gravity is at such a distance from the fulcrum, that a moderate motion of the arms may bring it into the vertical line. This motion is measured by an index on a graduated arc, which gives the instrument a considerable range; and where expedition is particularly desired, it may often be used with advantage; but if the weights to be determined are large, the scale becomes very much contracted, and the instrument requires to be levelled with great accuracy. A counterpoise acting on a spiral or conical barrel, has also been applied to a similar purpose; it is capable of a scale somewhat more extended than a bent lever balance, but it is less simple, and scarcely more accurate. (Plate IX. Fig. 113.)

A spring, which is usually of a spiral form, being made to support a hook by the intervention of a graduated bar, the divisions of this bar, which are drawn out beyond the fixed point, indicate the weight supported by the hook. This instrument is called a spring steelyard. Mr. Hanin's spring steelyard has a long index, which revolves on a centre, and shows at once the weight according to the standards of different countries. The divisions of the scales in moderate flexures of the spring are nearly equal: hence it may be inferred, that the space through which a spring is bent, and consequently its curvature, or change of curvature, is simply proportional to the force which acts on it, and that the vibrations of a weight supported by a spring, must, like those of a cycloidal pendulum, be performed in equal times, whatever may be their magnitude. The strength of all springs is somewhat diminished by heat, and for each degree of Fahrenheit that the temperature is raised, we must deduct about one part in five thousand from the apparent weight indicated by the spring steelyard. (Plate IX. Fig. 114.)

The spring steelyard affords us the most convenient method of measuring the immediate intensity of the forces exerted by animals of different kinds, in the labour which they perform. When it is adapted for this purpose, it is sometimes called the dynamometer. We may also estimate the force of an animal, which is employed in drawing a distant boat or carriage, by the inclination of the rope or chain to the horizon, compared with the weight of that portion of it which the animal supports, that is, of the part which extends to the point where the curve becomes horizontal.

All animal actions, or, at least, all the external actions of animals, are ultimately dependent on the contractions and relaxations of the fleshy parts, which are called muscles. The operation of the particular muscles belongs properly to the science of physiology; but their mechanism may in general be understood from the properties of the lever and of the centre of gravity. The bones are the levers, the joints the fulcrums, and the force is applied by the muscles, which are usually attached to the bones by the intervention of tendinous cords. When a muscle contracts in the direction of its fibres, it becomes at the same time thicker, and its total bulk is little if at all diminished: when it relaxes itself, it is merely passive, for the fibres, being extremely flexible, can have little or no effect in separating the parts to which they are attached; this separation is generally performed by the action of other muscles, which are called the antagonists of the first, but sometimes by elastic ligaments, or by other means. The bone forms a lever of the second kind, where the two forces opposing each other are on the same side of the fulcrum. In general the insertion of a muscle is much nearer to the fulcrum than the point of action, and the obliquity of its direction gives it a still greater mechanical disadvantage with regard to rotatory power; but it is more convenient in the animal economy to produce a great contractile force, than a great extent in the original motion. For instance, when the arm is raised by the exertion of the deltoid muscle of the shoulder, a very strong contraction takes place in the muscle, but the action is only continued through a short space; had the contractile power been weaker and more extensive, the shoulder must have been made higher, in order to give it sufficient purchase, and the projection would have been inconvenient.

Borelli has calculated that the immediate force of the biceps, or double-headed muscle which bends the arm, is equivalent to about 300 pounds, and that of the muscles which raise the lower jaw, above 500 in man, but in beasts of prey far greater. It is obvious that in muscles of the same kind, the strength must be as the number of fibres, or as the extent of the surface which would be formed by cutting the muscle across; and it is not improbable that the contractile force of the muscles of a healthy man is equivalent to about 500 pounds for each square inch of their section. The weakest man can lift with his hands about 125 pounds, a strong man 400. Topham, a carpenter, mentioned by Desaguliers, could lift 800 pounds. He rolled up a strong pewter dish with his fingers; he lifted with his teeth and knees a table six feet long, with a half hundred weight at the end. He bent a poker, three inches in circumference, to a right angle, by striking it upon his left fore arm: another he bent and unbent about his neck; and snapped a hempen rope two inches in circumference. A few years ago there was a person at Oxford who could hold his arm extended for half a minute, with half a hundred weight hanging on his little finger. A young gentleman, who has distinguished himself as a pedestrian by going 90 miles in 19 hours, has also lifted two hundred weights, one in each hand, and made them meet over his head.

Sometimes feats of strength apparently extraordinary have been exhibited by men who have not really been possessed of any material superiority. Desaguliers relates, that one of them used to withstand the force of two horses drawing at a girdle passed round his middle, while his feet acted on a firm obstacle. By falling suddenly backwards, in an oblique position, he broke a rope which was fixed a little before his feet. He supported one or two men by forming his body into an arch; and by a harness fitted to his hips, he sustained a cannon, weighing two or three thousand pounds. In all these cases the muscles principally employed are the extensors of the legs and thighs, but the passive strength of the bones is more concerned than the active force of the muscles. In the instance, mentioned by Lahire, of a young man who raised an ass from the ground, by cords tied to the hair of his head, the sensibility of the nerves of the skin must have been diminished by habit, so as to allow the hair to be thus forcibly extended, without immoderate pain.

The application of animal force is usually performed by means of a progressive motion. The muscles employed in this process are in general, if not always, the strongest of the body, both by nature, and by habit; so that when force alone is required, it is most advantageously obtained from their exertions. In walking, the centre of gravity is moved forwards with a velocity nearly uniform. If the legs were perfectly inflexible, the centre of gravity would describe, in succession, portions of circles, of which each leg would alternately be the radius: but if the velocity were great enough to create a centrifugal force more than equivalent to the force of gravity, the pressure would be removed from each leg after the first instant of its touching the ground; the path would become parabolic instead of circular, and the walking would be converted into running: for the difference between walking and running is this, that in running, one foot is removed from the ground before the other touches it; while in walking, the hindmost foot is only raised after the foremost has touched the ground. Now supposing the length of the inflexible leg three feet, the centrifugal force would become equal to the weight, with a velocity which would be acquired by a heavy body in falling through a foot and a half, that is, near 10 feet in a second, or 7 miles an hour; and this is the utmost velocity with which it would be mechanically possible to walk with inflexible legs. But the flexibility of the legs makes the progressive motion much more uniform, by softening the angles of the path, which the centre of gravity describes, and rendering it either more or less curved at pleasure; so that it becomes mechanically if not physically possible, to walk with a velocity somewhat greater than 7 miles an hour, and to run or dance with as small a velocity as we please, since we may make the path of the centre of gravity somewhat less, or much more curved, than a circle described on the point of the foot as a centre. (Plate IX. Fig. 115, 116.)

The flexions and extensions of the legs are also almost the only means by which an impulse is given to the body; if the legs were perfectly inflexible, it would be extremely difficult, although not absolutely impossible, to obtain a progressive motion. The centre of gravity is principally impelled forwards in the beginning of the ascending part of the curve which it describes, in walking, by the action of the leg which is left behind, but in running or hopping, by that of the only foot which touches the ground at any one time. When we thrust against any obstacle, or draw a rope in a horizontal, or in a descending direction, the body is inclined forwards, and in some cases its action is limited by the effect of the weight of the body reduced to the direction of the line of draught: but we much more usually draw or pull in an ascending direction, so that our whole muscular force may be exerted without any limit of this kind.

It happens, however, very frequently, that we have occasion for motions of such a nature as to be more conveniently performed by the hands and arms than by the action of walking or running; and where delicacy is required rather than strength, the form of the hand and fingers gives the human species a great superiority over all other animals, although by no means, as some authors have supposed, an advantage equivalent to that of the higher perfection of the intellectual powers. It is true, as we may observe in the manufactories of this country, that machinery has been invented by which a power of any kind may be converted to purposes seemingly the most intricate and refined; and after all that has been done by a Watt and an Arkwright, it is difficult to determine a positive limit to the ingenuity of mechanical invention.

It is necessary to consider, in examining the different sources of motion, not only the immediate magnitude of the forces which they produce, but also the velocity with which they are capable of acting, and the time for which that action can be continued. The daily work of a labouring man, of middle age, and in good health, will serve as a convenient unit for the comparison of moving powers of all kinds. It may be most easily remembered in this form: a man can raise a weight of 10 pounds to the height of 10 feet in a second, and can continue this labour for 10 hours a day. The actual velocity of the man's motion must vary according to the mode in which his force is applied; but we suppose that velocity to be such as to give the greatest effect under the circumstances of the machine. This is a moderate estimate of the work of a labourer, without any deduction for friction. Desaguliers states the performance of a man working at a winch, with the assistance of a fly, as considerably greater, but he does not allege any correct experiments in support of his estimate. Professor Robison, however, mentions a hydraulic machine in which the effect was actually more than one tenth greater, without making any allowance for friction; so that it is probable, considering the loss both from friction and from the momentum with which the water must have been disengaged, that the immediate performance was at least one third more than this unit: the machine was worked by a light man carrying a weight, and walking backwards and forwards on a lever. According to Mr. Buchanan's experiments, an action like that of ringing bells produced an effect about one third greater than turning a winch, and the action of rowing, an effect four ninths greater; but it does not appear that these experiments were continued for a whole day; and the greatest number of observations make the daily performance of workmen considerably less. It is indeed seldom that the muscles employed in progressive motion are so much exerted as in the arrangement described by Professor Robison. A Chinese, in the operation called sculling, is said to beat a European at his oar.

For a short time a much greater effect than this may be produced by a great exertion: thus a man weighing above 160 pounds can ascend by means of steps, at the rate of more than three feet in a second, for a quarter, or perhaps half a minute; and this is an effort five times as great as that which can be continued for a day. Usually, however, where the hands are chiefly employed, whether in turning a winch, or in pumping, it is only possible to exert a double, or at most a triple action, for a minute or two: thus, although a machine may only enable a man to raise a hogshead of water in a minute to the height of ten feet for a whole day, yet it is easy to work it so rapidly for a single minute as to raise double the quantity, or to raise a single hogshead to a height of twenty feet. The whole exertion of force must be a little greater than that which is thus estimated, because a certain degree of superfluous momentum must be generated in removing weights from one situation to another: but this loss is usually inconsiderable.

The action of carrying a load horizontally requires an exertion of a different kind, and admits of no direct comparison with the application of a constant force to overcome the gravitation of a weight, or any other immediate resistance. The work of a labourer thus employed is however confined within moderate limits. A strong porter can carry 200 pounds at the rate of three miles an hour; and, for a short distance, even 300 pounds: a chairman carries 150 pounds, and walks four miles an hour; and in Turkey it is said that there are porters, who, by stooping forwards, and placing the weight very low on their backs, are enabled to carry from 700 to 900 pounds. The subjects of Mr. Coulomb's experiments appear to have been either weaker, or more inactive, than the generality of porters in this country: he calculates that the most advantageous load for a man of common strength is about a hundred weight; or, if he is to return without a burden, 135 pounds.

The daily work of a horse is equal to that of five or six men: its immediate force is something greater, but it cannot support the labour of more than 8 hours a day, when drawing with a force of 200 pounds, or of 6 hours when with a force of 240, walking two miles and a half an hour. It is generally supposed, that in drawing up a steep ascent a horse is only equivalent to 3 or 4 men, and the employment of horses in walking wheels, where the action is similar to that of ascending a hill, has for this reason been condemned. For men, on the contrary, an ascent of any kind appears to afford a favourable mode of exertion. But, perhaps, the weight of the carriage, and of the horse itself, has not always been sufficiently considered in the comparison. The strength of a mule is equal to that of three or four men. The expense of keeping a horse is in general about twice or three times as great as the hire of a day labourer; so that the force of horses may be reckoned about half as expensive as that of men. The horse Childers is said, although, perhaps, without sufficient authority, to have run an English mile in a single minute; his velocity must in this case have been 88 feet in a second, which would have been sufficient to carry him on an inclined plane without friction, or in a very long sling, to the perpendicular height of 120 feet.

A large windmill, on which Mr. Coulomb made many experiments, was capable, on an average, of working eight hours a day; its whole performance was equivalent to our estimate of the daily labour of 34 men; 25 square feet of the sails doing the work of one labourer. The expense of the machinery, with its repairs, would probably amount to less than half the expense of a number of horses capable of exerting the same force. Where a stream of water can be procured, its force is generally more convenient, because more regular, than that of the wind.

A steam engine of the best construction, with a thirty inch cylinder has the force of 40 horses; and, since it acts without intermission, will perform the work of 120 horses, or of 600 men, each square inch of the piston being nearly equivalent to a labourer. According to Mr. Boulton, the consumption of a bushel, or 84 pounds of coals, will raise 48000 cubic feet of water 10 feet high, which is equivalent to the daily labour of 8 $\frac{1}{3}$ men, or perhaps more: the value of this quantity of coals is seldom more than that of the work of a single labourer for a day; but the expense of the machinery generally renders a steam engine somewhat more than half as expensive as the number of horses for which it is substituted. According to other accounts, a 24 inch cylinder, being equivalent to about 72 horses, requires only a chaldron of coals in a day, each bushel doing the work of ten men.

The force of gunpowder is employed with advantage where a very powerful action is required for a short space, as in dividing rocks, or in generating a great velocity in a projectile. As a source of momentum or energy only, this power is by no means economical, the daily labour of a man being equivalent to the effect of about 40 pounds of powder; but the advantage of artillery consists in having the force communicated by means of an elastic fluid extremely rare, which is capable of generating a very great velocity in the ball only, without any waste of power in producing a useless momentum in any other substance.

The comparative force of different kinds of gunpowder is determined by an eprouvette, or powder proof: the effect is measured by the angular motion of a little wheel, a projecting part of which is impelled by the explosion of a small quantity of the powder, while the friction of a spring or a weight creates a resistance which may be varied if it be required. The absolute force of a given quantity of powder may be ascertained either by suspending a cannon as a pendulum, and measuring its angular recoil; or by shooting into a large block, and finding the velocity which is imparted to it by the ball.

For measuring very small attractive or repulsive forces, with great accuracy, the most convenient test is furnished by the effects of twisting. An arm or beam is suspended horizontally by a long wire, and the force required to cause the beam to make one or more revolutions being ascertained, we may divide the circle described by its extremities into as many parts as we think proper, and the force required to bring the beam into any position will always be proportional, without a sensible error, to the magnitude of the part of the circle intercepted between the given position, and that in which the arm would naturally rest. When the force is of such a nature as to be capable of producing a vibration, the body on which it acts being suspended by the thread of a silkworm, or of a spider, we may compare its magnitude with that of gravitation, by observing the time required for each vibration, and determining the operation of the force according to the laws of pendulums. It is in this manner that the forces concerned in the effects of electricity and of magnetism have been measured by Mr. Coulomb.

LECTURE XIII.

ON PASSIVE STRENGTH AND FRICTION.

THE passive strength of the materials employed in the mechanical arts depends on the cohesive and repulsive forces of their particles, and on the rigidity of their structure. The consideration of the intimate nature of these forces belongs to the discussion of the physical properties of matter; but the estimation of their magnitude, and of their relative value in various circumstances, is of undeniable importance to practical mechanics, and requires to be examined as a continuation of the subject of statics. The retarding force of friction is very nearly allied to some kinds of passive strength, and may be in great measure explained from similar considerations.

The principal effects of any force, acting on a solid body, may be reduced to seven denominations; extension, compression, detrusion, flexure, torsion, alteration, and fracture. When a weight is suspended below a fixed point, the suspending substance is extended, or stretched, and retains its form by its cohesion, assisted by its rigidity: when the weight is supported by a block, or pillar, placed below it, the block is compressed, and resists primarily by a repulsive force, but secondarily also by its rigidity. The effect here called detrusion, is produced when a transverse force is applied close to a fixed point, in the same manner as the blades of a pair of scissors act on the pin, and the force which resists this operation is principally the rigidity, or lateral adhesion of the strata of the substance, but it could scarcely be effectual without some degree of cohesive and repulsive force. When three or more forces are applied to different parts of any substance, they produce flexure, that is, they bend it, some of its parts being extended, and others compressed. In torsion, or twisting, the central particles remain in their natural state, while those which are in opposite parts of the circumference are detruded, or displaced, in opposite directions. The operation of forces applied in any of these ways may produce a permanent alteration, or change of figure, in substances sufficiently soft, and perhaps, in a certain degree, in all substances: this change is sometimes called by workmen settling, or taking a set. But the limit of all these effects is fracture, which is the consequence of the application of any force capable of overcoming the strength of the substance, and to which the generality of writers on mechanics have hitherto confined their attention.

The forces, by which the form of any substance is changed, may also be divided into two kinds, simple pressure, and impulse; but it is only with regard to fracture that it will be necessary to take the force of impulse into consideration.

Extension and compression follow so nearly the same laws, that they may be best understood by comparison with each other. The cohesive and repulsive forces, which resist these effects, depend almost as much on the solidity, or rigidity of the substances, as on the attractions and repulsions which are their immediate causes: for a substance perfectly liquid, although its particles are in full possession of their attractive and repulsive powers, may be extended or compressed by the smallest force that can be applied to it. It is not indeed certain that the actual distances of the particles of all bodies are increased when they are extended, or diminished when they are compressed: for these changes are generally accompanied by contrary changes in other parts of the same substance, although probably in a smaller degree. We may easily observe, that if we compress a piece of elastic gum in any direction, it extends itself in other directions; and if we extend it in length, its breadth and thickness are diminished.

If the rigidity of a body were infinite, and all lateral motions of its particles were prevented, the direct cohesion alone would be the measure of the force required to produce extension, and the direct repulsion of the force required to produce compression; in this respect indeed, the actual rigidity of some substances may be considered as infinite, wherever the extension or compression is moderate, and no permanent alteration of form is produced; and within these limits these substances may be called perfectly elastic. If

2

the cohesion and repulsion were infinite, and the rigidity limited, the only effect of force would be to produce alteration of form: and such bodies would be perfectly inelastic, but they would be harder or softer according to the degree of rigidity.

It is found by experiment, that the measure of the extension and compression of uniform elastic bodies is simply proportional to the force which occasions it; at least when the forces are comparatively small. Thus if a weight of 100 pounds lengthened a rod of steel one hundredth of an inch, a weight of 200 would lengthen it very nearly two hundredths, and a weight of 300 pounds three hundredths. The same weights acting in a contrary direction would also shorten it one, two, or three hundredths respectively. The former part of this law was discovered by Dr. Hooke, and the effects appear to be perfectly analogous to those which are more easily observable in elastic fluids.

According to this analogy, we may express the elasticity of any substance by the weight of a certain column of the same substance, which may be denominated the modulus of its elasticity, and of which the weight is such, that any addition to it would increase it in the same proportion, as the weight added would shorten, by its pressure, a portion of the substance of equal diameter. Thus if a rod of any kind, 100 inches long, were compressed 1 inch by a weight of 1000 pounds, the weight of the modulus of its elasticity would be 100 thousand pounds, or more accurately 99000, which is to 100000 in the same proportion as 99 to 100. In the same manner, we must suppose that the subtraction of any weight from that of the modulus will also diminish it, in the same ratio that the equivalent force would extend any portion of the substance. The height of the modulus is the same, for the same substance, whatever its breadth and thickness may be: for atmospheric air, it is about 5 miles, and for steel nearly 1500. This supposition is sufficiently confirmed by experiments, to be considered at least as a good approximation: it follows that the weight of the modulus must always exceed the utmost cohesive strength of the substance, and that the compression produced by such a weight must reduce its dimensions to one half: and I have found that a force capable of compressing a piece of elastic gum to half its length will usually extend it to many times that length, and then break or tear it; and also that a force capable of extending it to twice its length will only compress it to two thirds. In this substance, and others of a similar nature, the resistance appears to be much diminished by the facility by which a contrary change is produced in a different direction; so that the cohesion and repulsion thus estimated appears to be very weak, unless when the rigidity is increased by a great degree of cold. It would be easy to ascertain the specific gravity of such a substance in different states of tension and compression, and some light might be thrown, by the comparison, on the nature and operation of the forces which are concerned.

It is difficult to compare the lateral adhesion, or the force which resists the detrusion of the parts of a solid, with any form of direct cohesion. This force constitutes the rigidity or hardness of a solid body, and is wholly absent from liquids, although their immediate cohesion appears to be nearly equal to that of solids. Some experiments have been made on the fracture of bodies by means of detrusion, but it does not appear that the force necessary to produce a temporary derangement of this kind has ever been examined: it may be inferred, however, from the properties of twisted substances, that the force varies in the simple ratio of the distance of the particles from their natural position, and it must also be simply proportional to the magnitude of the surface to which it is applied.

The most usual, as well as the most important effect, produced by the application of force, is flexure. When a force acts on a straight column in the direction of its axis, it can only compress or extend it equally through its whole substance; but if the direction of the force be only parallel to the axis, and applied to some point more or less remote from it, the compression or extension will obviously be partial: it may be shown that in a rectangular column, when the compressing force is applied to a point more distant from the axis than one sixth of the depth, the remoter surface will no longer be compressed but extended, and it may be demonstrated that the distance of the neutral point from the axis is inversely as that of the point to which the force is applied. From the effect of this partial compression, the column must necessarily become curved; and the curvature of the axis at any point will always be proportional to its distance from the line of direction of the force, not only while the column remains nearly straight, but also when it is bent in any degree that the nature of the substance will allow. If the column was originally bent, any force, however small, applied to the extremities of the axis, will increase the curvature according to the same law, but if the column was originally straight, it cannot be kept in a state of flexure by any longitudinal force acting precisely on the axis, unless it be greater than a certain determinate force, which varies according to the dimensions of the column. It is not however true, as some authors have asserted, that every column pressed by such a force must necessarily be bent; its state when it is straight, and submitted to the operation of such a force, will resemble a tottering equilibrium, in which a body may remain at rest until some external cause disturbs it. The figure of a column naturally straight, but bent a little by a longitudinal force, will coincide with that of the harmonic curve, in which the curvature is as the distance from the basis. (Plate IX. Fig. 117 . . 121.)

Considerable irregularities may be observed in all the experiments which have been made on the flexure of columns and rafters exposed to longitudinal forces; and there is no doubt but that some of them were occasioned by the difficulty of applying the force precisely at the extremities of the axis, and others by the accidental inequalities of the substances, of which the fibres must often have been in such directions as to constitute originally rather bent than straight columns.

When a rod, not very flexible, is fixed at one end in a horizontal position, the curvature produced by its own weight is every where as the square of the distance from the other end: and if a rod be simply supported at each end, its curvature at any point will be proportional to the product of the two parts into which that point divides it. But when the weights are supposed to be applied to any given points of the rod only, the curvature always decreases uniformly between these points and the points of support. (Plate IX. Fig. 122, 123.)

The stiffness of any substance is measured by the force required to cause it to recede through a given small space in the direction of the force. It is only necessary to consider this property with regard to forces applied transversely. In such cases the stiffness is directly as the breadth and the cube of the depth of the beam, and inversely as the cube of its length. Thus if we have a beam which is twice as long as another, we must make it, in order to obtain an equal stiffness, either twice as deep, or eight times as broad. The property of stiffness is fully as useful in many works of art as the ultimate strength with which a body resists fracture: thus for a shelf, a lintel, or a chimney piece, a great degree of flexure would be almost as inconvenient as a rupture of the substance.

When a beam is supported at both ends, its stiffness is twice as great as that of a beam of half the length firmly fixed at one end; and if both ends are firmly fixed, the stiffness is again quadrupled. For if the whole beam were inverted and supported by a fulcrum in the middle, each half would resemble a separate beam fixed at one end, and the fulcrum would bear the sum of two equal weights placed at the extremities, disregarding that of the beam; and consequently the same flexure will be produced by placing a double weight on the middle of the beam in an inverted position. If both ends were firmly fixed, the curvature would be every where as the distance from the middle of each half, the whole being in the same state as four separate beams fixed at their extremities: each of these beams would be eight times as stiff as beams of twice the length, and the whole beam, in this state, would be eight times as stiff as if the ends were simply supported. It is, however, difficult to fix the ends of a beam so firmly as to increase its resistance in this proportion, unless it be continued both ways considerably beyond the supports.

It is evident that a tube, or hollow beam, of any kind, must be much stiffer than the same quantity of matter in a solid form: the stiffness is indeed increased nearly in proportion to the square of the diameter, since the cohesion and repulsion are equally exerted with a smaller curvature, and act also on a longer lever.

Torsion, or twisting, consists in the lateral displacement, or detrusion, of the opposite parts of a solid, in opposite directions, the central particles only remaining in their natural state. We might consider a wire as composed of a great number of minute threads, extending through its length, and closely connected together; if we twisted such a wire, the external threads would be extended, and, in order to preserve the equilibrium, the internal ones would be contracted; and it may be shown that the whole wire would be shortened one fourth as much as the external fibres would be extended if the length remained undiminished; and that the force would vary as the cube of the angle through which the wire is twisted. But the force of torsion, as it is determined by experiment, varies simply as the angle of torsion; it cannot, therefore, be explained by the action of longitudinal fibres only; but it appears rather to depend principally, if not intirely, on the rigidity, or lateral adhesion, which resists the detrusion of the particles. If a wire be twice as thick as another of the same length, it will require sixteen times as much force to twist it once round; the stiffness varying as the fourth power of the diameter, that is, as the square of its square. But if the length vary, it is obvious that the resistance to the force of torsion will be inversely as the length.

A permanent alteration of form is most perceptible in such substances as are most destitute of rigidity, and approach most to the nature of fluids. It limits the strength of materials with regard to practical purposes, almost as much as fracture, since in general the force which is capable of producing this effect, is sufficient, with a small addition, to increase it till fracture takes place. A smaller force than that which has first produced an alteration of form, is seldom capable either of increasing, or of removing it, a circumstance which gives such materials, as are susceptible of an alteration of this kind, a great advantage for many purposes of convenience and of art. The more capable a body is of a permanent alteration of form, the more ductile it is said to be; pure gold and silver, lead, annealed iron and copper, wax when warm, glass when red hot, and clay when moist, possess considerable ductility. Wood admits of little permanent change of form, except in a green state, although it sometimes settles a little, when it has been exposed to pressure. Even stone will become permanently bent in the course of years, as we may observe in old marble chimney pieces. But the most ductile of all solid substances appears to be a spider's web. Mr. Bennet twisted a thread of this kind many thousand times, and shortened it more than a fourth of its length, yet it showed no disposition to untwist.

A ductile substance acquires the same cohesive and repulsive powers with regard to its new form, as it possessed in its original state; and when the alteration of form has once commenced, those powers are neither increased nor diminished by continuing the operation: the degree of flexure or torsion, required for producing a further alteration, appears also to be little varied: thus if the spider's web could at first be twisted only one half round, so as to retain the power of returning to its original state, without any permanent alteration of form, it would never acquire the power of returning more than half a revolution, however it might be twisted. From a want of attention to this consideration, a late respectable author has called in question, without sufficient reason, the accuracy of Mr. Bennet's experiments.

A variation of ductility, in any substance, does not appear to depend on any change in the magnitude of the ultimate powers of cohesion and repulsion. Steel, whether perfectly hard, or of the softest temper, resists flexure with equal force, when the deviations from the natural state are small: but at a certain point the steel, if soft, begins to undergo an alteration of form; at another point it breaks if much hardened; but when the hardness is moderate, it is capable of a much greater curvature without either permanent alteration or fracture; and this quality, which is valuable for the purposes of springs, is called toughness, and is opposed to rigidity and brittleness on the one side, and to ductility on the other. There may, however, be an apparent difference in the stiffness of some substances in different states, arising from the greater facility with which their dimensions are extended in one direction while they are contracted in another: thus elastic gum appears to possess a much greater degree of stiffness when its hardness is increased by cold than when it is at a more elevated temperature; but the change produced in this case by heat is not an increase of that ductility which facilitates a permanent alteration of form, but rather of the toughness which allows a temporary change of figure, continuing only while the force is applied. The effect of forging and of wiredrawing tends to lessen the ductility of metals, and to render them tough, and even rigid: so that in hammering copper and brass, and in drawing wire, it is necessary to anneal the metals more than once by fire, in order to restore their ductility, which is lessened by the operation. The corrosion of the surface of a metal by an acid is also said to render it brittle; but it is not impossible that this apparent brittleness may be occasioned by some irregularity in the action of the acid.

The last effect of force on solid materials is their fracture, which, as well as the former changes, may be produced either by impulse, or by pressure alone. The action which resists pressure is called strength, and that which resists impulse may properly be termed resilience. The strength of every body is in the joint ratio of its immediate cohesion and repulsion, or elasticity, and of its toughness, or the degree in which it may be extended, compressed, or otherwise deranged, without a separation of its parts. The resilience is jointly proportional to its strength and its toughness, and is measured by the product of the mass and the square of the velocity of a body capable of breaking it, or of the mass and the height from which it must fall in order to acquire that velocity; while the strength is merely measured by the greatest pressure that it can support in a state of rest.

The simplest way in which a body can be broken, is by tearing it asunder. The cohesive force continues to be increased as long as the tenacity of the substance allows the particles to be separated from each other without a permanent alteration of form; when this has been produced, the same force, if its action is continued, is generally capable of causing a total solution of continuity; and sometimes a separation takes place without any previous alteration of this kind that can be observed.

It follows from the nature of resilience, that a body of a pound weight, falling from the height of a yard, will produce the same effect in breaking any substance, as a body of three pounds falling from the height of a foot; so that here, as well as in the estimation of mechanical power, it is the energy, and not the momentum, that is to be considered as the measure of the effect. If we know the strength of any substance, and the degree in which it is capable of extension, we may easily determine its resilience from a consideration of the laws of pendulums. For the same weight which would break it by pressure, will acquire a sufficient impulse for breaking it, if it fall from a height equal to half the space through which the substance may be extended, supposing the direction of the stroke to be horizontal, so that its effect may not be increased by the force of gravity. Thus if the pressure of a weight of 100 pounds broke a given substance, after extending it through the space of an inch, the same weight would break it by striking it with the velocity that would be acquired by the fall of a heavy body from the height of half an inch, and a weight of one pound would break it by falling from a height of 50 inches.

It is obvious that the cohesive strength, as well as the resilience, of any substance must be simply proportional to the magnitude of its transverse section, that is, of the surface of fracture. Some experiments appear to show that it increases in a greater proportion than this surface, others that it increases in a smaller proportion; but it is probable that in both cases some accidental irregularities must have interfered, and that a wire two inches in diameter is exactly four times as strong as a wire one inch in diameter. The length has no effect either in increasing or in diminishing the cohesive strength; but the resilience is proportional to the length, since a similar extension of a longer fibre produces a greater elongation.

There is however a limit beyond which the velocity of a body striking another cannot be increased without overcoming its resilience, and breaking it, however small the bulk of the first body may be, and this limit depends on the inertia of the parts of the second body, which must not be disregarded when they are impelled with a considerable velocity. For it is demonstrable that there is a certain velocity, dependent on the nature of a substance, with which the effect of any impulse or pressure is transmitted through it; a certain portion of time, which is shorter accordingly as the body is more elastic, being required for the propagation of the force through any part of it; and if the actual velocity of any impulse be in a greater proportion to this velocity than the extension or compression, of which the substance is capable, is to its whole length, it is obvious that a separation must be produced, since no parts can be extended or compressed which are not yet affected by the impulse, and the length of the portion affected at any instant is not sufficient to allow the required extension or compression. Thus if the velocity with which an impression is transmitted by a certain kind of wood be 15000 feet in a second, and it be susceptible of compression to the extent of $\frac{1}{200}$ of its length, the greatest velocity that it can resist will be 75 feet in a second, which is equal to that of a body falling from a height of about 90 feet. And by a similar comparison we may determine the velocity which will be sufficient to penetrate or to break off a substance in any other manner; if we calculate the velocity required to convey the impulse from one part of the substance to the other, and ascertain the degree in which it can have its dimensions altered without fracture.

It is easy to understand, from this statement, the different qualities of natural bodies with respect to hardness, softness, toughness, and brittleness. A column of chalk, capable of supporting only a pound, will perhaps be compressed by it only a thousandth part of its length; a column of elastic gum, capable of suspending a pound, may be extended to more than twice its length, the elastic gum will therefore resist the energy of an impulse incomparably greater than the chalk. A diamond, so hard as to resist an enormous pressure, may be broken, by a moderate blow, with a small hammer. A weight of 1000 pounds, moving with a velocity of one foot in a second, and acting on a small surface of a board, may possess sufficient energy to break or to penetrate it; with a velocity of 100 feet in a second, a weight of $\frac{1}{10}$ of a pound will possess the same energy, and produce the same effect, if it act on a similar surface; but if the wood be so constituted, as to be wholly incapable of resisting a velocity of 100 feet in a second, it may be penetrated by a weight of $\frac{1}{100}$ of a pound as well as by one tenth, and by a moderately soft body as well as by a harder one. The whole board, however, if at liberty, would receive a much greater momentum from the impulse of the large weight, than from that of the small one, its action being continued for a much longer time. And it is for this reason that a ball shot by a pistol will perforate a sheet of paper standing upright on a table, without overturning it.

The strength, or rather hardness, of a substance, exposed to the action of a force that tends to compress it, must not be confounded with its resistance to a force applied longitudinally and tending to produce flexure. A slender rod of wood, when it yields to a longitudinal pressure, commonly bends before it breaks, and gives way at last to the force by a transverse fracture; but a column of stone or brick, and even a thick pillar of wood, is crushed without bending, and generally by a smaller force than that which would produce or continue a flexure. In this case the parts slide away laterally, and in a rectangular pillar; if the texture of the substance is uniform, and not fibrous, the surfaces of fracture will make nearly a right angle with each other, supposing the resistance arising from the lateral adhesion, in the direction of any surface or section, to be simply proportional to that section: but if this force, like that of friction, is increased by a pressure which tends to bring the parts into closer contact, the angle left after fracture must be more acute. (Plate X. Fig. 124.)

The power of the force of lateral adhesion, in resisting fracture, is considered by Mr. Coulomb as nearly equal to that of the direct cohesion of the same substance, or a little greater; while Professor Robison makes it twice as great. If, however, this force be supposed to be simply equal to the direct cohesion, it may be inferred that the strength of a square bar in resisting compression is twice as great as its cohesive strength, allowing that the fracture takes place in the surface of least resistance. It is, however, seldom that the strength, with which a body resists compression, is in so great a proportion as this to its cohesive strength; and where the substance is in any degree composed of fibres, they must naturally produce great irregularities by their flexure. The strength in resisting compression, must, according to this statement, be simply proportional to the magnitude of the section of the substance, although some experiments on freestone appear to indicate that when the section is increased, the strength is increased in a greater proportion; and there is no reason to suppose that it can be influenced either way by the length. A cylindrical or prismatic form is therefore the best that can be given to materials of a given bulk, in order to enable them to resist a force which tends to crush them, except that the additional pressure of their own weight on the lower parts, requires that those parts should be a little stronger than the upper parts. It appears also that something is gained by making the outline a little convex externally: for it may be demonstrated, that for a column or upright beam, to be cut out of a slab of equable thickness, supposing the strength to be independent of pressure, the strongest form is a circle. (Plate IX. Fig. 126, 127.)

When a body is broken by a transverse force applied very near to a fixed point, its lateral adhesion is overpowered by the effect which we have called detrusion, and its strength in this case is therefore generally somewhat greater than its direct cohesive strength. But when the part to which the force is immediately applied is at a distance from the fixed point greater than about one sixth of the depth, the fracture is no longer the immediate consequence of detrusion, but of flexure.

2

Flexure is the most usual manner in which fracture is produced; the superficial parts on the convex side are most extended, and usually give way first; except in soft fibrous substances, such as moist or green wood, which is more easily crushed than torn; and in this case the concave side fails first, and becomes crippled, and the piece still remains suspended by the cohesion of the fibres. After the convex surface has been cracked, the whole substance is usually separated, but not always; for example, a triangular beam, with one of the edges uppermost, may be charged with such a weight that the upper edge may be divided and the lower part may remain intire.

When a column or rafter is broken by the operation of a longitudinal pressure, the stiffness of the column being once overcome, a small addition of force is usually sufficient to produce fracture, unless the pressure has been applied to a part more or less distant from the axis; for in this case a moderate force may produce a moderate flexure, and a much greater force may be required to break the column. But in general, the stiffness of columns is of more consequence, than their strength in resisting transverse fracture.

The strength of beams of the same kind, and fixed in the same manner, in resisting a transverse force, is simply as their breadth, as the square of their depth, and inversely as their length. Thus if a beam be twice as broad as another, it will also be twice as strong, but if it be twice as deep, it will be four times as strong: for the increase of depth not only doubles the number of the resisting particles, but also gives each of them a double power, by increasing the length of the levers on which they act. The increase of the length of a beam must also obviously weaken it, by giving a mechanical advantage to the power which tends to break it: and some experiments appear to show, that the strength is diminished in a proportion somewhat greater than that in which the length is increased.

The strength of a beam supported at both ends, like its stiffness, is twice as great as that of a single beam of half the length, which is fixed at one end; and the strength of the whole beam is again doubled if both the ends are firmly fixed.

The resilience of a prismatic beam, resisting a transverse impulse, follows a law very different from that which determines its strength, for it is simply proportional to the bulk or weight of the beam, whether it be shorter or longer, narrower or wider, shallower or deeper, solid or hollow. Thus a beam ten feet long will support but half as great a pressure, without breaking, as a beam of the same breadth and depth, which is only five feet in length; but it will bear the impulse of a double weight striking against it with a given velocity, and will require that a given body should fall from a double height in order to break it.

It is therefore of great consequence in the determination of the form and quantity of the materials to be employed for any mechanical purpose, that we should consider the nature as well as the magnitude of the forces which are to be resisted. Stiffness, strength, or resilience, may be separately or jointly required in various degrees. For a cieling, stiffness would be principally desirable; for a door, strength; for the floor of a ball room, resilience; for a coach spring, resilience and flexibility, that is, resilience without stiffness. An observatory should be as stiff as possible, a ship as strong as possible, a cable as resilient as possible.

It is a common remark that a floor which shakes is the strongest; and, improbable as it appears at first sight, it may perhaps be founded in truth: for if the absolute strength of a stiff and a shaking floor were equal, the shaking floor would bear the effects of motion with the least injury. It is possible that a stiff floor, which would support a numerous assembly, might give way at a ball; while a more resilient one, which would be suited for dancing, might be destroyed by a crowded concert.

A coach spring, divided into plates, has the same power of resisting, without being broken, the momentum of the carriage, arising from sudden elevations and depressions, as it would possess if it formed one entire mass, while its greater flexibility allows it to regulate these motions in a much more gradual and gentle manner. A single piece of timber may perhaps sometimes have too much of the flexibility of a coach spring, its strata sliding in some degree on each other: in such a case its stiffness and strength may be increased by binding it very firmly with hoops.

The transverse strength of a perfectly elastic substance, fixed at one end, is to its direct cohesive strength as the depth of the substance to six times its length. This proportion is equally applicable to such substances as resist compression more strongly than extension: for their immediate repulsive force is probably not greater than their cohesive force, when their dimensions are equally changed, so that the middle of the beam is always in its natural state; and when the curvature is sufficient to overcome the cohesive force, the whole beam must give way. When, however, the substance is less capable of resisting compression than extension, the concave surface gives way first, and the strength depends immediately on the repulsive strength of the substance. This is perhaps the reason, that, in experiments on beams of oak, the transverse strength has seldom been found in a greater ratio to the whole cohesive strength than that of the depth to nine times the length.

It may be inferred from the consideration of the nature of the different kinds of resistance which have been explained, that if we have a cylindrical tree a foot in diameter, which is to be formed into a prismatic beam by flattening its sides, we shall gain the greatest stiffness by making the breadth or thickness 6 inches, and the depth $10\frac{1}{2}$, the greatest strength by making the breadth 7 inches and the depth $9\frac{1}{4}$, and the greatest resilience by making the beam square. The stiffness and the strength of the beam may be much increased by cutting the tree into four pieces, turning their edges outwards, and uniting them so as to make a hollow beam: but it will require great strength of union, to make the whole act as one piece, and the resilience of the beam will be rather diminished than increased by the operation.

The adoption of the hollow masts and beams which an ingenious mechanic has lately introduced, requires, therefore, some caution. For where an impulse is to be resisted, such a mast is no stronger than a solid mast of the same weight, and much weaker than a solid mast of the same diameter. The force of the wind is, however, rather to be considered as constituting a pressure than a finite impulse, except when a sudden squall carries a loose sail before it with considerable velocity. A similar caution may also be extended to some other attempts to make improvements in naval architecture: it is a common opinion, and perhaps a well founded one, that flexibility is of great advantage to a ship's sailing; if therefore we sacrifice too much resilience to strength, and too much of both to stiffness, we may perhaps create greater evils than those which we wish to avoid.

We have hitherto supposed the beams, of which the strength has been compared, to be prismatic, that is, of equal breadth and thickness throughout, which is not only the simplest form in theory, but the most generally useful in practice. If however we have the power of giving any form that we please to materials of a certain weight, which may often be done where several smaller pieces are to be cut out of a larger one, or a larger one to be composed of several smaller ones, or where the materials are either ductile or fusible, it is frequently possible to determine a more advantageous form than that of an equable beam or column. For since the extension which the parts of the substance admit, without giving way, is the limit of their strength, if the depth of a beam be everywhere equal, and the curvature unequal, the fracture will first take place where the curvature is greatest, and the superfluous strength of the other parts will be lost; so that, in order to have the greatest strength that a given quantity of materials is capable of affording in a beam of given length, the form must be such that the strength may be everywhere equal, the tension of the surface being equal throughout; and the depth must be as much smaller as the curvature is greater. It is also necessary to consider whether the substance is likely to be crushed, and whether it is liable to be broken by detrusion, rather than by flexure. Sometimes the depth of the beam may be limited, and sometimes its breadth; or it may be required that the breadth and depth may be always equal or proportional to each other, and the force may be either applied at one end of the beam, or it may be equally divided throughout its length; it may also principally depend on the weight of the substance itself; and the strongest form will be different, according to the different conditions of its application. In the most common cases, the outline must be either triangular, or parabolic, as if the point of the triangle were rounded off; but the curves required are sometimes of much more difficult investigation. (Plate X. Fig. 128 . . 147.)

The strength of bodies is sometimes employed in resisting torsion, as in the case of the axles of wheels and pinions, rudders of ships, and screws of all kinds: but there is seldom occasion to determine their absolute strength in resisting a force thus applied: if they are sufficiently stiff, their parts are not often separated by any violent efforts.

In order to investigate the strength of the various substances employed for the purposes of the mechanical arts, it is most convenient to use a machine furnished with proper supports, and gripes, or vices, for holding the materials, and with steelyards for ascertaining the magnitude of the force applied, while the extension or compression is produced by a screw or a winch, with the intervention of a wire, a chain, or a cord: provision ought also to be made for varying the direction of the force, when the flexure of the materials renders such a change necessary. (Plate XI. Fig. 148.)

According to the experiments of various authors, the cohesive strength of a square inch of razor steel is about 150 thousand pounds, of soft steel 120, of wrought iron 80, of cast iron 50, of good rope 20, of oak, beech, and willow wood, in the direction of their fibres 12, of fir 8, and of lead about 3 thousand pounds: the cohesive strength of a square inch of brick 300, and of freestone 200. Teak wood, the tectona grandis, is said to be still stronger than oak.

The weight of the modulus of the elasticity of a square inch of steel, or that weight which would be capable of compressing it to half its dimensions, is about 3 million pounds; hence it follows, that when a square inch of steel is torn asunder by a weight of 150000 pounds, its length is first increased to one twentieth more than its natural dimensions.

The strength of different materials, in resisting compression, is liable to great variation. In steel, and in willow wood, the cohesive and repulsive strength appear to be nearly equal. Oak will suspend much more than fir; but fir will support twice as much as oak; probably on account of the curvature of the fibres of oak. Freestone has been found to support about 2000 pounds for each square inch, oak in some practical cases more than 4000.

The strongest wood of each tree is neither at the centre nor at the circumference, but in the middle between both; and in Europe it is generally thicker and firmer on the south east side of the tree. Although iron is much stronger than wood, yet it is more liable to accidental imperfections; and when it fails, it gives no warning of its approaching fracture. The equable quality of steel may be ascertained by corrosion in an acid; but there is no easy mode of detecting internal flaws in a bar of iron, and we can only rely on the honesty of the workman for its soundness. Wood, when it is crippled, complains, or emits a sound, and after this, although it is much weakened, it may still retain strength enough to be of service. Stone sometimes throws off small splinters when it is beginning to give way: it is said to be capable of supporting by much the greatest weight when it is placed in that position, with respect to the horizon, in which it has been found in the quarry.

It is obvious that when the bulk of the substance employed becomes very considerable, its weight may bear so great a proportion to its strength as to add materially to the load to be supported. In most cases the weight increases more rapidly than the strength, and causes a practical limitation of the magnitude of our machines and edifices. We see also a similar limit in nature: a tree never grows to the height of 100 yards; an animal is never strong enough to overset a mountain. It has been observed that whales are often larger than any land animals, because their weight is more supported by the pressure of the medium in which they swim.

The force of friction, which resists the sliding of different bodies on each other, seems to be intimately connected with that lateral adhesion, or rigidity, which is opposed to the internal displacement of the parts of a single body, by the effect which we have denominated detrusion: and when the friction is considered as resisting pressure rather than motion, it approaches still more nearly to the same force. It is probably derived in great measure from the strength of the protuberant particles, which must be broken, bent, or compressed by the motion of the bodies on each other: but it is not always that the existence of such particles can be asserted, much less can they be made perceptible to the senses, and we can only examine the effects which they may be supposed to produce, by immediate experiments on the forces required to counteract them. Such experiments have been made on a very extensive scale by Musschenbroek and Coulomb, and many of their results have been confirmed by Mr. Vince, in a simple and elegant manner.

With a few exceptions, the friction of all solid bodies is, either perfectly, or very nearly, a uniformly retarding force, neither increasing nor diminishing when the relative velocity of the bodies concerned is changed. The friction of some rough substances is a little increased with the velocity, but, as they become more polished, this variation disappears. When, however, the motion is wholly extinct, and the bodies remain in contact with each other, their adhesion is usually greater than the friction, and by a continuation of the contact, it may become twice or even thrice as great, especially where the surfaces are large, and the substances but moderately hard.

The truth of the assertion, that friction is a uniformly retarding force, may be shown very conveniently by means of Atwood's machine for experiments on accelerated motion. By suffering the axis of the pulley to rest on the surface of any fixed substance, we may subject it to a friction of which the magnitude may be varied by different methods; and we shall find that the motions of the boxes still indicate the action of a uniformly accelerating force, the spaces described being always proportional to the squares of the times of descent; it follows, therefore, that since the operation of gravity is uniform, that of friction which is deducted from it at each instant, must also be uniform, in order that the remaining acceleration may follow the same law.

The uniformity of the force of friction may also be shown by the descent of a flat substance on an inclined plane: if the body be caused to begin its descent with a certain velocity, it will be retarded, when the resistance is greater than the relative force of gravity: in this case the retardation will continue until it is wholly stopped, the resistance not diminishing with the velocity. If, on the contrary, the relative weight overpowers the resistance at first, the motion will be continually accelerated, the resistance not being increased by the increase of the velocity. But since every experiment of this kind must be performed in the presence of the air, the resistance of this fluid, which follows another law, will in the end prevent the acceleration.

It may in general be asserted, with some exceptions, that the force of friction is simply proportional to the weight or pressure that brings the substances concerned into contact, independently of the magnitude of their surfaces: but Mr. Coulomb has observed that in many cases there is, besides this force, another resistance, amounting to several pounds for each square foot of the surface, which is independent of the pressure; and by calculating these forces separately, we may probably always ascertain the whole resistance with sufficient accuracy. This constant portion is usually much smaller than that which varies with the weight, and in all common cases it may be safely neglected, and the friction of stone on stone may be called equal to one half of the pressure, that of wood on wood one third, and that of metal on metal one fourth; and this may serve as an estimate sufficiently accurate for calculating the effects of machines; although, if their parts were perfectly adjusted to each other, and all the surfaces well polished, the friction would not in general exceed one eighth of the pressure, whatever might be the nature of the materials. The application of unctuous substances lessens the friction in the first instance; but unless they are frequently renewed, they sometimes tend rather to increase it.

The simplest mode of ascertaining the magnitude of the friction of two bodies, is to incline their common surface to the horizon until the one begins to slide on the other: this point determines the magnitude of their adhesion; but in order to find that of their friction when they are in motion, they must be first separated, and then allowed to move on each other, while the whole apparatus is gently agitated. The friction will then be to the pressure, as the height of the inclined plane to its horizontal length, when the inclination is barely such as to allow the continuance of any motion which is imparted to the substance placed on the plane.

It follows from the doctrine of the resolution of force, that when any body is to be drawn along a horizontal surface, which produces a resistance proportionate to the pressure, a part of the force may be advantageously employed in diminishing the pressure produced by the weight of the body; hence, in order for the most advantageous application of the force, its direction must be inclined to the horizon, and it may be demonstrated, that the inclination must be the same with that of a plane on which the relative weight of the body is precisely equal to the friction. Thus if we can determine the inclination of a road which is barely sufficient for a carriage to descend on it by its own weight, the same inclination will be the best possible for the application of any force by which the carriage is to be drawn along a horizontal road of the same materials.

It is obvious that an inclined plane, on which a weight rests by means of an adhesion proportionate to the pressure, can never be forced backwards by any increase of that pressure, since the resistance increases in the same proportion, and continues always sufficient to prevent the relative motion of the weight and the inclined plane. Two such planes, put together, would constitute a wedge, which would be equally incapable of giving way to a pressure applied to its opposite surfaces, each of them possessing similar properties with respect to friction. Thus, if the friction or adhesion were exactly one eighth of the pressure, the height of the inclined plane would be one eighth of its length, and the back of the wedge one fourth. Such a wedge would therefore possess a perfect stability with respect to any forces acting on its inclined surfaces. But the effects of agitation, and the minute tremors produced by percussion, have a great tendency to diminish the force of adhesion, by interrupting the intimacy of contact; and where a pin, a nail, or a screw is required to retain its situation with firmness, the inclination of the surfaces must be smaller than the angle of such a wedge as is barely capable of affording a sufficient resistance in theory.

It appears, therefore, that the force of lateral adhesion, acting between two bodies in contact, is of great importance in all mechanical arts; the firmness of architecture and of carpentry depends in great measure on it. This kind of resistance being equally powerful, when the force is applied in the direction of the surface, to whatever part of the surface it may tend, it follows that any body which is subjected to friction on all sides, will retain its situation with the same force, that was used in overcoming the friction, in order to bring it into that situation, or rather with a greater force, since the lateral adhesion is generally a little greater than the friction: so that a cylindrical wire cannot be withdrawn from a perforation in a board, by any direct force less than that which was employed in introducing it; and this kind of stability, together with that of a wedge or nail resisting a lateral pressure, constitutes the security of the lighter structures of carpentry, while those of architecture receive a great part of their firmness from the accumulation of weight, which makes the resistance of their lower parts to any lateral motion almost insuperable.

When a hard body penetrates another, or when a substance is ground away by the attrition of another, the force, which opposes the motion, is to be considered, like the force of friction, as a uniformly retarding force. There is no reason for imagining the stiffness of a bar, whether longer or shorter, to depend on the velocity of the body that bends it, and the space through which it may be bent, without breaking, is also limited only by the toughness of the materials. In the same manner, when the internal parts of a solid are broken and displaced by the penetration of another, or its external parts abraded by its attrition, the resistance is the same, whatever the velocity may be, and the space described by the body, before its velocity is destroyed, is always proportional to the square of that velocity, or to the energy which results from a combination of the proportions of the velocity and the momentum.

LECTURE XIV.

ON ARCHITECTURE AND CARPENTRY.

THE subjects, which we have lately examined, are to be considered as preliminary to the particular departments of practical mechanics. The first division of these is to consist of such as are employed in resisting forces of various kinds, but they may almost all be referred, without inconvenience, to the general heads of architecture and carpentry, of which the principal business is, to resist the force of gravitation. Architecture, in its most extensive sense, may be understood as comprehending carpentry, but the term is more usually applied to the employment of those materials, which are only required to resist the effects of a force tending principally to produce compression, while the materials used by carpenters are frequently subjected to the operation of a force which tends to extend or to bend them: the works of architects being commonly executed in stone or brick, and those of carpenters in wood, besides the occasional use of iron and other metals, in both cases.

The simplest problem in mechanical architecture appears to be, to determine the most eligible form for a column. The length and weight being supposed to be given, it is of importance to investigate the form which affords the greatest possible strength; but it is somewhat difficult to ascertain the precise nature and direction of all the forces which are to be resisted. If we considered the column as a beam fixed in the ground, and impelled by a transverse force, it ought to be much tapered, and reduced almost to a point at its extremity; but it is seldom that any force of this kind can be powerful enough to do more than overcome the weight alone of the column, and it is only necessary to regard the load which presses vertically on it; and whether we consider the force as tending to bend or to crush it, the forms commonly employed will appear to be sufficiently eligible. Mr. Lagrange seems to have been misled by some intricacies of mathematical investigation, too remote from physical accuracy, when he calculated that a cylinder was the strongest form for resisting flexure; that form approaches in reality much more nearly to an oblong spheroid, of which the outline is elliptical. The consideration of the flexure of a column is, however, of little practical importance in architecture, for upon a rough estimate of the properties of the materials usually employed, it may be computed that a column of stone must be about forty times as high as it is thick, in order to be capable of being bent by any weight which will not crush it; although a bar of wood or of iron may be bent by a longitudinal force, if its length exceed about twelve times its thickness. The force may therefore be considered as tending only to crush the column; and since the inferior parts must support the weight of the superior parts, in addition to the load which presses on the whole column, their thickness ought to be somewhat increased; and it appears from a consideration of the direction in which the fracture is most easily effected, that the outline ought to be made a little convex externally, and more curved above than below, which is the usual, although not the universal practice; an elliptic arc is perhaps the most eligible outline, or a curve formed by bending a ruler fixed at the summit of the column; sometimes the form is made to differ little from a cone, but such a figure is very inelegant. The diminution of the thickness amounts in general to about one sixth or one seventh of the whole, and sometimes to one fourth. (Plate XI. Fig. 149.)

For a light house, where a great force of wind and water was to be resisted, Mr. Smeaton chose a curve with its concavity turned outwards. If we calculated what would be the best form for a wooden pillar, intended to remain always immersed in the water to a certain depth, we should find that a cone or pyramid would possess the greatest possible strength for supporting the motion of the water; and a cone more acute than this would be equally capable of resisting the force of the wind, supposing it to be less active than that of the water; the part below the water might, therefore, be widened so as to become a portion of a more obtuse cone, the upper part remaining more slender; and the greatest agitation of the sea being near its surface, the basis of the pillar might be a little contracted, so as to have the outline of the lower part a little convex outwards, if the depth of the water were considerable. But in the case of a building of stone, the strength often depends as much on the weight of the materials as on their cohesive power: and the lateral adhesion, which is materially influenced by the weight, constitutes a very important part of the strength. For resisting a force which tends to overset the building, the form in which the weight gives the greatest strength is that of a conoid, or a solid of which the outline is a parabola, concave towards the axis: and for procuring, by means of the weight, a lateral adhesion which is every where proportional to the force, the form must be cylindrical. So that in a building circumstanced as we have supposed the pillar to be, there appears to be no reason for making either portion of the outline taken separately, convex towards the axis, although the angular junction of the two portions of cones might very properly be rounded off; and the upper parts might be a little enlarged if it were desirable to reduce the thickness of the walls. But the Eddystone light house is completely above the level of the sea, although in stormy weather every part of it is exposed to the action of the waves, the water being sometimes thrown up to a much greater height than that of the light house: so that it may be considered as exposed to the force of a fluid more and more powerful as it is nearer to the foundation; and in this point of view its form differs but little from that which the most accurate theory would point out; but it is probably a little weaker about the middle of its height, or somewhat lower, than in any other part. (Plate XI. Fig. 150.)

A wall must be reduced in thickness as it rises, for the same reason as a column is diminished; and if the wall is a part of a house, it must be reduced in a still greater degree, since the load, which is to be supported by it at different parts of its height, is usually much varied, by the weight of the floors, and of the contents of the apartments. But sometimes the obliquity of the surface of the wall may become inconvenient, by promoting the growth of moss and weeds. In building a wall, the first precaution that is required, is to dig deep enough to ascertain the nature of the ground; the next, to lay a sufficiently extensive and firm foundation; and it has been very properly recommended that where a well is wanted, it should be dug before the foundations of the house are laid, in order to examine the qualities of the different strata which are to support them. The disposition of the stones, or bricks, is not a matter of indifference; the strength is obviously greatest when all the surfaces are either horizontal or vertical; for if they are oblique, they must have a tendency to slide away laterally, and the wall must be very liable to crack: hence the reticulated walls, sometimes employed by the ancients, of which all the joints were oblique, possessed but little durability. If the materials are thrown together without order, they press on the parts in contact with them; but occasionally, as in the case of piers, or quays, this circumstance may be of some advantage in opposing external pressure; or at least the effect of such a pressure may remove the inconvenience which would otherwise arise from the irregularity of the structure.

In some cases it is necessary to unite the stones of a building mechanically, either by cramps of iron, fixed by means of melted lead, or by other methods, similar to those which are more usually employed in carpentry. Mr. Smeaton was obliged to fix the stones of his light house to the rock and to each other, by dovetail joints, and to connect each horizontal tier with the tier below it, by pins of wood passing through the stones, with wedges driven in at each end, to make them expand, and tie the stones fast together. But, in general, it is sufficient to employ mortar, made of lime or terras, and sand, of which the utility depends principally on the firmness and cohesive strength that it acquires in consequence of its chemical properties. Sometimes the whole structure is composed of a mass which is at first soft, but hardens as it dries; in this manner mud walls are built; and the materials called pisé are of a similar nature. (Plate XI. Fig. 151.)

The wall or column, when raised, must in general help to support a single lintel or beam, an arch, a dome, or a roof of carpentry. The strength of the lintel depends more on the nature of the substance, than on any art employed in forming it, excepting the precaution to give it as much depth as is convenient, especially towards the middle, if the depth be any where unequal; but the construction of an arch affords considerable scope for the exertion of mechanical science.

The simplest theory of the arch, supporting itself in equilibrium, is that of Dr. Hooke, the greatest of all philosophical mechanics. The arch, when it has only its own weight to bear, may be considered as the inversion of a chain suspended at each end; for the chain hangs in such a form, that the weight of each link is held in equilibrium by the result of the two forces acting at its extremities; and these forces or tensions are produced, the one by the weight of the portion of the chain below the link, the other by the same weight increased by that of the link; both of them acting originally in a vertical direction. Now supposing the chain inverted, so as to constitute an arch of the same form and weight, the relative situations of all the lines, indicating the directions of the forces, will remain the same, the forces acting only in contrary directions, so that they are compounded in a similar manner, and balance each other on the same conditions, but with this difference, that the equilibrium of the chain is stable, and that of the arch tottering. This property of the equilibrium renders an accurate experimental proof of the proposition somewhat difficult; but it may be shown that a slight degree of friction is sufficient for retaining in equilibrium an arch formed by the inversion of a chain of beads. The figure is called a catenaria, when the links are supposed to be infinitely small, and the curvature is greatest at the middle of the chain. It is not at all necessary to the experiment that the links of the chain be equal; the same method may be applied to the determination of the form requisite for an equilibrium, whatever may be the length or weight of the constituent parts of the arch; and when the arch is to be loaded unequally in different parts, we may introduce this circumstance into the experiment, by suspending proportional weights from different parts of the chain. Thus we may employ wires or other chains to represent the pressure, and adjusting them by degrees, till their extremities hang in a given line, we may find the form which will best support the weight of the materials, the upper surface, or extrados, of the arch being represented by the same line in an inverted position, while the original chain shows the form of the intrados, or of the curve required for the arch stones themselves. In common cases, the form thus determined will differ little from a circular arc, of the extent of about one third of a whole circle, rising from the abutments with an inclination of 30° to the vertical line, and it never acquires a direction much more nearly perpendicular to the horizon. It usually becomes more curved at some distance below the summit, and then again less curved. (Plate XI. Fig. 152 . . 154.)

But the supposition of an arch resisting a weight, which acts only in a vertical direction, is by no means perfectly applicable to cases which generally occur in practice. The pressure of loose stones and earth, moistened as they frequently are by rain, is exerted very nearly in the same manner as the pressure of fluids, which act equally in all directions: and even if they were united into a mass, they would constitute a kind of wedge, and would thus produce a pressure of a similar nature, notwithstanding the precaution recommended by some authors, of making the surfaces of the arch stones vertical and horizontal only. This precaution is, however, in all respects unnecessary, because the effect which it is intended to obviate, is productive of no inconvenience, except that of exercising the skill of the architect. The effect of such a pressure only requires a greater curvature near the abutments, reducing the form nearly to that of an ellipsis, and allowing the arch to rise at first in a vertical direction.

A bridge must also be so calculated as to support itself without being in danger of falling by the defect of the lateral adhesion of its parts, and in order that it may in this respect be of equal strength throughout, its depth at each point must be proportional to the weight of the parts beyond it. This property belongs to the curve denominated logarithmic, the length corresponding to the logarithm of the depth. If the strength were afforded by the arch stones only, this condition might be fulfilled by giving them the requisite thickness, independently of the general form of the arch : but the whole of the materials employed in the construction of the bridge, must be considered as adding to the strength, and the magnitude of the adhesion as depending in great measure on the general outline.

We must examine in the next place what is the most advantageous form for supporting any weight which may occasionally be placed on the bridge, in particular at its weakest part, which is usually the middle. Supposing the depth at the summit of the arch and at the abutments to be given, it may be reduced considerably in the intermediate parts, without impairing the strength, and the outline may be composed of parabolic arcs, having their convexity turned towards each other. This remark also would be only applicable to the arch stones, if they afforded the whole strength of the bridge, but it must be extended in some measure to the whole of the materials forming it.

If therefore we combine together the curve best calculated for resisting the pressure of a fluid, which is nearly elliptical, the logarithmic, and the parabolic curves, allowing to each its due proportion of influence, we may estimate, from the comparison, which is the fittest form for an arch intended to support a road. And in general, whether the road be horizontal, or a little inclined, we may infer that an ellipsis, not differing much from a circle, is the best calculated to comply as much as possible with all the conditions. (Plate XI. Fig. 155.)

The tier of bricks cut obliquely, which is usually placed over a window or a door, is a real arch, but so flat as to allow the apparent outline to be horizontal. Mr. Coulomb observes, that the greatest strength is obtained by causing all the joints to tend to a single point: but little dependence can be placed on so flat an arch, since it produces a lateral thrust which may easily overpower the resistance of the wall. For the horizontal force, required to support each end of any arch, is equal to the weight of a quantity of the materials which are supported by its summit, supposed to be continued, of their actual depth, to the length of a semidiameter of the circle of which the summit of the arch is a portion. This simple calculation will enable an architect to avoid such accidents, as have too often happened to bridges for want of sufficient firmness in the abutments. The equilibrium of a bridge, so far as it depends only on the form of the arch, is naturally tottering, and the smallest force which is capable of deranging it, may completely destroy the structure; but when the stones or blocks composing it have flat surfaces in contact with each other, it is necessary that the line expressing the direction of the pressure be so much disturbed, as to exceed at some part the limits of these surfaces, before the blocks can be displaced. When this curve, indicating the general pressure which results from the effect of a disturbing force, combined with the original thrust, becomes more remote from the centre of the blocks than one sixth of their depth, the joints will begin to open on the convex side, but the arch may still stand, while the curve remains within the limits of the blocks.

It is desirable that the piers of bridges should be so firm, as to be able not only to support the weight of half of each adjoining arch, but also to sustain, in case of the failure of one of those arches, the horizontal thrust of the other; and the same condition is obviously necessary for the stability of walls of any kind which support an arched or vaulted roof, wherever there is no opportunity of assisting the strength by ties or chains of any kind. There are two ways in which such a pier or wall may give way: it may either be overset, or caused to slide away horizontally; but since the friction or adhesion which resists the horizontal motion is usually greater than one third of the pressure, it seldom happens that the whole thrust of the arch is so oblique as not to produce a sufficient vertical pressure for securing the stability in this respect; and it is only necessary to make the pier heavy enough to resist the force which tends to overset it. It is not, however, the weight of the pier only, but that of the half of the arch which rests on it, that resists any effort to overset it, and in order that the pier may stand, the sum of these weights, acting on the end of a lever equal to half the thickness of the pier, must be more than equivalent to the horizontal thrust, acting on the whole height of the pier. The pier may also be simply considered as forming a continuation of the arch, and the stability will be preserved as long as the curve, indicating the direction of the pressure, remains within its substance.

The arches of Black Friars bridge are of an oval form, composed of circular arcs, and differing but little from ellipses; the arch stones are so large that the pressure in any direction might be very greatly increased without causing the general result to exceed the limits of their magnitude, or even to approach very near to their surfaces. (Plate XII. Fig. 156.)

The construction of a dome is less difficult than that of an arch, since the tendency of each part to fall is counteracted, not only by the pressure of the parts above and below, but also by the resistance of those which are situated on each side. A dome may therefore be erected without any temporary support, like the centre which is required for the construction of an arch, and it may at last be left open at the summit, without standing in need of a keystone, since the pressure of the lower parts is sufficiently resisted, by the collateral parts of the same horizontal tier, to prevent the possibility of their falling in, or of their forcing out the upper parts. The weight of the dome may however force out its lower parts, if it rises in a direction too nearly vertical; and supposing its form spherical, and its thickness equable, it will require to be confined by a hoop or chain as soon as the span becomes eleven fourteenths of the whole diameter. But if the thickness of the dome be diminished as it rises, it will not require to be bound so high: thus, if the increase of thickness in descending begin at about 30 degrees from the summit, and be continued until, at about 60 degrees, the dome becomes a little more than twice as thick as at first, the equilibrium will be so far secure; and at this distance it would be proper to employ either a chain, or some external pressure, to preserve the stability, since the weight itself would require to be increased without limit, if it were the only source of pressure on the lower parts. (Plate XII. Fig. 157.)

The dome of St. Paul's cathedral is elliptical, and is built of wood, and confined by strong chains, consisting of iron bars; that of the Pantheon at Rome is nearly circular, and its lower parts are so much thicker than its upper parts, as to afford sufficient resistance to their pressure: they are supported by walls of great thickness, and furnished with many projections which answer the purpose of abutments and buttresses. (Plate XII. Fig. 158, 159.)

A knowledge of the parts and proportions usually assigned to columns, and to buildings in general, and of their technical names and divisions, belongs rather to the subject of ornamental than to that of useful architecture; and the consideration of symmetry and elegance is in great measure foreign to that of the mechanical properties of bodies, which it is our present business to investigate. The five orders of ancient architecture are found to differ considerably in their proportions, in the different remains of Greek and Roman edifices; but there always remain some characteristic distinctions: the Tuscan is known by its strength and simplicity, without any peculiar ornament; the Doric by its triglyphs, or triangular grooves, above each column, imagined to represent the ends of beams; the Ionic by the large volutes, and the Corinthian by the foliage, respectively enveloping their capitals; and the Composite usually by the combination of both these characters; each order being lighter than the preceding, and being sometimes employed with it in the upper parts of the same building. In general, the length of the Tuscan column, with its capital, is equal to about seven diameters of the base, that of the Doric eight, of the Ionic nine, and of the Corinthian and Composite ten diameters. (Plate XII. Fig. 160 . . 164.)

The Gothic architects appear to have been superior to the Greeks in the mechanical arrangement of the parts of their edifices, so as to produce the most advantageous effect in preserving the general equilibrium. They made every essential member of their buildings a constituent part of their system of ornament, and even those embellishments, which, by a superficial observer, might be deemed useless or prejudicial, are frequently calculated, either by their strength, or by their weight, to serve some beneficial purposes. The pointed arch is not in all cases well calculated for equilibrium, but when it has a pillar resting on its summit, it is exceedingly strong. The most celebrated of modern architects have sometimes been less successful than those of the middle ages; and for want of paying sufficient attention to mechanical principles, have committed such errors in their attempts to procure an equilibrium, as have been followed by the most mischievous consequences. Examples of this might be pointed out in the bridges of our own country, and the churches of others; but if we are masters of the true theory of pressure, we shall be able to avoid similar errors, without examining the particular circumstances which have occasioned these accidents. (Plate XII. Fig. 165.)

The principles of equilibrium, which are employed in architecture, are equally applicable to many cases in carpentry; and where the work is principally calculated to withstand a thrust, there is little difference in the operation of the forces concerned; but where a tie is introduced, that is, a piece which resists principally by its cohesive strength, the parts often require to be arranged in a different manner. The general principle, that three forces, in order to retain each other in equilibrium, must be proportional to the sides of a triangle corresponding to their directions, is sufficient for determining the distribution of pressure in almost all cases that can occur. The conclusions which have been drawn from this principle, and from other similar considerations, respecting the strength of materials, will also be of great use in directing us how to determine the best forms for beams, rafters, and timbers of all kinds, and how to arrange and connect them in the best manner with each other.

The employment of the cohesive strength of materials in carpentry introduces a difficulty which scarcely exists in architecture. Two blocks, placed on each other, resist the force of a weight compressing them, as effectually as if they formed but one piece: but they have no sensible cohesion to enable them to withstand a force tending to separate them, and if they are required to co-operate by their cohesive strength, some mode of uniting them must be found. For this purpose, it is generally necessary to sacrifice a considerable portion of the strength of the materials employed. The most usual mode is to place the ends of the pieces side by side, first reducing their dimensions, where a regular outline is required; and to procure a firm adhesion between them by means of external pressure, or to employ the natural adhesion of some parts which are made to project beyond the rest in each piece, and receive in their interstices the corresponding projections of the other piece.

Where the adhesion is produced by external pressure only, it is of advantage to subdivide the joints into a considerable number of parts, as is usually done in the masts of ships, and to make the junction of any two pieces, following each other in the same line, as distant as possible from any other junction; for in this manner, the loss of strength may be diminished almost without limit, provided that the distance between the joints be great enough to afford a firm adhesion to each part. The junction may also be formed by an oblique line; but the obliquity must be so great that any lateral pressure may increase the stability of the wedge, the length being in a greater proportion to the depth than the pressure to the adhesion that it occasions; and the pieces must be pressed together very forcibly by means of hoops or bolts. (Plate XIII. Fig. 166 . . 168.)

Where the natural adhesion of some projecting parts in each piece is employed, the projections must be sufficiently long to secure their strength, and they must be as little prominent as possible, partly because the contiguous piece must be excavated for their reception, and partly because their strength is diminished when they project more than one sixth of their length. A beam united to another in this manner is said to be scarfed. (Plate XIII. Fig. 169.)

In order to preserve the strength of a compound beam, intended to resist a transverse action in a particular direction, it is necessary to avoid, as much as possible, reducing the depth of the beam in that direction, and to secure the union with the greatest care on the convex side of the beam, which is stretched by the operation of the force. Where no inconvenience can result from the projection of a piece on one side, it is easy to preserve the strength unimpaired, by splicing or fishing it on the convex side; and if the depth of the piece added be only half as great as that of the original beam, the strength will be somewhat increased by the operation, supposing the two ends to meet each other without any connexion. Such pieces require, however, to be firmly united, either by pins passing through them, or by blocks or joggles let in to a certain depth, in order to prevent their sliding on each other; and this mode of union is stronger than scarfing them, because it does not diminish the depth. (Plate XIII. Fig. 170, 171.)

Where the pieces to be connected together are in different directions, the end of one of them is usually reduced in its size, and becomes a tenon, while a mortise is cut in the other for its reception, and the joint is also often secured still more firmly by a strap of iron. If a joist be let into a beam, at its upper edge, and made very tight by wedges, the strength of the beam will not be materially diminished; but the vicissitudes of moisture and dryness may very much impair the firmness of the union, and the end of the joist may fail in dry weather to afford sufficient resistance to the flexure of the beam: so that in some cases it might be more adviseable to cut the mortise near the middle of the depth of the beam. If two pieces meet obliquely, and one of them exerts a thrust against the other, the simplest mode of opposing this thrust is to bind them together by a strap of iron fixed to the second piece; this strap renders it impossible for the first to advance without having its extremity crushed; it is also common to make a mortise in the second piece, a part of which serves as an abutment for the first: and for this purpose the piece must be continued far enough beyond the abutment to give the projection sufficient force of adhesion, a condition which is the more easily fulfilled when the action of the strap produces a pressure on it. The assistance of a strap is still more indispensable where the pieces are perpendicular to each other, and the force tends to draw one of them away from the other: in this case the mortise may be made a little wider at the remoter part, and the end of the tenon may be made to fit it by driving in wedges, in the same manner as Mr. Smeaton united his blocks of stone; but a large mortise would weaken the beam too much, and a strong strap or hoop is usually required for additional security. Such a strap ought always to be as straight as possible, so as to act only in the direction of the force to be resisted: it has been too customary to accommodate the strap to the form of the beams, or to make it deviate in other ways from a right line: but wherever a strap is bent in any direction, to a distance from a right line equal only to its depth in that direction, its strength is so reduced, as not to exceed one seventh of what it would have been, if it had remained straight. (Plate XIV. Fig. 172 . . 174.)

It is equally necessary in all other cases which occur in carpentry, to avoid as much as possible a transverse strain, the disadvantage of which is obvious from the great inferiority of the strength of any substance, resisting a transverse force, to its primitive cohesive or repulsive strength.. For similar reasons, it is proper to avoid employing a very open angle at a point where a load is supported, the great obliquity of the two pieces forming the angle requiring them to exert a great force in order to oppose a much smaller one. Allowance must also be made for the contraction of the timber, and care must be taken that it do not so alter the arrangement of the parts, as to bring a disproportionate strain on a point not calculated to support it. If the two pieces forming an obtuse angle consisted, either wholly or partly, of wood cut across the grain, and the piece joining their extremities were cut in the usual manner, the oblique pieces would contract considerably more as they became drier, and the angle would become more obtuse, so that the strain, produced by a given weight, would be greater than in the original state of the triangle. Sometimes the work is liable to be deranged by the operation of a lateral force, which may have appeared too trifling to produce any considerable effect, but which may still destroy the greater part of the strength, by causing the resistances to deviate from the plane of the forces which they are intended to oppose.

The framing of a roof is one of the most common and most important subjects for the employment of the theory of carpentry. If the rafters were simply to abut on the walls, they would force them outwards; a tie beam is therefore necessary, to counteract the thrust. In order to enable the tie beam to support a weight, a king post is suspended from the rafters; and frequently braces are again erected from the bottom of the king post, to support the middle of the rafters. Sometimes a flat or less inclined portion is placed in the middle, forming a kirb or mansard roof, somewhat resembling an arch; this form has the advantage, when it is properly proportioned, of lessening the transverse strain on the rafters, by making them shorter; but this purpose is answered equally well by the addition of the braces which have been already mentioned. A kirb roof affords, however, a greater space within, than a plain roof of the same height, and produces also somewhat less strain on the tie beam or on the abutments: the tie beam may be suspended from it by a king post and two queen posts, descending perpendicularly from the joints; and the place of the king post may be supplied by a cross beam uniting the heads of the queen posts and keeping them at a proper distance; this beam may also be suspended by a shorter king post from the summit. Such a roof appears to be more advantageous than it has been commonly supposed. (Plate XIII. Fig. 175 . . 177.)

The angle of inclination of a roof to the horizon usually varies in different climates: in Italy the height is generally less than one fourth of the breadth; in England it was formerly three fourths, but it now commonly approaches much more to the Italian proportion. In northern climates, a steep roof is required on account of falls of snow, which greatly increase the lateral thrust of the rafters; for the horizontal force exerted by a roof is always proportional to the length of a line perpendioular to the rafter, descending from its extremity till it meets another similar line drawn from the opposite rafter; and this perpendicular is obviously much increased when the roof becomes very flat. But for bearing the transverse strain, which tends to break the rafters themselves, a low roof is stronger than a high one, supposing the number of braces and queen posts equal on both: for if we have to support a given weight by a beam or rafter, whether it be placed in the middle, or equally divided throughout the length, we neither gain nor lose force by lengthening the beam and raising it higher, while the horizontal span continues the same, since the obliquity lessens the effect of the weight precisely in the same ratio that the length of the beam diminishes its strength; but by lengthening the beam we also add to the weight which is to be supported, and we thus diminish the strength of the roof. It must be observed, in calculating the strength of a rafter, that the slight flexure, produced by the transverse strain, has a material effect in diminishing its strength in resisting a longitudinal force; and this diminution must be determined according to the principles that have been laid down respecting the equilibrium of elastic substances.

Wooden bridges, and the temporary centres on which arches of stone are supported during their construction, depend nearly on the same principles as roofs: the external parts usually support a thrust, and the internal act as ties; but the abutments are generally capable of withstanding a horizontal thrust without inconvenience, so that by their assistance the strain on the ties is considerably diminished. Great strength may also be obtained, where it is practicable to support each part of the centre by two beams, in the direction of chords, bearing immediately on the abutments. (Plate XIV. Fig. 178, 179.)

The various articles of household furniture belong to subordinate branches of carpentry, but their form is in general more accommodated to convenience and elegance than to strength and durability. Yet even in making a chair, there is room for error and for improvement; the same principles that direct us in framing a roof, are capable of application here; but if they were implicitly followed, they would lead us to the employment of bars crossing each other in an inelegant manner. Doors, gates, locks, and hinges, are either parts of the carpenter's employment, or appendages to his works; and it is possible that, by attentive consideration, improvements might be made in all of them. Mr. Parker has devoted much time and labour to the subject of gates, with their hinges and fastenings, and has presented to the Royal Institution a very useful collection of models, which show the result of his investigations.

LECTURE XV.

ON MACHINERY.

HAVING taken a general view of those branches of practical mechanics in which forces are to be resisted, we are next to consider the modifications of forces and of motions; and in the first place the modes of applying forces, of changing their direction and intensity, and of communicating them to different parts of our machines by the intervention of rods, joints, cranks, wheelwork, ropes, or other flexible substances; in the second place, the structure of these substances, and the methods by which the union of flexible fibres in general may be effected; and in the third place, the regulation and equalisation of motion, by means of clocks and watches.

The modes of applying mechanical forces are almost as various as the machines that are constructed, and the purposes for which they are employed: but in general, the strength of men is applied by means of levers, or winches, or by walking wheels, which slide beneath them as they attempt to ascend; and that of other animals, by a horizontal arm projecting from a vertical axis, to which they are harnessed, and sometimes also by causing them to walk on or in a moveable wheel. Many of these arrangements may however be very conveniently considered as belonging to the particular objects for which each machine is constructed, especially to the modes of raising weights by cranes, and of grinding substances by mills.

When motion is simply communicated to a substance placed before the moving body, such materials must be employed as are capable of exerting a repulsive force, or a thrust; and these are generally of the same kind as are sometimes concerned in the operations of architecture, but more commonly in those of carpentry, particularly metal and wood. But when the body to be moved is behind the moving power, and is pulled along by it, chains or ropes are sometimes more convenient. In the union of wood for moveable machinery, it is generally advisable to avoid employing pins or bolts of metal; for these, by their superior weight and hardness, sometimes injure the wood in contact with them, and become loose.

When the direction of the motion communicated is also to be changed, levers or cranks may be employed, united by joints or hinges of various kinds. Sometimes a long series of connected rods is suspended by other rods or chains, so as to convey the effect of the force to a considerable distance; in this case the motion is generally alternate, when, for example, pumps are worked by means of a waterwheel at a distance from the shaft in which the pumps are placed. In this arrangement, there is no necessary loss on account of the alternation of the motion of the rods; for if they are suspended at equal distances from a number of fixed points, they will move backwards and forwards, in the manner of a single pendulum; but the magnitude of the friction is the principal inconvenience produced by the weight of the series. Where a lever is employed for changing the direction of a great force, its strength may be increased by the addition of a frame projecting in the direction of its depth; and if the lever is bent, a cross piece uniting its arms is still more requisite. (Plate XIV. Fig. 180 . . 182.)

For the communication of a rotatory motion, Dr. Hooke's universal joint is sometimes of use, especially when the inclination is not required to be materially changed; but if the obliquity is great, the rotation is not communicated equably to the new axis at all points of its revolution. This joint is formed by a cross, making the diameters of two semicircles, one of which is fixed at the end of each axis. (Plate XIV. Fig. 183.)

The best mode of connecting a rotatory motion with an alternate one, is, in all common cases, to employ a crank, acting on one end of a long rod, which has a joint at the other. If the rotatory motion of the crank be equable, the progressive motion of the rod will be gradually accelerated and retarded, and for a considerable part of the revolution the force exerted will be nearly uniform: but if we attempted to communicate at once to the rod its whole velocity in each direction, as has sometimes been done by inclined planes, or by wheelwork, the motion would become extremely

3

irregular, and the machinery would be destroyed by the strain. (Plate XIV. Fig. 184.)

On the other hand it must be observed, that the force applied to a machine may, in general, be divided into two portions; the one employed in opposing another force, so as to produce equilibrium only, the other in generating momentum. With respect to the first portion, a single crank has the inconvenience of changing continually the mechanical advantage of the machine; with respect to the second, its motion in the second quarter of its revolution is accelerated, instead of being retarded, by the inertia which this portion of the force is intended to overcome: and from a combination of both these causes, the motion must necessarily be rendered very irregular. They may, however, be completely removed by employing always cranks in pairs, one of them being fixed so as to make a right angle with the other, which is also the best position for two winches to be turned by two labourers; since the point of the circle, in which a man can exert his greatest strength, is nearly at the distance of a right angle, or a little more, from the point at which his force is smallest.

An alternate motion may be comunicated to a rod, so that the force may be either uniformly exerted, or varied according to any given law, by means of an inclined surface, formed into a proper curve, and acting on a friction wheel fixed to the rod; and a single plane surface, placed obliquely, would answer sufficiently well for this purpose. But in such cases, as well as when a crank is used, it is necessary to employ other means for supporting the rod in its proper situation; this may either be done by additional friction wheels, or in a more elegant manner, by such an arrangement of jointed rods, as will cause the extremity of one of them to move in a curve which does not sensibly differ from a right line. If we fix two pins in a beam, so as to connect to it two equal rods, of which the extremities are joined by a third, and the end of this third rod which is nearest to the centre of the beam be connected to a second beam of a proper length, the opposite end of the rod will initially describe a right line; and for this purpose the length of the second beam must be to the distance of the nearest pin from the centre as that distance is to the distance of the pins from each other. The same effect may also be produced by means of a frame, made of two pieces, each a

2

yard long, united by joints to each other, and to two other pieces of a foot each; one of the first pieces being fixed, if the shorter piece opposite to it be produced to the length of four feet, its extremity will move at first in a right line. The proportions of the rods may also be made more convenient than these, and others may be added to them, if it be required, which may make a line move so as to remain always in parallel directions. (Plate XIV. Fig. 185 . . 188.)

But of all the modes of communicating motion, the most extensively useful is the employment of wheelwork, which is capable of varying its direction and its velocity without any limit.

Wheels are sometimes turned by simple contact with each other; sometimes by the intervention of cords, straps, or chains, passing over them; and in these cases the minute protuberances of the surfaces, or whatever else may be the cause of friction, prevents their sliding on each other. Where a broad strap runs on a wheel, it is usually confined to its situation, not by causing the margin of the wheel to project, but, on the contrary, by making the middle prominent: the reason of this may be understood by examining the manner in which a tight strap running on a cone would tend to run towards its thickest part. Sometimes also pins are fixed in the wheels, and admitted into perforations in the straps; a mode only practicable where the motion is slow and steady. A smooth motion may also be obtained, with considerable force, by forming the surfaces of the wheels into brushes of hair. (Plate XV. Fig. 189.)

More commonly, however, the circumferences of the contiguous wheels are formed into teeth, impelling each other, as with the extremities of so many levers, either exactly or nearly in the common direction of the circumferences; and sometimes an endless screw is substituted for one of the wheels. In forming the teeth of wheels, it is of consequence to determine the curvature which will procure an equable communication of motion, with the least possible friction. For the equable communication of motion, two methods have been recommended; one, that the lower part of the face of each tooth should be a straight line in the direction of the radius, and the upper a portion of an epicycloid, that is, of a curve described by a point of a circle rolling on the wheel, of which the diameter must be half that of the opposite wheel; and in this case it is demonstrable that the plane surface of each tooth will act on the curved surface of the opposite tooth so as to produce an equable angular motion in both wheels: the other method is, to form all the surfaces into portions of the involutes of circles, or the curves described by a point of a thread which has been wound round the wheel, while it is uncoiled; and this method appears to answer the purpose in an easier and simpler manner than the former. It may be experimentally demonstrated, that an equable motion is produced by the action of these curves on each other: if we cut two boards into forms terminated by them, divide the surfaces by lines into equal or proportional angular portions, and fix them on any two centres, we shall find that as they revolve, whatever parts of the surfaces may be in contact, the corresponding lines will always meet each other. (Plate XV. Fig. 190 . . 192.)

Both of these methods may be derived from the general principle, that the teeth of the one wheel must be of such a form, that their outline may be described by the revolution of a curve upon a given circle, while the outline of the teeth of the other wheel is described by the same curve revolving within the circle. It has been supposed by some of the best authors that the epicycloidal tooth has also the advantage of completely avoiding friction; this is however by no means true, and it is even impracticable to invent any form for the teeth of a wheel, which will enable them to act on other teeth without friction. In order to diminish it as much as possible, the teeth must be as small and as numerous as is consistent with strength and durability; for the effect of friction always increases with the distance of the point of contact from the line joining the centres of the wheels. In calculating the quantity of the friction, the velocity with which the parts slide over each other has generally been taken for its measure: this is a slight inaccuracy of conception, for, as we have already seen, the actual resistance is not at all increased by increasing the relative velocity; but the effect of that resistance, in retarding the motion of the wheels, may be shown, from the general laws of mechanics, to be proportional to the relative velocity thus ascertained. When it is possible to make one wheel act on teeth fixed in the concave surface of another, the friction may be thus diminished in the proportion of the difference of the diameters to their sum. If the face of the teeth, where they are in contact, is too much inclined to the radius, their mutual friction is not much affected, but a great pressure on their axes is produced; and this occasions a strain on the machinery, as well as an increase of the friction on the axes.

If it is desired to produce a great angular velocity with the smallest possible quantity of wheelwork, the diameter of each wheel must be between three and four times as great as that of the pinion on which it acts. Where the pinion impels the wheel, it is sometimes made with three or four teeth only; but it is much better in general to have at least six or eight; and considering the additional labour of increasing the number of wheels, it may be advisable to allot more teeth to each of them than the number resulting from the calculation; so that we may allow 30 or 40 teeth to a wheel acting on a pinion of 6 or 8. In works which do not require a great degree of strength, the wheels have sometimes a much greater number of teeth than this; and on the other hand, an endless screw or a spiral acts as a pinion of one tooth, since it propels the wheel through the breadth of one tooth only in each revolution. For a pinion of six teeth, it would be better to have a wheel of 35 or 37 than 36; for each tooth of the wheel would thus act in turn upon each tooth of the pinion, and the work would be more equally worn than if the same teeth continued to meet in each revolution. The teeth of the pinion should also be somewhat stronger than those of the wheel, in order to support the more frequent recurrence of friction. It has been proposed, for the coarser kinds of wheelwork, to divide the distance between the middle points of two adjoining teeth into 30 parts, and to allot 16 to the tooth of the pinion, and 13 to that of the wheel, allowing 1 for freedom of motion.

The wheel and pinion may either be situated in the same plane, both being commonly of the kind denominated spur wheels, or their planes may form an angle: in this case one of them may be a crown or contrate wheel, or both of them may be bevilled, the teeth being cut obliquely. According to the relative magnitude of the wheels, the angle of the bevil must be different, so that the velocities of the wheels may be in the same proportion at both ends of their oblique faces: for this purpose, the faces of all the teeth must be directed to the point where the axes would meet. (Plate XV. Fig. 193, 194.)

In cases where a motion not quite equable is required, as it sometimes happens in the construction of clocks, but more frequently in orreries, the wheels may either be divided a little unequally, or the axis may be placed a little out of the centre; and these eccentric wheels may either act on other eccentric wheels, or, if they are made as contrate wheels, upon a lengthened pinion. (Plate XV. Fig. 195, 196.)

An arrangement is sometimes made for separating wheels which are intended to turn each other, and for replacing them at pleasure; the wheels are said to be thrown by these operations out of gear and into gear again.

When a wheel revolves round another, and is so fixed as to remain nearly in a parallel direction, and to cause the central wheel to turn round its axis, the apparatus is called a sun and planet wheel. In this case, the circumference of the central wheel moves as fast as that of the revolving wheel, each point of which describes a circle equal in diameter to the distance of the centres of the two wheels: consequently, when the wheels are equal, the central wheel makes two revolutions, every time that the exterior wheel travels round it. If the central wheel be fixed, and the exterior wheel be caused to turn on its own centre during its revolution, by the effect of the contact of the teeth, it will make in every revolution one turn more with respect to the surrounding objects, than it would make, if its centre were at rest, during one turn of the wheel which is fixed: and this circumstance must be recollected when such wheels are employed in planetariums.

Wheels are usually made of wood, of iron, either cast or wrought, of steel, or of brass. The teeth of wheels of metal are generally cut by means of a machine; the wheel is fixed on an axis, which also carries a plate furnished with a variety of circles, divided into different numbers of equal parts, marked by small excavations; these are brought in succession under the point of a spring, which holds the axis firm, while the intervals between the teeth are expeditiously cut out by a revolving saw of steel. The teeth are afterwards finished by a file; and a machine has also been invented for holding and working the file. (Plate XV. Fig. 197.)

It is frequently necessary in machinery to protract the time of application of a given force, or to reserve a part of it for future use. This is generally effected by suffering a weight to descend, which has been previously raised, or a spring to unbend itself from a state of forcible flexure, as is exemplified in the weights and springs of clocks and watches. The common kitchen jack is also employed for protracting and equalising the operation of a weight: in the patent jack the same effect is produced by an alternate motion, the axis being impelled backwards and forwards, as in clocks and watches, by means of an escapement, and the place of a balance spring being supplied by the twisting and untwisting of a cord.

In these machines, as well as in many others of greater magnitude, the fly wheel is a very important part, its velocity being increased by the operation of any part of the force which happens to be superfluous, and its rotatory power serving to continue the motion when the force is diminished or withdrawn. Thus, when a man turns a winch, he can exert twice as much force in some positions as in others, and a fly enables him in some cases to do nearly one third more work. In the pile engine, also, without the help of the fly, the horses would fall for want of a counterpoise, as soon as the weight is disengaged. Such a fly ought to be heavy, and its motion must not be too rapid, otherwise the resistance of the air will destroy too much of the motion; but in the kitchen jack, as well as in the striking part of a clock, where the superfluous force is purposely destroyed, the fly is made light, and strikes the air with a broad surface. An effect similar to that of a fly and a spring is sometimes produced in hydraulic machines by the introduction of an air vessel, the air contained in which is compressed more or less according to the intensity of the force, and exerts a more uniform pressure in expelling the fluid which is forced irregularly into it.

LECTURE XVI.

ON THE UNION OF FLEXIBLE FIBRES.

THE strength of cordage, and of other substances which are employed in the communication of motion, where flexibility is required, as well as the utility of other flexible materials which serve for furniture or for clothing, depends principally upon the lateral adhesion produced by twisting, or by the intermixture of fibres. The union of flexible fibres, therefore, being frequently subservient to the communication of motion, and the machinery, usually employed for producing it, belonging immediately to the subject of the modification of motion, we may with propriety consider at present, as far as our plan will allow us, those important branches of the mechanical arts, of which the object is to effect a union of this kind.

When a chain is made of wire, each link is separately bent, and remains united with the neighbouring links in virtue of its rigidity: but the fibres of vegetable and of animal substances must be united by other means. For this purpose we have recourse to the force of friction, or rather of lateral adhesion, and the fibres are so disposed, that besides the mutual pressure which their own elasticity causes them to exert, any additional force applied in the direction of the length of the aggregate, tends to bring the parts into closer contact, and to augment the adhesion, in the same manner as we have already seen that a wedge and a screw may be retained in their situations. The simple art of tying a knot, and the more complicated processes of spinning, ropemaking, weaving, and felting, derive their utility from this principle.

When a line is coiled round a cylinder, for instance, in letting down a weight, by means of a rope which slides on a post, or on such a grooved cylinder as is sometimes employed to enable a person to lower himself from a window in cases of fire, the pressure on the whole circumference is to the weight, as twice the circumference to the diameter; supposing, for example, that the friction of rope on metal were one tenth of the pressure, then a single coil of rope round a cylinder of metal would support about two thirds of the weight; or if the weights acting on the different ends are different, the adhesion may be a little greater or less than in this proportion, according to the manner in which the rope is applied. If such a rope made two or three coils, it would be impossible to apply a force sufficient to cause it to slide in the grooves.

From considering the effect of a force which is counteracted by other forces acting obliquely, we may understand both the effect of twisting, in binding the parts of a rope together, and its inconvenience, in causing the strength of the fibres to act with a mechanical disadvantage. The greater the obliquity of the fibres, the greater will be their adhesion, but the greater also will be their immediate tension, in consequence of the action of a given force in the direction of the rope: so that after employing as much obliquity and as much tension, as is sufficient to connect the fibres firmly, in all cases of relaxation and of flexure, and to prevent in some measure the penetration of moisture, all that is superfluously added tends to overpower the primitive cohesion of the fibres in the direction of their length.

The mechanism of simple spinning is easily understood; care is taken, where the hand is employed, to intermix the fibres sufficiently, and to engage their extremities as much as possible in the centre; for it is obvious that if any fibre were wholly external to the rest, it could not be retained in the yarn; in general, however, the materials are previously in such a state of intermixture as to render this precaution unnecessary. Where we have a number of single continuous fibres, as in reeled silk, they are sufficiently connected by twisting, and we have no need of spinning. In both cases such machinery has been invented for performing the necessary operations, as is both honourable and lucrative to the British nation.

A single thread or yarn, consisting of fibres twisted together, has a tendency to untwist itself; the external parts are the most strained in the operation, and at first shorten the thread, until the internal parts have no longer room for spreading out laterally, as they must necessarily do when their length is diminished; the elasticity of all the parts, therefore, resists, and tends to restore the thread to its natural state. But if two such threads are retained in contact at a given point of the circumference of each, this point is rendered stationary by the opposition of the equal forces acting in contrary directions, and becomes the centre, round which both threads are carried by the remaining forces, so that they continue to twist round each other till the new combination causes a tension, capable of counterbalancing the remaining tension of the original threads. Three, four, or more threads may be united nearly in the same manner: a strand consists of a considerable number of yarns thus twisted together, generally from sixteen to twenty five, a hawser of three strands, a shroud of four, and a cable of three hawsers or shrouds. Shroud laid cordage has the disadvantage of being hollow in the centre, or of requiring a greater change of form in the strands to fill up the vacuity, and in undergoing this change, the cordage stretches, and is unequally strained. The relative position and the comparative tension of all the fibres in these complicated combinations are not very easily determined by calculation; but it is found by experience to be most advantageous to the strength of the ropes to twist the strands, when they are to be compounded, in such a direction as to untwist the yarns of which they are formed; that is, to increase the twist of the strands themselves: and probably the greatest strength is obtained when the ultimate obliquity of the constituent fibres is the least, and the most equable. This advantage is obtained in a considerable degree by Mr. Huddart's method of adjusting the length of the strand to its position in the rope, and his registered cordage appears to derive a decided superiority from this arrangement of the strands. A very strong rope may also be made by twisting five or six strands round a seventh as an axis: the central strand, or heart, is found after much use to be chafed to oakum; it should be more twisted than the rest, in order to allow it to extend a little; such ropes are, however, unfit for running rigging, or for any use in which they are liable to be frequently bent.

Ropes are most commonly made of hemp, but various other vegetables are occasionally employed; the Chinese even use woody fibres, and the barks of trees furnish cordage to other nations; we have indeed in this country an example of the use of the bark of the lime tree, which is employed for garden matting. The finest hemp is imported from Riga and St. Petersburg. The

2

male and female flowers of hemp are on different plants; the male plants are soonest ripe, and require to be first pulled. They are prepared for dressing by being exposed to the air, and the fibrous part is separated from the dry pulp by beating and hackling. In spinning the yarn, the hemp is fastened round the waist; the wheel is turned by an assistant, and the spinner, walking backwards, draws out the fibres with his hands. When one length of the walk has been spun, it is immediately reeled, to prevent its untwisting. The machines employed in continuing the process of ropemaking are of simple construction, but both skill and attention are required in applying them so as to produce an equable texture in every part of the rope. The tendency of two strands to twist, in consequence of the tension arising from the original twist of the yarns, is not sufficient to procure an equilibrium, because of the friction and rigidity to be overcome; hence it is necessary to employ force in order to assist this tendency, and the strands or ropes afterwards retain spontaneously the form which has thus been given them: the largest ropes even require external force in order to make them twist at all.

The constituent ropes of a common cable, when separate, are stronger than the cable, in the proportion of about 4 to 3; and a rope worked up from yarns 180 yards in length to 135 yards, has been found to be stronger than when reduced to 120 yards, in the ratio of 6 to 5. The difference is owing partly to the obliquity of the fibres, and partly to the unequal tension produced by twisting. Mr. Huddart's ropes of 100 yarns lose but about one eighth of the whole strength of the yarns; and his experiments appear to show that similar ropes made in the common manner retain only one half of their original strength. The tarring of ropes, although sometimes necessary for their preservation from decay, is found to lessen their strength, probably because it produces partial adhesions between some of the fibres, which cause them to be disproportionally strained. A rope is also said to be weaker when wet than when dry, perhaps because the water enables the fibres to slide more readily on each other, or because the presence of water is in general favourable to separation of any kind. A good hempen rope will support, without danger, one fifth as many tons as the square of its circumference contains inches.

Flax is weaker than hemp, but not less extensively useful. Its growth considerably exhausts the strength of the soil which produces it; its cultivation is encouraged in this country by a bounty from government, and a large quantity is also imported from the north of Europe. The plant, while green, is laid in water for ten days, and undergoes a chemical change, which softens the pulpy part, without injuring the strength of the fibres, and renders it more easy, when it has been dried and exposed to the air for a fortnight, to separate the two substances in the process of dressing it. This is performed by beating it with the edge of a flat piece of wood, the stroke being oblique, and nearly in the direction of the fibres, and afterwards combing it, in order to reduce the fibres into regular order, and to prepare them for spinning. The refuse, consisting of the shorter fibres, is tow.

Cotton is a fine fibrous substance, that envelopes the seeds of a plant. The best is brought from the isle of Bourbon; but by far the greatest quantity from the West Indies, although the Turkish dominions as well as the East Indies furnish us with a considerable supply. It is usually white, but there is a yellow kind, which is used for nankeens. It is separated from the seeds by means of rollers, between which it passes, and leaves the seeds behind. It is then beaten, on a flake, or a stool covered with a texture of cord. Next, it is carded, either by hand, the fibres being drawn into regular order by cards, that is, by brushes of fine pointed wire; or, more commonly, by machinery, the cards being disposed in cylinders which revolve nearly in contact with each other. The drawing or roving machine then draws it into long flakes, a state preparatory to its being spun by Sir Richard Arkwright's machines or jennies, which form at once forty threads by the labour of one person.

The silkworm is bred in the greatest abundance in Italy and in Asia; it has lately been introduced very successfully into the British possessions in the East Indies. The principal food of the caterpillar is the white mulberry tree, which is too delicate to thrive well in northern climates: in Italy the trees are planted in beds, like willows, and the foliage is cut as it is wanted. The room in which the worms are fed, is kept at the temperature of 80 degrees of Fahrenheit. The eggs of a former year are hatched either by animal heat, or by that of the sun; at the age of six weeks, the caterpillars begin to spin, first a light external texture, which is carded and spun for coarse silk, and afterwards a compact oval pod or cocoon, of one continued thread. The threads of several cocoons are reeled off at the same time: for this purpose they are generally put into warm water, which kills the chrysalis; but when it is preserved, it soon turns to a moth, which lives but a few days, taking no food, and dies after producing eggs for the next season.

The silk is either yellow or white, but the white is an accidental variety only. By repeated washings, the yellow silk is bleached, and that which is originally white, acquires a more perfect whiteness. Soap is also used for removing a gummy substance that accompanies the silk of the cocoons.

Wool is distinguished into two principal varieties, long and short wool. The longest is from Lincolnshire; it is combed, by means of instruments furnished with a double row of long and sharp teeth of iron or steel; it is repeatedly drawn from one comb to the other, heat being used in the process, and also a little oil. The fleeces of long wool are generally heavier than those of short wool, but less valuable, on account of their coarseness; they are used for worsteds, and for cloths in which the separate threads remain visible, as stuffs, shalloons, serges, and tammies. Short wool, on the contrary, is carded, and is used for cloths in which the individual threads are concealed by the projecting fibres.

The principal use of thread and yarn, when spun, is for the purpose of weaving. The same force of lateral adhesion that retains the twisted fibres of each thread in their situations, is here also employed in giving firmness to the cloth; and this adhesion is generally increased by the action of any external force, tending to strain the whole texture.

The first step in weaving is to form a warp, which consists of threads placed side by side, continued through the length of the piece, and sufficient in number to constitute its breadth. This being wound on a beam or roller, in the loom, the threads are drawn through a harness, consisting of loops formed by twine fixed to bars or frames, which elevates and depresses the threads in succession by means of treadles, moved by the feet, in an order which is different, according to the different nature of the intended work; the cross thread or woof, being thrown between them at each alternation, by means of a shuttle, and forced into its place by a batten or comb, made of wire or reeds, while the piece, in proportion as it is completed, is rolled upon a second beam, opposite to the first.

Crape is composed of threads which are so strongly twisted, as to have a disposition to curl, and in weaving it, moisture is sometimes employed, in order to obviate this tendency during the process. Woollen cloth, when woven, is rendered stronger and more compact by means of the fulling mill, in which it is beaten by heavy hammers of wood, at the same time that fullers' earth, or alcaline substances of animal origin, are applied in order to cleanse it. In this operation, both its length and breadth are diminished, and it is reduced to a texture approaching to that of felt. The reason of the contraction is probably this, that all the fibres are bent by the operation of the hammer, but not all equally, and those which have been the most bent are prevented by their adhesion to the neighbouring fibres from returning to their original length. After fulling, the cloth is roughened by means of teasels, which are cultivated for the purpose; and the most projecting fibres are cut away by the operation of shearing.

The lateral adhesion of fibres of various kinds gives strength also to felted substances, assisted, as some assert, by minute barbs, with which the fibres of furs are said to be furnished. The whole strength is, however, much inferior to that of cloth; partly because the fibres are in general much shorter, and partly because their arrangement is less accurately adjusted.

The materials commonly used for felting, are the furs of rabbits and beavers, mixed with each other, and with sheep's wool, in various proportions, according to the quality required. A very fine fur has lately been discovered on the skin of a species of seal, mixed with its hair, and it has been employed not only for felting, but also for spinning and weaving into a cloth resembling the shawls of the East Indies. The fur of the rabbit is also mixed with a coarser hair, which is separated from it, by being first pulled off from the skins, with a sharper knife. The materials to be felted are intimately mixed by the operation of bowing, which depends on the vibrations of an elastic string; the rapid alternations of its motion being peculiarly well adapted to remove all irregular knots and adhesions among the fibres, and to dispose them in a very light and uniform arrangement. This texture, when pressed under cloths and leather, readily unites into a mass of some firmness; this mass is dipped into a liquor containing a little sulfuric acid, and when intended for a hat, is moulded into a large conical figure, which is reduced in its dimensions by working it with the hands, and is formed into a flat surface, with several concentric folds, which are still more compacted in order to make the brim, and the circular part of the crown, and forced on a block, which serves as a mould for the cylindrical part. The black dye is composed of logwood, sulfate of iron, and a little acetite of copper, or verdigris; and the stiffening is a thin glue.

The texture of paper is scarcely different from that of felt, except that its fibres are less visible to the naked eye. To make white paper, linen rags are ground with warm water in a mill, into a paste of the consistence of cream: a portion of the paste is taken up in a wire sieve, which is passed obliquely through it, and this, being a little shaken, subsides into a sheet, which is turned out on a piece of flannel; a number of sheets being thus formed, they are then pressed, first with the interposition of flannel, and afterwards alone, while they are still moist. For thick paper, two or more sheets are laid on each other before the first pressing. To fill up the pores of the paper, and to increase its strength, a size is employed, which is generally made by boiling shreds of parchment or untanned leather. Sometimes the size is added after printing on the paper, but this is only done in works of inferior elegance, and in this country not at all.

Such are the principal cases of the union of flexible fibres, for the different purposes of strength or of convenience. Their importance is such that they might be esteemed worthy of a more detailed consideration; but we are not likely to make any material improvements in these departments of mechanical art by the application of theoretical refinements.

LECTURE XVII.

ON TIMEKEEPERS.

THE measurement of time by clocks and watches is a very important and interesting department of practical mechanics. The subject is intimately connected with the consideration of astronomical instruments, but it is not essentially dependent on astronomical principles.

Time is measured by motion; but in order that motion may be a true measure of time, it must be equable. Now a motion perfectly free and undisturbed, and consequently uniform, is rendered unattainable by the resistances inseparable from the actual constitution of material substances. It becomes therefore necessary to inquire for some mode of approximating to such a motion. Astronomical determinations of time, which are the most accurate, can only be made under particular circumstances, and even then they assist us but little in dividing time into small portions.

The first timekeepers somewhat resembled the hour glasses which are still occasionally employed; they measured the escape of a certain quantity, not of sand, but of water, through a small aperture. In these clepsydrae, it appears from Vitruvius's account that wheelwork was employed, and the hour was shown on a graduated scale; the graduations were also probably so adjusted as to correct the error arising from the inequality of the velocity occasioned by the variation of the height of the water in the reservoir. This inconvenience was however sometimes wholly avoided, by means of a constant steam, which kept the vessel full, or still more elegantly, by the siphon of Hero, which was a bent tube supported by a float, so that its lower orifice, at which the water was discharged, was always at a certain distance below the surface. Dr. Hooke proposed to keep the reservoir full, by means of a semicylindrical counterpoise, so that the time might be determined either from the measure or weight of the quantity of water discharged, or from the position of the counterpoise. Various other modes might also be devised for making cheap and simple timekeepers on similar principles, dependent on the motion of various liquids or elastic fluids; but great accuracy could scarcely be expected from them. A candle sometimes serves as a coarse measure of time; and by burning a thread which passes through it, it may easily be made to answer the purpose of an alarm.

Clocks and watches are machines in which wheelwork is employed for the measurement of time, being driven by a weight or by a spring, and regulated by a pendulum or a balance. Watches differ from clocks, in being portable, and this condition excludes the pendulum and the weight from their construction.

It is conjectured that the Saracens had clocks which were moved by weights, as early as the eleventh century. Trithemius mentions an orrery, moved by a weight, and keeping time, which was sent, in 1232, by the Sultan of Egypt, as a present to the Emperor Frederic II. Wallingford, in 1326, had made a clock which was regulated by a fly. The use of such a fly in equalising motion depends on the resistance of the air, which increases rapidly when the velocity is increased, and therefore prevents any great inequality in the motion, as long as the moving power varies but little; and if the action of the weight were transmitted with perfect regularity by the wheels, and the specific gravity of the air remained unaltered by pressure or by temperature, a fly clock might be a perfect machine, the weight being always exactly counterbalanced by the resistance of the air, attending a certain velocity of the fly; and it might even be possible to regulate the inequalities of the action of the weight, by causing the fly to open and shut, or to turn on an axis, by means of a spring, according to the magnitude of the resistance. The unequal density of the air would however still remain uncompensated; and in this respect a liquid would be a better medium than an elastic fluid. For experiments which are but of short duration, and which require great precision, a chronometer regulated by a simple fly is still a useful instrument. Mr. Whitehurst's apparatus for measuring the time occupied in the descent of heavy bodies, is governed by a fly; the index is stopped by the machinery, and points out the time elapsed without an error of the hundredth part of a second.

The alternate motion of a balance, thrown backwards and forwards by the successive actions of a wheel impelling its pallets, is also capable of producing a degree of uniformity in the motion of the wheel; for the force operating on the pallet is consumed in destroying a velocity in one direction, and in generating a velocity in the contrary direction; and the space in which it acts being nearly the same in all cases, the velocity generated will also be nearly the same at all times, as long as the force remains the same. The addition of a balance to a clock was made soon after the year 1400, for Arnault, who died in 1465, describes a planisphere, constructed by his master De Fondeur, which had a balance with a scapement like that of a common watch, but without a spring. Such a balance vibrates much more slowly than a balance provided with a spring; if the balance spring of a common watch be removed, the hands will pass over the space of about twenty eight minutes in an hour.

It is said that before the pendulum was used, a balance wheel was sometimes suspended in a horizontal position by a thread passing through its axis, which coiled round it, and caused it to rise and fall as it oscillated backwards and forwards. This mode of regulation differed but little in principle from the modern pendulums, but it was more complicated and less accurate. Huygens, in somewhat later times, constructed a clock with a revolving weight, which rose higher, and increased the resistance, whenever an augmentation of the force increased the velocity; and he caused the thread, which supported the weight, to bend round a curve of such a form as to preserve the equality of the revolutions.

A chronometer may be constructed on this principle for measuring small portions of time, which appears to be capable of greater accuracy than Mr. Whitehurst's apparatus, and by means of which an interval of a thousandth part of a second may possibly be rendered sensible. If two revolving pendulums be connected with a vertical axis, in such a manner, as to move two weights backwards and forwards accordingly as they fly off to a greater or smaller distance, the weights sliding, during their revolution, on a fixed surface, a small increase of velocity will considerably increase the distance of the weights from the axis, and consequently the effect of their friction, so that the machine will be immediately retarded, and its motion may thus be made extremely regular. It may be turned by a string coiled round the upper part, and this string may serve as a support to a barrel, sliding on a square part of the axis, which will consequently descend as it revolves. Its surface, being smooth, may be covered either with paper or with wax, and a pencil or a point of metal may be pressed against it by a fine spring, so as to describe always a spiral line on the barrel, except when the spring is forced a little on one side by touching it slightly, either with the hand, or by means of any body of which the motion is to be examined, whether it be a falling weight, a vibrating chord or rod, or any other moving substance. In this manner, supposing a barrel a foot in circumference to revolve in two seconds, each hundredth of an inch would correspond to the six hundredth part of a second; and the scale might be still further enlarged if it were necessary. (Plate XV. Fig. 198.)

By means of this instrument we may measure, without difficulty, the frequency of the vibrations of sounding bodies, by connecting them with a point, which will describe an undulated path on the roller. These vibrations may also serve in a very simple manner for the measurement of the minutest intervals of time; for if a body, of which the vibrations are of a certain degree of frequency, be caused to vibrate during the revolution of an axis, and to mark its vibrations on a roller, the traces will serve as a correct index of the time occupied by any part of a revolution, and the motion of any other body may be very accurately compared with the number of alternations marked, in the same time, by the vibrating body. For many purposes, the machine, if heavy enough, might be turned by a handle only, care being taken to keep the balls in a proper position, and it would be convenient to have the descent of the barrel regulated by the action of a screw, and capable of being suspended at pleasure.

But for the general purposes of timekeepers, all other inventions have been almost universally superseded by the pendulum and the balance spring, or pendulum spring. About the year 1000, Ibn Junis, and the other Arabian astronomers were in the habit of measuring time, during their observations, by the vibrations of pendulums; but they never connected them with machinery. The equality of the times occupied by these vibrations, whether larger or smaller, was known to Galileo in 1600, and some time before 1633, he proposed that they should be applied to the regulation of clocks. But Sanctorius, in his commentary on Avicenna, describes an instrument to which he had himself applied the pendulum in 1612. Huygens made the same application only in 1658, which is the date of his work on the subject. In the same year, Hooke applied a spring to the balance of a watch; and soon after, he conceived the idea of improving timekeepers sufficiently for ascertaining the longitude at sea, but he was interrupted in the pursuit of his plan. Hooke was also probably the first that employed for a clock a heavy weight vibrating in a small arc; an arrangement from which the peculiar advantages of a pendulum are principally derived.

The objects which require the greatest attention in the construction of timekeepers, are these; to preserve the moving power, or sustaining force, as equable as possible, to apply this force to the pendulum or balance in the most eligible manner, and to employ a pendulum or balance of which the vibrations are in their nature as nearly isochronous as possible. In clocks, the sustaining force, being generally derived from a weight, is already sufficiently equable, provided that care be taken that the line by which it is suspended may be of equal thickness throughout, and may act on a perfect cylinder. But in some clocks, and in all watches, the moving power is a spring. One of the first clock springs is said to have been an old sword blade; a clock with such a spring was lately preserved at Brussels: the spring which is at present used, is a thin elastic plate of steel, coiled into a spiral form. Every spring exerts the more force as it is more bent; in order to correct this inequality, the chain or cord by which it acts on the work is wound on a spiral fusee; so that, in proportion as the force is lessened, it is applied to a larger cylinder, or a longer lever. The general outline of the fusee must be nearly such, that its thickness at any part may diminish in the same proportion as it becomes more distant from the point at which the force would cease altogether, the curve being that which is denominated a hyperbola; but the workmen have in general no other rule than a habitual estimation. (Plate XV. Fig. 199.)

Notwithstanding all possible precautions in the immediate application of the weight or spring, the irregular action of the teeth of the wheels, the increasing tenacity of the oil usually employed, and other accidental disturbances, make it still desirable to procure a further equalisation of the force, which is sometimes obtained in clocks, by raising the loaded arm of a lever to a given height, whence it may descend; and in watches, by bending a spring into a given position, from which it may return, so as to limit with great precision the propelling force employed in each vibration. The necessity of applying oil is sometimes in great measure removed by jewelling the holes in which the axes or verges run; a perforation being made in a plate of ruby, and a diamond applied upon this, in contact with the end of the axis; the hardness and high polish of these stones tending very considerably to diminish the friction.

There are also different methods of continuing the action of the force, while the clock or watch is wound up: a spring is interposed between the fusee and the wheel impelled by it, a little inferior in force to the original weight or spring, so as to remain always bent, until, when the pressure of the main spring is removed, it begins to act upon a fixed point on one side, and upon the wheel of the fusee on the other, so that it propels the work for a short time with a force nearly equal to that of the main spring. Sometimes also the spring is wound up by causing a small wheel to revolve round the centre of the fusee, having its teeth engaged on one side in those of a wheel which makes a part of the fusee, and on the other side with the internal teeth of a hoop connected with the work, so that the same pressure which winds up the spring tends also to turn the hoop round, and to continue the motion. (Plate XVI. Fig. 200.)

The scapement, by which the sustaining force is communicated to the pendulum or balance, demands a greater exertion of skill and accuracy than any other part of a timekeeper. Sometimes the alternate motion of the pendulum has been produced by the action of a crank, but this construction subjects it too much to the irregularities of the wheelwork, and is liable to several other objections. A crank cannot properly be called a scapement, for according to the etymology of the term, the pendulum must escape for a time from the action of the wheelwork, and in general, the more independent its motion is rendered, the better is the effect of the machine. The simplest forms in common use are the crutch scapement for a clock; and the pallets with a vertical wheel, for a watch; the dead beat scapement, and the cylinder with a horizontal wheel, are improvements on these; and the detached scapement is a still further refinement.

The crutch scapement, called by the French the anchor scapement, is an arch in the plane of the scape wheel, and parallel to that in which the pendulum vibrates, supporting at each extremity a pallet, of which the face is a plane, and which is impelled in its turn by the teeth of the scape wheel. The faces are so inclined, that the pallets are alternately forced, by the action of the teeth, to retire from the centre of the wheel: and great care is taken in making the teeth exactly at equal distances, so that they may fall regularly on the pallet, immediately after the disengagement of the teeth on the other side from the opposite pallet. (Plate XVI. Fig. 201.)

In the common watch, the axis of the balance is parallel to the plane of the scape wheel, which is a contrate or crown wheel, and the flat pallets are fixed on the axis of the balance, at the opposite parts of the circumference of the scape wheel. (Plate XVI. Fig. 202.)

In both these cases, the impulse given to one pallet carries the opposite pallet with some force against the approaching tooth, and drives the wheel a little backwards, with a visible recoil. Here the sustaining power, being applied principally at the extremities of the vibrations, disturbs their isochronism, or the equality of the times in which they are performed, by partially increasing the force. We may recollect that, in order that all vibrations, of whatever magnitude, may be performed in equal times, the force must be exactly proportional to the distance from a given point, consequently, if an additional force be applied near the extremities of the vibration only, the longer vibrations will occupy less time than the shorter; and we may observe that, by adding to the force of the spring of a common watch with the key, we may accelerate its motion, at the same time that the angular magnitude of the vibration is increased. The motion of the balance also, being slowest at the extremities of its vibration, where the sustaining force is applied, is more affected by the inequalities of this force than if it were subjected to its action through an equal space in the middle of the vibration. Yet a good clock on this construction may keep time without an error of the ten thousandth part of the whole, and a watch within a two thousandth. In the common watch scapement, there is little friction, for the force acts almost perpendicularly on the pallet; it appears to have been the oldest scapement, and was employed before the application of springs to balances: it requires a considerable extent of motion in the balance, and cannot therefore well be applied to clocks with such pendulums as vibrate in small arcs. The crutch scapement, on the contrary, cannot be applied immediately to a vibration in a very large arc; but by the interposition of a lever with a roller, or of a part of a wheel with a pinion, it may be adapted to the balance of a watch; and some watches thus constructed by Emery, Letherland, and others, appear to have succeeded very well.

To avoid the inconveniences of the recoiling scapements, Mr. Graham invented or introduced the dead beat for the clock, and the cylinder for the watch. In both of these, the tooth of the scape wheel rests, during the greater part of the vibration, on a cylindrical surface, and acts on the inclined plane for a short time only, in the middle of each vibration; so that a change of the sustaining power scarcely produces a sensible derangement of the isochronism; for which ever way we turn the key of a horizontal watch, as long as it continues to go, the frequency of its vibrations is scarcely affected. A good horizontal watch will keep time within about a ten thousandth part, especially if a little oil be frequently applied to it, or if the cylinder be made of a ruby: and the timekeeper in the observatory at Greenwich, with a dead beat scapement, made by Graham, varies from true time only two parts in a million. (Plate XVI. Fig. 203, 204.)

Still, however, the friction of the teeth of the scape wheel on the cylinder or pallet, and the tenacity of the oil, where it is employed, may interfere in a slight degree with the time of vibration, especially by the irregularities to which they are liable. If the friction were perfectly uniform, it would scarcely disturb the isochronism, but friction is always increased by an increase of pressure; hence, therefore, the effect of any addition to the sustaining force must tend in some degree to retard the vibrations; and to obviate this, the surfaces, on which the teeth rest, have sometimes been so formed as to create a slight recoil; but this construction does not appear to have been very successful in practice. The friction may, however, be considerably diminished by the duplex scapement, apparently so called from the double series of teeth employed. The teeth of the more prominent series are detained on a cylinder so small as to be unfit for receiving an impulse from them, the balance is therefore impelled by the other series of teeth, acting on a pallet at a greater distance from its axis. The French have sometimes employed a construction somewhat similar, which they call the comma scapement, the teeth first resting on a small arch of repose, and then impelling the curved surface of a pallet extending to a considerable distance beyond it. In both these cases the single pallet, which is impelled by a tooth of a simple form, requires less labour in the execution than a number of a larger teeth, each of which is to be finished with great accuracy: but watches on these constructions, especially those with the comma scapement, are too liable to be stopped by any sudden motion, although the duplex scapement begins to be frequently employed for pocket timekeepers. (Plate XVI. Fig. 205.)

Mr. Harrison avoided all friction on the pallet, by connecting it with the pendulum by means of a slender spring, so flexible as to follow the motion of the scape wheel to a sufficient extent without sliding on its teeth. But the construction which is most usually employed where the greatest accuracy is required, is the detached scapement; in which the teeth of the scape wheel always rest on a detent, excepting a short interval, when it is unlocked in order to impel the pallets. Mr. Mudge employed a detached scapement, actuated by a subsidiary spring, of which the force is scarcely liable to any variation; the detent being unlocked by the motion of the balance. Mr. Haley has refined still further on this construction, by causing the subsidiary spring to unlock the wheel in its return, so that the balance is relieved from this action, which may sometimes produce a slight irregularity. These constructions are, however, much too delicate for common pocket watches. In a clock, Mr. Cumming has employed a detached scapement, in which a lever is raised to a certain height by each tooth of the scape wheel, and acts immediately on the pendulum in its descent in the middle of the vibration. The scape wheel is unlocked by the pendulum during its ascent, and a variation of the pressure may, therefore, produce a very slight inequality in the motion of the pendulum. Mr. Nicholson has attempted to remove this cause of error, by a construction in which the scape wheel only assists the pendulum in raising the lever; but it depends on the degree of force applied, to determine what part of the weight the scape wheel shall sustain; this scapement cannot, therefore, by any means be considered as detached. It is, however, easy to remove the defect of Mr. Cumming's scapement, if it can be called a defect, by a method similar to that which Mr. Haley has applied to watches; each tooth of the wheel being unlocked by the descent of the lever on the opposite side, at the moment that it ceases to act on the pendulum, and remaining inactive until the pendulum meets it. (Plate XVI. Fig. 206, 207.)

The detents of the scapements of Mudge and Cumming are parts of the pallet, but in the timekeepers now commonly made by Arnold, Earnshaw, and others, the tooth is detained by a pallet or pin projecting from a lever, the point of which is forced back by the balance, at the moment that the pallet presents itself to another of the teeth. Mr. Arnold employs an epicycloidal tooth, acting on a single point of the pallet; Mr. Earnshaw makes a flat surface of the tooth first act on the point of the pallet, and then the point of the tooth on a flat surface of the pallet. In other respects there is little difference in these scapements; and both the artists have been judged worthy of a public reward for their success. (Plate XVI. Fig. 208, 209.)

The last of the three principal objects which require the attention of the watchmaker, is to employ a pendulum or balance of which the vibrations are in their nature perfectly isochronous. For this purpose the weight of the pendulum ought to move in a cycloidal arc, but the difficulty of producing such a motion in practice is much greater than the advantage derived from it, and a circular vibration, confined to a small arc, is sufficiently isochronous for all practical purposes. The error of such a vibration is nearly proportional to the square of the arc described by the pendulum, and amounts to a second and a half, in a day of 24 hours, for a single degree on each side the point of rest; so that a pendulum keeping true time in an arc of three degrees, would gain $13\frac{1}{2}$ seconds if the arc were very much contracted or made cycloidal, and would lose $10\frac{1}{2}$ seconds by having the vibration extended to an arc of four degrees. In order to avoid the friction which would be occasioned by the motion of the pendulum on an axis, it is usually suspended by a flexible spring, which is wholly free from friction.

The elasticity of this spring adds a minute force to the power of gravitation, which acts on the pendulum, and this force must be considered when the length of a simple pendulum is compared with the frequency of its vibrations. It does not, however, interfere with the equality of the vibrations among each other; for in all springs, Dr. Hooke's general law, that the force increases as the degree of flexure, is found for moderate oscillations to be perfectly accurate; such a force, therefore, accelerates the larger and the smaller vibrations precisely in the same degree. But in balances, it is desirable to have the velocity, and the extent of the vibration, as great as possible, in order that the motion may be the less influenced by the inequalities of the sustaining power; and in large excursions, Dr. Hooke's law is not so precisely true; there must also necessarily be some inaccuracy from the loss of a certain portion of the force in generating the momentum of the spring itself, which, when the form is spiral, introduces great intricacy into the calculation of the properties of the vibration. Yet it has been found by experiment that a certain length may be determined for almost every spring, which will afford vibrations either perfectly or very nearly isochronous. In order that the weight or inertia of the spring may interfere the less with the regularity of its motion, it is sometimes tapered, and made thinner at the extremity: it is now also usual in the best watches to employ a spring coiled into a cylindrical form, like that of the spring of a bell, of which the motion appears to be somewhat more regular than that of a flat spiral. This was indeed the original construction, but was probably laid aside on account of the space which it required. The balance springs are made of the finest steel, and the best are manufactured in this country, although the French are said to have the art of making their main springs of a better temper than ours. Sometimes the balance spring is made of an alloy of gold and copper; these springs are very elastic, but they are too liable to break. Mr. Earnshaw observes, that the strength of a spring always diminishes a little as it wears; and endeavours to derive a compensation for this diminution of strength, by employing a spring of such a form, that the vibrations in small arcs may be a little more frequent than in larger ones, in order that when the presence of dust and the tenacity of the oil contract the extent of the vibrations, this contraction may tend to produce an acceleration which compensates for the diminished force of the spring. But it is perhaps more eligible to make every compensation, as far as possible, independent of circumstances foreign to the cause of the error. The strength of the spring is found to be less impaired by use when it is hardened than when the steel is softer. It sometimes happens, that from a sudden motion, or from some other accidental circumstance, the balance of a timekeeper may be thrown beyond the point at which the pallets are impelled by the scape wheels, and the whole motion may from this cause be interrupted. To prevent this accident, a small bar or pin is usually fixed on the balance spring, which is carried outwards when the vibration begins to be extended too far, and stops the further progress of the balance, by intercepting a pin which projects from it. This arrangement is called banking the balance.

We have already seen that the squares of the times of vibration of two pendulums are proportional to their lengths; so that if we add to a pendulum one hundredth part of its length, we increase the time of its vibration very nearly one two hundredth. But since all bodies are expanded by heat, the variable temperature of the atmosphere must necessarily produce changes of this kind in the motions of pendulums, and it may be observed that a clock goes somewhat more slowly in summer than in winter. The same expansion has a similar effect in the motion of a balance, and the increase of temperature produces also a diminution of the elastic force of the spring itself. There is, however, a great difference in the expansibilities of various substances; dry deal is one of the least expansible, and is therefore often used for the rods of pendulums. Brass expands one part in a hundred thousand for every degree of Fahrenheit, or a little more or less than this, accordingly as it contains more or less zinc. Glass and platina are less than half as expansible as brass, iron about two thirds, and mercury three times as much. A pendulum of brass would therefore make one vibration in ten thousand less at 70° than at 50°, and would lose $8\frac{1}{4}$ seconds in a day; a balance regulated by a spring would lose much more; for I have observed that vibrations governed by the elasticity of steel have lost in frequency as much as one ten thousandth part for a single degree of Fahrenheit; and Berthoud informs us, that where a clock, probably with a pendulum of steel, loses 20 seconds by heat, a watch loses eight minutes.

Mr. Graham appears to have been the first that attempted to compensate for the effects of temperature by the different expansibilities of various substances. He employed, for a pendulum, a tube partly filled with mercury; when the tube expanded by the effect of heat, the mercury expanded much more; so that its surface rose a little more than the end of the pendulum was depressed, and the centre of oscillation remained stationary. This mode of compensation is still sometimes practised with success; but the gridiron pendulum is more commonly used: it was the invention of Harrison, who combined seven bars, of iron or steel, and of brass, in such a manner, that the bars of brass raised the weight as much as the bars of iron depressed it. At present five bars only are usually employed, two of them being of a mixture of zinc and silver, and three of steel. Mr. Ellicott suspended a pendulum at the extremity of a lever, which was supported by a pillar of brass, much nearer to the fulcrum; as the pendulum expanded, the end of the lever was raised in the same degree, and the weight remained at its original distance from the point of suspension, which was determined by a fixed plate, transmitting the slender spring, as usual, between two opposite edges. The same effect is produced more simply by suspending the pendulum from the summit of a bar nearly parallel to it, and of the same substance with itself, resting on a fixed support, and either of the same length with the pendulum, or a little longer, accordingly as the distance of the fixed plate from the point of support of the bar, is determined by materials which may be considered as nearly of an invariable length, or as liable to a certain degree of expansion. (Plate XVI. Fig. 210.)

All these methods of compensation are peculiar to clocks; for watches, it is usual to unite together two metals which differ in expansibility, so as to form a compound plate; one side of the plate is commonly of steel, the other of brass, and it is obvious that any increase of temperature, by causing the brass to expand more than the steel, must bend the whole plate. Such a plate is variously applied; the most accurate method, which is employed by Arnold and other modern artists, is to make it a part of the balance itself, fixing a weight on its extremity, which is brought nearer to the centre, by the increase of curvature of the plate, whenever the expansion of the arms of the balance tends to remove it further off. The best way of making the plate appears to be to turn a ring of steel, and to immerse it in melted brass, and then to turn away what is superfluous of the brass. The magnitude of the weight, and the length of the plate, may easily be so regulated, as to compensate not only for the expansion produced by heat, but also for the diminution of the elasticity of the spring. Sometimes also a plate has been applied in such a way as to shorten the spring when the temperature is increased, by an operation similar to that which serves to regulate a common watch, the clip, that determines the effective length of the spring, being moved backwards and forwards; and a similar effect has also been produced by dividing this clip into two parts, one of which is fixed to a compound plate, and is made to approach the other so as to confine the spring more narrowly, and thus diminish its length, upon an increase of temperature. (Plate XVI. Fig. 211.)

The flexure of a compound plate has also been applied in a simple and elegant manner by Mr. Nicholson to the pendulum of a clock, by causing it to support the upper extremity of the pendulum. The plate is placed horizontally, the brass being uppermost, and carries the pendulum in the middle, while the ends rest on two fixed points, of which the distance may be adjusted with great accuracy, so that when the temperature is increased, the curvature of the plate may raise the rod of the pendulum, enough to keep the weight or bob at a constant distance below the fixed point, which determines its upper extremity. (Plate XVI. Fig. 212.)

The resistance, opposed to the motion of a pendulum by the air, affects in some degree its velocity, and the variation of the density of the atmosphere must therefore also produce some irregularities in timekeepers: they are, however, too small to be sensible. Derham found that the resistance of the air accelerated the motion of a half second pendulum about four vibrations in an hour, by diminishing the arc in which it vibrated: and when the vibrations were restored to their original magnitude, the resistance of the air produced a retardation of eight vibrations in the same time. But a heavy pendulum, vibrating in a small arc, is very little affected by this resistance.

Besides these more essential parts of the watchmaker's art, there are several subordinate considerations which require his attention; the striking part in particular occupies, in clocks, and in repeating watches, no inconsiderable portion of the bulk of the machine. But the apparatus employed on these occasions requires neither refinement of invention nor delicacy of execution.

In old clocks, the number of hours struck is usually determined by the revolution of a certain portion of a wheel, which supports an arm, and allows the hammer to strike, until at a proper time it falls into a notch. In watches, and in more modern clocks, the same effect is produced by means of a spiral of 12 teeth, revolving once in 12 hours.

It is of considerable importance to the accurate performance of a good clock, that it should be firmly fixed to a solid support. Any unsteadiness in the support causes the point of suspension to follow the motion of the pendulum, and enlarges the diameter of the circle of which the pendulum describes an arc; it must, therefore, tend in general to retard the motion of the clock. Sometimes, however, an unsteady support may be of such a nature as to accelerate the motion; and an observation of this kind, made by Berthoud, has suggested to Bernoulli a theory of compound vibrations, which may perhaps be true in some cases, but is by no means universally applicable to every case. On account of some circumstances of this kind, it happens that when two clocks are placed near each other, and rest in some degree on the same support, they have often a remarkable effect on each other's vibrations, so as to continue going for several days, without varying a single second, even when they would have differed considerably if otherwise situated: and it sometimes happens that the clock which goes the more slowly of the two will set the other in motion, and then stop itself; a circumstance which has been explained from the greater frequency of the vibrations of a circular pendulum when confined to a smaller arc, the tendency of the pendulums to vibrate in the same time causing the shorter to describe an arc continually larger and larger, and the longer to contract its vibrations, until at last its motion entirely ceases. This sympathy has some resemblance to the alternate vibrations of two scales hanging on the same beam, one of which may often be observed to stop its vibrations when the other begins to move, and to resume its motion when its companion is at rest; but it is still more analogous to the mutual influence of two strings, or even two organ pipes, which, though not separately tuned to a perfect unison, still influence each other's vibrations in such a manner as to produce exactly the same note when they sound together.

LECTURE XVIII.

ON RAISING AND REMOVING WEIGHTS.

THE methodical arrangement of our subject leads us, after having considered the modifications of force, to those machines which are intended for counteracting it, or for producing motion in opposition to an existing force. The simplest of the forces to be counteracted, is gravitation, and it is one of the most common employments of mechanical powers to raise a weight from a lower to a higher situation. This operation is also intimately connected with the modes of overcoming the corpuscular force of friction or adhesion, which constitutes the principal difficulty in removing bodies horizontally from place to place; for if we had only to produce motion in an unresisting mass of matter, a loaded waggon might in time be drawn along by a silk worm's thread. The raising and removing of weights, therefore, together with the modes of avoiding friction in general, constitute the first part of the subject of the counteraction of forces, and the remaining part relates to the machinery intended for overcoming the other corpuscular powers of bodies, by such operations as are calculated to change their external forms.

Machines for raising weights, which involve only the mechanics of solid bodies, are principally levers, capstans, wheels, pullies, inclined planes, screws, and their various combinations, in the form of cranes.

A lever is a very simple instrument, but of most extensive utility in raising weights to a small height. We may recollect that levers are distinguished into two principal kinds, accordingly as the power and weight are on different sides, or on the same side of the fulcrum; the forces counteracting each other being in the one case in the same direction, in the other, in opposite directions. Thus, when a man lifts a stone by means of a lever of the first kind, resting on a fulcrum between himself and the stone, he presses down the end of the lever, and the utmost force that he can apply is equal to the whole weight of his body: but when he thrusts the lever under the stone, so that its extremity bears on the ground, it becomes a lever of the second kind, and in order to raise the stone, he must now draw the end of the lever upwards. In this direction, a strong man can exert a force equivalent to twice his weight; consequently the second kind of lever possesses here a temporary advantage over the first; although, if the operation were continued, the workman would be more fatigued by raising even the same weight by this method, than if he could conveniently apply his weight to a lever of the first kind; and for this purpose, cross bars have sometimes been added to levers, in order to enable several workmen to stand on them with advantage at once. A bent lever operates precisely with the same power as a straight one, provided that the forces be applied in a similar manner with respect to its arms: and in all cases, the forces capable of balancing each other are inversely as the distances of the points of action from the fulcrum. Some addition of force is necessary for overcoming the equilibrium, and producing motion, but the velocity of the motion being seldom of much consequence, a small preponderance is usually sufficient.

The principal inconvenience of the lever is the short extent of its action: this may, however, be obviated by means of the invention of Perrault, in which two pins are fixed in the lever, at a short distance from each other, sliding in two pairs of vertical grooves, provided with ratchets, so that when the long arm of the lever is pulled by means of a rope, the nearer pin serves as a fulcrum, and the more distant one is elevated at the same time with the weight, and is detained in its place by the click; but when the rope is slackened, the weight sinks a little, and raises the pin, which first served as a fulcrum, to a higher place in its groove. The same effects may also be produced by catches or clicks resting upon ratchets on the opposite sides of a single upright bar, which passes through a perforation in the lever. There must, however, be a considerable loss of force, from the continual intermission of the motion. (Plate XVII. Fig. 213.)

An axis with a winch, that is, a lever bent at the end, is known from the common machine for raising a bucket out of a well. A vertical or upright axis, with two or more levers inserted into it, becomes a capstan. In these cases, if we wish to estimate the force with accuracy, we must add to the radius of the axis half the thickness of the rope, when we compare it with the arm of the lever.

Sometimes the weight of a reservoir or bucket of water is employed for raising another bucket, filled with coals or other materials, by means of a rope or chain, coiled round a cylinder or drum, or two drums of different sizes. This machine is called a water whimsey: when the bucket of water has reached the bottom, a valve is opened by striking against a pin, and lets out the water. In a machine of this kind, employed in the Duke of Bridgwater's coal works, the water descends thirty yards, and raises a smaller quantity of coals from a depth of sixty. In such cases, supposing the action to be single, and the stream of water to be unemployed during the descent of the reservoir, a considerable preponderance may be advantageously employed in giving velocity to the weights, provided that the machinery be not liable to injury from their impulse.

An erect axis or drum, turned by the force of horses walking in a circle, is used for raising coals and other weights, and is called a gin, probably by corruption from engine: the buckets being attached to the opposite ends of a rope which passes round the drum, and which is drawn by means of its adhesion to the drum. One of the buckets descends empty, while the other is drawn up full, and when the motions of the buckets are to be changed, the horses are turned, or the wheels are made to impel the axis in a contrary direction, when any other moving power is employed.

When a ship's anchor is weighed, the cable itself would be too large to be bent round the capstan; it is therefore connected with it by means of an endless rope, called the messenger. As the messenger is coiled round the lower part of the capstan, it quits the upper part; so that its place becomes lower and lower, till at last it has no longer room on the capstan; it is therefore necessary to force it up from time to time: this is called surging the messenger; it is commonly done by beating it, and to facilitate the operation, the capstan is made somewhat conical. It has been proposed to employ lifters in different parts of the circumference, which are raised once in each revolution, by passing over an inclined plane, with the interposition of friction wheels; a patent has been taken out for the invention, and it has already been introduced in the navy. Some experienced judges, however, are of opinion, that it would be better and more simple to employ a capstan so much tapered, that the tension of the rope itself, guided only by a pulley, might always be sufficient to bring the messenger into its place.

The capstan, which consists of two cylinders of different sizes on the same axis, with a rope passing from the smaller one over a pulley, which is connected with the weight, and returning to be wound up by the larger one, is very powerful in its operation; but it requires a great length of rope for a small extent of motion. (Plate IV. Fig. 51.)

Wheelwork is employed in a variety of ways for raising weights: its powers are in all cases derived from the same principles as the actions of levers, each wheel and pinion being considered as composed of a series of bent levers, of which the axis is the common fulcrum, and which act in succession on the teeth of the next wheel. The simplest combination of wheelwork used for this purpose constitutes a jack; a bar which is furnished with teeth on one side, being raised by the last pinion. Such instruments were not unknown even to the ancients; the barulcus described by Hero was a machine of this nature.

A series of buckets connected by ropes, and passing over a wheel, is often employed for raising water to a small height; and sometimes even for solid substances in the state of powder, in particular for raising flour, in a corn mill; and in this case the flour must be brought within reach of the buckets by means of a revolving spiral, which pushes it gradually forwards. When a weight of any kind is raised in buckets distributed through the circumference of a wheel, the force, required for retaining the weight in equilibrium, is as much less than the weight, as the diameter of a circle is less than half the circumference, the remainder of the weight being supported by the axis of the wheel.

Pullies, and their combinations in blocks, are universally employed on board of ships. They are very convenient where only a moderate increase of power is required; but in order to procure a very great advantage, the number of separate pullies or sheaves must be very much multiplied; a great length of rope must also be employed; and it is said that in a pair of blocks with five pullies in each, two thirds of the force are lost by the friction and the rigidity of the ropes. The inconvenience resulting from a large number of pullies, may, however, as we have already seen, be considerably lessened when they are arranged in Mr. Smeaton's manner, the acting rope being introduced in the middle, so as to cause no obliquity in the block. Tackles, or combinations of pullies for raising weights, are most conveniently supported on shore by means of shears, which consist of three rods, or poles, resting on the ground, and meeting each other in the point of suspension. For raising stones in building, two poles are employed, with a rope fixed to their summit, which keeps them in a proper position; their lower ends are usually connected by a third pole, which serves as an axis. (Plate IV. Fig. 56. Plate XVII. Fig. 214.)

Sometimes a pulley is drawn horizontally along a frame, setting out from the point where the rope is fixed, so that while the bucket is raised, it is also transferred diagonally to the opposite end of the scaffolding. This apparatus is used in some of the Cornish stream works, in which the earth of a whole valley is raised, in order to be washed for the separation of tin ore. (Plate XVII. Fig. 215.)

A fixed inclined plane is often of use in assisting the elevation of great weights, by means of other machinery. It is supposed that in all the edifices of remote antiquity, where great masses of stone were employed, as in the pyramids of Egypt, and the druidical temples of this country, these vast blocks were elevated on inclined planes of earth, or of scaffolding, with the assistance also of levers and rollers. Inclined planes are frequently used for drawing boats out of one canal into another; and sometimes the local circumstances are such that this may be done with great convenience, merely by allowing a loaded boat to descend, and to turn the axis which raises an empty one. An example of this may be seen, on a large scale, in the Duke of Bridgwater's canal. This canal is extended, above ground, for forty miles on one level; an underground navigation twelve miles long joins it at Worsley, leading to the coal mines under Walkden moor. At a height of 35½ yards above this, is another subterraneous portion, nearly six miles in length. The connection between these levels is formed by an inclined plane: the boats are let down loaded, and proceed three miles along the tunnel into the open canal. The inclined plane is fixed in a stratum of stone, which fortunately has the most eligible inclination of 1 in 4, and is 33 yards in thickness, affording the most advantageous means of fixing every part of the machinery with perfect security. The whole length of the plane is 151 yards, besides a lock of 18 yards at the upper end. (Plate XVII. Fig. 216.)

Inclined planes are also universally employed for facilitating the ascent of heights, by men or by animals; they may either be uniform, as roads, or the general inclination of the surface may be superseded by the formation of separate steps or stairs. The inclination of the surface may be governed by the proportion of the strength of the animal to its weight, the force required to support any weight on a plane being to the whole weight as the height of the plane to its length; and if the plane be a little less inclined than the exact equilibrium would require, the animal will be able to acquire a sufficient velocity at first to carry it easily up the ascent with a motion nearly equable. The strength of a labourer may be advantageously employed in ascending a given height by a flight of steps, and placing himself on a stage which may raise a weight by its descent; but it appears that the force of other animals is less calculated for exertions of this kind.

The screw is not often immediately applied to the elevation of weights; although sometimes a number of screws has been used for raising by slow degrees a large and unmanageable weight, for instance, that of an obelisc: and a perpetual screw is frequently employed in giving motion to wheelwork. Such machines possess a considerable mechanical advantage, but they are subject to much friction, and are deficient in durability. Mr. Hunter's double screw might be applied with advantage, if the extent of the motion required were extremely small; but this limitation confines its utility within very narrow bounds.

A crane is a machine for raising weights by means of a rope or chain, descending from an arm, which is capable of horizontal motion, and passing over a pulley to be wound up on an axis. The axis is turned, either immediately, or with the interposition of wheelwork, by a winch, by the horizontal bars of a windlass, or by a walking wheel, and sometimes by the force of wind, of water, or of steam. A walking wheel is an advantageous mode of employing the strength of a labourer, but the bulk of the machine is sometimes inconvenient and detrimental: when, however, the man walks upon the wheel, and not within it, this objection is in great measure obviated. A walking wheel requires to be provided with some method of preventing the dangerous consequences of the rapid descent of the weight, in case of an accidental fall of the labourer: for this purpose, a catch is usually employed, to prevent any retrograde motion; a bar has also sometimes been suspended from the axis of the wheel, on which the man may support himself with his hands, and other similar precautions have been adopted. Sometimes the plane of a walking wheel is but little inclined to the horizon, and the man walks on its flat surface. In either case the labour of horses, asses, or oxen, may be substituted for that of men: but for cranes this substitution would be very disadvantageous, since much force would be lost in stopping frequently so bulky a machine as would be required. The employment of a turnspit dog is an humble example of the same operation, and even goats appear to have been sometimes made to climb in a similar manner. In a walking wheel used for raising water at Carisbrook Castle, in the Isle of Wight, the work was performed by the same individual ass for the whole of forty five years preceding 1771. Walking wheels have also been invented, on which horses were to act externally with their fore feet or hind feet only; but they have seldom, if ever, been applied to practical purposes. In general it is advisable that walking wheels for quadrupeds should present to them a path as little elevated as possible; and it might probably be of advantage to harness them either to a fixed point, or to a spring or weight, which would enable them to exert a considerable force even in a horizontal direction; but probably after all they might be more advantageously employed in a circular mill walk. (Plate XVII. Fig. 217.)

Mr. White's crane affords a good specimen of an oblique walking wheel; the force may be varied accordingly as the labourer stands at a point more or less distant from the centre; and in order to avoid accidents, a break is always acting on the axis of the wheel by its friction, except when it is removed by the pressure of the man's hand on a lever, upon which he leans as he walks. The force is also varied in some cranes by changing the pinion, which acts on the principal wheel, and an expanding drum has been contrived for the same purpose, consisting of a number of bars moveable in spiral grooves, so as to form a greater or smaller cylinder at pleasure. In order to place the weight in any situation that may be required, the pulley may be made to slide horizontally on the gib or arm. (Plate XVII. Fig. 218.)

A model of a crane was exhibited some years ago to the Royal Society, in which a large wheel fixed to a short axis was made to roll round on a plane, while the lower end of its axis was connected by a joint with another axis in a vertical position: then the wheel, having to describe a circumference somewhat larger than its own, was turned slowly, and therefore powerfully, round its axis, and the motion was communicated to the fixed axis. The machine, however, appears to be more curious than useful.

Sometimes a steelyard has been combined with a crane, for weighing goods at the same time that they are raised by it. A small crane, fixed in a carriage, is convenient for loading and unloading goods. In France, the carts used on the wharfs are generally so long as to reach the ground behind when depressed, and to furnish an inclined plane, along which the goods are raised by a lever and axis, or a kind of capstan, fixed in front.

For taking hold of stones which are to be raised by means of a rope, a hole is sometimes formed in them, wider within than at its opening, and in this a lewis is inserted, consisting of two inverted wedges, separated by a plug, to which they are fastened by a pin. (Plate XVII. Fig. 219.)

When a rope or chain, which is to raise a weight, is so long as to require a counterpoise, the effect of this may be varied according to the length of the rope, which is unbent, by hanging it on a second rope or chain, which acts on a spiral fusee, slowly turned by a wheel and pinion.

The use of cranes is so extensive and so indispensable, that their forms have been often multiplied on account of local circumstances, or even from caprice; but the constructions which have been described appear to be of the most general utility, and from them it will be easy to judge of others.

When weights of any kind are simply to be removed from one situation to another, the most natural and obvious method, if they are portable, is to carry them. There is, however, some scope for theory even in this common operation, and we have seen that calculations have been made in order to determine the most advantageous burden for a porter to carry, but the experience of a few trials would in general be a better guide. Some carry weights on their heads, others on their shoulders, others low down on their backs: and according to the situation of the burden, they bend forwards or backwards, so that the common centre of gravity of the weight and the body comes immediately or very nearly over some part of the ground between their feet. The difficulty of carrying a weight at the extremity of a long rod is easily understood from the properties of the lever, and the same principles will enable us to determine the distribution of a load between two porters, in whatever way they may carry it. Supposing the weight to be placed on a porter's horse, or hand barrow, and at equal distances from both extremities, each of the men will support an equal portion of it; but if it be nearer to the one than to the other, the load will be distributed in the same proportion as the poles are divided by the centre of the burden. For instance, if the weight were 300 pounds, and it were one foot distant from the one, and two from the other, the first would have to carry 200 pounds, and the second 100. If the porters ascend a hill, or a flight of steps, the distribution of the load will remain the same, provided that the centre of the weight lie in the plane of the poles. But if the weight consists of a large body placed on that plane, the centre of gravity being above it, the effect of an inclination to the horizon may materially change the distribution of the load, since the pressure will always be determined by the distance of the ends of the poles from the line passing perpendicularly through the centre of gravity; so that, if the elevation were sufficient, the whole burden might rest on the lower porter. And in the same manner, if the weight were suspended below the poles, the inclination would cause a greater proportion of the load to be borne by the upper porter. The force is, however, only thus distributed as long as the arms of the porters continue parallel to each other; but the inequality would naturally be lessened by a change of the directions in which they would act; it would only be necessary that those directions should meet in some part of the vertical line passing through the centre of gravity; the magnitude of each force would then be determined by the length of the side of a triangle corresponding to its direction, and the load might be either equally or unequally divided, according to the positions of the arms. (Plate XVII. Fig. 220, 221.)

A man can carry in general a weight four or five times as great as that which he can raise continually in a vertical direction with the same velocity: so that we may consider the resistance to be overcome as a kind of friction which amounts to about a fourth or a fifth of the weight. If we attempted to draw a weight along a horizontal surface, the resistance of the surface would often not only impede the motion, but also injure the texture of the substance to be moved. This injury may, however, be avoided by the interposition of a simple frame or dray, and the dray may be armed with a substance subject to little friction, as with iron: the friction may also be somewhat further diminished by making the outline of the dray a little convex below so that a slight agitation may be continually produced during its motion. Sometimes the simple expedient of placing a load on two poles of elastic wood, the thickest ends of which are supported by the horse, and the thinner drag on the ground, is of use both in diminishing the friction, by confining it to a smaller and smoother surface, and in equalising the motion, by the flexibility of the poles.

It often happens that agitation of any kind enables us to lessen considerably the friction between two bodies, especially when they are elastic. If we wish, for instance, to draw a ring along an iron rod, by a thread which is nearly perpendicular to it, we may exert all our strength in vain if we apply it by slow degrees, since the increase of force continues to increase the adhesion. But if we pull the ring suddenly, and then slacken the thread, it rebounds from the rod by its elasticity, and in this manner it slides readily along, by a continuance of alternations. In such a case, however, it would be more natural, if the thread were sufficiently heavy, to give it a serpentine motion, which would draw the ring in a more oblique direction. It is said that when a screw is fixed very firmly in a piece of iron, it may be extricated much more easily while the iron is filed in some neighbouring part. The agitation thus produced probably operates in a manner somewhat similar to that of the rod.

Friction may in general be considerably diminished by the interposition of oily substances, where the surfaces are of such a nature as to admit of their application. Thus common oil, tallow, or tar, are usually interposed between metals which work on each other. It is necessary to attend to the chemical properties of the oil, and to take care that it be not of such a nature as to corrode the metals employed, especially where the work requires great accuracy. Tallow is liable to lose its lubricating quality, unless it be frequently renewed. Between surfaces of wood, soap is sometimes applied, but more commonly black lead, which becomes highly polished. The advantages of canals, and of navigation in general, are principally derived from the facility with which the particles of fluids make way for the motion of bodies floating on them.

The interposition of rollers or of balls bears some resemblance to the application of fluids. Supposing the surfaces to be flat and parallel, a roller moves between them without any friction: but it has still to overcome the resistance occasioned by the depression which it produces in the substance on which it moves, and which is greater or less according to the softness and want of elasticity of the substance. If the substance were perfectly elastic, the temporary depression would produce no resistance, because the tendency to rise behind the roller would be exactly equivalent to the force opposing its progress before; and the actual resistance only arises from a greater or smaller want of elasticity in the materials concerned. The continued change of place of the rollers is often a material objection to their employment; their action may in some cases be prolonged by fixing wheels on their extremities, as well as by some other arrangements; but these methods are too complicated to afford much practical utility. Rollers may also be placed between two cylinders, the one convex and the other concave, and the friction may in this manner be wholly removed, whatever may be the magnitude of the rollers. (Plate XVII. Fig. 222, 223.)

The effect of friction in any machine being always diminished, in proportion as the velocity of the parts sliding on each other is diminished, it is obvious that by reducing the dimensions of the axis of a wheel as much as possible, we also reduce the friction. When the pressure on the axis is derived principally from the weight of the wheel itself, the friction may be lessened by placing the wheel in a horizontal position, and making the axis vertical:

for in this manner the weight may be supported on an axis ending in a very small surface, and the effect of the friction on this surface will be about one third less than if it acted at the circumference. The velocity of the parts sliding on each other may be still more reduced, by placing each extremity of the axis on another wheel, or between two wheels, on which the axis rolls as they turn round, so that the friction is transferred to the axis of these wheels, of which the motion is very slow. But when a great weight is to be supported, it is necessary that the friction wheels be very strong, and very accurately formed; for if their surface were irregular, they might stand still, and their use would be destroyed. (Plate XVIII. Fig. 224.)

Perrault attempted to avoid all friction by supporting the axis of a wheel in the coil of a rope, which allowed it to turn while the whole wheel ascended and descended; but the stiffness of a rope occasions in general even a greater resistance than the friction for which it is substituted.

The wheels of carriages owe a great part of their utility to the diminution of friction, which is as much less in a carriage than in a dray, as the diameter of the axle is less than that of the wheel, even supposing the dray to slide on a greased surface of iron. The wheels also assist us in drawing the carriage over an obstacle, for the path which the axis of the wheel describes, is always smoother and less abrupt than the surface of a rough road on which the wheel rolls. It is obvious that both these advantages are more completely attained by large wheels than by smaller ones; the dimensions of the axis not being increased in the same proportion with those of the wheel, and the path of the axis, to which that of the centre of gravity is similar, consisting of portions of larger circles, and consequently being less curved; and if the wheels are elastic, and rebound from an obstacle, the difference is still increased. It is, however, barely possible, that the curvature of the obstacle to be overcome may be intermediate between those of a larger and of a smaller wheel; and in this case the higher wheel will touch a remoter part of the obstacle, so that the path of the axis will form an abrupt angle, while the smaller wheel follows the curve, and produces a more equable motion; this, however, is a case of rare occurrence, and an advantage of little importance. (Plate XVIII. Fig. 225, 226.)

The greater part of the resistance to the motion of a carriage very frequently arises from the continual displacement of a portion of the materials of the road, which do not react on the wheels with perfect elasticity, but undergo a permanent change of form proportional to the loss of force. Hence, in a soft sand, although the axles of the wheels may move in a direction perfectly horizontal, the draught becomes extremely heavy. The more the wheel sinks, the greater is the resistance, and if we suppose the degree of elasticity of the materials, and their immediate resistance at different depths to be known, we may calculate the effect of their reaction in retarding the motion of the carriage. Thus, if the materials were perfectly inelastic, acting only on the preceding half of the immersed portion of the wheel, and their immediate pressure or resistance were simply proportional to the depth, like that of fluids, or of elastic substances, the horizontal resistance would be to the weight nearly as the depth of the part immersed to two thirds of its length; but if the pressure increased as the square of the depth, which is a more probable supposition, the resistance would be to the weight as the depth to about four fifths of the length; the pressure may even vary still more rapidly, and we may consider the proportion of the resistance to the weight as no greater than that of the depth of the part immersed to its length, or of half this length to the diameter of the wheel; and if the materials are in any degree elastic, the resistance will be lessened accordingly. But on any of these suppositions, it may be shown that the resistance may be reduced to one half, either by making a wheel a little less than three times as high, or about eight times as broad as the given wheel. This consideration is of particular consequence in soft and boggy soils, as well as in sandy countries; thus, in moving timber in a moist situation, it becomes extremely advantageous to employ very high wheels, and they have the additional convenience that the timber may be suspended from the axles by chains, without the labour of raising it so high as would be necessary for placing it upon a carriage of any kind. (Plate XVIII. Fig. 227.)

But the magnitude of wheels is practically limited, by the strength or the weight of the materials of which they are made, by the danger of overturning when the centre of gravity is raised too high, and in the case of the first pair of wheels of a four wheeled carriage, by the inconvenience that would arise, in turning a corner, with a wheel which might interfere with the body of the carriage. It is also of advantage that the draught of a horse should be in a direction somewhat ascending, partly on account of the shape of the horse's shoulder, and partly because the principal force that he exerts is in the direction of a line passing through the point of contact of his hind feet with the ground. But a reason equally strong, for having the draught in this direction, is, that a part of the force may always be advantageously employed in lessening the pressure on the ground; and to answer this purpose the most effectually, the inclination of the traces or shafts ought to be the same with that of a road on which the carriage would begin or continue to descend by its own weight only. In order to apply the force in this manner to both pairs of wheels, where there are four, the line of draught ought to be directed to a point half way between them, or rather to a point immediately under the centre of gravity of the carriage; and such a line would always pass above the axis of the fore wheels. If the line of draught pass immediately through this axis, the pressure on the hind wheels will remain unaltered; and if the traces or shafts be fixed still lower, the pressure on the hind wheels will even be somewhat increased by the draught. It is evident, therefore, that this advantage cannot be obtained if the fore wheels are very high; we may also understand that in some cases the common opinion of the eligibility of placing a load over the fore wheels, rather than the hind wheels, may have some foundation in truth. When several horses are employed, the draught of all but the last must be nearly horizontal; in this case the flexure of the chain brings it into a position somewhat more favourable for the action of the horses; but the same cause makes the direction of its attachment to the waggon unfavourable; further than this there is no absolute loss of force, but it appears to be advisable to cause the shaft horse to draw in a direction as much elevated as possible; and on the whole it is probable that horses drawing singly have a material advantage, when they do not require additional attendance from the drivers.

The practice of making broad wheels conical has obviously the disadvantageous effect of producing a friction at each edge of the wheel, when the carriage is moving in a straight line; for such a wheel, if it moved alone, would always describe a circle round the vertex of the cone to which it belongs. When the wheels are narrow, a slight inclination of the spokes appears to be of use in keeping them more steady on the axles than if they were exactly vertical; and when, by an inclination of the body of the carriage, a greater proportion of the load is thrown on the lower wheel, its spokes, being then in a vertical position, are able to exert all their strength with advantage. The axles being a little conical, in order that they may not become loose, or may easily be tightened as they wear, it is necessary that they should be bent down, so that their lower surfaces may be horizontal, otherwise the wheels would press too much on the linch pin. For this reason, the distance between the wheels should be a little greater above than below, and their surfaces of course slightly conical. (Plate XVIII. Fig. 228.)

It has been proposed to fix the wheels to their respective axles, to continue the axles as far as the middle of the carriage only, and to cause them to turn on friction wheels or rollers; a plan which may succeed if the apparatus is not too complicated for use; but in fact the immediate friction on the axles is not great enough to render this refinement necessary. If both opposite wheels were fixed to a single axis, one of them would be dragged backwards and the other forwards, whenever the motion deviated from a straight line; and a similar effect actually takes place in those carriages which are supported on a single roller.

The effect of the suspension of a carriage on springs is to equalise its motion, by causing every change to be more gradually communicated to it, by means of the flexibility of the springs, and by consuming a certain portion of every sudden impulse in generating a degree of rotatory motion. This rotatory motion depends on the oblique position of the straps suspending the carriage, which prevents its swinging in a parallel direction; such a vibration as would take place if the straps were parallel, would be too extensive, unless they were very short, and then the motion would be somewhat rougher. The obliquity of the straps tends also in some measure to retain the carriage in a horizontal position: for if they were parallel, both being vertical, the lower one would have to support the greater portion of the weight, at least according to the common mode of fixing them to the bottom of the carriage, the spring, therefore, being flexible, it would be still further depressed. But when the straps are oblique, the upper one assumes always the more vertical position, and consequently bears more of the load; for when a body of any kind is supported by two oblique forces, their horizontal thrusts must be equal, otherwise the body would move laterally; and in order that the horizontal portions of the forces may be equal, the more inclined to the horizon must be the greater: the upper spring will, therefore, be a little depressed, and the carriage will remain more nearly horizontal than if the springs were parallel. The reason for dividing the springs into separate plates has already been explained: the beam of the carriage, that unites the wheels, supplies the strength necessary for forming the communication between the axles: if the body of the carriage itself were to perform this office, the springs would require to be so strong that they could have little or no effect in equalising the motion, and we should have a waggon instead of a coach. The ease with which a carriage moves, depends not only on the elasticity of the springs, but also on the small degree of stability of the equilibrium, of which we may judge in some measure, by tracing the path which the centre of gravity must describe, when the carriage swings. (Plate XVIII. Fig. 229.)

The modes of attaching horses and oxen to carriages are different in different countries, nor is it easy to determine the most eligible method. When horses are harnessed to draw side by side, they are usually attached to the opposite ends of a bar or lever; and if their strength is very unequal, the bar is sometimes unequally divided by the fulcrum, the weaker horse being made to act on the longer bar, and being thus enabled to counteract the greater force of his companion. But even without this inequality, a compensation takes place, for the centre on which the bar moves is always considerably behind the points of attachment of the horses; and when one of them falls back a little, the effective arm of the lever becomes more perpendicular to the direction of his force, and gives him a greater power, while the opposite arm becomes more oblique, and causes the other horse to act at a disadvantage: so that there is a kind of stability in the equilibrium. If the fulcrum were further forwards than the extremity of the bar, the two horses could never draw together with convenience. (Plate XVIII. Fig. 230.)

In mining countries, and in collieries, it is usual, for facilitating the motion of the carriages employed in moving the ore or the coals, to lay wheelways of wood or iron along the road on which they are to pass; and this practice has of late been extended in some cases as a substitute for the construction of navigable canals. Where there is a turning, the carriages are usually received on a frame, supported by a pivot, which allows them to be turned with great ease. In particular situations, these waggons are loaded by little carts, rolling without direction down inclined planes, and emptying themselves; they are also provided with similar contrivances for being readily unloaded, when they arrive at the place of their destination. The carriages used for drawing loaded boats over inclined planes, where they have to ascend and again to descend, are made to preserve their level by having at one end four wheels instead of two, on the same transverse line; the outer ones as much higher than the pair at the other end, as the inner ones are lower; and the wheelway being so laid, that either the largest or the smallest act on it, accordingly as the corresponding part of the plane is lower or higher than the opposite end. It is possible that roads paved with iron may hereafter be employed for the purpose of expeditious travelling, since there is scarcely any resistance to be overcome, except that of the air, and such roads would allow the velocity to be increased almost without limit.

For removing earth from one situation to another, a series of baskets has sometimes been hung on two endless ropes, moving on pullies of such a form, as to suffer the bars supporting the baskets to pass freely over them; the baskets being moved by means of a winch, acting on the rope by a wheel like one of the pullies. Sometimes also a series of little carts has been connected by ropes, and drawn in a circle or oval up and down an inclined plane. These methods may be adopted in making roads, where a hill is to be levelled, and the materials are to be employed in filling up the valley below: but in such cases two carts, connected by a cylinder or windlass, are generally sufficient; and they may be arranged in the same manner as the carriages for removing boats on an inclined plane.

LECTURE XIX.

ON MODES OF CHANGING THE FORMS OF BODIES.

THE corpuscular forces by which bodies retain their peculiar forms of aggregation, require in many cases to be counteracted or modified by mechanical processes: thus we have frequent occasion to compress bodies into a smaller space, to augment their dimensions in a particular direction, to divide their substance, either partially or totally, in given lines or surfaces, or to destroy their general form, by reducing them into more minute portions; and we may consider these subjects as principally referable to the effects of compression, extension, penetration, division, attrition, digging, boring, agitation, trituration and demolition. The two first of these articles depend on such a change as we have examined, in considering the strength of materials, under the name of alteration, the remainder on fracture.

The instruments peculiarly intended for compression are in general of the description of presses; and the most common act by means of a screw. The friction on the screw interferes considerably with the power of the machine; but it is of use in keeping the press fixed in a situation into which it has been brought by force. The screw is always turned by a lever; for without this assistance, however powerful it might be, the friction would render it almost useless. When great force is required, the screw is made as close as is consistent with the strength of its spires. Mr. Hunter's double screw may also be used with advantage, where only a small extent of motion is required. The screw of a printing press, or of a stamping press, is, on the contrary, open, and it is caused to descend with considerable momentum, the handle being loaded with a weight. Wherever a force is so employed as to produce an impulse which acts on any body, the momentum, which is the result of the action of the force for a certain time, is usually much more powerful than the simple pressure; the degree of its efficacy depends, however, on the degree of compressibility of the substance. Thus, if a heavy body fall from a certain height, so as to acquire a momentum, in consequence of the force of gravity, it will ultimately exert on the substance upon which it falls, a force about as much greater than its weight, as the space, through which the surface of the substance struck is depressed, by means of the impulse, is less than twice the height from which the body has fallen; and unless either the substance is very compressible, or the height very small, this force must be incomparably greater than the pressure of the weight only.

For a printing press, a single heavy roller is sometimes made to pass over the paper, when it has been laid on the types; and since the whole action of such a roller is confined to a small part, at any one time, it is said to exert sufficient force, and to perform its work more equably than a common press; but its operation must be comparatively slow. A common mangle for linen acts nearly in a similar manner. In calendering mills, the force of a spring is employed, for exerting a pressure on the block, with which the materials are glazed.

The copper plate printing press, and the machine for copying letters, are composed of two rollers, parallel to each other, pressing on the substance which is interposed, and which is brought into its situation partly by the friction of the surface of the roller, and partly by external force.

The rollers, by which sugar canes are pressed, are in general situated vertically, the middle one of three being turned by horses, by mules, or by water, and the canes being made to return round it, so as to pass through both interstices in succession. It appears to be of some advantage in presses of this kind, that all the rollers should be turned independently of their action on the materials interposed, since the friction of two rollers may tend to draw the materials into the space between them, with more regularity and greater force, than the action of a single roller would do. For this reason, it may be advisable to retain the toothed wheels turning the rollers, even when their axes are not firmly fixed, but held together by an elastic hoop. (Plate XVIII. Fig. 231.)

In oil mills, a still greater momentum is applied to the purpose of compression than in the printing press: hammers, or long wooden beams, placed vertically, are raised by a water wheel, and suffered to fall on wedges, which act very forcibly on the materials contained in bags on each side.

Compression is also sometimes performed by the operation of hammering: thus, cast brass is generally hammered before it is used, in order to increase its strength; the hammer renders it so much stiffer, that if it is necessary to preserve its ductility, it must be frequently annealed by exposure to heat. Anvils and vices are necessary appendages to the hammer; their use depends principally on their firmness, which is chiefly derived from weight in the one case, and from strength in the other; and pincers may be considered as portable vices.

For the purpose of producing a continued pressure on such substances as have a tendency to contract their dimensions, under the operation of a press, a spring has been interposed between the press and the materials, which is capable of pursuing them with a certain degree of force: the utility of such an arrangement must, however, be extremely limited. Mr. Bramah has applied a well known law of hydrostatics to the construction of a very useful press, which is simple, powerful, and portable.

Extension is seldom performed by forces that tend immediately to increase the dimensions of the substance only: it is generally procured by reducing the magnitude of the substance in another direction, sometimes by means of pressure, but more effectually by percussion. The rollers of the press employed for laminating metals are turned by machinery, and are capable of being moved backwards and forwards, in order to repeat the operation on the same substance; their distance is adjusted by screws, which are turned at once by pinions fixed on the same axis, in order that they may be always parallel. In this manner lead, copper, and silver, are rolled into plates, and a thin plate of silver being soldered to a thicker one of copper, the compound plate is sub mitted again to the action of the press, and made so thin as to be afforded at a moderate expense. The glazier's vice is a machine of the same nature, for forming window lead: the softness of the lead enables it to assume the required shape, in consequence of the pressure of the rollers or wheels; and the circumference of these wheels is indented, in order to draw the lead along by the corresponding elevations. (Plate XVIII. Fig. 232.)

In drawing wire, the force is originally applied in the direction of the extension, but it produces a much stronger lateral compression, by means of the conical apertures through which the wire is successively drawn. For holding the large wire, pincers are at first used, which embrace it strongly while they pull, and open when they advance to a new position, the interruption being perhaps of use, by enabling the pincers to acquire a certain momentum before they begin to extend the wire; but afterwards, when the wire is finer, it is simply drawn through the aperture from one wheel or drum to another. During the operation, it requires frequent annealing, which causes a scale to form on its surface; and this must be removed by rolling it in a barrel with proper materials; for the application of an acid is said to injure the temper of the metal. Copper is sometimes drawn into wire so large as to serve for the bolts used in shipbuilding, especially for sheathing ship's bottoms. Silver wire, thinly covered with gold, is rendered extremely fine, and then flattened, in order to be fit for making gold thread: the thickness of the gold is inconceivably small, much less than the millionth part of an inch, and sometimes only a ten millionth.

In order to form the handles of vessels of earthenware, the clay is forced through a hole of a proper shape in an iron box. The operation of the potter's wheel consists in great measure of compression and extension, performed by the hands; the vessels are finished, when they are partly dry, in a lathe, or by other instruments; some kinds of earthenware are formed in a mould only.

When a thread or a plate of glass is extended in a semifluid state, it has a tendency to preserve an equable thickness throughout: this is derived from the effect of the air in cooling it, the thinnest parts becoming immediately a little colder than the rest, and consequently harder, so that they retain their thickness, until the neighbouring parts are brought into a similar state.

Extension is performed by means of percussion, in forges, and in the common operation of the smith's hammer. In forges, the hammers are raised by machinery, and thrown forcibly against a spring, so as to recoil with great velocity. With the help of this spring, the hammer sometimes makes 500 strokes in a minute, its force being many times greater than the weight of the hammer. Such forges are used in making malleable iron, in forming copper plates, and in manufacturing steel. (Plate XVIII. Fig. 233.)

Gold is beaten between the intestines of animals, on a marble anvil; for this purpose it is alloyed with copper or silver. It is reduced to the thickness of little more than the three hundred thousandth of an inch. Silver leaf is about the hundred and sixty thousandth: it is made of silver without alloy.

The operation of coining depends also principally on an extension of the metal into the recesses of the die; it is performed by a strong pressure, united with a considerable impulse, communicated by a screw like that of a printing press; and sometimes the impression is formed by the repeated blows of a hammer only.

Thin plates of silvered copper are moulded into any figure that may be required, by being placed between two corresponding stamps, of which the one is fixed, and the other attached to the bottom of a heavy hammer. The hammer is raised and suffered to fall in a right line, by means of pincers, which open when they have acquired a certain height. Sometimes the contact, produced by the forcible impulse of a die, is sufficiently intimate to cause a thin plate of silver to cohere permanently with a surface of iron; and this mode of uniting metals is actually employed in some manufactures.

The operations of perforating, cutting, turning, boring, digging, sawing, grinding, and polishing, resemble each other, in great measure, with respect to the minute actions of the particles of bodies which they have to overcome. Penetration is generally performed in the first instance by the effect which we have called detrusion, where the magnitude of the penetrating substance is considerable: but when a fine point or edge is employed, it probably first tears the surface where it is most depressed, and then acts like a wedge on the portions of the substance left on each side, with a force so much the greater as the edge is thinner. The resistance opposed by a solid, or even by a soft substance, to the motion of a body tending to penetrate it, appears to resemble in some measure the force of friction, which is nearly uniform, whether the motion be slow or rapid, destroying a certain quantity of momentum in a certain time, whatever the whole velocity may be, or whatever may be the space described. Hence arises the advantage of giving a great velocity to a body which is to penetrate another, the distance to which a body penetrates being as the square of its velocity, or as its energy; and a certain degree of energy being required in order to make it even penetrate at all. It is true that when we exchange a slow motion for a more rapid one, by the immediate action of any mechanical power, we can only obtain the same energy from the same power, for we must diminish the mass in the same proportion as the square of the velocity is increased; but a very small part of the force, which is consumed in the operation of a machine of any kind, is employed in generating momentum; by much the greatest part is spent in overcoming resistances which vary but little with the velocity; a small portion only of the resistance increasing in proportion to the square of the velocity; so that by applying a triple force, we may obtain more than a double velocity, and more than a quadruple effect: and besides it has already been observed that when the velocity begins to exceed a certain limit, the effect is increased in a much greater proportion than that of its square. The same work is also performed with less pressure, and less strain on the machinery, where a great velocity is employed. It is on account of the efficacy of velocity, in facilitating penetration, that soft substances, moving very swiftly, will readily perforate much harder ones; and for the same reason a gunshot wound, and even the loss of a limb, takes place with so little disturbance of the neighbouring parts, that it is sometimes scarcely felt. The advantage of an impulse, however inconsiderable, above a pressure, however great, may be easily understood from the ease with which a moderate blow of a hammer causes a nail to penetrate a substance, into which the whole force of the arm could not have thrust it.

In the engine for driving the piles, or upright beams, used for the foundations of buildings in water, or in soft ground, the weight is raised slowly to a considerable height, in order that, in falling, it may acquire sufficient energy to propel the pile with efficacy. The same force, if applied by very powerful machinery immediately to the pile, would perhaps produce an equal ffect in driving it, but it would be absolutely impossible in practice to construct machinery strong enough for the purpose, and if it were possible, there would be an immense loss of force from the friction. For example, supposing a weight of 500 pounds, falling from a height of 50 feet, to drive the pile 2 inches at each stroke; then, if the resistance be considered as nearly uniform, its magnitude must be about 150 thousand pounds, and the same moving power, with a mechanical advantage of 300 to 1, would perform the work in the same time. But for this purpose some parts of the machinery must be able to support a strain equivalent to the draught of 600 horses. In the pile driving engine, the forceps, or tongs, sometimes called the monkey, or follower, is opened as soon as the weight arrives at its greatest height; and at the same time a lever detaches the drum, employed for raising the weight, from the axis or windlass, at which the horses are drawing; the follower then descends after the weight, uncoiling the rope from the drum, and the force of the horses is employed in turning a fly wheel, until the connexion with the weight is again restored. (Plate XVIII. Fig 234.)

When we throw a stone, or a missile weapon of any kind, with the hand, the stone can acquire no greater velocity than the hand itself, accompanied by the neighbouring part of the arm: so that the whole velocity must be produced in a mass of matter comparatively very large. A sling enables us to throw a stone or a ball much further; for here the stone may be moved with a velocity far greater than the hand that impels it, although the action of the force on the stone is indirect, and the resistance of the air considerable. An elastic bow, furnished with a strong and light string, enables us to apply to an arrow or to a ball the whole force of our arms, unencumbered with any considerable portion of matter, that requires to be moved with the arrow; hence a very great velocity may be obtained in this manner. An air gun possesses the same advantage in a still greater degree, and the force of fired gunpowder excels perhaps all others, from its concentrating an immense force in the form of an inconceivably light elastic fluid; of course a ball impelled by this force, becomes a most effectual instrument in penetrating the most refractory substances. We may easily calculate the velocity of an arrow, by comparing its motion with that of a pendulum, if we know the proportion of its weight to the force that bends the bow; including in the weight a small addition for the inertia of the bow and bowstring; the height to which the arrow will rise, being about as much greater than the space through which the bowstring acts on it, as the greatest force applied in drawing the bow is greater than twice the weight to be moved.

The action of a whip, either on the air, or on a solid body, depends on the increase of velocity, occasioned by the successive transmission of the motion from a thicker to a thinner portion of its flexible substance, so that at last, the energy of the lash, and of its knots, gives it a sufficient capability of exciting sound, or of inflicting pain.

The instruments generally employed for the division of solid bodies, are wedges, chisels, knives, and scissors; they sometimes act by pressure only, but they are more powerful when impulse is added to it. Hatchets, planes, saws, and files, always act with some rapidity. Cutting instruments are in general very thin wedges, but the edge itself is usually much more obtuse; Mr. Nicholson has estimated the angle, formed ultimately by the surfaces constituting the finest edge, at about 56 degrees. Knives are sometimes fixed on wheels, so as to revolve in a direction oblique to their edges, as in some machines for cutting chaff, where the straw is also drawn forwards, through a space variable at pleasure, during each revolution of the knife. An instrument of a similar nature has also been invented for the purpose of cutting weeds under water.

For the edges of all cutting instruments, steel is principally employed. After being hardened, by plunging it when red hot into cold water, it is tempered, by laying it on a heated iron, or more accurately, by Mr. Stodart's method, of immersing it in a metallic composition in the state of fusion. When its surface has acquired a yellow tinge, it is fit for edge tools, and the degree of heat proper for watch springs is indicated by a blue colour. The backs of knives are often made of iron, which is less brittle than steel: these substances are generally welded together, by hammering them when red hot; but sometimes, in large instruments, a back of iron is only rivetted on.

The iron employed for making nails, and other small articles, is first rolled into flat bars, and then cut into narrow rods, by causing it to pass between the cylinders of the slitting mill, the surfaces of which are formed into rectangular grooves, and which are placed close to each other, so that the prominent parts of the one are opposed to the depressions of the other, and the bars are divided by the pressure of the opposite forces, acting transversely at the same points, so as to separate them by the effect which we have already considered under the name detrusion. The same machinery also generally works a pair of large shears, for cutting bars of any kind. (Plate XVIII. Fig. 235.)

The lathe is an elegant instrument, in which a considerable relative velocity is produced between the tool and the substance to be cut, by the revolution of this substance on an axis, while the tool is supported by a rest. Ornamental lathes admit of a great variety of mechanical contrivance, but they are of little practical use, except for amusement. Picture frames are, however, sometimes turned in oval lathes; and in the manufacture of buttons, machines of a similar nature are occasionally employed. The effect of every lathe of a complicated construction depends on a certain degree of motion of which its axis is capable: if this motion be governed by a screw, a screw of any diameter may be turned by its assistance; if by a frame producing an elliptic curve, any number of ovals, having the same centre, may be described at once; and if a moveable point connected with the work, be pressed by a strong spring against a pattern of any kind, placed at one end of the axis, a copy, of the same form, may be made at the other end of the axis.

The process of boring is a combination of penetration and division, and sometimes of attrition. Awls, gimlets, screws, augers, and centrebits, are various forms of borers. The drill has the advantage of a rapid motion, communicated by the drill bow, which turns it round by means of a little wheel or pulley. In boring cannon, the tool is at rest, while the cannon revolves, and by this arrangement the bore of the cannon is formed with much more accuracy than according to the old method of putting the borer in motion; perhaps because the inertia of so large a mass of matter, as constitutes the cannon, assists in defining the axis of revolution with more accuracy. The borer is pressed against the cannon by a weight, hung on the arm of a bent spring, and during the operation, the outside is also turned into its intended shape by the application of proper instruments. Cylinders for steam engines are cast hollow, and afterwards bored; but in this case the borer revolves, and the cylinder remains at rest.

Ploughs, spades, pickaxes, mattocks, harrows, and other agricultural instruments, resemble in their operation the chisel and the wedge: the numerous diversities in their form and the complications of their structure, are determined more by the various modifications of their action, required for particular purposes, than by any material difference in the mode of application of the principles on which they depend. (Plate XVIII. Fig. 236.)

The process of mining is a combination of boring and digging. Shafts are sunk, levels are driven, and drains are carried off, by the help of picks or pickaxes, wedges, and hammers, the rocks being also sometimes loosened by blasting with gunpowder. In searching for coal, a shaft is sunk through the uppermost soft strata, and the rock is then bored, by striking it continually with an iron borer, terminating in an edge of steel, which is in the mean time turned partly round; and at proper intervals a scoop is let down, to draw up the loose fragments. In this manner a perforation is sometimes made for more than a hundred fathoms, the borer being lengthened by pieces screwed on to it; it is then partly supported by a counterpoise, and is worked by machinery; if it happens to break, the piece is raised by a rod furnished with a hollow cone, like an extinguisher, which is driven down on it. Sometimes the borer is furnished with knives, which are made to act on any part at pleasure, and to scrape off a portion of the surrounding substance, which is collected in a proper receptacle.

For sawing wood on a large scale, sawing mills are very advantageously employed, being usually driven by water. Several saws are generally fixed in a frame, parallel to each other; they are worked up and down by a crank, and at every alternation a wheel is drawn round a little, by a catch, or click, and moves forwards the frame which supports the timber. When the machine is employed for cutting the fellies which form the circumference of wheels, the frame supporting the timber is made to turn round a centre. A circular saw is used in the construction of blocks and pullies; and in order to make the motion more secure from the effect of accidental irregularities, the wheels are made to turn each other by contact only, without teeth. The machinery for making blocks, in the Royal dock yard at Portsmouth, has been lately much improved and enlarged; it is worked by a steam engine, the action of which is applied to a great variety of purposes. The advantage of a saw which revolves continually, appears to be very considerable, since a much greater velocity may be given to it than can be obtained when the motion is alternate. Such a saw has also sometimes been applied to cutting off piles under water.

In mills for sawing marble into slabs, the saws are drawn backwards and forwards horizontally: they are made of soft iron, without teeth; and sand being applied to them, with water, during the operation, the sand is partly imbedded in the iron, and grinds away the marble.

Granite is worked by driving a number of thin wedges very gradually into it, at various parts of the section desired; and sometimes wedges of wood are employed, which being moistened by water, their expansion separates the parts from each other. It is also said that many stones may be divided by drawing lines on them with oil, and then exposing them to heat. Perhaps some processes of this kind might be performed with advantage under water; it is well known that glass may be cut in a rough manner under water, without much difficulty, by a common pair of scissors.

For reducing the magnitude of a substance in a particular part, instruments of attrition are used; rasps, files, grindstones, and hones; and of all these the immediate actions appear to resemble those of chisels and saws. The hatches of files are cut with a hard chisel while the steel is soft, and the files are afterwards hardened. In using the grindstone, water is applied, in order to avoid the inconvenience produced by too much heat; and sometimes tallow is substituted for water with equal advantage: but oil is not found to answer the same purpose; and it has been conjectured that the cold continually occasioned by the melting of the tallow at the point of friction, serves as a substitute for the cooling effect of the evaporation of the water. For grinding and polishing steel, the grindstones are made to revolve, either vertically or horizontally, with a velocity so great as to describe sometimes as much as 60 feet in a second. The steel is also in some cases drawn backwards and forwards horizontally on a circular surface, and in order that the action may be equally divided throughout the surface, it is allowed to revolve on an axis by means of the friction; its motion being confined to one direction by the action of a catch.

Various substances, chiefly of mineral origin, are also used, on account of their hardness, as intermediate materials, for grinding and polishing others. These are diamond dust, corundum, emery, tripoli, putty, glass, sand, flint, red oxid of iron, or crocus martis, and prepared chalk; they are sometimes applied in loose powder, and sometimes fixed on leather, wood, or paper. Cuttle fish bone, and seal skin, are furnished by the animal kingdom, and Dutch rushes by the vegetable; these are employed chiefly in polishing wood or ivory.

Marble is made smooth by rubbing one piece on another, with the interposition of sand; the polishing blocks are sometimes caused to revolve by machinery in a trough, in which the marble is placed under water, and are drawn at the same time gradually to and from the centre; or the slab itself, with the frame on which it rests, is drawn slowly backwards and forwards, while the blocks are working on it. Granite is polished with iron rubbers, by means of sand, emery, and putty; it is necessary to take care during the operation that the water, which trickles down from the rubbers, and carries with it some of the iron, may not collect below the columns, and stain them; but this inconvenience may be wholly avoided by employing rubbers of glass.

Optical lenses are fixed on blocks by means of a cement, and ground with emery, by a tool of proper convexity or concavity: if they are small, a large number is fixed on the blocks at the same time. The tool is sometimes first turned round its axis by machinery, and when the lenses are to be finished, a compound motion is given to it by means of a crank; and in order to make it more smooth, the wheels turn each other by brushes instead of cogs. The point of the lens where its two surfaces are parallel, is determined by looking through it at a minute object, while it is fixed on a wheel with a tubular axis, and shifting it, until the object no longer appears to move; a circle is then described, as it revolves, in order to mark its outline.

4

Machines for trituration, by means of which the larger masses of matter are crushed, broken, or ground, into smaller parts, are in general comprehended under the denomination of mills. After the pestle and mortar, the simplest machine of this kind appears to be the stamping mill; the stampers resemble the hammers of the mill employed in the extraction of oils from seeds, and the machine is used for reducing to powder the ores of metals, and sometimes also barks, and linseed; the surface of the stampers being armed with iron or steel. But barks and seeds are more usually ground by the repeated pressure of two wheels of stone, rolling on an axis which is forced in a horizontal direction round a fixed point. A nobleman of distinguished rank and talents has lately employed for a mortar mill, a wheel of cast iron, formed of two portions of cones, joined at their bases: after thirty revolutions, the mortar being sufficiently ground, a bell rings, and the horse stops.

The materials for making gunpowder are also ground by a wheel revolving in a trough: in order to corn them, they are moistened, and put into boxes with a number of holes in their bottoms, and these boxes being placed side by side, in a circular frame, suspended by cords, the frame is agitated by a crank revolving horizontally, and the paste shaken through the holes: the corns are polished by causing them to revolve rapidly within a barrel.

A revolving barrel is used for forming and polishing small round bodies of different kinds, and it is often employed in agriculture as a churn for making butter. The purpose of agitation is perhaps more effectually answered by an alternate motion, which has sometimes been produced in a barrel churn, by means of a cord attached to a heavy pendulum.

Threshing machines are of two kinds; the one consists of a number of flails, beating the corn nearly in the same manner as they are used by labourers; in the other, which is more commonly employed in this country, the corn is drawn along by two revolving rollers, and caused to pass between a cylinder, and its concave cover, while a number of blocks, projecting from the surface of the cylinder, beat or rub out the grains very effectually from the ears; the corn falls out at the lower part, and is winnowed by a fan which the machine turns at the same time. In this manner it is said that a horse will thresh about 100

2

bushels of corn in a day. It is commonly reckoned the work of a labourer to thresh about six bushels in a day. (Plate XVIII. Fig. 237.)

Some kinds of grain are occasionally ground in mills of iron or steel, which consist of a solid cylinder or cone turning within a hollow one, both the surfaces being cut obliquely into teeth. But the common mill for grinding corn is composed of two circular stones of silicious grit, placed horizontally; the upper one revolves with considerable velocity, and is supported by an axis passing through the lower one, at a distance variable at pleasure: When the diameter is five feet, the stone usually makes about 90 revolutions in a minute; if the velocity were greater, the flour would be too much heated. The corn is shaken out of a funnel, or hopper, by means of projections from the revolving axis, which strike against the orifice; it passes through the middle of the upper millstone, and is readily admitted between the stones; the lower stone is slightly convex, and the upper one somewhat more concave, so that the corn passes over more than half the radius of the stone before it begins to be ground: after being reduced to powder, it is discharged at the circumference, its escape being favoured by the convexity of the lower stone, as well as by the centrifugal force. The surface of the stones is cut into grooves, in order to make them act more readily and effectually on the corn. The resistance, in grinding wheat, has been estimated at about a thirty fifth of the weight of the millstone. The stones have sometimes been placed vertically, and the axis supported on friction wheels: but the common position appears to be more eligible for mills on a large scale. It is said that a man and a boy can grind by a hand mill a bushel of wheat in an hour; in a watermill, the grinding and dressing of a bushel of wheat is equivalent to the effect of 20160 pounds of water falling through a height of 10 feet, which is about as much as the work of a labourer for a little more than half an hour. In a windmill, when the velocity is increased by the irregular action of the wind, the corn is sometimes forced rapidly through the mill, without being sufficiently ground. There is an elegant method of preventing this, by means of the centrifugal force of two balls, which fly out as soon as the velocity is augmented, and as they rise in the arc of a circle, allow the end of a lever to rise with them, while the opposite end of the lever descends with the upper millstone, and brings it a little nearer to the lower one. The bran or husk is separated from the flour, by sifting it in the bolting mill, which consists of a cylindrical sieve, placed in an inclined position, and turned by machinery. (Plate XVIII. Fig 238.)

When the flour is made into bread, the dough requires to be kneaded: for this purpose a machine is sometimes used, in which four or more bars, parallel to the axis of motion, are turned round, by means of a walking wheel. The dough is placed in a circular trough, in which the bars revolve not quite in the middle, so as to approach in each revolution to one of its sides, and thus the dough is perpetually compelled to change its form.

A machine of nearly the same construction is employed for levigating flints, after they have first been made red hot, and plunged into cold water, in order to render them friable. They are mixed, when it is necessary, with other large stones, and the water, in which the process is performed, carries off the powder, and deposits its coarser parts in a short time, while the finer remain much longer suspended, and are thus separated from the rest.

When a mechanical structure is to be demolished, or a natural substance to be broken into smaller parts, we have often occasion to employ the collected force of men, the powers of machinery, or the expansive force of chemical agents. Battering rams, or wooden beams, suspended by ropes, and armed with iron, which were used by the warriors of antiquity in besieging a town, are now generally superseded by the introduction of artillery, although they may perhaps still afford, in some cases, a more economical and equally powerful mode of operation. The same momentum, and the same energy, may be given to a battering ram at a less expense than to a cannon ball; but it is probable that the efficacy of a cannon ball is chiefly owing to the augmentation of its velocity beyond that limit, which is the utmost that the substance to be destroyed can sustain without giving way, independently of the mass of the body which strikes it.

For demolishing smaller aggregates, pincers, hammers, and crows, are generally sufficient; to these sometimes more complicated instruments are added. Thus, for example, several machines have been invented for drawing out ship's bolts. A hook which grapples like the common instrument for drawing teeth, has been applied for holding them firmly, and sometimes a screw, turned by means of wheelwork, has been used for gaining a force sufficient to overcome their adhesion. In all such cases, however, the effect of percussion has a considerable advantage; and even if other means are employed, it is of use to begin with lessening the firmness of the adhesion by the blows of a hammer; and in this manner a screw may be extracted, which is so firmly attached by its rust, as to be immoveable by other means.

The expansive force of heat is frequently of great service in dividing rocks, or in destroying old buildings. This is sometimes done simply by the application of fire, as in the mine of Rammelsberg, in the Hartz, where the stratum containing the ore is of such a nature, partly, perhaps, on account of the combustible matter which enters into its composition, that, by the effect of a large quantity of fuel, which is burnt in the vast excavation, of which it forms the side, it is rendered so friable as to be worked with ease. More commonly, however, the force of gunpowder is employed, and rocks are generally blasted with great convenience by an explosion of this powerful agent. A hole being bored to the depth of three or four feet, the powder is placed at the bottom, and a wire being introduced, small stones and sand are rammed round it, and the wire is withdrawn, leaving a communication for firing the powder, by means of a train of sufficient length to insure the safety of the workman. It is said that the explosion is more efficacious when the powder does not fill the whole of the cavity; this, however, appears to require confirmation. The chemical powers, which are the ultimate causes of the operation of gunpowder, belong to a department of philosophy which it is not our business to investigate: but the elasticity of the gases and vapours which are extricated, as modified by the heat which accompanies their production, will be considered and explained in the subsequent divisions of this Course of Lectures.

LECTURE XX.

ON THE HISTORY OF MECHANICS.

THE order which we have pursued, in considering the various departments of mechanical science, has been in great measure synthetical, dictated by the plan of proceeding logically from the most simple principles to their more complicated combinations, so as to build at every step on foundations which had been firmly laid before: and this method is unquestionably the best adapted for the expeditious progress of a student in sciences with which he is unacquainted. But having once acquired a certain degree of knowledge, he is anxious to be informed by what steps that knowledge was originally obtained, and to what individuals mankind is indebted for each improvement that has been successively made. Hence, although we cannot attempt to enter into a complete history of mechanics, it may still be satisfactory to take a short retrospect of a few of the most remarkable eras in mechanical philosophy, and in those parts of mathematics on which it immediately depends.

It is universally allowed that the Greeks derived the elements of mathematical, mechanical, and astronomical learning from Egypt and from the East. Diogenes Laertius, who appears to be very desirous of claiming, for his countrymen, the merit of originality, does not deny that Thales and Pythagoras acquired much of their knowledge in their travels. Thales of Miletus is the first that can be supposed to have introduced these studies into Greece. Moeris, who was probably a king of Egypt, and Theuth or Thoth, a native of the same country, are mentioned as having laid the foundations of geometry; but the science could scarcely have extended, in those ages, further than was barely necessary for the measurement of land: since Thales, or even a later philosopher, is said to have first discovered that two lines drawn from the extremities of the diameter of a circle, and meeting in any other part of its circumference form with each other a right angle. Thales was one of the seven whom antiquity distinguished by the appellation of wise men; he flourished about 600 years before the Christian era, and he was the father of the Ionian school, the members of which, in subsequent times, devoted themselves more particularly to the study of moral than of natural philosophy.

The Italian school, on the contrary, which was founded by Pythagoras, appears to have been more inclined to the study of nature and of its laws; although none of the departments of human knowledge were excluded from the pursuits of either of these principal divisions of the Grecian sages, until Socrates introduced, into the Ionian school, a taste for metaphysical speculations, which excluded almost all disposition to reason coolly and clearly on natural causes and effects. To Pythagoras, philosophy is indebted for the name which it bears; his predecessors had been in the habit of calling themselves wise; he chose to be denominated a lover of wisdom only. He had studied under Pherecydes, and Pherecydes under Pittacus: but with respect to mathematical and mechanical researches, it does not appear that either of his teachers had made any improvements. On his return from his travels in Egypt and the East, in the time of the last Tarquin, about 500 years before Christ, he found his native country Samos under the dominion of the tyrant Polycrates, and went as a voluntary exile to seek a tranquil retreat in a corner of Italy. At Croto, says Ovid, he studied and taught the laws of nature.

> " From human view what erst had lain concealed
> His piercing mind to open light revealed ;
> To patient toil his ardent soul constrained,
> Of Nature's richest stores possession gained:
> And thence, with glowing heart and liberal hand,
> He dealt her treasures o'er the listening land.
> The wondering crowd the laws of nature hears,
> And each great truth in silent awe reveres."

However erroneous the opinion may be, that Pythagoras was acquainted with the laws of gravitation, it is certain that he made considerable improvements both in mathematics and in mechanics, and in particular that he discovered the well known relation between the hypotenuse and the sides of a right angled triangle, and demonstrated that the square of the hypotenuse is always equal to the sum of the squares of the sides. This theorem is more essential to the perfection of geometry than any other proposition that can be named: and if we may judge by the story of his having sacrificed a hecatomb to the Muses, on occasion of the discovery, he seems to have had a foresight of the magnificence of the edifice, that was in subsequent times to be built on this foundation.

Democritus of Abdera lived about a century after Pythagoras, whose works he studied, and whose principles he adopted. He appears to have been possessed of very extensive knowledge and profound learning; but little remains of his works, excepting their titles. Some have attributed to him the invention of the method of arranging stones so as to form an arch. Seneca thinks that so simple an invention must have been practised in earlier ages: but Mr. King has endeavoured to show that its general introduction in building was of much later date. Architecture, and other mechanical arts had however been considerably advanced some time before this period, if it is true that Ctesiphon or Chersiphron, who built the temple of Ephesus, was cotemporary with Croesus and Thales. It is uncertain at what time bridges of stone were first built; and it is doubtful whether the art of building bridges of wood was very well understood in those ages: for according to Herodotus, it was commonly believed, that Thales avoided the necessity of procuring a passage over the Halys for the army of Croesus, by encamping them on its banks, and cutting a channel for the river in their rear, although the historian himself is of opinion, that they passed over bridges which already existed. Curtius speaks of a bridge of stone over the Euphrates at Babylon, which appears to have been built long before the time of Alexander, whose expedition he relates; and it is scarcely probable that a stone bridge could have withstood the impulse of so rapid a river, if it had been supported by columns only, without arches. We are informed by Pliny that Ctesiphon lowered his large blocks of stone by placing them on heaps of sand bags, and letting out the sand by degrees; it does not appear how he raised them, but the inclined plane seems to afford the simplest and most obvious method.

Archytas of Tarentum, and Eudoxus of Cnidus were also Pythagoreans. They were the first that attempted to make the mathematical sciences familiar by popular illustrations; and Archytas is said by some to have invented the pulley and the screw. They lived nearly 150 years after Pythagoras, and geometry had made in the mean time very rapid advances, for the properties of the conic sections were well known to these philosophers. "The first persons," says Plutarch, "that cultivated the method of organic geometry, were of the school of Eudoxus and Archytas. These philosophers introduced elegance and variety into science, by illustrations derived from sensible objects, and made use of mechanical contrivances for expediting and familiarising the solutions of problems, which, if more mathematically treated, are complicated and difficult: each of them invented a method of determining in this manner the magnitude of two mean proportionals between two given lines, by the assistance of certain curves and sections. Plato by no means approved of their mode of proceeding, and reprehended them severely, as giving up and perverting the most essential advantages of geometry, and causing the science to revert from pure and incorporeal forms to the qualities of sensible bodies, subjected to narrow and servile restraints. It was for this reason that practical mechanics were separated from geometry, and were long neglected by philosophers, being considered as a department only of the art of war."

Aristotle, who was almost the last of the Ionian school, flourished a little less than half a century after Archytas; he was perhaps the author of no original discoveries relating to the principles of mechanics, but we find, in his treatise on this science, the law of the composition of motion very distinctly laid down; he makes, however, some mistakes respecting the properties of levers. His general merit in elegant literature, as well as in natural history and natural philosophy, is too well known to require encomium.

The foundation of Alexandria commences a period memorable for science in general, but more particularly for mathematics and astronomy. Dinocrates was the architect whom Alexander employed in laying out and in building this celebrated city. Among those who studied in this school, the sciences are indebted to none more than to Euclid, who lived about 300 years before our era. It is uncertain how much of his Elements may have been derived from his own investigations; but the masterly manner in which this well known work is arranged, and the precision and accuracy which reign in every part of it, demand almost as great a share of praise as is due to original discovery.

Epicurus was a cotemporary of Euclid, and is considered as the last of the Pythagorean or Italian philosophers. The penetration that he discovered in assigning the true causes to many mechanical phenomena, his explanations of which are copied by Lucretius, is sufficient to induce us to look forwards with impatience to the publication of such of his works, as have lately been discovered among the manuscripts of Herculaneum. Apollonius of Perga lived about half a century later; the elegance and extent of his investigations of the most abstruse properties of the conic sections left but little to be added to them by more modern geometricians. The architect Philo appears to have been more ancient than Apollonius: but he is not the Philo whose essay on warlike engines is published in the collection of the Ancient mathematicians; since this author was a pupil of Ctesibius.

For the demonstration of the fundamental properties of the lever and of the centre of gravity; for the discovery of the laws of hydrostatics, and of the modes of determining the specific gravities of bodies; for the construction of the first cranes, and of the first planetarium; and for those improvements of the methods of mathematical investigation which have been the basis of every modern refinement in analytical calculation; for all these additions to our knowledge and our powers, we are indebted to Archimedes. On a character so conspicuous, we can with pleasure dwell long enough to attend to some particulars of his history, which are related by Plutarch, in his account of the siege of Syracuse; omitting, however, such details as are evidently fabulous. "Archimedes," says Plutarch, "armed with his own inventions only, made light of the splendour of the Roman preparations, and of the glory of the name of Marcellus. And these were inventions that he even considered as of subordinate value, as geometrical playthings, which had been the amusements of his leisure hours. It was king Hiero that first induced him to transfer a portion of his science from intellectual to material objects, and to condescend in some degree to the comprehension of the multitude, by giving a sensible form to those truths, which in their abstract state are discoverable only to the reasoning faculty. Archimedes, who was a friend and a relation of Hiero, had asserted that any weight whatever might be moved by any given power: and depending on the validity of his arguments, had given scope to his imagination, and boasted that if he had another earth to which he could step over, he would draw the whole of the present globe out of its place. Hiero, surprised at the boldness of his assertion, requested him to give some substantial proof of its truth, by moving a great weight with a small power: upon this Archimedes procured a ship, which was with great labour drawn up on the shore, and having completely manned and freighted her, he seated himself at a distance, and by lightly touching the first movement of a machine, he drew her along as smoothly and as safely as if she had been sailing in the deepest water. Hiero, full of astonishment, and admiring the powers of mechanical art, prevailed on Archimedes to construct such engines both of defence and of offence, as might be of use to him in case of a siege: for these, however, Hiero, who lived a life of peace and prosperity, was not so unfortunate as to have occasion; but they now became highly valuable to the Syracusans, and they were of the more advantage, as their inventor was present, to direct their use. And in fact the whole people of Syracuse constituted but a part of Archimedes's corporeal machinery, and he was the soul that moved and governed the whole. All other arms were deserted, and they employed his engines alone, both for their own defence, and for the annoyance of the enemy. In short, the Romans soon became so terrified, that if they saw a stick or a rope upon the walls, they cried out that it was some machine of Archimedes, and immediately fled; so that Marcellus at last determined to desist from attempting to take the place by assault, and resolved to blockade it only.

"Archimedes, however, had such depth of intellect, and such sublimity of mind, that notwithstanding he had obtained, by these inventions, the credit and glory of an intelligence rather divine than human, he thought it unworthy of him to leave any written treatise on the subject, considering practical mechanics, and every art that is concerned in satisfying the wants of life, as ignoble and sordid; and resting all his hopes of fame on those works, in which the magnificent and the elegant are exhibited, uncontaminated by the imperfections of the material world: works that are comparable to nothing else that the mind of man has produced; in which the subject only contends with the mode of treating it, the magnitude and beauty of the one being rivalled by the accuracy and vigour of the other. It is impossible that propositions more difficult and important should be deduced from simpler and purer elements. Some attribute this excellence to his natural genius, others to his indefatigable application, which has given to every thing that he has attempted the appearance of having been performed with ease. For we might ourselves search in vain for a demonstration of his propositions; but so smooth and direct is the way by which he leads us, that when we have once passed it, we fancy that we could readily have found it without assistance. We may, therefore, easily give credit to what is said of him, that being as it were fascinated by this domestic syren, that bore him company, he often neglected his food and his clothing; that when sometimes brought by compulsion to the baths, he used to draw his figures in the ashes of the fire places, and to make his calculations upon the cosmetics that were employed by the attendants; deriving, like a true votary of the muses, every pleasure from an intellectual origin. Among all his beautiful discoveries, he is said to have chosen that of the proportion of the sphere and cylinder for his sepulchral honours; requesting of his friends that they would place on his tomb a cylinder containing a sphere, and inscribe on it the ratio which he had first determined.

"By artifice, and through the thoughtlessness and security of a day of festivity, the Romans at length obtained possession of Syracuse, and in the pillage, although orders had been issued that the life of Archimedes should be spared, he was killed by a private soldier. His death is variously related, but all accounts agree, that Marcellus was deeply concerned for his loss, that he held his assassin in abhorrence, and conferred distinguished favours on his surviving relations." This event is supposed to have happened about 212 years before the birth of Christ: and the cultivation of mechanical philosophy, which had been continued for four hundred years with increasing success, was almost wholly interrupted for eighteen centuries.

There lived, however, in the mean time, some mathematicians and mechanics of considerable merit. A work on warlike machines, addressed to Marcellus by Athenaeus, is still extant, and may be found in the splendid collection of writers on military mechanics entitled Mathematici Veteres. Ctesibius of Alexandria was about a century later than Archimedes; he enriched hydraulics with several valuable machines; although he contributed little to the advancement of theoretical investigation. Hero was of the same school, and his pursuits were similar; some of his treatises on hydraulics, pneumatics, and mechanics, are published in the collection of Ancient mathematicians, and some others are still extant in manuscript. We are informed by Pappus, that Hero and Philo had referred the properties of the lever, the wheel and axis, the pulley, the wedge, and the screw, to the same fundamental principle; so that the theory of the mechanical powers began at that time to be extremely well understood. The treatises of Hero on pneumatics and on automatons contain many very ingenious inventions, but they are rather calculated for amusement than for utility; among them is a cupping instrument, which operates nearly in the manner of an air pump. A work of Bito, on warlike machinery, addressed to king Attalus, is included in the same collection.

Vitruvius was an author of great general knowledge: he lived under one of the earliest of the Caesars, and the greatest part of our information respecting the mechanics of antiquity has been derived from his works. Apollodorus was employed by Trajan, in building a bridge over the Danube, in the year 102; he has left a treatise on besieging a town, which is to be found among the Ancient mathematicians. Diophantus, Pappus, and Proclus, were mathematicians of eminence; Diophantus confined himself in great measure to arithmetic and pure geometry; but the last book of Pappus's collections is devoted to mechanics, and Proclus wrote a treatise on motion, which is still extant. The rudiments of algebraical notation and calculation may be found in the works of Diophantus; but the Arabians appear to have first practised the method of denoting quantities in general by literal characters; they made, however, no considerable advances, and mathematics in general remained nearly stationary until the time of the revival of letters.

During the long interval, in which learning and science were involved in the darkness of the middle ages, the arts subservient to the convenience of life were also in great measure neglected. It is evident from many remains of antiquity, that various manufactures had attained, in Greece and at Rome, a high degree of perfection; but the irruptions of the barbarians were as effectual in suppressing the refinements of civilisation, as in checking the pursuit of literary acquirements: our own country was not the earliest in recovering the arts which had been lost, but it has always received with open arms those who have excelled in them; and the improvements which have been made, within a few centuries, in the British manufactures, have obtained for them a celebrity unrivalled by those of any other nation. The ancient Britons are supposed to have made, in common with the other Celtic nations, coarse cloths and felts of wool, and perhaps some articles of linen; their chariots of war, which are mentioned by Caesar, could not have been executed without some skill in the arts of the carpenter and the smith. The Romans introduced a certain degree of civilisation into England, but it appears to have been in great measure forgotten soon after they left the country. In the seventh century, several architects and workmen were brought from the continent by Wilfrid and Biscop; they restored the practice of building with stone, which had been generally superseded by wood, and laid the foundation for other improvements. In the time of king Alfred, the English goldsmiths began to excel, and before the conquest, the woollen manufactures had acquired a considerable degree of perfection. The paper now in use was introduced about the year 1100; it was probably imported from the continent, since the linen manufacture was little advanced in England till 150 years later; but embroidery was much practised, although in the 12th century silks were principally woven in Sicily. The manufactory of cloth was considerably improved, in the 14th century, by the establishment of Kempe and other Flemish weavers in England: and many of the arts were benefited, about the same time, by the invention of the method of drawing wire, which was first introduced at Nuremberg. In the succeeding century, the increasing number of hands employed in various manufactures, suggested to some mind of superior penetration the great principle of the division of labour, by which each individual is enabled to acquire so high a degree of perfection in a very limited branch of each manufacture, that the whole work is performed much more perfectly, as well as more expeditiously, than if it had been begun and completed by any one person, even of greater abilities and experience. The invention of the modern spinning wheel is attributed to Jürgen of Brunswick, and the year 1530 is assigned as its date: England soon profited by the improvement; many manufacturers took refuge in this country from the Duke of Alva's persecutions in Flanders, and before the end of the century a new modification of the art of weaving was introduced by Lee of Cambridge, who invented the stocking loom, imitating the texture of the knit stockings, which were first manufactured in Spain about the year 1550. Mills for drawing wire and for slitting iron were also first erected in the sixteenth century; Birmingham and Sheffield were even at that time, according to Camden, celebrated for their manufactures; and the machinery which has been since introduced at different periods in those places, affords a facility and expedition which astonish every unexperienced spectator. The names of Watt and of Boulton have acquired a just celebrity from their refined improvements; but many other mechanics of inferior rank have exhibited a degree of ingenuity which would have done honour to the most distinguished talents. The manufactures of Manchester are also of considerable antiquity; but they are very greatly indebted to the inventions of Arkwright and his followers, which have also been introduced in many other parts of the united kingdom. The importance of these improvements may be estimated from the quantity of cotton which is annually imported into Great Britain; in 1787, it amounted to 23 million pounds, and gave employment to 420 thousand manufacturers; in 1791, it was increased to 32 millions: about one half is consumed in white goods, one fourth in fustians, and the remainder in hosiery, mixtures, and candle wicks. But the woollen manufactory affords a subsistence to above a million persons, who receive annually for their work about nine millions sterling, and employ as much wool as is worth about three.

In architecture, the Anglonorman stile prevailed in this country from the conquest to the beginning of the thirteenth century; the arch was frequently employed, and its form was semicircular. The Gothic architecture, distinguished by its pointed arches, which is said to have originated from the Saracens, was first introduced into England about the year 1170, and was more and more generally adopted for about three centuries. Of the architects of this school, two of the most celebrated were William of Sens, and Walter of Coventry: the most elegant specimen of its performances is, perhaps, King's College Chapel at Cambridge, which was founded by Henry the Sixth, and begun in the year 1441. The Cathedral of Lincoln appears to have been one of the earliest Gothic edifices; Westminster Abbey was finished about 1285, the Minster of York was begun a few years afterwards; and it is difficult to determine which of these three buildings most deserves the attention of the antiquary and the architect, or whether the Cathedral at Canterbury may not be equal to either of them.

In the midst of an age of darkness, an insulated individual arrests our attention by merits of no ordinary kind. Roger Bacon was born at Ilchester, in the year 1214; it is well known that his experiments had led him to a discovery of the properties of gunpowder, although he humanely concealed the nature of its composition from the public, and described it only in an enigma: the extent of his optical knowledge has been variously estimated, but it was unquestionably much greater than that of the ancient philosophers. He appears, however, to have had some companions in his mechanical pursuits; he declares that he had seen chariots which could move with incredible rapidity, without the help of animals; he describes a diving bell; and he says that he had been informed, on good authority, that machines had been made, by the assistance of which men might fly through the air. Cimabue, who first began to revive the long neglected art of painting, was cotemporary with Bacon. The use of oil in painting is commonly supposed to have been introduced by Van Eyck, but there are traces, in the records of this country, of its employment as early as the year 1239.

The clepsydrae, or water timekeepers of the ancients appear to have been gradually transformed, in the middle ages, into the clocks of the Saracens and of the Arabians; and these were introduced into Europe in the thirteenth century. About the year 1290, turret clocks were erected at Westminster and at Canterbury. The first clock, of which we know the construction, is that which was made by Wallingford in 1326, and which was regulated by a fly; and the second that of Defondeur, or Fusorius, with a simple balance, made about 1400. But it appears that some portable watches had been constructed in the beginning of the fourteenth century; and about the year 1460, several clock makers are said to have come to England from Flanders.

The art of engraving on metal, and of printing with the rolling press, is supposed to have been invented in the year 1423. Some attribute the art of printing with types, to Laurentius Coster of Haerlem, who, as they say, in 1430, employed for the purpose separate blocks of wood, tied together with thread. Gensfleich, one of his workmen, went to Mentz, and was there assisted by Gutenberg, who invented types of metal. But the best authors appear to disbelieve this story; and Gutenberg, in partnership with Fust and

Schaeffer, is the first that is universally allowed to have practised the art. It was introduced into this country by William Caxton.

Leonardo da Vinci, the most accomplished man of his age, was born about the year 1443, and excelled not only in painting and poetry, but also in architecture, mathematics, and mechanics. The state of practical mechanics in this and the subsequent centuries may be estimated from Ramelli's collection of machines, which contains several curious and useful inventions; some of them long since forgotten, and even lately proposed again as new.

The works of Bacon, Lord Verulam, although not immediately tending to the advancement of mathematics or of mechanics, are universally allowed to have conduced very materially to the improvement of every branch of science, by the introduction of a correct and conclusive method of philosophical argument and inquiry. Guido Ubaldi published, in 1577, a treatise on mechanics, not wholly exempt from inaccuracies, and in the following year a valuable commentary on the works of Archimedes: some of the properties of projectiles were about the same time rather imagined than demonstrated by Tartalea: Benedetti soon after began to reason correctly respecting the principles of mechanics; but it was reserved for Galileo to lay the foundations of the discoveries, which have succeeded each other with increasing rapidity for more than two centuries. He investigated, in the year 1589, the laws of accelerating forces, and showed the nature of the curve which is described by a projectile: he inferred from observation the isochronism of the vibrations of a pendulum, and the principle was soon after applied by Sanctorius to the regulation of timekeepers. Stevinus, a Dutchman, was the first that clearly stated the important law by which the equilibrium of any three forces is determined: and the properties of the centre of gravity were successively investigated by Lucas Valerius, Lafaille, and Guldinus, who made some additions to the elegant propositions of Archimedes which relate to it.

The application of the more intricate parts of the mathematics, to practical purposes of all kinds, has become incomparably easier and more convenient since the invention of logarithms. This important improvement was made by Baron Napier; his tables were published in 1614: and they were reduced to a still more useful form by the labours of Briggs and of Gunter. Descartes, about the same time, was making considerable additions to the science of algebra, and the mathematics were soon after enriched by Cavalleri's invention of the method of indivisibles. This method was founded on the principles introduced by Archimedes, it was further improved by Wallis, and it led to the still more valuable invention of the fluxional analysis.

The laws of collision were investigated nearly at the same time in England by Wren and Wallis, and in France by Huygens. After the discoveries of Archimedes and of Galileo, those of Huygens hold the third place, in the order of time, among the greatest benefits that have been conferred on science. His theory of cycloidal pendulums, and his doctrine of central forces were the immediate foundations of Newton's improvements.

Hooke was as great in mechanical practice, and in ingenious contrivance, as Huygens was in more philosophical theory; he was the first that applied the balance spring to watches, and he improved the mode of employing pendulums in clocks; the quadrant, the telescope, and the microscope, were materially indebted to him; he had the earliest suspicions of the true nature of the cause that retains the planets in their orbits; and the multitude of his inventions is far too great to be enumerated in a brief history of the progress of science.

The composition of motion, and several other mechanical and optical subjects, are elegantly treated in the lectures published by the learned Dr. Barrow. He was professor of mathematics at Cambridge, and voluntarily resigned his chair to make way for his successor, the pride of his country, and the ornament of mankind. Sir Isaac Newton was born at Woolsthorpe in Lincolnshire, on Christmas day in 1642, the year of Galileo's death: At the age of 12 he was sent to school at Grantham, and at 18 to Cambridge. He made some important improvements in algebraical analysis, and laid the foundation of his admirable method of fluxions, before he was 24 years old; but his modesty prevented him from immediately publishing any work on these subjects. His first optical experiments were also made in the year 1666, and they were communicated to the Royal Society, then in its infancy, on his admission as a member, in 1672. The theory of gravitation, and the mechanics of the universe, are developed in his Mathematical Principles of Natural Philosophy, first published in 1687. The following year he was chosen representative of the university of Cambridge, in parliament, and in 1696, he was placed, upon the recommendation of the Earl of Halifax, in a lucrative situation in the Mint. From 1703 until his death in 1727, he continued president of the Royal Society, and enjoyed, to the age of 80, an uninterrupted state of good health. He was knighted by Queen Anne, in 1705, and died possessed of a considerable fortune. "He had the singular happiness," says Mr. Fontenelle, "of obtaining, during his life, all the credit and consideration to which his sublime researches and his fortunate discoveries entitled him. All men of science, in a country which produces so many, placed Newton, by a kind of acclamation, at their head; they acknowledged him for their chief and their master; no opponent, nor even a cool admirer, dared to appear. His philosophy was adopted throughout England, and it is supported in the Royal Society, and in all the excellent productions of the members of that Society, with as much confidence, as if it had been consecrated by the respect of a long course of ages." A remarkable instance of the extent and refinement of Newton's mathematical acquirements may be found in a paper of a celebrated modern mathematician, on the subject of atmospherical refraction; Mr. Kramp observes, with a mixture of surprise and doubt, that Newton appears to have been acquainted with those methods of algebraical calculation which he had himself pursued; at the same time he says that this is almost incredible, since "he must have discovered certain improvements in the higher analysis which were unknown even to Euler, and to every other mathematician before Laplace."

Although Newton was unquestionably the first inventor of the method of fluxions, yet Leibnitz, whether he had received any hints of Newton's ideas, as there is some reason to suspect, or whether his investigations were wholly independent of those of Newton, was the first that published any work on the subject; and he extended its application to many important problems, earlier, perhaps, than any English mathematician. James and John Bernoulli also pursued the same studies with considerable success, and the general laws of mechanics were very elegantly investigated, and successfully applied by these three contemporary philosophers on the continent, while Machin, Cotes, Halley, and Demoivre, were applying themselves to similar pursuits in this country. Perrault, Lahire, Amontons, and Parent, members of the Parisian academy of sciences, were the authors of many useful investigations relating to practical mechanics; but few of them were made public till after the year 1700; some of their inventions made their appearance much later, in the valuable collection of machines approved by the academy, and some of them have been inserted in the useful work published by Leupold, at Leipzig, under the title of a Theatrum Machinarum. Throughout the last century, the transactions of various societies, established for the promotion of science, became every year more numerous, and the publication of the literary journals of Leipzig and of Paris formed a mode of communication, which was extremely serviceable in facilitating the dissemination of all new discoveries.

The philosophy of Newton assumed also a more popular and attractive form in the writings of Clarke, Pemberton, Maclaurin, and Musschenbroek, and the lectures of S'Gravesande and Desaguliers; at the same time that its more refined investigations were pursued with success in this country by Maclaurin and Simpson, and on the continent by Hermann, Daniel Bernoulli, Leonard Euler, and Clairaut. Maclaurin, Bernoulli, and Euler, had the honour of sharing with each other the prize, proposed by the academy of sciences at Paris, for the best essay on the intricate subject of the tides; but a premature death prevented Maclaurin from long pursuing the career which he began so successfully. Bernoulli and Euler continued for many years to vie with each other, for the elegance and extent of their researches: Euler appears to have been the more profound mathematician, and Bernoulli the more accurate philosopher.

The latter half of the eighteenth century was in many respects extremely auspicious to the progress of the sciences; the names of Dalembert, Landen, Waring, Frisi, Robison, Lagrange, and Laplace, deserve to be enumerated in the first class of mathematicians and theoretical mechanics; those of Smeaton, Wedgwood, and Watt are no less distinguished for their success in improving the practice of the useful arts and manufactures. The union of all these objects, into one system of knowledge, was effected, on a magnificent scale, in the Encyclopédie, a work which does as much honour to the labour and genius of some of its authors, as it reflects disgrace on the principles and politics of others. The Society for the encouragement of arts, manufactures, and commerce, was established in London about the same time that the Encyclopédie began to appear at Paris, and its premiums and publications have, without doubt, excited a degree of attention to the subjects of practical mechanics, and agricultural, as well as commercial improvements, which must have been beneficial both to individuals and to the public. The academy of Paris began to print, in 1762, a collection of the descriptions of arts and trades of all kinds, on a still more extended scale than had been attempted in the Encyclopédie; the work was carried to a very considerable length, but it by no means comprehends all the articles which were intended to compose it.

The construction of watches has been so much improved, by the artists both of this country and of France, that they have been rendered capable of affording very essential service to navigation, especially since the astronomical methods of determining a ship's place have been brought to such a degree of perfection, as greatly to facilitate the frequent correction of the accidental errors of the timekeeper. The first artist that constructed watches, sufficiently accurate for the determination of the longitude, was William Harrison, who was indebted to himself alone for his education and his inventions; in 1765 he received for his labours, from the Board of Longitude, the promised reward of ten thousand pounds.

There has scarcely been a period, in any age of the world, in which the sciences, and literature in general, have been so rapidly promoted, and so universally disseminated, as within the last forty years. This advancement has partly been the cause, and partly the effect, of the great multiplication of scientific journals, cyclopaedias, and encyclopaedias, which have been annually increasing since the beginning of the Journal de Physique in 1773; supported by the interest which they have derived, in great measure, from the new and amusing discoveries and improvements, which have been made in chemistry and natural history: some of the most copious of these works have had a sale, unprecedented even for books of more moderate extent.

The charter of the Royal Institution is dated in 1799; its foundation will not perhaps make an era in the history of the refinements of science; but if it be hereafter found to have given notoriety to what is useful, and popularity to what is elegant, the purposes of those who established it will not have been frustrated.

After all that has been effected by the united labours and talents of the philosophers who have been mentioned, and of many more, who, though less fortunate, have yet been highly meritorious, there is still ample opportunity for the employment of genius and industry in following their steps. To suppose that little or nothing remains to be done, betrays a want either of knowledge, or of courage. The experimental researches of some of the greatest philosophers have been very imperfectly conducted, and the most interesting results may be expected from repeating and diversifying them. Whatever advances our neighbours may have made beyond us, in intricate calculations and combinations, we are still able to vie with them, and shall probably long remain so, in the accuracy of our instruments, and in the art of using them with precaution and with success.

When, however, we contemplate the astonishing magnitude to which a collection of books in any department of science may even at present be extended, and the miscellaneous nature of the works in which many of the most valuable disquisitions have been communicated to the public, together with the natural disposition to indolence, which a high degree of civilisation too frequently encourages, there is the greatest reason to apprehend, that from the continual multiplication of new essays, which are merely repetitions of others that have been forgotten, the sciences will shortly be overwhelmed by their own unwieldy bulk, that the pile will begin to totter under its own weight, and that all the additional matter that we heap on it, will only tend to add to the extent of the basis, without increasing the elevation and dignity of the fabric. Having been impressed, from continued experience, with the truth of this observation, I have employed no small portion of time and labour, in order to obtain an effectual remedy for the evil; and I trust that, in future, every one who is desirous of enlarging the sphere of our knowledge, with respect to any branch of science, connected with the subject of these Lectures, will find it easy, by consulting the authors who will be quoted in my catalogue of references, to collect that previous knowledge of all that has been already done with the same view, which, in justice to himself, he ought to acquire before he enters on the pursuit, or at any rate, in justice to the public, before he calls on the world at large to participate in his improvements and discoveries.

CHRONOLOGY OF MATHEMATICIANS AND MECHANICS.

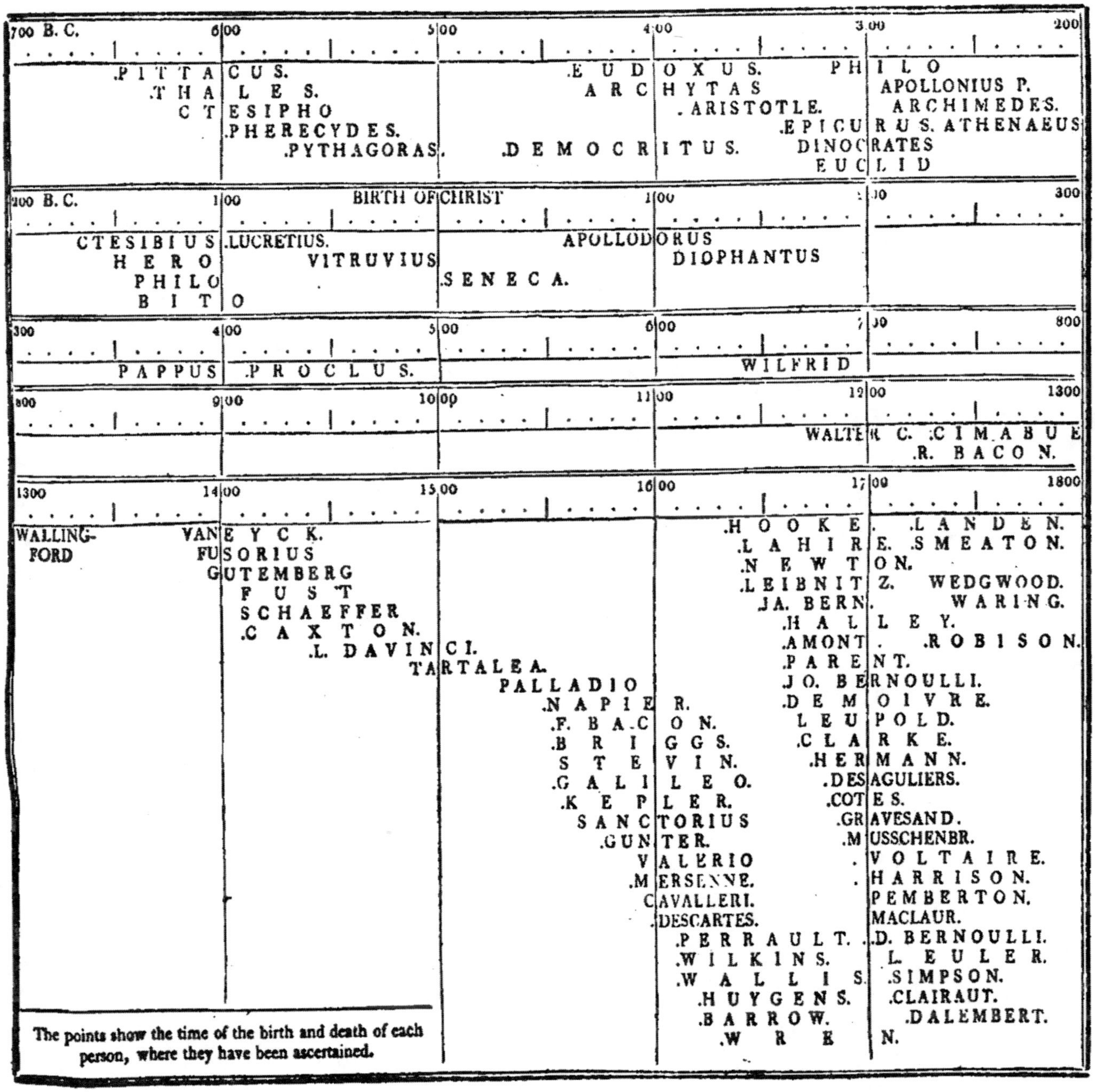

The points show the time of the birth and death of each person, where they have been ascertained.

A

COURSE OF LECTURES

ON

NATURAL PHILOSOPHY

AND THE

MECHANICAL ARTS.

PART II.

HYDRODYNAMICS.

A

COURSE OF LECTURES

ON

NATURAL PHILOSOPHY

AND THE

MECHANICAL ARTS.

LECTURE XXI.

ON HYDROSTATICS.

THE mechanical properties and affections of fluids, and the laws and phenomena of their motions, are to be the subjects of the second division of this Course of Lectures. Although these properties are in reality derived from the same fundamental principles as the doctrines of pure mechanics, they are yet in great measure incapable of being referred, in a demonstrative and accurate manner, to the operation of simple and general causes. We are therefore frequently under the necessity of calling in the assistance of experimental determinations; and for this reason, as well as others, the science of hydrodynamics may with propriety hold a middle rank, between mathematical mechanics and descriptive physics. In treating of the mechanics of solid bodies, we are able to begin with axioms, or self evident truths, almost inseparable from the constitution of the human mind; to deduce from them the general laws of motion, and to apply these laws, with little chance of error, to every combination of circumstances in which we have occasion to examine their consequences; and it requires only a certain degree of attention and of mathematical knowledge, to be perfectly convinced of the justice of all our conclusions, without any reference to experimental proof. But here our abstract reasonings begin to fail; and whether from the imperfection of our modes of considering the mechanical actions of the particles of fluids on each other, or from the deficiencies of our analytical calculations, or, as there is more reason to suppose, from a combination of both these causes, all attempts to reduce the affections of fluids to a perfect mechanical theory have been hitherto unsuccessful. At the same time it will appear, that by a proper mixture of calculation with experiment, we may obtain sufficient foundations for all such determinations as are likely to be of any practical utility.

The whole of the subjects, which will be classed under the denomination Hydrodynamics, may be divided into three general heads; Hydraulics, Acustics, and Optics; terms which are sufficiently understood, as relating to the common properties of fluids, to sound, and to light; but which do not allow of a very strict definition, without a still further division. The first subdivision which we shall consider, will relate to the laws of the equilibrium of fluids, or of the opposition of forces acting on them, without producing actual motion, comprehending hydrostatics, or the doctrine of the equilibrium of liquids, either within themselves, or with moveable bodies; and pneumatostatics, or the equilibrium of elastic fluids. The actual motions of fluids will be considered in the second subdivision: and the third will relate to the instruments and machines in which the principles of hydrostatics, hydraulics, and pneumatics, are applied to the purposes of the arts or of domestic convenience. The science of hydraulics must be allowed to be of as great importance to civil life, and especially to a maritime nation, as any department of practical mechanics. Let us only reflect for a moment to what the metropolis of England would be reduced, if deprived of pipes for the conveyance of water, of pumps, and of fire engines; and how much the commerce of the whole kingdom has been facilitated by the formation of navigable canals, and we shall soon be convinced of the obligations that we owe to the art of modifying the motion of water, and to the principles of hydraulics, on which that art depends.

The facts concerned in acustics and harmonics, or the doctrine of sound, and the science of music, are not exclusively dependent on the characteristic properties of fluids. In these departments, although we can by no means explain with precision the manner in which every appearance is produced, we shall still find a variety of very beautiful phenomena, which have indeed been too generally neglected, and supposed to be of the most abstruse and unintelligible nature; but which, when carefully examined, will appear to be much more within the reach of calculation, than the simplest doctrines of hydraulics. We may also apply some of these phenomena to a very complete explanation of an extensive class of facts in optics, which, in whatever other way they are considered, are inextricably obscure. Whether this explanation may or may not be admitted as satisfactory, it deserves at least a fair examination; it would, therefore, be impossible to assign to the science of optics an earlier place in the order of the system, even if we agree with those, who imagine that all the phenomena of light depend on causes wholly deducible from the mechanics of solid bodies.

We must commence the subject of hydrostatics, or the doctrine of the equilibrium of liquids, with a definition of the essential characteristics of a fluid substance. The most eligible definition appears to be, that a fluid is a collection of material particles, which may be considered as infinitely small, and as moving freely on each other in every direction, without friction. Some have defined a fluid as a substance which communicates pressure equally in all directions; but this appears to be a description of a property derivable from the former assumption, which is certainly more simple; and although it may be somewhat difficult to deduce it mathematically, in a manner strictly demonstrative, yet we may obtain from mathematical considerations a sufficient conviction of its truth, without assuming it as a fundamental or axiomatic character. A fluid which has no immediate tendency to expand when at liberty, is commonly considered as a liquid: thus water, oil, and mercury, are liquids; air and steam are fluids, but not liquids.

We shall for the present consider a liquid as without either compressibility or expansibility: and we must neglect some other physical properties essential to liquids, such as cohesion and capillary attraction; although in reality the particles of liquids are found, by very nice experiments, to be subject to the same laws of elasticity which we have already examined with regard to solids, and are possessed also of cohesive powers, which essentially distinguish them from elastic fluids, and which resist any force tending directly to separate the particles from each other, while they admit any lateral motion with perfect facility. In treating of hydrostatics, therefore, we suppose the fluids concerned to be of uniform density throughout; and as far as elastic fluids agree with this description, they are subject to the same laws with liquids; on the other hand, all fluids, as far as they are compressible, possess properties similar to those which will hereafter be examined, when we investigate the subjeot of pneumatic equilibrium.

The first law of hydrostatics which arrests our attention, is this, that the surface of every homogeneous gravitating fluid, when at rest, is horizontal. If any part of the surface were inclined to the horizon, the superficial particles would necessarily tend towards its lowest part, in the same manner as if they moved without friction on the inclined surface of a solid. And if any two portions of the surface of the fluid are separated, as in two branches of a tube or pipe, however they may be situated, the fluid cannot remain at rest, unless the surfaces be in the same level plane: for if we imagine such a tube, containing water, to be made of ice, and to be immersed in a large reservoir of water, and then thawed, the water will make a part of the general contents of the reservoir, and consequently will remain at rest, if its surfaces are level with that of the reservoir: and it is obvious that the tube has acquired no new power of supporting it from being thawed: consequently, the water would have remained in equilibrium at the same height in the original state of the solid tube. The experimental proof of this proposition is easy and obvious, and the property affords one of the most usual modes of determining a horizontal surface. But when we compare the heights of fluids occupying tubes of different magnitudes, it is necessary, if the tubes are small, to apply a slight correction on account of the actions of the tubes on the fluids which they contain, which are more apparent, as their diameters are smaller. The same cause produces also a curvature in each separate surface, which is always visible at the point of contact with the tube or vessel. (Plate XIX. Fig 239.)

If several separate fluids of different kinds be contained in the same vessel, they will never remain at rest unless all the surfaces intervening between them be horizontal; and this is in fact the state of the surface of common liquids, which is exposed to the pressure of the atmosphere.

The power of gravitation, strictly speaking, does not act precisely in parallel lines, so that the surface of lakes, instead of being perfectly plane, becomes, like that of the earth, a little convex. It is obvious that the surface of a fluid must always be perpendicular to the direction of the joint results of all the forces which act on it; and since the earth turns round on its axis, the centrifugal force resulting from its motion is combined with that of gravity, in determining the position of the general surface of the ocean.

A similar combination of a centrifugal force with gravitation may be observed when a bucket is suspended by a rope, and caused to turn round on its axis by twisting the rope: the direction of the joint forces is such that the surface, in order to be perpendicular to it, must assume a parabolic form. When also any number of different fluids are made to revolve in the same manner, or when they are inclosed in a glass globe and turned by means of the whirling table, the surfaces which separate them acquire always the forms of parabolic conoids, when the axis remains in a vertical position: but if the axis be in any other position, the situation of the surface will be of more difficult determination. (Plate XX. Fig. 240.)

In all these cases the equilibrium is stable; for if any part of the fluid be raised above the surface, it will immediately tend to return to its level. But if a heavier fluid were contained in a bent tube or siphon, with its legs or branches opening downwards, and immersed in a lighter fluid, the equilibrium would be tottering, since, if it were once disturbed, it would never be restored. (Plate XIX. Fig 241.)

From these principles, we may infer, that the pressure of a fluid on every particle of the vessel containing it, or of any other surface, real or imaginary, in contact with it, is equal to the weight of a column of the fluid, of which the base is equal to that particle, and the height to its depth below the surface of the fluid. Thus, if we have a vessel of water one foot deep, each square foot of the bottom will sustain the pressure of a cubic foot of water, or nearly 1000 ounces: if we have a vessel of mercury an inch in depth, each square foot will sustain a pressure of one twelfth part of a cubic foot of mercury, or 1130 ounces; the atmosphere presses on each square foot of the earth's surface with a force of about 34000 ounces, which is equivalent to the pressure of a column of mercury 30 inches high. The pressure of the water on a small portion of the lowest part of the side of the vessel containing it, is also equal to the weight supported by an equal portion of the bottom; but we cannot estimate the force sustained by any large portion of the side, without considering the different depths below the surface, at which its different parts are situated.

It is obvious that if we conceive a fluid to be divided by an imaginary surface of any kind, the particles contiguous to it are urged on either side by equal forces, the fluid below resisting them, and pressing them upwards, with as much force as the fluid above presses them downwards, their own weight being comparatively inconsiderable, for without this equality of pressures, they could not possibly remain at rest. And if we employ a vessel of such a form as to occupy the place of any superior portion of the fluid, the pressure against that part of the vessel which is thus substituted will be the same that before supported the weight of the fluid removed; and in order that all may remain in equilibrium, the vessel must itself exert an equal pressure on the fluid below it; so that the pressure on the bottom will be the same as if the vessel had remained in its original state, and were filled to the same height with the fluid. (Plate XIX. Fig. 242.)

In order to understand this the more readily, we may suppose the portion of the fluid, instead of being removed, to have been congealed into a solid mass of equal density; it is obvious that this congelation of the fluid would not have altered the quantity of its pressure; it would, therefore, have remained in equilibrium with the water below; the mass might also be united with the sides of the vessel, so as to form a part of it, without increasing or diminishing any of the pressures concerned: and we should thus obtain a vessel similar to that which was the subject of our investigation, the pressure on the bottom being always the same, as if the mass, supposed to be congealed, had remained fluid. Thus, the pressure on the base of a conical or pyramidical vessel, full of water, is three times as great as the weight of the water, since its content is one third of that of a column of the same height, and standing on the same base. (Plate XIX. Fig. 243.)

In this manner the smallest given quantity of any fluid contained in a pipe may be made to produce a pressure equivalent to any given weight, however large, which rests on the cover of a close vessel communicating with the pipe, and this may be done either by diminishing the diameter of the pipe, and increasing its height, while the weight is supported by a surface of a certain extent, or by increasing the magnitude of this surface, without adding to the height of the pipe; for in either case the ultimate force of the fluid, in supporting the weight, will be equal to the weight of a column of the same height, standing on the whole surface which is subjected to its action. And if the effect of the column be increased by any additional pressure, independent of its weight, that pressure may be represented by supposing the height of the column to be augmented; and the effect of the additional pressure will also be increased in proportion to the magnitude of the surface which supports the weight. It is on this principle that the pressure of water has been applied, by Mr. Bramah, to the construction of a very convenient press. (Plate XIX. Fig. 244.)

Although this property of fluids is the cause of some results which would scarcely be expected by a person not accustomed to reflect on the subject, and has, therefore, not improperly, been called the hydrostatic paradox, yet it depends wholly on the general and acknowledged principles of mechanical forces; nor can we agree with those authors, who have asserted, that a very small quantity of a fluid may, " without acting at any mechanical advantage" whatever, be made to balance a weight of any assignable magnitude: for the immediate operation of the force very much resembles, in the most common cases, the effect of a wedge, or of a moveable inclined plane; thus, a wedge remains in equilibrium, when the forces acting on each side are in proportion to its length, like the hydrostatic pressure on a vessel of a similar form. The conditions of the equilibrium of fluids may also be determined, in all cases, from the general law of the descent of the centre of gravity to the lowest point. Thus, it is easy to show that even when two branches of a tube are of unequal diameter, a fluid must stand at the same height in both of them, in order to remain in equilibrium: for if any portion be supposed to stand, in either leg, above the surface of the fluid in the other leg, it is obvious that its centre of gravity may be lowered, by removing so much of it as will raise the fluid in the opposite leg to its own level, the situation of the fluid below remaining unaltered: consequently the centre of gravity of the whole fluid can never acquire its lowest situation, unless both the surfaces are in the same level.

The air, and all other elastic fluids, are equally subject with liquids to this general law. Thus, a much greater force is required, in order to produce a blast of a given intensity, with a large pair of bellows, than with a smaller pair; and for the same reason, it is much easier to a glassblower, when he uses a blowpipe, to employ the muscles of his mouth and lips, than those of his chest, although these are much more powerful. If we estimate the section of the chest at a foot square, it will require a force of seventy pounds to raise a column of mercury an inch high, by means of the muscles of respiration, but the section of the mouth is scarcely more than eight or nine square inches, and a pressure of the same intensity may here be produced by a force of about four pounds. The glassblower obtains, besides, the advantage of being able to continue to breathe during the operation, the communication of the chest with the nostrils remaining open, while the root of the tongue is pressed against the palate.

It is obvious that the pressure on each square inch of the side of a vessel, or on each square foot of the bank of a river, continually increases in descending towards the bottom. If we wish to know the sum of the pressures on all the parts of the side or bank, we must take some mean depth by which we can estimate it; and this must be the depth of the point which would be the centre of gravity of the surface, if it were possessed of weight. Thus, if we had a hollow cube filled with water, the centre of gravity of each side being in its middle point, the pressure on each of the upright sides would be half as great as the pressure on the bottom, that is, it would be equal to half the weight of the water contained in the cube.

If, however, we wished to support the side of the cube externally by a force applied at a single point, that point must be at the distance of one third of the height only from the bottom. For the pressure at each point may be represented by a line equal in length to its depth below the surface, and a series of such lines may be supposed to constitute a triangle, of which the centre of gravity will indicate the place of the centre of pressure of the surface; and the height of the centre of gravity will always be one third of that of the triangle. It is easily inferred, from this representation, that the whole pressure on the side of a vessel, or on a bank, of a given length, is proportional to the square of the depth, below the water, to which it extends. (Plate XIX. Fig. 245.)

The magnitude of the whole pressure on a concave or convex surface may also be determined by the position of its centre of gravity; but such a determination is of no practical utility, since the portions of the forces, which act in different directions, must always destroy each other. Thus, the perpendicular pressure on the whole internal surface of a sphere filled with a fluid, is three times as great as the weight of the fluid; but the force tending to burst the sphere, in the circumference of any vertical circle, is only three fourths of that weight.

If two fluids are of different specific gravities, that is, if equal bulks of them have different weights, their opposite pressures will counterbalance each other, when their heights above the common surface are inversely as their specific gravities; for it is obvious that the greater density of the one will precisely compensate for its deficiency in height. Thus, a column of mercury, standing at the height of 30 inches, in a tube, will support the pressure of a column of water, in another branch of the tube, exactly 34 feet high: since the weight of 30 cubic inches of mercury is equal to that of 408 cubic inches of water. (Plate XIX. Fig. 246.)

We have hitherto considered the properties of fluids in contact with solids which are immoveable, and of invariable form; but it often happens that they act on substances which are moveable; and they are sometimes contained in vessels of which the form is susceptible of variation; in these cases, other considerations are necessary for the determination of the equilibrium of fluids and solids with each other; and in the first place the properties of floating bodies require to be investigated.

When a solid body floats in a fluid, it displaces a quantity of the fluid, equal to itself in weight; and every solid, which is incapable of doing this, must sink. For in order that the solid may remain at rest, the pressure of the fluid below it, reduced to a vertical direction, must be precisely equal to its weight; but before the body was immersed, the same pressure was exerted on the portion of the fluid which is now displaced, and was exactly counterbalanced by its weight; consequently that weight was equal to the weight of the floating body.

Since the force, which supports the weight of a floating body, is the pressure of the fluid immediately below it, if this pressure be removed or diminished, the body may remain at rest below the surface of the fluid, even when it is specifically lighter. Thus a piece of very smooth wood will remain, for some time, in contact with the flat bottom of a vessel of water, until the water insinuates itself beneath it; and it will continue at the bottom of a vessel of mercury, without any tendency to rise, since the mercury has no disposition to penetrate, like water, into any minute interstices which may be capable of admitting it. And, for a similar reason, if the pressure of the incumbent fluid be removed from the upper surface of a solid substance, wholly immersed in it, the solid may remain suspended, although heavier than an equal bulk of the fluid. Thus, if a tube or vessel of any kind, open above and below, have a bottom of metal, ground so as to come into perfect contact with it, without being fixed, the bottom will appear to adhere to the vessel, when it is immersed to a sufficient depth in water, the vessel remaining empty.

In order that a floating body may remain in equilibrium, it is also necessary that its centre of gravity be in the same vertical line with the centre of gravity of the fluid displaced; otherwise the weight of the solid will not be completely counteracted by the pressure of the fluid. The nature of the equilibrium, with respect to stability, is determined by the position of the metacentre, or centre of pressure, which may be considered as a fixed point of suspension, or support, for the solid body. It is obvious that when the lower surface of the body is spherical or cylindrical, the metacentre must coincide with the centre of the figure, since the height of this point, as well as the form of the portion of the fluid displaced, must remain invariable in all circumstances, and the nature of the equilibrium will depend on the distance of the centre of gravity above or below the centre of the sphere or cylinder. And the place of the metacentre may always be determined from the form and extent of the surface of the displaced portion of the fluid, compared with its bulk, and with the situation of its centre of gravity. For example, if a rectangular beam be floating on its flat surface, the height of the metacentre above the centre of gravity will be to the breadth of the beam, as the breadth to twelve times the depth of the part immersed. Hence, if the beam be square, it will float securely when either the part immersed or the part above the surface is less than $\frac{21}{100}$ of the whole; but when it is less unequally divided by the surface of the fluid, it will overset. If, however, the breadth be so increased as to be nearly one fourth greater than the depth, it will possess a certain degree of stability whatever its density may be. (Plate XIX. Fig. 247.)

When the equilibrium of a floating body is stable, it may oscillate backwards and forwards in the neighbourhood of the quiescent position: and the oscillations will be the more rapid in proportion as the stability is greater in comparison with the bulk of the body. Such oscillations may also be combined with others which take place in a transverse direction: a ship, for example, may roll on an axis in the direction of her length, and may heel, at the same time, upon a second axis in the direction of the beams. Besides these rotatory vibrations, a floating body which is suffered to fall into a fluid, will commonly rise and sink several times by its own weight; and in all these cases, the vibrations of any one kind, when they are small, are performed nearly in equal times: but various and intricate combinations may sometimes arise, from the difference of the times, in which the vibrations of different kinds are performed.

When a solid body is wholly immersed in a fluid, and is retained in its situation by an external force, it loses as much of its weight as is equivalent to an equal bulk of the fluid. For, conceiving the fluid, which is displaced by the body, to have been converted into a solid by congelation, it is obvious that it would retain its situation, and the difference of the pressures of the fluid on its various parts would be exactly sufficient to support its weight. But these pressures will be the same if a body of any other kind be substituted for the congealed fluid; their buoyant effect may, therefore, be always estimated by the weight of a portion of the fluid equal in bulk to the solid.

Thus, when a little figure, containing a bubble of air, is immersed in a jar of water, which is so covered by a bladder that it may be compressed by the hand, the bulk of the figure with its bubble is diminished by the pressure, it is, therefore, less supported by the water, and it begins to sink: and when the hand is removed, it immediately rises again. (Plate XIX. Fig. 248.)

While a body is actually rising or sinking in a fluid, with an accelerated motion, the force of gravity being partly employed in generating momentum, either in the fluid or in the solid, the whole pressure on the bottom of the vessel is necessarily somewhat lessened. Hence the apparent weight of a jar of water will suffer a slight diminution, while a bullet is descending in it, or while bubbles of air are rising in it, but the difference can seldom be great enough to be rendered easily discoverable to the senses.

It sometimes happens that a solid body is partly supported by a fluid, and partly by another solid; of this we have an example in one of Dr. Hooke's ingenious inventions for keeping a vessel always full. A half cylinder, or a hemisphere, being partly supported on an axis, which is in the plane of the surface of the fluid, its weight is so adjusted, as to be equal to that of a portion of the fluid of half its magnitude: when the vessel is full, it is half immersed, and exerts no pressure on the axis: it descends as the fluid is exhausted, and its tendency to turn round its axis can only be counteracted, by the pressure of the fluid on its flat side, as long as the surface of the remaining portion of the fluid retains its original level. (Plate XIX. Fig. 249.)

When a fluid is contained in a vessel of a flexible nature, the sides of the vessel will always become curved, in consequence of the pressure, and the more, in proportion as the pressure is greater; the form of the curved surface will also be such that the common centre of gravity of the fluid and the vessel may descend to the lowest point that the circumstances of the case allow; this form is generally of too intricate a nature to be determined by calculation: no mathematician has hitherto been able to investigate, for example, the curvature which a square or rectangular bag of leather will assume when filled with water or with corn. When, indeed, one dimension only of a vessel is considered, for instance, when the bottom of a cistern is supposed to be flexible, and to be fixed at two opposite sides, while the ends are simply in-contact with upright walls, without allowing the water to run out, the nature of the curve may be determined with tolerable facility, whether the weight of the bottom itself be considered or not. If the weight be exactly equal to that of the water, the form of a semicircle will agree with the conditions of equilibrium, as Bernoulli has demonstrated, supposing the fixed points at the distance of its diameter; but if the weight of the bottom be neglected, the curvature will be every where proportional to the distance below the surface, the form being the same as that of an elastic rod, bent by two forces in the direction of the surface. The same principles, with a slight difference in the calculations, will serve to determine the forms adapted to the equilibrium of arches, intended for supporting the weight of superincumbent fluids, or of such soft materials as approach nearly in their operation to more perfect fluids. (Plate XIX. Fig. 250.)

LECTURE XXII.

ON PNEUMATIC EQUILIBRIUM.

THE laws of the pressure and equilibrium of liquids, which are the peculiar subjects of hydrostatics, are also applicable in general to fluids of all kinds, as far as they are compatible with the compressibility of those fluids, or with their tendency to expand.

Elastic fluids are distinguished from liquids by the absence of all cohesive force, or by their immediate tendency to expand when they are at liberty. Such are atmospheric air, steam, and gases of various kinds; and the consideration of these fluids, in the state of rest, constitutes the doctrine of pneumatostatics, or of the equilibrium of elastic fluids.

That the air is a material substance, capable of resisting pressure, is easily shown, by inverting an empty jar in water; and by the operation of transferring airs and gases from vessel to vessel, in the pneumatic apparatus used by chemists. The tendency of the air to expand is shown by the experiment in which a flaccid bladder becomes distended, and shrivelled fruit recovers its full size, as soon as the external pressure is removed from it, by the operation of the air pump: and the magnitude of this expansive force is more distinctly seen, when a portion of air is inclosed in a glass vessel, together with some mercury, in which the mouth of a tube is immersed, while the other end is open, and without the vessel; so that when the whole apparatus is inclosed in a very long jar, and the air of the jar is exhausted, the column of mercury becomes the measure of the expansive force of the air. (Plate XIX. Fig. 251.)

If the diameter of the tube, in an apparatus of this kind, were very small in comparison with the bulk of the air confined, the column of mercury would be raised, in the ordinary circumstances of the atmosphere, to the height of nearly 30 inches. But supposing the magnitude of the tube such, that the portion of air must expand to twice its natural bulk, before the mercury acquired a height sufficient to counterpoise it, this height would be 15 inches only. For it appears to be a general law of all elastic fluids, that their pressure on any given surface is diminished exactly in the same proportion as their bulk is increased. If, therefore, the column of mercury in the vacuum of the air pump were 60 inches high, the air would be reduced to half its natural bulk; and for the same reason, the pressure of a column of 30 inches of mercury in the open air will reduce any portion of air to half its bulk, since the natural pressure of the atmosphere, which is equal to that of about 30 inches of mercury, is doubled by the addition of an equal pressure. In the same manner the density of the air in a diving bell is doubled at the depth of 34 feet below the surface of the water, and tripled at the depth of 68 feet. This law was discovered by Dr. Hooke; he found, however, that when a very great pressure had been applied, so that the density became many times greater than in the natural state, the elasticity appeared to be somewhat less increased than the density; but this exception to the general law has not been confirmed by later and more accurate experiments.

Not only the common air of the atmosphere, and other permanently elastic gases, but also steams and vapours of all kinds, appear to be equally subject to this universal law: they must, however, be examined at temperatures sufficient to preserve them in a state of elasticity; for example, if we wished to determine the force of steam twice as dense as that which is usually produced, we should be obliged to employ a heat 30 or 40 degrees above that of boiling water: we should then find that steam of such a density as to support, when confined in a dry vessel, the pressure of a column of 30 inches of mercury, would be reduced to half its bulk by the pressure of a column of 60 inches. But if we increased the pressure much beyond this, the steam would be converted into water, and the experiment would be at an end.

That the air which surrounds us is subjected to the power of gravitation, and possesses weight, may be shown by weighing a vessel which has been exhausted by means of the air pump, and then allowing the air to enter, and weighing it a second time. In this manner we may ascertain the specific gravity of the air, even if the exhaustion is only partial, provided that we know the proportion of the air left in the vessel to that which it originally contained. The pressure derived from the weight of the air is also the cause of the ascent of hydrogen gas, or of another portion of air which is rarefied by heat, and carries with it the smoke of a fire; and the effect is made more conspicuous, when either the hydrogen gas, or the heated air, is confined in a balloon. The diminution of the apparent weight of a body, by means of the pressure of the surrounding air, is also shown by the destruction of the equilibrium between two bodies of different densities, upon their removal from the open air into the vacuum of an air pump. For this purpose, a light hollow bulb of glass may be exactly counterpoised in the air by a much smaller weight of brass, with an index, which shows, on a graduated scale, the degree in which the large ball is made to preponderate in the receiver of the air pump, by the rarefaction of the air, lessening the buoyant power which helps to support its weight. (Plate XIX. Fig. 252.)

From this combination of weight and elasticity in the atmosphere, it follows, that its upper parts must be much more rare than those which are nearer to the earth, since the density is every where proportional to the whole of the superincumbent weight. The weight of a column of air one foot in height is one twenty eight thousandth of the whole pressure; consequently that pressure is increased one twenty eight thousandth by the addition of the weight of one foot, and the next foot will be denser in the same proportion, since the density is always proportionate to the pressure; the pressure thus increased will therefore still be equal to twenty eight thousand times the weight of the next foot. The same reasoning may be continued without limit, and it may be shown, that while we suppose the height to vary by any uniform steps, as by distances of a foot or a mile, the pressures and densities will increase in continual proportion; thus, at the height of about 3000 fathoms, the density will be about half as great as at the earth's surface; at the height of 6000, one fourth; at 9000, one eighth as great. Hence it is inferred that the height in fathoms may be readily found from the logarithm of the number expressing the density of the air: for the logarithm of the number 2, multiplied by 10000, is 3010, the logarithm of 4, 6020, and that of 8, 9031; the logarithms of numbers always increasing in continual proportion, when the numbers are taken larger and larger by equal steps. (Plate XIX. Fig. 253.)

Hence we obtain an easy method of determining the heights of mountains with tolerable accuracy: for if a bottle of air were closely stopped on the summit of a mountain, and, being brought in this state into the plain below, its mouth were inserted into a vessel of water or of mercury, a certain portion of the liquid would enter the bottle; this being weighed, if it were found to be one half of the quantity that the whole bottle would contain, it might be concluded that the air on the mountain possessed only half of the natural density, and that its height was 3000 fathoms. It appears also, from this statement, that the height of a column of equal density with any part of the atmosphere, equivalent to the pressure to which that part is subjected, is every where equal to about 28000 feet.

Many corrections are, however, necessary for ascertaining the heights of mountains with all the precision that the nature of this kind of measurement admits; and they involve several determinations, which require a previous knowledge of the effects of heat, and of the nature of the ascent of vapours, which cannot be examined with propriety at present.

We may easily ascertain, on the same principles, the height to which a balloon will ascend, if we are acquainted with its bulk and with its weight: thus, supposing its weight 500 pounds, and its bulk such as to enable it to raise 300 pounds more, its specific gravity must be five eighths as great as that of the air, and it will continue to rise, until it reach the height, at which the air is of the same density: but the logarithm of eight fifths, multiplied by 10000, is 2040; and this is the number of fathoms contained in the height, which will, therefore, be a little more than two miles and a quarter. It may be found, by pursuing the calculation, that at the distance of the earth's semidiameter, or nearly 4000 miles, above its surface, the air, if it existed, would become so rare, that a cubic inch would occupy a space equal to the sphere of Saturn's orbit: and on the other hand, if there were a mine about 42 miles deep, the air would become as dense as quicksilver at the bottom of it.

It appears, therefore, that all bodies existing on or near the earth's surface may be considered as subjected to the pressure of a column of air, 28000 feet high, supposing its density every where equal to that which it possesses at the earth's surface, and which is usually such, that 100 wine gallons weigh a pound avoirdupois, creating a pressure equal to that of 30 inches of mercury, or 34 feet of water, and which amounts to $14\frac{3}{4}$ pounds for each square inch. This pressure acts in all directions on every substance which is exposed to it: but being counterbalanced by the natural elasticity of these substances, it produces in common no apparent effects; when, however, by means of the air pump, or otherwise, the pressure of the air is removed from one side of a body, while it continues to act on the other, its operation becomes extremely evident. Thus, when two hollow hemispheres, in contact with each other, are exhausted of air, they are made to cohere with great force; they are named Magdeburg hemispheres, because Otto von Guerike, of Magdeburg, constructed two such hemispheres, of sufficient magnitude to withstand the draught of the emperor's six coach horses, pulling with all their force to separate them. By a similar pressure, a thin square bottle may be crushed when it is sufficiently exhausted, and a bladder may be torn with a loud noise: and the hand being placed on the mouth of a vessel which is connected with the air pump, it is fixed to it very forcibly, when the exhaustion is performed, by the pressure of the air on the back of the hand; the fluids also, which circulate in the bloodvessels of the hand, are forced towards its lower surface, and the effect which is called suction is produced in a very striking manner. It is on the same principle that cupping glasses are employed, a partial exhaustion being procured by means of the flame of tow, which heats the air, and expels a great part of it: so that the remainder, when it cools, is considerably rarefied.

It was Galileo that first explained the nature of suction from the effects of the pressure of the atmosphere; and his pupil Torricelli confirmed his doctrines by employing a column of mercury, of sufficient height to overcome the whole pressure of the atmosphere, and to produce a vacuum in the upper part of the tube or vessel containing it. In the operation of sucking up a fluid through a pipe, with the mouth or otherwise, the pressure of the air is but partially removed from the upper surface of the fluid, and it becomes capable of ascending to a height which is determined by the difference of the densities of the air within and without the cavity concerned: thus, an exhaustion of one fourth of the air of the cavity would enable us to raise water to the height of $8\frac{1}{2}$ feet, and mercury to $7\frac{1}{2}$ inches, above the level of the reservoir from which it rises. We can draw up a much higher column of mercury by sucking with the muscles of the mouth only, than by inspiring with the chest, and the difference is much more marked than the difference in the forces with which we can blow: for in sucking, the cavity of the mouth is very much contracted by the pressure of the external air, and the same force, exerted on a smaller surface, is capable of counteracting a much greater hydrostatic or pneumatic pressure.

When a tube of glass, about three feet long, closed at one end and open at the other, is filled with mercury, and then immersed in a bason of the same fluid, the pressure of the atmosphere is wholly removed from the upper surface of the mercury in the tube, while it continues to act on the mercury in the bason, and by its means on the lower surface of the column in the tube. If such a tube be placed under the receiver of an air pump, the mercury will subside in the tube, accordingly as the pressure of the atmosphere is diminished; and if the exhaustion be rendered very perfect, it will descend very nearly to the level of the open bason or reservoir. When the air is readmitted, the mercury usually rises, on the level of the sea, to the height of about 30 inches; but the air being lighter at some times than at others, the height varies between the limits of 27 and 31 inches. This well known instrument, from its use in measuring the weight of the air, is called a barometer. In the same manner a column of water from 30 to 35 feet in height may be sustained in the pipe of a pump; but if the pipe were longer than this, a vacuum would be produced in the upper part of it, and the pump would be incapable of acting.

In order to observe the height of the mercury in the barometer with greater convenience and accuracy, the scale has sometimes been amplified by various methods; either by bending the upper part of the tube into an oblique position, as in the diagonal barometer, or by making the lower part horizontal, and of much smaller diameter than the upper, or by making the whole tube straight, and narrow, and slightly conical, or by placing a float on the surface of the mercury in the reservoir, and causing an axis, which carries an index, to revolve by its motion. But a good simple barometer, about one third of an inch in diameter, furnished with a vernier, is perhaps fully as accurate as any of these more complicated instruments. In order to exclude the air the more completely from the tube, the mercury must at least be shaken in it for a considerable time, the tube being held in an inverted position; and where great accuracy is required, the mercury must be boiled in the tube. The reservoir most commonly employed is a flat wooden box, with a bottom of leather; the cover, which is unscrewed at pleasure, being cemented to the tube. Sometimes a screw is made to act on the leather, by means of which the surface of the mercury is always brought to a certain level, indicated by a float, whatever portion of it may be contained in the tube; but the necessity of this adjustment may be easily avoided, by allowing the mercury to play freely between two horizontal surfaces of wood, of moderate extent, and at the distance of one seventh of an inch: the height may then be always measured from the upper surface, without sensible error. But if the surfaces were closer than this, the mercury would stand too high in the tube. (Plate XIX. Fig. 254.)

The same method which is employed for determining the relation between the heights and densities of elastic fluids, may be extended to all bodies which are in any degree compressible, and of which the elasticity is subjected to laws similar to those which are discoverable in the air and in other gases: and it is not improbable that these laws are generally applicable to all bodies in nature, as far as their texture will allow them to submit to the operation of pressure, without wholly losing their form. Water, for example, has been observed by Canton to be compressed one twenty two thousandth of its bulk by a force equal to that of the pressure of the atmosphere; consequently this force may be represented by that of a column of water 750 thousand feet in height; the density of the water at the bottom of a lake, or of the sea, will be increased by the pressure of the superincumbent fluid; and supposing the law of compression to resemble that of the air, it may be inferred that at the depth of 100 miles, its density would be doubled; and that at 200 it would be quadrupled. The same measures would also be applicable to the elasticity of mercury. But there is reason to suppose that they are in both cases a little too small.

LECTURE XXIII.

ON THE THEORY OF HYDRAULICS.

HAVING considered the principal cases of the equilibrium of fluids, both liquid and aeriform, we proceed to examine the theory of their motions. Notwithstanding the difficulties attending the mathematical theory of hydraulics, so much has already been done, by the assistance of practical investigations, that we may in general, by comparing the results of former experiments with our calculations, predict the effect of any proposed arrangement, without an error of more than one fifth, or perhaps one tenth of the whole: and this is a degree of accuracy fully sufficient for practice, and which indeed could scarcely have been expected from the state of the science at the beginning of the last century. Many of these improvements have been derived from an examination of the nature and magnitude of the friction of fluids, which, although at first sight it might be supposed to be very inconsiderable, is found to be of so much importance in the application of the theory of hydraulics to practical cases, and to affect the modes of calculation so materially, that it will require to be discussed, hereafter, in a separate lecture.

There is a general principle of mechanical action, which was first distinctly stated by Huygens, and which has been made by Daniel Bernoulli the basis of his most elegant calculations in hydrodynamics. Supposing that no force is lost in the communication of motion between different bodies, considered as belonging to any system, they always acquire such velocities in descending through any space, that the centre of gravity of the system is capable of ascending to a height equal to that from which it descended, notwithstanding any mutual actions between the bodies. The truth of this principle may easily be inferred from the laws of collision, compared with the properties of accelerating and retarding forces. Thus, if an elastic ball, weighing 10 ounces, and descending from a height of 1 foot, be caused to act in any manner on a similar ball of one ounce, so as to lose the whole of its motion, the smaller ball will acquire a velocity capable of carrying it to the height of 10 feet. It is true that some other suppositions must be made, in applying this law to the determination of the motions of fluids, and that in many cases it becomes necessary to suppose that a certain portion of ascending force or energy is lost, in consequence of the internal motions of the particles of the fluid. But still, with proper restrictions and corrections, the principle affords us a ready method of obtaining solutions of problems, which, without some such assistance, it would be almost impossible to investigate. The principal hypothesis which is assumed by Bernoulli, without either demonstration, or even the appearance of perfect accuracy, is this, that all the particles of a fluid in motion, contained in any one transverse section of the vessels or pipes through which it runs, must always move with equal velocities; thus, if water be descending through a vessel of any form, either regular or irregular, he supposes the particles at the same height to move with the same velocity; so that the velocity of every particle in every part of a cylindrical vessel 10 inches in diameter, through which a fluid is moving, must be one hundredth part as great as in passing through a circular orifice, an inch in diameter, made in its bottom. It is evident that this cannot possibly be true of the portions of the fluid nearest the bottom of the vessel, since the particles most distant from the orifice must be nearly at rest, while those which are immediately over the orifice are in rapid motion; but still the calculations founded on the hypothesis agree tolerably well with experiments. In this case the actual descent, in any instant, may be estimated by the removal of the quantity discharged, from the surface of the fluid to the orifice, since the intermediate space remains always occupied. The ascending force thus obtained is to be distributed throughout the fluid, according to the respective velocities of its different portions; and it may easily be shown, that when the orifice is small, the part which belongs to the fluid in the vessel is wholly inconsiderable in comparison with the ascending force required for the escape of the small portion which is flowing through the orifice, and the whole ascending force may, therefore, be supposed to be employed in the motion of this portion; so that it will acquire the velocity of a body falling from the whole height of the surface of the reservoir, or the velocity due to that height. It appears also that very nearly the same velocity is acquired by almost the first particles that escape from the orifice, so that no sensible time elapses before the jet flows with its utmost velocity.

This velocity may be found, as we have already seen, by multiplying the square root of the height of the reservoir, expressed in feet, by 8, or more correctly, by $8\frac{1}{44}$; thus, if the height be 4 feet, the velocity will be sixteen feet in a second; if the height be 9 feet, the velocity will be 24, the squares of 2 and 3 being 4 and 9; and if the height were 14 feet, the velocity would be 30 feet in a second, and a circular orifice an inch in diameter would discharge exactly an ale gallon in a second. In the same manner, the pressure of the atmosphere being equal to that which would be produced by a column of air of uniform density 28000 feet high, the air would rush into a vacuum with a velocity of more than 1300 feet in a second.

The velocity is also equal, whatever may be the direction of the stream; for since the pressure of fluids acts equally in all directions, at equal depths, the cause being the same, the effect must also be the same. And if the motion be occasioned by a pressure derived from a force of any other kind, the effect may be found by calculating the height of a column of the fluid, which would be capable of producing an equal pressure. When also the force arises from the difference of two pressures, the velocity may be determined in a similar manner. Thus, the pressure of a column of water, 1 foot in height, would force the air through a small orifice, with a velocity of 230 feet in a second, corresponding to the height of 830 feet; a column of mercury 1 inch high, would produce the same effect as a reservoir of water more than thirteen times as high, and the force of the air confined in a closed bottle under the receiver of the air pump, will cause a jet to rise to the same height as a column of mercury which measures the difference of the elasticities of the air in the bottle and in the receiver.

But these calculations are only confirmed by experiment in cases when the ajutage through which the fluid runs is particularly constructed; that is, when it is formed by a short tube, of which the sides are so curved that the particles of the fluid may glide along them for some distance, and escape in a direction parallel to the axis of the stream. A short cylindrical pipe is found to answer this purpose in some measure; but the end may be more completely obtained by a tube nearly conical, but with its sides a little convex inwards, so as to imitate the shape which a stream or vein of water spontaneously assumes when it runs through an orifice in a thin plate: for in such cases the stream contracts itself, after it has passed the orifice, for the distance of about half its diameter, so that at this point its thickness is only four fifths as great as at its passage; and the quantity discharged is only five eighths as great as that which the whole orifice would furnish, according to the preceding calculation: instead, therefore, of multiplying the square root of the height by 8, we may employ the multiplier 5 for determining the actual discharge. But the velocity, where the stream is most contracted, is only one thirtieth less than that which is due to the whole height; and when the jet is discharged in a direction nearly perpendicular, it rises almost as high as the surface of the fluid in the reservoir.

This contraction of the stream, and the consequent diminution of the discharge, is unquestionably owing to the interference of the particles of the fluid coming from the parts on each side of the orifice, with those which are moving directly towards it; and the effect is more perceptible when the orifice is made by a pipe projecting within the reservoir, so that some of the particles approaching it must acquire in their path a motion contrary to that of the stream. It would be possible to obtain an approximate calculation of the magnitude of this contraction, from the equilibrium which must subsist between the centrifugal forces of the particles, as they pass out of the orifice, describing various curves, according to their various situations, and the pressure required for the contraction of the internal parts of the stream, which obliges the particles to move more rapidly as they proceed, and which must be proportional to the height required for producing this acceleration. (Plate XX. Fig. 255.)

When a short cylindrical tube is added to the orifice, it is probable that the motion of the fluid within the tube is still in some measure similar: but the vessel must now be supposed to be prolonged, and to have a new orifice at the end of the tube, at which the particles cannot arrive by any lateral motions, and which will, therefore, not be liable to a second contraction: the discharge may, therefore, be estimated nearly according to the true measure of this orifice; the original pressure of the fluid continuing to act until the stream escapes.

The effect of a short pipe, in increasing the discharge, ceases when the water separates from its sides, so that it is no longer filled by the stream: since there is then nothing to distinguish its motion from that of a stream passing through a simple orifice: but the increase is not owing merely to the cohesion of the water to the sides of the pipe; for the effect, as I have found by experiment, is nearly the same in the motion of air as in that of water. The contraction caused by the motion of the water at the entrance of the short pipe, may be considered simply as a contraction in the pipe itself, and the subsequent part of the pipe either as cylindrical or as nearly conical: for in this case it follows, from the general law on which Bernoulli's calculations are founded, that as long as the fluid remains in one mass, the discharge will be nearly the same, as if the mouth of the pipe were the only orifice, supposing that no force is lost: and the exceptions which Bernoulli has made to the general application of the principle in such cases, although partly supported by experiments, have been extended somewhat further, both by himself and by other authors, than those experiments have warranted. In the case of a diverging conical pipe, or of a pipe with a conical termination, the discharge is found to be considerably greater than that which a cylindrical pipe would produce, but not quite so great as would be produced on the supposition that no force is lost. (Plate XX. Fig. 256.)

This analogy between the effects of a cylindrical and conical pipe is strongly supported by the experiments of Venturi, compared with those of Bernoulli. Bernoulli found that when a small tube was inserted into any part of a conical pipe, in which the water was flowing towards the wider end, not only none of the water escaped through the tube, but the water of a vessel, placed at a considerable distance below, was drawn up by it; Venturi observed the same, when the tube was inserted into the side of a cylindrical pipe, near to its origin; and in both cases air was absorbed, as well as water, so that cohesion could not be in any manner concerned. But the pressure of the atmosphere is generally necessary for all effects of this kind, and both Venturi and Dr. Matthew Young have observed, that a short pipe has no effect, in increasing the discharge through an orifice, in the vacuum of an air pump: but even if the difference were sometimes found to exist in the absence of atmospherical pressure, it might be produced by an accidental cohesion, like that which sometimes causes a column of mercury to remain suspended in similar circumstances. (Plate XX. Fig. 257.)

The effect of ajutages of different kinds, on the quantity of water discharged through an orifice of a given magnitude, may be most conveniently exhibited by placing them side by side at the same height in a reservoir, and suffering the water to begin to flow at the same moment through any two of them; the quantities discharged in a given time will then obviously indicate the respective velocities. If a very long pipe were employed, some time would be required before the velocity became uniform; but in such cases the retardation arising from friction is so considerable, as to cause a still greater deviation from the quantity which would be discharged by a shorter pipe in the same time.

When the aperture, through which a fluid is discharged, instead of being every way of inconsiderable magnitude, is continued throughout the height of the vessel, and is every where of equal breadth, the velocity must be materially different at different parts of its height; but we may find the quantity of the discharge, by supposing the whole velocity equal to two thirds of the velocity at the lowest point. And we may find the quantity discharged by an orifice not continued to the surface, but still of considerable height, by subtracting from the whole discharge of an orifice so continued, that which would have been produced by such a portion of it, as must be shut up, in order to form the orifice actually existing. But in this case, the result will seldom differ materially, from that which is found by considering the pressure, on the whole orifice, as derived from the height of the fluid above its centre.

When a cylindrical vessel empties itself by a minute orifice, the velocity of the surface, which is always in the same proportion to the velocity of the fluid in the orifice, is, therefore, uniformly retarded, and follows, in its descent the same law as a heavy body, projected upwards, in its ascent; consequently the space actually described, in the whole time of descent, is equal to half of that which would have been described, if the initial motion had been uniformly continued; and in the time that such a vessel occupies, in emptying itself, twice the quantity of the fluid would be discharged if it were kept full by a new supply. This may be easily shown, by filling two cylindrical vessels, having equal orifices in their bottoms, and while the one is left to empty itself, pouring into the other the contents of two other equal vessels, in succession, so as to keep it constantly full; for it will be seen that both operations will terminate at the same instant.

A similar law may be applied to the filling of a lock, from a reservoir of constant height; for in all such cases, twice as long a time is required for the effect, as would be necessary if the initial velocity were continued. The immersion of the orifice in a large reservoir has been found to make no difference in the magnitude of the discharge, so that the pressure may always be estimated by the difference of the levels of the two surfaces. Thus, when a number of reservoirs communicate with each other by orifices of any dimensions, the velocity of the fluid flowing through each orifice being inversely as the magnitude of the orifice, and being produced by the difference of the heights of the fluid in the contiguous reservoirs, this difference must be every where as the square of the corresponding velocity. But if the reservoirs were small, and the orifices opposite and near to each other, a much smaller difference in the heights of the surfaces would be sufficient for producing the required velocity. The same circumstances must be considered, in determining the velocity of a fluid, forced through a vessel divided by several partitions, with an orifice in each; if the orifices are small in proportion to their distance from each other, and if they are turned in different directions, each orifice will require an additional pressure, equivalent to the whole velocity produced in it: but if the partitions occupy a small part only of the vessel, and are placed near to each other, the retardation will be much less considerable. Cases of this kind occur very frequently in the passage of water through the pipes and valves of pumps, and it is, therefore, of consequence to avoid all unnecessary expansions, as well as contractions, in pipes and in canals, since there is always a useless expense of force in restoring the velocity which is lost in the wider parts.

When a siphon, or bent tube, is filled with a fluid, and its extremities are immersed in fluids of the same kind, contained in different vessels, if both their surfaces are on the same level, the whole remains at rest; but if otherwise, the longer column in the siphon preponderates, and the pressure of the atmosphere forces up the fluid from the higher vessel, until the equilibrium is restored; provided, however, that this pressure be sufficiently powerful: for if the height of the tube were more than 34 feet for water, or than 30 inches for mercury, the pressure of the atmosphere would be incapable of forcing up the fluid to its highest part, and this part remaining empty, the fluid could no longer continue to run. (Plate XX. Fig. 258.)

If the lower vessel be allowed to empty itself, the siphon will continue running as long as it is supplied from the upper, with a velocity nearly corresponding to the height of that portion of the fluid in the longer leg, which is not counterbalanced by the fluid in the shorter; that is, to the height of the surface of the upper vessel above that of the lower one, or above the end of the siphon, when it is no longer immersed; for the height of the pipe is in all cases to be considered as constituting a part of that height which produces the pressure. Thus the discharge of a pipe, descending from the side or bottom of a vessel, is nearly the same as from a similar horizontal pipe, inserted into a reservoir of the whole height of the descending pipe and of the fluid above it; and this is true even when the depth of the vessel is inconsiderable, in comparison with the length of the pipe, if its capacity is sufficient to keep the pipe running full. It appears at first sight extremely paradoxical, that the whole water discharged, each particle of which is subjected to the action of gravitation in a pipe 16 feet long, for half a second only, should acquire the velocity of 32 feet in a second, which would require, in common circumstances, the action of the same force of gravitation for a whole second, and this fact may be considered as favourable to the opinion of those, who wish to estimate the magnitude of a force, rather by the space through which it is continued, than by the time during which it acts; but if we attend to the nature of hydrostatical pressure, we shall find that the effect of the column on the atmosphere is such, as to produce, or to develope, a portion of accelerating force which is actually greater than the weight of the particles immediately concerned. If a doubt could be entertained of the truth of this theory, it might be easily removed by recurring to the general law of ascending force, since it follows from that law, that each particle, which descends in any manner through the space of 16 feet, must acquire, either for itself or for some other particles, a power of ascending to the same height; and on the other hand, the event of the experiment confirms the general law. For if we fix a shallow funnel on a vertical pipe, and pour water into it, so as to keep it constantly full, while the pipe discharges itself into a reservoir, out of which the water runs through a second pipe, placed horizontally, of exactly the same dimensions with the first, the height, at which the water in the reservoir becomes stationary, will be very nearly equal to the height of the funnel above its surface, so that the same height produces the same velocity in both cases. (Plate XX. Fig. 259.)

We may understand the action of the forces immediately concerned in this experiment, by attending to the mutual effects of the water and of the atmosphere. The water entering the orifice must immediately acquire a velocity equal to that of the whole water in the pipe, otherwise there would be a vacuum in the upper part of the pipe, which the pressure of the atmosphere will not permit; and this pressure, considered as a hydrostatic force, is equal to that which would be derived in any other way from a column of the same height with the pipe, since the weight of the water in the pipe is wholly employed in diminishing the counterpressure of the atmosphere below, not only in the beginning, when it is at rest, but also while it is in motion; for that motion being uniform throughout its descent, the power of gravitation is expended in producing pressure only; so that the pressure of the atmosphere on the water in the funnel becomes completely analogous to the pressure of a reservoir of water, of the same height with the pipe. The circumstance, which causes the appearance of paradox in this experiment, exists also in the simplest case of the discharge of water; for it may be shown, that the portion of accelerating force actually employed in generating the velocity with which a stream is discharged through a small orifice, is twice as great as the pressure of the fluid on a part of the vessel equal in extent to the orifice; and in the same manner the quantity of force exerted by the atmosphere on the water in the funnel, as well as that with which the descending fluid impels the air below, is equal to twice the weight of the quantity existing at any time in the pipe.

There is, however, a limit, which the mean velocity in such a pipe can never exceed, and which is derived from the magnitude of the pressure of the atmosphere. For the water cannot enter the pipe with a greater velocity than that with which it would enter an exhausted pipe, and which is produced by the whole pressure of the atmosphere; and this pressure being equivalent to that of a column of water 34 feet high, the velocity derived from it is about 47 feet in a second: so that if the vertical pipe were more than 34 feet long, there would be a vacuum in a part of it near the funnel.

Wherever a pipe of considerable length descends from a funnel, if the supply of the fluid be scanty, and especially if it approach the orifice obliquely, the pressure of the atmosphere, and the centrifugal force of the particles which must necessarily revolve round the orifice, will unite in producing a vacuity in the centre; and when this happens, the discharge is considerably diminished.

In order that a siphon may run, it is obvious that it must first be filled; and when it is once filled, it will continue to run till the reservoir is exhausted, as far as the level of its upper orifice. And from this circumstance, the phenomena of some intermitting springs have been explained, which only begin to run, when the reservoirs from which they originate have been filled by continued rains, and then go on to exhaust them, even though the weather may be dry. From a combination of several such siphons and reservoirs, a great number of alternations may sometimes be produced. (Plate XX. Fig. 260.)

Since the velocity of a stream or jet issuing in any direction, out of a simple orifice, or a converging one, is nearly equal to that of a heavy body falling from the height of the reservoir, it will rise, if directed upwards, very nearly to the same height, excepting a slight difference occasioned by the resistance of the air, and by the force which is lost, in producing the velocity with which the particles must escape laterally, before they begin to descend. The truth of this conclusion is easily confirmed by experiment. (Plate XX. Fig. 261.)

If a jet issue in an oblique or in a horizontal direction, its form will be parabolic, since every particle tends, as a separate projectile, to describe the same parabola in its range: and it may be demonstrated, that if it be emitted horizontally from any part of the side of a vessel, standing on a horizontal plane, and a circle be described, having the whole height of the fluid for its diameter, the jet will reach the plane, at a distance from the vessel twice as great as the distance of that point of the circle, through which it would have passed, if it had continued to move horizontally. And if the jet rise in any angle from the bottom of the vessel, the utmost height of its ascent will be equal to that of the point in which it would meet the same semicircle, if it continued to move in a right line, and the horizontal range will be equal to four times the distance, intercepted between the same point and the side of the vessel. This law is equally true with regard to simple projectiles: but the experiment is most conveniently exhibited in the motion of a jet. (Plate XX. Fig. 262.)

We have hitherto considered the motions of fluids as continued principally in the same direction; but they are frequently subjected to alternations of motion, which bear a considerable analogy to the vibrations of pendulums; thus, if a long tube be immersed in a fluid, in a vertical direction, and the surface of the fluid within the tube be elevated a very little, by some external cause, the whole contents of the fluid will be urged downwards by a force, which decreases in proportion to the elevation of the surface above the general level of the vessel, and when both surfaces have acquired the same level, the motion will be continued by the inertia of the particles of the fluid, until it be destroyed by the difference of pressures, which now tends to retard it; and this alternation will continue, until the motion be destroyed by friction and by other resistances. It is also obvious, that since any two vibrations, in which the forces are proportional to the spaces to be described, are performed in equal times, these alternations will require exactly the same time for their completion, as the vibrations of a pendulum, of which the length is equal to that of the whole tube; for the relative force in the tube is to the whole force of gravity as the elevation or depression is to the whole length of the tube. Hence it follows, that if two such tubes were united below, so as to form a single bent tube, the vibrations might take place in the whole compound tube, in the same manner, and in the same time, as in each of the separate tubes; nor would the effects be materially altered if any part of the middle of the tube were in a horizontal or in an oblique direction, provided that the whole length remained unaltered. In such a tube also, all vibrations, even if of considerable extent, would be performed in the same time, and would long remain nearly of the same magnitude; but in a single tube, open below, the vibrations would continually become less extensive, and their duration would also be altered as well as their extent; besides the unavoidable resistances, which would in both cases interfere with the regularity of the effects.

But it does not appear that the laws of the vibrations of fluids in pipes will at all serve to elucidate the phenomena of waves. Sir Isaac Newton has supposed that each wave may be compared with the fluid oscillating in a bent pipe; but the analogy is by far too distant to allow us to found any demonstration on it. The motions of waves have been investigated in a new and improved manner by Mr. Lagrange; and I have given a concise demonstration of a theorem similar to his, but perhaps still more general and explicit. It appears from these determinations, that supposing the fluids concerned to be infinitely elastic, that is, absolutely incompressible, and free from friction of all kinds, any small impulse, communicated to a fluid, would be transmitted every way along its surface, with a velocity equal to that which a heavy body would acquire in falling through half the depth of the fluid; and I have reason to believe, from observation and experiment, that where the elevation or depression of the surface is considerably extensive in proportion to the depth, the velocity approaches nearly to that which is thus determined, being frequently deficient one eighth or one tenth only of the whole; in other cases, where a number of small waves follow each other at intervals considerably less than the depth, I have endeavoured to calculate the retardation which must be occasioned by the imperfect elasticity or compressibility of the fluid; but it seems probable that the motion of small waves is still much slower than this calculation appears to indicate.

Whatever corrections these determinations of the velocity of waves may be found to require, the laws of their propagation may still be safely inferred from the investigation. Thus, it may be shown, supposing the waves to flow in a narrow canal of equable depth, that, whatever the initial figure of the waves may be, every part of the surface of the fluid will assume in succession the same form, except that the original elevations and depressions, extending their influence in both directions, will produce effects only half as great on each side, and those effects will then be continued until they are destroyed by resistances of various kinds. It may also be inferred, that the surface of a fluid thus agitated by any series of impressions, will receive the effects of another series, in the same manner as a horizontal surface, and that the undulations, thus crossing each other, will proceed without any interruption, the motion of each particle being always the sum or difference of the motions belonging to the separate series.

Supposing two equal and similar series of waves to meet each other in such a canal, in opposite directions, the point in which their similar parts meet must be free from all horizontal motion, so that any fixed obstacle in an upright position would have the same effect on the motions of the fluid on either side as the opposition of a similar series; and this effect constitutes the reflection of a series of waves, which is easily observed, when they strike against a steep wall or bank; and when this reflection is sufficiently regular, it is easy to show, that the combination of the direct with the reflected motions must constitute a vibration, of such a nature, that the whole surface is divided into portions, which appear to vibrate alternately upwards and downwards, without any progressive motion, while the points which separate the portions remain always in their natural level. (Plate XX. Fig. 263.)

But those series of waves which are usually observable in any broad surface, and which constitute a number of concentric circles, are usually reflected in such a manner as to appear to diverge after reflection from a centre beyond the surface which reflects them, and to be subject to all those laws, which are more commonly noticed in the phenomena of reflected light; but as these laws are of more practical importance in their application to optics, than to hydraulics, it is unnecessary at present to examine their consequences in detail. It may, however, be easily understood, that a new series of waves, proceeding from a centre at the same distance behind the reflecting surface, as the centre of the original series is before it, would produce precisely the same effect as a fixed obstacle; consequently the law of reflection at equal angles is a very simple inference from this mode of reasoning. (Plate XX. Fig. 264.)

When a series of waves proceeds in an equable canal, it is obvious that the centre of gravity of the whole fluid neither rises nor falls; from this analogy, as well as from the general application of the law of ascending force, it is probable that in all cases of the propagation of waves, the place of the centre of gravity remains unaltered; so that when a circular wave spreads further and further from its centre, its height is not diminished in the same ratio as its diameter is increased, but the square of its height only varies in this proportion; that is, a wave which is a yard in diameter, and an inch high, will retain a height of half an inch, when its diameter is increased to four yards.

Many of the phenomena of waves may be very conveniently exhibited, by means of a wide and shallow vessel, with a bottom of glass, surrounded by sides inclined to the horizon, in order to avoid the confusion which would arise from the continual reflections produced by perpendicular surfaces. The waves may be excited by the vibrations of an elastic rod or wire, loaded with a weight, by means of which its motions may be made more or less rapid at pleasure; and the form and progress of the waves may be easily observed, by placing a light under the vessel, so that their shadows may fall on a white surface, extended in an inclined position above. In this manner the minutest inflections of the surface of the water may be made perfectly conspicuous. (Plate XX. Fig. 265.)

By means of this apparatus, we may examine the manner in which a wave diverges, when a portion of it has been intercepted on either side or on both sides. Thus, if a wave is admitted, by an aperture which is very narrow in proportion to its own breadth, into the surface of a part of the water which is at rest, it diverges from the aperture as from a new centre; but when the aperture is considerably wider than the wave, the wave confines its motion in great measure to its original direction, with some small divergence, while it is joined on each side by fainter circular portions, spreading from the angles only. (Plate XX. Fig. 266.)

When two equal series of circular waves, proceeding from centres near each other, begin their motions at the same time, they must so cross each other, in some parts of their progress, that the elevations of the one series tend to fill up the depressions of the other; and this effect may be actually observed, by throwing two stones of equal size into a pond at the same instant; for we may easily distinguish, in favourable circumstances, the series of points in which this effect takes place, forming continued curves, in which the water remains smooth, while it is strongly agitated in the intermediate parts. These curves are of the kind denominated hyperbolas, each point of the curve being so situated with respect to its foci, as to be nearer to one than the other by a certain constant distance. (Plate XX. Fig. 267.)

The subject of waves is of less immediate importance for any practical application than some other parts of hydraulics; but besides that it is intimately connected with the phenomena of the tides, it affords an elegant employment for speculative investigation, and furnishes us with a sensible and undeniable evidence of the truth of some facts, which are capable of being applied to the explanation of some of the most interesting phenomena of acustics and optics.

It may be shown, by steps nearly similar to those by which the velocity of the motions of waves is investigated, that a fluid which is contained in an elastic pipe, and which receives an impulse at any part of the pipe, will transmit its effects, with the same velocity, as a wave would have in a reservoir, of that depth which measures the elasticity of the pipe, that is, with half the velocity which a body would acquire, in falling from the height at which a portion of the fluid, connected with the contents of the pipe, would stand in a vertical tube. It is in this manner that the blood is transmitted, by means of the impulse given to it by the heart through the bloodvessels; the pulse moves on with great rapidity, the elastic force of the vessels being considerably assisted by the temporary actions of the muscular coats of the arteries, which cause a contraction more rapid than the dilatation; while the whole mass of the arterial blood continues, at the same time, to advance with a much smaller velocity; like the slow stream of a river, on the surface of which undulations are continually propelled, with motions independent of its own.

LECTURE XXIV

ON THE FRICTION OF FLUIDS.

WE have hitherto considered the motions of fluids independently of the resistance which they undergo from the vessels containing them, and from the surfaces in contact with them, as well as from the interference of the neighbouring particles with each other; there is, however, a variety of cases of very common occurrence, in which these frictions most materially affect the results of our calculations; so that before this subject was laboriously and judiciously investigated by the Chevalier de Buat, it was almost impossible to apply any part of our theoretical knowledge of hydraulics to practical purposes.

The effect of friction is particularly exemplified by the motions of rivers, in which almost the whole force of gravity is employed in overcoming it. When the inclination and the dimensions of a river continue uniform, the velocity is also every where equal; for otherwise the depth would become unequal: here, therefore, the force of gravitation must be an exact counterpoise to the resistance which is to be overcome, in order that the water may flow with its actual velocity; this velocity having been originally derived from the effect of a greater inclination near the origin of the river. When the river is thus proceeding, with an equable motion, it is said to be in train; and it is obvious that no increase of its length will produce any alteration in its velocity. There is, therefore, a very material difference between the course of a river, and the descent of a body, with an accelerated motion, along an inclined surface. For when a solid body is placed on an inclined plane, the force of friction is either great enough to overpower its relative weight, and to retain it at rest, or else the friction is constantly less than the gravitation, and the motion is always accelerated. But the resistance to the motions of fluids arises principally from different causes; not from the tenacity of the fluids, which, where it exists, is a force nearly uniform, like that of friction, but principally from the irregular motions and mutual collisions of their particles; and in this case, according to the laws of mechanics, it must vary nearly in proportion to the square of the velocity. For when a body is moving in a line of a certain curvature, the centrifugal force is always as the square of the velocity; and the particles of water in contact with the sides and bottom of a river or pipe, must be deflected, in consequence of the minute irregularities of the surfaces on which they slide, into nearly the same curvilinear paths, whatever their velocity may be, so that the resistance, which is in great measure occasioned by this centrifugal force, must also vary as the square of the velocity. Thus also the curvature assumed by the outline of a stream of water issuing from a simple orifice, which constitutes the contraction already described, is very nearly the same, whatever the velocity may be: nor does the friction increase with the pressure, as is demonstrated by an experiment of Professor Robison on the oscillations of a fluid through a bent tube, terminated by two bulbs, which were performed in the same time, whether the tube was in a horizontal or in a vertical position. Mr. Coulomb has also proved the same fact by experiments on the vibrations of bodies immersed in fluids, and suspended by twisted wires; he finds that precisely at the surface, the friction is somewhat greater than at any depth below it: he also considers a certain part of the friction as simply proportional to the velocity, and a very small portion only, in common fluids, as perfectly independent of it.

It is obvious that wherever the friction varies as the square of the velocity, or even when it increases in any degree with the velocity, there must always be a limit, which the velocity can never exceed, by means of any constant force, and this limit must be the velocity at which the resistance would become equal to the force. It is for this reason that a light body, descending through the air, soon acquires a velocity nearly uniform; and if it be caused, by any external force, to move for a time more rapidly, it will again be speedily retarded, until its velocity be restored very nearly to its original state. In the same manner the weight of the water in a river, which has once acquired a stationary velocity, is wholly employed in overcoming the friction produced by the bottom and the banks.

From considering the effect of the magnitude of the surface exposed to the friction of the water, in comparison with the whole quantity contained in the river, together with the degree in which the river is inclined to the horizon, we may determine, by following the methods adopted by Mr. Buat, the velocity of any river of which we know the dimensions and the inclination. Supposing the whole quantity of water to be spread on a horizontal surface, equal in extent to the bottom and sides of the river, the height, at which it would stand, is called the hydraulic mean depth; and it may be shown that the square of the velocity must be jointly proportional to the hydraulic mean depth, and to the fall in a given length. If we measure the inclination by the fall in 2800 yards, the square of the velocity in a second will be nearly equal to the product of this fall multiplied by the hydraulic mean depth. For example, in the Ganges, and in some other great rivers, the mean depth being about 30 feet, and the fall 4 inches in a mile, the fall in 2800 yards will be about $6\frac{1}{2}$ inches, which, multiplied by 360 inches, gives 2340 inches for the square of the mean velocity, and $48\frac{1}{3}$ inches, or about four feet, for the mean velocity in a second, that is, not quite three miles an hour, which is the usual velocity of rivers moderately rapid. If, however, great precision were required in the determination, some further corrections would be necessary, on account of the deviation of the resistance from the exact proportion of the squares of the velocities; since the friction, as we have already seen, does not increase quite so fast as this.

It is obvious that the friction of a fluid, moving on the surface of a solid alone, would not produce any material retardation of its motion, if the particles of the fluid themselves were capable of moving on each other, without the least resistance; for in this case a small portion of the fluid, in immediate contact with the solid, might remain at rest, and the remaining mass of the fluid might slide over this portion without any retardation. It appears, however, that the water in contact with the bottom of a river moves with a very considerable velocity, and the water next above this only a little faster, so that the velocity increases almost uniformly as we ascend towards the surface. It follows, therefore, that the resistance must be much greater where the particles of water slide on each other, than where they glide along the surface of a solid. This internal friction operates gradually throughout the water; the surface being retarded by the particles immediately below it, those particles by the next inferior stratum, and each stratum being actuated, besides its own relative weight, by the friction of the water above, tending to draw it forwards, and by that of the water below, tending still more to retard it; the retardation being communicated, from below upwards, in such a manner as to be every where equivalent to the relative weight of the water above the part considered. It appears from observation, that when we have determined the mean velocity in English inches, we may find the superficial velocity, very nearly, by adding to it its square root, and the velocity at the bottom, by subtracting from it the same number: thus the square root of $48\frac{1}{3}$ being nearly 7, the superficial velocity of the Ganges will be about 55 inches, or 4 feet 7 inches in a second, and the velocity at the bottom $41\frac{2}{3}$. There are, however, frequent irregularities in the proportions of the velocities at different depths, and it has sometimes been observed, perhaps on account of the resistance of the air, that the velocity is a little less, immediately at the surface, than a few inches below it.

For similar reasons, the velocity of a river is also generally greater in the middle than at the sides; and the motion of the particles in the middle must be retarded, not only by those which are below them, but also by those on each side, while these, on the contrary, are dragged on by the water in the middle: the middle parts tend, therefore, to draw the sides towards them, which they cannot do, without lowering the surface of the fluid next to the banks, in such a degree as to make the difference of level an equivalent to this tendency to approach the middle. This appears to be the reason, that the surface of a large river may generally be observed to be slightly convex, or a little elevated in the middle.

The course of a river is sometimes interrupted by a were or a fall, natural or artificial; in such cases the velocity may be calculated in the same manner as when a fluid is discharged from a reservoir through an aperture of considerable height: supposing the whole section of the were to be such an aperture, in a vessel so much higher, that the velocity of a fluid issuing from it at the upper part of the aperture would be precisely equal to the actual velocity of the river. The extent of the swell caused by a were, or by any partial elevation thrown across the bed of a river, may also be found by first determining the height at which the surface must stand immediately above the were, and then calculating the inclination of the surface which will be required for producing the actual velocity, in the river thus made deeper; which of course will determine the situation of the surface where the water approaches the were; and this surface, which is more nearly horizontal than the general surface of the river, will be so joined to it as to have a curvature nearly uniform throughout.

It appears from calculations of the effects of various changes in the dimensions of rivers, as well as from immediate observation, that a considerable diminution of the breadth of a river at a particular place, will often produce but a small elevation of its surface. The velocity, however, may sometimes be considerably increased by such a change, and where the bottom is of a loose nature, its particles may be carried away by means of the increased velocity, and the bed of the river may be deepened.

Where a river bends in a considerable degree, it is generally remarked that the velocity of the water is greater near the concave than the convex side of the flexure, that is, at the greatest distance from the centre of its curvature. This effect is probably occasioned by the centrifugal force, which accumulates the water on that side; so that the banks are undermined, and the channel is deepened by its friction. Some authors have been led to expect that the velocity would be greater nearest to the convex bank, because the inclination of the surface must be a little greater there; but the effect of the accelerating force, in any short distance, is inconsiderable, and it is more than compensated by the want of depth. It may easily be understood, that all angles and flexures must diminish the general velocity of the river's motion, and the more as they are more abrupt.

It has sometimes been imagined, that because the pressure of fluids is propagated equally in all directions, their motions ought also to diverge in a similar manner; but this opinion is by no means well founded, even with respect to those particles which receive their motions in an unlimited reservoir from the impulse of a stream which enters it. An experiment, which sets this fact in a clear point of view, was made long ago by Hauksbee. He produced a very rapid current of air, by means of a vessel, into which three or four times as much air as it naturally contained had been condensed by means of a syringe, and causing the current to pass through a small box, in which the bason of a barometer was placed, the mercury was depressed more than two inches, in consequence of the rarefaction which the current produced in the air of the box. (Plate XXI. Fig 268.)

Professor Venturi has also made several experiments of a similar nature on the motion of water: he observes that not only the water in contact with a stream is drawn along by it, but that the air in the neighbourhood of a jet is also made to partake of its motion. When the mouth of a pipe, through which a stream of water is discharged, is introduced into a vessel a little below the surface of the water which it contains, and is allowed to escape by ascending an inclined surface placed opposite to the pipe, and leading over the side of the vessel, the stream not only ascends this surface without leaving any portion of itself behind, but carries also with it the whole of the water of the vessel, until its surface becomes level with the lowest part of the stream. (Plate XXI. Fig. 269.)

The effect of a jet of water, in drawing towards it a current of air, is in some measure illustrated by an experiment which is often exhibited among the amusements of hydraulics. A ball of cork, or even an egg, being placed in the middle of a jet, which throws up a pretty large stream to a moderate height, the ball, instead of falling, or being thrown off, as it might naturally have been expected to do, remains either nearly stationary, or playing up and down, as long as the experiment is continued. Besides the current of air which Venturi has noticed, and which tends to support the ball in a stable equilibrium, the adhesion of the water, combined with its centrifugal force in turning round the ball, assists in drawing it back, when it has declined a little on either side, so that the stream has been principally in contact with the other side. A similar effect may be observed in the motions of the air only, as I have shown by some experiments of which an account is published in the Philosophical Transactions. Thus, if we bend a long plate of metal into the form of the letter S, and suspend it in the middle by a thread, so that it may move freely on its centre, and if we then blow on its convex surface with a tube directed obliquely towards the extremity, instead of retreating before the blast, it will on the contrary appear to be attracted; the pressure of the atmosphere being diminished by the centrifugal force of the current, which glides along the convex surface, because it finds a readier passage in the neighbourhood of the solid, towards which it is urged by the impulse of the particles of the air approaching it on one side, and by the defect of pressure on the other side, occasioned by the removal of a certain portion of the air which it carries with it. (Plate XXI. Fig. 270, 271.)

From considerations similar to those by which the velocity of a river is determined, we may calculate the quantity of water discharged from a pipe of any given dimensions, and in any position. The same expressions will serve for estimating the magnitude of the friction in both cases; the pipe being considered as a small river, of which the mean depth is one fourth of its diameter: but a part only of the force of gravity is now expended in overcoming the friction, the rest being employed in producing the momentum of the water. We may obtain a sufficiently accurate determination of the velocity, by supposing the height of the reservoir above the orifice of the pipe to be diminished in the same proportion as the diameter of the pipe would be increased by adding to it one fiftieth part of the length, and finding the whole velocity corresponding to four fifths of this height. Thus, if the diameter of the pipe were one inch, and its length 100 inches, we must suppose the effective height to be reduced to one third by the friction, and the discharge must be calculated from a height four fifths as great as this, which may be considered as a reduction derived from the interference of the particles, entering the pipe, with each other's motions. If the diameter of the pipe had been two inches, the height must only have been supposed to be reduced to one half by the friction; such a pipe would, therefore, discharge about five times as much water as the former, although of only twice the diameter; and this circumstance requires the attention of all those who are concerned in regulating the distribution of water by pipes for domestic use, or for any other purpose.

In such cases it becomes also frequently necessary to attend to the angle in which a small pipe is inserted into a larger; whenever a pipe is bent, there is a loss of force according to the degree of flexure, and to the velocity of the water, which may be calculated, if it be required; but if a pipe be fixed into another through which the water is moving very rapidly, in a direction contrary to that of the stream, its discharge will not only be much smaller than if the directions more nearly coincided, but sometimes such a pipe will discharge nothing at all; on the contrary, like the air in Hauksbee's experiment, the water which it contains may be dragged after the stream in the larger pipe.

The bad effect of unnecessary dilatations, as well as contractions, in aqueducts and in pipes, may be understood from what has been already said of the loss of force attendant on every change of velocity; a circumstance of a similar nature sometimes happens in the animal economy. When an artery is dilated so as to form an aneurism, it has been observed that the artery is usually distended above the cavity; and this effect is easily understood from the actual increase of resistance which the aneurism produces, united perhaps with the previous debility of the artery.

Mr. Gerstner, has found by some very accurate observations on the motion of water in very small pipes, that the resistance is considerably affected by the temperature at which the experiment is performed; but in the cases of rivers, and of such pipes as are commonly used in practice, no variations of temperature, to which they can be liable, will produce any sensible effects. His experiments indicate a resistance, where the tubes are very small, which follows a law so different from that which is observed in more common cases, that it appears to be owing to some other cause: this cause is perhaps the capillary attraction of the open end of the tube, and it is the more probable that the resistance depends on some such circumstance, as there is reason to think that the irregularity may be in great measure removed by placing the tube wholly under water.

LECTURE XXV.

ON HYDRAULIC PRESSURE.

THE mutual effects of fluids and moveable solids on each other depend principally on the laws of hydraulic pressure, and of the resistance of fluids, which have been considered by Bernoulli as constituting a separate department of hydrodynamics, under the name of hydraulicostatics, and which are of the utmost practical importance, since the application of the powers of wind or water to the working of mills, and to the navigation of ships, are wholly dependent on them. The impulse of a fluid differs very materially from that of a solid, for in the motions of solids, the least possible finite momentum must overpower the strongest possible pressure; but since the particles of fluids are supposed to be infinitely small, the momentum of a fluid stream may always be balanced by a certain determinate pressure, without producing motion in the solid opposed to it; so that this division of the subject of hydraulics has nothing analogous to it in simple mechanics. It is true that when a certain quantity of a fluid is made to concentrate its action almost instantaneously, its effect is nearly similar to that of a solid, for here the essential distinction, derived from the successive action of the particles, no longer exists. Thus, when a stream of fluid filling a pipe acts suddenly on an obstacle at the end of it, it requires to be resisted by a force far greater than that which originally caused its motion, unless the action of the force be continued through a considerable space: and for this reason the strength of the pipe ought to be so calculated as to be able to resist this action; its intensity may, however, be easily diminished by means of an air vessel communicating with the pipe, which will allow the motion to be changed in a less abrupt manner. But in the principal cases which we are about to consider, the action of the fluid on the solid is supposed to be confined to such of its particles as are nearly in contact with the surface.

When a part of the weight of any fluid is expended in producing a motion in any direction, an equal force is deducted from its pressure on the vessel in that direction: for the gravitation, employed in generating velocity, cannot at the same time be causing pressure; and when the motion produced is in any other direction than a vertical one, its obliquity must be immediately derived from the reaction of the vessel, or of some fixed obstacle; for it is obvious that a vertical force, like that of gravity, cannot of itself produce an oblique or a horizontal motion.

If a small stream descends from the bottom of a vessel, the weight expended in producing its motion is equal to that of a column of the fluid standing on a base equal to the contracted orifice, and of twice the height of the vessel. Thus, if the vessel be 16 feet high, the velocity of the stream will be 32 feet in a second, and a column 32 feet in length will pass through the orifice in each second, with the whole velocity derivable from its weight acting for the same time; so much, therefore, of the pressure of the fluid in the reservoir must be expended in producing this motion, and must of course be deducted from the whole force with which the fluid acts on the bottom of the reservoir; in the same manner as when two unequal weights are connected by means of a thread passing over a pulley, and one of them begins to descend, the pressure on the pulley is diminished, by a quantity, which is as much less than the sum of the weights, as the velocity of their common centre of gravity is less than the velocity of a body falling freely. If the stream issue from the vessel in any other direction, the effect of the diminution of the pressure in that direction will be nearly the same as if the vessel were subjected to an equal pressure of any other kind in a contrary direction; and if the vessel be moveable, it will receive a progressive or rotatory motion in that direction. Thus, when a vessel or pipe is fixed on a centre, and a stream of water is discharged from it by a lateral orifice, the vessel turns round at first with an accelerated motion, but on account of the force consumed in producing the rotatory motion, in successive portions of the water, the velocity soon becomes nearly stationary. (Plate XXI. Fig. 272.)

From similar reasoning it appears, that the effect of a detached jet on a plane surface perpendicular to it must be equivalent to the weight of a portion of the same stream equal in length to twice the height which is capable of producing the velocity. And this result is confirmed by experiments: but it is necessary, that the diameter of the plane be at least four times as great as that of the jet, in order that the full effect may be produced. When also a stream acts on an obstacle in a channel sufficiently closed, on all sides, to prevent the escape of any considerable portion of water, its effect is nearly the same as that of a jet playing on a large surface. But if the plane, opposed to the jet, be only equal to it in diameter, or if it be placed in an unlimited stream, the whole velocity of the fluid column will not be destroyed, it will only be divided and diverted from its course, its parts continuing to move on, in oblique directions; in such cases the pressure is usually found to be simply equivalent to the weight of a column equal in height to the reservoir, the surface being subjected to a pressure nearly similar to that which acts on a part of the bottom of a vessel, while a stream is descending through a large aperture in another part of it. (Plate XXI. Fig. 273.)

It is obvious that, in all these cases, the pressure varies as the square of the velocity, since the height required to produce any velocity is proportional to its square. This inference was first made in a more simple manner, from comparing the impulse of a fluid on a solid with that of a number of separate particles, striking the surface of the body, each of which would produce an effect proportional to its velocity, while the whole number of particles, acting in a given time, would also vary in the same ratio. If the solid were in motion, and the fluid either in motion or at rest, it is obvious that the relative velocity of the solid and the fluid, with regard to each other, would be the only cause of their mutual effects, and that the hydraulic pressure or resistance must be dependent on this velocity alone, except so far as the limited dimensions of the reservoir, containing the fluid, might produce a difference in the internal motions of its particles in different cases. Thus, where the fluid is so confined, that the whole of the stream acts on a succession of planes, each portion into which it is divided may be considered as an inelastic solid, striking on the surface exposed to it with a certain velocity: and in this case the force must be considered as simply proportional to the relative velocity, and not to its square. For want of this consideration, the effects of water wheels have frequently been very erroneously stated.

When a jet strikes a plane surface obliquely, its force, in impelling the body forwards, in its own direction, is found to be very nearly proportional to the height to which the jet would rise, if it were similarly inclined to the horizon. But when a plane is situated thus obliquely with respect to a wide stream, the force impelling it in the direction of the stream is somewhat less diminished by the obliquity, at least if we make allowance for its intercepting a smaller portion of the stream: thus, if the anterior part of a solid be terminated by a wedge more or less acute, the resistance, according to the simplest theory of the resolution of forces, might be found by describing a circle on half the base of the wedge as a diameter, which would cut off a part from the oblique side of the wedge that would be the measure of the resistance, the whole side representing the resistance to the same solid without the wedge: but the resistance is always somewhat more than this, and the portion to be added may be found, very nearly, by adding to the fraction thus found one ten millionth of the cube of the number of degrees contained in the external angle of the wedge. (Plate XXI. Fig. 274.)

The pressure of a fluid, striking perpendicularly on a plane surface, has been found to be very different at different parts of the surface; being greatest at the centre, and least towards the edges; so that if an aperture be made in the centre of a circular plane, covering the mouth of a bent tube, the fluid within it will rise half as high again as if the whole mouth were open. It is also observable, that two bodies, equal and similar in the form of the part meeting the fluid, undergo very different degrees of resistance according to the forms of their posterior terminations, and that a thin circular plate is much more retarded than a long cylinder of the same diameter. These circumstances are utterly inexplicable upon the vague approximation of supposing the resistance produced by the immediate impulse of separate particles of the fluid on the solid; but they are no longer surprising, when we consider the true mode of action of continuous fluids, since all the motion which is produced by the fluid in the solid or by the solid in the fluid is communicated much more by means of pressure than by immediate impulse. The minute operations of this pressure are too intricate to be accurately developed, but we may observe in general, that when a body moves along the surface of a resisting medium at rest, or when an obstacle at rest is opposed to a fluid in equable motion, the pressure is increased before the moving substance, and diminished behind it; so that the surface is elevated at the one part, and depressed at the other, and the more as the velocity is greater. Now it is obvious that the pressure must be greatest where the elevation is greatest, and hence a perforation at the centre of the surface indicates a greater pressure than at the circumference. Behind the body, this pressure becomes negative, and has sometimes been called nonpressure; hence it happens that a tube, opening in the centre of the posterior surface, exhibits the fluid within it depressed below the level of the general surface of the water. Thus, if we suppose the velocity of a body, terminated by perpendicular surfaces, to be 8 feet in a second, it will require the pressure of about a foot, to produce such a velocity, and we may, therefore, expect an elevation of about a foot before the body, and an equal depression behind it: consequently an equivalent difference must be found in the pressure of the water at any equal depths on the anterior and posterior surfaces of the body. The water elevated before the body escapes continually towards each side, and the deficiency behind is also filled up in some measure by the particles rushing in and following the body: but there is in both cases, a certain quantity of water which moves forwards, and constitutes what is called the dead water: before, where it is usually most observable, it forms an irregular triangle, of which the sides are convex inwards. If the posterior part of the body be formed like a wedge, the water on each side will be advancing to fill up the vacuity, even while it remains in contact with the sides, and the negative pressure will be considerably diminished. For this reason, the bottoms of ships are made to terminate behind in a shape somewhat resembling a wedge; and the same economy may be observed in the forms of fishes, calculated by nature for following their prey with the greatest possible rapidity. In general, fishes, as well as ships, are of a more obtuse form before than behind, but it is not certain that there would be any material difference in the resistance in a contrary direction, although some experiments seem to favour such an opinion. Perhaps if the natural form of the dead water, moving before an obtuse body, were ascertained, it might serve to indicate a solid calculated to move through the water with the least resistance; for the water must naturally assume such a form for its own motions, and the friction of fluids on solids being less than that of fluids moving within themselves, the resistance would be diminished by substituting a solid of the same form for a fluid. (Plate XXI. Fig. 275.)

Supposing a body to move through a fluid at a considerable depth below its surface, there will still be an elevation before and a depression behind it, the less in height, and the greater in extent, as the depth at which the body is situated is greater. Such an elevation appears to be in some measure analogous to the effect of a low were thrown across a river, which raises its surface, and produces a swell.

If two or more bodies, differently formed, the resistances to the motions of which had been ascertained, were caused to move through a fluid in contact with each other, it is obvious that the paths described by the particles of the fluid, in gliding by them, must be very materially altered by their junction; and it seems natural to expect that the joint disturbance produced in the motions of the fluid, when the surfaces are so united as to form a convex outline, would be somewhat less than if each surface were considered separately. Accordingly it is found that no calculation, deduced from experiments on the resistance opposed to oblique plane surfaces, will determine with accuracy the resistance to a curved surface. It appears from experiment that the resistance to the motion of a sphere is usually about two fifths of the resistance to a flat circular substance of an equal diameter. The resistance to the motion of a concave surface is greater than to a plane, and it is easily understood, that since the direction, in which the particles of the fluid recede from the solid, must be materially influenced by the form of the solid exposed to their action, their motion in this case must be partly retrograde, when they glide along towards the edges of the concave surface, and a greater portion of force must have been employed, than when they escape with a smaller deviation from their original direction. (Plate XXI. Fig. 276.)

For some reason which is not well understood, the hydraulic pressure of the air appears to be somewhat greater, in proportion to its density, than that of water. It has been found that the perpendicular impulse of the air, on a plane surface, is more than equivalent to the weight of a column of air of a height corresponding to the velocity, and the excess is said by some to amount to one third, by others to two thirds of that weight. The resistance appears also to be a little greater for a large surface, than for a number of smaller ones, which are together of equal extent.

The resistance or impulse of the air, on each square foot of a surface directly opposed to it, may in general be found, with tolerable accuracy, in pounds, by dividing the square of the velocity in a second, expressed in feet, by 500. Thus, if the velocity were 100 feet in a second, the pressure on each square foot would be 20 pounds; if 1000 feet, 2000 pounds. For a sphere of a foot in diameter, we may divide the square of the velocity by 1600. We may also find, in a similar manner, the utmost velocity that a given body can acquire or retain in falling through the air; for the velocity at which the resistance is equal to the weight must be its limit. Thus, if a sphere one foot in diameter weighed 100 pounds, the square of its utmost velocity would be 160000, and the velocity itself 400 feet in a second; if a stone of such dimensions entered the atmosphere with a greater velocity, its motion would very soon be reduced to this limit; and a lighter or a smaller body would move still more slowly. The weight of Mr. Garnerin's parachute, with its whole load, was about a quarter of a pound for each square foot, the square of its greatest velocity must, therefore, have been about 125, and the velocity 11 feet in a second, which is no greater than that with which a person would descend, in leaping from a height of two feet, without stooping. Mr. Garnerin found the velocity even less than this, and it is not improbable that the concave form of the parachute might considerably increase the resistance. Thus, Mr. Edgeworth found that a plate 9 inches long, when bent into an arc of which the chord was $7\frac{1}{2}$, had the resistance increased more than one seventh. The diminution of the resistance of the air by the obliquity of the surface is still less than that of the resistance of water: thus, the resistance on the oblique surfaces of a wedge is not quite so much less than the resistance on its base, as its breadth is less than the length of those surfaces.

When the velocity of a body moving through an elastic fluid is very great, the resistance is increased in a much greater proportion than the square of the velocity: thus, the retardation of a cannon ball moving with a velocity of 1000 feet in a second, or a little more, becomes suddenly much greater than the calculation indicates. The reason of this change appears to be, that the condensation of the air before the ball is necessarily confined to a smaller portion, which is very intensely compressed, because the effect of the impulse can only spread through the air with a certain velocity, which is not much greater than that of the ball; and this smaller portion of air must necessarily be much more condensed than a larger portion would have been. Thus, when a cannon ball moves slowly, its effect at any instant is in some degree divided throughout all that part of the atmosphere, which the sound of the report has reached; and if the ball follows the sound very speedily, it is obvious that the portion of the air before the ball, which partakes of the effect, must be very small. The sound is observed to be propagated with a velocity of about 1130 feet in a second, and a cannon ball may be discharged with a velocity of 2000; but one half of this is very speedily lost, so as to be wholly useless with regard to the effect of the ball. If, therefore, we wish to increase the range of a cannon ball, we must increase its weight; for the resistance increases only in proportion to the surface of the ball, while the weight is determined by its solid content.

It is not easy to explain, in a manner perfectly satisfactory, the reflection of a cannon ball, or of a stone, which strikes the surface of the sea, or of a piece of water, in an oblique direction. We may, however, assign some causes which appear to be materially concerned in this effect. In the first place the surface of the water, acting at first for some time on the lower part of the ball, produces, by its friction, a degree of rotatory motion, by means of which the ball, as it proceeds, acts upon the mass of water which is heaped up before it, and is obliged by a similar friction to roll upwards, so that it mounts again to a much greater height than it could possibly have attained by the mere hydrostatic pressure of the water at a depth so inconsiderable. But a more powerful cause than this appears to be the continual succession of new surfaces which are to be depressed, and which may be supposed to react on the ball, so as to produce the same effect, as a more intense pressure would have done, if it had continued stationary; and the mutual action of the water and the ball may be compared to the impulse of an oblique stream, moving with the velocity of the ball, which would impel it much more powerfully than the simple hydrostatic pressure at a much greater depth. It happens in this case, as in many others, that the effects which appear to be the most familiar to us, do not by any means admit the clearest and simplest explanation.

LECTURE XXVI.

ON HYDROSTATIC INSTRUMENTS, AND HYDRAULIC ARCHITECTURE.

WE have now examined the fundamental laws of the principal departments of hydrodynamics, which may be considered as constituting the theory of the science: we are next to proceed to the application of this theory to a variety of practical purposes. Following the same general order as we have observed in mechanics, our first division will be analogous to the subject of statics, and will relate to hydrostatic instruments; the second to architecture, containing some particulars respecting canals and embankments; the third to machinery, comprehending the modification and application of the force of fluids considered as inelastic; the fourth and the fifth to the methods of raising and removing weights, in which the principal hydraulic and pneumatic machines will be respectively explained, and, as a part of this subject, the application of pneumatic force will also be examined.

The principles of hydrostatics are very frequently applied to the determination of the specific gravities of the various productions of nature or of art. The diminution of the apparent weight of a solid body, upon immersion into a fluid, affords an easy method of comparing its density with that of the fluid. For the weight of the solid being previously determined, if we examine how much that weight is diminished by plunging the body in pure water, we shall have the weight of an equal bulk of water; and thence we may immediately obtain the proportion of the specific gravity of the body to that of water, which is the usual standard of comparison. And if we weigh a solid of given magnitude, for instance, a ball of glass, first in water, and then in any other fluid, the quantities of weight lost in each case will be in the same proportion as the specific gravities of the two fluids. A balance adapted for such examinations is called a hydrostatic balance; on one side it has a scale as usual, and on the other a loop of fine wire, or of horse hair, for holding the solid to be weighed, which may be changed occasionally for a ball of glass, suspended in a similar manner: sometimes also a dish is added for holding any loose substances which will sink in water, proper counterpoises being used as equivalents for the weight of the dish either in air or in water; and when a body lighter than water is examined, a weight of known magnitude and density is employed for sinking it. (Plate XXI. Fig. 277.)

The specific gravities of any substances, and in particular of such as are lighter than water, may also be very conveniently determined by means of a common balance, employing a phial with a conical ground stopple, filling it first with water, and then either with a given fluid, or with a portion of the solid of which the weight has been ascertained, together with as much water as is sufficient to exclude all the air.

For the speedy examination of a variety of fluids, differing but little in specific gravity from some known standard, a hydrometer may be very conveniently employed. This instrument is said to have been invented by Archimedes: it consists of a hollow ball, with a weight below it, and a slender stem above, so graduated as to express the specific gravity of the fluid by the degree to which it sinks. Sometimes the instrument is sunk to a certain mark, by means of weights placed in a dish at the end of the stem; or different weights are fixed to it below, while the graduations of the scale are still observed; and it may even be applied to finding the specific gravities of solids, the solid being first placed in the dish at the end of the stem, and then in a second dish which is suspended from the bulb below the water. (Plate XXI. Fig. 278.)

Another mode of ascertaining the specific gravities of fluids differing but little from each other in density, is to have a series of globules of glass, so loaded as to correspond to the specific gravities indicated by as many numbers, which are marked on them; and, throwing several of them together into the fluid, to observe which of them remains nearly stationary, without either rising to the surface or sinking. This method, though not expeditious, appears to be very secure from error: the globules are sold by patent, adapted for the measurement of the strength of spirituous liquors.

In whatever manner we compare the specific gravities of bodies with that of water, it is necessary, for very accurate experiments, either that the water be employed at the temperature of the air when moderately warm, or that a proper correction should be made for its change of bulk at different temperatures. Platina, the densest known substance, is 23 times as heavy as distilled water, gold $19\frac{1}{2}$, mercury $13\frac{1}{2}$, lead $11\frac{1}{4}$, silver 11, copper 9, iron and steel $7\frac{3}{4}$, stony substances usually about $2\frac{1}{2}$, rectified spirits $\frac{5}{6}$, naphtha, the lightest liquid $\frac{7}{10}$, cork about $\frac{1}{3}$, common air $\frac{1}{830}$, steam $\frac{1}{1500}$, and pure hydrogen gas $\frac{1}{12000}$. From this comparison, the weight of a cubic foot of any of these substances may be easily determined; since a cubic foot of water weighs nearly 1000 ounces avoirdupois, or more nearly 998; thus a cubic foot of gold would weigh about 195 000 ounces, and be worth above 60 000 pounds sterling; a cubic foot of iron weighs 7750 ounces, and a cubic foot of common stone about 2500.

The method of measuring the bulk of solid bodies by immersing them in a fluid was applied, by its inventor Archimedes, to the detection of a fraud in the composition of a mixed metal: and at present the principal use of hydrometers is for ascertaining, by the specific gravity of a compound of alcohol and water, the proportional quantities of its ingredients. But in all experiments of this kind, it is necessary to be aware, that a considerable change of the joint bulk of two substances is often produced by their mixture: and that in general their dimensions are considerably contracted. Thus, 18 gallons of water, and 18 of alcohol, instead of 36 gallons, make only 35, consequently the specific gravity of the compound is one 35th greater than the mean of the specific gravities of the ingredients. And in some cases the whole dimensions of a single substance may even be contracted by the addition of another substance: thus iron, by the addition of one eighth of its bulk of platina, becomes contracted one fortieth of that bulk.

The use of the spirit level depends on the tendency of all fluids to preserve a horizontal surface, and the freedom, with which the particles of fluids move on each other, renders it an instrument capable of the greatest delicacy. A tube, which is very slightly curved, being nearly filled with alcohol or ether, and then perfectly closed, the bubble will always rise to the highest part of the tube, and will never be stationary at the point which is marked as its proper place, unless the instrument be very accurately horizontal, or in the same position in which the mark was adjusted. The surface of the bubble, especially when it is small, cannot, in a strict sense, be called perfectly horizontal, since its form approaches nearly to that of a sphere; but in order that the centre of gravity of the water may attain the lowest possible situation, the bubble must necessarily occupy the highest point of the tube. (Plate XXI. Fig. 279.)

The principles of hydrostatics have been employed in various ways for supplying lamps with oil. It is found that a lamp will burn, without consuming any considerable portion of its wick, as long as it is amply supplied with oil; hence it becomes desirable that it should always be level with the surface of the reservoir, and this may be effected sufficiently well by placing the wick at the edge of a very large vessel, or at the end of a tube projecting from such a vessel, or from a vessel closed above, and opening only by an orifice below, which lets in the air as the oil escapes through it. But all these methods are often attended with inconveniences of various kinds, especially where the lamp is to be employed like a candle, and placed on a table. A French artist has applied a little pump, which is worked by means of a spring, for raising the oil from a vessel under the lamp; but this refinement is too complicated to be practically useful. Mr. Keir's lamp contains a divided cavity, one part of which is filled with oil, and the other with a saline or saccharine fluid of greater density, so that when the oil contained in the upper part of the tube is exhausted, its place is partly supplied by a fresh portion, which is forced up in consequence of the descent of the denser fluid in a much larger vessel. Still, however, the surface must be lowered by degrees; but by combining the invention with Dr. Hooke's semicylindrical counterpoise, a little modified, the height of this fluid may be so regulated, that the surface of the oil may remain almost invariable, until the reservoir is quite exhausted. For this purpose, the centre of gravity of the counterpoise must be a little higher than the line which bisects it; and its specific gravity must be about three fourths as great as that of the fluid; and in this manner it may be made to raise the surface of the heavier fluid, in proportion as a greater quantity of it escapes, to supply the place of the oil; and to keep it always at a sufficient height above the surface which separates it from the oil, so that the wick may be amply and almost uniformly supplied. (Plate XXI. Fig. 280.)

The art of embankment is a branch of architecture entirely dependent on hydrostatical and hydraulic principles. In Holland, and in some parts of Germany, this art is indispensable to the existence of large tracts of country; and even in this island, it has been of extensive utility, in gaining and securing ground on the sea coast. The construction of canals, and the management of rivers and harbours, are also dependent on the same principles; and these important subjects have been discussed by various writers, in many copious treatises, expressly devoted to hydraulic architecture.

When a bank or dike is to be constructed, it must be composed of materials capable of resisting, by their weight, the effort of the fluid to overturn them; by their lateral adhesion, the force tending to thrust them aside horizontally; and by their density and tenacity, the penetration of the water into their substance. If the water be in motion, they must also be able to resist its friction, without being carried away by it, and they must be arranged in such a form, as to be least liable to be undermined. For many of these reasons, the surface of the bank exposed to the water must be inclined to the horizon: the line expressing the general direction of the pressure of the water ought to be confined entirely within its substance, so that no force thus applied may be able to overturn it as a whole; and this condition will always be fulfilled, when the sides of the bank make an angle with each other not less than a right angle. The pressure acting on a bank thus inclined will also tend to condense the materials, and to increase their lateral adhesion, and the particles will become less liable to crumble away by their weight, than if the surface were more nearly vertical. For embankments opposed to the sea, a bank much inclined has also the additional advantage of breaking the force of the waves very effectually. An embankment of this kind is usually furnished with drains, formed by wooden pipes or by brickwork, closed by falling doors, or valves, which allow the water to flow out at low water, but do not permit the tide to enter. To prevent the penetration of the water, clay is often used, either mixed with gravel, or sunk in a deep trench cut on each side of the canal or reservoir. (Plate XXI. Fig. 281.)

The greater or less velocity of a river must determine what substances are capable of withstanding its tendency to disturb them; some are carried away by a velocity of a few inches in a second, others remain at rest when the velocity amounts to several feet. But in general, the velocity of a river is sufficient to produce a gradual transfer of the particles of its bed, which are shifted slowly downwards, towards the sea, being occasionally deposited in those parts where the water has least motion, and serving at last to form the new land, which is always advancing into the sea, on each side of the mouth of a large river. It has been recommended, as a good form for a navigable river or canal, to make the breadth of the horizontal bottom one fifth of that of the surface, and the depth three tenths. (Plate XXI. Fig. 282.)

If a canal or a reservoir were confined by a perpendicular surface of boards, and it were required to support it by a single prop, the prop should be placed, as we have already seen, at the distance of one third of the whole height from the bottom; but it would be always more convenient in practice to fix the side of the reservoir at the bottom, than to allow the whole pressure to be supported by the prop, and it might also be strengthened by means of ribs, thicker below than above, so as to produce an equal strength throughout, wherever the prop might be placed: but if the side were formed of a single plank, of uniform thickness, the strain would be most equally divided by placing the prop very near the middle of its height.

The strength of the materials employed for flood gates and sluices requires to be determined according to the principles, which have been laid down, in treating of the passive strength of substances used for purposes simply mechanical; but the calculations become in this case much more intricate. Thus, if we have a circular plate or plank, of a uniform elastic substance, constituting the bottom of a pipe or cistern, and simply supported at the circumference, a very complicated calculation is required for determining the proportion of its strength to that of a square plate of the same breadth, supported only at two opposite ends, since at each point of the circular piece, there are two curvatures which require to be considered. The square plate will support a column of fluid twice as heavy as the weight which would break it, if placed at its centre; and if I have been correct in the calculation, a circular plate will support a height of water nearly sixteen sevenths as great as a square plate. But for ordinary purposes, it will be sufficient to consider the strength as derived only from the resistance opposed to the flexure in one direction, since the additional strength, obtained from the lateral supports, may very properly be neglected, as only assisting in affording that additional security which is always necessary, to compensate for any accidental defects of the materials. It has been asserted that the strength of a square plate is doubled when it is supported on both sides; but this appears to be a mistake.

We may, therefore, be contented with determining the strain on the materials in that direction in which they afford the greatest resistance, either from the shorter distance between the supports, or by the disposition of the fibres; and it will be always most eligible to combine these circumstances, so that the fibres of the wood may be arranged in the direction of the shortest dimensions of the sluice. If a sluice be supported above and below only, the greatest strain will be at the distance of about three sevenths of its height from the bottom; and it is at this point that the greatest strength is required. But if the boards forming the sluice be fixed across it, in horizontal directions, their strength must be greatest at the bottom. (Plate XXI. Fig. 283.)

In the construction of flood gates, the principles of carpentry must be applied in a manner nearly similar to that which serves for the determination of the best forms of roofs. The flood gates, if they are double, without a solid obstacle between them, must meet at an angle: and when this angle is very open, the thrust against the walls or hinges must necessarily be very great. If, however, the angle were too acute, the flood gates would require to be lengthened, and in this case their strength would be far more diminished than that of a roof similarly elevated, since the hydrostatic pressure acts always with full force in a perpendicular direction. The thickness required for each flood gate may be determined in the same manner as the thickness of a sluice.

Where a sluice board of considerable dimensions is to be occasionally raised, it may be necessary to ascertain the force which will be required for overcoming its friction; this friction is nearly proportional to the whole pressure of the water, and may be found, with sufficient accuracy, in pounds, by multiplying the square of the depth of the sluice, in feet, by 10. Thus, if the depth be 3 feet, the friction or adhesion will be about 90 pounds for each foot of the breadth.

If the side of a canal gives way, it is sometimes of consequence to prevent, as much as possible, the escape of the water. For this purpose it is usual to have doors or valves in various parts of the canal, which, when the water is at rest, lie nearly flat at the bottom; but when it begins to run over them, with a considerable velocity, they are raised by its force, and put a stop to its motion.

The utility of the introduction of canals into a commercial country may be estimated in some measure by the effect of the same labour, employed in removing weights by land carriage and by water. Thus, a single horse can scarcely draw more than a ton weight on the best road, but on a canal, the same horse can draw a boat of 30 tons at the same rate.

The construction of piers and quays, and the management of harbours, are also important departments of hydraulic architecture; it often happens that besides the application of the general principles of mechanics and hydrostatics to these purposes, the peculiar circumstances of the case may indicate to an ingenious artist a mode of performing the required work in an effectual and economical manner. We may find a good example of such an arrangement, in the account given, by Mr. Smeaton, of the method which he adopted for the improvement of the port of Ramsgate, and which indeed resembles some that had been before employed in similar cases: by forming a large excavation, which is furnished with flood gates, and is constantly filled at high water, he has procured a number of artificial torrents, which escape through the sluices, and become powerful agents for carrying away the matter deposited by the sea, and tending to impede the navigation of the harbour.

LECTURE XXVII.

ON THE REGULATION OF HYDRAULIC FORCES.

THOSE modifications of the motions of fluids which are employed either for conducting them from place to place, or for applying their powers to the production of mechanical effects, may be considered as constituting a separate division of practical hydraulics, which is analogous to the subject of general machinery in practical mechanics.

A supply of water may be obtained from a reservoir, situated above the level at which it is wanted, whatever its distance may be, either by means of open canals, or aqueducts, or of closed pipes. Where an uninterrupted declivity cannot be obtained, it is necessary to employ pipes, which may be bent upwards or downwards at pleasure, provided that no part of them be more than thirty feet above the reservoir, and when the pipe is once filled, the water will continue to flow from the lower orifice; but it is best in all such cases to avoid unnecessary angles; for when the pipe rises and falls again, a portion of the air, which is always contained in water, is frequently collected in the angle, and very materially impedes the progress of the water through the pipe. When the bent part is wholly below the orifices of the pipe, this air may be discharged by various methods. The ancients used small upright pipes, called columnaria, rising from the convexity of the principal pipe, to the level of the reservoir, and suffering the air to escape without wasting any of the water. It may however frequently be inconvenient or impossible to apply a pipe of this kind; and the same purpose may be answered, by fixing on the pipe a box containing a small valve, which opens downwards, and is supported by a float, so as to remain shut while the box is full of water, and to fall open when any air is collected in it. (Plate XXI. Fig. 288.)

If the pipe were formed into a siphon, having its flexure above both orifices, it would be necessary to bend it upwards at the extremities, in order to keep it always full: but in this case the accumulation of the air would be extremely inconvenient, since it would collect so much the more copiously, as the water in the upper part of the pipe would be more free from pressure, and neither of the methods which have been mentioned would be of any use in extricating it. It has been usual in such cases to force a quantity of water violently through the pipe, in order to carry the air with it; but perhaps the same effect might be produced much more easily, by making a small airtight valve in the upper part of the pipe, opening outwards, and a stopcock immediately before it: the stopcock being suddenly turned as often as might be necessary, the momentum of the water in the pipe would probably carry it forwards with sufficient force to throw out the air; or, if it were necessary external pressure might be added, and the air might even in this manner be discharged by the valve much more readily than without it. But it might be still simpler to have a pretty large vessel of water screwed on to the pipe, which would not be filled with air for a considerable time; and which, when full, might be taken off and replenished with water. (Plate XXI. Fig. 285.)

The diameter of a pipe, required for conveying a given quantity of water to a given distance, may be calculated from the experiments of Mr. Buat, which have been already mentioned. Pipes are usually made of wood, of lead, or of cast iron; but most commonly of lead; and of late tinned copper has been employed with considerable advantage. A pipe of lead will bear the pressure of a column of water 100 feet high, if its thickness be one hundredth of its diameter, or even less than this; but when any alternation of motion is produced, a much stronger pipe is required, and it is usual to make leaden pipes of all kinds far thicker than in this proportion.

The form and construction of stopcocks and valves are very various, according to their various situations and uses. Stopcocks usually consist of a cylindrical or conical part, perforated in a particular direction, and capable of being turned in a socket formed in the pipe, so as to open or shut the passage of the fluid, and sometimes to form a communication with either of two or more vessels at pleasure. A valve is employed where the fluid is to be allowed to pass in one direction only, and not to return. For water, those valves are the best which interrupt the passage least; and none appears to fulfil this condition better than the common clack valve of leather, which is generally either single, or divided into two parts; but it is sometimes composed of four parts, united so as to form a pyramid, nearly resembling the double and triple valves which are formed by nature in the hearts of animals. A board, or a round flat piece of metal, divided unequally by an axis on which it moves, makes also a very good simple valve. Where a valve is intended to intercept the passage of steam, it must be of metal; such a valve is generally a flat plate, with its edge ground a little conically, and guided in its motion by a wire or pin. For air, valves are commonly made of oiled silk, supported by a perforated plate or grating. (Plate XXI. Fig. 286, 287.)

Before we consider the application of the force of fluids in motion to practical purposes, we must attend to the methods of measuring the velocity of their motions. This may be done either by a comparison with linear measures, or by instruments founded on the laws of hydraulic pressure. One of the best of such instruments is the tube invented by Pitot, and improved by Buat. A funnel is presented to the stream, and the water in a vertical tube connected with it is elevated above the level of the river, nearly to the height corresponding to the velocity: but it is said that the result will be less liable to error, if the funnel be covered by a plate with a small orifice in its centre, the elevation being in this case always half as great again as the height due to the velocity. Other instruments, intended for the same purpose, require some previous experiments for determining the degree in which they are affected by different velocities; in this manner the hydrometrical fly is adjusted; the impulse of the water on two inclined planes turning an axis to which they are fixed, and by its means a series of wheels, with an index, which expresses the space described during the time of observation. Instruments similar to these have also sometimes been employed, for measuring the relative velocity, with which a ship under way passes through the water; and an apparatus, resembling Pitot's, has been adapted to this purpose by Captain Hamilton, with the addition of a tube inserted into it on a level with the surface of the water, which continually discharges a small stream into a reservoir with a velocity regulated by the pressure, and consequently equal or proportional to that of the ship itself. In this manner he obtains an accurate register of the whole distance described, including the effect of all the variations of the velocity. If the orifice be small, it will be necessary to attend to the temperature of the water, since the discharge is considerably retarded by any considerable degree of cold. But when the aperture, which determines the magnitude of the discharge, is wholly under water, as Captain Hamilton has placed it, this source of error is probably much diminished. (Plate XXII. Fig. 288, 289.)

The motions of the air may also be measured by instruments similar to those which are employed for determining the velocity of streams of water. The direction of the wind is sometimes indicated by a wind dial, consisting simply of an index, connected by wheels with a common vane or weather-cock. Its velocity may be found by means of wind gages of different kinds: these are sometimes constructed by opposing a flat surface to the wind, the pressure being measured by the flexure of a spring, or by the winding up of a weight on a spiral barrel; and sometimes by receiving the stream in the mouth of a funnel, so as to raise a column of water, in a vertical tube, to a height equivalent to the pressure, or to condense a quantity of air inclosed in a cavity, to a degree which is indicated by the place of a small portion of mercury, moving in a horizontal tube, which leads to the cavity. A little windmill, like the hydrometrical fly, may also be employed for measuring the velocity of the wind, with the assistance of a watch.

The principal methods of applying the force of fluids to useful purposes are to employ their weight, their impulse, or their pressure. The weight of water may be applied, by collecting it in a reservoir, which alternately ascends and descends, by causing it to act within a pipe on a moveable piston, or by conducting it into the buckets of a revolving wheel; its impulse may be directed either perpendicularly or obliquely against a moveable surface; and its pressure may be obtained, without any immediate impulse, by causing a stream to flow horizontally out of a moveable pipe which revolves round an axis. The force of the air can only be applied by means of its impulse, and this may be employed either perpendicularly or obliquely.

When water is collected in a single reservoir, which serves to work a pump or to raise a weight, the mode of its operation may be determined from mechanical considerations only; and it is obvious that if we are desirous of preserving the whole force of the water, we must employ a second reservoir to be filled during the descent of the first, which may either descend in its turn, or empty itself into the first when it has ascended again to its original situation. The action of a column of water, inclosed in a pipe, is of a nature nearly similar to that of such a reservoir, excepting that the apparatus is more liable to friction; the arrangement of its parts is nearly similar, although in an inverted position, to that which is more commonly employed for raising water by means of pumps. But both these methods of employing the weight of water, are in great measure confined to those cases in which it is to be procured in a small quantity, and may be allowed to descend through a considerable height, and when the circumstances do not allow us to employ machines which require a greater space.

We have seen that in order to determine the effect of any force employed in machinery, we must consider not only its magnitude, but also the velocity with which it can be brought into action, and we must estimate the ultimate value of the power, by the joint ratio, or the product, of the force and the velocity. Thus, if we had a corn mill, for example, in which we wished the millstone to revolve with a certain velocity, and to overcome a given resistance, and supposing that this effect could be obtained by means of a certain train of wheels from a given source of motion; if the velocity of the motion at its source be reduced to one half, we must double the diameter of one of the wheels by which the force is communicated, in order to give the millstone the desired velocity, and thus we must introduce a mechanical disadvantage, which can only be compensated by a double intensity in the force at its origin.

If we apply this estimation of effect to the motion of an overshot wheel, we shall find that the velocity of the wheel, and consequently its breadth, and the magnitude of its buckets, is perfectly indifferent with respect to the value of its operation: for supposing the stream to enter the buckets with the uniform velocity of the wheel, the quantity of water in the wheel at any one time, and consequently the pressure, must be inversely as the velocity, so that the product of the force into the velocity will be the same, however they may separately vary. If, however, the velocity were to become very considerable, it would be necessary to sacrifice a material part of the fall, in order that the water might acquire this velocity before its arrival at the wheel; but a fall of one foot, or even less, is sufficient for producing any velocity that would be practically convenient: and it is obvious, on the other hand, that a certain velocity may be procured from a wheel moving rapidly, with less machinery than from another which moves more slowly. In general the velocity of the surface of the wheel is between two and six feet in a second; and whether it be greater or smaller, the force actually applied will always be equal in effect to the weight of a portion of the stream employed, equal in length to the height of the wheel. In order to avoid the resistance which might be occasioned by the stagnant water below the wheel, it is a good practice to turn the stream backwards upon its nearer half, so that the water, when discharged, may run off in the general direction of its motion. (Plate XXII. Fig. 290.)

If we suffer the stream of water to acquire the utmost velocity that the whole fall can produce, and to strike horizontally against the floatboards of an undershot wheel, or if we wish to employ the force of a river running in a direction nearly horizontal, the wheel must move, in order to produce the greatest effect, with half the velocity of the stream. For the whole quantity of water impelling the floatboards is nearly the same, whatever may be the velocity, especially if the wheel is properly inclosed in a narrow channel, and hence it is easy to calculate that the greatest possible effect will be produced when the relative velocity of the stream, striking the floatboards, is equal to the velocity of the wheel itself. The pressure on the floatboards is equal to that of a stream containing the same quantity of water, and striking a fixed obstacle with half the velocity, that is, such a stream as escapes from the wheel, which must be twice as deep or twice as wide as the original stream, since its motion is only one half as rapid; and a column of such a stream, of twice the height due to its velocity, that is, of half the height of the fall, being, as we have already seen, the measure of the hydraulic pressure, this force will be precisely half as great as that of a similar column, acting on an overshot wheel, which moves with the same velocity. But the stream thus retarded will not retain the other half of its mechanical power; since its greatest effect will be in the same proportion to that of an equal stream acting on an overshot wheel with one fourth of the fall of the former: and the remaining fourth of the power is lost in producing the change of form of the water and in overcoming its friction. In whatever way we apply the force of water, we shall find that the mechanical power which it possesses must be measured by the product of the quantity multiplied by the height from which it descends: for example, a hogshead of water capable of descending from a height of 10 feet, possesses the same power as 10 hogsheads descending from a height of one foot; and a cistern filled to the height of 10 feet above its orifice possesses 100 times as much power as the same cistern filled to the height of one foot only.

When, therefore, the fall is sufficiently great, an overshot wheel is far preferable to an undershot wheel, and where the fall is too small for an overshot wheel, it is most advisable to employ a breast wheel, which partakes of its properties; its floatboards consisting of two portions meeting at an angle, so as to approach to the nature of buckets, and the water being also in some measure confined within them by the assistance of a sweep or arched channel which follows the curve of the wheel, without coming too nearly into contact with it, so as to produce unnecessary friction. When the circumstances do not admit even of a breast wheel, we must be contented with an undershot wheel: it is recommended, for such a wheel, that the floatboards be so placed as to be perpendicular to the surface of the water at the time that they rise out of it: that only one half of each should ever be below the surface, and that from three to five should be immersed at once, according to the magnitude of the wheel. Sometimes, however, it has been thought eligible to employ a much smaller number: thus the water wheel which propels Mr. Symington's steam boat has only six floatboards in its whole circumference. (Plate XXII. Fig. 291, 292.)

Since the water escaping from an undershot wheel still retains a part of its velocity, it is obvious that this may be employed for turning a second wheel, if it be desirable to preserve as much as possible of the force. In this case, by causing the first wheel to move with two thirds of the velocity of the stream, the whole effect of both will be one third greater than that of a single wheel placed in the same stream; but it must be considered that the expense of the machinery will also be materially increased.

Considerable errors have frequently been made by mathematicians and practical mechanics in the estimation of the force of the wind or the water on oblique surfaces: they have generally arisen from inattention to the distinction between pressure and mechanical power. It may be demonstrated that the greatest possible pressure of the wind or water, on a given oblique surface at rest, tending to turn it in a direction perpendicular to that of the wind, is obtained when the surface forms an angle of about 55° with the wind; but that the mechanical power of such a pressure, which is to be estimated from a combination of its intensity with the velocity of the surface, may be increased without limit by increasing the angle of inclination, and consequently the velocity. The utmost effect that could be thus obtained would be equal to that of the same wind or stream acting on the floatboards of an undershot wheel: but since in all practical cases the velocity is limited, the effect will be somewhat smaller than this: for example, if the mean velocity of the sails or floatboards be supposed equal to that of the wind, the mechanical power will be more than four fifths as great as that of an undershot wheel, that is, in the case of a windmill, more than four fifths of the utmost effect that can be obtained from the wind. In such a case Maclaurin has shown that the sails ought to make an angle of 74° with the direction of the wind: but in practice it is found most advantageous to make the angle somewhat greater than this, the velocity of the extremities of the sails being usually, according to Mr. Smeaton, more than twice as great as that of the wind. It appears, therefore, that the oblique sails of the common windmill are in their nature almost as well calculated to make the best use of any hydraulic force as an undershot wheel; and since they act without intermission throughout their whole revolution, they have a decided advantage over such machines as require the sails or fans to be exposed to a more limited stream of the wind, during one half only of their motion, which is necessary in the horizontal windmill, where a screen is employed for covering them while they are moving in a direction contrary to that of the wind: and such machines, according to Smeaton, are found to perform little more than one tenth of the work of those which are more usually employed. The sails of a common windmill are frequently made to change their situation according to the direction of the wind, by means of a small wheel, with sails of the same kind, which turns round whenever the wind strikes on either side of it, and drives a pinion turning the whole machinery; the sails are sometimes made to furl or unfurl themselves, according to the velocity of the wind, by means of a revolving pendulum, which rises to a greater or less height, in order to prevent the injury which the flour would suffer from too great a rapidity in the motion, or any other accidents which might happen in a mill of a different nature. The inclination of the axis of a windmill to the horizon is principally intended to allow room for the action of the wind at the lower part, where it would be weakened if the sails came too nearly in contact with the building, as they must do if they were perfectly upright. When it is necessary to stop the motion of a windmill, a break is applied to the surface of a large wheel, so that its friction operates with a considerable mechanical advantage. Water wheels with oblique floatboards are sometimes used with good effect in China and in the south of France: for tide wheels, such floatboards have the advantage that they may be easily made to turn on a hinge with the stream, so as to impel the wheel in the same direction whether the tide be flowing or ebbing. (Plate XXII. Fig. 293.)

A smoke jack is a windmill in miniature; a kite affords a very familiar example of the effect of the oblique impulse of the air, of which the action first causes a pressure perpendicular to the surface of the kite, and this force, combined with the resistance of the string, produces a vertical result capable of counteracting the weight of the kite. (Plate XXII. Fig. 294.)

The counterpressure of the water, occasioned by the escape of a stream from a moveable reservoir, was applied by Parent to the purpose of turning a millstone, and various other authors have described machines of a similar nature: they may be constructed with little or no wheel work, and it does not appear to be necessary that much of the force of the water should be lost in their operation; but they have never been practically employed with success, nor have they perhaps ever had a fair trial.

The art of seamanship depends almost entirely on the management of the forces and resistances of air and water, and if the laws of hydraulic pressure, with respect to oblique and curved surfaces, were more completely ascertained, we might calculate not only what the motions of a ship would be under any imaginable circumstances, but we might also determine precisely what would be the best possible form of a ship, and what the best arrangement of her rigging.

When a ship is sailing immediately before the wind, little or no art is required in setting her sails, and her velocity is only limited by that of the wind, and by the resistance of the water: but for sailing with a side wind, it becomes necessary that the immediate force of the wind should be considerably modified.

If we had a circular vessel or tub, with a single mast, and a sail perfectly flat, and if the sail were placed in a direction deviating but little from that of the wind, the tub would begin to move in a direction nearly at right angles to that of the wind, since the impulse of the wind acts almost entirely in a direction perpendicular to that of the sail: but the slightest inequality of the dimensions of the sail, or of the force of the wind, would immediately disturb the position of the vessel; and in order to avoid this inconvenience, it would be necessary to have a moveable body projecting into the water, so as to create a resistance by means of which the vessel might be steered, and the sail confined to its proper place: and this might be done more effectually by changing the form of the vessel from round to oval; it would then also have the advantage of moving much more easily through the water in the direction of its length than a circular vessel of equal size, and of creating still more resistance in a transverse direction, so that when urged by an oblique force, it would move in some measure obliquely, but always much more nearly in the direction of its length than of its breadth. The angular deviation from the track of the ship is called its lee way, and if we know the direction of the sails, and the actual proportions of the resistances opposed to the ship's motion in different directions, we may calculate from these resistances the magnitude of the angular deviation or lee way: but hitherto such calculations have generally indicated a lee way three or four times as great as that which has been observed. The use of the keel is not only to assist in confining the motion of the ship to its proper direction, but also to diminish the disposition to vibrate from side to side, which would interfere with the effect of the sails, and produce many other inconveniences. When the principal force of the wind is applied to the anterior part of the ship, her head would be naturally turned from the wind if the rudder were not made to project from the stern in a contrary direction, and to present the surface of an inclined plane to the water which glides along the keel, so as to preserve the ship, by means of the pressure which it receives, in any direction that may be required for her manoeuvres. Commonly, however, although the sails may be so arranged that the principal force of the wind appears to be on the fore part of the ship, the curvature of the sails, or some other cause, throws the pressure further backwards, and the action of the rudder is necessary to prevent the ship's head turning towards the wind. (Plate XXII. Fig 295.)

When a ship is steering in this manner on a side wind, the effect of the wind has a natural tendency to overset her, and if she is too crank, that is, deficient in stability, she cannot sail well, otherwise than directly before the wind. The place of the centre of gravity, compared with that of the metacentre, or imaginary centre of pressure, determines the degree of stability, and the most general way of increasing it is to lessen the weight of the upper part, and of the rigging of the vessel, to diminish her height, or to increase her breadth, and to stow the ballast as low as possible in the hold. Too little attention has frequently been paid to this subject, as well as to many other departments of naval architecture; and although mere theoretical investigations have hitherto been but of little service to the actual practice of seamanship, yet it cannot be doubted that an attention to what has already been discovered of the laws of hydrodynamics, as well as to the principles of mechanics in general, must be of great advantage to the navigator, in enabling him to derive from his own experience all the benefits, which a correct mode of reasoning is capable of procuring him.

LECTURE XXVIII.

ON HYDRAULIC MACHINES.

WE shall apply the denomination of hydraulic machines to such only, as are intended for counteracting the gravity of water, that is, for raising it from a lower situation to a higher. The simplest of these are buckets, bucket wheels, and friction ropes; moveable pipes are the next in order; and pumps of various kinds constitute the most extensive and the most important part of the subject. Besides these and some other similar machines, hydraulic air vessels and artificial fountains will also require to be examined.

A series of earthen pitchers, connected by ropes, and turned by trundles or pinions, over which they pass, has long been used in Spain, under the name of noria: in this country, buckets of wood are sometimes employed in a similar manner. A bucket wheel is the reverse of an overshot waterwheel, and the water may be raised by buckets nearly similar to those which are calculated for receiving it in its descent: sometimes the buckets are hung on pins, so as to remain full during the whole ascent; but these wheels are liable to be frequently out of repair. Sometimes the reverse of an undershot wheel or rather of a breast wheel, is employed as a throwing wheel, either in a vertical or in an inclined position. Such wheels are frequently used for draining fens, and are turned by windmills; the floatboards are not placed in the direction which would be best for an undershot wheel, but on the same principle, so as to be perpendicular to the surface when they rise out of it, in order that the water may the more easily flow off them. (Plate XXII. Fig. 296 . . 298.)

Instead of a series of buckets connected by ropes or chains, a similar effect is sometimes produced by a simple rope, or a bundle of ropes, passing over a wheel above, and a pulley below, moving with a velocity of about 8 or 10 feet in a second, and drawing a certain quantity of water up by its friction. It is probable that the water commonly ascends with about half the velocity of the rope, and on this supposition we might calculate its depth on the rope by comparing its relative motion with that of a little river: but the rules, which serve for calculating the velocity of rivers, do not perfectly agree in this case with the results of direct experiments; for the friction required for elevating the quantity raised by such a machine, appears from calculation to correspond to a velocity about twice as great as the actual relative velocity. While the water is principally supported by the friction of the rope, its own cohesion is amply sufficient to prevent its wholly falling, or being scattered, by any accidental inequality of the motion. (Plate XXII. Fig. 299.)

The lateral friction of water has been applied in a very simple manner by Venturi to the draining of land, by means of a stream which runs through it, allowing the stream to acquire sufficient velocity to carry it over an inclined surface, and to drag with it a certain portion of water from the lowest part of this surface: but the quantity of water raised in this manner must be very inconsiderable, and the loss of force by friction very great.

A system of spiral pipes may be placed in the plane of a wheel, receiving the water at its circumference, and raising it by degrees, as the wheel turns, towards the axis, where it is discharged; the motion of the wheel being usually derived from the same stream which supplies the pipes: but the height to which the water is raised by this machine is very small in proportion to its bulk. A single pipe wound spirally round a cylinder which revolves on an axis in an oblique situation, has been denominated the screw of Archimedes, and is called in Germany the water snail. Its operation, like that of the flat spiral, may be easily conceived by imagining a flexible pipe to be laid on an inclined plane, and its lower part to be gradually elevated, so that the fluid in the angle or bend of the pipe may be forced to rise; or by supposing a tube, formed into a hoop, to be rolled up the same plane, the fluid being forced by the elevation of the tube behind it to run as it were up hill. This instrument is sometimes made by fixing a spiral partition round a cylinder, and covering it with an external coating, either of wood or of metal; it should be so placed with respect to the surface of the water as to fill in each turn one half of a convolution; for when the orifice remains always immersed, its effect is much diminished. It is generally inclined to the horizon in an angle of between 45 and 60 degrees; hence it is obvious that its utility is limited to those cases in which the water is only to be raised to a moderate height. The spiral is seldom single, but usually consists of three or four separate coils, forming a screw which rises slowly round the cylinder. (Plate XXII. Fig. 300, 301.)

An instrument of a similar nature is called by the Germans a water screw; it consists of a cylinder with its spiral projections detached from the external cylinder or coating, within which it revolves. This machine might not improperly be considered as a pump, but its operation is precisely similar to that of the screw of Archimedes. It is evident that some loss must here be occasioned by the want of perfect contact between the screw and its cover; in general, at least one third of the water runs back, and the machine cannot be placed at a greater elevation than 30°; it is also very easily clogged by accidental impurities of the water: yet it has been found to raise more water than the screw of Archimedes, when the lower ends of both are immersed to a considerable depth; so that if the height of the surface of the water to be raised were liable to any great variations, the water screw might be preferable to the screw of Archimedes. (Plate XXII. Fig. 302.)

When a spiral pipe, consisting of many convolutions, arranged either in a single plane, or in a cylindrical or conical surface, and revolving round a horizontal axis, is connected at one end by a watertight joint with an ascending pipe, while the other end receives during each revolution nearly equal quantities of air and water, the machine is called a spiral pump. It was invented about 1746, by Andrew Wirtz, a pewterer at Zurich, and it is said to have been used with great success at Florence and in Russia: it has also been employed in this country by Lord Stanhope, and I have made trial of it for raising water to a height of forty feet. The end of the pipe is furnished with a spoon, containing as much water as will fill half a coil, which enters the pipe a little before the spoon has arrived at its highest situation, the other half remaining full of air, which communicates the pressure of the column of water to the preceding portion, and in this manner the effect of nearly all the water in the wheel is united, and becomes equivalent to that of the column of water, or of water mixed with air, in the ascending pipe. The air nearest the joint is compressed into a space much smaller than that which it occupied at its entrance, so that where the height is considerable, it becomes advisable to admit a larger portion of air than would naturally fill half the coil, and this lessens the quantity of water raised, but it lessens also the force required to turn the machine. The joint ought to be conical, in order that it may be tightened when it becomes loose, and the pressure ought to be removed from it as much as possible. The loss of power, supposing the machine well constructed, arises only from the friction of the water on the pipe, and the friction of the wheel on its axis; and where a large quantity of water is to be raised to a moderate height, both of these resistances may be rendered inconsiderable. But when the height is very great, the length of the spiral must be much increased, so that the weight of the pipe becomes extremely cumbersome, and causes a great friction on the axis, as well as a strain on the machinery: thus, for a height of 40 feet, I found that the wheel required above 100 feet of a pipe which was three quarters of an inch in diameter; and more than one half of the pipe being always full of water, we have to overcome the friction of about 80 feet of such a pipe, which will require 24 times as much excess of pressure to produce a given velocity, as if there were no friction. The centrifugal force of the water in the wheel would also materially impede its ascent if the velocity were considerable, since it would be always possible to turn it so rapidly as to throw the whole water back into the spoon. The machine which I had erected being out of repair, I thought it more eligible to substitute for it a common forcing pump, than to attempt to make any further improvement in it, under circumstances so unfavourable. But if the wheel with its pipes were entirely made of wood, it might in many cases succeed better: or the pipes might be made of tinned copper, or even of earthenware, which might be cheaper and lighter than lead. (Plate XXII. Fig. 303.)

The centrifugal force, which is an impediment to the operation of Wirtz's machines, has sometimes been employed, together with the pressure of the atmosphere, as an immediate agent in raising water, by means of the rotatory pump. This machine consists of a vertical pipe, caused to revolve round its axis, and connected above with a horizontal pipe, which is open at one or at both ends, the whole being furnished with proper valves to prevent the escape of the water when the machine is at rest. As soon as the rotation becomes sufficiently rapid, the centrifugal force of the water in the horizontal pipe causes it to be discharged at the end, its place being supplied by means of the pressure of the atmosphere on the reservoir below, which forces the water to ascend through the vertical pipe. It has also been proposed to turn a machine of this kind by the counterpressure of another portion of water, in the manner of Parent's mill, where there is fall enough to carry it off. This machine may be so arranged that, according to theory, little of the force applied may be lost; but it has failed of producing in practice a very advantageous effect. (Plate XXIII. Fig. 304.)

A pump is a machine so well known, and so generally used, that the denomination has not uncommonly been extended to hydraulic machines of all kinds; but the term, in its strictest sense, is to be understood of those machines, in which the water is raised by the motion of one solid within another, and this motion is usually alternate, but sometimes continued so as to constitute a rotation. In all the pumps most commonly used, a cavity is enlarged and contracted by turns, the water being admitted into it through one valve, and discharged through another.

One of the simplest pumps, for raising a large quantity of water to a small height, is made by fitting two upright beams or plungers, of equal thickness throughout, into cavities nearly of the same size, allowing them only room to move without friction, and connecting the plungers by a horizontal beam moving on a pivot. The water being admitted, during the ascent of each plunger, by a large valve in the bottom of the cavity, it is forced, when the plunger descends, to escape through a second valve in the side of the cavity, and to ascend by a wide pipe to the level of the beam. The plungers ought not to be in any degree tapered, because of the great force which would be unnecessarily consumed, in continually throwing out the water, with great velocity, as they descend, from the interstice formed by their elevation. This pump may be worked by a labourer, walking backwards and forwards, either on the beam or on a board suspended below it. By means of an apparatus of this kind, described by Professor Robison, an active man, loaded with a weight of thirty pounds, has been able to raise 580 pounds of water every minute, to a height of $11\frac{1}{2}$ feet, for ten hours a day, without fatigue; this is the greatest effect produced by a labourer that has ever been correctly stated by any author; it is equivalent to somewhat more than 11 pounds raised through 10 feet in a second, instead of 10 pounds, which is a fair estimate of the usual force of a man, without any deduction for friction. (Plate XXIII. Fig. 305.)

It is obvious that if the plungers were so well fitted to the cavity as to prevent the escape of any water between them, the ascending pipe might convey the water to any required height; the machine would then become a forcing pump, and the plungers might be shortened at pleasure, so as to assume the form of a piston sliding within a barrel. The piston might also be situated above the level of the reservoir, and in this case the water would be forced up after it by the pressure of the atmosphere to the height of about 30 feet, but not much further: and even this height would be somewhat too great for practice, because the water might sometimes follow the piston in its ascent too slowly. Such a pump, partaking of the nature of a forcing and a sucking pump, is sometimes called a mixed pump. In Delahire's pump, the same piston is made to serve a double purpose, the rod working in a collar of leathers, and the water being admitted and expelled in a similar manner, above and below the piston, by means of a double apparatus of valves and pipes. (Plate XXIII. Fig. 306.)

For forcing pumps of all kinds, the common piston, with a collar of loose and elastic leather, is preferable to those of a more complicated structure: the pressure of the water on the inside of the leather makes it sufficiently tight, and the friction is inconsiderable. In some pumps the leather is omitted, for the sake of simplicity, the loss of water being compensated by the greater durability of the pump; and this loss will be the smaller in proportion as the motion of the piston is more rapid. (Plate XXIII. Fig. 307.)

Mr. Bramah has very ingeniously applied a forcing pump, by means of the well known properties of hydrostatic pressure, to the construction of a convenient and powerful press. The water is forced, by a small pump, into a barrel in which it acts on a much larger piston; consequently this piston is urged by a force as much greater than that which acts on the first pump rod, as its surface is greater than that of the small one. (Plate XXIII. Fig. 308.)

In the common sucking pump, the valve through which the water escapes is placed within the piston itself, so that the same barrel serves for the ascent of the water, which rises in one continued line, while the piston is raised, and rests on the fixed valve while it is depressed. The velocity of the stroke ought never to be less than 4 inches in a second, nor greater than two or three feet; the stroke should also be as long as possible, in order to avoid unnecessary loss of water during the descent of the valves. The diameter of the pipe, through which the water rises to the barrel, ought not to be less than two thirds of the diameter of the barrel itself. (Plate XXIII. Fig. 309.)

A bag of leather has also been employed for connecting the piston of a pump with the barrel, and in this manner nearly avoiding all friction: but it is probable that the want of durability would be a great objection to such a machine. (Plate XXIII. Fig. 310.)

Where the height, through which the water is to be raised, is considerable, some inconvenience might arise from the length of the barrel through which the piston rod of a sucking pump would have to descend, in order that the piston might remain within the limits of atmospheric pressure. This may be avoided by placing the moveable valve below the fixed valve, and introducing the piston at the bottom of the barrel. Such a machine is called a lifting pump: in common with other forcing pumps, it has the disadvantage of thrusting the piston before the rod, and thus tending to bend the rod, and produce an unequal friction on the piston, while, in the sucking pump, the principal force always tends to straighten the rod. (Plate XXIII. Fig. 311.)

The rod of a sucking pump may also be made to work in a collar of leather, and the water may be forced through a valve into an ascending pipe. By applying an air vessel to this, or to any other forcing pump, its motion may be equalised, and its performance improved; for if the orifice of the air vessel be sufficiently large, the water may be forced into it, during the stroke of the pump, with any velocity that may be required, and with little resistance from friction, while the loss of force, from the frequent accelerations and retardations of the whole body of water, in a long pipe, must always be considerable. The condensed air, reacting on the water, expels it more gradually, and in a continual stream, so that the air vessel has an effect analogous to that of a fly wheel in mechanics. (Plate XXIII. Fig. 312.)

If, instead of forcing the water to a certain height through a pipe, we cause it to form a detached jet, we convert the forcing pump into a fire engine; and in general two barrels, acting alternately, are connected, for this purpose, with the same air vessel; so that the discharge is thus rendered very nearly uniform. The form of the ajutage, or orifice of the pipe, is by no means indifferent to the effect of the machine, since the height of the jet may be much increased by making it moderately contracted, and a little conical rather than cylindrical. When the air vessel is half filled with water, the height of such a jet will be about 30 feet, when two thirds filled, about 60, the height being always nearly proportional to the degree of condensation of the air, or to the excess of its density above that of the surrounding atmosphere. Sometimes a double forcing pump, or fire engine, is formed by the alternate rotatory motion of a flat piston within a cylindrical barrel; the axis of its motion coinciding with that of the barrel, and the barrel being divided by a partition into two cavities, which are filled and emptied in the same way as the separate barrels of the common fire engine. The mechanical advantage of this machine is nearly the same as that of the more usual constructions, but it appears to be somewhat more simple than a common engine of equal force. The partition may be extended throughout the diameter of the cylinder, the opposite pairs of cavities being made to communicate with each other, and thus both sides of the piston may be employed at once. (Plate XXIII. Fig. 313.)

A piston placed in a similar manner has sometimes been made to revolve continually, and to force the water through a pipe by means of a slider or a spring, which intercepts its passage in any other direction. Machines of this kind have been invented and reinvented, by Ramelli, Cavalleri, Amontons, Prince Rupert, Dr. Hooke, Mr. Bramah, and Mr. Gwynn. Mr. Gwynn's engine, which has been employed in many cases with considerable success, consists of a piston or roller nearly elliptical, well fitted to the cylinder within which it revolves, with a valve pressed lightly against it by a spring, which causes a considerable part of the water contained in the cylinder to be forced in each revolution into the pipe: the whole machine is made of brass; the spring requires very little force, for the pressure of the water on the valve keeps it always close to the roller, and the friction arising from this cause is even an objection to the machine. The stream, although never wholly intermitted, is, however, by no means uniform in its velocity. (Plate XXIII. Fig. 314 . . 317.)

The pipes, through which water is raised, by pumps of any kind, ought to be as short and as straight as possible; thus, if we had to raise water to a height of 20 feet, and to carry it to a horizontal distance of 100 by means of a forcing pump, it would be more advantageous to raise it first vertically into a cistern 20 feet above the reservoir, and then to let it run along horizontally, or find its level in a bent pipe, than to connect the pump immediately with a single pipe carried to the place of its destination. And for the same reason a sucking pump should be placed as nearly over the well as possible, in order to avoid a loss of force in working it. If very small pipes are used, they will much increase the resistance, by the friction which they occasion.

Water has been sometimes raised by stuffed cushions, or by oval blocks of wood, connected with an endless rope, and caused by means of two wheels or drums, to rise in succession in the same barrel, carrying the water in a continual stream before them; but the magnitude of the friction of the cushions appears to be an objection to this method. From the resemblance of the apparatus to a string of beads, it has been called a bead pump, or a paternoster work. When flat boards are united by chains, and employed instead of these cushions, the machine may be denominated a cellular pump; and in this case the barrel is usually square, and placed in an inclined position, but there is a considerable loss from the facility with which the water runs back. The chain pump generally used in the navy is a pump of this kind, with an upright barrel, through which leathers, strung on a chain, are drawn in constant succession; these pumps are only employed, when a large quantity of water is to be raised, and they must be worked with considerable velocity in order to produce any effect at all. Mr. Cole has improved the construction of the chain pump, so as materially to increase the quantity of water raised by it. (Plate XXIII. Fig. 318.)

It is frequently necessary to procure alternate motion in pumps by means of wheelwork, and for this purpose the application of a crank is the most usual and perhaps the best method. Provided that the bar by which it acts be sufficiently long, very little will be lost by the obliquity of its situation, and it is easy, by means of rollers, or of a compound frame, to confine the head of the pump rod to a rectilinear motion. When any other mode is employed, it must be remembered that the motion of the pump rod ought always to be slower at the beginning of each alternation, since a considerable part of the force is consumed in setting the water in motion, especially where the pipe is long, and the velocity considerable. But it may happen that, from the nature of hydraulic pressure under other circumstances, the resistance may be nearly equal throughout the stroke: for example, when the motion of the piston is slow in comparison of that of the water in the pipe, or when the force employed in producing velocity is inconsiderable, in comparison with that which is required for counteracting the pressure. In such cases it may sometimes be eligible to employ inclined surfaces, of such forms as are best adapted to communicate the most advantageous velocity to the pump rod by their pressure on a roller, which may be confined to its proper direction by the same means as when a crank is used. (Plate XIV. Fig. 184 . . 187.)

The Chinese work their cellular pumps, or bead pumps, by walking on bars which project from the axis of the wheel or drum that drives them, and whatever objection may be made to the choice of the machine, the mode of communicating motion to it must be allowed to be advantageous.

Pumps have sometimes been worked by means of the weight of water acting within a barrel, which resembles a second pump placed in an inverted position. The only objection to the machine appears to be the magnitude of the friction, and even this inconvenience may perhaps be inconsiderable. The invention is by no means modern, but it is best known in Germany under the name of Höll's machine, and it has been introduced into this country by Mr. Westgarth and Mr. Trevithick. A chain pump, or a series of buckets, may also be applied, in a manner nearly similar, to the working of machinery of any kind. (Plate XXIII. Fig. 319.)

The mediation of a portion of air is employed for raising water, not only in the spiral pump, but also in the air vessels of Schemnitz. A column of water, descending through a pipe into a closed reservoir, full of air, obliges the air to act, by means of a pipe, leading from the upper part of the reservoir or air vessel, on the water in a second reservoir, at any distance either below or above it, and forces this water to ascend through a third pipe to any height less than that of the first column. The air vessel is then emptied, and the second reservoir filled, and the whole operation is repeated. The air must, however, acquire a density equivalent to the pressure, before it can begin to act; so that if the height of the columns were 34 feet, it must be reduced to half its dimensions before any water would be raised; and thus half of the force would be lost; in the same manner, if the height were 68 feet, two thirds of the force would be lost. But where the height is small, the force lost in this manner is not greater than that which is usually spent in overcoming friction and other imperfections of the machinery employed; for the quantity of water, actually raised by any machine, is not often greater than half the power which is consumed. The force of the tide, or of a river rising and falling with the tide, might easily be applied by a machine of this kind, to the purposes of irrigation. (Plate XXIII. Fig. 320, 321.)

The fountain of Hero precisely resembles in its operation the hydraulic vessels of Schemnitz, which were probably suggested to their inventor by the construction of this fountain. The first reservoir of the fountain is lower than the orifice of the jet; a pipe descends from it to the air vessel, which is at some distance below, and the pressure of the air is communicated, by an ascending tube, to a third cavity, containing the water which supplies the jet. Many other hydraulic and pneumatic instruments, intended for amusement only, and some of them of much more complicated structure, are also described in the works of Hero. (Plate XXIII. Fig. 322.)

The spontaneous vicissitudes of the pressure of the air, occasioned by changes in the weight and temperature of the atmosphere, have been applied, by means of a series of reservoirs, furnished with proper valves, to the purpose of raising water by degrees to a moderate height. But it seldom happens that such changes are capable of producing an elevation in the water of each reservoir of more than a few inches, or at most a foot or two, in a day: and the whole quantity raised must, therefore, be very inconsiderable.

The momentum of a stream of water, flowing through a long pipe, has also been employed for raising a small quantity of water to a considerable height.

The passage of the pipe being stopped by a valve, which is raised by the stream, as soon as its motion becomes sufficiently rapid, the whole column of fluid must necessarily concentrate its action almost instantaneously on the valve; and in this manner it loses, as we have before observed, the characteristic property of hydraulic pressure, and acts as if it were a single solid; so that, supposing the pipe to be perfectly elastic, and inextensible, the impulse must overcome any pressure, however great, that might be opposed to it, and if the valve open into a pipe leading to an air vessel, a certain quantity of the water will be forced in, so as to condense the air, more or less rapidly, to the degree that may be required, for raising a portion of the water contained in it, to any given height. Mr. Whitehurst appears to have been the first that employed this method: it was afterwards improved by Mr. Boulton; and the same machine has lately attracted much attention in France under the denomination of the hydraulic ram of Mr. Montgolfier. (Plate XXIII. Fig. 323.)

LECTURE XXIX.

ON PNEUMATIC MACHINES.

PNEUMATIC machines are such as are principally dependent, in their operation, upon the properties of elastic fluids; they may be calculated either for diminishing or increasing their density and pressure, as air pumps and condensers; or for directing and applying their force, as bellows, ventilators, steam engines, and guns.

The density and pressure of the air may be diminished, or the air may be perfectly or very nearly withdrawn from a given space, either by means of a column of mercury, or by the air pump. The ancients sometimes exhausted a vessel imperfectly by the repeated action of the mouth, and preserved the rarefaction by the assistance of a stopcock. The Torricellian vacuum, obtained by inverting a receiver filled with mercury, and furnished with a descending tube at least 30 inches long, is the most perfect that can be procured; but there is generally a portion of air adhering to the vessels, and mixed with the mercury, which may often be considerably diminished by agitation, but can only be completely expelled by boiling the mercury for some time in the vessel and its tube, previously to their inversion. (Plate XXIV. Fig. 324.)

The construction of an air pump greatly resembles that of a common sucking pump for raising water; but the difference in the operation to be performed requires a difference in several particular arrangements. The objects are, to rarefy or exhaust the air as completely, as expeditiously, and as easily, as possible. In order that the exhaustion may be complete, it is necessary that no air remain in the barrel when the valve is opened, and that the process be very long continued. For, supposing all the parts of an air pump to be perfectly well fitted, and the exhaustion to be carried on for any length of time, the limit of its perfection will be a rarefaction expressed by the proportion of the air remaining in the barrel, when the piston is down, to the whole air that the barrel is capable of containing; for such will be the rarity of the air in the barrel when the piston is raised. It becomes, therefore, of consequence to lessen the quantity of this residual air as much as possible; and at the same time to take care that the valve may be capable of being accurately closed and easily opened, or that a stopcock may be occasionally substituted for it, which may be opened and shut by external force, when the elasticity of the air remaining is too small to lift the valve. In pumping water from a well, we raise an equal quantity at each stroke, but in the air pump, we withdraw at most only equal bulks of the air differently rarefied, so that the quantity extracted is continually diminished as the operation proceeds. Thus, if one tenth of the air were exhausted by the first stroke, only nine tenths as much, that is, one tenth of the remainder, would be drawn out by the second; hence, in order that the process may be expeditious, it is of importance to have the barrel as large as possible in proportion to the receiver. In cases where the presence of aqueous vapour would be of no consequence, the exhaustion might be made very rapidly by filling the whole apparatus with water, which was the method first employed by Otto von Guericke, the inventor of the modern air pump.

In order to lessen the labour of the operation, two barrels may be employed, and so connected as to work alternately; in this manner the pressure of the atmosphere, acting on both pistons at once, opposes no resistance to their motion in either direction. In Smeaton's pump a single barrel has nearly the same advantage, the rod of the piston working in a collar of leathers with oil, and the air being excluded from the upper part of the barrel by a valve, through which the air passes when the piston is raised near to the top; so that in the descent of the piston there is a vacuum above it, and the air below opens the valve much earlier, and passes more completely through it, than in the common air pump; and the piston is only exposed to the whole pressure of the atmosphere during the discharge of the air through the upper valve. (Plate XXIV. Fig. 325.)

That the air is really removed by the operation of the air pump, may be demonstrated by various experiments, which show the absence of its resistance, of its buoyant effect, and of its pressure; such are the descent of a guinea and a feather at the same time, the equal duration of the motion of two fly wheels, with their plates placed in different directions, the preponderance of the larger of two bodies which balance each other in the open air, the descent of mercury or of water in a barometrical tube, the playing of a fountain urged by the expansion of a portion of confined air, and the ebullition of ether, or of water moderately warm. (Plate XXIV. Fig. 326, 327.)

The degree of perfection of the vacuum formed by the air pump, or the rarity of the air remaining in the receiver, is measured by gages of different kinds. The simplest gage is a short tube filled with mercury, and inverted in a bason of the same fluid: in this the mercury begins to descend when the elasticity of the air becomes diminished in the proportion of the height of the gage to that of the barometer; but on account of the capillary attraction of the particles of mercury for each other, there is a depression within the tube, differing in quantity according to its magnitude, which renders it difficult to observe the exact situation of the surface when the height of the column is very small, although, if that height were correctly ascertained, the allowance to be made for the depression might easily be calculated. It is, however, more usual to employ the long barometer gage, in which the pressure is removed from the upper surface of the column of mercury in proportion as the exhaustion proceeds, and the height to which it is raised by the pressure of the external atmosphere, is compared with that of a common barometer, the difference always indicating the density of the air left in the receiver. Sometimes also a bent tube is employed instead of the short gage, the difference of the height in its two branches indicating the pressure; and this instrument has the advantage of requiring no correction on account of capillary attraction, since the depressions of the two columns exactly counterbalance each other. But in all these cases the mercury must be well boiled in the tubes; and in the bent tube, or siphon gage, the operation is somewhat difficult.

The pressure indicated by a gage of any kind depends on the elasticity of the whole of the fluid remaining in the receiver; but this fluid is not always atmospheric air alone. In all common temperatures, water, and many other liquids, have the property of emitting a vapour which possesses a very sensible degree of elasticity; so that if either water, or any moist substance, be present under the receiver, it will be impossible to procure a total absence of pressure, the short mercurial gage commonly standing at the height of at least half an inch, in the best pumps. Hence, the vacuum may be made more perfect when the receiver is ground to the plate of the pump, with the interposition of an unctuous substance, than when it is placed on wet leather, as it has sometimes been usual to do. The quantity of atmospherical or incondensable air actually existing in the receiver, whether mixed with vapour or alone, is measured by means of Smeaton's pear gage, which is left open under the receiver during the exhaustion, and having its orifice then plunged, by means of a wire passing through a collar of leather, into a bason of mercury, receives, upon the readmission of the air, as much of the mercury as is sufficient to fill it, leaving only in a tube rising from the neck of the gage, the small quantity of air which had before filled the whole cavity, so that from the space occupied by this air, compared, by means of previous measurements, with the capacity of the gage, the degree of exhaustion of the pump with respect to air may be estimated. It is said that in an air pump of Cuthbertson's construction, such a rarefaction has been procured that the air sustained but one hundredth part of an inch of mercury, that is, it was expanded to nearly 3000 times its original bulk. The pear gage often indicates a much more complete exhaustion, but this measurement relates only to the quantity of dry air present. (Plate XXIV. Fig. 328.)

A condenser is the reverse of an air pump; and sometimes the same machine is made to serve for both purposes; but the condenser requires more strength than the air pump, and less delicacy. The gage for measuring the degree of condensation is a small portion of air contained in a graduated cylindrical tube, the space that it occupies being indicated by a drop of mercury which confines it. (Plate XXIV. Fig. 329.)

Diving bells were formerly supplied with air by means of barrels let down continually from the surface of the water, and taken into the bell by the divers; but it is now more usual to force down a constant stream by means of a pump resembling a condenser in its construction and operation: the heated air is suffered to escape by a stopcock at the upper part of the bell. When proper care is taken to lower the machine gradually, the diver can support the pressure of an atmosphere of twice or thrice the natural density. It would be advisable that every diver should be provided with a float of cork, or with a hollow ball of metal, which might be sufficient to raise him slowly to the surface, in case of any accident happening to the bell; for want of a precaution of this kind, several lives have been lost from confusion in the signals. (Plate XXIV. Fig. 330.)

Bellows are commonly made of boards connected by leather, so as to allow of alternately increasing and diminishing the magnitude of their cavities, the air being supplied from without by a valve. The blast must be intermitted while the cavity is replenished; and in order to avoid this inconvenience, a second cavity is sometimes added, and loaded with a weight, which preserves the continuity of the stream. If great uniformity be required in the blast, it will be necessary to take care that the cavity be so formed as to be equally diminished while the weight descends through equal spaces; but notwithstanding this precaution, there must always be an additional velocity while the new supply of air is entering from the first cavity. Sometimes the construction of the bellows resembles that of a forcing pump; and then, if the barrel is single, a second barrel, loaded with a weight, must be provided, in order to equalise the blast: or a vessel inverted in water, and either loaded or fixed, may supply the place of the second barrel. The first cavity may also be formed of a similar inverted vessel, suspended to a beam, so as to be moved up and down in the water, and such a machine is much used, in large founderies, under the name of hydraulic bellows. The quantity of water employed may be much diminished, and the operation expedited, by introducing, in the centre of the inverted vessel, a fixed solid, or an internal inverted vessel, capable of nearly filling up the cavity of the moveable vessel when it is in its lowest position, so that the water only occupies a part of the interstice between the vessels. (Plate XXIV. Fig. 331.)

The gasometer differs little from the hydraulic bellows, except that it is provided with stopcocks instead of valves, and the moveable cylinder is supported by a counterpoise, which, in the best kind, acts on a spiral fusee, calculated to correct the difference of pressure arising from the greater or less immersion of the cylinder. (Plate XXIV. Fig. 332.)

A shower of water, or even an irregular stream, being conveyed through a descending pipe, plunged into the water of a reservoir, a large quantity of air is carried down with the water, and rises to the upper part of an inverted vessel which surrounds the pipe, whence it may be conveyed through another pipe, in a rapid stream, for any required purpose; and the water escapes at the bottom of the air vessel into the general reservoir, from the surface of which it runs off. The quantity of air supplied by these shower bellows is, however, small. (Plate XXIV. Fig. 333.)

The velocity of the blast produced by any pressure, forcing the air through a pipe of moderate dimensions, may readily be determined from the height of a column of air equivalent to the pressure. Thus, if the hydraulic bellows were worked with a constant pressure of 4 feet of water, the velocity would correspond to a height of about 3300 feet, and the air would move through a space of about 460 feet in a second. But in this calculation no allowance is made for any of the causes which diminish in all cases the discharge of fluids, and the velocity actually observed is only five eighths as great as that which corresponds to the height; that is, in the example here given, 285 feet in a second, when the air escapes through a small orifice; but when it moves in a pipe, about three fourths, or 345 feet. If the pipe were of considerable length, there would also be a diminution of velocity on account of friction. In some bellows actually employed, a pressure equivalent to 9 feet of water is applied, and in this case the velocity must be about 500 feet in a second.

Bellows may be used for the ventilation of a mine, either by forcing air into it, or by drawing it out through a pipe connected with the valve. The wind may also be received by the mouth of a tube a little conical, and may be made to cause a current where it is conveyed; such an instrument is sometimes called a windsail, or a horse head. It has been proposed to draw the air up through a pipe by the lateral friction of a current of air received by such a funnel, but the effect would probably be too small to be of much practical utility.

A corn fan is turned by the hand, or by machinery; its simplest operation is to cause a portion of air to revolve with it, and to create a wind in the direction of its circumference. But when a small fan is made to revolve with great rapidity, as in Papin's Hessian bellows, the centrifugal force causes the air admitted at the centre to rush towards the circumference, and to pass with great velocity through a pipe inserted there. The common ventilator placed in windows, which revolves in the same manner as a smoke jack, in consequence of the impulse of a current of air, serves only to retard a little the entrance of that current, to disperse it in some measure in different directions, and to prevent any sudden increase of the intensity of the draught; but it has little or no power of acting on the air, so as to prevent the decrease of the velocity of the current. (Plate XXIV. Fig. 334.)

The operation of heat affords us also a very effectual mode of ventilation. Its action upon air at common temperatures occasions an expansion of about $\frac{1}{500}$ for every degree that Fahrenheit's thermometer is raised; the air becomes in the same proportion lighter, and the fluid below it is consequently relieved from a part of its weight: the pressure of the surrounding atmosphere, therefore, preponderates, and the lighter column is forced upwards. When the shaft of a mine communicates with the external air at two different heights, there is generally a sufficient ventilation from the difference of the temperatures of the air in the shaft, and of the surrounding atmosphere: for the temperature of the earth is nearly invariable, it therefore causes the air in the shaft to be warmer in winter than the external air, and colder in summer; so that there is a current upwards in winter, and downwards in summer; and in the more temperate seasons, the alternations take place in the course of the day and night. For a similar reason, there is often a current down a common chimney in summer; but when the fire is burning, the whole air of the chimney is heated, and ascends the more rapidly as the height is greater. It would be easy, from the principles of hydraulics, if the length of the chimney, and the mean temperature of the air in it were given, to calculate the velocity of the draught; thus, if the height of the chimney were 50 feet, and the air contained in it 10 degrees hotter than the external air, the expansion would be one fiftieth, and the pressure of the whole column being diminished one fiftieth, the difference would be equivalent to a column of one foot in height, and such a column would represent the pressure causing the draught, which might, therefore, be expected to have a velocity of 6 feet in a second. If the room were perfectly closed, the air contained in it would by degrees become so much lighter than the external air, as would be equivalent to one foot of the height of the column causing the pressure, and the current would then stop; if fresh air were gradually admitted by a small orifice, the current would again go on, but the air in the room would always remain somewhat rarer than the external atmosphere, unless a fresh supply were admitted through ample openings.

The object of a chimney is not so much to ventilate the room, as to provide a sufficiently rapid supply of air for maintaining the process of combustion, and to carry off the products of that process: hence, it is desirable to allow as little air as possible to enter the chimney without passing through the fire; and this is the best general mode of avoiding smoky chimnies. For wind furnaces, the flue should be as equable as possible, throughout its height, or widened rather than contracted in its ascent, and free from any considerable angles.

The ascent of a balloon is an effect of the same kind as that of air in a chimney, and arises sometimes from the same cause, when the air within it is expanded by heat; but more commonly from the greater rarity of hydrógen gas, with which the balloon is filled, and which, when pure, is only one thirteenth as heavy as atmospherical air, but as it is commonly used, about one fifth or one sixth.

The steam engine is perhaps the most magnificent effort of mechanical power; it has undergone successive changes, and it appears to have been brought very near to perfection by the improvements of Mr. Watt. The pressure of steam was first applied by the Marquis of Worcester, and afterwards by Savery, to act immediately on the surface of water contained in a close vessel, and this water was forced, by the elasticity of the steam, to ascend through a pipe. But a great degree of heat was required for raising water to any considerable height by this machine: for, in order that steam may be made capable of supporting, in addition to the atmospherical pressure, a column of 34 feet of water, its temperature must be raised to 248° of Fahren-

4

heit, and for a column of 68 feet, to 271°; such a pressure, also, acting on the internal surface of the vessels, made it necessary that they should be extremely strong; and the height to which water could be drawn up from below, when the steam was condensed, was limited to 33 or 34 feet. A still greater objection was, however, the great quantity of steam necessarily wasted, on account of its coming into contact with the cold water and the receiver, the surfaces of which required to be heated to its own temperature, before the water could be expelled; hence a tenth or a twentieth part only of the steam produced could be effective; and there would probably have been a still greater loss, but for the difficulty with which heat is conducted downwards in fluids. These inconveniences were in great measure avoided in Newcomen's engine, where the steam was gradually introduced into a cylinder, and suddenly condensed by a jet of water, so that the piston was forced down with great violence by the pressure of the atmosphere, which produced the effective stroke: this effect was, however, partly employed in raising a counterpoise, which descended upon the readmission of the steam, and worked a forcing pump in its return, when water was to be raised. The condensation, although rapid, was, however, neither instantaneous, nor complete, for the water injected into the cylinder had its temperature considerably raised by the heat emitted by the steam during its condensation; it could only reduce the remaining steam to its own temperature, and at this temperature it might still retain a certain degree of elasticity; thus, at the temperature of 180° steam is found to be capable of sustaining about half the pressure of the atmosphere, so that the depression of the piston must have been considerably retarded by the remaining elasticity of the steam, when the water was much heated. The water of the jet was let off when the piston was lowest, and was afterwards pumped up to serve the boiler, as it had the advantage of being already hot. This engine, with Beighton's apparatus for turning the cocks, was until lately in general use, and it is still very frequently employed. In this, as well as in other steam engines, the boiler is furnished with a safety valve, which is raised when the force of the steam becomes a little greater than that of the atmospheric pressure; and it is supplied with water by means of another valve, which is opened, when the surface of the water within it falls too low, by the depression of a block of stone, which is partly supported by the water. (Plate XXIV. Fig. 335, 336.)

The cylinder of Beighton's machine is necessarily much cooled by the admission of the jet, and by exposure to the air. Mr. Watt has avoided this inconvenience by performing the condensation in a separate vessel, into which a small jet is flowing without intermission; and by introducing the steam alternately above and below the piston, the external air is wholly excluded; the piston rod working in a collar of leathers, so that the machine has a double action, somewhat resembling that of Lahire's double pump; and the stroke being equally effectual in each direction, the same cylinder, by means of an increased quantity of steam, performs twice as much work as in the common engine. We might also employ, if we thought proper, a lower temperature than that at which water usually boils, and work in this manner with a smaller quantity of steam; but there would be some difficulty in completely preventing the insinuation of the common air. On the other hand, we may raise the fire so as to furnish steam at 220° or more, and thus obtain a power somewhat greater than that of the atmospheric pressure; and this is found to be the most advantageous mode of working the engine; but the excess of the force above the atmospheric pressure cannot be greater than that which is equivalent to the column of water descending to supply the boiler, since the water could not be regularly admitted in opposition to such a pressure. The steam might also be allowed to expand itself within the cylinder for some time after its admission, and in this manner it appears from calculation that much more force might be obtained from it than if it were condensed in the usual manner as soon as its admission ceases; but the force of steam thus expanding is much diminished by the cold which always accompanies such an expansion, and this method would be liable to several other practical inconveniences.

The peculiarities of Mr. Watt's construction require also some other additional arrangements; thus, it is necessary to have a pump, to raise not only the water out of the condenser, but also the air, which is always extricated from the water during the process of boiling. If the water employed has been obtained from deep wells or mines, it contains more air than usual, and ought to be exposed for some time in an open reservoir before it is used; for it appears that the quantity of air, which can be contained in water, is nearly in proportion to the pressure to which it is subjected. The admission of the steam into the cylinder is regulated by the action of a double revolving pendulum. The piston is preserved in a situation very nearly vertical by means of a moveable parallelogram, fixed on the beam, which corrects its curvilinear motion by a contrary curvature. In the old engines, a chain working on an arch was sufficient, because there was no thrust upwards. When a rotatory motion is required, it may be obtained either by means of a crank, or of a sun and planet wheel, with the assistance of a fly wheel; this machinery is generally applied to the opposite end of the beam; but it is sometimes immediately connected with the piston, and the beam is not employed. The cylinder is usually inclosed within a case, and the interval is filled with steam, which serves to confine the heat very effectually. (Plate XXIV. Fig. 337.)

The steam engines of Messrs. Boulton and Watt are said to save three fourths of the fuel formerly used; and it appears that only one fourth of the whole force of the steam is wasted. Such a machine, with a thirty inch cylinder, performs the work of 120 horses, working 8 hours each in the day.

When the water producing the condensation is to be raised from a great depth, a considerable force is sometimes lost in pumping it up. Hence, Mr. Trevithick has attempted, as Mr. Watt had indeed long before proposed, to avoid entirely the necessity of condensation, by employing steam at a very high temperature, and allowing it to escape, when its elasticity is so reduced by expansion, as only to equal that of the atmosphere: the air pump is also unnecessary in this construction, and for a small machine, it may perhaps succeed tolerably well. But there must always be a very considerable loss of steam, and although the expense of fuel may not be increased quite in the same proportion as the elasticity of the steam, yet the difference is probably inconsiderable. A great number of less essential alterations have also been made in Mr. Watt's arrangements by various engineers, but they have generally been calculated either for obtaining some subordinate purpose of convenience, or for imposing on the public by a fallacious appearance of novelty. (Plate XXIV. Fig. 338.)

The force of steam, or of heated vapour, is probably also the immediate agent in the astonishing effects produced by the explosion of gunpowder. The initial elasticity of the fluid by which a cannon ball is impelled, appears, from Bernoulli's calculation, to be at least equal to ten thousand times the pressure of the atmosphere, and upon the most moderate computation, from Count Rumford's experiments, to be more than three times as great as this. The quantity of moisture, or of water of crystallization, contained in the powder, is certainly too small to furnish steam enough for so great an effect. We have no reason to suppose that the elasticity of a given quantity of any aeriform fluid or vapour is increased more than about one five hundreth for each degree of Fahrenheit that its temperature is elevated; and if we suppose the heat to be raised to more than 5000 degrees, the force of each grain of water converted into steam will only be increased tenfold; so that if the elasticity were 40 thousand times as great, the density must be 4 thousand times as great as that of ordinary steam, and the whole space must be filled with an aqueous vapour almost twice as dense as water itself. It is, therefore, probable that some other parts of the materials assume, together with the water, the state of vapour, and possess in this form a much greater elasticity than that of the steam: for the quantity of fluids permanently elastic, which are extricated, must be allowed to be wholly inadequate to the effect.

The force of fired gunpowder is found to be very nearly proportional to the quantity employed; consequently, if we neglect the consideration of the resistance of the atmosphere, the square of the velocity of the ball, the height to which it will rise, and its greatest horizontal range, must be directly as the quantity of powder, and inversely as the weight of the ball. Count Rumford, however, found that the same quantity of powder exerted somewhat more force on a large ball than on a smaller one.

The essential properties of a gun are to confine the elastic fluid as completely as possible, and to direct the motion of the bullet in a rectilinear path; and hence arises the necessity of an accurate bore. The advantage of a rifle barrel is principally derived from the more perfect contact of the bullet with its cavity; it is also supposed to produce a rotation round an axis in the direction of its motion, which renders it less liable to deviations from its path on account of irregularities in the resistance of the air. The usual charge of powder is one fifth or one sixth of the weight of the ball, and for battering one third. When a 24 pounder is fired with two thirds of its weight of powder, it may be thrown almost four miles, the resistance of the air reducing the distance to about one fifth of that which it would describe in a vacuum.

Bullets of all kinds are usually cast in separate moulds: shot are granulated by allowing the lead, melted with a little arsenic, to pass through perforations in the bottom of a vessel, and to drop in a shower into water. The patent shot fall in this process through a height of 120 feet: the roundest are separated by rolling them down an inclined plane slightly grooved, those which are of an irregular form falling off at the sides.

Condensed air may also be employed for propelling a bullet by means of an air gun, an instrument of considerable antiquity, but of little utility. It is obvious that no human force can so far increase the density of air as to make its elasticity at all comparable to that of the fluid evolved by fired gunpowder, and even if it were reduced to such a state, its effects would still be far inferior to those of gunpowder; for the utmost velocity, with which it could expand itself, would not exceed 1300 feet in a second, and it would, therefore, be incapable of imparting to a ball a velocity even as great as this, while the vapour of gunpowder impels a heavy ball with a velocity of more than 2000 feet in a second. When, however, it is considered that by far the greatest part of such a velocity as this is uselessly employed, and that the mechanical power which is practically obtained from gunpowder is much more expensive than an equivalent exertion of any of the ordinary sources of motion, it must be allowed that the force of condensed air may possibly be applied in some cases, with advantage, as a substitute for that of gunpowder. (Plate XXIV. Fig. 339.)

LECTURE XXX.

ON THE HISTORY OF HYDRAULICS AND PNEUMATICS.

NOTWITHSTANDING a few observations and experiments made by Aristotle and his predecessors, the properties of fluids had scarcely been the subjects of much accurate investigation before the time of Archimedes. The progress, which the science of hydrostatics in particular made under this eminent mathematician, does the highest honour to his genius and penetration. His treatise on floating bodies, although the theorems which it contains are not so general as they have been rendered since the late improvements in the methods of calculation, still affords us instances of very ingenious determinations of the equilibrium of floating bodies of different forms, grounded on the true principles of the opposition of the general directions of the weight of the body and of the pressure of the fluid ; and in this manner he has shown in what cases the equilibrium of conical and conoidal solids will be stable, and in what cases unstable. Archimedes was the inventor of the mode of measuring the bulk of a solid by immersing it in a fluid: to us, indeed, there appears to have been little difficulty in the discovery, but the ancients thought otherwise. Vitruvius observes that this invention indicates a degree of ingenuity almost incredible. The philosopher himself is said to have valued it is so highly, that when it first occurred to him, in a public bath, he hastened home in an ecstasy, without recollecting to clothe himself, in order to apply it to the determination of the specific gravity of Hiero's crown, and to the detection of the fraud of the maker, who had returned the crown equal in weight to the gold that was given him, but had adulterated it with silver, and imagined, that on account of the complicated form of the work, which rendered it almost impossible to determine its bulk by calculation, he must infallibly escape conviction. The hydrometer, which has sometimes been attributed to Hypatia, a learned Greek lady of

Constantinople, is mentioned by Fannius, an early writer on weights and measures, and is ascribed by him to Archimedes.

The forcing pump, or rather the fire engine, was the invention of Ctesibius of Alexandria, the greatest mechanic of antiquity after Archimedes. He is also said to have invented the clepsydra, for the hydraulic measurement of time, and Philo informs us that he constructed an air gun, for propelling a stone, or rather a ball, by means of air, previously condensed by a syringe. The ball was not immediately exposed to the action of the air, but was impelled by the longer end of a lever, while the air acted on the shorter. Ctesibius is said to have been the son of a barber, and to have had his attention turned to mechanics and pneumatics, by being employed to fit a shutter, with a counterpoise sliding in a wooden pipe, for his father's shop window.

Hero was a cotemporary, and a scholar of Ctesibius; he describes, in his treatise on pneumatics, a number of very ingenious inventions, a few of which are calculated for utility, but the greater part for amusement only; they are principally siphons variously concealed and combined, fountains, and water organs, besides the syringe and the fire engine. The description of this engine agrees precisely with the construction which is at this day the most usual; it consists of two barrels, discharging the water alternately into an air vessel; and it appears from Vitruvius, that this was the original form in which Ctesibius invented the pump. Hero supposes the possibility of a vacuum in the intervals of the particles of bodies, observing that without it no body could be compressible; but he imagines that a vacuum cannot exist throughout a perceptible space, and thence derives the principle of suction. The air contained in a given cavity may be rarefied, he says, by sucking out a part of it, and he describes a cupping instrument, which approaches very nearly to the nature of an imperfect air pump. (Plate XXIV. Fig. 324.)

After the time of Ctesibius and Hero, the science of hydraulics made little further progress, until the revival of letters. The Romans had water mills in the time of Julius Caesar, which are described by Vitruvius; and it appears that their aqueducts were well built, and their waterpipes well arranged. Pipes of lead were, however, less frequent than at present, from an apprehension of the poisonous quality of the metal, which was not wholly without foundation. Some say that the ancients had no chimnies, but whatever may be the authorities, the opinion is extremely improbable.

It was in the middle ages that navigable canals began to be considerably multiplied, first in China, and afterwards in other parts of the world. The canal from the Trent to the Witham, which is the oldest in England, is said to have been dug in 1134. The date of the earliest windmills has been referred to the year 1299. The invention of gunpowder possesses perhaps an equal claim with the art of printing, to the honour of being considered as constituting the most marked feature, that distinguishes the character of ancient from that of modern times; its introduction must necessarily have tended to produce material alterations, and perhaps improvements, in the habits of nations and of individuals. It is said to have been known long since to the Chinese, and our countryman Roger Bacon was evidently acquainted with its properties; but it was not actually employed in Europe or in its neighbourhood till about the year 1330; and the earliest artillery appears to have been that which was used by the Moors, at the siege of Algesiras, in 1334. King Edward had four pieces of cannon at the memorable battle of Cressy, in 1346.

About the year 1600, Galileo made the important discovery of the effects of the weight and pressure of the atmosphere, in the operation of suction, and in various other phenomena. Before his time, it was generally supposed that water was raised by a sucking pump, on account of the impossibility of the existence of a vacuum: if, however, a vacuum had been impossible in nature, the water would have followed the piston to all heights, however great, but Galileo found that the height of its ascent was limited to about 34 feet; and concluded that the weight of a column of this height was the measure of the magnitude of the atmospherical pressure. His pupil Torricelli afterwards confirmed the explanation, by showing that a column of mercury was only supported when its weight was equal to that of a column of water standing on the same base; hence the vacuum obtained by means of mercury is often called the Torricellian vacuum. Torricelli corrected also, in 1644, the mistake of Castelli respecting the quantities of water discharged by equal orifices, at different distances below the surface of the water in the reservoir. Castelli's experiments, made about 1640, were the first of the kind, and some of them really tended to the improvement of the science of hydraulics, but others appeared to show, that a double height of the head of water produced a double discharge. Torricelli's more accurate observations proved that a quadruple height was required in order to produce a double velocity; and his assertions were afterwards fully confirmed by Mariotte and by Guglielmini.

A little before the year 1654, Otto von Guericke, of Magdeburg, first constructed a machine similar to the air pump, by inserting the barrel of a fire engine into a cask of water, so that when the water was drawn out by the operation of the piston, the cavity of the cask remained nearly void of all material substance. But finding that the air rushed in between or through the staves of the cask, he inclosed a smaller cask in a larger one, and made the vacuum in the internal one more complete, while the intervening space remained filled with water; yet still he found that the water was forced into the inner cask through the pores of the wood. He then procured a sphere of copper, about two feet in diameter, and was exhausting it in the same way, when the pressure of the air crushed it, with a loud noise. This machine was more properly a water pump, than an air pump, but the inventor soon after improved his apparatus, and made all the experiments which are to this day the most usually exhibited with the air pump, such as the apparent cohesion of two exhausted hemispheres, the playing of a jet by means of the expansion of a quantity of air inclosed in a jar, the determination of the air's weight, and others of a similar nature. He also observed, that for very accurate experiments, the valve of the pump might be raised at each stroke by external force; and he particularly noticed the perpetual production of air, from the water that he generally employed, which caused an imperfection in the vacuum. An account of his experiments was first published in different works, by Caspar Schott, and afterwards by himself, in his book intitled Experimenta nova Magdeburgica, printed in 1672 at Amsterdam.

In the year 1658, Hooke finished an air pump for Boyle, in whose laboratory he was an assistant: it was more convenient than Guericke's, but the vacuum was not so perfect; yet Boyle's numerous and judicious experiments gave, to the exhausted receiver of the air pump, the name of the Boylean vacuum, by which it was long known in the greatest part of Europe. Hooke's air pump had two barrels, and with some improvements by Hauksbee, it remained in common use, until the introduction of Smeaton's pump, which, however, has not wholly superseded it. The theory of pneumatics was also considerably indebted to Hooke's important experiments on the elasticity of the air, which were afterwards confirmed and extended by Mariotte and Amontons, in France, by Hales in this country, and by Richmann at Petersburg.

About the same time the first steam engine was constructed, by the celebrated Marquis of Worcester. Hints of the possibility of such a machine had been given a hundred years before, by Matthesius, in a collection of sermons intitled Sarepta, and at a subsequent period by Brunau; but the Marquis of Worcester professes to have carried the project into full effect, as we are informed by his account of what he called a fire water work, which is one of his Century of Inventions, first published in 1663, and which is thus described: "I have taken a piece of a whole cannon, whereof the end was burst, and filled it three quarters full of water, stopping and screwing up the broken end, as also the touch hole; and making a constant fire under it, within 24 hours it burst, and made a great crack: so that having a way to make my vessels, so that they are strengthened by the force within them, and the one to fill after the other, I have seen the water run like a constant fountain stream forty foot high. One vessel of water, rarefied by fire, driveth up forty of cold water; and a man that tends the work is but to turn two cocks, that one vessel of water being consumed, another begins to force and refill with cold water, and so successively, the fire being tended and kept constant, which the self same person may likewise abundantly perform in the interim between the necessity of turning the said cocks." The machine was, however, not at that time practically introduced, and it was soon forgotten; Savery's engines were constructed in a manner precisely similar, some time before 1700; and it is uncertain whether he adopted the Marquis of Worcester's ideas, or reinvented a similar machine. About 1710, the piston and cylinder were invented by Newcomen, and with Beighton's apparatus for turning the cocks by its own motion, the engine remained nearly stationary for many years.

As early as the year 1667, the pressure of fluids in motion, and the resistance opposed by fluids at rest to the motion of solid bodies, were experimentally examined by Huygens, and some other members of the Parisian Academy. Pardies, whose works were published in 1673, attempted to determine, although upon some inaccurate suppositions, the effects of the wind on a ship's sails, under different circumstances. His principles were adopted by Renaud, who published a work on the subject in 1689; their imperfections were, however, soon after pointed out by Huygens, and by James Bernoulli; and in 1714, John Bernoulli published an extensive treatise on the mánoeuvres of ships, which at last compelled Renaud to submit to so many united authorities.

It must be confessed, that the labours of Newton added fewer improvements to the doctrines of hydraulics and pneumatics, than to many other departments of science; yet some praise is undeniably due both to his computations and to his experiments relating to these subjects. No person before Newton had theoretically investigated the velocity with which fluids are discharged, and although his first attempt was unsuccessful, and the method which he substituted for it in his second edition is by no means free from objections, yet either of the determinations may be considered in some cases as a convenient approximation; and the observation of the contraction of a stream passing through a simple orifice, which was then new, serves to reconcile them in some measure with each other. His modes of considering the resistance of fluids are far from being perfectly just, yet they have led to results which, with proper corrections, are tolerably accurate; and his determination of the oscillations of fluids, in bent tubes, was a good beginning of the investigation of their alternate motions in general.

The accurate experiments of Poleni were published in 1718: he has the merit of having first distinctly observed that the quantity of water, discharged by a short pipe, is greater than by a simple orifice of the same diameter;

although there is some reason to suppose that Newton was before acquainted with the circumstance.

In 1727, Mr. Bouguer received a prize from the academy of Paris for his essay on the masts of ships, which is said to be ingenious, but by no means practically useful. He was probably tempted by this encouragement to continue his application to similar studies, and about twenty years afterwards he published his valuable essay on the construction and manoeuvres of ships, which appears to have superseded all that had been done before respecting the subjects of his investigation.

The first researches of Daniel Bernoulli, concerning the properties and motions of fluids, bear also the date of 1727. This justly celebrated man was as happy in his application of mathematics to natural philosophy, as he was ready and skilful in his calculations. The greatest part of his hydraulic theorems are founded on the principle first assumed by Huygens, and called by Leibnitz the law of living or ascending force, which is confessedly only true where there is no loss of velocity, from the imperfection of the elasticity of the bodies concerned; for it is only with this limitation, that the motions of any system of bodies are always necessarily such, as to be capable of carrying the common centre of gravity to the height, from which it has descended, while the bodies have been acquiring their motions. This law of ascending force is of considerable utility in facilitating the solution of a great variety of problems; it is certain that mechanical power is always to be estimated by the product of the mass of a body into the height to which it is capable of ascending; and whatever objections may have been made to the employment of this product as the measure of the force of a body in motion, which is indeed an expression inconsistent with a correct definition of the term force, yet it must be confessed, on the other hand, that some of the best English mathematicians have fallen into material errors for want of paying sufficient attention to the general principle. Bernoulli estimates very justly in this manner the mechanical power of a variety of natural and artificial agents, and among the rest, he examines that of gunpowder; but from an accidental combination of errors, he states the force of a pound of gunpowder, as equivalent to the daily labour of 100 men, while in fact the effect which is actually obtained from two tons of powder is no greater than that which is here attributed to a pound. His calculations of the motions of fluids, in some very intricate cases, are very ingenious and satisfactory, and they are in general sufficiently confirmed by well imagined experiments. He examines the force of the wind acting on the sails of a windmill, but by another mistake in calculation, which Maclaurin has detected, of two angles which answer the conditions of the determination, he has taken the wrong one, and assigned that position of the sail as the most effectual, which produces absolutely no effect at all.

It may be objected to Bernoulli's calculations, that some of the circumstances which are necessarily neglected in them, produce a very material effect in the results of all experiments; but it must be allowed that the corrections required, on account of this unavoidable omission, may easily be deduced from simple experiments, and then applied to the most complicated cases. It is, however, a more material objection, that the fundamental law of the preservation of ascending force can only be adopted with certain limitations; thus, when a small stream passes through a large reservoir, Bernoulli is obliged to suppose the whole of its force consumed by the resistance which it meets. The immediate mode, in which the accelerating forces must be supposed to act, remains also wholly undetermined; and it was principally for this reason, that John Bernoulli attempted to substitute, for his son's calculations, a method of deducing the motions of fluids more immediately from the gravitation of their different parts. The peculiarity of John Bernoulli's mode of investigation consists in his imagining the weight of each individual particle to be transferred to the surface of the fluid, causing there a pressure in the direction of gravity; and he examines the manner in which this force must operate, in order to produce every acceleration which is required for the motion of fluids, in vessels of all imaginable forms.

Maclaurin, in his treatise of fluxions, investigated several of the properties of fluids in his usual concise and elegant manner. His remarks on the positions of the sails of windmills and of ships are peculiarly interesting: he added much to what had been done respecting the effects of the wind, and showed the possibility of arranging the sails of a ship in such a manner as to make her advance with a greater velocity than that of the wind itself. At that time, however, the science of hydraulics had been too little assisted by experiments to be capable of affording determinations of all questions which are of very frequent occurrence in practice. An application was made to Maclaurin, and at the same time to Desaguliers, a man of considerable eminence in the mechanical sciences, respecting the quantity of water that might be brought, by a train of pipes of certain dimensions, to the city of Edinburgh. The project was executed with a confidence founded on their opinions, but the quantity actually obtained was only about one sixth of Desaguliers's calculation, and one eleventh of Maclaurin's. At a still later period, the French Academicians were consulted respecting a great undertaking of a similar nature; and their report was such as to dissuade the projectors from making the attempt, which was consequently at the point of being abandoned, till a celebrated practical architect insisted, from a rough estimation, deduced from his general experience, that more than double the quantity assigned by the Academicians might be obtained; and the event justified his assertion.

The experiments and calculations of Robins, respecting the resistance of the air, and the operation of gunpowder, deserve to be mentioned with commendation on account of their practical utility; but he appears to have been less successful in his theoretical investigations than Daniel Bernoulli had been a few years before.

Dalembert attempted, in his treatise on the motions of fluids, which was published in 1744, to substitute, for the suppositions of John Bernoulli, a more general law, relating to all changes produced in the motions of a system of bodies, by their mutual actions on each other; but his calculations are more intricate, and less easily understood, than some others, which are capable of an application equally extensive. The late Professor Kaestner of Gottingen has defended Bernoulli against Dalembert's objections with some success, and has in many instances facilitated and extended Bernoulli's theory; but there is often a singular mixture of acuteness and prolixity in this author's works. By the side of an intricate and difficult fluxional calculation, he inserts a long string of logarithms for performing a simple multiplication; and in a work which comprehends the whole range of the mathematical sciences, he does not venture to determine the square root of 10 without quoting an authority.

About the same time, the profound Leonard Euler applied himself, with some success, to the examination of the motions of fluids, particularly as they are connected with the subjects of seamanship and naval architecture; but the investigations of Euler are in general more remarkable for mathematical address, than for philosophical accuracy and practical application; although his calculation of the resistance of the air to the motions of projectiles may be employed with considerable advantage by the gunner.

The beginning of the modern experimental improvements in hydraulics may perhaps be dated from the investigations of Smeaton respecting the effects of wind and water, which were published in the Philosophical Transactions for 1759. His observations are of material importance, as far as they are capable of immediate application to practice, but he has done little to illustrate their connexion with the general principles of mechanics. It was Mr. Borda that first derived from a just theory, about 10 years after, the same results, respecting the effects of undershot water wheels, as Smeaton had obtained from his experiments. Before this time, the best essay on the subject of water wheels was that of Elvius, published in 1742; his calculations are accurate and extensive; but they are founded, in great measure, on the imperfect suppositions respecting the impulse of a stream of water, which were then generally adopted.

Our countryman Mr. Watt obtained, in 1769, a patent for his improvements of the steam engine, which includes almost every essential change that has been made since the time of Beighton. On a subject so important, it cannot be superfluous to insert the words of the inventor, whose admirable application of the sciences, to practical purposes, most justly entitles him to a rank among philosophical mechanics, not inferior to that of Ctesibius and Dr. Hooke.

"My method of lessening the consumption of steam, and consequently fuel, in fire engines," says Mr. Watt, in his specification of his patent, "consists of the following principles. First, that vessel in which the powers of steam are to be employed, to work the engine, which is called the cylinder in common fire engines, and which I call the steam vessel, must, during the whole time the engine is at work, be kept as hot as the steam that enters it; first, by inclosing it in a case of wood, or any other materials that transmit heat slowly; secondly, by surrounding it with steam, or other heated bodies; and thirdly, by suffering neither water, nor any other substance colder than the steam, to enter or touch it during that time. Secondly, in engines that are to be worked wholly or partially by condensation of steam, the steam is to be condensed in vessels distinct from the steam vessels, or cylinders, although occasionally communicating with them; these vessels I call condensers; and, whilst the engines are working, these condensers ought at least to be kept as cold as the air in the neighbourhood of the engines, by application of water, or other cold bodies. Thirdly, whatever air or other elastic vapour is not condensed by the cold of the condenser, and may impede the working of the engine, is to be drawn out of the steam vessels, or condensers, by means of pumps, wrought by the engines themselves, or otherwise. Fourthly, I intend, in many cases, to employ the expansive force of steam to press on the pistons, or whatever may be used instead of them, in the same manner as the pressure of the atmosphere is now employed in common fire engines: in cases where cold water cannot be had in plenty, the engines may be wrought by this force of steam only, by discharging the steam into the open air after it has done its office. Fifthly, where motions round an axis are required, I make the steam vessels in form of hollow rings, or circular channels, with proper inlets and outlets for the steam, mounted on horizontal axles, like the wheels of a water mill; within them are placed a number of valves, that suffer any body to go round the channel in one direction only; in these steam vessels are placed weights, so fitted to them as entirely to fill up a part or portion of their channels, yet capable of moving freely in them by the means herein after mentioned or specified. When the steam is admitted in these engines, between the weights and the valves, it acts equally on both, so as to raise the weight to one side of the wheel, and, by the reaction of the valves, successively, to give a circular motion to the wheel, the valves opening in the direction in which the weights are pressed, but not in the contrary; as the steam vessel moves round, it is supplied with steam from the boiler, and that which has performed its office may either be discharged by means of condensers, or into the open air. Sixthly, I intend, in some cases, to apply a degree of cold, not capable of reducing the steam to water, but of contracting it considerably, so that the engines may be worked by the alternate expansion and contraction of the steam. Lastly, instead of using water to render the piston or other parts of the engines air and steam tight, I employ oils, wax, resinous bodies, fat of animals, quicksilver, and other metals, in their fluid state."

It is probable that the rotatory engines described by Mr. Watt, although they appear to produce some advantage in theory, will never be generally introduced, on account of the difficulty of constructing steam vessels so large, and of so complicated a form, as would be necessary, in order to give full effect to the machine. The term of this patent was prolonged by act of parliament until the year 1799; but although the legal privilege of the original manufacturers is expired, yet the superiority of their workmanship still gives their engines a decided preference.

Much of the labour of the later writers on hydraulics has been employed on the determination of the resistance of fluids to bodies of different forms, which move through them; a subject which derives great importance from its immediate application to the manoeuvres of ships. The most extensive experiments on these subjects were made by Bossut, and some other members of the Academy of Sciences. About the same time Don George Juan, a gentleman who had enjoyed the best possible opportunity for actual observation and practical study in serving with Ulloa, published at Madrid his Examen Maritimo, which appears to be the most ingenious and useful treatise on the theory and practice of seamanship that has yet appeared. But unfortunately his deductions, however refined and diversified, are principally founded on a mistaken theory respecting the effects of hydraulic pressure; since he tacitly assumes, in his fundamental proposition on the subject, that a double force, acting in a given small space, will produce a double velocity; while it is well known that in such circumstances a quadruple force would be required. Hence he derives some conclusions which indicate that the resistance must vary very materially at different depths below the surface of the water, and alleges in support of the assertion a few imperfect experiments of Mariotte and of his own, in which some accidental circumstances not noticed may easily have caused great irregularities. Mr. Prony, in his Architecture Hydraulique, appears to have followed Juan; and Professor Robison very justly observes, in speaking of this work, that if the pressure of the water alters the magnitude of the resistance at different depths, that of the atmosphere ought by no means to be omitted in the calculation. But if a more correct mathematician and mechanic would take the pains to model Juan's book anew, to correct his errors, and to adapt his modes of calculation to the laws of resistance previously deduced from accurate experiments rather than from theory, there is no doubt but that the work thus modified might essentially improve the science of seamanship. He alleges indeed that the results of his calculations are in almost every instance rigidly conformable to observation and experience, but it is probable that where such a coincidence really exists, it must be owing to some combination of errors compensating each other; and it is indeed very possible that his calculations, with all their errors, may approach nearer to the truth than the imperfect approximations which had been before employed. Juan has generally made use of the English weights and measures, on account of their convenience in computations respecting the descent of falling bodies and the impulse of water.

The works of Chapman and of Romme, upon various departments of seamanship, possess also considerable merit. These authors appear to have avoided the errors of Juan, but without entering so minutely into the detail of nautical operations as he has done.

The accurate experiments of Dr. Hutton and of Count Rumford on the force of fired gunpowder, and the resistance of the air, deserve to be mentioned as affording valuable materials to the speculative investigator, and useful information to the practical gunner. Robins had very erroneously supposed that the whole of the effects of gunpowder might be derived from the expansive force of fluids permanently elastic; but Vandelli soon after maintained a contrary opinion in the commentaries of Bologna, and Count Rumford has very satisfactorily shown the insufficiency of the agents considered by Robins, although he has been unsuccessful in attempting to deduce the whole force from the elasticity of aqueous vapour alone.

The theory of practical hydraulics, as affected by friction, may be considered as having been begun and completed by the highly meritorious labours of the Chevalier du Buat. He had some assistance in expressing the results of his experiments by means of general rules or formulae, and these, al-

4

though they agree sufficiently well with the experiments, have not always been reduced to the simplest and most convenient forms; nor have they been much improved either by Langsdorf or Eytelwein in Germany, or by Robison in this country, who have gone over nearly the same ground with each other, and have shown the way in which the results of Buat's investigations may be applied to a variety of cases, which occur in hydraulic architecture.

One of the latest inventions, which require to be mentioned in speaking of the history of pneumatics, is that of the aerostatic globe or air balloon. The suggestions of Lohmeier, of Albertus, and of Wilkins, respecting the various modes of passing through the air, had long remained disregarded as idle speculations; and Rosnier, who, in the 17th century, descended obliquely over some houses, by means of wings, was wholly unable to employ them in ascending. Dr. Black had exhibited in his lectures a bladder filled with hydrogen gas, and floating in the air by means of its smaller specific gravity, many years before Montgolfier conceived the idea of applying a similar machine to the elevation of human beings into the aerial regions. It was in 1783 that this project was first executed, and persons of a warm imagination were disposed to believe that the discovery would be of great importance to the convenience of mankind. But if we coolly consider the magnitude of the force with which the wind unavoidably impels a surface so large as that of a balloon, we shall be convinced of the absolute impossibility of counteracting it, in such a manner, as to direct the balloon in any course, materially different from that of the wind which happens to blow. With this limitation, the invention may still in some cases be capable of utility, wherever we are only desirous of ascending to a great height, without regarding the place in which we are to descend: or where we wish to attain only a height so moderate that the machine may be kept by ropes in the situation which is desired. In France the balloon has lately been employed with considerable success as a meteorological observatory; Mr. Biot and Mr. Gay Lussac having ascended to a height of above four miles, for the laudable purpose of ascertaining some facts relating to the constitution of the atmosphere, and to the magnetic properties of the earth.

CHRONOLOGY OF AUTHORS ON HYDRODYNAMICS.

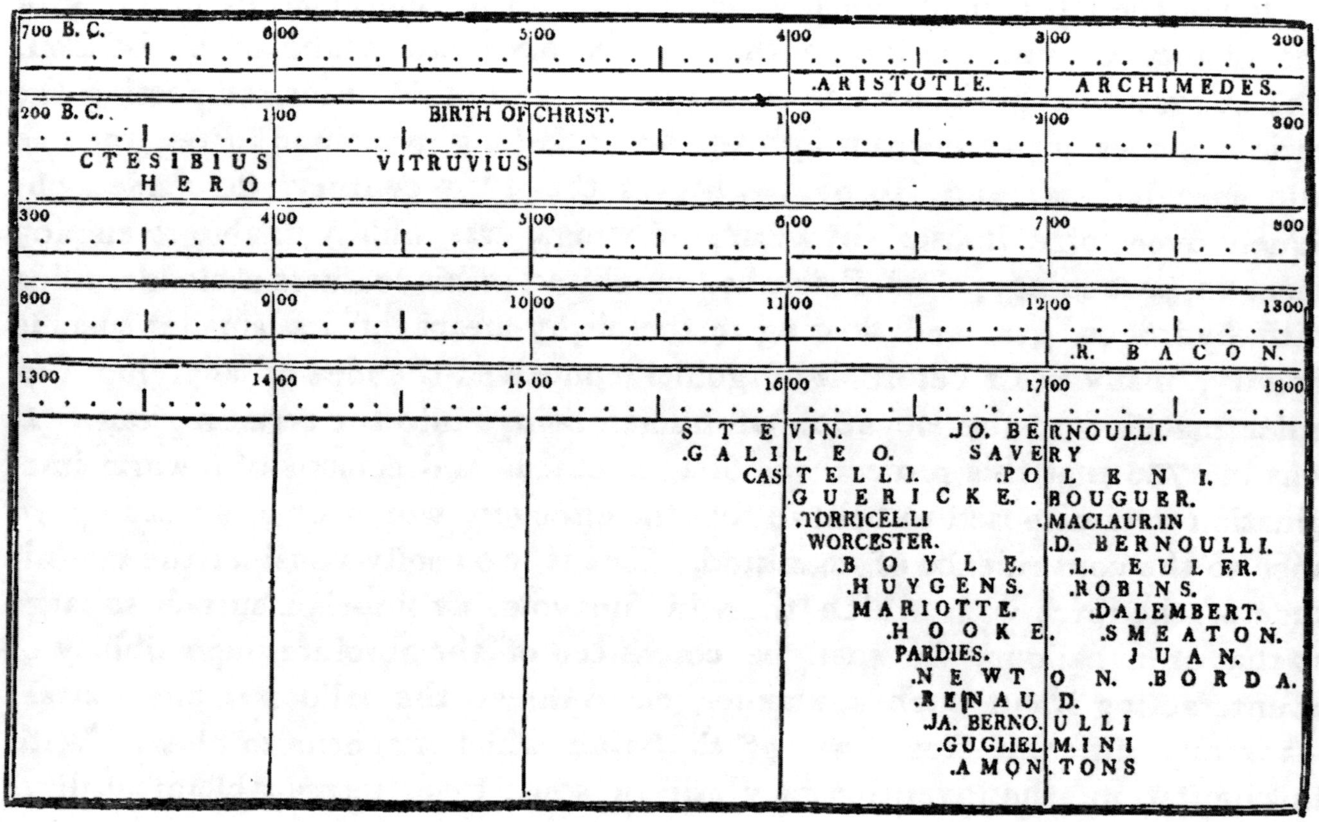

LECTURE XXXI.

ON THE PROPAGATION OF SOUND.

THE theory of sound, which constitutes the science of acustics, is on many accounts deserving of particular attention, for it not only involves many interesting properties of the motions of elastic substances, but it also affords us considerable assistance in our physiological inquiries respecting the nature and operation of the senses. The subject has usually been considered as exceedingly abstruse and intricate, but the difficulty has in some measure originated from the errors which were committed in the first inquiries respecting it; and many of the phenomena belonging to it are so remarkable, and so amusing, as amply to repay the labour of examining them by the entertainment that they afford. We shall consider first the nature and propagation of sound in general, secondly, the origin of particular sounds, and the effects of single sounds; thirdly, the consequences of the combinations of sounds variously related, constituting the doctrine of harmonics, and fourthly, the construction of musical instruments, and the history of the science of acustics.

Sound is a motion capable of affecting the ear with the sensation peculiar to the organ. It is not simply a vibration or undulation of the air, as it is sometimes called; for there are many sounds in which the air is not concerned, as when a tuning fork or any other sounding body is held by the teeth: nor is sound always a vibration or alternation of any kind; for every noise is a sound, and a noise, as distinguished from a continued sound, consists of a single impulse in one direction only, sometimes without any alternation; while a continued sound is a succession of such impulses, which, in the organ of hearing at least, cannot but be alternate. If these successive impulses form a connected series, following each other too rapidly to be separately distinguished, they constitute a continued sound, like that of the voice in speaking; and if they are equal among themselves in duration, they produce a musical or equable sound, as that of a vibrating chord or string, or of the voice in singing. Thus, a quill striking against a piece of wood causes a noise, but, striking against the teeth of a wheel or of a comb, a continued sound; and if the teeth of the wheel are at equal distances, and the velocity of the motion is constant, a musical note.

Sounds of all kinds are most usually conveyed through the medium of the air; and the necessity of the presence of this or of some other material substance for its transmission is easily shown by means of the air pump; for the sound of a bell struck in an exhausted receiver is scarcely perceptible. The experiment is most conveniently performed in a moveable receiver or transferrer, which may be shaken at pleasure, the frame which suspends the bell being supported by some very soft substance, such as cork or wool. As the air is gradually admitted, the sound becomes stronger and stronger, although it is still much weakened by the interposition of the glass: not that glass is in itself a bad conductor of sound, but the change of the medium of communication from air to glass, and again from glass to air, occasions a great diminution of its intensity. It is perhaps on account of the apparent facility with which sound is transmitted by air, that the doctrine of acustics has been usually considered as immediately dependent on pneumatics, although it belongs as much to the theory of the mechanics of solid bodies as to that of hydrodynamics.

A certain time is always required for the transmission of an impulse through a material substance, even through such substances as appear to be the hardest and the least compressible. It is demonstrable that all minute impulses are conveyed through any homogeneous elastic medium, whether solid or fluid, with a uniform velocity, which is always equal to that which a heavy body would acquire by falling through half the height of the modulus of elasticity, that is, in the case of the air, half the height of the atmosphere, supposed to be of equal density; so that the velocity of sound passing through an atmosphere of a uniform elastic fluid must be the same with that of a wave moving on its surface. In order to form a distinct idea of the manner in which sound is propagated through an elastic substance, we must first consider the motion of a single particle, which, in the case of a noise, is pushed forwards, and then either remains stationary, or returns back to its original situation; but in the case of a musical sound, is continually moved backwards and forwards, with a velocity always varying, and varying by different degrees, according to the nature or quality of the tone; for instance, differently in the notes of a bell and of a trumpet. We may first suppose for the sake of simplicity, a single series of particles to be placed only in the same line with the direction of the motion. It is obvious that if these particles were absolutely incompressible, or infinitely elastic, and were also retained in contact with each other by an infinite force of cohesion or of compression, the whole series must move precisely at the same time, as well as in the same manner. But in a substance which is both compressible and extensible or expansible, the motion must occupy a certain time in being propagated to the successive particles on either side, by means of the impulse of the first particle on those which are before it, and by means of the diminution of its pressure on those which are behind; so that when the sound consists of a series of alternations, the motion of some of the particles will be always in a less advanced state than that of others nearer to its source; while at a greater distance forwards, the particles will be in the opposite stage of the undulation, and still further on, they will again be moving in the same manner with the first particle, in consequence of the effect of a former vibration.

The situation of a particle at any time may be represented by supposing it to mark its path, on a surface sliding uniformly along in a transverse direction. Thus, if we fix a small pencil in a vibrating rod, and draw a sheet of paper along, against the point of the pencil, an undulated line will be marked on the paper, and will correctly represent the progress of the vibration. Whatever the nature of the sound transmitted through any medium may be, it may be shown that the path thus described will also indicate the situation of the different particles at any one time. The simplest case of the motion of the particles is that in which they observe the same law with the vibration of a pendulum, which is always found opposite to a point supposed to move uniformly in a circle: in this case the path described will be the figure denominated a harmonic curve; and it may be demonstrated that the force, impelling any particle backwards or forwards, will always be represented by the distance of the particle before or behind its natural place; the greatest condensation and the greatest direct velocity, as well as the greatest rarefaction and retrograde velocity, happening at the instant when it passes through its natural place.

We are ready to imagine that very hard bodies transmit motion instantaneously, because we have no easy means of measuring the interval of time that elapses between the action of pushing the end of a rod, and the protrusion of an obstacle at the other end, or between the instant of pulling a bell rope, and that of the ringing of the bell. But it is demonstrable that in order to transmit an impulse in a time infinitely small, the hardness of the substance must be infinitely great, and it must be absolutely incompressible and inextensible by any force, which is a property not discoverable in any natural bodies: the hardest steel and the most brittle glass being very susceptible both of extension and compression.

The least elastic, substance that has been examined, is perhaps carbonic acid gas, or fixed air, which is considerably denser than atmospheric air exposed to an equal degree of pressure. The height of the atmosphere, supposed to be homogeneous, is in ordinary circumstances, and at the sea side, about 28 000 feet, and in falling through half this height a heavy body would acquire a velocity of 946 feet in a second. But from a comparison of the accurate experiments of Derham, made in the day time, with those of the French Academicians, made chiefly at night, it appears that the true velocity of sound is about 1130 feet in a second, which agrees very nearly with some observations made with great care by Professor Pictet. This difference between calculation and experiment has long occupied the attention of natural philosophers, but the difficulty appears to have been in great measure removed by the happy suggestion of Laplace, who has attributed the effect to the elevation of temperature, which is always found to accompany the action of condensation, and to the depression produced by rarefaction. It is true that a greater change of temperature would be required than Mr. Dalton's experiments on the compression of air appear to indicate; but those experiments do not perfectly agree among themselves; and the observation which has been made in France, that a heat, sufficient to set tow on fire, may be produced by the operation of a condensing syringe, seems to show that Mr. Dalton's results are somewhat below the truth. In this manner the theory may be completely reconciled with experiments; we may estimate the modulus of the air's effective elasticity, which is the measure of its immediate force, from the velocity which is thus observed, and its height will appear to be 39 800 feet, instead of 27 800, which is the supposed height of the atmosphere. This velocity remains unchanged by any alternation of pressure indicated by the barometer, but it may be affected by a change of temperature. For when an elastic fluid is compressed, its elasticity is increased in the same ratio as its density; and the height of a homogeneous atmosphere equivalent to the pressure, remains the same, consequently the velocity calculated from that height remains unaltered; but the identity of the acceleration, from the effect of heat which has been mentioned, can only be inferred from observation: this identity may, however, be satisfactorily shown, by means of experiments on the sounds of organ pipes, which are intimately connected with the velocity of the transmission of sound through the air, and which are found to remain precisely the same, however the air may be rarefied or condensed. The Academicians del Cimento inclosed an organ pipe, with bellows worked by a spring, in the receiver of an air pump and of a condenser, and they found that, as long as the sound was audible, its pitch remained unchanged. Papin screwed a whistle on the orifice which admits the air into the receiver of the air pump, and I have fixed an organ pipe in the same manner; and the result agreed with the experiment of the academicians. But if the density of the air is changed, while its elasticity remains unaltered, which happens when it is expanded by heat, or condensed by cold, the height of the column, and consequently the velocity, will also be altered; so that for each degree of Fahrenheit's thermometer the velocity will vary about one part in a thousand. Bianconi has actually observed this difference of velocity according to the different heights of the thermometer, and it may be shown less directly by means of the sounds of pipes; but it has not been accurately determined whether or no the correction on account of the effect of compression in causing heat, remains unaltered, although Bianconi's experiments agree very well with the supposition that no material change takes place in this respect. The velocity of sound must also be in some measure influenced by the quantity of moisture contained in the atmosphere: it must be a little diminished by cold fogs, which add to the density, without augmenting the elasticity, and increased by warm vapours, which tend to make the air lighter; and these two opposite states are probably often produced in succession in wind instruments blown by the mouth, the air within them being at first cold and damp, and afterwards warm and moist.

In pure hydrogen gas, the velocity of sound ought, from calculation, to be more than three times as great as in common air, but the difference does not appear to have been so great in any experiment hitherto made on the sounds of pipes in gases of different kinds. For such experiments, the comparative specific gravity of the gas may be most conveniently ascertained by Mr. Leslie's method of observing the time employed in emptying a vessel through a small orifice, by means of the pressure of an equal column of water; according to the simple theory, the velocities of the gas thus discharged ought to be in the same proportion as the respective velocities with which sounds would be transmitted by them: and if any variation from this proportion were discovered, it must be attributed to the different degrees of heat produced by condensation in the different fluids. Steam, at the temperature of boiling water, is only one third as heavy as common air; consequently the velocity of sound in steam must be nearly three fourths greater than in air.

It does not appear that any direct experiments have been made on the velocity with which an impulse is transmitted through a liquid, although it is well known that liquids are capable of conveying sound without difficulty; Professor Robison informs us, for example, that he heard the sound of a bell transmitted by water at the distance of 1200 feet. It is, however, easy to calculate the velocity with which sound must be propagated in any liquid of which the compressibility has been measured. Mr. Canton has ascertained that the elasticity of water is about 22 000 times as great as that of air; it is, therefore, measured by the height of a column which is in the same proportion to 34 feet, that is 750 thousand feet, and the velocity corresponding to half this height is 4900 feet in a second. In mercury, also, it appears from Mr. Canton's experiments, that the velocity must be nearly the same as in water, in spirit of wine a little smaller. These experiments were made by filling the bulb of a thermometer with water, and observing the effects of placing it in an exhausted receiver, and in condensed air; taking care to avoid changes of temperature, and other sources of error:

the fluid rose in the tube when the pressure was removed, and subsided when it was increased. A slight correction is, however, required on account of the expansion and contraction of the glass, which must have tended to make the elasticity of the fluids appear somewhat greater than it really was.

It is also well known that solid bodies in general are good conductors of sound: thus, any agitation communicated to one end of a beam is readily conveyed to the ear applied to the other end of it. The motion of a troop of cavalry is said to be perceived at a greater distance by listening with the head in contact with the ground, than by attending to the sound conveyed through the air; and we may frequently observe that some parts of the furniture of a house are a little agitated by the approach of a wagon, before we hear the noise which it immediately occasions. The velocity, with which impulses are transmitted by solids, is in general considerably greater than that with which they are conveyed by the air: Mr. Wunsch has ascertained this by direct observations on a series of deal rods closely united together, which appeared to transmit a sound instantaneously, while a sensible interval was required for its passing through the air: I have also found that the blow of a hammer on a wall, at the upper part of a high house, is heard as if double by a person standing near it on the ground, the first sound descending through the wall, the second through the air. It appears from experiments on the flexure of solid bodies of all kinds, that their elasticity, compared with their density, is much greater than that of the air: thus, the height of the modulus of elasticity of fir wood, is found, by means of such experiments, to be about 9 500 000 feet, whence the velocity of an impulse conveyed through it must be 17 400 feet, or more than three miles, in a second. It is obvious, therefore, that in all common experiments such a transmission must appear perfectly instantaneous. There are various methods of ascertaining this velocity from the sounds produced under different circumstances by the substances to be examined, and Professor Chladni has in this manner compared the properties of a variety of natural and artificial productions.

We have hitherto considered the propagation of sound in a single right line, or in parallel lines only; but it usually happens, at least when a sound is transmitted through a fluid, that the impulse spreads in every direction, so as to occupy at any one time nearly the whole of a spherical surface. But it is impossible that the whole of this surface should be affected in a similar manner by any sound, originating from a vibration confined to a certain direction, since the particles behind the sounding body must be moving towards the centre, whenever the particles before it are retreating from the centre; so that in one half of the surface, the motions may be called retrograde or negative, while in the other they are direct or positive; consequently at the sides, where these portions join, the motions can be neither positive nor negative, and the particles must remain at rest; the motions must also become gradually less and less sensible as they approach to the limit between the two hemispheres. And this statement may be confirmed by an experiment on the vibration of a body of which the motion is limited to a certain direction, the sound being scarcely audible when the ear is in a direction precisely perpendicular to that of the vibration.

The sound thus diverging must always be spread through a part of a spherical surface, because its velocity must be equal in every direction, so that the impulse will always move forwards in a straight line passing through the centre of the sphere, or the vibrating body. But when a hemispherical pulse arrives at the surface of a plane solid obstacle, it is reflected, precisely in the same manner as we have already seen that a wave of water is reflected, and assumes the form of a pulse proceeding from a centre at an equal distance on the opposite side of the surface. This reflection, when it returns back perpendicularly, constitutes what is commonly called an echo: but in order that the echo may be heard distinctly, it is necessary that the reflecting object be at a distance moderately great, otherwise the returning sound will be confused with the original one; and it must either have a smooth surface, or consist of a number of surfaces arranged in a suitable form; thus there is an echo not only from a distant wall or rock, but frequently from the trees in a wood, and sometimes, as it is said, even from a cloud.

If a sound or a wave be reflected from a curved surface, the new direction which it will assume may be determined, either from the condition that the velocity with which the impulse is transmitted must remain unaltered, or from the law of reflection, which requires that the direction of the reflected pulse or wave be such as to form an angle with the surface, equal to that which the incident pulse before formed with it. Thus, if a sound or wave proceed from one focus of an ellipsis, and be reflected at its circumference, it will be directed from every part of the circumference towards the other focus; since the distance which every portion of the pulse has to pass over in the same time, in following this path, is the same, the sum of the lines drawn from the foci to any part of the curve being the same; and it may also be demonstrated that these lines form always equal angles with the curve on each side. The truth of this proposition may be easily shown by means of the apparatus already described for exhibiting the motions of the waves of water; we may also confirm it by a simple experiment on a dish of tea: the curvature of a circle differs so little from that of an ellipsis of small eccentricity, that if we let a drop fall into the cup near its centre, the little wave which is excited will be made to converge to a point at an equal distance on the other side of the centre. (Plate XXV. Fig. 340, 341.)

If an ellipsis be prolonged without limit, it will become a parabola: hence a parabola is the proper form of the section of a tube, calculated for collecting a sound which proceeds from a great distance, into a single point, or for carrying a sound nearly in parallel directions to a very distant place. It appears, therefore, that a parabolic conoid is the best form for a hearing trumpet, and for a speaking trumpet; but for both purposes the parabola ought to be much elongated, and to consist of a portion of the conoid remote from the vertex; for it is requisite, in order to avoid confusion, that the sound should enter the ear in directions confined within certain limits: the voice proceeds also from the mouth without any very considerable divergence, so that the parts of the curve behind the focus would in both cases be wholly useless. A trumpet of such a shape does not very materially differ from a part of a cone; and conical instruments are found to answer sufficiently well for practice; it appears, however, unnecessary to suppose, as Mr. Lambert has done, that they differ essentially in principle from parabolic trumpets. It is not yet perfectly decided whether or no a speaking trumpet has any immediate effect in strengthening the voice, independently of the reflection of sound. (Plate XXV. Fig. 342.)

An umbrella, held in a proper position over the head, may serve to collect the force of a distant sound by reflection, in the manner of a hearing trumpet; but its substance is too slight to reflect any sound very perfectly, unless the sound fall on it in a very oblique direction. The whispering gallery at St. Paul's produces an effect nearly similar, by a continued repetition of reflections. Mr. Charles's paradoxical exhibition of the Invisible Girl has also been said to depend on the reflection of sound; but the deception is really performed by conveying the sound through pipes, artfully concealed, and opening opposite to the mouth of the trumpet, from which it seems to proceed.

When a portion of a pulse of sound is separated by any means from the rest of the spherical or hemispherical surface to which it belongs, and proceeds through a wide space, without being supported on either side, there is a certain degree of divergence, by means of which it sometimes becomes audible in every part of the medium transmitting it: but the sound thus diverging is comparatively very faint; and more so indeed than the effect of a wave of water, admitted under similar circumstances, into a wide reservoir, which we have already examined. Hence, in order that a speaking trumpet may produce its full effect, it must be directed in a right line towards the hearer: and the sound collected into the focus of a concave mirror is far more powerful than at a little distance from it, which could not happen if, as some have erroneously supposed, sound in all cases tended to spread equally in all directions. The sounds that enter a room, in which there is an open window, are generally heard by a mixture of this faint divergence with the reflection from various parts of the window and of the room, and with the effect of the impulse transmitted through the walls. This diverging portion, however faint, probably assists in preserving the rectilinear motion of the principal sound, and gradually gains a little additional strength at the expense of this portion.

The decay of sound is the natural consequence of its distribution throughout a larger and larger quantity of matter, as it proceeds to diverge every way from its centre. The actual velocity of the particles of the medium transmitting it appears to diminish simply in the same proportion as the distance from the centre increases; consequently their energy, which is to be considered as the measure of the strength of sound, must vary as the square of the distance; so that, at the distance of ten feet from the sounding body, the velocity of the particles of the medium becomes one tenth as great as at the distance of one foot, and their energy, or the strength of the sound, only one hundredth as great.

LECTURE XXXII.

ON THE SOURCES AND EFFECTS OF SOUND.

THE examination of the origin of sound might naturally be deemed anterior to the inquiry respecting its propagation; but it will appear, that the properties of many of the most usual sources of sound depend immediately on the velocity, with which an impulse of any kind is transmitted through an elastic medium; it was therefore necessary to consider this velocity, before the production of sound in general could be discussed.

The origin of a simple sound, without any alternation, requires very little investigation: it appears that the only condition necessary for its production is a sufficient degree of velocity in the motion or impulse which occasions it. A very moderate velocity must be sufficient for producing an impression on the ear; there is reason to believe, that, when the sound is continued, it may remain audible with a velocity of no more than one hundredth of an inch in a second, and perhaps even with a velocity much smaller than this: but, at its origin, it is probable that the velocity of the motion, constituting a sound, must always be considerably greater.

A continued sound may be produced by a repetition of separate impulses independent of each other, as when a wheel strikes in rapid succession the teeth of a pinion, so as to force out a portion of air from between them; when a pipe, through which air is passing, is alternately opened and shut, either wholly or partially, by the revolution of a stopcock or valve; or when a number of parallel surfaces is placed at equal distances in a line nearly perpendicular to them, and a noise of any kind is reflected from each of them in succession; a circumstance which may often be observed when we are walking near an iron railing, an acute sound being heard, which is composed of such reflections from the surfaces of the palisades.

Musical sounds are, however, most frequently produced by the alternate motions of substances naturally capable of isochronous vibrations, and these substances may be either fluids or solids, or instruments composed of a combination of fluids with solids. The resonance of a room or passage is one of the simplest sources of a musical sound; the walls being parallel, the impulse is reflected backwards and forwards continually, at equal intervals of time, so as to agree with the definition, and to produce the effect, of a musical sound. When we blow obliquely and uniformly into a cylindrical pipe closed at one end, it is probable that the impulse or condensation must travel to the bottom and back, before the resistance is increased; the current of our breath will then be diverted from the mouth of the pipe, for an equal time, which will be required for the diminution of the resistance by the discharge of the condensed air, so that the whole time of a vibration will be equal to the time occupied by an impulse of any kind in passing through four times the length of the pipe. An open pipe may be considered nearly as if it consisted of two such pipes, united at their closed ends, the portions of air contained by them being agitated by contrary motions, so as always to afford each other a resistance similar to that which the bottom of the stopped pipe would have furnished. It is probable that when an open pipe is once filled with air a little condensed, the oblique current is diverted, until the effect of the discharge, beginning at the remoter end, has returned to the inflated orifice, and allowed the current to reenter the pipe. Where the diameter of the pipe is different at different parts of its length, the investigation of the sound becomes much more intricate; but it has been pursued by Daniel Bernoulli with considerable success, although upon suppositions not strictly consistent with the actual state of the motions concerned.

In the same manner as an open pipe is divided by an imaginary basis, so as to produce the same sound with a stopped pipe of half the length, a pipe of any kind is capable of being subdivided into any number of such pipes, supposed to meet each other's corresponding ends only; and in general the more violently the pipe is inflated, the greater is the number of parts into which it subdivides itself, the frequency of the vibrations being always proportional to that number. Thus, an open pipe may be divided not only into two, but also into four, six, eight, or more portions, producing the same sounds as a pipe of one half, one third, one fourth, or any other aliquot part of the length; but a stopped pipe cannot be divided into any even number of similar parts; its secondary sounds being only those of a pipe of which the proportion is determined by the odd numbers, its length being, for example, one third, one fifth, or one seventh of the original length. These secondary notes are sometimes called harmonics; they are not only produced in succession from the same pipe, but they are also often faintly heard together, while the fundamental note of the pipe continues to sound. When the pipe has a large cavity connected with it, or consists principally of such a cavity, with a small opening, its vibrations are usually much less frequent, and it is generally incapable of producing a regular series of harmonics.

It is obvious from this statement of the analogy between the velocity of sound and the vibrations of the air in pipes, that they must be affected in a similar manner by all alterations of temperature. Thus, the frequency of the vibrations of a pipe must be increased nearly in the ratio of 33 to 34 by an elevation of 30 degrees of Fahrenheit's thermometer; and if this change be accompanied by a transition from dampness to simple moisture, the sound will be still more altered.

Dr. Chladni has discovered that solids of all kinds, when of a proper form, are capable of longitudinal vibrations, exactly resembling in their nature those of the air in an organ pipe, having also their secondary or harmonic notes related to them in a similar manner. These vibrations are always far more frequent than those of a column of air of equal length, the velocity, with which an impulse is transmitted by a solid of any kind, being usually from 5 to 16 times as great as the velocity of sound in air; so that the longitudinal sounds are always extremely acute, when they are produced by substances of moderate length. These sounds afford perhaps the most accurate mode of determining the velocity of the transmission of an impulse through any elastic substance, and of obtaining from that velocity the exact measure of its elasticity: they may be easily exhibited by holding a long bar or wire of iron or brass in the middle, and striking it at one end with a small hammer, in the direction of its length.

The vibrations by which solid bodies most usually produce sound are, however, not longitudinal, but lateral, and they are governed either by a tension, derived from the operation of a weight, or of some other external force, or by the natural elasticity of the substance. The vibrations of extended substances resemble most in their properties those of elastic fluids, and they occur the most frequently in practice, although the vibrations produced by the elasticity of the substance may be considered as the most natural.

Vibrations derived from tension are either those of chords or musical strings, or those of membranes; but the vibrations of membranes afford little variety, and have not hitherto been very accurately investigated, the drum being almost the only instrument in which they are concerned; they do not however appear to differ materially in their properties from the vibrations of strings. A musical string or chord is supposed to be perfectly flexible, and of uniform thickness, to be stretched between two fixed points by a force incomparably greater than its own weight, and to vibrate in a single plane through a minute space on each side of its natural position. Its motions may then be traced through all their stages, by comparing the chord to a portion of an elastic medium of the same length, contained between two bodies capable of reflecting any impulse at each end; for example, to a portion of air situated between two walls, or inclosed in a pipe stopped at both ends; for the vibration of such a medium will be performed in the time occupied by any impulse in travelling through twice its length; and the vibration of the chord will be performed in the same time, supposing the height or depth of the medium equal to the length of a portion of the chord, of which the weight is equivalent to the force applied to stretch it, and which may be called with propriety the modulus of the tension. If the chord be at first bent into a figure of any kind, and then set at liberty, the place of any part of it at every subsequent time will be such, that it will always be in a right line between two points moving along the figure each way with the appropriate velocity; but in order to pursue this determination, we must repeat the figure of the chord on each side of the fixed points in an inverted position, changing the ends as well as the sides. Hence it appears that, at the end of a single vibration, the whole chord will assume a similar figure on the opposite side of its natural place, but in an inverted position, and after a complete or double vibration, it will return precisely to the form which it had in the beginning. The truth of this result is easily shown by inflecting any long chord near one of its ends, having first drawn a line under its natural position, and it will then be evident that the chord returns in each vibration nearly to the point of inflection, and passes at that end but to a much shorter distance on the opposite side of the line, while at the other end its excursions are greatest on the opposite side of the line. The result of the calculation of the frequency of vibration agrees also perfectly with experiment, nor is the coincidence materially affected by the inflexibility or elasticity of the string, by the resistance of the air, nor by the slight increase of the tension occasioned by the elongation of the string when it is inflected: thus, if the weight or force causing the tension of a string were equal or equivalent to the weight of 200 feet of the same string, that is, if the modulus of tension were 200 feet long, the velocity corresponding to half this height would be 80 feet in a second; and every impulse would be conveyed with this velocity from one end of the string to the other, so that if the string were 1 foot long, it would vibrate 40 times in a second, if 6 inches, 80 times, and if it were 40 feet long, only once in a second. Hence, it is obvious that the time of vibration of any chord is simply proportional to the length; and this may be shown either by means of such vibrations as are slow enough to be reckoned, or by a comparison with the sounds of pipes, or with other musical sounds. But if the tension of a chord of given length were changed, it would require to be quadrupled in order to double the frequency of vibration; and if the tension and length remained unaltered, and the weight of the chord were caused to vary, it would also be necessary to make the weight four times as great in order to reduce the frequency of vibration to one half.

It appears from the mode of tracing the progress of a vibration, which has already been laid down, that every chord vibrates in the same manner as if it were a part of a longer chord, composed of any number of such chords, continually repeated in positions alternately inverted; consequently if a long chord be initially divided into any number of such equal portions, its parts will continue to vibrate in the same manner as if they were separate chords; the points of division only remaining always at rest. Such subordinate sounds are called harmonics: they are often produced in violins by lightly touching one of the points of division with the finger, when the bow is applied, and in all such cases it may be shown, by putting small feathers or pieces of paper on the string, that the remaining points of division are also quiescent, while the intervening portions are in motion. (Plate XXV. Fig. 343.)

These harmonic sounds are also generally heard together with the fundamental sound of the chord, and it is, therefore, necessary, in such cases, to consider the subordinate vibrations as combined with a general one. It is not, however, universally true that the fundamental sound must always be accompanied by all the harmonics of which the chord is susceptible; for I have found that by inflecting the chord exactly at any point in which the chord might be divided into a number of equal parts, and then suffering it to vibrate, we lose the effect of the corresponding harmonic. There is some difficulty in explaining the reason of the distinct production of these sounds, in cases where the theory appears to indicate a single and simple vibration only; but it appears to be most probable that they usually become audible in consequence of some imperceptible irregularity in the form or weight of the chord, which is just sufficient to derange the perfect coincidence of the actual motions with those which the theory indicates, without producing a discordance capable of offending the ear. That a chord irregularly loaded may have the relations of its harmonics disturbed, may easily be understood by considering the effect of a small weight placed at one of the points of division, which will obviously retard the principal vibration, without materially affecting that of the portions terminated by it. An abrupt and irregular agitation appears also in many cases to make the secondary notes more audible, while a gradual and delicate impulse, like that of the wind on the strings of an Aeolian harp, produces a sound almost entirely free from subordinate harmonics.

It usually happens that the vibration of a chord deviates from the plane of its first direction, and becomes a rotation or revolution, which may be considered as composed of various vibrations in different planes, and which is often exceedingly complicated. These vibrations may be combined in the first instance in a manner similar to that which has been already explained respecting the vibrations of pendulums; and if the motion of the chord be supposed to follow the same law as that of a pendulum, the result of two entire vibrations thus united, may be either a vibration in an intermediate direction, or a revolution, in a circle or in an ellipsis. But besides these compound vibrations of the whole chord, it is also frequently agitated by subordinate vibrations, which constitute harmonic notes of different kinds, so that the whole effect becomes very intricate; as we may observe by a microscopic inspection of any luminous point on the surface of the chord, for instance the reflection of a candle in the coil of a fine wire wound round it. The velocity of the motion is such that the path of the luminous point is marked by a line of light, in the same manner as when a burning coal is whirled round; and the figures, thus described, are not only different at different parts of the same chord, but they often pass through an amusing variety of forms during the progress of the vibration; they also vary considerably according to the mode in which that vibration is excited. (Plate XXV. Fig. 344, 345.)

The vibrations immediately dependent on elasticity are those of rods, plates, rings, and vessels. These admit of much greater variety, and are of more difficult investigation than the vibrations of chords. A rod may be either wholly loose, or fixed at one end only, or at both; and it may either be loosely fixed, in situation only, or firmly fixed, in direction as well as in situation; and these conditions may be variously combined with each other; the rod may also have a variety of secondary vibrations besides the principal or fundamental sound. All these cases have been examined by various mathematicians: the subject was begun by Daniel Bernoulli, and much extended by Euler, some of whose conclusions have been corrected by Riccati; and Chladni has compared them all with experiment. The sounds produced by the same rod, either under different circumstances, or as harmonics which may be heard at the same time, are scarcely ever related to each other in any simple proportion, except that when a rod is loosely fixed at both ends, the frequency of the vibrations of the subordinate notes is expressed by the series of the squares of the natural numbers, as 1, 4, 9, and 16. But the times occupied by any similar vibrations of rods, similarly circumstanced, are always directly as the squares of their lengths, and inversely as their depths. When the rod is wholly at liberty, two at least of its points must be at rest, and these are at the distance of about one fifth of its length from either end: in the next sound of the same rod, the middle point is at rest, with two others near the ends. There is by no means the same regularity in the progress of the vibrations of rods of different kinds as in those of chords; it can only happen in particular cases that the rod will return after a complete vibration to its original state, and these cases are probably such as seldom occur in nature.

The vibrations of plates differ from those of rods in the same manner as the vibrations of membranes differ from those of chords, the vibrations which cause the plate to bend in different directions being combined with each other, and sometimes occasioning singular modifications. These vibrations may be traced through wonderful varieties by Professor Chladni's method of strewing dry sand on the plates, which, when they are caused to vibrate by the operation of a bow, is collected into such lines as indicate those parts, which remain either perfectly or very nearly at rest during the vibrations. Dr. Hooke had employed a similar method, for showing the nature of the vibrations of a bell, and it has sometimes been usual, in military mining, to strew sand on a drum, and to judge, by the form in which it arranges itself, of the quarter from which the tremors produced by countermining proceed. (Plate XXV. Fig. 346 . . 348.)

The vibrations of rings and of vessels are nearly connected with those of plates, but they are modified in a manner which has not yet been sufficiently investigated. A glass, or a bell, divides in general into four portions vibrating separately, and sometimes into six or eight; they may readily be distinguished by means of the agitations excited by them in a fluid contained in the glass. It is almost unnecessary to observe, that the fluid thus applied, by adding to the mass of matter to be moved, makes the vibration slower, and the sound more grave.

In some cases the vibrations of fluids and solids are jointly concerned in the production of sound: thus, in most of the pipes of an organ denominated reed pipes, the length of a tongue of metal is so adjusted, as to be capable of vibrating in the same time with the air contained in the pipe. Sometimes, however, the air only serves to excite the motion of the solid, as in some other organ pipes, which are usually much shorter than would be required for producing the proper note alone, and probably in the glottis, or organ of the voice, of animals. On the other hand, the alternate opening and shutting of the lips, in blowing the trumpet or French horn, can scarcely be called a vibration, and the pitch of the sound is here determined by the properties of the air in the pipe only. The vibrations of a solid may be excited by an undulation propagated through a fluid; thus, when a loud sound strikes against a chord, capable of vibrating, either accurately, or very nearly, with the same frequency, it causes a sympathetic tone, resembling that from which it originated; and the chord may produce such a sound either by vibrating as a whole, or by dividing itself into any number of equal parts. Thus, if the damper be raised from any of the strings of a harpsichord, it may be made to vibrate, by striking or singing any note, of which the sound corresponds either to that of the whole string, or to that of any of its aliquot parts. Sometimes also two chords that are very nearly alike, appear, when sounding together, to produce precisely the same note, which differs a little from each of those which the chords would produce separately; and a similar circumstance has been observed with respect to two organ pipes placed near each other. In these cases the vibrating substances must affect each other through the medium of the air; nearly in the same manner as two clocks, which rest on the same support, have been found to modify each other's motions, so as to exhibit a perfect coincidence in all of them.

It is uncertain whether any fibres in the ear are thus sympathetically agitated in the process of hearing, but if there are any such vibrating fibres, their motions must necessarily be of short duration, otherwise there would be a perpetual ringing in our ears, and we should never be able to judge accurately of the termination of a sound. Besides, a sympathetic vibration may be excited not only by a sound producing vibrations of equal frequency, but also by a sound, of which every alternate, or every third or fourth vibration, coincides with its motions: it would, therefore, be difficult to distinguish such sounds from each other, if hearing depended simply on the excitation of sympathetic vibrations. It is true that we generally distinguish, in listening to a loud and deep sound, precisely such notes as would be thus produced; but it is only when the sounding body is capable of affording them from the nature of its vibrations; for we may listen for them in vain in the sound of a bell or of a humming top. There is, however, no doubt that the muscles, with which the different parts of the ear are furnished, are concerned in accommodating the tension of some of them to the better transmission of sound;

and it is equally certain that their operation is not absolutely necessary in the process.

The external ear serves in some measure for collecting the undulations of sounds transmitted through the air, and reflecting them into the auditory passage, at the bottom of which they strike against the membrane of the tympanum or drum, which, being larger and more moveable than some of the subsequent parts, is capable of transmitting a stronger impulse than they would immediately receive. In the same manner we may often feel the tremors produced in a sheet of thick paper, held in the hand, by the agitation of the air, derived from a loud sound, which would not otherwise have affected the organ of touch. The impulse received by the membrane of the tympanum is conveyed by the hammer and anvil, two small bones, which together constitute a kind of bent lever, through a third minute flattened bone, to a fourth called the stirrup, which serves merely as a handle to its basis, a plate covering the orifice of a cavity called the vestibule, and communicating the impulse to the mucous fluid which fills this cavity. The fluid of the vestibule, thus agitated, acts immediately on the terminations of the nerves, which form a loose membranous tissue, almost floating in it, while another portion of them is distributed on the surface of three semicircular tubes or canals, opening at both ends into the cavity, and a third portion supplies the cochlea, a detached channel, which appears to be arranged with singular art as a micrometer of sound. It resembles the spiral convolutions of a snail shell, and if uncoiled, would constitute two long conical tubes connected at their summits, the base of one opening into the vestibule, that of the other being covered by a membrane only, which separates the fluid from the air contained in the general cavity of the ear, or the tympanum. It is evident from the properties of fluids moving in conical pipes, that the velocity of any impulse affecting the fluid at the base of the cone must be extremely increased at its vertex, while the flexibility of the membrane at the base of the second channel allows this motion to be effected without difficulty. It has also been supposed that a series of fibres are arranged along the cochlea, which are susceptible of sympathetic vibrations of different frequency according to the nature of the sound, which acts on them; and, with some limitations, the opinion does not appear to be wholly improbable. We must, however, reason with great caution respecting the functions of every part of the ear, since its structure varies so much in different animals, that we cannot pronounce with certainty respecting the indispensable necessity of any one arrangement for the perfection of the sense. And even in the case of the human ear, many of these parts may be spared without great inconvenience; thus, we hear very perfectly, by means of impressions communicated to the teeth, and through them to the large bones of the head; and even when the membrane of the tympanum, and all the small bones of the ear have been destroyed by disease, the undulations of the air still continue to affect the organ in the usual manner. (Plate XXV. Fig. 349 . . 351.)

Such is the delicacy of the organs of hearing in their perfect state, that we readily distinguish not only the frequency of the vibrations of a sound, whether constant or variable, and its loudness or softness, but also the quality of tone, depending on the law which governs each separate vibration, and which constitutes the difference between instruments of different kinds, or different instruments of the same kind, or even the same instrument differently employed. Thus, we can distinguish very accurately the voices of our friends, even when they whisper, and those modifications of the same voice which constitute the various vowels and semivowels, and which, with the initial and final noises denominated consonants, compose the words of a language. We judge also, without an error of many degrees, of the exact direction in which the sound approaches us; but respecting the manner in which the ear is enabled to make this discrimination, we cannot reason upon any satisfactory grounds.

LECTURE XXXIII.

ON HARMONICS.

THE philosophical theory of harmonics, or of the combinations of sounds, was considered by the ancients as affording one of the most refined employments of mathematical speculation; nor has it been neglected in modern times, but it has been in general either treated in a very abstruse and confused manner, or connected entirely with the practice of music, and habitually associated with ideas of mere amusement. We shall, however, find the difficulties by no means insuperable, and the subject will appear to be worthy of attention, not only on its own account, but also for the sake of its analogy with many other departments of science.

It appears both from theory and from experience, that the transmission of one sound does not at all impede the passage of another through the same medium. The ear too is capable of distinguishing, without difficulty, not only two sounds at once, but also a much greater number. The motions produced by one series of undulations being wholly indifferent with respect to the effect of another series, and each particle of the medium being necessarily agitated by both sounds, its ultimate motion must always be the result of the motions which would have been produced in it by the separate sounds, combined according to the general laws of the composition of motion, which are the foundation of the principal doctrines of mechanics. When the two sounds, thus propagated together, coincide very nearly in direction, the motions belonging to each sound may be resolved into two parts, the one in the common or intermediate direction, and the other transverse to it; the latter portions will obviously be very small; they will sometimes destroy each other, and may always be neglected in determining the effect of the combination, since the ear is incapable of distinguishing a difference in the directions of sounds which amounts to a very few degrees.

only. Thus, when two equal undulations, of equal frequency, coincide in this manner, and when the particular motions are directed the same way at the same time, the velocities in each direction are added together, and the joint effect is doubled, or perhaps quadrupled, since it appears that the strength of sound ought to be estimated from the squares of the velocities of the particles: but when the particular motions of the two sounds counteract each other, both their effects are wholly destroyed. These combinations resemble the effects of the waves of water in similar circumstances, which we have already examined, and they may be illustrated by drawing two curved lines representing the motions which constitute the sounds, in the same manner as we have already supposed them to be described, by a vibrating particle, on a surface moving uniformly in a transverse direction; these figures being placed side by side, the joint effect may be represented by a third curve drawn in such a direction as to be always in the middle between the corresponding points of the first two. A similar result, but still more strongly marked, may be obtained mechanically, by cutting two boards or plates of any kind into the form of the curves, and then dividing one of them into a number of thin pieces or sliders, by lines perpendicular to the general direction of the curve, or to the termination of the plate which is parallel to it: the bottom of these sliders being then placed on the other curve, their general outline will represent the effect of the combination. We may assume for this purpose the form of the harmonic curve, which represents the motions of a body vibrating like a pendulum, and which probably agrees very nearly with the purest and simplest sounds. (Plate XXV. Fig. 352.)

If the two undulations differ a little from each other in frequency, they alternately tend to destroy each other, and to acquire a double or perhaps a quadruple force, and the sound gradually increases and diminishes in continued succession at equal intervals. This intension and remission is called a beat, and furnishes us with a very accurate mode of determining the proportional frequency of the vibrations, when the absolute frequency of one of them is known, or the absolute frequency of both, when their proportion is known; since the beats are usually slow enough to be reckoned, although the vibrations themselves can never be distinguished. Thus, if one sound consisted of 100 vibrations in a second, and produced with another acuter sound a single beat in every second, it is obvious that the second sound must consist of 101 vibrations in a second. Again, if we have two portions of a similar chord equally stretched, or two simple pipes, of which the lengths are in the proportion of 15 to 16, they will produce a beat in 15 vibrations of the longer; and if we count the number of beats in 15 seconds, we shall find the number of vibrations in a single second. The easiest way of procuring two such strings or pipes, in practice, is to tune them by a third, so that they may be respectively $\frac{4}{5}$ and $\frac{3}{4}$ of its length; the vibrations of the third pipe in a second will also be equal to the number of beats of the first two in 12 seconds. (Plate XXV. Fig. 353.)

When the beats of two sounds are too frequent to be heard as distinct augmentations of their force, they have the same effect as any other impulses which recur in regular succession, and produce a musical note, which has been denominated a grave harmonic. Thus, two sounds in the proportion of 4 to 5, produce, when they are both very low or grave, an audible succession of beats; but when they are higher or more acute, a grave harmonic, which may be separately distinguished as a third sound by an attentive ear. Those combinations of sounds, which produce beats distinctly audible, have in general a harsh and coarse effect, and are called discords; but those of which the vibrations are so related, as to have a common period after a few alternations, and which may be observed to produce a third sound, constitute concords, which are in themselves the more perfect as the common periods are shorter. (Plate XXV. Fig. 353.)

The natural association of the secondary sounds, which generally accompany almost all musical notes, serves in some measure as a foundation for the science of harmonics, the same sounds, as are thus frequently connected in nature, being found to be agreeable when united by art. But it appears to depend still more immediately on a love of order, and a predilection for a regular recurrence of sensations, primitively implanted in the human mind. Hence, when two sounds are heard together, those proportions are the most satisfactory to the ear, which exhibit a recurrence of a more or less perfect coincidence at the shortest intervals, expressed by the smallest numbers of the separate vibrations: for though we cannot immediately estimate the magnitude of the vibrations, yet the general effect of a regular or irregular succession necessarily produces the impression of sweetness or harshness. The same sounds, as form the best accompaniment for each other, are also in general the most agreeable for melodies, consisting of a succession of single notes; their intervals are, however, too large to be sufficient for the purposes of music, and they require to be mixed with other sounds which are related to them in a manner nearly similar.

The same constitution of the human mind, which fits it for the perception of harmony, appears also to be the cause of the love of rhythm, or of a regular succession of any impressions whatever, at equal intervals of time. Even the attachment to the persons and places to which we are accustomed, and to habits of every kind, bears a considerable resemblance to the same principle. The most barbarous nations have a pleasure in dancing; and in this case, a great part of the amusement, as far as sentiment and grace are not concerned, is derived from the recurrence of sensations and actions at regular periods of time. Hence not only the elementary parts of music, or the single notes, are more pleasing than any irregular noise, but the whole of a composition is governed by a rhythm, or a recurrence of periods of greater or less extent, generally distinguished by bars, which are also the constituent parts of larger periods, and are themselves subdivided into smaller. An interruption of the rhythm is indeed occasionally introduced, but merely for the sake of contrast; nearly in the same manner as, in all modern pieces of music, discords are occasionally mixed with concords, in order to obtain an agreeable variety of expression.

In a simple composition, all the intervals are referred to a single fundamental or key note. Thus, any air which can be played on a trumpet or on a bugle horn, must consist of the harmonics of a single sound only: and when an accompaniment is performed by a French horn, the length of the instrument is first adjusted to the principal note, and all the sounds which it is to produce are selected from this natural series. But the notes constituting the most natural scale are not, without exception, comprehended among the harmonics; they are, however, all immediately dependent on a similar relation. A sound of which the vibrations are of equal frequency with those of another, is called a unison; if two vibrations occur for every one of the fundamental note, the sound is called its superior octave, being the eighth of those which are commonly considered as filling up the scale; and on account of its great resemblance to the fundamental note, it is described by the same letter of the alphabet, or by the same syllable; so that all audible sounds are considered as repetitions of a series contained within the interval of an octave. One third part of the string or pipe gives the fifth above the octave; one fourth the double octave, and one fifth of the string its third. Thus we obtain the common accord or chord, or the harmonic triad, consisting of the fundamental note, with its third and fifth, which produces the most perfect harmony, and which also contains the constituent parts of the most simple and natural melodies. But we are still in want of intermediate steps for the scale; these are supplied by completing first, the triad of the fifth, which gives us the second, and the seventh, of which 9 and 15 vibrations correspond respectively to 8 of the fundamental, and which may also be found in the ascending series of natural harmonics; and in the second place, by adding the fourth and sixth in such proportions as to make up another perfect triad with the octave; the respective notes consisting of 4 and 5 vibrations, while the fundamental note makes 3, and being no where found among the natural harmonics. The complete scale is, therefore, formed by these harmonic triads contiguous to and connected with each other; the middle one being the triad of the key note, the superior one that of its fifth, which is sometimes called the dominant or governing note, and the inferior one that of the fourth, or subdominant. This scale is derived from principles so simple, that it may properly be considered as a natural arrangement, and it appears to be found with little variation in barbarous as well as in civilised countries. (Plate XXV. Fig. 354.)

A long piece would, however, be too monotonous, unless the fundamental note were sometimes changed; we may, therefore, take at pleasure one of the auxiliary triads for the principal harmony, and we may continue the modulation or progression, until every note of the scale becomes in succession a key note. But, in order to fill up the intervals of these several scales in just proportion, it becomes necessary to add several new notes to the first series, for instance, if we take the seventh for a key note, we shall want five new sounds within the octave, making twelve in the whole, which is the number usually employed in music The interval between any two adjoining sounds of the twelve is called a semitone or half note, two semitones making a tone or note; these terms are, however, sometimes employed with various subordinate distinctions and limitations. The five additional sounds have no separate names, but they are denominated from the neighbouring notes on either side, with the addition of the term sharp or flat, accordingly as they are a semitone higher or lower than the notes of which they bear the names.

For still further variety, we sometimes change the place of the middle note of each triad, placing the minor third, or the interval expressed by the ratio of 5 to 6, below the major, which is in the ratio of 4 to 5; and the scale thus formed is called the scale of the minor mode, in contradistinction to the major, the three principal thirds being depressed a semitone. Sometimes, however, the alteration is made in the third of the key note only, especially in ascending, in order to retain the seventh of the major scale, which leads so naturally into the octave, as to be sometimes called the characteristic semitone of the key; and it is for this reason, that the triad, in which it is found, is called the accord of the dominant, which, in all regular compositions, immediately precedes the termination in the key note.

The major and minor triads, with the discord of the flat seventh, may be considered as constituting the foundation of all essential harmonies. The flat seventh is principally used with the major triad, in transitions from the fundamental key into its fourth, to which that seventh naturally belongs as a concord; so that it serves to introduce the new key, by strongly marking the particular note in which it differs from the old one; and in such cases the flat seventh always descends into, or is followed by, the third of the new key, and the third of the first triad ascends into the new key note. Other discords are also sometimes introduced, but they are in general either partial continuations of a preceding, or anticipations of a following accord. Two different parts of a harmony are never allowed, in regular and serious compositions, to accompany each other in successive octaves or fifths, since such a succession is found to produce a disagreeable monotony of effect, except when a series of octaves is continue for some time, so as to be considered as a repetition of the same part.

These are almost the only principles, upon which the art of accompaniment, as well as the general theory of practical music, is founded. Many prolix treatises have been written on the subject, but they only contain particular illustrations of the application of these principles, together with a few refinements upon them. The art of composition, however, depends much more on a good taste, formed by habitual attention to the best models, and aided, perhaps, by some little natural predisposition, than upon all the precepts of science, which teach us only how to avoid what is faulty, without instructing us in the mode of attaining what is beautiful or sublime.

It is impossible to assign any such proportions for the twelve sounds thus employed, that they may be perfectly appropriate to all the capacities in which they are used; their number is, therefore, sometimes considerably increased; and in some instruments they may be varied without limit, at the performer's pleasure, as in the voice, in instruments with finger boards, and in some wind instruments; but in many cases this is impracticable, nor could any imaginable alteration make all the intervals perfect, unless every note were varied, whenever we returned to it by steps different from those by which we had left it. The simplest mode of arranging the twelve sounds, is to divide the octave into twelve equal intervals, all the notes being in the same proportion to those which immediately follow them: this is called the equal temperament, because the imperfection is equal in all keys. In this system of temperament, the fifths, which consist of seven semitones, are a little too flat, that is, the interval is a little too small; the minor thirds, consisting of three semitones, are also too flat, and the major thirds too sharp. But it has generally been esteemed best to preserve some keys more free from error; partly for variety, and partly because some are more frequently used than others: this cannot, however, be done without making some of the scales more imperfect than in the equal temperament. A good practical mode of performing it, is to make six perfect fifths, in descending from the key note of the natural scale, and six ascending fifths equally imperfect among themselves. We thus retain a slight imperfection in the scales most commonly used, and make the keys which are most remote from them considerably less perfect. Another method, which is perhaps somewhat more easily executed, is to make the fifth and third of the natural scale perfectly correct, to interpose between their octaves, the second and sixth, so as to make three fifths equally tempered, and to descend from the key note by seven perfect fifths, which will complete the scale. Any of these modes of temperament may be actually executed, either by the estimation of a good ear, or, still more accurately, by counting the frequency of the beats which the notes make with each other.

For denoting precisely the absolute as well as the relative frequency of the sounds of the different octaves, we employ the first seven letters of the alphabet; A being the key note of the minor mode, in the scale of natural notes, and C of the major. The peculiar characters used in music are generally disposed on five or more lines, with their intervening spaces, each implying a separate step in the scale, setting out from any line at pleasure, which is marked with an ill formed G, a C, or an F; a sharp or a flat implying that all the notes written on the line, or in the space, to which it belongs, are to be raised or depressed a semitone, and a natural restoring the note to its original value. The actual frequency of the vibration of any note, according to the pitch most usually employed, may be found, if we recollect to call a noise, recurring every second, the first C, then the C denoted by the mark of the tenor cliff will be the ninth, consisting of 256 vibrations in a second. The fifth, consisting of sixteen vibrations, will be nearly the lowest audible note, and the fourteenth the highest note used in music, but the sixteenth, consisting of above 30 000 vibrations in a second, may perhaps be an audible sound. The frequency of the vibrations of the other notes may easily be calculated from the known relations which they bear to the note thus determined. (Plate XXV. Fig. 355.)

LECTURE XXXIV.

ON MUSICAL INSTRUMENTS.

THE application of the theory of harmonics to practice depends on the construction of musical instruments of different kinds: of these we shall only be able to take a cursory view, and we shall afterwards attend to the historical order of the most remarkable steps, by which both the theory and practice of music have been advanced to a high degree of refinement.

Musical instruments may be most conveniently arranged, accordingly as they are principally calculated for exciting sound by the vibrations of chords, of membranes, of elastic plates, or of the air; or by the joint effects of the air and a solid body vibrating together. The essential varieties of stringed instruments are found in the harp, the harpsichord, the pianoforte, the clavichord, the guitar, the violin, the vielle or monochord, and the aeolian harp. In all these, the immediate force of the sound of the strings is increased by means of a sounding board, which appears to be agitated by their motion, and to act more powerfully on the air than the strings could do alone.

In the harp, the sound is produced by inflecting the string with the finger, and suffering it to return to its place. The lyre of the ancients differed from the harp only in its form and compass, except that the performer sometimes used a plectrum, which was a small instrument, made of ivory, or some other substance, for striking the strings. Each note in the harp has a separate string; and in the Welch harp there are two strings to each note of the principal scale, with an intermediate row for the semitones; but in the pedal harp, the half notes are formed by pressing pins against the strings, so as to shorten their effective length. Instead of this method, an attempt has lately been made to produce the semitones by changing the tension of the strings, which is said to have succeeded tolerably well although it appears at first sight somewhat unpromising.

In the harpsichord, and in the spinet, which is a small harpsichord, the quill acts like the finger in the harp, or the plectrum in the lyre, and it is fixed to the jack by a joint with a spring, allowing it without difficulty to repass the string, which is here of metal. Sometimes leather is used instead of quills; and this serves to make the tone more mellow, but less powerful. Besides two strings in unison, for each note, the harpsichord has generally a third which is an octave above them. Different modifications of the tone are sometimes produced by striking the wire in different parts, by bringing soft leather loosely into contact with its fixed extremity, and by some other means. When the finger is removed from the key, a damper of cloth falls on the string, and destroys its motion. In all instruments of this kind, the perfection of the tone depends much on the construction and situation of the sounding board: it is usually made of thin deal wood, strengthened at different parts by thicker pieces fixed below it.

In the pianoforte, the sound is produced by a blow of a hammer, raised by a lever, which is as much detached from it as possible. The dulcimer, or hackbrett of the Germans, is also made to sound by the percussion of hammers, but they are simply held in the hand of the performer.

The clavichord, the clavier of the Germans, differs from other keyed instruments in the manner in which the length of the string is determined; it is attached at one end to a bridge, and at the other to a pin or screw as usual; but the effective length is terminated on one side by the bridge, and on the other by a flat wire projecting from the end of the key, which strikes the string, and at the same time serves as a temporary bridge as long as the sound continues: the remaining portion of the string is prevented from sounding by being in contact with a strip of cloth, which also stops the whole vibration as soon as the hammer falls. The instrument is capable of great delicacy and neatness of expression, but it is deficient in force. The guitar is generally played with the fingers, like a harp; but each string is made to serve for several notes, by means of frets, or cross wires, fixed to the finger board, on which it is pressed down by the other hand. But in the pianoforte guitar, hammers are interposed between the fingers and the strings, acting like those of the pianoforte. The mandoline and lute are species of the guitar: and the arch lute was a very powerful instrument of the same kind, formerly much used in full pieces.

In the violin, and in other instruments resembling it, all the strings are capable of having their length altered at pleasure, by being pressed down on the finger board. The sound is produced by the friction of the bow, rubbed with resin: the string is carried forwards by its adhesion to the bow, and when its resistance has overcome this adhesion, it begins to return in opposition to the friction; for the friction of bodies in motion is generally less than their adhesion when they are at rest with respect to each other, besides that the contact of the string with the bow is usually in great measure interrupted by subordinate vibrations, which may be distinguished, by the assistance of a microscope, in the manner already described; but when the string changes once more the direction of its motion, it adheres again to the bow, and is accelerated by it as before. The original instrument appears to have been the viola or tenor, its diminutive the violino, its intensitive, expressing a greater bulk, the violone or double bass, and the diminutive of this, the violoncello, or common bass. The viola di gamba had one or more long strings separate from the finger board, serving as an occasional accompaniment.

The vielle, or monochord, commonly called the hurdy gurdy, has frets which are raised by the action of the fingers on a row of keys; and instead of a bow, the string is made to vibrate by the motion of a wooden wheel: there is a second string serving as a drone, producing always the same sound; this is furnished with a bridge loosely fixed, which strikes continually against the sounding board, and produces a peculiar nasal effect. The trumpet marine, or trumpet Marigni, was a string of the same kind, which was lightly touched at proper points, so as to produce harmonic notes only; it was impelled by a bow. The aeolian harp, when agitated by the wind, affords a very smooth and delicate tone, frequently changing from one to another of the harmonics of the string, accordingly as the force of the wind varies, and as it acts more or less unequally on different parts of the string. (Plate XXV. Fig. 356.)

The human voice depends principally on the vibrations of the membranes of the glottis, excited by a current of air, which they alternately intercept and suffer to pass; the sounds being also modified in their subsequent progress through the mouth. Perhaps the interception of the air by these membranes is only partial; or it may be more or less completely intercepted in sounds of different kinds: the operation of the organs concerned is not indeed perfectly understood, but from a knowledge of their structure, we may judge in some measure of the manner in which they are employed.

The trachea, or windpipe, conveys the air from the chest, which serves for bellows: hence, it enters the larynx, which is principally composed of five elastic cartilages. The lowest of these is the cricoid cartilage, a strong ring, which forms the basis of the rest: to this are fixed, before, the thyreoid cartilage, and behind, the two arytaenoid cartilages, composing together the cavity of the glottis, over which the epiglottis inclines backwards, as it ascends from its origin at the upper part of the thyreoid cartilage. Within the glottis are extended its ligaments, contiguous to each other before, where they are inserted into the thyreoid cartilage, but capable of diverging considerably behind whenever the arytaenoid cartilages separate. These ligaments, as they vary their tension, in consequence of the motions of the arytaenoid cartilages, are susceptible of vibrations of various frequency, and as they vibrate, produce a continuous sound. Properly speaking, there are two ligaments on each side; but it is not fully understood how they operate; probably one pair only performs the vibrations, and the other assists, by means of the little cavity interposed, in enabling the air to act readily on them, and in communicating the vibrations again to the air. (Plate XXVI. Fig. 357, 358.)

The vowels and semivowels are continuous sounds, chiefly formed by this apparatus in the glottis, and modified either in their origin or in their progress by the various arrangements of the different parts of the mouth. Of simple vowels sixteen or eighteen may be enumerated in different languages: in the French nasal vowels the sound is in part transmitted through the nostrils, by means of the depression of the soft palate: the perfect semivowels differ from the vowels only in the greater resistance which the air undergoes in its passage through the mouth; there are also nasal and seminasal semivowels. The perfect consonants may be either explosive, susurrant, or mute; the explosive consonants begin or end with a sound formed in the larynx, the others are either whispers, or mere noises, without any vocal sound. By attending to the various positions of the organ, and by making experiments on the effects of pipes of different forms, it is possible to construct a machine which shall imitate very accurately many of the sounds of the human voice; and this has indeed been actually performed by Kratzenstein and by Kempelen. (Plate XXVI. Fig. 359.)

Although the vibrating ligaments of the glottis may be anatomically denominated membranes, yet their tension is probably confined to the direction of their length, and their action is, therefore, the same with that of a simple string or chord. But in the case of a tambourine and a drum, the membrane is stretched in every direction, and the force of tension consequently acts in a different manner. The principal character of such instruments is their loudness, derived from the magnitude of the surface which strikes the air, and the short duration of the sound, on account of the great resistance necessarily produced by the air's reaction.

Musical instruments which produce sounds, by means of vibrations depending on the elasticity of solid bodies, are less frequently employed than others; they have a peculiar character of tone, which is by no means unpleasant, but which renders them less fit to be mixed with other instruments, since their secondary harmonics are in different proportions. Such is the stacada, a series of cylinders of glass, or of metal, struck either immediately with hammers, or by means of keys; the tuning fork, the gong, the cymbal, and the bell. Bells and other similar instruments are usually made of a mixture of copper, and tin, with a little brass or zinc, which is more highly elastic than either of the component parts taken separately. The harmonica consists of a series of vessels of glass, either placed side by side, or fixed on a common axis, and made to sound by the friction of the fingers, and sometimes by that of rubbers of cork. The vibrations of an elastic plate, agitated by a current of air, which it continually admits and excludes, constitute the sound of the vox humana and regal organ pipes, resembling the human voice as much in their effects as in the mechanism on which they depend. (Plate XXVI. Fig 360 . . 362.)

Of simple wind instruments, in which the quality of the sound is determined by the vibrations of the air, the principal are the syrinx, the flute, the flageolet, the diapason organ pipe, whether open, stopped, or with a chimney, the humming top, and the cavity of the mouth in whistling, or in playing on the jew's harp. The pipes of the syrinx are adjusted to their respective notes by cutting them, or filling them up, until they are reduced to a proper length; and the effective length of the flute and flageolet is altered at pleasure by opening or shutting the holes made at proper distances in them; the opening a hole at any part having the same effect as if the pipe were cut off a little beyond it, and the elevation of the tone being somewhat greater as the hole is larger. The instruments differ little except in the mechanism, by which the breath is directed, in such a manner as to excite a sound; and the flageolet, when furnished with bellows, becomes a bagpipe. The tongue of the jew's harp is an elastic plate, but the sound, which it immediately produces, serves only as a drone; its vibration, however, appears to act like the motion of the bow of a violin in exciting another sound: this sound, although faint, is still sufficiently musical, and appears to be determined by the magnitude of the cavity of the mouth, nearly in the same manner as that of the humming top, or as the sound of the same cavity produced in whistling, by a current of air which is forced through it. (Plate XXVI. Fig. 363 . . 367.)

In mixed wind instruments, the vibrations or alternations of solid bodies are made to cooperate with the vibrations of a given portion of air. Thus, in the trumpet, and in horns of various kinds, the force of inflation, and perhaps the degree of tension of the lips, determines the number of parts into which the tube is divided, and the harmonic which is produced. In the serpent, the lips cooperate with a tube, of which the effective length may be varied by opening or shutting holes, and the instrument which has been called an organized trumpet appears to act in a similar manner; the trombone has a tube which slides in and out at pleasure, and changes the actual length of the whole instrument. The hautboy, and clarinet have mouth pieces of different forms, made of reeds or canes, and the reed pipes of an organ, of various constructions, are furnished with an elastic plate of metal, which vibrates in unison with the column of air that they contain. An organ generally consists of a number of different series of such pipes, so arranged, that by means of registers, the air proceeding from the bellows may be admitted to supply each series, or excluded from it, at pleasure, and a valve is opened, when the proper key is touched, which causes all the pipes belonging to the note, in those series of which the registers are open, to sound at once. These pipes are not only such as are in unison, but frequently also one or more octaves above and below the principal note, and sometimes also twelfths and seventeenths, imitating the series of natural harmonics. But these subordinate sounds ought to be comparatively faint, otherwise their irregular interference would often occasion an intolerable discord, instead of the grand and sublime effect which this instrument is capable of producing, when it is judiciously constructed and skilfully employed. (Plate XXVI. Fig. 368.)

The practice of music appears to be of earlier origin than either its theory, or any attention to the nature and general phenomena of sound. The first lyre, with three strings, is said to have been invented in Egypt by Hermes, under Osiris, between the years 1800 and 1500 before Christ; but a tradition so remote, concerning a personage so enveloped in fable, can scarcely be considered as constituting historical evidence: we cannot, therefore, expect to ascertain with any certainty the proportions of these strings to each other; some suppose that they were successive notes of the natural scale, others that they contained the most perfect concords; perhaps in reality each performer adjusted them in the manner which best suited his own fancy. The trumpet is said to have been employed about the same time; its natural harmonics might easily have furnished notes for the extension of the scale of the lyre, but it does not appear that the ancients ever adopted this method of regulating the scale. The lyre with seven strings is attributed to Terpander, about 700 years before our era, and two centuries afterwards, either Pythagoras, or Simonides, completed the octave, which consisted of intervals differing very little from the modern scale, the key note being nearly in the middle of the series. In subsequent times the number of the strings was much increased; the modulations, and the relations of the intervals, became very intricate, and were greatly diversified in a variety of modes or scales, which must have afforded an inexhaustible supply of original and striking melodies, but which could scarcely admit so many pleasing combinations, as our more modern systems. Although it is certain that the ancients had frequent accompaniments in perfect harmony with the principal part, yet they had no regular art of counterpoint, or of performing different melodies together; nor does it appear that they ever employed discords. The tibia of the ancients resembled a hautboy or clarinet, for it had a reed mouth piece, about three inches long; the same performer generally played on two of these instruments at once. There were, however, several varieties of the tibia; and it is not improbable that some of them may have had the simple mouth piece of the flageolet.

The first philosophical observer of the phenomena of sound, after Pythagoras, appears to have been Aristotle; he notices a great variety of curious facts in harmonics among his mechanical problems; and he entertained a very correct idea of the true nature of the motions of the air constituting sound. He knew that a pipe or a chord of a double length produced a sound of which the vibrations occupied a double time; and that the properties of concords depended on the proportions of the times occupied by the vibrations of the separate sounds. It is not indeed improbable that at least as much as this was known to Pythagoras, since he established correctly the numerical ratios between various sounds; but so little justice has been done to his discoveries by the imperfect accounts of them which have been preserved, that we cannot expect to be able to ascertain his opinions on any subject with accuracy.

The invention of the organ, by Ctesibius of Alexandria, about 2000 years ago, forms a remarkable epoch in harmonics. The larger instruments of this kind were furnished with hydraulic bellows, the smaller with bellows of leather only; and they had keys which were depressed, like those of the modern organs, by the fingers of the performer, and which opened valves communicating with the pipes.

The modern system of music is one of the few sciences, if such it can be called, which owe their improvement to the middle ages. The old ecclesiastical music was probably founded in great measure on that of the Greeks; its peculiar character consisted in the adoption of any note of the scale at pleasure for a key note, without altering materially the other intervals; and in this manner they obtained a variety much resembling that of the modes or kinds of music in use among the ancients. Pope Gregory, about the year 600, distinguished the notes by literal characters; the rules of counterpoint were formed by degrees from the experience of the ecclesiastical musicians; and early in the eleventh century, Guido of Arezzo, otherwise called Aretin the monk, introduced, together with some improvements in the theory and practice of music, a new method of naming the notes by syllables.

Some curious experiments on sound may be found in the works of Bacon, but they added very little to the true theory of acustics, and some of them are not perfectly accurate. Galileo rediscovered what was well known to Aristotle, respecting the nature of sound; for the words of Aristotle had been so much misunderstood and misinterpreted, that he could have profited but little by them. His cotemporaries Mersenne and Kircher made a variety of very ingenious experiments and observations, on sound and on sounding bodies, many of them unknown to authors of later date. The theory of the ancient music was very accurately investigated, in the middle of the 17th century, by Meibomius: our countryman Wallis, also, besides employing much learning and penetration in the illustration of the ancient music, observed some insulated facts in harmonics which were new and interesting.

Sir Isaac Newton's propositions respecting the velocity of the propagation of sound were the beginning of all the more accurate investigations relating to acustics. It must not be denied that these propositions contain some very inconclusive reasoning respecting the nature of the motions constituting sound, by which the determination of a particular case is erroneously extended into a general solution of the problem. The velocity is, however, truly calculated, because it is in fact independent of the particular nature of the vibration, and all that is wanting to generalise the proposition is the remark, that if the velocity of sound is the same in all cases, it must be such as the calculation indicates. An error nearly similar was committed by Brook Taylor, who in the year 1714 investigated the time occupied by the vibration of a string or chord upon a particular supposition, which he considered as a necessary condition, but which in fact confined the inquiry to a limited case. It happens, however, that the same determination of the frequency of vibration is equally true in all possible cases. Sauveur obtained, about the same time a similar conclusion from reasoning still less accurate: his merits with respect to the theory of acustics in general are, however, by no means contemptible. Lagrange and Euler have corrected and much extended the investigations of Newton, and of Taylor; and Bernoulli and Dalembert have also materially contributed to the complete examination and discussion of the subject.

About the year 1750, Daniel Bernoulli succeeded in obtaining a solution of a problem still more difficult than those which relate to the motions of chords: he determined the frequency of the vibrations of an elastic rod fixed at one end, as well as the relations of its subordinate sounds. The solution is not indeed absolutely general, but it may perhaps be adapted to all possible cases, by considering the effect of a combination of various sounds produced at the same time. Euler has also great merit in extending and facilitating the mathematical part of this investigation, although he has committed several mistakes respecting the mechanical application of it, some of which he has himself corrected, and others have been noticed by Riccati and Chladni.

The grave harmonics produced by the combination of two acute sounds were noticed about the same time by Romieu and by Tartini, but first by Romieu: their existence is not only remarkable in itself, but particularly as it leads to some interesting consequences respecting the nature of sound and hearing in general.

Bernoulli has also investigated, in a very ingenious manner, the sounds produced by the air in pipes of various forms, although confessedly on suppositions deviating in some measure from the truth: the results of his computations have, however, been amply confirmed by the experiments of Lambert on the sounds of flutes.

Dr. Chladni's method of examining the sounds of plates has afforded a very interesting addition to our knowledge of the nature of vibrations; his discovery of the longitudinal sounds of solids is of considerable importance, and he is said to be engaged in an extensive work on the subject of acustics in general. Some remarks which I have made in the Philosophical Transactions may perhaps also be considered as tending to illustrate the vibrations of chords. The latest improvement which deserves to be mentioned, with respect to the theory of sound, is Laplace's explanation of the increase of its velocity on account of the effect of heat, which appears to afford a satisfactory explanation of a difficulty so much the more important, as it tended to lessen our confidence in every part of a theory, which differed so widely from the most accurate and best established observations.

CHRONOLOGY OF ACUSTICS.

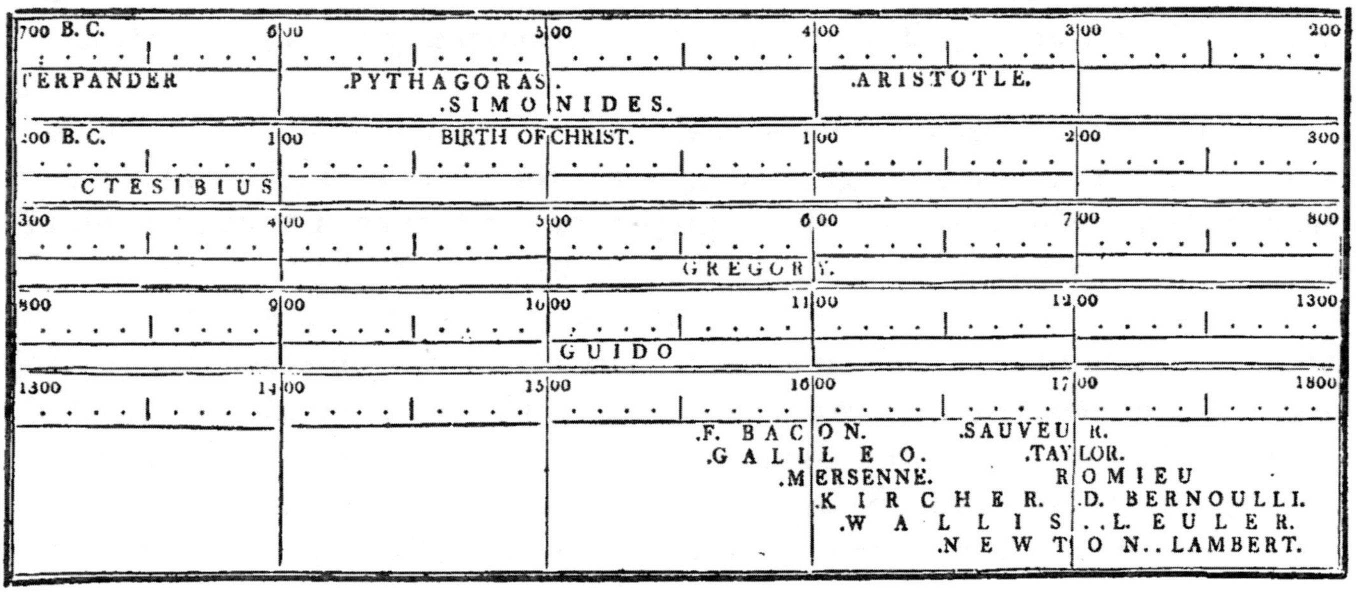

LECTURE XXXV.

ON THE THEORY OF OPTICS.

THE science of optics is one of the most elegant, and the most important branches of natural and mechanical philosophy. It presents us with experiments attractive by their beauty and variety, with investigations affording an ample scope for mathematical refinements, and with instruments of extensive utility both in the pursuit of other sciences, and in the common employments of life; nor is there any department of the study of nature in which an unprejudiced observer is more convincingly impressed with the characteristic marks of the perfect works of a supremely intelligent Artist.

We shall first consider the essential properties which we discover in light, and which are the basis of our calculations, together with the conclusions immediately deducible from those properties; and next, the application of these laws to practical purposes, in the construction of optical instruments. We shall afterwards proceed to examine the more complicated phenomena, which are derived from the same laws, and which are observed as well in natural as in artificial circumstances, constituting the subdivision of physical optics. The description of the eye, and the explanation of the sense of vision, by means of which all these effects are connected with the human mind, is properly a continuation of the subject of physical optics: the intimate nature of light will be the next subject of investigation, and a historical sketch of the progress of the science of optics will conclude the second part of this course of lectures.

In order to avoid all hypothesis in the beginning, it will be necessary to define light from its sensible qualities. The sensation of light is sometimes produced by external pressure on the eye; we must exclude this sensation from the definition of light, and must therefore call light an influence capable of entering eye, and of affecting it with a sense of vision. A body, from which this influence appears to originate, is called a luminous body. We do not include in this definition of the term light the invisible influences which occasion heat only, or blacken the salts of silver, although they both appear to differ from light in no other respects than as one kind of light differs from another; and they might probably have served the purpose of light, if our organs had been differently constituted.

A ray of light is considered as an infinitely narrow portion of a stream of light, and a pencil as a small detached stream, composed of a collection of such rays accompanying each other. As a mathematical line is sometimes conceived to be described by the motion of a mathematical point, so a ray of light may be imagined to be described by the motion of a point of light. We cannot exhibit to the senses a single mathematical line, except as the boundary of two surfaces; in the same manner, we cannot exhibit a single ray of light, except as the confine between light and darkness, or as the lateral limit of a pencil of light.

When light passes through a space free from all material substances, it moves, with great velocity, in a direction perfectly rectilinear; when also it passes through a material substance perfectly uniform in its structure, it probably always moves in a similar manner. But in many cases its motions are much interrupted. Those substances through which light passes the most freely, and in straight lines, are called homogeneous transparent mediums. Perhaps no medium is, strictly speaking, absolutely transparent; for even in the air, a considerable portion of light is intercepted. It has been estimated that of the horizontal sunbeams, passing through about 200 miles of air, one two thousandth part only reaches us; and that no sensible light can penetrate more than 700 feet deep into the sea; a length of seven feet of water having been found to intercept one half of the light which enters it.

It is possible that mediums, not in other respects identical, may be homogeneous with respect to the transmission of light; for example, a glass may be filled, with a fluid of such a density, that the light may pass uninterruptedly through their common surface; but it generally happens, that whenever the nature of the medium is changed, the path of light deviates from a straight line: thus, the apparent places of the sun and stars are changed by the effect of the atmosphere, because the light, by which we judge of their situations, is deflected, in its passage out of the empty space beyond the atmosphere, first into the rarer and then into the denser air. In the same manner, when we view a distant object over a fire or a chimney, it appears to dance and quiver, because the rays of light, by which it is seen, are perpetually thrown into new situations, by the different changes of the density of the air in consequence of the action of heat.

When rays of light arrive at a surface, which is the boundary of two mediums not homogeneous, they continue their progress without deviating from those planes, in which their former paths lay, and which are perpendicular to the surface of the mediums; but they no longer retain the same direction, a part of them, and sometimes nearly the whole, is reflected back from the surface, while the remaining part is transmitted and refracted, or bent. The name refraction is derived from the distortion which it occasions in the appearance of an object viewed in part only by refracted light: thus an oar, partially immersed in water, appears to be bent, on account of the refraction of the light by which its lower part is seen, in its passage out of the water into the air.

There is no instance of an abrupt change of the density of a medium, without a partial reflection of the light, passing either into the denser or into the rarer medium; and the more obliquely the light falls on the surface, the greater, in general, is the reflected portion. No body is so black as to reflect no light at all, and to be perfectly invisible in a strong light; although at the surface separating two very rare bodies, as two kinds of gas, the reflection is too faint to be perceptible; but in this case the separation is seldom perfectly abrupt.

The angles of incidence and reflection are the angles made by a ray of light, before and after its reflection, with a line perpendicular to the reflecting surface; and these angles are always equal to each other; consequently the inclination of the rays to the surface remains also the same. The quantity of light reflected, when other circumstances are equal, appears to be always greatest when the difference of the optical or refractive density of the two substances is greatest. Thus the reflection from the common surface of glass and water is much weaker, than from a surface of glass exposed to the air. Metals in general reflect a great proportion of the light falling on them, and even the reflection from the common surface of glass and mercury appears to be but little weaker than the reflection from the surface of mercury immediately exposed to the air, so that the optical density of the metals must be exceedingly great.

It appears also that a portion of the light falling on a reflecting surface is always transmitted, at least to a certain depth, notwithstanding the apparent opacity of any large masses of the substance. Thus, if we cover a small hole of a window shutter with the thinnest leaf gold, we shall find that it transmits a greenish light, which must have passed the reflecting surface, but which, if the gold had been but one ten thousandth of an inch in thickness, would have been wholly intercepted, and probably almost in the same manner as by passing through 700 feet of water. In transparent substances, however, the greater part of the light penetrates to all distances with little interruption, and all rays of the same kind, thus transmitted by the same surface, form with the perpendicular an angle of refraction which is ultimately in a certain constant proportion to the angle of incidence; that is, for instance, one half, three fourths, or two thirds, according to the nature of the surface. Thus, if the refractive properties of the substance were such, that an incident ray, making an angle of one degree with the perpendicular, would be so refracted as to make an angle of only half a degree with the same line, another ray, incident at an angle of two degrees, would be refracted, without sensible error, into an angle of one degree. But when the angles are larger, they vary from this ratio, their sines only preserving the proportion with accuracy: for example, if the angle of incidence at the supposed surface were increased to 90°, the angle of refraction would be 30° only, instead of 45°. Rays of the same kind are in general distinguished by the same colour, although some rays, which differ from each other in refrangibility, have scarcely a discernible difference of colour; and it is possible, on the other hand, to find a surface at which the ratio of the angles is the same for rays of all kinds. (Plate XXVI. Fig. 369, 370.)

In order to obtain the effects of regular reflection and transmission, we must have perfectly smooth and polished substances; for all rough bodies, and sometimes even such as to the touch seem tolerably smooth, have their surfaces divided into innumerable eminences and depressions, constituting, in reality, as many separate surfaces, disposed in all imaginable directions, so that from the equality of the angles of incidence and reflection, with respect to each of these surfaces, the light must be scattered every way, and no regularity can be observed in its direction. It is true that by continuing the mechanical operation of polishing, we only render these surfaces more minute and more numerous; but when they are so much reduced in magnitude as not to be elevated or depressed more than about the millionth part of an inch, they appear to become, for some physical reason, incapable of acting separately, and only to conspire in the general effect.

In all cases of refraction, as well as of reflection, if the ray of light returned directly backwards in the same line to the surface, it would proceed, after a second refraction or reflection, in the direction precisely opposite to that in which it first was incident, so that the same lines would mark its path in both cases. Thus, if we stand before a looking glass, with one eye shut, and cover its place on the glass with a finger, the same finger will hide the other eye as soon as it is shut, and the first is opened in its place; and a similar effect might be observed, if the glass were under water, or behind any other refracting substance. (Plate XXVI. Fig. 371.)

The medium, in which the rays of light are caused to approach nearest to the line perpendicular to its surface, is said to have the greatest refractive density. In general there is a considerable analogy between this refractive density and the specific gravity of the substance: thus water is more refractive than air, and glass than water. But inflammable bodies are usually more refractive than bodies of the same specific gravity, which are not inflammable; and it is well known that from the high refractive power of the diamond, in proportion to its actual density, Sir Isaac Newton most ingeniously conjectured that it was combustible, as more modern experiments have actually shown it to be. It is still more singular that he also imagined, from the same analogy, that water consists of a combination of oily or inflammable particles, with others earthy or not inflammable. In the order of refractive density, beginning from the lowest, or a vacuum, we have airs and gases of different rarities, water, which is the least refractive of all liquids, and which is still less refractive when frozen into ice: alcohol, oils, glass, and lastly the diamond; but probably some metallic substances are much more refractive than even the diamond.

The refractive powers of different substances, are usually estimated by a comparison of the refractions produced at their surfaces in contact with the air, which, in all common experiments, has the same sensible effect as a vacuum or an empty space; the ratio of the angles of refraction and incidence, when small, and that of their sines, in all cases, being expressed by the ratio of 1 to a certain number, which is called the index of the refractive density of the medium. Thus, when a ray of light passes out of air into water, the sines of the angles are in the ratio of 3 to 4, or of 1 to $\frac{4}{3}$, which is, therefore, the index of the refractive density of water. In the same manner, for crown glass, the ratio is that of 2 to 3, and the index $1\frac{1}{2}$; but for flint glass it is somewhat greater, the ratio being nearly that of 5 to 8.

It may easily be shown that a refractive substance, limited by parallel surfaces, must transmit a ray of light, after a second refraction at its posterior surface, in a direction parallel to that in which it first passed through the air. It is also found by experiment that such a substance, interposed between any two mediums of different kinds, produces no alteration in the whole angular deviation of a ray passing from one of them into the other. Hence it may be inferred, that the index of refraction at the common surface of any two mediums is the quotient of their respective indices. For instance, a plate of crown glass being interposed between water on one side and air on the other, it produces no change in the direction of a ray of light entering the water; and the index of refraction at the common surface of glass and water is $\frac{9}{8}$. (Plate XXVI. Fig. 372, 373.)

There is one remarkable consequence of the general law, by which the angles of incidence and refraction are related, that when the angle of incidence exceeds a certain magnitude, the refraction may become impossible; and in this case the ray of light is wholly reflected, in an angle equal to the angle of incidence.

Thus, if the law of refraction required the sine of the angle of refraction to be twice as great as that of incidence, this condition could not take place if the angle of incidence were greater than 30°, so that when a ray passing within a dense medium falls very obliquely on its surface, it must be wholly reflected; and the greater the density of the medium, the more frequently will the light be totally reflected. This reflection is more perfect than any other; the diamond owes much of its brilliancy to it: the great refractive density of this substance not only giving a lustre to its anterior surface, but also facilitating the total reflection of such rays as fall obliquely on its posterior surface. If we hold a prism, near a window, in a proper position, we may observe that its lower surface appears to be divided into two parts, the one much brighter than the other; the common partial reflection taking place in one, and the total reflection in the other. The two surfaces are separated by a coloured arch: it is coloured, because the total reflection commences at different angles for the rays of different colours; and it is curved, because the points, at which the light passing to the eye forms a given angle with the surface, do not lie in a straight line; and if we throw a light on a wall by a reflection of this kind, we may easily observe, as we turn the prism, the point at which the brightness of the image is very conspicuously increased. (Plate XXVI. Fig. 374.)

Such are the principal properties which we discover in light. Before we consider their immediate application to optical instruments, we must examine the general theory of refraction and reflection, at surfaces of different kinds, or the doctrines of dioptrics and catoptrics.

The rays, which constitute a pencil of light, are sometimes parallel to each other, sometimes divergent from a point, and sometimes convergent to a point. The intersection of the directions of any two or more rays of light is called their focus; and the focus is either actual or virtual, accordingly as they either meet in it, or only tend to or from it. Thus, a small luminous object may represent an actual focus of diverging rays, since the light spreads from it in all directions; and the small surface, into which the image of such an object, or of the sun, is collected by a lens or mirror, may represent the actual focus of converging rays. It was to such an image of the sun that the term focus, meaning a fire place, was first applied. But if the rays tending to this focus be intercepted, and made to diverge, the point will then be their virtual focus, since they will never actually arrive at it, being made to diverge as if they proceeded from a new point, which will also be a virtual focus. When the divergence or convergence of rays of light is altered by refraction or reflection at any surface, the foci of the incident and refracted or reflected rays are called conjugate to each other: the new focus is also called the image of the former focus. Thus, in the case already mentioned, where the convergence of the rays to one focus is converted into divergence from another, the two virtual foci are conjugate to each other; and the original focus of the lens or mirror is conjugate to the place of the sun, or of the luminous object. If the object had been put in the place of its image, the image would then have occupied that of the object; a property which follows from the direct return of every ray of light through the path by which it has arrived, and which may easily be illustrated by experimental confirmation. (Plate XXVII. Fig. 375.)

Whenever light is reflected by a plane surface, the conjugate foci are at equal distances from it, and in the same perpendicular. Thus, every point of an image in a looking glass is perpendicularly opposite to the corresponding point of the object, and is at the same distance behind the looking glass, as the point of the object is before it. (Plate XXVII. Fig. 376.)

The focus into which parallel rays are collected, or from which they are made to diverge, is called the principal focus of a surface or substance. The sun is so distant, that the rays, proceeding from any point of his surface, affect our senses as if they were perfectly parallel, and the principal focal distance of a surface or substance may often be practically determined by measuring the distance of the image of the sun, or of any other remote object, which is formed by it.

In order that the rays of light, proceeding from or towards any one point, may be made to converge by reflection towards another, the form of the surface must be elliptical, parabolic, or hyperbolic; there are also curves of still more intricate forms, which possess the same property with respect to refraction. A small portion, however, of any of these curves, differs very little from a circle; and a spherical surface is almost universally substituted in practice for all of them, except that the mirrors of large reflecting telescopes are sometimes made parabolical.

The principal focus of a spherical reflecting surface, whether convex or concave, is half way between the surface and its centre. If a luminous point be placed in the centre of a concave mirror, the rays will all return to the same point; if the point be beyond the centre, the image will be between the centre and the principal focus, its distance from that focus being always inversely as that of the radiant point. Such a focus is never absolutely perfect, for the rays are never collected from the whole surface of the mirror into the same point, except when both the point and its image are in the centre: but, provided that the surface be only a small portion of that of the whole sphere, the aberration will be too small to be easily observed: and the same is true of the foci produced by refracting surfaces. (Plate XXVII. Fig. 377, 378.)

When a ray of light passes through two surfaces forming an angle with each other, including a denser medium, as in the case of a prism of glass, it is always deflected from the angle in which the two surfaces meet. A greater number of surfaces, placed in different directions, constitute what is sometimes called a multiplying glass, each of them bending the rays of light into a different direction. (Plate XXVII. Fig. 379, 380.)

A lens is a detached portion of a transparent substance, of which the opposite sides are regular polished surfaces, of such forms as may be described by lines revolving round a common axis. These lines may be portions of circles, of ellipses, hyperbolas, or of any other curves, or they may be right lines. But in general, one of the sides is a portion of a spherical surface, and the other either a portion of a spherical surface or a plane; whence we have double convex, double concave, planoconvex, planoconcave, and meniscus lenses. The figures of all these are sufficiently described by their names, except that the term meniscus, which properly implies a little moon or crescent, is applied in general to all lenses which are convex on the one side, and concave on the other, although they may be thicker at the edges than in the middle. Sometimes, however, a lens of this kind is distinguished by the term concavoconvex. A lens is generally supposed, in simple calculations, to be infinitely thin, and to be denser than the surrounding medium. (Plate XXVII. Fig. 381.)

The general effect of a lens may be understood, from conceiving its surface to coincide at any given point with that of a prism; for if the angle of the prism be external, as it must be when the lens is convex, the rays will be inflected towards the axis; but if the base of the prism be external, and the lens concave, the rays will be deflected from the axis: so that a convex lens either causes all rays to converge, or lessens their divergence, and a concave lens either causes them to diverge, or lessens their convergence. (Plate XXVII. Fig. 382.)

The principal focus of a double convex or double concave lens, of crown glass, is at the distance of the common radius of its surfaces; and the focal length of a planoconvex lens is equal to the diameter of the convex surface. If the radii of the surfaces are unequal, their effect will be the same as if they were each equal to the harmonic mean between them, which is found by dividing the product by half the sum; or, in the meniscus, by half the difference. Thus, if one of the radii were two inches, and the other six, the effect would be the same as that of a lens of three inches radius; and if it were a meniscus, the same as that of a lens of six inches. (Plate XXVII. Fig. 383, 384.)

The focal length of a lens of flint glass, of water, or of any other substance, may be found, by dividing that of an equal lens of crown glass by twice the excess of the index of refraction above unity. Thus, the index for water being $1\frac{1}{3}$, we must divide the radius by $\frac{2}{3}$, or increase it one half, for the principal focal distance of a double convex or double concave lens of water.

When a radiant point is at twice the distance of the principal focus from a convex lens, the image is at an equal distance on the other side; when the radiant point is nearer than this, the image is more remote, the distance of the image from the principal focus nearest to it being always inversely as the distance of the object from the principal focus on the opposite side. (Plate XXVII. Fig. 385.)

The joint focus of two lenses, in contact with each other, is also found by multiplying together their separate focal lengths, and dividing the product by their sum or difference, accordingly as they agree or differ with respect to convexity and concavity.

We have hitherto considered the place of the focus only in relation to a single point, placed in the axis of the lens or mirror; but it is equally necessary to attend to other points, out of the principal axis; for, in order to form a picture, the rays from a great number of such points must be collected into as many distinct points of the image. Some of the rays proceeding from every radiant point must be considerably bent, in order to be collected into a common focus; others remain nearly straight; and if we can discover which of the rays are ultimately either in the same line with their original direction, or in a direction parallel to it, we may determine the line in which the image of the point in question is to be found. For this purpose, we employ the property of the optical centre, which is a point so situated, that all rays which pass through it, or tend towards it, while they are within the lens, must ultimately acquire a direction parallel to their original direction. In some cases, the optical centre may be without the lens, but no practical inconvenience results from supposing it to be always situated within the lens, especially when its thickness is inconsiderable; so that all rays which pass through the middle point of the lens must proceed, without sensible error, in the same straight line, and the image of any radiant point must consequently be found somewhere in this line: but in the case of a mirror, the centre of its figure is also the optical centre. Now when any radiant point is removed a little from the axis of a lens or mirror, the distance of its image is in general a little diminished, but the difference is too small to be observable in common cases. We may, therefore, suppose it to be at the same distance as if the point remained in the axis, or even to be in a plane crossing the axis perpendicularly at that distance, so as to form part of a flat image, of which the magnitude is determined by straight lines drawn from the extremities of the object through the centre of the lens. This is, however, an approximation which is only admitted for the greater convenience of computation and representation, the image being almost always in reality considerably curved. (Plate XXVII. Fig. 386.)

LECTURE XXXVI.

ON OPTICAL INSTRUMENTS.

AMONG the great variety of instruments depending on optical principles, it is more consistent with our plan to attend first to those which may be denominated optical measures, which are calculated either for the determination of the quantity or intensity of light itself, or for the examination of the properties of various material substances with respect to light. Reflecting quadrants and circles, which are often used in astronomical and nautical observations, although they derive their utility in some measure from optical laws, may most properly be considered as belonging to the subject of practical astronomy.

It is a problem of frequent occurrence in economical investigations, to compare the intensity of the light afforded by any two luminous objects. For this purpose, it is necessary to assume as a principle, that the same quantity of light, diverging in all directions from a luminous body, remains undiminished at all distances from the centre of divergence. Thus, we must suppose that the quantity of light falling on every body is the same as would have fallen on the place occupied by its shadow: and if there were any doubt of the truth of the supposition, it might be confirmed by some simple experiments. It follows that since the shadow of a square inch of any surface, occupies, at twice the distance of the surface from the luminous point, the space of four square inches, the intensity of the light diminishes as the square of the distance increases. We can judge with tolerable accuracy of the equality of two lights by the estimation of the eye, but we cannot form any idea of the proportions of lights of different intensities: if, however, we remove two sources of light to such distances from an object, that they may illuminate it in equal degrees, we may conclude that their original intensities are inversely as the squares of their distances. Count Rumford's photometer performs this very conveniently, by casting two shadows of a given object near each other, on the same surface, the lights being removed to such distances that the shadows appear equally dark. (Plate XXVII Fig. 387, 388.)

For determining the refractive density of solids, it has been usual to form them into a prism, and to measure the angular deviations which they produce; and for fluids, to inclose them either in a hollow prism, or between two meniscus lenses, and to measure the angular deviations produced by the prisms, and the focal distances of the lenses. But in most cases, Dr. Wollaston's apparatus is far preferable to both these methods: it is arranged for ascertaining the angle at which light, moving within a certain dense transparent substance, begins to be totally reflected from the common surface of that substance and the solid or fluid which is to be examined. Thus, if we first measure the angle, at which light begins to be totally reflected from the posterior surface of a prism of glass, in contact with air, we may readily determine its refractive power; and then, having caused a drop of a fluid to adhere to that surface, or fixed a solid to it by a small portion of some fluid denser than itself, we may observe, as we turn the prism round its axis, at what angle the drop or spot begins to disappear, and may thence calculate the refractive density of the substance; and even without actual measurement of the angle, we may readily compare the disappearance of the drop or spot with that of others placed near it, of which the properties are known. Dr. Wollaston has, however, rendered the process still easier and more simple, by employing a rectangular prism of glass, with sights fixed to a jointed frame, of such a construction as to enable him to read off, by a vernier, without any calculation, the index of the refractive power of any substance less dense than glass. (Plate XXVII. Fig. 389.)

All instruments strictly optical are employed for forming an image of an external object: the simplest are mirrors and lenses, which form a single image only, either actual or virtual, and sometimes depict it on a surface calculated for receiving and exhibiting it. Other instruments repeat the image once or more under several forms, in general enlarging it continually; and these are either microscopes or telescopes, which present us with great diversity in their arrangements, and in the appurtenances subservient to their uses.

It is a general rule, that when an image of an actual object is formed by any lens or speculum, if the rays converge to an actual focus, the image is inverted; but erect, if they diverge from a virtual focus, and the object and image subtend equal angles at the centre of the lens or speculum. Hence, a convex lens and a concave mirror form an inverted image, smaller than the object, whenever the object is at a greater distance than twice the principal focal length; but larger, when the object is within this distance; and when it is within the principal focal distance, the magnified image is virtual and erect, and may be seen by looking into the concave mirror, or by looking through the lens towards the object. But a concave lens and a convex mirror always form a virtual image of a real object, which is erect, and smaller than the object. (Plate XXVII. Fig. 390 . . 394.)

When the object is precisely in the principal focus of a convex lens or a concave mirror, the virtual image becomes infinitely distant; so that from whatever point in the neighbourhood of the lens it may be viewed, it must subtend the same angle, which is always equal to that which the object subtends at the centre of the lens: and since this angle may easily be much greater than that under which the object can be conveniently viewed by the naked eye, such a lens or mirror is often used as a simple microscope; and its magnifying power may be ascertained from a comparison of the angles which the object and image subtend. Thus, if a person cannot see a minute object with the naked eye at a distance less than eight inches, a lens of half an inch focal length will represent it to him in an angle 16 times as great: but if he can see it without the lens at the distance of four inches, the lens will magnify it to his eye but eight times. Supposing, however, the eye to be applied close to the lens, the object may be viewed a little within the focal distance, and its apparent angular magnitude may be increased 17 times instead of 16, and 9 times instead of 8. (Plate XXVII. Fig. 395, 396.)

Since the magnifying power of a lens is the greater, the smaller its focus, it is usual ot employ the minutest lenses that can be ground, and sometimes a small globule is formed by fusion in a lamp. Even a drop of water, placed in the perforation of a plate, makes a tolerable magnifier; and it has been proposed to substitute for water a transparent varnish, which is less liable to evaporate.

Supposing the whole light that proceeds from a distant object, and falls on a lens or speculum, to be collected in the image, its intensity must be increased in the ratio of the surface of the lens or speculum to that of the image. The image is greater in proportion as the object is greater; consequently the degree of condensation produced by any lens is the greater as the object is smaller, thus if the diameter of a lens were an inch, and the image of the sun formed by it were also an inch in diameter, the density of the light would be unaltered; but the image of a star would be infinitely brighter than the direct light of the star falling on the lens. The illumination of any image formed by a lens or mirror, supposing no light to be lost, is always the same as would be produced by the direct light of the surface of the lens or mirror, if it were equally luminous with the surface of the object which emits the light. It may also be shown, that when two lenses are of similar forms, their focal lengths being proportional to their diameters, they must produce the same degree of illumination in the image: but as far as the heat excited may be supposed to be a measure of the quantity of light, this conclusion is not confirmed by experiment: it is probable, however, that the greater heat. produced by a larger lens, is only derived from the greater extent of surface exposed at once to the solar rays.

Lenses are most commonly made of glass, but sometimes of rock crystal, or of other transparent substances. It is difficult to find glass, especially flint glass, for large lenses, sufficiently free from veins: it has been proposed to suffer the melted glass to cool without agitation, and to cut the lens out of any of its strata taken in a horizontal direction; but this method appears to be liable to several practical objections. Mirrors are made either of glass, coated with an amalgam of mercury and tin, or of metal, as of platina, of silver, or of an alloy of copper and tin, to which a little arsenic and silver are sometimes added. Mirrors of metal are more perfect than those of glass, because they are free from the inconvenience of a double reflection; but they are more expensive, and are liable to tarnish. Where a large mirror is required, with a weak reflection only, we may employ a single surface of glass, the back of the piece being covered with a black coating of some substance differing little from glass in its refractive density, by means of which the second reflection is avoided.

When the image formed by a lens or mirror is received on a smooth but unpolished surface, which is capable of irregular reflection, it is visible in every direction. Such an image is exhibited in the camera obscura, the solar microscope, and the magic lantern, or lucernal microscope.

The general effect of the camera obscura is the same as may often be observed in a dark room, where there is a small hole in the window shutter: the great masses of light and shade, before the windows, being represented in an inverted position, in the parts of the room diametrically opposite to them, which are illuminated in different degrees, according to the quantity of light which can reach them in straight lines from the external objects. A lens, of a focal length somewhat smaller than the distance of the surface on which the picture is projected, renders the images much more distinct; but some of them are unavoidably imperfect and ill defined, unless the objects happen to be situated at the same distance from the aperture; for the focus of the lens can never be adjusted at once to nearer and more remote objects; nor would the picture be rendered more natural by such an adjustment, for it would present to the eye at one view, with equal distinctness, objects which never can be seen at once without some degree of confusion. Sometimes the picture is intercepted, by a speculum placed obliquely, and is thrown upwards on the surface of a plate of ground glass, upon which its outline may be traced with a black lead pencil, and an impression may be taken from it on moist paper, which will represent the natural situation of the objects without inversion. Another arrangement is, to place the lens horizontally, with the speculum above it, which throws the image through the lens, upon a flat surface placed below, on which the objects may be delineated in their natural position, but not without some impediment from the interception of the light by the hand and the instrument employed. Such a surface, however, ought not to be perfectly flat, in order to afford the most distinct image, although by means of a meniscus lens, with a cover admitting the light only through a small aperture near its centre, on the principle of Dr. Wollaston's periscopic spectacles, an image nearly flat might be obtained; but in this case too much of the light would be excluded. It has been usual to consider the image of a very distant object, formed by a convex lens, as a portion of a spherical surface, of which every part is equally distant from the centre of the lens; but this estimate is extremely erroneous, for the effect of the obliquity of the different pencils of rays materially increases the curvature of the image. In fact no pencil of rays, falling obliquely on a spherical surface, can be collected any where to a perfect focus: the image of a circle would become most distinct at one distance, and that of its diameter at another; but for both these images, the surface ought to be much more curved than that which has been usually considered, and the mean of the curvatures required for them, which must be the best form for the ground or bottom of a camera obscura, is equal to that of a sphere of which the radius is three eighths of the focal distance, when a double convex lens of crown glass is employed. (Plate XXVII. Fig. 397 . . 399.)

In the solar microscope, an image is formed on a wall or screen, by means of a lens of small focal length, near to which the object is placed, so that the image is very much magnified. For this purpose the room must be darkened, and the object strongly illuminated by the sun's light, which is condensed by means of a large lens, and sometimes by two or more lenses placed at a distance from each other; but care must be taken to avoid burning the object by bringing it exactly into the focus; and, on the other hand, if it be much beyond the focus, the light will be thrown upon a small part of the image only; the best arrangement appears to be, to bring the focus of the condensing lenses very near to the small lens; and in order to adjust the instrument in the most convenient manner, the distances of all the lenses ought to be moveable at pleasure: the want of this precaution is a material defect in the usual construction of the instrument. The speculum which first receives the light must be capable of motion in all angular directions, in order to allow us to accommodate its position to the changeable place of the sun; and the adjustment has sometimes been performed by means of a heliostate, an instrument calculated for turning the speculum by clockwork, into such a position as always to reflect the sun's light in the required direction. An easier method would be to employ two speculums, the one moveable round an axis parallel to that of the earth, and reflecting the sun's light into the direction of its axis, the other fixed, and changing this direction into any other that might be required. When an opaque object is to be examined, the light may be thrown on it either by a plane mirror placed obliquely, or by a perforated concave mirror; and if the object is small, the concave mirror appears to be the more eligible. (Plate XXVIII. Fig. 400.)

By night, a lamp, with a large lens before it, may supply the place of the sun's light, and the instrument will become a lucernal microscope, which, when painted glass sliders are employed as objects for the amusement of children, is called a magic lantern: and this, exhibited on a larger scale, and projecting an image on a semitransparent screen of taffetas, instead of a wall, has of late been the source of much entertainment under the name of the phantasmagoria, a term which implies the raising of spectres. In order to favour the deception, the sliders are made perfectly opaque, except where the figures are introduced, the glass being covered, in the light parts, with a more or less transparent tint, according to the effect required. Several pieces of glass may also be occasionally placed behind each other, and may be made capable of such motions as will nearly imitate the natural motions of the objects which they represent. The figures may also be drawn with water colours on thin paper, and afterwards varnished. By removing the lantern to different distances, and altering at the same time more or less the position of the lens, the image may be made to increase or diminish, and to become more or less distinct at pleasure, so that to a person unaccustomed to the effects of optical instruments, the figures may appear actually to advance and retire. In reality, however, these figures become much brighter as they are rendered smaller, while in nature the imperfect transparency of the air causes them to appear fainter when they are remote than when they are near: this imperfection might be easily remedied by the interposition of some semiopaque substance, which might gradually be caused to admit more light as the figure became larger, or by uncovering a larger or a smaller portion of the lamp, or of its lens. Sometimes, by throwing a strong light upon an actual opaque object, or on a living person, its image is formed on the curtain, retaining its natural motions: but in this case the object must be considerably distant, otherwise the images of its nearer and remoter parts will never be sufficiently distinct at once, the refraction being either too great for the remoter, or too small for the nearer parts: and there must also be a second lens, placed at a sufficient distance from the first to allow an inverted image to be formed between them, and to throw a second picture of this image on the screen, in its natural erect position, unless the object be of such a nature that it can be inverted without inconvenience. This effect was very well exhibited at Paris by Robertson; he also combined with his pictures the shadows of living objects, which imitate tolerably well the appearance of such objects in a dark night, or by moonshine: and while the room was in complete darkness, concealed screens were probably let down in various parts of it, on which some of the images were projected; for they were sometimes actually situated over the heads of the audience. (Plate XXVIII. Fig. 401.)

In almost all telescopes and compound microscopes, the image formed by one lens or mirror stands in the place of a new object for another. The operation of such instruments may be illustrated by placing a screen of fine gauze at the place of the image, which receives enough light to make the image visible in all directions, and yet transmits enough to form the subsequent image. The simplest of such instruments is the astronomical telescope. Here the object glass first forms an actual inverted image, nearly in the principal focus of the eye glass, through which this image is viewed as by a simple microscope, and therefore still remains apparently inverted. In order to find the angular magnifying power, we must divide the focal length of the object glass by that of the eye glass: this quotient is consequently the greater as the focal length of the object glass is greater, and as that of the eye glass is smaller; but the power of the instrument cannot be increased at pleasure by lessening the focal length of the eye glass, because the object glass would not furnish light enough to render the view distinct, if the magnifying power were too great. (Plate XXVIII. Fig. 402.)

The double or compound microscope resembles in its construction the astronomical telescope, except that the distance of the lenses much exceeds their joint focal length; and the angular magnitude is greater than when the same object is viewed through the eye glass alone, in proportion as the first image is further from the object glass than the object itself. (Plate XXVIII. Fig. 403.)

In the Galilean telescope, or opera glass, a concave eye glass is placed so near the object glass, that the first image would be formed beyond it, and near its principal focus; and the second image, formed by the eye glass, which is the virtual image viewed by the eye, being on the opposite side of the centre, is inverted with respect to the first image, and erect with respect to the object. In this case also the magnifying power is indicated by the quotient of the numbers expressing the focal lengths of the glasses. (Plate XXVIII. Fig. 404.)

The inverted image of the astronomical telescope may be made erect by means of an additional eye glass. In the common day telescope of Rheita, two such eye glasses are employed, of nearly equal focus, which have the advantage of procuring a greater extent in the field of view; they are usually so placed as to have little or no effect on the magnifying power. (Plate XXVIII. Fig. 405.)

Dr. Herschel's reflecting telescopes resemble, in their effects, the simple astronomical telescope; a concave speculum, or mirror, being substituted for the object glass, and the eye glass being so placed as to magnify the image formed by the speculum. But since the speculum, if it received the principal rays perpendicularly, would send them back in the same direction, it is necessary, in this construction, to have them reflected somewhat obliquely, the speculum being a little inclined to the axis of the telescope, in order that the light may have free access to it. An arrangement of this kind was proposed long ago by Maire, but it has been very little employed before Dr. Herschel's time. This excellent philosopher and mechanic has carried the perfection of his telescopes to a degree far exceeding all that could have been expected from the labours of former opticians. His instruments allow him to extend the linear dimensions of his objects several thousand times:

5

but he commonly finds it more eligible to employ only powers of 5 or 600, which afford a much stronger illumination. (Plate XXVIII. Fig. 406.)

The Newtonian reflector has a plane speculum placed in its axis, at the inclination of half a right angle, which intercepts the rays about to form the image, and throws them into the focus of an eye glass fixed in the side of the tube. The plane speculum which he employed was the posterior surface of a rectangular prism of glass, which produces a total reflection: but Dr. Herschel has found that the sources of error are diminished by wholly omitting this speculum. (Plate XXVIII. Fig. 407.)

In the Gregorian telescope, the object speculum is perforated, and the image formed by it is received into the focus of a smaller concave speculum, which returns it to be viewed through the aperture by the eye glasses. It has been objected to this form of the reflecting telescope, which is the first that was invented, that the best part of the speculum is sacrificed by the perforation. But Dr. Herschel has found that the image formed by the external part of a speculum is in general more perfect than that which is formed by the central part. (Plate XXVIII. Fig. 408.)

For the smaller concave speculum of Gregory, Mr. Cassegrain substituted a convex one, placing it within the focal distance of the large speculum, so as to form the first actual image nearly in the same place as the second image of the Gregorian telescope; but this image is inverted. The instrument has some advantage in theory, with respect to the perfection of the focus; but it is little used. (Plate XXVIII. Fig. 409.)

Dr. Smith's reflecting microscope resembles Cassegrain's telescope, but the rays of light are first admitted through a perforation in the small speculum, that part of them which tends to fall immediately on the eye being intercepted by a screen. The convexity of the one mirror is nearly equal to the concavity of the other; and the instrument, although seldom employed, is said to succeed extremely well. (Plate XXVIII. Fig. 410.)

The image of a very distant object, formed by a speculum of any kind, is considerably less curved than that which is depicted by a lens of equal focal length. There is a similar imperfection in the nature of the focus of oblique pencils, but it is confined within narrower limits, the remotest part of the image in which any radiating lines would be most distinctly represented, being a flat surface, and the nearest, in which circles would become most distinct, being a part of a sphere touching the speculum: so that the radius of the mean curvature is equal to the focal distance. (Plate XXVIII. Fig. 411.)

The magnifying power of a refracting telescope may often be measured, by comparing the diameter of the object glass with that of the narrowest space, into which the beam of light is contracted beyond the eye glass, provided that none of the light has been intercepted in its passage through the telescope: for the object will be viewed through the telescope in an angle as much greater than that which it naturally subtends, as the diameter of the object glass is greater than that of this contracted pencil, which may be considered as an image of the object glass. But in the Galilean telescope, this method cannot be employed, since no such image is formed. The field of view, in a simple telescope, or the angular magnitude of that part of an object which can be seen through it at once, is nearly equal to the magnitude of the eye glass as seen from the object glass.

If a lens be added to any refracting telescope at the place of the first image, it will have no effect either on the place or on the magnitude of any subsequent image, but it will enlarge the field of view, by throwing more pencils of light on the original eye glass. If, however, the image fell exactly on such a lens, it would be liable to be impaired by any accidental impurities of its substance or on its surface, every opaque particle intercepting the whole of the light belonging to one of its points, which would not happen if the image were at a small distance from the lens. A field glass is, therefore, usually placed, both in telescopes, and in the common compound microscope, a little nearer to the object glass than the place of the first image. The best places for the various lenses, in an eye piece, are partly determined from similar considerations, but they require also in general to be adjusted by experiment, for several circumstances are concerned in the performance of a telescope, which are almost too intricate for practical calculation, although some assistance may certainly be obtained from theory with regard to the most important of them. The curvature of the image produced by any lens has already been mentioned: it may be in some measure remedied by Mr. Ramsden's method of placing a planoconvex lens a little beyond the image, with its flat side turned towards it. Mr. Ramsden also employs an eye piece constructed on this principle instead of a simple microscope, under the name of a double magnifier. The aberration of the different parts of any single pencil of rays, from the corresponding point of the image, requires also to be considered in the construction of telescopes: its magnitude is such, in the case of a double convex lens of crown glass, that those parts of a pencil of parallel rays which fall on it near the circumference meet each other in a point, which is within the true focus, by a distance a little more than half as great again as the thickness of the lens. In an image formed by a concave speculum, of equal focal length, this aberration would be only $\frac{1}{13}$ as great; it may, however, be almost entirely corrected, in refracting telescopes, by employing proper proportions in the dimensions of the various lenses. (Plate XXVIII. Fig. 412, 413.)

A still more important aberration, from which reflecting telescopes are also wholly free, is that which arises from the different refrangibilities of the rays of light of different colours, which form an infinite number of images, neither agreeing perfectly in situation nor in magnitude, so that the objects are rendered indistinct by an appearance of colours at their edges: this imperfection, however, Mr. Dollond has in great measure obviated, by his achromatic object glasses: the construction of which depends on the important discovery, that some kinds of glass separate the rays of different colours from each other much more than others, while the whole deviation produced in the pencil of light is the same. Mr. Dollond combined, therefore, a concave lens of flint glass with a convex lens of crown glass, and sometimes with two such lenses; the concave lens of flint glass being sufficiently powerful to correct the whole dispersion of coloured light produced by the crown glass, but not enough to destroy the effect of its refraction, which was still sufficient to collect the rays of light into a distant focus. For this purpose, it is necessary that the focal lengths of the two lenses should be in the same proportion as the dispersive powers of the respective substances, when the mean deviations of the pencils are equal; that is, in the case of the kinds of glass commonly used, nearly in the ratio of 7 to 10. Sometimes also the chromatic aberration, that is, the error arising from the different refrangibilities of the different rays, is partially corrected in an eye piece, by placing a field glass in such a manner, as considerably to contract the dimensions of the image formed by the least refrangible rays, which is nearest to the eye glass, and to cause it to subtend an equal angle with the image formed by the most refrangible rays, this image being little affected by the glass. (Plate XXVIII. Fig. 414, 415.)

The apparent magnitude of an object, viewed through a telescope, may be measured, with great accuracy, by a scale or by wires, introduced at the place of the last image, reducing afterwards the angle thus ascertained according to the magnifying power. Care must, however, be taken to avoid as much as possible the distortion which usually accompanies any curvature of the image; and the wires, one of which is sometimes made moveable by means of a micrometer screw, must be sufficiently illuminated to be distinctly visible. Sometimes a scale is introduced, which, from the apparent magnitude of a known object, such as that of a man of ordinary height, or of a portion of a wall built with bricks of the usual size, enables us at once to read off its actual distance, which is expressed on the scale in hundreds of yards. The angular magnitude of an object, seen through a telescope, may also be found, by viewing at the same time, with the other eye, either a scale, or any other object of known dimensions, placed at a given distance: the lucid disc micrometer of Dr. Herschel is employed in this manner for judging of the magnitude of the celestial bodies. The divided object glass micrometer affords another mode of measurement: the object glass being divided into two semicircular portions, one of which slides on the other; each portion acts as a separate lens, and two images of every part of the object being formed, the angular distance of any two points is determined by bringing their images together, and measuring the displacement of the moveable portion of the object glass, which is required for procuring the coincidence. Sometimes also a similar purpose is answered by inserting a divided glass in the eye piece, which acts nearly on the same principle, and which seems to be somewhat less liable to error. In a reflecting telescope of Cassegrain's construction, Mr. Ramsden has also produced the same effect by dividing the convex speculum, and causing a part of it to turn round an axis. All these arrangements particularly deserve the attention of those who are employed in practical astronomy and in geography, since the advancement of these sciences much depends on the accuracy of the telescopic and microscopic measures, which are performed by means of optical instruments. (Plate XXVIII. Fig. 416, 417.)

LECTURE XXXVII.

ON PHYSICAL OPTICS.

HAVING examined the general theory of optics, and the construction of optical instruments, we are now to consider those properties and affectiots of light, which rather belong to its natural history, than to its mechanical effects; to trace its relations to the particular phenomena of nature; to investigate the manner in which it is connected with our sensations, and to inquire on what intimate mode of action the various effects of light depend. All these subjects may be properly comprehended under the denomination of physical optics, but we shall find it convenient to reserve each of the two last for a separate examination. The sources of light, the velocity of its motion, its interception and extinction, its dispersion into different colours; the manner in which it is affected by the variable density of the atmosphere, the meteorological appearances in which it is concerned, and the singular properties of particular substances with regard to it, will be the first subjects of our investigation.

The sources, from which light is commonly derived, are either the sun or stars, or such terrestrial bodies as are undergoing those changes which constitute combustion. The process of combustion implies a change in which a considerable emission of light and heat is produced; but it is not capable of a very correct definition: in general it requires an absorption, or at least a transfer, of a portion of oxygen; but there appear to be some exceptions to the universality of this distinction; and it has been observed that both heat and light are often produced where no transfer of oxygen takes place, and sometimes by the effect of a mixture which cannot be called combustion.

Light is also afforded, without any sensible heat, by a number of vegetable and animal substances, which appear to be undergoing a slow decomposition, not wholly unlike combustion. Thus decayed wood, and animal substances slightly salted, often afford spontaneously a faint light, without any elevation of temperature; and it is not improbable that the light of the ignis fatuus may proceed from a vapour of a similar nature.

The effects, which are commonly attributed to the motions of the electrical fluid, are often attended by the production of light; and violent or rapid friction frequently seems to be the immediate cause of its appearance. But it is difficult to ascertain whether friction may not be partly concerned in the luminous phenomena attributedto electricity, or electricity in the apparent effects of friction. Light is sometimes produced by friction with a much lower degree of heat than is required for combustion, and even when it is accompanied by combustion, the heat produced by the union of these causes may be very moderate: thus it is usual in some coal mines, to obtain a train of light by the continual collision of flint and steel, effected by the machine called a fire wheel, in order to avoid setting fire to the inflammable gas emitted by the coal, which would be made to explode if it came near the flame of a candle.

There is a remarkable property, which some substances possess in an eminent degree, and of which few, except metals and water, are entirely destitute. These substances are denominated solar phosphori; besides the light which they reflect and refract, they appear to retain a certain portion, and to emit it again by degrees till it is exhausted, or till its emission is interrupted by cold. The Bolognan phosphorus was one of the first of these substances that attracted notice; it is a sulfate of barytes, found in the state of a stone; it is prepared by exposure to heat, and is afterwards made up into cakes: these, when first placed in a beam of the sun's light, and viewed afterwards in a dark room, have nearly the appearance of a burning coal, or a red hot iron. Burnt oyster shells, and muriate of lime have also the same property, and some specimens of the diamond possess it in a considerable degree. From the different results of experiments apparently accurate, made by different persons, there is reason to conclude that some of these phosphori emit only the same kind of light as they have received, while others exhibit the same appearances, to whatever kind of light they may have been exposed. Sometimes it has even been found that light of a particular colour has been most efficacious in exciting in a diamond the appearance of another kind of light, which it was naturally most disposed to exhibit. The application of heat to solar phosphori in general expedites the extrication of the light which they have borrowed, and hastens its exhaustion; it also produces, in many substances, which are not remarkable for their power of imbibing light, a temporary scintillation, or flashing, at a heat much below ignition: the most remarkable of these are fluor spar in powder, and some other crystallized substances. It appears that luminous bodies in general emit light equally in every direction, not from each point of any of their surfaces, as some have supposed, but from the whole surface taken together, so that the surface, when viewed obliquely, appears neither more nor less bright than when viewed directly.

However light of any kind may have at first originated, there is reason to believe that the velocity with which it passes through a given medium is always the same. It has been ascertained by the astronomical observations of Roemer and of Bradley, that each ray of light, emitted by the sun, arrives at the earth in eight minutes and one eighth, when the earth is at its mean distance of about 95 millions of miles. Roemer deduced this velocity from observations on the eclipses of the satellites of Jupiter, and Bradley confirmed it by his discovery of the cause of the apparent aberration of the fixed stars.

This aberration is produced by the effect of the revolution of the earth in its orbit, combined with that of the progressive motion of light. Since light proceeds always in right lines, when its motion is perfectly undisturbed, if a fine tube were placed so as to receive a ray of light, passing exactly through its axis when at rest, and then, remaining in the same direction, were moved transversely with great velocity, it is evident that the side of the tube would strike against the ray of light in its passage, and that in order to retain it in the axis, the tube must be inclined, in the same manner as if the light, instead of coming in its actual direction, had also a transverse motion in a contrary direction to that of the tube. The axis of a telescope, or even of the eye, may be considered as resembling such a tube, the passage of the light through the refracting substances not altering the necessary inclination of the axis. In various parts of the earth's orbit, the aberration of any one star must be different in quantity and in direction; it never exceeds 20 seconds each way, and must, therefore, in common observations, be wholly insensible. (Plate XXIX. Fig. 418.)

The quantity of light, which is reflected by a substance of any kind, depends not only on the nature of the substance, but also on the obliquity of its incidence: and it sometimes happens, that a surface, which reflects a smaller portion of direct light than another, reflects a greater portion when the light falls very obliquely on its surface. Bouguer found that the surface of water reflected only one fifty fifth part of the light falling perpendicularly on it, that of glass one fortieth, and that of quicksilver more than two thirds: but when the obliquity was as great as possible, the water reflected nearly three fourths of the incident light, and the glass about two thirds only.

Of the light which passes by a dense substance of any kind, the greatest part pursues its course undisturbed, but there is always a certain divergence, which has been called by Grimaldi diffraction, and by Newton inflection. This effect is usually attended by the production of colours, and will therefore require to be more particularly considered hereafter.

The separation of colours by refraction is one of the most striking of all optical phenomena. It was discovered by Newton that white light is a compound of rays of different kinds, mixed in a certain proportion, that these rays differ in colour and in refrangibility, that they constitute a series, which proceeds by gradual changes from red to violet, and that those substances which appear coloured when placed in white light, derive their colours only from the property of reflecting some kind of rays most abundantly, and of transmitting or extinguishing the rest. Dr. Herschel has added to this series rays of heat less refrangible than the red, and Ritter and Dr. Wollaston have discovered, beyond the violet, other still more refrangible rays, which blacken the salts of silver.

It has generally been supposed, since the time of Newton, that when the rays of light are separated as completely as possible by means of refraction, they exhibit seven varieties of colour, related to each other with respect to the extent that they occupy, in ratios nearly analogous to those of the ascending scale of the minor mode in music. The observations were, however, imperfect, and the analogy was wholly imaginary. Dr. Wollaston has determined the division of the coloured image or spectrum, in a much more accurate manner than had been done before: by looking through a prism, at a narrow line of light, he produces a more effectual separation of the colours, than can be obtained by the common method of throwing the sun's image on a wall. The spectrum formed in this manner consists of four colours only, red, green, blue, and violet, which occupy spaces in the proportion of 16, 23, 36, and 25, respectively, making together 100 for the whole length; the red being nearly one sixth, the green and the violet each about one fourth, and the blue more than one third of the length. The colours differ scarcely at all in quality within their respective limits, but they vary in brightness; the greatest intensity of light being in that part of the green which is nearest to the red. A narrow line of yellow is generally visible at the limit of the red and green, but its breadth scarcely exceeds that of the aperture by which the light is admitted, and Dr. Wollaston attributes it to the mixture of the red with the green light. There are also several dark lines crossing the spectrum within the blue portion and in its neighbourhood, in which the continuity of the light seems to be interrupted. This distribution of the spectrum Dr. Wollaston has found to be the same, whatever refracting substance may have been employed for its formation; and he attributes the difference, which has sometimes been observed in the proportions, to accidental variations of the obliquity of the rays. The angular extent of the spectrum formed by a prism of crown glass is one 27th of the deviation of the red rays; by a prism of flint glass, one 19th. (Plate XXIX. Fig. 419.)

In light produced by the combustion of terrestrial substances, the spectrum is sometimes still more interrupted; thus, the bluish light of the lower part of the flame of a candle is separated by refraction into five parcels of various colours; the light of burning spirits, which appears perfectly blue, is chiefly composed of green and violet rays; and the light of a candle into which salt is thrown abounds with a pure yellow, inclining to green, but not separable by refraction. The electrical spark furnishes also a light which is differently divided in different circumstances. (Plate XXIX. Fig. 420.)

If the breadth of the aperture viewed through a prism is somewhat increased, the space occupied by each variety of light in the spectrum is augmented in the same proportion, and each portion encroaches on the neighbouring colours, and is mixed with them: so that the red is succeeded by orange, yellow, and yellowish green, and the blue is mixed on the one side with the green, and on the other with the violet; and it is in this state that the prismatic spectrum is commonly exhibited. (Plate XXIX. Fig. 421.)

When the beam of light is so much enlarged as to exceed the angular magnitude of the spectrum, it retains its whiteness in the centre, and is terminated by two different series of colours at the different ends. These series are still divided by well marked lines: on the one hand the red remains unmixed; the space belonging to the green and blue becomes a greenish yellow, nearly uniform throughout, and here the appearance of colour ends, the place of the violet being scarcely distinguishable from the neighbouring white light: on the other hand, the space belonging to the red, green, and blue, of the simple spectrum appears of a bluish green, becoming more and more blue till it meets the violet, which retains its place without alteration. This second series is also the same that accompanies the limit of total reflection at the posterior surface of a prism. (Plate XXIX. Fig. 422.)

Sir Isaac Newton observed that the effect of white light on the sense of sight might be imitated by a mixture of colours taken from different parts of the spectrum, notwithstanding the omission of some of the rays naturally belonging to white light. Thus, if we intercept one half of each of the four principal portions into which the spectrum is divided, the remaining halves will still preserve, when mixed together, the appearance of whiteness; so that it is probable, that the different parts of those portions of the spectrum, which appear of one colour, have precisely the same effect on the eye. It is certain that the perfect sensations of yellow and of blue are produced respectively, by mixtures of red and green, and of green and violet light, and there is reason to suspect that those sensations are always compounded of the separate sensations combined: at least this supposition simplifies the theory of colours; it may, therefore, be adopted with advantage, until it be found inconsistent with any of the phenomena; and we may consider white light as composed of a mixture of red, green, and violet, only, in the proportion of about two parts red, four green, and one violet, with respect to the quantity or intensity of the sensations produced.

If we mix together, in proper proportions, any substances exhibiting these colours in their greatest purity, and place the mixture in a light sufficiently strong, we obtain the appearance of perfect whiteness; but in a fainter light the mixture is grey, or of that hue which arises from a combination of white and black; black bodies being such as reflect white light but in a very scanty proportion. For the same reason, green and red substances mixed together usually make rather a brown than a yellow colour, and many yellow colours, when laid on very thickly, or mixed with black, become brown. The sensations of various kinds of light may also be combined in a still more satisfactory manner by painting the surface of a circle with different colours, in any way that may be desired, and causing it to revolve with such rapidity, that the whole may assume the appearance of a single tint, or of a combination of tints, resulting from the mixture of the colours. (Plate XXIX. Fig. 423 . . 426.)

From three simple sensations, with their combinations, we obtain seven primitive distinctions of colours; but the different proportions, in which they may be combined, afford a variety of tints beyond all calculation. The three simple sensations being red, green, and violet, the three binary combinations are yellow, consisting of red and green; crimson, of red and violet; and blue, of green and violet; and the seventh in order is white light, composed by all the three united. But the blue thus produced, by combining the whole of the green and violet rays, is not the blue of the spectrum, for four parts of green and one of violet make a blue differing very little from green; while the blue of the spectrum appears to contain as much violet as green: and it is for this reason that red and blue usually make a purple, deriving its hue from the predominance of the violet.

It would be possible to exhibit at once to the eye the combinations of any three colours in all imaginable varieties. Two of them might be laid down on a revolving surface, in the form of triangles placed in opposite directions, and the third on projections perpendicular to the surface, which, while the eye remained at rest in any one point, obliquely situated, would

2

exhibit more or less of their painted sides, as they passed through their different angular positions: and the only further alteration, that could be produced in any of the tints, would be derived from the different degrees of light only. The same effect may also be exhibited by mixing the colours in different proportions, by means of the pencil, beginning from three equidistant points as the centres of the respective colo nrs, (Plate XXIX. Fig. 427.)

The ordinary atmospherical refraction cannot be determined in the usual manner from the knowledge of its density, and of the angular direction of the incident or refracted light, since the constitution of the atmosphere is such, that its density varies every where with its height, and the curvature of the earth's surface causes the inclination of the strata through which the ray passes to be perpetually changed; the difference of temperature at different elevations increases also the difficulty of an exact calculation, and it is only very lately that Mr. Laplace, by a comparison of astronomical with meteorological observations, has given a satisfactory solution of the problem in all its extent. But for practical uses, the refraction may be determined with sufficient accuracy by an approximation which is easily remembered; the deviation being at all altitudes one sixth part as great as the refracted ray would undergo, at the horizontal surface of a medium six times as dense as the air. When a celestial object appears exactly in the horizon, it is actually more than half a degree below it, since the refraction amounts to 33 minutes, when the barometer stands at $29\frac{6}{10}$ inches, and Fahrenheit's thermometer at 50°.

The accidental variations of the temperature of the air, at different parts, produce, however, great irregularities in its refraction, especially near the horizon. The most remarkable of these is occasioned by the rarefaction of the air in the neighbourhood of the surface of water, of a building, or of the earth itself, in consequence of which a distant object appears to be depressed instead of being elevated, and is sometimes seen at once both depressed and elevated, so as to appear double, one of the images being generally in an inverted position, as if the surface possessed a reflective power; and there seems indeed to be a considerable analogy between this kind of refraction and the total reflection which happens within a denser medium. These effects are known by the appellations looming, mirage, and Fata Morgana:

they may be very completely imitated, as Dr. Wollaston has shown, by looking at a distant object along a red hot poker, or through a saline or saccharine solution with water and spirit of wine floating on it. The effect of refraction on the apparent places of terrestrial objects must be frequently disturbed by circumstances of this kind; but its magnitude is usually about one tenth of the angular distance of the object, considered as a part of the earth's circumference. (Plate XXIX. Fig. 428, 429.)

The atmospherical phenomena of rainbows and halos present us with examples of the spontaneous separation of colours by refraction. The rainbow is universally attributed to the refraction and reflection of the sun's rays in the minute drops of falling rain or dew, and the halos, usually appearing in frosty atmospheres, are in all probability produced by the refraction of small triangular or hexagonal crystals of snow. It is only necessary, for the formation of a rainbow, that the sun should shine on a dense cloud, or a shower of rain, in a proper situation, or even on a number of minute drops of water, scattered by a brush or by a syringe, so that the light may reach the eye after having undergone a certain angular deviation, by means of various refractions and reflections; and the drops so situated must necessarily be found somewhere in a conical surface, of which the eye is the vertex, and must present the appearance of an arch. The light, which is reflected by the external surface of a sphere, is scattered almost equally in all directions, setting aside the difference arising from the greater efficacy of oblique reflection; but when it first enters the drop, and is there reflected by its posterior surface, its deviation never exceeds a certain angle, which depends on the degree of refrangibility, and is, therefore, different for light of different colours: and the density of the light being the greatest at the angle of greatest deviation, the appearance of a luminous arch is produced by the rays of each colour at its appropriate distance. The rays which never enter the drops produce noother effect, than to cause a bright ness, or haziness round the sun, where the reflection is the most oblique: those which are once reflected within the drop exhibit the common internal or primary rainbow, at the distance of about 41 degrees from the point opposite to the sun: those which are twice reflected, the external or secondary rainbow, of 52°: and if the effect of the light, three times reflected, were sufficiently powerful, it would appear at the distance of about 42 degrees trom the sun. The colours of both rainbows encroach considerably on each other; for each point of the sun may be considered as affording a distinct arch of each colour, and the whole disc as producing an arch about half a degree in breadth for each kind of light; so that the arrangement nearly resembles that of the common mixed spectrum. There is, however, another cause of a further mixture of the colours: the arch of any single colour, which belongs to any point of the sun, is accurately defined on one side only, while on the other it becomes gradually fainter, the breadth of the first minute containing about five times as much light as a minute at the distance of a quarter of a degree: the abrupt termination is on the side of the red, that is, without the inner bow, and within the outer, so that, for this reason, the order of colours partakes, in some degree, of the nature of the red termination of a broad beam of light seen through a prism; but it is more or less affected by this cause, on account of some circumstances, which will be explained when we examine the supernumerary rainbows, which sometimes accompany the bows more commonly observed. A lunar rainbow is much more rarely seen than a solar one, but its colours differ little, except in intensity, from those of the common rainbow. (Plate XXIX. Fig. 430.)

In the highest northern latitudes, where the air is commonly loaded with frozen particles, the sun and moon usually appear surrounded by halos or coloured circles, at the distances of about 22 and 46 degrees from their centres; this appearance is also frequently observed in other climates, especially in the colder months, and in the light clouds which float in the highest regions of the air. The halos are usually attended by a horizontal white circle, with brighter spots, or parhelia, near their intersections with this circle, and with portions of inverted arches of various curvatures: the horizontal circle has also sometimes anthelia, or bright spots nearly opposite to the sun. These phenomena have usually been attributed to the effect of spherical particles of hail, each having a central opaque portion of a certain magnitude, mixed with oblong particles, of a determinate form, and floating with a certain constant obliquity to the horizon. But all these arbitrary suppositions, which were imagined by Huygens, are in themselves extremely complicated and improbable, and are wholly unauthorised by observation. A much simpler, and more natural, as well as more accurate explanation, which was suggested at an earlier period by Mariotte, had long been wholly forgotten, until the same idea occurred to me, without any previous knowledge of what Mariotte had done. The natural tendency of water to crystallize, in freezing, at an angle of 60 degrees, is sufficiently established, to allow us to assume this as the constant angle of the elementary crystals of snow, which are probably either triangular or hexagonal prisms: the deviation produced by such a prism differs very little from the observed angle at which the first circle is usually seen; and all the principal phenomena, which attend this circle, may be explained, by supposing the axis of the crystals to assume a vertical or a horizontal position, in consequence of the operation of gravity: thus the parhelia, which are sometimes a little more distant from the sun than the halo, are attributed by Mariotte to the refraction of the prisms which are situated vertically, and produce a greater deviation, on account of the obliquity of the rays of light with respect to their axes. The horizontal circle may be deduced from the reflection, or even the repeated refractions of the vertical facets; the anthelia from two refractions with an intermediate reflection, and the inverted arch from the increase of the deviation, in the light passing obliquely, through prisms lying in a horizontal position. The external circle may be attributed either to two successive refractions through different prisms, or with greater probability, as Mr. Cavendish has suggested to me, to the effect of the rectangular terminations of the single crystals. The appearance of colours, in halos, is nearly the same as in rainbows, but less distinct; the red being nearest to the luminary, and the whole halo being externally very ill defined. (Plate XXIX. Fig. 431, 432.)

From the observed magnitude of these halos, I had concluded that the refractive power of ice must be materially less than that of water, although some authors had asserted that it was greater: and Dr. Wollaston afterwards fully confirmed this conclusion by means of the very accurate instrument which has already been described: his measurement agreeing precisely with the mean of the best observations on these halos; so that ice must be considered as the least refractive of any known substances not aeriform.

Sometimes the figures of halos and parhelia are so extremely complicated, as to defy all attempts to account for the formation of their different parts: but if we examine the representations which have been given, by various authors, of the multiplicity of capricious forms frequently assumed by the flakes of snow, we shall see no reason to think them inadequate to the production of all these appearances. (Plate XXIX. Fig. 433, 434.)

2

The most singular of all the phenomena of refraction is perhaps the property of some natural substances, which have a double effect on the light transmitted through them, as if two mediums of different densities freely pervaded each other, the one only acting on some of the rays of light, the other on the remaining portion. These substances are usually crystallized stones, and their refractions have sometimes no further peculiarity; but the rhomboidal crystals of calcarious spar, commonly called Iceland crystals, possess the remarkable property of separating such pencils of light, as fall perpendicularly on them, into two parts, one of them only being transmitted in the usual manner, the other being deflected towards the greater angle of the crystal. It appears from the experiments of Huygens, confirmed and extended by Dr. Wollaston, that the medium, which causes the unusual refraction, has a different refractive power, according to the direction in which the light passes through it, and that if an oblate or flattened spheroid be described within a crystal, its axis being in the middle of one of the obtuse solid angles, and its principal diameters in the proportion of 9 to 10, the refractive power, with respect to light passing in any direction, will always be inversely as the diameter of the spheroid which is parallel to it; and where it is greatest, will be equal to that of the medium which produces the usual refraction, of which the index is $\frac{5}{3}$. A ray of light, falling perpendicularly on any surface of the spar, its point of incidence being considered as the centre of the spheroid, will meet the surface of the spheroid at the point where it is parallel to that of the spar; and a ray incident on the same surface in any other direction, will preserve a relation to the perpendicular ray, which is nearly the same as in ordinary refraction. (Plate XXIX. Fig. 435.)

It is also remarkable, that the two portions of light, thus separated, will not be further subdivided by a transmission through a second piece, provided that this piece be in a position parallel to that of the first; but if it be placed in a transverse direction, each of the two pencils will be divided into two others; a circumstance which appears to be the most unintelligible of any that has been discovered respecting the phenomena of double refraction.

The appearances of colours, which are produced by transparent plates of different thicknesses, and of those which are seen in light variously diffracted or inflected, will be more conveniently examined, when we investigate the intimate nature of light, since the general explanation of these colours, which will then be given, will enable us to follow them through all their varieties, with much more ease than could be done at present, without the help of some theory respecting their origin.

LECTURE XXXVIII.

ON VISION.

THE medium of communication, by which we become acquainted with all the objects that we have been lately considering, is the eye; an organ that exhibits, to an attentive observer, an arrangement of various substances, so correctly and delicately adapted to the purposes of the sense of vision, that we cannot help admiring, at every step, the wisdom by which each part is adjusted to the rest, and made to conspire in effects, so remote from what the mere external appearance promises, that we have only been able to understand, by means of a laborious investigation, the nature and operations of this wonderful structure, while its whole mechanism still remains far beyond all rivalship of human art.

The eye is an irregular spheroid, not very widely differing from a sphere; it is principally composed of transparent substances, of various refractive densities, calculated to collect the rays of light, which diverge from each point of an object, to a focus on its posterior surface, which is capable of transmitting to the mind the impression of the colour and intensity of the light, together with a distinction of the situation of the focal point, as determined by the angular place of the object. (Plate XXX. Fig. 436.)

The first refraction happens at the surface of the cornea, or that transparent coat which projects forwards from the ball of the eye; but the cornea, being very nearly of equable thickness, has little effect by its own refractive power, and serves only to give a proper form to the aqueous humour, which fills its concavity, and distends it. This humour is partially divided by the uvea or iris, which is of different colours in different persons, having a perforation in its centre, called the pupil. Immediately behind the uvea, and closely connected to its base, are the ciliary processes, the summits of which hang, like a short fringe, before the crystalline lens, a substance much more refractive than the aqueous humour, and increasing in density towards its centre. The remaining cavity is filled by an aqueous fluid, lodged in a cellular texture of extremely fine membrane, and called the vitreous humour. The retina lines the whole posterior part of this cavity; it is semitransparent, and is supported by the choroid or chorioid coat, a very opaque black or brown membrane, continued from the uvea and ciliary processes: but immediately where the retina is connected with the optic nerve, thechoroid is necessarily perforated; and at this part a small portion of the retina is nearly insensible. The whole is surrounded by an opaque continuation of the cornea, called the sclerotica.

The rays of light, which have entered the cornea, and passed through the pupil, being rendered still more convergent by the crystalline lens, are collected into foci on the retina, and form there an image, which, according to the common laws of refraction, is inverted, since the central rays of each pencil cross each other a little behind the pupil; and the image may easily be seen in a dead eye, by laying bare the posterior surface of the retina. (Plate XXX. Fig. 437.)

By means of this arrangement of the various refracting substances, many peculiar advantages are procured. The surface of the cornea only, if it had been more convex, could not have collected the lateral rays of a direct pencil to a perfect focus, without a different curvature near its edges; and then the oblique pencils would have been subjected to greater aberration, nor could they have been made to converge to any focus on the retina. A second refraction performs both these offices much more completely, and has also the advantage of admitting a greater quantity of light. If also the surfaces of the crystalline lens, thus interposed, had been abrupt, there would have been a reflection at each, and an apparent haziness would have interfered with the distinct view of every luminous object; but this inconvenience is avoided by the gradual increase of density in approaching the centre, which also makes the crystalline equivalent to a much more refractive substance of equal magnitude; while, at the same time, the smaller density of the lateral parts prevents the usual aberration of spherical surfaces, occasioned by the too great refraction of the lateral rays of direct pencils, and causes also the focus of each oblique pencil to fall either accurately or very nearly on the concave surface of the retina, throughout its extent.

Opticians have often puzzled themselves, without the least necessity, in order to account for our seeing objects in their natural erect position, while the image on the retina is in reality inverted: but surely the situation of a focal point at the upper part of the eye could be no reason for supposing the object corresponding to it to be actually elevated. We call that the lower end of an object which is next the ground; and the image of the trunk of a tree being in contact with the image of the ground on the retina, we may naturally suppose the trunk itself to be in contact with the actual ground: the image of the branches being more remote from that of the ground, we necessarily infer that the branches are higher and the trunk lower: and it is much simpler that we should compare the image of the floor with the image of our feet, with which it is in contact, than with the actual situation of our forehead, to which the image of the floor on the retina is only accidentally near, and with which indeed it would perhaps be impossible to compare it, as far as we judge by the immediate sensations only.

We might indeed call in experience to our assistance, and habitually correct the errors of one sense by a comparison with the perceptions of another. But it appears that some philosophers have been too hasty in supposing, that the use of all our senses is derived from experience alone, and in disbelieving the existence of instinct independent of it. Without any other authority than that of their own imaginations, they have denied the observation recorded by Galen, on the instincts of a kid, which is sufficiently credible to counterbalance much more than bare assertion. The instant after its birth, accompanied by the loss of its mother, the little animal ran to some green vegetables, and having first smelt them, chewed and swallowed them. The kid could have been taught by no experience to be tempted by the sight, to act with the proper muscles of locomotion, to go near and smell, and to be induced by the smell to masticate, and by the taste to swallow and digest its food, had it not been provided with some fundamental instinct, by the same intelligence, which so calculated the adjustments of the eye, that the lens should be able to produce a perfect image of every object, and that the retina should be of that precise form, which is exactly suited to the reception of the image to be depicted on it.

The whole surface of the retina appears to be usually occupied by such an image, but it is not all of equal sensibility; a certain portion only, near the axis, is capable of conveying distinct impressions of minute objects. But the perfection of this limited distinctness is a far greater advantage to us, than a more extensive field of moderately accurate vision would have been; for by means of the external muscles, we can easily so change the position of the eye, that the image of any object before us may be made to fall on the most sensible part of the retina. We may readily observe the want of sensation at the entrance of the optic nerve, by placing two candles so that the distance of each from the eye may be about four times their distance from each other: then if we direct our right eye to to the left hand candle, the right hand candle will be lost in a confused mass of faint light, its image on the retina falling on the point at which its sensibility is deficient.

When the attention is not directed to any particular object of sight, the refractive powers of the eye are adapted to the formation of an image of objects at a certain distance only, which is different in different individuals, and also generally increases with increasing age. Thus, if we open our eyelids suddenly, without particular preparation, we find that distant objects only appear as distinct as we are able to make them; but by an exertion of the will, the eye may be accommodated to the distinct perception of nearer objects, yet not of objects within certain limits. Between the ages of 40 and and 50, the refractive powers of the eye usually begin to diminish, but it sometimes happens that where they are already too great, the defect continues unaltered to an advanced age. It appears also that after 50 or 60, the power of changing the focus of the eye is always much impaired, and sometimes wholly lost.

The mode, in which the accommodation of the eye to different distances is effected, has long been a subject of investigation and dispute among opticians and physiologists, but I apprehend that at present there is little further room for doubting, that the change is produced by an increase of the convexity of the crystalline lens, arising from an internal cause. The arguments in favour of this conclusion are of two kinds; some of them are negative; derived from the impossibility of imagining any other mode of performing the accommodation, without exceeding the limits of the actual dimensions of the eye, and from the examination of the eye in its different states by several tests, capable of detecting any other changes if they had existed: for example, by the application of water to the cornea, which completely removes the effect of its convexity, without impairing the power of altering the focus, and by holding the whole eye, when turned inwards, in such a manner as to render any material alteration of its length utterly impossible. Other arguments are deduced from positive evidence of the change of form of the crystalline, furnished by the particular effects of refraction and aberration which are observable in the different states of the eye; effects which furnish a direct proof that the figure of the lens must vary; its surfaces, which are nearly spherical in the quiescent form of the lens, assuming a different determinable curvature when it is called into exertion. The objections which have been made to this conclusion are founded only on the appearance of a slight alteration of focal length in an eye from which the crystalline had been extracted; but the fact is neither sufficiently ascertained, nor was the apparent change at all considerable: and even if it were proved that an eye without the lens is capable of a certain small alteration, it would by no means follow that it could undergo a change five times or ten times as great.

The iris serves, by its variable magnitude, to exclude more or less of the light falling on the cornea, when its intensity would otherwise be too great; hence the pupil is usually smallest by day, and its increased magnitude at night sometimes gives the eye a greater apparent lustre. The iris also intercepts such rays as would fall on parts incapable of refracting them regularly; and by its contraction when a nearer object is viewed, it lessens the confusion which would arise, in such eyes as cannot accommodate themselves sufficiently, from the magnitude of the imperfect focal points on the retina. Such a contraction almost always accompanies the diminution of the focal length, even in a perfect eye, and it may easily be rendered visible by walking gradually up to a looking glass, and observing the magnitude of the pupil as we approach nearer and nearer to our image. It would be difficult to assign a reason for this change of the state of the pupil within the limits of perfect vision, unless we allowed the irregularity of the form assumed by the marginal parts of the crystalline lens. The iris is also peculiarly useful in excluding such parts of lateral pencils of light as fall very obliquely on the cornea, and are too much refracted, while a smaller pencil only, which enters the eye more directly, is admitted into the pupil.

The refractive powers and properties of the eye may be very conveniently ascertained by means of an instrument to which I have given the name optometer, a term first employed in a sense nearly similar by Dr. Porterfield. If two or more separate parcels of the rays of the same pencil be admitted at distant parts of the pupil, they will only be reunited on the retina when the focus is perfect, so that if we look through two small perforations, or slits, at a minute object, to the distance of which the eye is not accommodated, it will appear as if double; and when the object is a line directed nearly towards the eye, each point of it will appear double, except that which is at the distance of perfect vision, and an image of two lines will be seen, crossing each other in this point; so that the measurement of the focal length of the eye is immediately performed by inspection of the optometer only. The scale may be extended by the addition of a lens, which enables us to produce the effect of a longer line, while the instrument still remains portable.

When the eye is possessed of too great a refractive power for the distinct perception of distant objects, the pupil is generally large, so that the confusion of the image is somewhat lessened by partially closing the eyelids; and from this habit an eye so formed is called myopic. In such cases, by the help of a concave lens, the divergence of the rays of light may be increased, and a virtual image may be formed, at a distance so much smaller than that of the object as to afford perfect vision. For a long sighted or presbyopic eye, on the contrary, a convex lens is required, in order to obtain a virtual image at a greater distance than the object; and it often happens that the rays must be made not only to diverge less than before, but even to converge towards a focus behind such an eye, in order to make its vision distinct. Presbyopic persons have in general a small pupil, and, therefore, seldom acquire the habit of covering any part of it with their eyelids.

When the images of the same object fall on certain corresponding points of the retina in each eye, they appear to the sense only as one; but if they fall on parts not corresponding, the object appears double; and in general, all objects at the same distance, in any one position of the eyes, appear alike either double or single. The optical axes, or the directions of the rays falling on the points of most perfect vision, naturally meet at a great distance; that is, they are nearly parallel to each other, and in looking at a nearer object we make them converge towards it, wherever it may be situated, by means of the external muscles of the eye; while in perfect eyes the refractive powers are altered, at the same time, by an involuntary sympathy, so as to form a distinct image of an object at the given distance. This correspondence of the situation of the axes with the focal length is in most cases unalterable; but some have perhaps a power of deranging it in a slight degree, and in others the adjustment is imperfect: but the eyes seem to be in most persons inseparably connected together with respect to the changes that their refractive powers undergo, although it sometimes happens that those powers are originally very different in the opposite eyes.

These motions enable us to judge pretty accurately, within certain limits, of the distance of an object; and beyond these limits, the degree of distinctness or confusion of the image still continues to assist the judgment. We estimate distances much less accurately with one eye than with both, since we are deprived of the assistance usually afforded by the relative situation of the optical axes; thus we seldom succeed at once in attempting to pass a finger or a hooked rod sideways through a ring, with one eye shut. Our idea of distance is also usually regulated by a knowledge of the real magnitude of an object, while we observe its angular magnitude; and on the other hand a knowledge of the real or imaginary distance of the object often directs our judgment of its actual magnitude. The quantity of light intercepted by the air interposed, and the intensity of the blue tint which it occasions, are also elements of our involuntary calculation: hence, in a mist, the obscurity increases the apparent distance, and consequently the supposed magnitude, of an unknown object. We naturally observe, in estimating a distance, the number and extent of the intervening objects; so that a distant church in a woody and hilly country appears more remote than if it were situated in a plain; and for a similar reason, the apparent distance of an object seen at sea, is smaller than its true distance. The city of London is unquestionably larger than Paris; but the difference appears at first sight much greater than it really is; and the smoke, produced by the coal fires of London, is probably the principal cause of the deception.

The sun, moon, and stars, are much less luminous when they are near the horizon, than when they are more elevated, on account of the greater quantity of their light that is intercepted, in its longer passage through the atmosphere: we also observe a much greater variety of nearer objects almost in the same direction: we cannot, therefore, help imagining them to be more distant, when they rise or set, than at other times; and since they subtend the same angle, they appear to be actually larger. For similar reasons the apparent figure of the starry heavens, even when free from clouds, is that of a flattened vault, its summit appearing to be much nearer to us than its horizontal parts, and any of the constellations seems to be considerably larger when it is near the horizon than when in the zenith. (Plate XXX. Fig. 438.)

The faculty of judging of the actual distance of objects is an impediment to the deception, which it is partly the business of a painter to produce. Some of the effects of objects at different distances may, however, be imitated in painting on a plane surface. Thus, supposing the eye to be accommodated to a given distance, objects at all other distances may be represented with a certain indistinctness of outline, which would accompany the images of the objects themselves on the retina: and this indistinctness is so generally necessary, that its absence has the disagreeable effect called hardness. The apparent magnitude of the subjects of our design, and the relative situations of the intervening objects, may be so imitated by the rules of geometrical perspective as to agree perfectly with nature, and we may still further improve the representation of distance by attending to the art of aerial perspective, which consists in a due observation of the loss of light, and the bluish tinge, occasioned by the interposition of a greater or less depth of air between us and the different parts of the scenery.

We cannot indeed so arrange the picture, that either the focal length of the eye, or the position of the optical axes, may be such as would be required by the actual objects: but we may place the picture at such a distance that neither of these criterions can have much power in detecting the fallacy; or, by the interposition of a large lens, we may produce nearly the same effects in the rays of light, as if they proceeded from a picture at any required distance. In the panorama, which has lately been exhibited in many parts of Europe, the effects of natural scenery are very closely imitated: the deception is favoured by the absence of all other visible objects, and by the faintness of the light, which assists in concealing the defects of the representation, and for which the eye is usually prepared, by being long detained in the dark winding passages, which lead to the place of exhibition.

The impressions of light on the retina appear to be always in a certain degree permanent, and the more so as the light is stronger; but it is uncertain whether the retina possesses this property merely as a solar phosphorus, or in consequence of its peculiar organization. The duration of the impression is generally from one hundredth of a second to half a second, or more; hence a luminous object revolving in a circle makes a lucid ring; and a shooting star leaves a train of light behind it, which is not always real. If the object is painfully bright, it generally produces a permanent spot, which continues to pass through various changes of colour for some time, without much regularity, and gradually vanishes: this may, however, be considered as a morbid effect.

When the eye has been fixed on a small object of a bright colour, and is then turned away to a white surface, a faint spot, resembling in form and magnitude the object first viewed, appears on the surface, of a colour opposite to the first, that is, of such a colour as would be produced by withdrawing it from white light; thus a red object produces a bluish green spot; and a bluish green object a red spot. The reason of this appearance is probably that the portion of the retina, or of the sensorium, that is affected, has lost a part of its sensibility to the light of that colour, with which it has been impressed, and is more strongly affected by the other constituent parts of the white light. A similar effect is also often produced, when a white, or grey, object is viewed on a coloured ground, even without altering the position of the eye: the whole retina being affected by sympathy nearly in the same manner as a part of it was affected in the former case. These appearances are most conveniently exhibited by means of the shadows of objects placed in coloured light: the shadow appearing of a colour opposite to that of the stronger light, even when it is in reality illuminated by a fainter light of the same colour. It seems that the eye cannot perfectly distinguish the intensity of a colour, either when the light is extremely faint, as that of many of the fixed stars, which Dr. Herschel has found to be strongly coloured, or when the light is excessively vivid; and that when a considerable part of the field of vision is occupied by coloured light, it appears to the eye either white, or less coloured than it is in reality: so that when a room is illuminated either by the yellow light of a candle, or by the red light of a fire, a sheet of writing paper still appears to retain its whiteness; and if from the light of the candle we take away some of the abundant yellow light, and leave or substitute a portion actually white, the effect is nearly the same as if we took away the yellow light from white, and substituted the indico which would be left: and we observe accordingly, that in comparison with the light of a candle, the common daylight appears of a purplish hue. (Plate XXX. Fig. 439 . . 441.)

LECTURE XXXIX.

ON THE NATURE OF LIGHT AND COLOURS.

THE nature of light is a subject of no material importance to the concerns of life or to the practice of the arts, but it is in many other respects extremely interesting, especially as it tends to assist our views both of the nature of our sensations, and of the constitution of the universe at large. The examination of the production of colours, in a variety of circumstances, is intimately connected with the theory of their essential properties, and their causes; and we shall find that many of these phenomena will afford us considerable assistance in forming our opinion respecting the nature and origin of light in general.

It is allowed on all sides, that light either consists in the emission of very minute particles from luminous substances, which are actually projected, and continue to move, with the velocity commonly attributed to light, or in the excitation of an undulatory motion, analogous to that which constitutes sound, in a highly light and elastic medium pervading the universe; but the judgments of philosophers of all ages have been much divided with respect to the preference of one or the other of these opinions. There are also some circumstances which induce those, who entertain the first hypothesis, either to believe, with Newton, that the emanation of the particles of light is always attended by the undulations of an etherial medium, accompanying it in its passage, or to suppose, with Boscovich, that the minute particles of light themselves receive, at the time of their emission, certain rotatory and vibratory motions, which they retain as long as their projectile motion continues. These additional suppositions, however necessary they may have been thought for explaining some particular phenomena, have never been very generally understood or admitted, although no attempt has been made to accommodate the theory in any other manner to those phenomena.

We shall proceed to examine in detail the manner in which the two principal hypotheses respecting light may be applied to its various properties and affections; and in the first place to the simple propagation of light in right lines through a vacuum, or a very rare homogeneous medium. In this circumstance there is nothing inconsistent with either hypothesis; but it undergoes some modifications, which require to be noticed, when a portion of light is admitted through an aperture, and spreads itself in a slight degree in every direction. In this case it is maintained by Newton that the margin of the aperture possesses an attractive force, which is capable of inflecting the rays: but there is some improbability in supposing that bodies of different forms and of various refractive powers should possess an equal force of inflection, as they appear to do in the production of these effects; and there is reason to conclude from experiments, that such a force, if it existed, must extend to a very considerable distance from the surfaces concerned, at least a quarter of an inch, and perhaps much more, which is a condition not easily reconciled with other phenomena. In the Huygenian system of undulation, this divergence or diffraction is illustrated by a comparison with the motions of waves of water and of sound, both of which diverge when they are admitted into a wide space through an aperture, so much indeed that it has usually been considered as an objection to this opinion, that the rays of light do not diverge in the degree that would be expected if they were analogous to the waves of water. But as it has been remarked by Newton, that the pulses of sound diverge less than the waves of water, so it may fairly be inferred, that in a still more highly elastic medium, the undulations, constituting light, must diverge much less considerably than either. (Plate XX. Fig. 266.)

With respect, however, to the transmission of light through perfectly transparent mediums of considerable density, the system of emanation labours under some difficulties. It is not to be supposed that the particles of ligh can perforate with freedom the ultimate atoms of matter, which compose a substance of any kind; they must, therefore, be admitted in all directions through the pores or interstices of those atoms: for if we allow such suppositions as Boscovich's, that matter itself is penetrable, that is, immaterial, it is almost useless to argue the question further. It is certain that some substances retain all their properties when they are reduced to the thickness of the ten millionth of an inch at most, and we cannot therefore suppose the distances of the atoms of matter in general to be so great as the hundred millionth of an inch. Now if ten feet of the most transparent water transmits, without interruption, one half of the light that enters it, each section or stratum of the thickness of one of these pores of matter must intercept only about one twenty thousand millionth, and so much must the space or area occupied by the particles be smaller than the interstices between them, and the diameter of each atom must be less than the hundred and forty thousandth part of its distance from the neighbouring particles: so that the whole space occupied by the substance must be as little filled, as the whole of England would be filled by a hundred men, placed at the distance of about thirty miles from each other. This astonishing degree of porosity is not indeed absolutely inadmissible, and there are many reasons for believing the statement to agree in some measure with the actual constitution of material substances; but the Huygenian hypothesis does not require the disproportion to be by any means so great, since the general direction and even the intensity of an undulation would be very little affected by the interposition of the atoms of matter, while these atoms may at the same time be supposed to assist in the transmission of the impulse, by propagating it through their own substance. Euler indeed imagined that the undulations of light might be transmitted through the gross substance of material bodies alone, precisely in the same manner as sound is propagated; but this supposition is for many reasons inadmissible.

A very striking circumstance, respecting the propagation of light, is the uniformity of its velocity in the same medium. According to the projectile hypothesis, the force employed in the free emission of light must be about a million million times as great as the force of gravity at the earth's surface; and it must either act with equal intensity on all the particles of light, or must impel some of them through a greater space than others, if its action be less powerful, since the velocity is the same in all cases; for example, if the projectile force is weaker with respect to red light than with respect to violet light, it must continue its action on the red rays to a greater distance than on the violet rays. There is no instance in nature besides of a simple projectile moving with a velocity uniform in all cases, whatever may be its cause, and it is extremely difficult to imagine that so immense a force of repulsion can reside in all substances capable of becoming luminous, so that the light of decaying wood, or of two pebbles rubbed together, may be projected precisely with the same velocity, as the light emitted by iron burning in oxygen gas, or by the reservoir of liquid fire on the surface of the sun. Another cause would also naturally interfere with the uniformity of the velocity of light, if it consisted merely in the motion of projected corpuscles of matter; Mr. Laplace has calculated, that if any of the stars were 250 times as great in diameter as the sun, its attraction would be so strong as to destroy the whole momentum of the corpuscles of light proceeding from it, and to render the star invisible at a great distance; and although there is no reason to imagine that any of the stars are actually of this magnitude, yet some of them are probably many times greater than our sun, and therefore large enough to produce such a retardation in the motion of their light as would materially alter its effects. It is almost unnecessary to observe that the uniformity of the velocity of light, in those spaces which are free from all material substances, is a necessary consequence of the Huygenian hypothesis, since the undulations of every homogeneous elastic medium are always propagated, like those of sound, with the same velocity, as long as the medium remains unaltered.

On either supposition, there is no difficulty in explaining the equality of the angles of incidence and reflection; for these angles are equal as well in the collision of common elastic bodies with others incomparably larger, as in the reflections of the waves of water and of the undulations of sound. And it is equally easy to demonstrate, that the sines of the angles of incidence and refraction must be always in the same proportion at the same surface, whether it be supposed to possess an attractive force, capable of acting on the particles of light, or to be the limit of a medium through which the undulations are propagated with a diminished velocity. There are, however, some cases of the production of colours, which lead us to suppose that the velocity of light must be smaller in a denser than in a rarer medium; and supposing this fact to be fully established, the existence of such an attractive force could no longer be allowed, nor could the system of emanation be maintained by any one.

The partial reflection from all refracting surfaces is supposed by Newton to arise from certain periodical retardations of the particles of light, caused by undulations, propagated in all cases through an ethereal medium. The mechanism of these supposed undulations is so complicated, and attended by so many difficulties, that the few who have examined them have been in general entirely dissatisfied with them: and the internal vibrations of the particles of light themselves, which Boscovich has imagined, appear scarcely to require a serious discussion. It may, therefore, safely be asserted, that in the projectile hypothesis this separation of the rays of light of the same kind by a partial reflection at every refracting surface, remains wholly unexplained. In the undulatory system, on the contrary, this separation follows as a necessary consequence. It is simplest to consider the ethereal medium which pervades any transparent substance, together with the material atoms of the substance, as constituting together a compound medium denser than the pure ether, but not more elastic; and by comparing the contiguous particles of the rarer and the denser medium with common elastic bodies of different dimensions, we may easily determine not only in what manner, but almost in what degree, this reflection must take place in different circumstances. Thus, if one of two equal bodies strikes the other, it communicates to it its whole motion without any reflection; but a smaller body striking a larger one is reflected, with the more force as the difference of their magnitude is greater; and a larger body, striking a smaller one, still proceeds with a diminished velocity; the remaining motion constituting, in the case of an undulation falling on a rarer medium, a part of a new series of motions which necessarily returns backwards with the appropriate velocity: and we may observe a circumstance nearly similar to this last in a portion of mercury spread out on a horizontal table; if a wave be excited at any part, it will be reflected from the termination of the mercury almost in the same manner as from a solid obstacle.

The total reflection of light, falling, with a certain obliquity, on the surface of a rarer medium, becomes, on both suppositions, a particular case of refraction. In the undulatory system, it is convenient to suppose the two mediums to be separated by a short space in which their densities approach by degrees to each other, in order that the undulation may be turned gradually round, so as to be reflected in an equal angle: but this supposition is not absolutely necessary, and the same effects may be expected at the surface of two mediums separated by an abrupt termination.

The chemical process of combustion may easily be imagined either to disengage the particles of light from their various combinations, or to agitate the elastic medium by the intestine motions attending it: but the operation of friction upon substances incapable of undergoing chemical changes, as well as the motions of the electric fluid through imperfect conductors, afford instances of the production of light in which there seems to be no easy way of supposing a decomposition of any kind. The phenomena of solar phosphori appear to resemble greatly the sympathetic sounds of musical instruments, which are agitated by other sounds conveyed to them through the air: it is difficult to understand in what state the corpuscles of light could be retained by these substances so as to be reemitted after a short space or time; and if it is true that diamonds are often found, which exhibit a red light after having received a violet light only, it seems impossible to explain this property, on the supposition of the retention and subsequent emission of the same corpuscles.

The phenomena of the aberration of light agree perfectly well with the system of emanation; and if the ethereal medium, supposed to pervade the earth and its atmosphere, were carried along before it, and partook materially in its motions, these phenomena could not easily be reconciled with the theory of undulation. But there is no kind of necessity for such a supposition: it will not be denied by the advocates of the Newtonian opinion that all material bodies are sufficiently porous to leave a medium pervading them almost absolutely at rest; and if this be granted, the effects of aberration will appear to be precisely the same in either hypothesis.

The unusual refraction of the Iceland spar has been most accurately and satisfactorily explained by Huygens, on the simple supposition that this crystal possesses the property of transmitting an impulse more rapidly in one direction than in another; whence he infers that the undulations constituting light must assume a spheroidical instead of a spherical form, and lays down such laws for the direction of its motion, as are incomparably more consistent with experiment than any attempts which have been made to accommodate the phenomena to other principles. It is true that nothing has yet been done to assist us in understanding the effects of a subsequent refraction by a second crystal, unless any person can be satisfied with the name of polarity assigned by Newton to a property which he attributes to the particles of light, and which he supposes to direct them in the species of refraction which they are to undergo: but on any hypothesis, until we discover the reason why a part of the light is at first refracted in the usual manner, and another part in the unusual manner, we have no right to expect that we should understand how these dispositions are continued or modified, when the process is repeated.

In order to explain, in the system of emanation, the dispersion of the rays of different colours by means of refraction, it is necessary to suppose that all refractive mediums have an elective attraction, acting more powerfully on the violet rays, in proportion to their mass, than on the red. But an elective attraction of this kind is a property foreign to mechanical philosophy, and when we use the term in chemistry, we only confess our incapacity to assign a mechanical cause for the effect, and refer to an analogy with other facts, of which the intimate nature is perfectly unknown to us. It is not indeed very easy to give a demonstrative theory of the dispersion of coloured light upon the supposition of undulatory motion; but we may derive a very satisfactory illustration from the well known effects of waves of different breadths. The simple calculation of the velocity of waves, propagated in a liquid perfectly elastic, or incompressible, and free from friction, assigns to them all precisely the same velocity, whatever their breadth may be: the compressibility of the fluids actually existing introduces, however, a necessity for a correction according to the breadth of the wave, and it is very easy to observe, in a river or a pond of considerable depth, that the wider waves proceed much more rapidly than the narrower. We may, therefore, consider the pure ethereal medium as analogous to an infinitely elastic fluid, in which undulations of all kinds move with equal velocity, and material transparent substances, on the contrary, as resembling those fluids, in which we see the large waves advance beyond the smaller; and by supposing the red light to consist of larger or wider undulations and the violet of smaller, we may sufficiently elucidate the greater refrangibility of the red than of the violet light.

It is not, however, merely on the ground of this analogy that we may be induced to suppose the undulations constituting red light to be larger than those of violet light: a very extensive class of phenomena leads us still more directly to the same conclusion; they consist chiefly of the production of colours by means of transparent plates, and by diffraction or inflection, none of which have been explained, upon the supposition of emanation, in a manner sufficiently minute or comprehensive to satisfy the most candid even of the advocates for the projectile system; while on the other hand all of them may be at once understood, from the effect of the interference of double lights, in a manner nearly similar to that which constitutes in sound the sensation of a beat, when two strings, forming an imperfect unison, are heard to vibrate together.

Supposing the light of any given colour to consist of undulations, of a given breadth, or of a given frequency, it follows that these undulations must be liable to those effects which we have already examined in the case of the waves of water, and the pulses of sound. It has been shown that two equal series of waves, proceeding from centres near each other, may be seen to destroy each other's effects at certain points, and other points at to redouble them; and the beating of two sounds has been explained from a similar interference. We are now to apply the same principles to the alternate union and extinction of colours. (Plate XX. Fig. 267.)

In order that the effects of two portions of light may be thus combined, it is necessary that they be derived from the same origin, and that they arrive at the same point by different paths, in directions not much deviating from each other. This deviation may be produced in one or both of the portions by diffraction, by reflection, by refraction, or by any of these effects combined; but the simplest case appears to be, when a beam of homogeneous light falls on a screen in which there are two very small holes or slits, which may be considered as centres of divergence, from whence the light is diffracted in every direction. In this case, when the two newly formed beams are received on a surface placed so as to intercept them, their light is divided by dark stripes into portions nearly equal, but becoming wider as the surface is more remote from the apertures, so as to subtend very nearly equal angles from the apertures at all distances, and wider also in the same proportion as the apertures are closer to each other. The middle of the two portions is always light, and the bright stripes on each side are at such distances, that the light, coming to them from one of the apertures, must have passed through a longer space than that which comes from the other, by an interval which is equal to the breadth of one, two, three, or more of the supposed undulations, while the intervening dark spaces correspond to a difference of half a supposed undulation, of one and a half, of two and a half, or more.

From a comparison of various experiments, it appears that the breadth of the undulations constituting the extreme red light must be supposed to be, in air, about one 36 thousandth of an inch, and those of the extreme violet about one 60 thousandth; the mean of the whole spectrum, with respect to the intensity of light, being about one 45 thousandth. From these dimensions it follows, calculating upon the known velocity of light, that almost 500 millions of millions of the slowest of such undulations must enter the eye in a single second. The combination of two portions of white or mixed light, when viewed at a great distance, exhibits a few white and black stripes, corresponding to this interval; although, upon closer inspection, the distinct effects of an infinite number of stripes of different breadths appear to be compounded together, so as to produce a beautiful diversity of tints, passing by degrees into each other. The central whiteness is first changed to a yellowish, and then to a tawny colour, succeeded by crimson, and by violet and blue, which together appear, when seen at a distance, as a dark stripe; after this a green light appears, and the dark space beyond it has a crimson hue; the subsequent lights are all more or less green, the dark spaces purple and reddish; and the red light appears so far to predominate in all these effects, that the red or purple stripes occupy nearly the same place in the mixed fringes as if their light were received separately.

The comparison of the results of this theory with experiments fully establishes their general coincidence; it indicates, however, a slight correction in some of the measures, on account of some unknown cause, perhaps connected with the intimate nature of diffraction, which uniformly occasions the portions of light, proceeding in a direction very nearly rectilinear, to be divided into stripes or fringes a little wider than the external stripes, formed by the light which is more bent. (Plate XXX. Fig. 442, 443.)

When the parallel slits are enlarged, and leave only the intervening substance to cast its shadow, the divergence from its opposite margins still continues to produce the same fringes as before, but they are not easily visible, except within the extent of its shadow, being overpowered in other parts by a stronger light; but if the light thus diffracted be allowed to fall on the eye, either within the shadow, or in its neighbourhood, the stripes will still appear; and in this manner the colours of small fibres are probably formed. Hence if a collection of equal fibres, for example a lock of wool, be held before the eye when we look at a luminous object, the series of stripes belonging to each fibre combine their effects, in such a manner, as to be converted into circular fringes or coronae. This is probably the origin of the coloured circles or coronae sometimes seen round the sun and moon, two or three of them appearing together, nearly at equal distances from each other and from the luminary, the internal ones being, however, like the stripes, a little dilated. It is only necessary that the air should be loaded with globules of moisture, nearly of equal size among themselves, not much exceeding one two thousandth of an inch in diameter, in order that a series of such coronae, at the distance of two or three degrees from each other, may be exhibited. (Plate XXX. Fig. 444.)

If, on the other hand, we remove the portion of the screen which separates the parallel slits from each other, their external margins will still continue to exhibit the effects of diffracted light in the shadow on each side; and the experiment will assume the form of those which were made by Newton on the light passing between the edges of two knives, brought very nearly into contact; although some of these experiments appear to show the influence of a portion of light reflected by a remoter part of the polished edge of the knives, which indeed must unavoidably constitute a part of the light concerned in the appearance of fringes, wherever their whole breadth exceeds that of the aperture, or of the shadow of the fibre.

The edges of two knives, placed very near each other, may represent the opposite margins of a minute furrow, cut in the surface of a polished substance of any kind, which, when viewed with different degrees of obliquity, present a series of colours nearly resembling those which are exhibited within the shadows of the knives: in this case, however, the paths of the two portions of light before their incidence are also to be considered, and the whole difference of these paths will be found to determine the appearance of colour in the usual manner; thus when the surface is so situated, that the image of the luminous point would be seen in it by regular reflection, the difference will vanish, and the light will remain perfectly white, but in other cases various colours will appear, according to the degree of obliquity. These colours may easily be seen, in an irregular form, by looking at any metal, coarsely polished, in the sunshine; but they become more distinct and conspicuous, when a number of fine lines of equal strength are drawn parallel to each other, so as to conspire in their effects.

It sometimes happens that an object, of which a shadow is formed in a beam of light, admitted through a small aperture, is not terminated by parallel sides; thus the two portions of light, which are diffracted from two sides of an object, at right angles with each other, frequently form a short series of curved fringes within the shadow, situated on each side of the diagonal, which were first observed by Grimaldi, and which are completely explicable from the general principle, of the interference of the two portions encroaching perpendicularly on the shadow. (Plate XXX. Fig. 445.)

But the most obvious of all the appearances of this kind is that of the fringes, which are usually seen beyond the termination of any shadow, formed in a beam of light, admitted through a small aperture: in white light three of these fringes are usually visible, and sometimes four; but in light of one colour only, their number is greater; and they are always much narrower as they are remoter from the shadow. Their origin is easily deduced from the interference of the direct light with a portion of light reflected from the margin of the object which produces them, the obliquity of its incidence causing a reflection so copious as to exhibit a visible effect, however narrow that margin may be; the fringes are, however, rendered more obvious as the quantity of this reflected light is greater. Upon this theory it follows that the distance of the first dark fringe from the shadow should be half as great as that of the fourth, the difference of the lengths of the different paths of the light being as the squares of those distances; and the experiment precisely confirms this calculation, with the same slight correction only as is required in all other cases; the distances of the first fringes being always a little increased. It may also be observed, that the extent of the shadow itself is always augmented, and nearly in an equal degree with that of the fringes: the reason of this circumstance appears to be the gradual loss of light at the edges of every separate beam, which is so strongly analogous to the phenomena visible in waves of water. The same cause may also perhaps have some effect in producing the general modification or correction of the place of the first fringes, although it appears to be scarcely sufficient for explaining the whole of it. (Plate XXX. Fig. 446.)

A still more common and convenient method, of exhibiting the effects of the mutual interference of light, is afforded us by the colours of the thin plates of transparent substances. The lights are here derived from the successive partial reflections produced by the upper and under surface of the plate, or when the plate is viewed by transmitted light, from the direct beam which is simply refracted, and that portion of it which is twice reflected within the plate. The appearance in the latter case is much less striking than in the former, because the light thus affected is only a small portion of the whole beam, with which it is mixed; while in the former the two reflected portions are nearly of equal intensity, and may be separated from all other light tending to overpower them. In both cases, when the plate is gradually reduced in thickness to an extremely thin edge, the order of colours may be precisely the same as in the stripes and coronae already described; their distance only varying when the surfaces of the plate, instead of being plane, are concave, as it frequently happens in such experiments. The scale of an oxid, which is often formed by the effect of heat on the surface of a metal, in particular of iron, affords us an example of such a series formed in reflected light: this scale is at first inconceivably thin, and destroys none of the light reflected, it soon, however, begins to be of a dull yellow, which changes to red, and then to crimson and blue, after which the effect is destroyed by the opacity which the oxid acquires. Usually, however, the series of colours produced in reflected light follows an order somewhat different: the scale of oxid is denser than the air, and the iron below than the oxid; but where the mediums above and below the plate are either both rarer or both denser than itself, the different natures of the reflections at its different surfaces appear to produce a modification in the state of the undulations, and the infinitely thin edge of the plate becomes black instead of white, one of the portions of light at once destroying the other, instead of cooperating with it. Thus when a film of soapy water is stretched over a wine glass, and placed in a vertical position, its upper edge becomes extremely thin, and appears nearly black, while the parts below are divided by horizontal lines into a series of coloured bands; and when two glasses, one of which is slightly convex, are pressed together with some force, the plate of air between them exhibits the appearance of coloured rings, beginning from a black spot at the centre, and becoming narrower and narrower, as the curved figure of the glass causes the thickness of the plate of air to increase more and more rapidly. The black is succeeded by a violet, so faint as to be scarcely perceptible; next to this is an orange yellow, and then crimson and blue. When water, or any other fluid, is substituted for the air between the glasses, the rings appear where the thickness is as much less than that of the plate of air, as the refractive density of the fluid is greater; a circumstance which necessarily follows from the proportion of the velocities with which light must, upon the Huygenian hypothesis, be supposed to move in different mediums. It is also a consequence equally necessary in this theory, and equally inconsistent with all others, that when the direction of the light is oblique, the effect of a thicker plate must be the same as that of a thinner plate, when the light falls perpendicularly upon it; the difference of the paths described by the different portions of light precisely corresponding with the observed phenomena. (Plate XXX. Fig. 447 . . 449.)

Sir Isaac Newton supposes the colours of natural bodies in general to be similar to these colours of thin plates, and to be governed by the magnitude of their particles. If this opinion were universally true, we might always separate the colours of natural bodies by refraction into a number of different portions, with dark spaces intervening; for every part of a thin plate, which exhibits the appearance of colour, affords such a divided spectrum, when viewed through a prism. There are accordingly many natural colours in which such a separation may be observed; one of the most remarkable of them is that of blue glass, probably coloured with cobalt, which becomes divided into seven distinct portions. It seems, however, impossible to suppose the production of natural colours perfectly identical with those of thin plates, on account of the known minuteness of the particles of colouring bodies, unless the refractive density of these particles be at least 20 or 30 times as great as that of glass or water; which is indeed not at all improbable with respect to the ultimate atoms of bodies, but difficult to believe with respect to any of their arrangements constituting the diversities of material substances.

The colours of mixed plates constitute a distinct variety of the colours of thin plates, which has not been commonly observed. They appear when the interstice between two glasses, nearly in contact, is filled with a great number of minute portions of two different substances, as water and air, oil and air, or oil and water: the light, which passes through one of the mediums, moving with a greater velocity, anticipates the light passing through the other; and their effects on the eye being confounded and combined, their interference produces an appearance of colours nearly similar to those of the colours of simple thin plates, seen by transmission; but at much greater thicknesses, depending on the difference of the refractive densities of the substances employed. The effect is observed by holding the glasses between the eye and the termination of a bright object, and it is most conspicuous in the portion which is seen on the dark part beyond the object, being produced by the light scattered irregularly from the surfaces of the fluid. Here, however, the effects are inverted, the colours resembling those of the common thin plates, seen by reflection; and the same considerations on the nature of the reflections are applicable to both cases. (Plate XXX. Fig. 450.)

The production of the supernumerary rainbows, which are sometimes seen within the primary and without the secondary bow, appears to be intimately connected with that of the colours of thin plates. We have already seen that the light producing the ordinary rainbow is double, its intensity being only greatest at its termination, where the common bow appears, while the whole light is extended much more widely. The two portions concerned in its production must divide this light into fringes; but unless almost all the drops of a shower happen to be of the same magnitude, the effects of these fringes must be confounded and destroyed: in general, however, they must at least cooperate more or less in producing one dark fringe, which must cut off the common rainbow much more abruptly than it would otherwise have been terminated, and consequently assist the distinctness of its colours. The magnitude of the drops of rain, required for producing such of these rainbows as are usually observed, is between the 50th and the 100th of an inch: they become gradually narrower as they are more remote from the common rainbows, nearly in the same proportions as the external fringes of a shadow, or the rings seen in a concave plate. (Plate XXX. Fig. 451.)

The last species of the colours of double lights, which it will be necessary to notice, constitutes those which have been denominated, from Newton's experiments, the colours of thick plates, but which may be called, with more propriety, the colours of concave mirrors. The anterior surface of a mirror of glass, or any other transparent surface placed before a speculum of metal, dissipates irregularly in every direction two portions of light, one before, and the other after its reflection. When the light falls obliquely on the mirror, being admitted through an aperture near the centre of its curvature, it is easy to show, from the laws of reflection, that the two portions, thus dissipated, will conspire in their effects, throughout the circumference of a circle, passing through the aperture; this circle will consequently be white, and it will be surrounded with circles of colours very nearly at equal distances, resembling the stripes produced by diffraction. The analogy between these colours and those of thin plates is by no means so close as Newton supposed it; since the effect of a plate of any considerable thickness must be absolutely lost in white light, after ten or twelve alternations of colours at most, while these effects would require the whole process to remain unaltered, or rather to be renewed, after many thousands or millions of changes. (Plate XXX. Fig. 452.)

It is presumed, that the accuracy, with which the general law of the interference of light has been shown to be applicable to so great a variety of facts, in circumstances the most dissimilar, will be allowed to establish its validity in the most satisfactory manner. The full confirmation or decided rejection of the theory, by which this law was first suggested, can be expected from time and experience alone; if it be confuted, our prospects will again be confined within their ancient limits, but if it be fully established, we may expect an ample extension of our views of the operations of nature, by means of our acquaintance with a medium, so powerful and so universal, as that to which the propagation of light must be attributed.

LECTURE XL

ON THE HISTORY OF OPTICS.

THE science of optics is not one of those which had been cultivated with the greatest diligence and success by the philosophers of antiquity: almost every refinement relating to it has originated in the course of about two centuries; and some of its greatest improvements have been made within these fifty years. The reflection of the rays of light is indeed an occurrence too frequent and too obvious to have escaped the notice even of the earliest observers: a river or a fountain was the first mirror; its effect was easily imitated by speculums of metal; and as soon as any philosophical attention was paid to the phenomenon, it was easy to collect the equality of the angles of incidence and reflection; but although it was well known that an oar, partially immersed in water, no longer appeared straight, it was long before any attempts were made to ascertain the relation between the angles of incidence and refraction. The Greeks were, however, acquainted with the properties of the burning glass, which was sold as a curiosity in the toy shops; for it is well known, that one of the personages, introduced by Aristophanes, proposes to destroy the papers of his adversary by the assistance of this instrument. The magnifying powers of lenses were, however, but little understood, although it is scarcely credible that they could have escaped the notice of a person in possession of a burning glass; it appears from Seneca that the Romans at least were informed of the effects of spherical refracting substances, and it is not improbable that some use was occasionally made of them in the arts.

Empedocles is perhaps the first person on record that wrote systematically on light. He maintained that it consisted of particles projected from luminous bodies, and that vision was performed both by the effect of these particles on the eye, and by means of a visual influence, emitted by the eye itself. Both of these doctrines were combated by Aristotle, who thought it absurd to suppose that a visual influence should be emitted by the eye, and that it should not enable us to see in the dark; and who considered it as more probable that light consisted in an impulse, propagated through a continuous medium, than in an emanation of distinct particles. Light, he says, is the action of a transparent substance; and if there were absolutely no medium between the eye and any visible object, it would be absolutely impossible that we should see it.

It is said that Archimedes made a compound burning mirror, of sufficient power to set on fire the Roman ships: in this form the story is scarcely probable, although the possibility of burning an object at a great distance by a collection of plane mirrors has been sufficiently shown by the experiments of Buffon. It is, however, not unlikely that Archimedes was acquainted with the properties of reflecting surfaces, and that he confirmed his theories by some experimental investigations. The work on catoptrics, attributed to Euclid, contains the determination of the effects of reflecting surfaces of different forms; but it is not supposed to be genuine. The existence and the magnitude of the atmospheric refraction were well known to Ptolemy, and a treatise of this astronomer on the subject is still extant in manuscript.

The mathematical theory of optics, or the science of dioptrics and catoptrics, made some advances in the middle ages from the labours of Alhazen and Vitellio. Alhazen was mistaken in some of his propositions respecting refraction; Vitellio, a native of Poland, gave a more correct theory of this subject, and constructed a table of refractive densities, showing the supposed proportions of the angles of incidence and refraction in the respective mediums.

The invention of the magic lantern is attributed to Roger Bacon, and the lens was soon afterwards commonly applied to the assistance of defective sight. It has been much disputed whether or no Bacon was acquainted with telescopes; the prevalent opinion is, that the passages, which have been alleged to prove it, are insufficient for the purpose; but there is reason to suspect, from the testimony of Recorde, who wrote in 1551, not only that Bacon had actually invented a telescope, but that Recorde himself knew something of its construction. Digges also, in a work published in 1571, has a passage of a similar nature, and from Bacon's own words it has been conjectured that an instrument resembling a telescope was even of much higher antiquity. But the first person, who is certainly known to have made a telescope, is Janson, a Dutchman, whose son, by accident, placing a concave and a convex spectacle glass at a little distance from each other, observed the increased apparent magnitude of an object seen through them; the father upon this fixed two such glasses in a tube a few inches long, and sold the instrument in this form. He also made some telescopes of greater powers, and one of his family discovered a satellite of Jupiter with them. Galileo had heard of the instrument, but had not been informed of the particulars of its construction, he reinvented it in 1609, and the following year rediscovered also the satellite which Janson had seen a little before.

It was, however, Kepler that first reduced the theory of the telescope to its true principles; he laid down the common rules for finding the focal lengths of simple lenses of glass; he showed how to determine the magnifying power of the telescope, and pointed out the construction of the simple astronomical telescope, which is more convenient for accurate observations than the Galilean telescope, since the micrometer may be more easily applied to it; a third glass, for recovering the erect position of the object, was afterwards added by Scheiner, and a fourth, for increasing the field of view, by Rheita. Kepler made also some good experiments on the nature of coloured bodies, and showed the inverted situation of the image formed on the retina of the eye. Maurolycus of Messina had demonstrated, in 1575, that the pencils of light are brought to focal points on the retina; Kepler's observations were thirty or forty years later.

The next great step in optics was made by De Dominis, who in 1611 first explained the cause of the interior or primary rainbow, and this was soon followed by a still more important discovery respecting the nature of refraction, first made by Snellius, who ascertained, about 1621, that the sines of the angles of incidence and refraction are always in the same proportion to each other at the same surface; he died, however, in 1626, without having made his discovery public. Descartes, is generally supposed to have seen Snellius's papers, although he published the law of refraction without acknowledging to whom he was indebted for it. Descartes also explained the formation of the secondary rainbow, and truly determined the angular magnitude of both the bows from mathematical principles; he did not, however, give a sufficient reason for the production of colours in either case. Descartes imagined light to consist in motion, or rather pressure, transmitted instantaneously through a medium infinitely elastic, and colours he attributed to a rotatory motion of the particles of this medium. He supposed that light passed more rapidly through a denser medium than through a rarer; other philosophers about the same time maintained a contrary opinion, without deciding with respect to any general theory of light: thus Fermat and Leibnitz deduced, on this supposition, the path of refracted light from the natural tendency of every body to attain its end by the shortest possible way; and Barrow derived the same law, in a more geometrical manner, from a similar hypothesis respecting the velocity of light, by considering a pencil of light as a collection of collateral rays influencing each other's motions. We are indebted to this learned mathematician for the first accurate investigation of the properties of refracting and reflecting surfaces, and for the most general determination of the situations of focal points.

The industrious Mr. Boyle had noticed with attention the phosphorescence of diamonds, the colours produced by the effect of scratches on the surfaces of polished metals, and the diversified tints which a bubble or a film of soapy water usually assumes. His assistant, Dr. Hooke, investigated these and other similar appearances with still greater accuracy, and proposed, in his Micrographia, which was published in 1665, a theory of light considerably resembling that of Descartes: he supposes that light is an impulse propagated through a medium highly, but not infinitely, elastic; that refraction is produced by the readier transmission of light through the denser medium, and that difference of colour consists in the different law of the particular impulse constituting coloured light, so that red and blue differ from each other in the same manner as the sound of a violin and of a flute. He explained the colours of thin plates from the interference of two such pulses, partially reflected from the upper and under surface; but the hypothesis which he assumed, respecting the nature of colours, renders this explanation wholly inadequate, nor were the phenomena at that time sufficiently investigated for a complete solution of the difficulties attending them.

It was still believed that every refraction actually produces colour, instead of separating the colours already existing in white light; but in the year 1666, Newton first made the important discovery of the actual existence of colours of all kinds in white light, which he showed to be no other than a compound of all possible colours, mixed in certain proportions with each other, and capable of being separated by refraction of any kind.

About the same time that Newton was making his earliest experiments on refraction, Grimaldi's treatise on light appeared; it contained many interesting experiments and ingenious remarks on the effects of diffraction, which is the name that he gave to the spreading of light in every direction upon its admission into a dark chamber, and on the colours which usually accompany these effects. He had even observed that in some instances the light of one pencil tended to extinguish that of another, but he had not inquired in what cases and according to what laws such an interference must be expected.

The discoveries of Newton were not received without some controversy either at home or abroad; the essential points of his theory were, however, soon established, but Dr. Hooke very warmly opposed the hypothesis which Newton had suggested respecting the nature and propagation of light. On this subject Newton professed himself by no means tenacious; he was not, however, convinced by Dr. Hooke, and disliked the dispute so much, that he deferred the publication of his treatise on optics till after Hooke's death in 1703. Very soon after his first communication to the Royal Society, in 1672, he had sent them a description of his reflecting telescope, which was perhaps the first that had been constructed with success, although Gregory had invented his instrument some years before, and a plan of a similar kind had been suggested by Eskinard as early as 1615. The principal parts of the treatise on optics had been communicated at different times to the Royal Society; besides the experiments on refraction and the theory of the rainbow, they consist of an elegant analysis of the colours of thin transparent substances, in which the phenomena are reduced to their simplest forms, and of a collection of miscellaneous experiments on the colours produced in cases of inflection or diffraction.

With respect to the nature of light, the theory which Newton adopted was materially different from the opinions of most of his predecessors. He considered indeed the operation of an ethereal medium as absolutely necessary to the production of the most remarkable effects of light, but he denied that the motions of such a medium actually constituted light; he asserted, on the contrary, that the essence of light consisted in the projection of minute particles of matter from the luminous body, and maintained that this projection was only accompanied by the vibration of a medium as an accidental circumstance, which was also renewed at the surface of every refractive or reflective substance.

In the mean time Bartholin had called the attention of naturalists and opticians to the singular properties of the Iceland crystal, and had hastily examined the laws of its unusual refraction. On this subject Huygens had been much more successful: his analysis of the phenomena of the double refraction is a happy combination of accurate experiment with elegant theory; it was published in 1690, making a part of his treatise on light, the fundamental doctrines of which he had communicated to the Academy of Paris in 1678. They scarcely differ in their essential parts from those of our countryman Dr. Hooke, but the subject of colours Huygens has left wholly untouched. Roemer had then lately made the discovery of the immense velocity with which light passes through the celestial regions, by observing the apparent irregularities of the eclipses of Jupiter's satellites; and Huygens readily admitted this property into his system; although Hooke, by a singular caprice, professed himself more ready to believe that the propagation of light might be absolutely instantaneous, than that its motion could be successive, and yet so inconceivably rapid. The merits of Huygens in the mathematical theory of optics were no less considerable than in the investigation of the nature of light; his determinations of the aberrations of lenses were the first refinement on the construction of telescopes.

In the year 1720 Dr. Bradley had the good fortune to discover both the existence and the cause of the aberration of the fixed stars. He had for some time observed an irregularity in the places of the stars, which he was wholly unable to explain, and the idea of attributing it to a combination of the effect of the earth's motion in its orbit, with the progressive motion of light, occurred to him first as he happened to observe the apparent direction of the wind on board of a boat which was moving in a transverse direction. He also determined with accuracy the magnitude of the atmospherical refraction, which had been theoretically investigated by Newton and by Taylor, but never before practically ascertained with sufficient precision. The formula, which Bradley appears to have deduced from observation only, agrees precisely with an approximation which was obtained by Simpson from calculation; but it cannot be considered as rigidly accurate.

The optics of Bouguer were first published in 1729, and an improved edition appeared thirty years afterwards; the merits of this author in the exmaination of the properties of a variety of substances, with respect to the transmission and reflection of light in different circumstances, and in the comparison of lights of different kinds, require to be mentioned with the highest commendation. Dr. Porterfield's investigations of the functions of the eye tended greatly to illustrate the economy of this admirable organ, and some valuable remarks of Dr. Jurin on the same subject were soon after published in Dr. Smith's elaborate treatise on optics, which contains all that had been done at that time with respect to the mathematical part of the science.

The invention of achromatic telescopes is with justice universally attributed to our countryman Mr. Dollond, but there is reason to believe that he was not absolutely the first author of the improvement. Mr. Hall, a gentleman of Worcestershire, is said to have discovered, about the year 1729, Sir Isaac Newton's mistake, in supposing that the rays of different colours must of necessity be equally separated by all surfaces which produce an equal mean refraction; and by combining the different dispersive properties of different kinds of glass, he constructed, in 1733, several compound object glasses, which were calculated not only for avoiding all appearance of colour, but also for correcting the imperfect refractions of the spherical surfaces of the separate lenses. He did not, however, make known the particulars of his investigations, and his invention was soon wholly forgotten. It was in consequence of a discussion with Euler, Klingenstierna, and some other mathematicians, that Mr. Dollond was led to make experiments on the refraction of different kinds of glass; these gentlemen had not questioned the general truth of Newton's opinion respecting the dispersion of the different colours, but Euler had asserted that the eye itself produced a refraction free from the appearance of colour, and Klingenstierna had shown the possibility of producing a deviation by refraction, without a separation of colour, according to the laws of refraction laid down by Newton himself. When Dollond had once discovered the material difference which exists between the dispersive properties of flint glass and of crown glass, it was easy to produce the combination required; but this ingenious artist was not satisfied with the advantage of freedom from colours only; he adjusted the forms and apertures of his lenses in the most skilful manner to the correction of aberrations of various kinds, and he was also particularly fortunate in being able to obtain, about the time of his discovery, a glass of a quality superior to any that has been since manufactured.

This opinion of Euler respecting the eye was, however, by no means well founded, for the eye acts very differently on rays of different colours, as we may easily observe by viewing a minute object in different parts of a beam of light, transmitted through a prism. It must be allowed that this great mathematician was less fortunate in his optical theories than in many other departments of science; his mathematical investigations of the effects of lenses are much more intricate and prolix than the subject actually requires, and with respect to the nature and propagation of light, he adopted several paradoxical opinions. Assuming the theory of Huygens, with the additional hypothesis respecting the nature of colours, which had been suggested by Newton, and maintained by Pardies and Malebranche, that is, that the difference of colours, like that of tones in music, depends on the different frequency of the vibrations constituting light; he imagined that opaque bodies are not seen by reflected light, but that their particles are agitated by the impulse of the light which falls on them, and that the vibrations of these particles render the bodies again visible in every direction; he also conceived that the undulations of light are simply propagated through the solid substances of transparent mediums, in the same manner as sound travels through the air. But on these suppositions, all bodies would have the properties of solar phosphori, and the refraction of the rarest of natural bodies would be incomparably greater than that of the densest is actually found to be: and on the whole, although the character of Euler has been so highly and so deservedly respected as to attach a certain degree of authority to all his opinions, so that in this instance the name of Huygens has been almost superseded by that of Euler, yet in fact he has added no argumentative evidence whatever to the theory, but, by inaccurate and injudicious reasoning, has done a real injury to the cause which he endeavoured to support.

The researches of Lambert may be considered as a continuation of those of Bouguer; they present us with many interesting observations on the natural history of light, and the properties of various bodies with regard to it. Mr. Lambert first ascertained that a luminous surface emits its light very nearly with equal intensity in all directions, so that any part of it appears almost equally brilliant to an eye placed in any direction, while the light thrown by each square inch or square foot of the surface in any direction differs according to the obliquity of that direction. The mathematical theory of optics is considerably indebted to the labours of Clairaut, Dalembert, and Boscovich; Jeaurat, Beguelin, Redern, and Klügel have also continued the investigation; their calculations may be of considerable utility to the practical optician, but it requires the ingenuity of a Dollond or a Ramsden to apply the whole of the results to any useful purposes.

The experiments of Mazéas on the colours of thin plates are mere repetitions of those of Newton under disadvantageous circumstances; Mr. Dutour has, however, considerably diversified and extended these experiments, as well as those on the colours which are produced in diffracted light, yet without obtaining any general results of importance. Comparetti's experiments on inflection have every appearance of accuracy, but they are much too intricate to be easily compared with each other, or with those of former observers.

The late Dr. Priestley rendered an essential service to the science of optics, considered as a subject for the amusement of the general reader, by an elegant and wellwritten account of the principal experiments and theories, which had been published before the year 1770. But this work is very deficient in mathematical accuracy, and the author was not sufficiently master of the science to distinguish the good from the indifferent.

Mr. Delaval's experiments on colours appear to show very satisfactorily, that all the colouring substances, in common use, owe their tints to rays, which are separated from white light, during its passage through them, and not, as Newton supposed, to the reflection of a particular colour from the first surface. It has been observed that Kepler and Zucchius had long ago made experiments nearly similar to those of Mr. Delaval. Dr. Robert Darwin's investigation of the effects of strong lights on the eye appears to comprehend almost all possible varieties of these ocular spectra, but it does not lead to any fundamental analogy, capable of explaining the most intricate of them.

The phenomena of the unusual atmospheric refraction, which frequently produces double or triple images of objects seen near a heated surface, have been successively illustrated by Mr. Huddart, Mr. Vince, and Dr. Wollaston, so that at present there appears to be little doubt remaining with respect to their origin. Dr. Wollaston's instrument, for the measurement of refractive densities, very much facilitates the examination of the optical properties of substances of various kinds: he has applied it very successfully to the confirmation of Huygens's theory of double refraction; he has corrected the common opinion respecting the division of the prismatic spectrum; he discovered, without being acquainted with the observations of Ritter, the dark rays which blacken the salts of silver; and he has remarked a singular property in some natural as well as artificial crystals, which appear of one colour when viewed in the direction of the axis, and of another when in a transverse direction.

To Dr. Herschel the sciences of optics and astronomy are equally indebted. He has carried the construction of the reflecting telescope to a degree of perfection, far exceeding all that had been before attempted, and the well known improvements, which astronomy has derived from his observations, are numerous and important. In the course of his researches for the attainment of his more immediate objects, he has also had the good fortune to discover the separation of the rays of heat from those of light by means of refraction; a fact which has been sufficiently established by the experiments of several other persons.

The investigations of Mr. Laplace, relating to atmospherical refraction, may be considered as the latest application of refined mathematics to the purposes of optics and of astronomy. I have myself attempted to attain a degree of certainty, in attributing the changes of the refractive powers of the eye to a variation in the form of the crystalline lens; I have discovered a general law of the mutual action of two portions of light interfering with each other, to which no exception has yet been shown; and by reviving a theory of light similar to that of Hooke and Huygens, with an improvement originally suggested by Newton, respecting the nature of colours, I have endeavoured to obtain a satisfactory explanation of many circumstances, which appear, upon a minute examination, to be in every other hypothesis difficulties absolutely insuperable. It cannot be expected that all objections to such a system will at once be silenced, but if a full and candid discussion only of the facts, which I have advanced, should be excited, I trust that the science of optics will be essentially benefited, even if the theory should be ultimately confuted.

CHRONOLOGY OF OPTICAL AUTHORS.

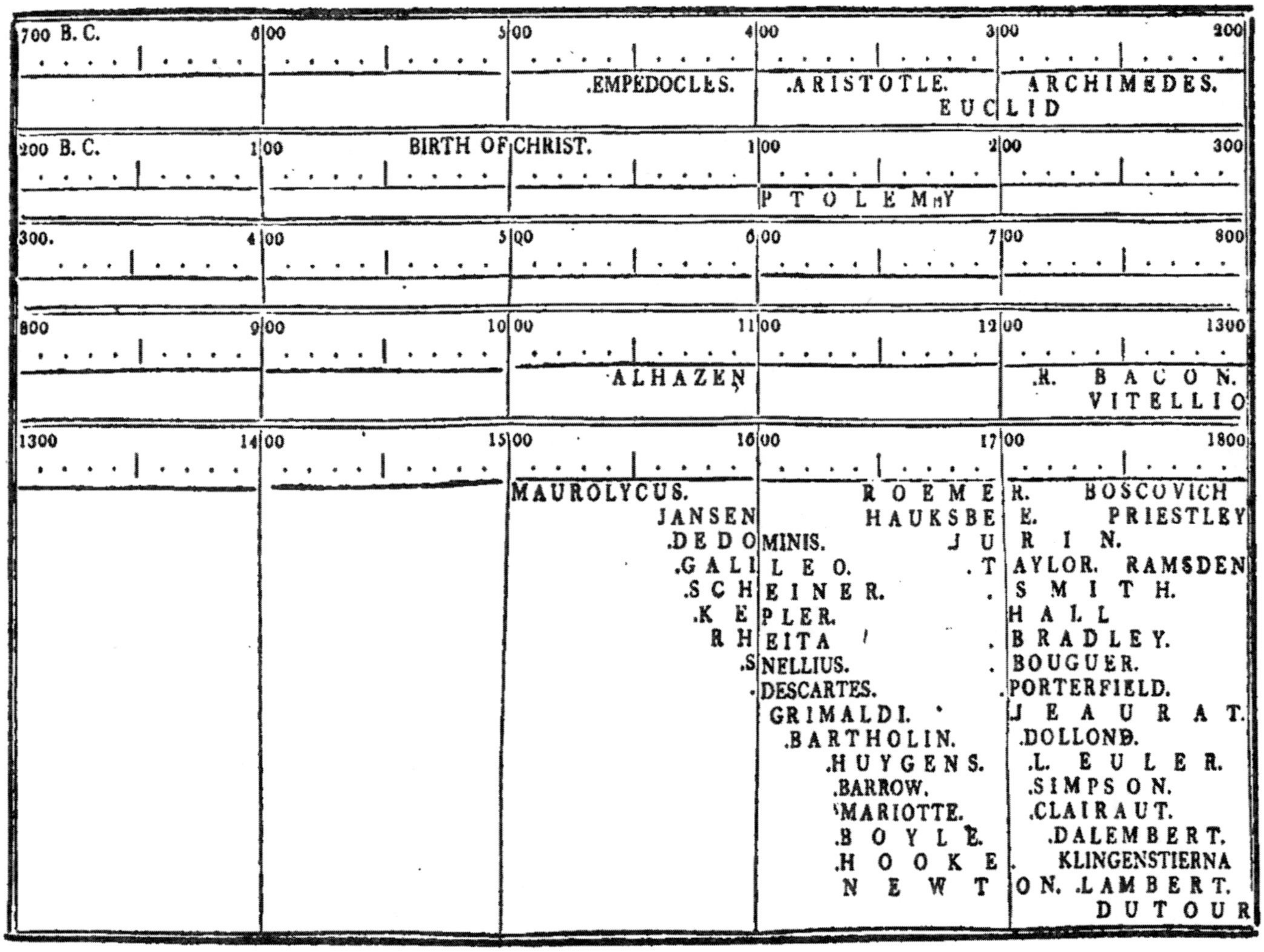

A

COURSE OF LECTURES

ON

NATURAL PHILOSOPHY

AND THE

MECHANICAL ARTS.

PART III.

PHYSICS.

A COURSE OF LECTURES ON NATURAL PHILOSOPHY AND THE MECHANICAL ARTS.

LECTURE XLI.

ON THE FIXED STARS.

THE departments of natural philosophy, which are to be the subjects of the third and last division of these lectures, are included in the description implied by the term physics, or the history of the particular phenomena of nature; and the account, which will be given of these phenomena, will be accompanied by as much of mechanical theory and analogical reasoning, as can be applied to them with sufficient certainty, and without too great intricacy of calculation.

The science of astronomy might, without any great impropriety, have been considered as a part of mechanics; but there are circumstances intimately connected with it, for the complete investigation of which, a knowledge of the motions of fluids in general, and also of optics, is absolutely necessary. It could not, therefore, hold any other place in a strict order of arrangement, than that which is here allotted to it; and, since it will not be in our power to enter completely into a mathematical examination of all the motions of the heavenly bodies, although we shall be able to pursue the detail of the most remarkable appearances which they exhibit, we may for this reason more properly consider such a view of astronomy as belonging to descriptive than to theoretical philosophy. This method of treating the subject is sometimes denominated plain astronomy, in contradistinction to the mechanical theory of the science, which is called physical astronomy; but it is obvious that in the sense which we are at present annexing to the word physics, that which is commonly called plain astronomy must be termed physical or descriptive, and what is usually called physical, must be denominated mathematical astronomy. We shall, therefore, confine ourselves in great measure to descriptive astronomy, and shall take only a general view of the laws of gravitation, as an illustration of the phenomena previously described. After having considered the magnificent objects of astronomy, which are scattered throughout the universe, we descend to geography, or the particular history of the terraqueous globe, and to the tides, produced by the influence of the celestial bodies on the ocean: and then, quitting the affections of the larger features of the matter, that constitutes the earth, we come naturally to the properties and powers of its individual particles, and to the phenomena of heat, electricity and magnetism, which are either qualities of matter, or dependent on substances differing in some respects from common matter; and in the next place, to the combination of all these substances and actions in meteorology, and in the phenomena of vegetable and animal life, a general view of which will complete our discussions on the subject of physics. The science of chemistry, or the doctrine of the qualities of particular kinds of matter, might be said to belong to the investigation of the properties of matter in general; but this science is of too great extent and importance to occupy a subordinate place in a system of natural philosophy, and must, therefore, be considered as requiring a separate course of study.

In our astronomical inquiries, we shall first examine the phenomena of the heavens and earth in their simplest form, not as they immediately appear to our observation, but as they are shown by unexceptionable proofs to be naturally arranged. The stars and sun, the planets and their satellites, and lastly the comets, will be severally described; the causes of the motions of these bodies will be superficially indicated; their sensible effects with respect to the inhabitants of the earth will be shown, and the practical modes of determining their situations and orbits will be explained.

When we begin to consider, on a large scale, the affections of matter and of space, we are impressed, at the first sight, with the inconceivable disproportion between the magnitude of space and of sensible matter: and we are naturally led to inquire if the apparently void expanse of the universe is wholly without all matter or all substance. The atmospheres of the planets cannot indeed be said absolutely to terminate at any given point, but they must become rare beyond all imagination at a very moderate distance. The substance which produces the sensation of light must, however, be every where found, at least without any sensible interval: for if an eye were placed in any point of the regions of unbounded space, wherever human investigation or fancy can penetrate them, some luminous object would at each instant be visible to it, and, in general, objects without number might be seen in every direction. Light, therefore, must be every where present, whether we suppose it to consist of separate projected corpuscles, or to be an affection of a highly elastic ether, pervading the universe in a state so rare, that although it constitutes a continuous medium, it suffers all bodies to move through it without sensible resistance, and is admitted even into their pores with perfect freedom; and if we follow Newton's opinion of the nature of light, we must suppose both such an ethereal medium, nearly at rest, and the particles of light also, moving swiftly through it, to exist together in all places: to say nothing of the possibility of the coexistence of a thousand other unseen and unknown substances, essences, and influences, in the same individual place, which may for ever set at defiance the pride of a presumptuous philosophy, that would aspire to comprehend, within its own contracted sphere, the whole extent of the mighty work of the creation.

The expanse of the universe is strewed, at immense distances, with detached portions of a substance, which we suppose to be matter, constituting stars, or suns, planets, and comets; bodies which certainly agree with each other in the power of emitting or reflecting light, and which, in all probability, have many other properties in common. Such of these, as emit their own light, are called fixed stars; and this appears to be the only criterion that we can apply to a star: for the word fixed is only to be understood in a comparative sense.

The stars must necessarily shine by their own light; for if we grant that they consist of gravitating matter, it must be allowed that no star could be near enough to another to be seen by reflected light, without a very sensible change of the places of both in consequence of their mutual gravitation, nor would it be possible, on account of their immense distance from us, to distinguish two such bodies from each other. It follows also, on the same supposition of the universality of the force of gravity, that the form of the stars must be nearly spherical.

The light of the stars appears to the naked eye to be generally white; being too faint to excite the idea of a particular colour; but when it is concentrated by Dr. Herschel's large speculums, it becomes in various stars of various hues; and indeed to the naked eye some of the stars appear a little redder and others a little bluer. The cause of the twinkling of the stars is not fully ascertained, but it is referred, with some probability, to changes which are perpetually taking place in the atmosphere, and which affect its refractive density. It is said that in some climates, where the air is remarkably serene, the stars have scarcely any appearance of twinkling.

Above two thousand stars are visible to the naked eye; and when a telescope is employed, their number appears to increase without any other limit than the imperfection of the instrument. Dr. Herschel has observed in the milky way above ten thousand stars in the space of a square degree. Lucretius and Dr. Halley have argued that their number must be absolutely infinite, in order that all of them may remain at rest by the opposition of attractions acting in every possible direction; but we are by no means certain that they do remain in perfect equilibrium.

Of the actual magnitude of the stars we can give no exact account; but they are divided into seven or more orders, according to the degrees of their apparent brightness. There is, however, reason to suppose, from the quantity of light emitted by the brightest stars, that some of them are much larger than the sun. Those stars which are below the sixth magnitude are scarcely visible without the help of telescopes. The distances of all the stars from us and from one another are so great, as not to be capable of being immediately compared with their diameters; for no star subtends an angle large enough to be ascertained by direct observation. The more perfect the instruments that we employ, the smaller are the apparent diameters of the fixed stars. Dr. Herschel found that one of the stars of the first magnitude, when viewed in his best telescopes, appeared to be about one third of a second in diameter. But there is always a limit to the perfection of the focus of the telescope and of the eye, and, however accurate both may be, the image of every radiant point will occupy on the retina a space of a certain magnitude, not depending on that of the object: so that it will perhaps be for ever impossible to measure any angle, which is only a very small fraction of a second. (Plate XXXI. Fig. 453, 454.)

There is, however, reason to suppose, that the angle subtended by the nearest stars is in reality more than a hundred times less than the angle measured by Dr. Herschel, for it may be conjectured that our distance from the nearest stars is about a hundred million million miles; taking about one third of a second for the annual parallax of the earth, that is, for the change of the apparent places of some of the fixed stars in consequence of the earth's annual motion. This seems to be nearly the utmost amount of an annual parallax that could wholly have escaped observation; for Dr. Herschel supposes that, by means of double stars, a parallax of one tenth of a second only might become sensible, and even this has never yet been discovered; on the other hand, if the parallax were really much smaller than this, it would be necessary to suppose the actual magnitude or splendour of the brightest stars to be incomparably greater than that of the sun; for at the distance of a hundred million million miles, our sun would appear, according to Lambert's calculations, but about one fourth as bright as Saturn, or like a star of the second or third magnitude only. Perhaps, indeed, the stars may differ as much from each other in magnitude as the planetary bodies, but it is somewhat more natural to imagine them more nearly equal, until we have some reason for supposing any material inequality in their dimensions. At any rate there is little doubt, that the diversity of their apparent magnitudes is principally owing to their different distances; perhaps none of them are much nearer to each other than the nearest to us; and there may still be a very great variety in their actual dimensions. There can be only twelve points on the surface of a sphere as far from each other as from the centre; in a sphere of twice the radius, there may be about 50 points at the same distance; in a sphere of three times the radius, more than 100: and it has been observed that these numbers do not greatly differ from the actual numbers of the stars of the first, second, and third magnitudes; although it is true that they are not by any means placed at equal angular distances from each other. But, from a comparison of the light of different stars, we may infer, that if their real magnitudes are nearly equal their distances must increase much faster than in this arithmetical progression; that is, that the stars of the second magnitude are more than twice as remote as those of the first, and those of the third more than three times as remote. Mr. Michell found the light of Sirius between 400 and 1000 times as great as that of a star of the sixth magnitude; consequently, supposing these stars actually equal, their distances must differ in the ratio of 1 to 20 or 30; since light always diminishes in proportion to the square of the distance of the luminous object. The light of stars of different magnitudes, situated near each other, may be compared by viewing them through two apertures of different sizes, cut in cards, one held before each eye, the apertures being reduced to such magnitudes, that the stars may appear equally bright; and the comparison may be extended to the light of the sun, by finding a star and a planet of equal brightness, and calculating what proportion of the sun's light must be reflected by the planet, upon the most probable supposition respecting the disposition of its surface to reflect more or less of the light which falls on it.

The stars are in general dispersed without any regular order, but we may observe in many parts of the heavens that a number of them are so much nearer together than to the rest, as to form a cluster or nebula. The ancients had noticed some of the most conspicuous nebulae, but Huygens first directed the attention of modern astronomers to the large one situated in the constellation Orion. Herschel has now given us catalogues of 2500 nebulae: many of them can be resolved by very high magnifying powers into separate stars; but others appear to consist of a luminous matter, spread uniformly in the neighbourhood of the several stars to which they seem to belong. (Plate XXI. Fig. 455 . . 463.)

It has been conjectured that all stars are disposed in nebulae, and that those, which appear to us to be more widely separated, are individual stars of that particular nebula in which we are placed, and of which the marginal parts may be observed, in the form of a lucid zone, which is called the milky way, being too distant to allow the single stars to be perceived by the naked eye. This opinion was first suggested by Professor Kant, the author of the system of metaphysics called the critical philosophy. The idea was adopted by Lambert, who considers the largest stars as constituting a distinct nebula placed among a multitude of others, which together produce the appearance of a continued zone; and Dr. Herschel has investigated very particularly the figure of a single nebula, which would be capable of being projected into the form of the milky way. We must not, however, suppose that each of Dr. Herschel's 2500 nebulae can be at all comparable in magnitude to this supposed nebula, since many of them are almost as much resolved by the telescope into single stars as the milky way itself; which would be utterly impossible, if the stars which they contain were equally numerous with those of the nebula to which the milky way belongs. Supposing all the stars of this nebula to be as remote from each other as the nearest of them are from the sun, it may be calculated that the most distant are about 500 times as far from us as the nearest, and that light, which is probably 15 or 20 years in travelling to us from Sirius, would be nearly twenty thousand in passing through the whole diameter of the milky way. A nebula of the same size as this, appearing like a diffused light of a degree in diameter, must be at such a distance, that its light would require a million years to reach us. (Plate XXXI. Fig. 464.)

The stars are not, properly speaking, absolutely fixed with respect to each other, for several of them have particular motions, which have been discovered by a comparison of accurate observations, made at very distant times. Arcturus, for instance, has a progressive motion, amounting to more than two seconds annually. Dr. Maskelyne found, that out of 36 stars, of which he ascertained the places with great precision, 35 had a proper motion. Mr. Michell and Dr. Herschel have conjectured, that some of the stars revolve round others which are apparently situated very near them; and perhaps even all the stars may in reality change their places more or less, although their re-

relative situations, and the directions of their paths may often render their motions imperceptible to us.

Respecting all these arrangements of stars into different systems, Dr. Herschel has lately entered into a very extensive field of observation and speculation, and has divided them into a number of classes, to each of which he has assigned a distinct character. Some he supposes, like our sun, to be insulated stars, beyond the reach of any sensible action of the gravitation of others; and around these alone he conceives that planets and comets revolve. Double stars, in general, he imagines to be much nearer to each other, so as to be materially affected by their mutual gravitation, and only to preserve their distance by means of the centrifugal force derived from a revolution round their common centre of inertia; an opinion which, he thinks, is strongly supported by his own observations of some changes in the positions of double stars. Others again he supposes to be united in triple, quadruple, and still more compound systems. A fourth class consists of nebulae like the milky way, the clusters of stars being rounded, and appearing brightest in the middle. Groups of stars Dr. Herschel distinguishes from these by a want of apparent condensation about a centre of attraction; and clusters by a still greater central compression. A seventh class includes such nebulae as have not yet been resolved into stars, some of which Dr. Herschel supposes to be so remote, that the light emitted by them must actually have been two millions of years in travelling to our system. The nebulae of another description resemble stars surrounded by a bur, or a faint disc of light: a diffused milky nebulosity, apparently produced by some cause distinct from the mmediate light of any stars, is the next in order: and Dr. Herschel has distinguished other more contracted nebulous appearances, in different states of condensation, into the classes of nebulous stars, and planetary nebulae, with and without bright central points. Many of these distinctions are perhaps too refined to be verified by common observers; but the discovery of the existence of double and triple stars, revolving round a common centre, will, if it be confirmed, add one more to the catalogue of Dr. Herschel's important improvements.

It is however fully ascertained, that some of the stars have periodical changes of brightness, which are supposed to arise either from the temporary interposition of opaque bodies revolving round them, or, still more probably, from a rotatory motion of their own, which brings at certain periodical times a less luminous part of the surface into our view. Thus, the star Algol, which is usually of the second magnitude, becomes, at intervals of 2 days and 21 hours each, of the fourth only, and occupies 7 hours in the gradual diminution and recovery of its light. A less probable conjecture respecting this change of brightness was advanced by Maupertuis, who imagined that the disc of the star might be greatly flattened by a rapid rotation, and its edge occasionally presented to us, in consequence of the disturbances produced by the attraction of planets revolving round the luminary. Other irregular variations may possibly be occasioned by the appearance and disappearance of spots, occurring, like the spots of the sun, without any determinate order or assignable cause; and many stars have in the course of ages wholly disappeared, and sometimes have been again recovered; others have made their appearance for a short time, where no star had before been seen. Such a temporary star was observed by Hipparchus, 120 years before our era, and the circumstance suggested to him the propriety of making an accurate catalogue of all the stars, with their respective situations, which is still extant, having been preserved by Ptolemy, who added 4 stars to the 1022 that it contained. In 1572, Cornelius Gemma discovered a new star in Cassiopeia, which was so bright as to be seen in the day time, and gradually disappeared in sixteen months. Kepler, in 1604, observed a new star in Serpentarius, more brilliant than any other star or planet, and changing perpetually into all the colours of the rainbow, except when it was near the horizon; it remained visible for about a year. Many other new stars have also been observed at different times.

For describing the particular fixed stars according to their relative situations, it is necessary to consider them as they are visible to the inhabitants of the earth. They have been divided, for the sake of convenience, into parcels, making up imaginary forms, denominated constellations. This division is of very remote antiquity, and though it may be useless, and sometimes even inconvenient, for the purposes of minute observation, yet for a general recollection of the great features of the heavens, these arbitrary names and associations cannot but greatly assist the memory. It is also usual to describe particular stars by their situation with respect to the imaginary figure to which they belong, or, more commonly, at present, by the letters of the Greek alphabet, which were first applied by Bayer in 1603, and in addition to these, by the Roman letters, and by the numbers of particular catalogues.

There are two principal modes of representing the stars; the one by delineating them on a globe, where each star occupies the spot in which it would appear to an eye placed in the centre of the globe, and where the situations are consequently reversed, when we look on them from without, in the same manner as a word appears reversed when seen from the back of the paper: the other mode is by charts, which are generally so arranged as to represent the stars in positions similar to their natural ones, or as they would appear on the internal concave surface of the globe. Sometimes also the stars have been delineated as they would be projected on imaginary surfaces, without any reference to a globe; for instance, on the surfaces of transparent cones or cylinders. The art of constructing all such projections belongs to the subject of perspective.

In describing the particular stars, it will be most convenient to begin with such as never set in our climates, and we may then refer the situations of others to their positions with respect to these.

The great bear is the most conspicuous of the constellations which never set; it consists of seven stars, placed like the four wheels of a waggon, and its three horses, except that the horses are fixed to one of the wheels. The two hind wheels are the pointers, which direct us to the pole star, in the extremity of the tail of the little bear: and further on, to the constellation Cassiopeia, which is situated in the milky way, where it is nearest to the pole, and which consists of several stars, nearly in the form of the letter W. The two northernmost wheels of the great bear, or wain, point at the bright star Capella, the goat, in Auriga. Descending along the milky way from Cassiopoia, if we go towards Capella, we come to Algenib, in Perseus; and a little further from the pole we find Algol, or Medusa's head: but if we take the opposite direction, we arrive at Cygnus, the swan; and beyond it, a little out of the milky way, is the bright star Lyra. The dragon consists of a chain of stars partly surrounding the little bear; and between Cassiopeia and the swan is the constellation Cepheus.

Near Algenib, and pointing directly towards it, are two stars of Andromeda, and a third is a little beyond them. A line drawn through the great bear and Capella passes to the Pleiades, and then, turning at a right angle towards the milky way, reaches Aldebaran, or the bull's eye, and the shoulders of Orion, who is known by his belt, consisting of three stars, placed in the middle of a quadrangle. Aldebaran, the Pleiades, and Algol, make the upper, and Menkar, or the whale's jaw, with Aries, the lower points of a W. In Aries we observe two principal stars, one of them with a smaller attendant.

A line drawn from the pole, midway between the great bear and Capella, passes to the twins and to Procyon; and then, in order to reach Sirius, it must bend across the milky way. Algol and the twins point at Regulus, the lion's heart, which is situated at one end of an arch, with Denebola at the other end.

The pole star and the middle horse of the wain direct us to Spica Virginis, considerably distant: the pole and the first horse nearly to Arcturus, in the waggoner, or Bootes. Much further southwards, and near the milky way, is Antares, in the scorpion, forming, with Arcturus and Spica, a triangle, within which are the two stars of Libra. The Northern crown is nearly in a line between Lyra and Arcturus, and the heads of Hercules and Serpentarius are between Lyra and Scorpio.

In the milky way, below the part nearest to Lyra, and on a line drawn from Arcturus through the head of Hercules, is Aquila, making with Lyra and Cygnus a conspicuous triangle. The last of the three principal stars in Andromeda makes, with three of Pegasus, a square, of which one of the sides points to Fomalhaut, situated at a considerable distance in the southern fish, and in the neighbourhood of the whale, which has already been mentioned.

By means of these allineations, all the principal stars that are ever visible in Britain may be easily recognised. Of those which never rise above our horizon, there are several of the first magnitude; Canopus, in the ship Argo, and Achernar, in the river Eridanus, are the most brilliant of them; the feet of the centaur, and the crosier are the next; and according to Humboldt's observations, perhaps some others may require to be admitted into the same class. (Plate XXXVI, XXXVII.)

LECTURE XLII.

ON THE SOLAR SYSTEM.

THE most conspicuous of all the celestial bodies, which we have been examining, is the sun, that magnificent luminary which occupies the centre of the system that comprehends our earth, together with a variety of other primary and secondary planets, and a still greater number of comets.

The sun agrees with the fixed stars in the property of emitting light continually, and in retaining constantly its relative situation with very little variation; it is probable also that these bodies have many other properties in common. The sun is, therefore, considered as a fixed star comparatively near us; and the stars as suns at immense distances from us: and we infer from the same analogy, that the stars are possessed of gravitation, and of the other general properties of matter; they are supposed to emit heat as well as light; and it has with reason been conjectured that they serve to cherish the inhabitants of a multitude of planetary bodies revolving round them.

The sun, like many other stars, has probably a progressive motion, which is supposed, from a comparison of the apparent motions of a great number of the stars, to be directed towards the constellation Hercules. It is beyond all question that many of the stars have motions peculiar to themselves, and it is not certain that any of them are without such motions: it is, therefore, in itself highly probable that the sun may have such a motion. But Dr. Herschel has confirmed this conjecture by arguments almost demonstrative. He observes that the apparent proper motions of 44 stars out of 56 are very nearly in the direction which would be the result of such a real motion of the solar system: and that the bright stars Arcturus and Sirius, which are probably the nearest to us, have, as they ought to have, the greatest apparent motions. Besides, the star Castor appears, when viewed with a telescope, to consist of two stars, of nearly equal magnitude; and though they have both a considerable apparent motion, they have never been found to change their distance a single second; a circumstance which is easily understood if both their apparent motions are supposed to arise from a real motion of the sun, but which is much less probable on the supposition of two separate and independent motions.

Besides this progressive motion, the sun is subjected to some small change of place, dependent on the situations of the planetary bodies, which was long inferred from theory only, but which has been actually demonstrated by modern observations. Supposing all the planets to be in conjunction, or nearly in the same direction from the sun, the common centre of inertia of the system is at the distance of about a diameter of the sun from his centre: and since the centre of inertia of the whole system must be undisturbed by any reciprocal actions or revolutions of the bodies composing it, the sun must describe an irregular orbit round this centre, his greatest distance from it being equal to his own diameter. We may form an idea of the magnitude of this orbit by a comparison with the orbit of the moon: a body revolving round the sun, in contact with his surface, must be nearly twice as remote from his centre as the moon is from the earth, and the sun's revolution round the common centre of gravity of the system must therefore be, where it is most remote, at four times the distance of the moon from the earth.

The sun revolves on his axis in 25 days 10 hours, with respect to the fixed stars: this axis is directed towards a point about half way between the pole star and Lyra, the plane of the rotation being inclined a little more than 7° to that in which the earth revolves. The direction of this motion is from west to east, terms which we can only define from our presupposed knowledge of the stars, by saying that the motion is such, that a point of the sun's surface at first opposite Aries, moves towards Taurus. Nor have we any better mode of describing north and south, or right and left: we can only say comparatively, that if we are placed with our heads northwards, and looking towards the centre, our right hands will be eastwards, and our left westwards. All the rotations of the different bodies which compose the solar system, as far as they have been ascertained, are in the same direction, and all their revolutions, excepting those of some of the comets, of which the motions are retrograde, and those of some of the satellites of the Georgian planet, which revolve in planes so distant from those of the other planetary motions, that the directions of their revolutions can scarcely be called either direct or retrograde.

The time and direction of the sun's rotation is ascertained by the change of the situation of the spots, which are usually visible on his disc, and which some astronomers suppose to be elevations, but others, apparently on better foundations, to be excavations or deficiencies in the luminous matter covering the sun's surface. These spots are frequently observed to appear and disappear, and they are in the mean time liable to great variations, but they are generally found about the same points of the sun's surface. Lalande imagines that they are parts of the solid body of the sun, which, by some agitations of the luminous ocean, with which he conceives the sun to be surrounded, are left nearly or entirely bare. Dr. Wilson and Dr. Herschel are disposed to consider this ocean as consisting rather of a flame than of a liquid substance, and Dr. Herschel attributes the spots to the emission of an aeriform fluid, not yet in combustion, which displaces the general luminous atmosphere, and which is afterwards to serve as fuel for supporting the process; hence he supposes the appearance of copious spots to be indicative of the approach of warm seasons on the surface of the earth, and he has attempted to maintain this opinion by historical evidence. The exterior luminous atmosphere has an appearance somewhat mottled, some parts of it, appearing brighter than others, have generally been called faculae; but Dr. Herschel distinguishes them by the names of ridges and nodules. The spots are usually surrounded by margins less dark than themselves, which Dr. Herschel calls shallows, and which he considers as parts of an inferior stratum consisting of opaque clouds, capable of protecting the immediate surface of the sun from the excessive heat produced by combustion in the superior stratum, and perhaps of rendering it habitable to animated beings. (Plate XXXI. Fig. 465 . . 469.)

But if we inquire into the intensity of the heat which must necessarily exist wherever this combustion is performed, we shall soon be convinced that no clouds, however dense, could impede its rapid transmission to the parts below. Besides, the diameter of the sun is 111 times as great as that of the earth; and at its surface, a heavy body would fall through no less than 450 feet in a single second: so that if every other circumstance permitted human beings to reside on it, their own weight would present an insuperable difficulty, since it would become nearly thirty times as great as upon the surface of the earth, and a man of moderate size would weigh above two tons. Some of the most celebrated astronomers have imagined, from the comparative light of different parts of the sun's disc, or apparent surface, that he is surrounded by a considerably dense and extensive atmosphere, imperfectly transparent; conceiving that, without such an atmosphere, the marginal parts, which are seen most obliquely, must appear considerably the brightest; but this opinion is wholly erroneous, and the inferences which have been drawn from it, respecting the sun's atmosphere, are consequently without foundation.

We are, however, assured, by direct observation, of the existence of some aerial substance in the neighbourhood of the sun, producing the appearance called the zodiacal light, which is sometimes seen, nearly in the plane of the sun's rotation on its axis, extending beyond the orbit of Mercury. It is said to have been first distinctly described in Childrey's Britannia Baconica, a work published in 1661, and it was afterwards more particularly observed by Cassini, Mairan, and others. In the torrid zone it is almost constantly visible; and in these climates, it may often be distinguished in the beginning of March, after the termination of twilight, exhibiting the appearance of a narrow triangle, somewhat rounded off, of a whiteness resembling the milky way, ascending from the sun as a base, like the projection or section of a very flat spheroid, and extending to a distance of more than 50° from the sun. The whole orbit of Venus never subtends so great an angle from the earth as 96°, consequently this substance must occasionally involve both Mercury and Venus; and if it were not extremely rare, it would produce some disturbance in their motions; while in fact it does not appear to impede the progress even of the tails of the comets, which are probably themselves of very inconsiderable density. It cannot be a continuous fluid atmosphere, revolving with the same velocity as the sun; for the gravitation of such an atmosphere would cause it to assume a form more nearly spherical; and the only probable manner in which it can be

4

supposed to retain its figure, is by means of a revolution much more rapid than the sun's rotation. Some persons have attributed the appearance to the refraction of the earth's atmosphere only; but if it arose from any such cause as this, its direction could scarcely be oblique with respect to the horizon, and it is highly improbable that it should always happen to coincide with the plane of the sun's rotation. (Plate XXXI. Fig. 470.)

The sun is accompanied in his progressive motion among the fixed stars by ten planetary bodies, of different magnitudes, revolving round him, from west to east, in orbits approaching to circles, and visible to us by means of the light which they receive from him. These are Mercury, Venus, the Earth, Mars, Juno, Pallas, Ceres, Jupiter, Saturn, and the Georgian planet. It is unnecessary to adduce at present any arguments to prove the actual existence or direction of any of these motions; their complete agreement with the visible phenomena of the heavens, and with the laws of gravitation, will hereafter appear to afford sufficient evidence of the accuracy of the received theory of the arrangement of the solar system. The motion of the earth is the most unanswerably proved by the apparent aberration of the fixed stars, derived from the different directions of this motion at different times, and corresponding precisely with the known velocity of light, deduced from observations of a very different kind. That the planets receive their light from the sun, is undeniably shown by the appearance of the discs of many of them, when viewed through a telescope, those parts of their surfaces only being luminous, on which the sun shines at the time of observation.

These planets are neither all in one plane, nor does any one of them remain precisely in the same plane at all times; but their deviations from their respective planes are inconsiderable, and they are commonly represented by supposing each planet to revolve in a plane passing through the sun, and the situation of this plane to be liable to slight variations. There is, however, a certain imaginary plane, determinable from the situations, the velocities, and the masses of the planets, which, like the centre of inertia, never changes its position on account of any mutual actions of the bodies of the system, and this plane of inertia is called the fixed ecliptic. Its situation is nearly half way between the orbits of Jupiter and of Saturn; and it is inclined in a small angle only to the plane of the earth's orbit, which is called the earth's ecliptic, or simply the ecliptic.

The ecliptic passes through the constellations denominated the signs of the zodiac, between Aries, the Pleiades, the twins, and Regulus, to the north, and Aldebaran, Spica, and Antares, to the south. Its position has varied slowly in the course of many ages, so that its northmost point is now more than one third of a degree more remote from the pole star than it was in the time of Eratosthenes, who observed its place 230 years before the birth of Christ. It appears from Lagrange's calculations, that the limit of its greatest possible variation is about 10 or 11 degrees. The ecliptic is supposed to be divided into twelve angular parts, or signs, each containing thirty degrees: they are named Aries, Taurus, Gemini, Cancer, Leo, Virgo, Libra, Scorpio, Sagittarius, Capricornus, Aquarius, Pisces. Those who prefer the cadence of a Latin distich, in order to assist the memory, may repeat them thus,

Sunt Aries, Taurus, Gemini, Cancer, Leo, Virgo,
Libraque, Scorpius, Arcitenens, Caper, Amphora, Pisces.

The planes of the orbits of the other primary planets, excepting the three minute planets lately discovered, intersect the ecliptic in small angles, and the lines of intersection are called lines of the nodes. The nodes of all the planets move very slowly, but not quite uniformly, from east to west, that is, with respect to the fixed stars. At present the inclinations of all the orbits appear to be somewhat diminishing: that of the orbit of Jupiter is less by 6 minutes than it was in the time of Ptolemy.

The orbit of each planet is very nearly an ellipsis, one of the foci of which coincides with the sun, or rather with the common centre of inertia of the sun and planet. The extremities of the greater axis, where the orbit is furthest from the sun and nearest to it, are called the upper and the lower apsis, or the aphelion and perihelion; the mean distance being at either end of the lesser axis; and the distance of the centre of the ellipsis from the sun is called the eccentricity. The slight deviations of the planets from these elliptic paths are expressed by considering the apsides as moveable, and this motion is direct, that is, from west towards east, in the case of all the planets except Venus, of which the aphelion has a retrograde motion, with respect to the fixed stars.

The elliptic motion of the planets was first discovered by Kepler; and he found that a right line, joining the sun and any planet, describes always equal areas in equal times. The observations, on which Kepler founded these important laws, were made principally on the planet Mars. He determined by calculation, upon the supposition which was then generally adopted, of a motion in an eccentric circle, what must be nearly the situation of the planet, with respect to the sun, that is, its heliocentric place, and observing its geocentric place, with respect to the earth, he was thus able to construct a triangle representing the situation of the three bodies; repeating this operation in various parts of the orbit, he discovered its form; and having done this, the velocity of the motion in different parts of the orbit was easily determined from the apparent change of place in a given time. (Plate XXXII. Fig. 471.)

The same astronomer also ascertained, that the squares of the times of revolution of the different planets are in proportion to the cubes of their mean distances from the sun. For example, if one planet were four times as distant as another, it would revolve in a period eight times as long, since the cube of 4 is equal to the square of 8; thus Mars is nearly four times as remote from the sun as Mercury, and the Georgian planet four times as remote as Jupiter, and their periods are nearly eight times as long respectively.

It is probable that all the planets have a rotatory motion from west to east, either perfectly or very nearly equable. This motion has been observed in Venus, the Earth, Mars, Jupiter, and Saturn; and from some phenomena of the satellites of the Georgian planet, Mr. Laplace thinks that it may also be assumed as nearly certain that this planet has also a rotatory motion. The figure of the planets is spheroidical; they are more or less flattened at the poles, as they revolve more or less rapidly on their axes. These axes retain, with a very slight deviation, a situation always parallel, in every part of the orbits.

But, in the course of time, the gradual change of the position of the axis produces a sensible effect. In the case of the earth, this effect is denominated the precession of the equinoxes. The equinoctial points are the intersections of the apparent ecliptic, or the path of the sun in the heavens, with the plane of the equinoctial, which is perpendicular to the earth's axis and which passes through the equator on the earth's surface; these points of intersection have a retrograde motion, from east to west, on the ecliptic. This motion was discovered by Hipparchus, in the year 128 before Christ, from a comparison of his own observations with those of Timocharis, made 155 years before; and since the time of Hipparchus, the equinoctial points have receded about $26\frac{1}{2}°$. Hence it happens that the constellations, called the signs of the zodiac, are now at a considerable distance from those divisions of the ecliptic which bear the same names.

The earth's axis has also a small periodical change of inclination, or a nutation, performed in about 19 years, and amounting in the whole to 18 seconds only. Its existence was determined by Newton from theory, although he failed in the attempt to ascertain its quantity with accuracy; it was first actually observed by Dr. Bradley, about the year 1747. The absolute direction of the axis in the heavens is also liable to some variation, in the course of many ages, but this change has not always been sufficiently distinguished from the change of the position of the ecliptic. The inclination of the equator to the ecliptic is now very nearly 23° 28'.

In order to retain in memory a general idea of the proportional distances of the primary planets from the sun, we may call that of the earth 10 and that of Saturn 100; the distance of Mercury will then be 4, to which we must add 3 for Venus, making 7; twice 3 or 6 for the earth, making 10; twice 6 or 12 for Mars, making 16; twice 12 or 24, making 28, for the three small planets, Juno, Pallas, and Ceres; twice 24 or 48, making 52, for Jupiter; twice 48 or 96 for Saturn, making 100; and twice 96 or 162, making 196, for the Georgian planet; and these sums will represent the distances, without any material exception, in the nearest integer numbers.

The planet Mercury is little more than one third as large as the earth in diameter. He performs his revolution in somewhat less than three months, at about two fifths of the distance of the earth. His orbit is more eccentric, and more inclined to the ecliptic, than those of any of the planets except the three small ones lately discovered; the eccentricity being one fifth of the mean distance, and the inclination 7°. Of his density and his rotation we know nothing but from conjecture.

Venus is very nearly as large as the earth; Dr. Herschel thinks her even a little larger. Her revolution occupies about 7 months, her distance from the sun being about seven tenths of that of the earth, and her orbit nearly circular, inclined in an angle of 3° 24′ to the ecliptic. Mr. Schroeter attributes to her mountains much higher than those of the earth, he has observed strong indications of an atmosphere surrounding her, and he assigns for her rotation on her axis the period of 23 hours 21 minutes. Her density has been estimated from the perturbations, occasioned by her attraction, in the motions of the other planets, and it has been supposed to be a little less than that of the earth.

The distance of the earth from the sun is about 95 million English miles; and this determination is generally supposed to be so far accurate, that there is no probability of an error of more than a million or two, at most, although some authors are still disposed to believe that the distance may be even greater than a hundred millions. The period of its revolution, with respect to the equinoctial points, which are the usual standard of comparison, since their situation determines the annual return of the seasons, is 365 days, 5 hours, 48 minutes, and 48 seconds; and this is called its tropical revolution; that of its absolute or sidereal revolution is 365 days, 6 hours, 9 minutes, and 8 seconds; the difference, which is 20 minutes and 20 seconds, being the time occupied in passing over the space, through which the equinoctial points have retreated in the course of the tropical year. By a day, we always understand the time which elapses during the rotation of the earth with respect to the sun; a sidereal day is about four minutes shorter.

At a distance from the sun exceeding that of the earth by one half, the planet Mars revolves, in about a year and seven eighths. He is of half the earth's linear dimensions: he has spots which change their form, and, therefore, probably, an atmosphere. Dr. Herschel found his rotation performed in 39 minutes more than a day; his equator inclined 28° 42′ to the plane of his orbit, and his figure so much flattened at the poles, that his axis is $\frac{1}{16}$th shorter than his equatorial diameter. From this form, compared with the time of his rotation, it may be inferred that his density must be very unequal in different parts: Laplace supposes it from calculation to be on the whole about three fourths as great as that of the earth.

In the interval between Mars and Jupiter, and nearly at the distance where, from a dependance on the regularity of the progression already mentioned, a number of astronomers had for some years been seeking for a primary planet, the observations of Mr. Piazzi, Dr. Olbers, and Mr. Harding have placed three very small bodies, differing but little in their mean distance and their periodical time. They have named them Ceres, Pallas, and Juno: none of them subtends an angle large enough to be measured by our best instruments; and all the circumstances of their motions are yet but imperfectly established. Juno, however, appears to be somewhat less remote than the other two; all their orbits are considerably inclined to the ecliptic, especially that of Pallas, which is also extremely eccentric. Dr. Herschel does not admit that they deserve the name of planets, and chooses to call them asteroids.

Jupiter is the largest of all the planets, his diameter being 11 times as great as that of the earth, and the force of gravitation at his surface being triple the terrestrial gravitation. He revolves in about 12 years, at a little more than five times the earth's distance from the sun. His rotation is performed in less than ten hours, his equator being inclined about three degrees to his ecliptic, which makes an angle of 1° 19′ with ours. His belts are supposed by many to be clouds in his atmosphere; they seem to have a rotation somewhat slower than that of the planet.

The diameter of Saturn is ten times as great as that of the earth, but, on account of the smaller density of his substance, the force of gravity at his surface scarcely exceeds its force at the surface of the earth. He revolves in 29 years and a half, in an orbit inclined $2\frac{1}{2}$° to the ecliptic, at the distance of $9\frac{1}{2}$ semidiameters of the earth's orbit: his rotation occupies only $10\frac{1}{4}$ hours, and his equator is inclined about 30° to our ecliptic. The most remarkable circumstance attending him is the appearance of a double ring, which is suspended over his equator, and revolves with a rapidity almost as great as that of the planet. His figure appears also, according to Dr. Herschel's observations, to be extremely singular; deviating very considerably from that of an elliptical spheroid, which is the form assumed by all the other planets that appear flattened, and approaching in some degree to a cylinder with its angles rounded off. Such a form can only be derived from some very great irregularities in the density of the internal parts of his substance.

The Georgian planet, discovered by Dr. Herschel in 1780, sometimes also called Herschel, and sometimes Uranus, revolves in $83\frac{3}{4}$ years, at a distance from the sun equal to 19 times that of the earth. Its diameter is a little more than 4 times that of the earth, and the weight of bodies at its surface a little less than here. Notwithstanding its dimensions are by no means comparatively small, it appears to us as a star of the sixth or seventh magnitude, and is seldom seen by the naked eye. Its orbit approaches very near to the ecliptic; its disc is said to be somewhat flattened, and it is supposed to revolve with considerable rapidity.

These ten planetary bodies are the only ones hitherto discovered which have any title to be considered as primary planets, that is, as bodies revolving round the sun, in orbits so nearly circular, as to remain always within the reach of our observation. It has been conjectured that the number of planets may in reality be much greater, that not only many small and perhaps invisible bodies may be revolving in the intervals of the planets with which we are acquainted, but that larger bodies also may belong to our system, which never approach within such a distance as to be seen by us. Some have even bestowed names, borrowed from the ancient mythology, on these imaginary planets; but the idea of such an appropriation of terms is rather to be regarded as belonging to the regions of poetical fiction than to those of solid philosophy.

The largest and the most remote of the primary planets have their attendant satellites, or secondary planets, accompanying them in their respective revolutions round the sun, and moving, at the same time, in subordinate orbits, round the primary planets. The earth is attended by the moon, Jupiter by four moons or satellites, Saturn by seven, besides his ring, and the Georgian planet by six moons. All these satellites move in the direct order of the signs, and in planes not very remote from the ecliptic, excepting those of the Georgian planet, which revolve in planes nearly perpendicular to the ecliptic. Each of these planets thus becomes the central luminary of a little system of its own, in which the motions and the periods observe the same general laws as prevail in the solar system at large. Of the 28 primary and secondary planets, we are indebted to Dr. Herschel for the knowledge of 9; the Georgian planet, with its six satellites, and the two innermost moons of Saturn.

The motions of some of these satellites, in particular of those of Jupiter, and of the moon, are of considerable importance for the assistance they afford us in determinations of time, and of the relative situations of places. They are subjected to considerable irregularities, but the united labours of various astronomers have enabled us to calculate all their motions with the greatest accuracy.

The moon performs a complete sidereal revolution in 27 days $7\frac{3}{4}$ hours, and a synodical revolution, during which she returns to the same position with respect to the earth and sun, in 29 days $12\frac{3}{4}$ hours; a period which constitutes a lunation, or a lunar month. Her orbit is inclined to the ecliptic in an angle of a little more than five degrees, but this inclination is liable to great variations: the place of its nodes is also continually changing, their motion being sometimes retrograde, and sometimes direct, but on the whole the retrograde motion prevails. The form of the moon's orbit is irregularly elliptic, and the velocity of its motion deviates considerably from the Keplerian law of the description of equal areas in equal times; the apsides, or the extremities of the greater axis of the ellipsis, which are called the apogee and perigee, have on the whole a direct motion. From a comparison of modern observations with the most ancient, the mean motion of the moon is found to be somewhat accelerated.

The moon revolves on her own axis with a very equable motion, and the period of her rotation is precisely equal to the mean period of her revolution round the earth; so that she always presents to us the same portion of her surface, excepting the apparent librations produced by her unequal velocities in her orbit, and by the position of her axis, which is inclined 1° 43′ to the ecliptic, and sometimes as much as 7° to her own orbit. Her distance from the earth is about 240 000 miles; her diameter $\frac{3}{11}$ of that of the earth, or 2160 miles; and the weight of bodies at her surface is supposed to be about one fifth of their weight at the surface of the earth.

The surface of the moon presents to us, when viewed with a telescope, a great diversity of light and shade, the principal features of which are visible even to the naked eye. Many of these inequalities resemble very strongly the effects of volcanos; several astronomers have imagined that they have seen volcanos actually burning in the unenlightened part of the planet; and Dr. Herschel's instruments have enabled him to obtain satisfactory evidence of the truth of the conjecture. The appearance of a perforation, which Ulloa supposed that he observed near the margin of the Moon's disc, in a solar eclipse, has been attributed by some to a volcano actually burning. Dr. Halley and Mr. Weidler have also observed flashes of light on the dark part of the moon, considerably resembling the effect of lightning. The height of the lunar mountains has been commonly supposed to exceed very considerably that of the mountains of the earth; but Dr. Herschel is of opinion that none of them are so much as two miles high. The names, which have been given by astronomers to various parts of the moon's surface, are of some utility in the observation of the progress of an eclipse.

Of the satellites of Jupiter, some are a little larger, and others smaller than the moon: they all revolve in planes inclined between $2\frac{1}{2}°$ and $3\frac{1}{2}°$ to the orbit of the planet, and they are therefore always seen nearly in the same line. It is inferred, from some periodical changes of light which they undergo, that, like our moon, they always present the same face to their primary planet.

The ring of Saturn is inclined 31 degrees to our ecliptic; of his seven satellites, six are nearly in the same plane with the ring; but the plane of the seventh or outermost satellite is but half as much inclined to the ecliptic. The ring has been observed by Dr. Herschel to revolve in $10\frac{1}{2}$ hours, which is considerably less than the time that would be occupied by the revolution of a satellite at the same distance. The planes of the six satellites of the

Georgian planet are nearly perpendicular to the ecliptic; and some of their revolutions are supposed to be rather retrograde than direct.

Besides the bodies which revolve completely round the sun, within the limits of our observation, there are others, of which we only conclude from analogy, that they perform such revolutions. These are the comets; they generally appear attended by a nebulous light, either surrounding them as a coma, or stretched out to a considerable length as a tail; and they sometimes seem to consist of such light only. Their orbits are so eccentric, that in their remoter situations the comets are no longer visible to us, although at other times they approach much nearer to the sun than any of the planets: for the comet of 1680, when in its perihelion, was at the distance of only one sixth of the sun's diameter from his surface. Their tails are often of great extent, appearing as a faint light, directed always towards a point nearly opposite to the sun: it is quite uncertain of what substance they consist; and it is difficult to determine which of the conjectures respecting them can be considered as the least improbable; it is possible that, on account of the intense cold, to which the comets are subjected in the greatest part of their revolutions, some substances, more light than any thing we can imagine on the earth, may be retained by them in a liquid, or even in a solid form, until they are disengaged by the effect of the sun's heat: but we are still equally at a loss to explain the rapidity of their ascent: for the buoyancy of the sun's atmosphere cannot possibly be supposed to be adequate to the effect; and on the whole there is, perhaps, reason to believe that the appearances are derived from some cause, bearing a considerable analogy to the fluid, supposed to be concerned in the effects of electricity. It is probable that the density of the nucleus, or the body of the comet itself, is comparatively small, and its attraction for the tail consequently weak, so that it has little tendency to reduce the tail, even if it consists of a material substance, to a spherical form: for since some comets have no visible nucleus at all, there is no difficulty in supposing the nucleus, when present, to be of very moderate density, and perhaps to consist of the same kind of substance as constitutes the tail or coma, in a state of somewhat greater condensation. If, therefore, it should ever happen to a planet to fall exactly in the way of a comet, of which there is but very little probability, it is to be supposed that the inconvenience

4

suffered by the inhabitants of the planet might be merely temporary and local: the chances are, however, much greater, that a comet might interfere in such a manner with a planet, as to deflect it a little from its course, and retire again without coming actually into contact with it.

Nearly 500 comets are recorded to have been seen at different times, and the orbits of about a hundred have been correctly ascertained: but we have no opportunity of observing a sufficient portion of the orbit of any comet, to determine with accuracy the whole of its form as an ellipsis, since the part which is within the limits of our observation does not sensibly differ from the parabola, which would be the result of an ellipsis prolonged without end.

Two comets at least, or perhaps three, have been recognised in their return. A comet appeared in 1770, which Prosperin suspected to move in an orbit materially different from a parabola: Mr. Lexell determined its period to be 5 years and 7 months, and its extreme distances to be between the orbits of Jupiter and of Mercury; but it does not appear that any subsequent observations have confirmed his theory. It has, however, been calculated, that supposing the theory correct, it must afterwards have approached so near to Jupiter as to have the form of its orbit entirely changed.

Dr. Halley foretold the return of a comet about 1758, which had appeared in 1531, in 1607, and in 1682, at intervals of about 75 years; and with Clairaut's further correction for the perturbations of Jupiter and Saturn, the time agreed within about a month. The mean distance of this comet from the sun must be less than that of the Georgian planet; so that by improving our telescopes still more highly, we may, perhaps, hereafter be able to convert some of the comets into planets, so far as their remaining always visible would entitle them to the appellation. Dr. Halley also supposed the comet of 1680 to have been seen in 1106, in 531, and in the year 44 before Christ, having a period of 575 years; and it has been suspected that the comets of 1556 and 1264 were the same, the interval being 292 years; a conjecture which will either be confirmed or confuted in the year 1848. Some persons have even doubted of the perfect coincidence of the orbits of any comets, seen at different times, with each other, and have been disposed to consider them as messengers forming a communication between the neighbouring systems of the sidereal world, and visiting a variety of stars in succession, so as to have their courses altered continually, by the attraction towards many different centres; but considering the coincidence of the calculation of Halley and Clairaut with the subsequent appearance of the comet of 1759, this opinion can scarcely be admitted to be in any degree probable with respect to the comets in general, however possible the supposition may be in some particular cases. (Plate XXXII. Fig. 472 . . 475. Plate XXXIII. Fig. 476 . . 485.)

LECTURE XLIII.

ON THE LAWS OF GRAVITATION.

It was first systematically demonstrated by Sir Isaac Newton, that all the motions of the heavenly bodies, which have been described, may be deduced from the effects of the same force of gravitation which causes a heavy body to fall to the earth; he has shown that in consequence of this universal property of matter, all bodies attract each other with forces decreasing as the squares of the distances increase; and of later years the same theory has been still more accurately applied to the most complicated phenomena. We are at present to take a general view of the operation of this law, in the same order in which the affections of the celestial bodies have been enumerated. It will not be possible to investigate mathematically the effects of gravity in each particular motion, but we may in some measure illustrate the subject, by considering in what manner astronomers have proceeded in their explanations and calculations, and we may enter sufficiently into the principles of the theory, to understand the possibility of its applications.

The bodies which exist in nature are never single gravitating points; and in order to determine the effects of their attraction, we must suppose the actions of an infinite number of such points to be combined. It was shown by Newton, that all the matter of a spherical body, or of a spherical surface, may be considered, in estimating its attractive force on other matter, as collected in the centre of the sphere. The steps of the demonstration are these: a particle of matter, placed at the summit of a given cone or pyramid, is attracted by a thin surface, composed also of attractive matter, occupying the base of the cone, with equal force, whatever may be the length of the cone, provided that its angular position remain unaltered: hence it is easily inferred that if a gravitating point be placed any where within a hollow sphere, it will remain in equilibrium, in consequence of the opposite and equal actions of the infinite number of minute surfaces, terminating the opposite pyramids into which the sphere may be divided: it is also demonstrable, by the assistance of a fluxional calculation, that a point, placed without the surface, is attracted by it, precisely in the same manner, as if the whole matter which it contains were collected in the centre; consequently the same is true of a solid sphere, which may be supposed to consist of an infinite number of such hollow spheres. If, however, the point were placed within a solid sphere, it would be urged towards the centre, by a force which is simply proportional to its distance from that centre. This proposition tends very much to facilitate all calculations of the attractions of the celestial bodies, since all of them are so nearly spherical, that their action on any distant bodies is the same, as if the whole of the matter of which they consist were condensed into their respective centres; but if the force of gravity varied according to any other law than that which is found to prevail, this simplification would no longer be admissible, even with respect to a sphere.

It can scarcely be doubted that the power of gravitation extends from one fixed star to another, although its effects may in this case be much too inconsiderable to be perceived by us. It may possibly influence the progressive motions of some of the stars; and if, as Dr. Herschel supposes, there are double and triple stars revolving round a common centre, they must be retained in their orbits by the force of gravity. Dr. Herschel also imagines that the motion of our sun is in some measure derived from the same cause, being directed nearly towards a point in which two strata of the milky way meet; the attraction of the stars, other things being equal, must, however, be proportional to their brightness, and that part of the heavens, to which the sun is probably moving, appears to afford less light than almost any other part, nor does the hemisphere, of which it is the centre, abound so much in bright stars as the opposite hemisphere. If Sirius is a million times as far from the sun as the earth, and if he should descend towards the sun by means of their mutual gravitation only, he would move, on a rough estimate, but about 40 feet in the first year, and in 1000 years only 8000 miles. It has been conjectured that the mutual gravitation of the stars of a nebula is sometimes the cause of the peculiar form of the aggregate, which somewhat resembles that of a drop of a liquid, held together by its cohesion: but unless the form of the nebula was originally spherical, it could scarcely have acquired that form from the operation of gravity, since the spherical form of a drop is owing as much to the elasticity as to the attractive force of the particles of water, and it would be necessary, in order to preserve the analogy, that the stars should also be floating in an incompressible fluid.

The sun's change of place, dependent on the relative situation of the planets, is so inconsiderable, that it escaped observation until its existence had been deduced from theory. Not but that this change would be sufficiently conspicuous if we had any means of detecting it, since it may amount in the whole to a distance equal to twice the sun's diameter, or seven times the distance of the moon from the earth; and this change is readily deducible from the general and unquestionable law of mechanics, that the place of the centre of inertia of a system cannot be changed by any reciprocal or mutual action of the bodies composing it, the action of gravity being found to be perfectly reciprocal. But the earth accompanies the sun in great measure in this aberration, and the other planets are also more or less affected by similar motions; so that the relative situations are much less disturbed than if the sun described this irregular orbit by the operation of a cause foreign to the rest of the system.

The simple revolution of a body, in a given plane, indicates, at first sight, the existence of an attractive force directed to some point within the orbit; and the Keplerian law of the equality of the areas described in equal times, by a line drawn from each planet to the sun, agrees precisely with what is demonstrable of the effects of central forces and points at once to the sun as the centre of attraction of the system. And since the orbits of the planets are elliptical, and the sun is placed in one of the foci of each, it may be mathematically proved that the force directed to the sun must increase in proportion as the square of the distance decreases.

The times of the revolutions of the planets are also in perfect conformity with the laws of gravitation, that is, the squares of the times are proportional to the cubes of the distances from the sun. It was easy to infer, from what Huygens had already demonstrated of centrifugal forces, that this Keplerian law must be true of bodies revolving in circles by the force of gravitation;

but Newton first demonstrated the same proportion with respect to elliptic orbits, and showed that the time of revolution in an ellipsis is equal to the time of revolution in a circle, of which the diameter is equal to the greater axis of the ellipsis, or the semidiameter to the mean distance of the planet.

The universality of the laws of gravitation, as applied to the different planets, shows also that the matter, of which they are composed, is equally subjected to its power; for if any of the planets contained a portion of an inert substance, requiring a force to put it in motion, and yet not liable to the force of gravitation, the motion of the planet would be materially different from that of any other planet similarly situated.

The deviations of each planet from the plane of its orbit, and the motions of its nodes, or the points in which the orbit intersects the plane of the ecliptic, as well as the motions of the aphelion, or the point where the orbit is remotest from the sun, have also been deduced from the attractions of the other planetary bodies; but the calculations of the exact quantities of these perturbations are extremely intricate. In general, each of the disturbing forces causes the nodes to have a slight degree of retrograde motion; but on account of the peculiar situation of the orbits of Jupiter and Saturn, it happens that the retrograde motion of Jupiter's node, on the plane of the orbit of Saturn, produces a direct motion on the ecliptic, so that the action of Saturn tends to lessen the effect of the other planets in causing a retrograde motion of Jupiter's nodes on the ecliptic.

The secular diminution of the obliquity of the ecliptic, or that slow variation of its position, which is only discovered by a comparison of very distant observations, is occasioned by the change of position of the earth's orbit, in consequence of the attractions of the other planets, especially of Jupiter. It has been calculated that this change may amount, in the course of many ages, to 10° or 11°, with respect to the fixed stars; but the obliquity of the ecliptic to the equator can never vary more than two or three degrees, since the equator will follow, in some measure, the motion of the ecliptic.

The mutual attraction of the particles of matter, composing the bulk of each planet, would naturally dispose them, if they were either wholly or partially fluid, to assume a spherical form: but their rotatory motion would require, for the preservation of this form, an excess of attraction in the equatorial parts, in order to balance the greater centrifugal force arising from the greater velocity of their motion: but since the attractive force of the sphere on the particles at an equal distance from its centre is every where equal, the equatorial parts would necessarily recede from the axis, until the greater number of particles, acting in the same column, compensated for the greater effect of the centrifugal force. The form would thus be changed from a sphere to an oblate or flattened spheroid; and the surface of a fluid, either wholly or partially covering a solid body, must assume the same figure, in order that it may remain at rest. The surface of the sea is therefore spheroidical, and that of the earth deviates so far only from a spheroidical figure, as it is above or below the general level of the sea. (Plate XXXIV. Fig. 486.)

The actions of the sun and moon, on the prominent parts about the earth's equator, produce a slight change of the situation of its axis, in the same manner as the attractions of the other planets occasion a deviation from the plane of its orbit. Hence arises the precession of the equinoxes, or the retrograde motion of the equinoctial points, amounting annually to about 50 seconds. The nutation of the earth's orbit is a small periodical change of the same kind, depending on the position of the moon's nodes; in consequence of which, according to Dr. Bradley's original observations, the pole of the equator describes in the heavens a little ellipsis, of which the diameters are 16 and 20 seconds. The same cause is also concerned in modifying the secular variation of the obliquity of the ecliptic; and on the other hand, this variation has a considerable effect on the apparent precession of the equinoxes. On account of the different quantity of the precession at different times, the actual length of the tropical year is subjected to a slight variation; it is now 4 or 5 seconds shorter than it was in the time of Hipparchus. The utmost change, that can happen from this cause, amounts to 43 seconds.

The exact computation of the moon's motion is one of the most difficult, as well as the most important problems in astronomy; but it is easy to understand, in general, how the difference in the quantity and direction of the sun's actions on the moon and earth, may cause such a derangement of the moon's gravitation towards the earth, that the inclination of the orbit must be variable, that the nodes must have a retrograde, and the apsides a direct motion; and that the velocity of the moon must often be different from that which she would have, according to the Keplerian law, in a simple elliptic orbit.

For, the sun's attraction as far as it acts equally on the earth and the moon, can have no effect in disturbing their relative position, being always employed in modifying their common annual revolution: but the difference of the forces, occasioned by the difference of distances, always tends to diminish the effect of their mutual attraction: since the sun acts more powerfully on the nearer than on the remoter of the two bodies. The difference of the directions, in which the sun acts on the earth and the moon, produces also a force, which tends, in some degree, to bring them nearer together; but this force is, on the whole, much smaller than the former; and the result of both these disturbing forces is always directed to some point in the line which joins the earth and the sun, on the same side of the earth with the moon. It is obvious that when the nodes are also in this line, the disturbing force can have no effect, either on their position, or on the inclination of the orbit, since it acts wholly in the plane of that orbit; but when they are in any other situation, the disturbing force must cause a deviation from the plane, towards the side on which the sun is situated, so that the inclination of the orbit increases and decreases continually and equally; but whatever may be the position of the nodes, it will appear that they must recede during the greater part of the moon's revolution, and advance during the smaller. (Plate XXXIV. Fig. 487.)

When the disturbing force tends to separate the earth and moon, it deducts from the gravitation of the moon towards the earth a portion which increases with the distance, and therefore causes the remaining force to decrease more rapidly than the square of the distance increases; and the reverse happens when the disturbing force tends to bring the earth and moon nearer together; but the former effect is considerably greater than the latter. Now in the simple ellipsis, when the body descends from the mean distance, the velocity continually prevails over the attractive force, so as to turn away the direction of the orbit more and more from the revolving radius, until, at a certain point, which is called the lower apsis, it becomes perpendicular to it: but if the central force increase in a greater proportion than is necessary for the description of the ellipsis, the point where the velocity prevails over it will be more remote than in the ellipsis; and this is expressed by saying that the apsis moves forwards. When, on the contrary, the force varies more slowly, the apsis has a retrograde motion. Since, therefore, the force attracting the moon towards the earth, increases, on the whole, a little more rapidly than the square of the distance decreases, the apsides must have, on the whole, a direct motion. And a similar theory is applicable to the mutual perturbations of the primary planets. (Plate XXXIV. Fig. 488.)

The secular acceleration of the moon's mean motion, which had long presented a difficulty amounting almost to an exception, against the sufficiency of the theory of gravitation, has at last been satisfactorily deduced by Mr. Laplace from the effect of the gradual change of the eccentricity of the earth's orbit, which is subject to a very slow periodical variation, and which causes a difference in the magnitude of the sun's action on the lunar revolution.

The perfect coincidence of the period of the moon's rotation, with that of a mean revolution, has been supposed to be in some degree an effect of the attraction exerted by the earth on a prominent part of her surface; there are however, many reasons to doubt of the sufficiency of the explanation. If the periods had originally been very nearly equal, we might imagine that the motion of the earth would have produced a libration or oscillation in the position of the moon, retaining it always within certain limits with respect to the earth: no libration is, however, observed, that can be derived from any inequality in the moon's rotation; and it has very properly been suggested that the same attraction towards the earth ought to have made the moon's axis precisely perpendicular to the plane of her orbit, instead of being a little inclined to it. At the same time the appearance of a similar coincidence, in the periods of the rotation and revolution of many other satellites, makes it probable that some general cause must have existed, which has produced the same effect in so many different cases.

The orbits of the comets afford no very remarkable singularity in the application of the laws of gravity, excepting the modifications which depend on their near approach to the parabolic form, and the great disturbance which their motions occasionally suffer, when they happen to pass through the neighbourhood of any of the larger planets. The velocity of a comet in its perihelion is such, that its square is twice as great as the square of the velocity of a body revolving in a circle at the same distance. It was determined by Halley and Clairaut, that the attractions of Jupiter and Saturn would delay the return of the comet of 1759 about 618 days; and the prediction was accomplished within the probable limits that they had assigned for the error of the calculation. The labours of Clairaut have indeed in many respects improved the science of mathematical astronomy; he was the first that obtained a complete determination of the effects of the mutual actions of three gravitating bodies, disturbing each other's motions; and his investigations, which were founded on those of Newton, led the way to still further improvements and refinements, which have been since made in succession by Euler, Lagrange, and Laplace.

LECTURE XLIV.

ON THE APPEARANCES OF THE CELESTIAL BODIES.

WE are next to proceed to examine the sensible effects produced by those motions which we have first considered in their simplest state, and afterwards with regard to their causes and their laws. Many authors have chosen rather to pursue a contrary method, and have attempted to imitate the original and gradual developement of the primitive motions from their apparent effects. But no conception is sufficiently clear, and no memory sufficiently strong, to comprehend and retain all these diversified appearances with accuracy and facility, unless assisted by some previous idea of the real changes which produce them, or by some temporary hypothesis respecting them, which may have been of use in its day for the better connexion of the phenomena, although it does not at present deserve to be employed for a similar purpose, in preference to simpler and better theories, which happen to be historically of a later date.

The proper motions of the fixed stars, as they are subjected to our observation, undergo two modifications; the one from the relative direction of the motion, by which it may be more or less concealed from our view; the other from the proper motion of the sun, and the planets attending him. This motion has indeed only been inferred from the apparent motions of a great number of stars, which are either partly or totally referable to it, and which could scarcely have agreed so correctly as they do, if they had arisen from the real and separate motion of each star.

Among the motions of the primary planets, that of the earth itself requires a principal share of our attention. The apparent places of the fixed stars are not sensibly affected by the earth's annual revolution: if any of them had been considerably less remote than they are, it is probable that this motion would have occasioned a sensible annual parallax, or a change of their relative situation, according to the earth's place in its orbit round the sun; for if this orbit, viewed from any of the stars, subtended an angle even of a single second, the place of that star might be observed to vary a second at different times of the year. Dr. Hooke supposed at one time that he had discovered such a parallax, but later observations have not confirmed those of Dr. Hooke. The stars have, however, a small aberration, in consequence of the progressive motion of the earth in its orbit, combined with the limited velocity of light; and the standard of comparison being the earth's axis, its nutation must also in some degree affect the apparent places of the stars. It was in endeavouring to ascertain the annual parallax, that Dr. Bradley discovered both the aberration of light and the nutation of the earth's axis.

The revolution of the earth, in its orbit round the sun, produces the apparent motion of the sun among the stars, by which he describes his annual path in the ecliptic, with an apparent angular velocity equal to the angular velocity of the earth, which varies considerably at various times. It required some investigation of the magnitudes and distances of the heavenly bodies, to be convinced that the sun and stars had not in reality the motion which a superficial inspection of the heavens would naturally lead a spectator to attribute to them; but it is at present perfectly unnecessary to enter into arguments to prove that the true cause of these apparent motions is the real motion of the earth. The effect of the earth's annual revolution is the change of place of the sun among the fixed stars: it is obvious that the sun will always appear, when viewed from the earth, in a place diametrically opposite to that in which the earth would appear, if seen from the sun: consequently, since the earth and sun remain in the same plane, the apparent path of the sun will mark the same circle among the stars as the earth would appear to describe, if viewed from the sun, that is, the ecliptic. If the light of the stars were much stronger, or that of the sun much weaker, we might see him pass by the stars in each part of the ecliptic, as we do the moon; but we are now obliged to observe what stars are in turn diametrically opposite to the sun, or at certain distances from him, and thus we obtain a correct knowledge of his path.

The sun's apparent diameter is larger by one thirtieth in January than in June; of course the earth is so much nearer to the sun in winter than in summer; and since the revolving radius of the earth's orbit describes equal areas in equal times, the angular motion must increase as the square of the distance diminishes, or about twice as fast as the distance itself diminishes; so that the whole variation of the apparent diurnal motion of the sun is one fifteenth of his mean motion: hence, the sun passes through the winter half of the ecliptic in a time 7 or 8 days shorter than the summer half. According to the different situations of the earth, with respect to the plane of the sun's equator, his rotation on his axis causes the paths of his spots to assume different forms; when the earth is in that plane, the paths appear straight, but in all other situations, elliptical.

The rotation of the earth on its axis produces the still more obvious vicissitudes of day and night; and, in combination with its annual motion, occasions the change of seasons. Since the axis remains always parallel to itself, and is inclined to the plane of the ecliptic in an angle of about $66\frac{1}{2}°$, the plane of the equator, which is perpendicular to the axis, must pass twice in the year through the sun. When this happens, the limit of illumination, or the circle which separates the dark portion of the earth from the enlightened part, will then pass through the poles; and as the earth turns on its axis, each point of its surface must remain for an equal length of time in light and in darkness. Hence the points of the ecliptic, in which the sun is situated at such times, are called the equinoctial points. At all other times, one pole of the earth is in the light, and the other in the shadow; and all the points of the earth nearest to the illuminated pole have their day longer than their night, while the parts on the opposite side of the equator, which are consequently nearer to the unenlightened pole, have their day shorter. The parts nearest to the poles have also one of their days and one of their nights protracted to a period of several common days, or even months, whenever they revolve entirely within the limit of illumination. (Plate XXXIV. Fig. 489.)

The sun appears to describe every day a circle in the heavens, more or less distant from the plane of the equator, according to the actual situation of the earth's axis; this distance being always the same as that of the poles from the limit of illumination, and never exceeding $23\frac{1}{2}°$; so that by determining the sun's path at the time of the equinoxes, or the apparent place of the equinoctial in the heavens, for any given point on the earth's surface, we may represent the sun's path at any other time by a smaller circle parallel to it. Speaking however, more correctly, the sun's apparent path is a spiral, formed by the continuation of these supposed circles into each other.

The effect of the centrifugal force, derived from the earth's rotation, is perceptible, at the equator, in the retardation of the vibrations of pendulums. The whole centrifugal force at the equator is found, by computation, to be $\frac{1}{289}$ of the force of gravity; but the diminution of the force of gravitation appears, by experiments on pendulums, to be $\frac{1}{179}$; this diminution being the sum of the centrifugal force, and of the decrease of gravity on account of the oblate figure of the earth, the equatorial parts being further removed from its centre, and the force of gravity being less powerful there. The changes of inclination in the earth's axis are observable in the places of the equinoctial points, and in the situation of the plane of the earth's equator with respect to the fixed stars; and the secular diminution of the obliquity of the ecliptic is discoverable by a comparison of distant observations on the sun's apparent motion, and on the places of the fixed stars with respect to the ecliptic.

For the phenomena of twilight, we are principally indebted to the light reflected by the atmosphere: when the sun is at a certain distance only below the horizon, he shines on some part of the air immediately visible to us, which affords us a portion of reflected light. The distance, at which this may happen, has been variously estimated, and it is perhaps actually different in different climates, being a little greater in countries near the poles than in those which are nearer the equator: there is also sometimes a secondary twilight, when the parts of the atmosphere, which reflect a faint light on the earth, are themselves indebted for this light to an earlier reflection. Some have assigned 18° as the limit of twilight, and on this supposition, allowing for refraction, the atmosphere must be capable of reflecting sensible light at the height of about 40 miles. Mr. Lambert, on the contrary, makes the limit only about $6\frac{1}{2}$°. The duration of twilight is greater or less as the sun moves more or less obliquely with respect to the horizon; it is, therefore, shortest near the time of the equinoxes, since the equinoctial intersects the horizon less obliquely than any lesser circle parallel to it. (Plate XXXIV. Fig. 490, 491.)

The revolutions of the primary planets, combined with that of the earth, necessarily produce the various relations, in which they are either in opposition or conjunction, with respect to each other or to the sun, and in which the apparent motion is direct or retrograde, or the planet is stationary, according to the directions and the comparative velocities of the real motions. If the earth were at rest, the inferior planets would appear to be stationary when they are at the greatest elongation or angular distance from the sun; but, on account of the effect of the earth's motion, Venus is stationary at an elongation of about 29°, while her greatest elongation is between 45° and 48°. The greatest elongation of Mercury, in each revolution, is from $28\frac{1}{3}$° to $17\frac{1}{2}$°, according to the position of his orbit, which is very eccentric. All these appearances are precisely the same as if the sun actually revolved round the earth, and the planets accompanied him in his orbit, performing at the same time their several revolutions round him; and the path which would thus be described in the heavens, and which is of a cycloidal nature, represents correctly the true positions of the planets with respect to the earth. The apparent angular deviation from the ecliptic, or the latitude of the planet, is also greater or less, accordingly as the earth is nearer or remoter to the planet, as well as according to the inclination of its orbit and its distance from the node. (Plate XXXIV. Fig. 492 . . 494.)

The various appearances of the illuminated discs, especially of the inferior planets, and the transits of these planets over the sun, depend on their positions in their orbits, and on the places of the nodes, with respect to the earth. Jupiter, Saturn, and the Georgian planet, are so remote in comparison of the earth's distance from the sun, that they appear always fully illuminated. Venus is brightest at an elongation of about 40° from the sun, in that part of her orbit which is nearest to the earth; she then appears like the moon when 5 days old, one fourth of her disc being illuminated; she casts a shadow, and may even be seen in the day time in our climates, if she happens to be far enough north: a circumstance which occurs once in about 8 years. In order that there may be a transit of Venus over the sun, she must be within the distance of $1\frac{1}{4}$° of her node at the time of conjunction, otherwise she will pass either to the north or to the south of the sun, instead of being immediately interposed between him and the earth.

The phases and eclipses of the moon are very obviously owing to the same causes; that part of the moon only, on which the sun shines, being strongly illuminated, although the remaining part is faintly visible, by means of the light reflected on it from the earth; it is, therefore, most easily seen near the time of the new moon, when the greatest part of the earth's surface turned towards the moon is illuminated. The parts of the moon which are immeaditely opposed to the earth, appear to undergo a libration, or change of situation, of two kinds, each amounting to about 7 degrees: the one arising from the inequality of the moon's velocity in her orbit at different times, the other from the inclination of the axis of her rotation to her orbit; besides these changes, the diurnal rotation of the earth may produce, to a spectator situated on some parts of it, a third kind of libration, or a change of almost two degrees in the appearance of the moon at her rising and setting. (Plate XXXIV. Fig. 495.)

When the moon passes the conjunction, or becomes new, near to the node, she eclipses the sun, and when she is full, or in opposition, in similar circumstances, she herself enters the earth's shadow. The earth's shadow consists of two parts, the true shadow, within which none of the sun's surface is visible, and the penumbra, which is deprived of a part only of the sun's light; the true shadow forms a cone terminating in a point at a little more than $3\frac{1}{4}$ times the mean distance of the moon; the penumbra, on the contrary, constitutes, together with the shadow, a portion of a cone diverging from the earth without limit; but the only effect of this imperfect shadow is, that it causes the beginning of a lunar eclipse to be incapable of very precise determination; for the limit of the darkened part of the moon, as it appears in the progress of the eclipse, is that of the true shadow, very little enlarged by the penumbra. The true shadow, where the moon crosses it, is about 80 minutes in diameter, as seen from the earth, while the moon herself is only 30. This shadow is not, however, wholly deprived of the sun's light; for the atmospheric refraction inflects the light passing nearest to the earth, in an angle of 66 minutes, and causes a great part of the shadow to be filled with light of a ruddy hue, by means of which the moon remains still visible to us, the cone of total darkness extending to somewhat less than two thirds of the moon's distance. But it has sometimes happened, probably from the effect of clouds occupying the greatest part of our atmosphere, that the moon has totally disappeared. (Plate XXXIV. Fig. 496.)

When the sun is eclipsed, it depends on the situations of the earth and moon in their orbits, whether the sun or moon subtends the greatest angle as seen from the earth; since at their mean distances their apparent diameters are each about half a degree. If the sun's apparent diameter is the greater, the eclipse, when the centres coincide, must be annular, the margin of the sun's disc being still visible in the form of a ring: when the moon's apparent diameter is greater than the sun's, the eclipse, if central, becomes total; but still a ring of pale light is seen round the disc, which has been attributed to the effect of the sun's atmosphere, since that of the moon is probably too inconsiderable to produce the appearance: a red streak is also sometimes observed at the margin, before the actual emersion of the sun. The degree of darkness depends on the situation of the place of observation within the shadow, on account of the greater or less illumination of the atmosphere within view: sometimes a considerable number of stars may be seen during a total eclipse of the sun.

It is obvious that, since the earth is much larger than the moon, the whole shadow of the moon will only pass over a part of the earth's surface: and that no solar eclipse can be visible in the whole of the hemisphere turned to the sun: while lunar eclipses, on the contrary, present the same appearance wherever the moon is visible. In the same manner, to a spectator on the moon, an eclipse of the earth, or a transit of the moon's shadow over the earth's disc, would have nearly the same appearance wherever he might be stationed; but an eclipse of the sun by the earth would be total to that part of the moon's surface only, which to us appears dark at the same time. (Plate XXXIV. Fig. 497 . . 499.)

The moon's nodes arrive very nearly at the same situation with respect to the earth after 223 lunations, or revolutions of the moon, which are performed in 18 years of 365 days each, 15 days, 7 hours, and 43¾ minutes; so that after a period of about 18 years, the series of eclipses recommences nearly in the same order. This circumstance was observed by the ancients, and is mentioned by Ptolemy and by Pliny. When the full moon happens within $7\frac{1}{3}°$ of the node, there must be a lunar eclipse and there may be an eclipse at the distance of 13° from the node. An eclipse, of the sun may happen when the moon changes, or comes into conj unction with the sun, at any distance within $17\frac{1}{3}°$ of the node. The mean number of eclipses which occur in a year is about 4; and there ase sometimes as many as 7: there must necessarily be two solar eclipses, but it is possible that there may not be even one lunar. In speaking of the magnitude of the part of the sun or moon eclipsed, it is usual to consider the whole diameter as divided into 12 parts, called digits, each of which contains 30 minutes: thus if one fifth part of the diameter were dark, the extent of the eclipse would be called 2 digits 12 minutes.

The moon travels through the heavens with a motion contrary to their apparent diurnal revolution. Hence she rises and sets, on an average, about three quarters of an hour later every day. The least possible difference between the time sof the moon's rising on two successive days, is, in London, 17 minutes; and this circumstance occurs once in about 19 years, which is nearly the period of the moon's nodes with respect to the heavens: the greatest possible difference is 1 hour 17 minutes. But it happens every month that the difference becomes greater and less by turns, and when the least difference is at the time of the full moon, it is usually called the harvest moon. In parts nearer to the poles, the moon often rises at the same hour on two succeeding days.

The eclipses of the satellites of Jupiter exhibit appearances extremely interesting for their utility in identifying the same instant of time in different places. On account of the small inclination of their orbits to the plane of Jupiter's orbit, the first three never pass the shadow without being plunged into it, and the fourth but seldom; while those of Saturn are much less frequently liable to be eclipsed, on account of their greater deviation from the plane of his ecliptic. These satellites are also frequently hidden behind the body of the planet, and this circumstance constitutes an occultation: hence it happens that we can never see both the immersion of the first satellite into the shadow of Jupiter, and its emersion from it; but both the immersion and emersion of the three outer satellites are sometimes observable. The ring of

Saturn exhibits a variety of forms according to its angular position: it disappears to common observation when either its edge or its dark side is presented to us: but to Dr. Herschel's telescopes it never becomes invisible; the light reflected from the planet being probably sufficient for illuminating in some measure the side not exposed to the sun's direct rays.

The comets are seen for a short time, and are again lost to our view; their tails are in general situated in the planes of their orbits, following them in their descent towards the sun, and preceding them in their ascent, with a slight degree of curvature in their forms; they must also appear to us more or less arched, and of different extent, according to their distances, and to the angular position of the orbits with respect to the ecliptic.

The proportion of the light afforded by the different heavenly bodies has been variously estimated by various authors; but there is little reason to call in question the accuracy of the experiments and calculations of Mr. Bouguer. He states the intensity of the moon's light as only one three hundred thousandth of that of the sun. These calculations have been extended by Euler and by Lambert; Euler considers the direct light of the sun as equal to that of 6560 candles of a moderate size, supposed to be placed at the distance of 1 foot from the object: that of the moon to the effect of 1 candle, at the distance of $7\frac{1}{2}$ feet; the light of Venus to a candle at 421 feet, and of Jupiter to a candle at 1620 feet; so that the sun would appear as bright only as Jupiter if he were removed to a distance 131 thousand times as great as his present distance. (Plate XXXIV. Fig. 500.)

When we reflect on the magnificence of the great picture of the universe, the outlines of which we have been considering, we are lost in the contemplation of the immensity of the prospect, and returning to the comparatively diminutive proportions of our individual persons, and of all the objects with which we are most immediately connected, we cannot help feeling our own insignificance in the material world. The mind, notwithstanding, endeavours to raise itself above the restraints which nature has imposed on the body, and to penetrate the abyss of space in search of congenial existences. But in speculations of this kind, reason and argument must give way to conjecture and imagination; and thus, from natural philosophy, our imaginations wander into the regions of poetry; and it must be confessed that the union of poetical embellishment with natural philosophy is seldom very happy. A poet has few facts to communicate, and these he wishes to expand and diversify; he dwells on a favourite idea, and repeats it in a thousand emblematical forms; his object is, to say a little, very elegantly, in very circuitous, and somewhat obscure terms. But the information, which the natural philosopher has to impart, is too copious to allow of prolixity in its detail; his subjects are too intricate to be compatible with digressions after amusement, which, besides interrupting, are too likely to enervate the mind; and if he is ever fortunate enough to entertain, it must be by gratifying the love of truth, and satisfying the thirst after knowledge. We have, however, a favourable specimen of highly ornamented philosophy in Fontenelle's Plurality of Worlds; a work which must be allowed to convey much information in a very interesting form, although somewhat tinctured with a certain frivolity which is not always agreeable. We need not attempt to accompany all the flights of Fontenelle's imagination; it will be sufficient for our purpose to pursue his ideas in a simple enumeration of the most remarkable phenomena, that would occur to a spectator placed on each of the planets.

Of Mercury we know little except the length of his year, which is shorter than three of our months. Supposing all our heat to come from the sun, it is probable that the mean heat on Mercury is above that of boiling quicksilver; and it is scarcely possible that there should be any point about his poles where water would not boil. The sun's diameter would appear, if viewed from Mercury, more than twice as great as to us on the earth.

Venus must have a climate far more temperate than Mercury, yet much too torrid for the existence of animals or vegetables, except in some circumpolar parts; her magnitude and diurnal rotation differ but little from those of the earth, and her year is only one third shorter; so that her seasons, and her day and night, must greatly resemble ours. The earth, when in opposition to the sun, must be about four times as bright as Venus ever appears to us, and must, therefore, always cast a shadow; it must be frequently, and perhaps generally, visible in the day; and together with the moon, must exhibit a very interesting object. The atmosphere of Venus is supposed to be nearly like our own, or somewhat more rare.

The inhabitants of the moon, if the moon is inhabited, must be capable of living with very little air, and less water: there is reason to think their atmosphere less than a mile high, and it is never clouded: so that the sun must shine without intermission for a whole fortnight on the same spot, without having his heat moderated by the interposition of air, or by the evaporation of moisture. The want of water in the moon is not, as some have supposed, the necessary consequence of the want of an atmosphere; but it is inferred partly from the total absence of clouds, and partly from the irregular appearance of the margin of the moon, as seen in a solar eclipse; no part of it being terminated by a line sufficiently regular to allow us to suppose it the surface of a fluid. The earth must always appear to occupy nearly the same part of the sky, or rather to describe a small oval orbit round a particular point, exposing a surface 13 times as great as that of the moon appears to us. This large surface, suspended, with phases continually changing, like those of the moon, must afford, especially when viewed with a telescope, an excellent timepiece; the continents and seas coming gradually and regularly into view, and affording a variety equally pleasing and useful. To us such a timepiece would be of inestimable value, as it would afford us an easy method of discovering the longitude of a place, by comparing its motion with the solar time: but in the moon, the relative position of the earth and sun, or of the earth and stars only, would be sufficient for determining the situation of any place in sight of the earth; if, however, there are no seas and no navigation, astronomical observations of this kind would be of very little utility. The assistance of the earth's phases in the measurement of time might, however, still be very useful, for many purposes, to the inhabitants of the nearer half of the moon; and probably the remoter part is much deserted, for in their long night of half a month, they must be extremely in want of the light reflected from the earth, unless the inhabitants have the faculty of sleeping through the whole of their dark fortnight. The surface of the moon appears to be very rocky and barren, and liable to frequent disturbances from volcanos. These have been supposed to project some of their contents within the reach of the earth's attraction, which they might easily do, if they could throw them out with a velocity of about eight thousand feet in a second, which is only four times as great as that of a cannon ball: and these stones, falling through the atmosphere, might very possibly generate so much heat, by compressing the air, as to cause the appearance of fiery meteors, and to fall in a state of ignition. The appearance of the moon, as viewed through a good telescope, is extremely well imitated by Mr. Russel's lunar globe, which is also capable of exhibiting, with great accuracy, the changes produced by its librations.

The climate of Mars is as much colder than ours, as that of Venus is warmer; in other respects there is no very striking difference: the inclination of his axis to his ecliptic being nearly the same as that of the earth's axis, the changes of seasons must be nearly like our own. Dr. Herschel has observed a constant appearance of two bright spots or circles near the poles of Mars, which he attributes to the ice and snow perpetually surrounding them. It is not, however, probable that water could remain fluid in any part of Mars, and even quicksilver and alcohol would, perhaps, be frozen in his temperate climates. It is pretty certain that Mars has an atmosphere, and his dark spots seem to be occasioned by clouds: this atmosphere may, perhaps, also be tne cause of the ruddy hue of his light.

It appears to be doubtful, whether either of the three little planets newly discovered can be sufficiently solid, to give a firm footing to any material beings: we should probably weigh only a few pounds each if transported there. According to Dr. Herschel's opinion, neither Ceres nor Pallas is much larger than a good Scotch estate, although they must, sometimes, appear to each other as planets of a most respectable size. The light reflected from Ceres is of a more ruddy hue than that of Pallas; both of these planets are attended by more or less of a nebulosity, proceeding, perhaps, from copious atmospheres; and in this respect, as well as in the great inclination of their orbits, they appear to have some affinity to comets. It is tolerably certain that neither of them is 200 miles in diameter; and Juno is also probably about the same size.

It is obvious that the most striking features of the heavens, when contemplated from Jupiter, would be the diversified positions and combinations of his satellites: their light must be faint, but yet of service; and to a traveller on the surface of this vast globe they must afford useful information, as well with respect to time as to place. Our little earth must probably be always invisible to a spectator situated on Jupiter, on account of its apparent proximity to the sun, in the same manner as a planet at half the distance of Mercury would be invisible to us. The year of Jupiter must contain nearly ten thousand of his days, and that of Saturn almost thirty thousand Saturnian days. Besides the vicissitudes of the seven satellites revolving round Saturn, his ring must afford, in different parts of his surface, very diversified appearances of magnificent luminous arches, stretched across the heavens, especially in that hemisphere which is on the same side of the ring with the sun.

From the Georgian planet the sun must be seen but as a little star, not one hundred and fiftieth part as bright as he appears to us. The axis of this planet being probably near to the plane of its ecliptic, it must be directed twice in the year towards the sun, and the limit of illumination must approach to the equator, so that almost every place on his surface must sometimes remain, for a great number of diurnal revolutions, in light and in darkness; the most moderate climates having one night, in their long year, equal in duration at least to several of our years: and it must be confessed that this planet would afford but a comfortless habitation to those accustomed to our summer sunshine, even if it were possible to colonise it. (Plate XXXIV. Fig. 501.)

On the whole, we are tempted, from an almost irresistible analogy, to conclude that the planets are all in some manner or other inhabited; but at the same time we can scarcely suppose that a single ecipses of terrestrial animals or even vegetables could exist in any of them; their minerals may, perhaps, resemble ours, and if the stones which Mr. Howard has analysed are really lunar productions, we have proofs that the moon at least contains some substances resembling those which compose the earth; but the seas and rivers of the other planets must consist of some fluid unknown to us, since almost all our liquids would either be frozen, or converted into vapour, in any of them.

LECTURE XLV.

ON PRACTICAL ASTRONOMY.

IT is generally most convenient in practical astronomy to neglect the real, and to consider only the apparent motions of the sun the stars, and planets, for the visible effects must be precisely the same, whether the sun or the earth perform a revolution in the plane of the ecliptic, and whether the earth actually move on its axis, or the whole of the celestial bodies move round it in a day. We may, therefore, suppose the sun to move, as he appears to do, from west to east in the ecliptic, so as to advance almost a degree in 24 hours, and from east to west, together with all the stars and planets, so as to perform a whole revolution in a day. Speaking more correctly, the sun appears to describe, in every sidereal day, a spiral, which differs a little from a circle, and is also about a degree shorter, so that about four minutes more are required for the return of the sun to the same part of the heavens, and the completion of a solar day.

In order to determine the place of any point in the heavens, it is usual to compare its situation either with the plane of the earth's equator, or with the ecliptic; its angular distance from the equator being called its declination, and from the ecliptic, its latitude; these distances must be measured in planes perpendicular to those of the equator or ecliptic, and the distances of these planes from their intersection, or from the equinoctial point in Aries, are called respectively the right ascension and the longitude of the point to be described. For the stars, the declination and right ascension are most usually laid down; but with respect to the sun and the planets, performing their revolutions in or near the ecliptic, it is most convenient to calculate their latitude and longitude.

The plane passing through the earth's axis and the place of a spectator is the plane of the meridian of that place; and a plane touching the earth in any point is its horizon. With respect to the appearances of the fixed stars, this plane may be considered as passing through the earth's centre in the same direction: and the difference is scarcely sensible with respect to the sun and the primary planets; but in observations of the moon's place, these planes must be carefully distinguished. (Plate XXXV. Fig. 502.)

The instruments requisite for astronomical observations are principally referable to geometrical or to optical apparatus, or to the measurement of time. Particular constructions and combinations are, however, necessary for the accommodation of quadrants, graduated circles, telescopes, and transit instruments, to the uses of observatories; and astronomical observations are as necessary to the correct determination of time, as artificial timekeepers are useful for other astronomical purposes.

The most accurate standard of time is the diurnal rotation of the earth on its axis, as ascertained by its situation with respect to the fixed stars. The time elapsing between two successive passages of any star over the same meridian, which constitutes a sidereal day, may be measured with great precision; and the star may for this purpose be observed, with almost equal accuracy, in any other situation, and sometimes with greater convenience. The length of the sidereal day may be considered as perfectly constant, the inequalities arising from the aberration of light, and from the nutation of the earth's axis, being too small to be easily discovered; but the correction for the aberration may be applied when it is necessary. For observations of this kind, it is usual to have a clock adjusted to sidereal time, which not only admits of a more ready comparison with the transits or passages of any one star over the meridian, but, by the difference of the times of any two transits, shows at once the difference of right ascension of the stars or planets, expressed in time instead of degrees.

The solar days are not only about four minutes longer than the sidereal days, but they are also unequal among themselves; and this inequality arises from two causes; the one, that even if the sun's motion in the ecliptic were uniform, his diurnal changes of right ascension would be different at different times, and the difference between his path in every sidereal day, and a whole circle, depending on this change, would also vary; the other that the sun's motion in the ecliptic is actually variable, consequently the diurnal change of right ascension is liable to adouble inequality. Hence it happens that the solar time agrees at four instants in the year only with the mean time, derived from supposing the whole 365 days to be divided into as many equal parts; the difference is called the equation of time, and amounts sometimes to as much as 16 minutes. The term equation is commonly applied in astronomy to any small quantity, which is to be added to, or subtracted from, another, quantity; thus it is usual, in calculating the place of a planet, to find from the tables of its motion, the mean place, in which it would be found if its orbit were circular, and thence to derive the true place, by means of various corrections called equations. In France the solar time is considered as the true time, and is used for all civil purposes, so that the clocks are sometimes embarrassed with a complicated apparatus, calculated forimitating the inequalities of the actual apparent motion of the sun. (Plate XXXV. Fig. 503.)

The art of dialling consists principally in projecting, on a given surface, such lines as will coincide with the shadow of an index or gnomon parallel to the earth's axis, at intervals corresponding to the different hours of the day: so that nothing more is necessary for the construction of a dial, than to determine the intersections of the surface on which the dial is to beconstructed, with planes passing through the edge of the gnomon, and situated at equal angular distances from each other: thus, supposing the plane of the dial perpendicular to the gnomon, and parallel to the equinoctial, the hour lines of the dial will be at equal distances from each other; but in other cases their distances will be unequal, and must be determined either by calculation or by construction. A point may also be used as a gnomon, as well as a line; but in this case the hour lines must cover a larger portion of the surface, in order that the shadow of the point may always fall among them. (Plate XXXV. Fig. 504 . . 506.)

The changes of the seasons depend on the return of the sun to the same position with respect to the equator, or on the length of the tropical year, so called from the tropics, which are circles supposed to be parallel to the equator, and between which the sun travels continually backwards and forwards, appearing to remain for some time, when he is near them, with very little change of declination; whence the time when the sun touches either tropic is called the solstice. The vicissitudes of light and darkness depending also on the solar day, it is necessary, for the regulation of the civil calendar, to establish the proportion between the periods of the solar day and the tropical year; and since the tropical year exceeds the time of 365 days, by 5 hours, 48 minutes and 48 seconds, it is usual to add to the common year an intercalary day once in about four years. The ancient Egyptians reckoned only 365 days in a year, and their nominal new year arrived continually earlier and earlier, so that after 1507 of their years, it would have happened successively on each of the 365 days, and returned to the original place: the same mode of computation was also adopted by the Greek astronomers. The Romans inserted intercalary days, at first without much regularity, according to the direction of their augurs, until the time of Julius Caesar; who, observing that the year was almost 6 hours longer than 365 days, added a day every fourth year to the calendar, by reckoning twice the day in February called sexto calendas Martias, whence this year of 366 days was denominated a bissextile year. The new moon immediately following the winter solstice, in the 707th year of Rome, was made the first of January of the first year of Caesar; the 25th of December in his 45th year is considered as the date of the Nativity of Christ, and Caesar's 46th year is reckoned the first of our era. The preceding year is commonly called by astronomers the year 0, but by chronologists the year 1 before Christ. The correction introduced by Caesar was, however, too great, the error being exactly 7 days in 900 years; so that in 1582 it amounted to about 12 days. This error was not wholly removed by Pope Gregory, who reformed the calendar; he omitted 10 days only of the usual reckoning, in order to bring back the course of the moveable feasts to the same state, in which they had been established by the Nicene council, in the fourth century. He determined at the same time that the last year of every century should be passed without an intercalary day, excepting that of every fourth century, which was still to be bissextile. Thus every year divisible by four, without a remainder, is in general a bissextile or leap year, but the last year of every century must be a common year, unless the number of the century be divisible by 4; the year 1800 being a common year, and 2000 a bissextile. In this manner 27 Julian bissextiles are omitted in 3600 years, while the true length of the year would require the omission of 28; but so small a difference can be of no material consequence. The Persians had introduced into their calendar, in the 11th century, an intercalation still more accurate; they make 8 bissextiles only every 33 years, reckoning four common years together instead of three, at the end of this period, so that in 132 years they have 32 leap years instead of 33; and the error is only a day in about five thousand years. If any change in the Gregorian calendar were thought necessary, it would be easy to make the last year of every fourth and fifth century alternately a bissextile, and this correction would be quite as accurate as it is possible for our calculations to render it. The adoption of the Gregorian calendar in this country was for some time delayed by religious prejudices; one of the best founded objections to it was, that 2 days of the real error was still uncorrected; but better arguments at last overcame these difficulties, and the new stile was introduced on the 14 September 1754, which would have been called, according to the old stile, the third.

Any tolerable approximation of this kind, when once generally established, appears to be more eligible than the mode which was lately adopted in France, where the republican year began at the instant of the midnight preceding the sun's arrival at the autumnal equinox. Mr. Lalande very judiciously observes, that there are several years, in which the sun will pass the equinox so near to midnight, that it is not at present in the power of calculation to determine on what day the republican year ought to begin; and perhaps these arguments have cooperated with others in facilitating the restoration of the ancient calendar.

The revolutions of the sun and moon are not very obviously commensurable, the solar year containing 12 lunations and almost 11 days; but Meto discovered, more than 2000 years ago, that 19 solar years contain exactly 235 lunations; and this determination is so accurate, that it makes the lunar month only about half a minute too long. Hence it happens, that in every period of 19 years, the moon's age is the same on the same day of the year. The number of the year, in the Metonic cycle, is called the golden number, the calendar of Meto having been ordered, at the celebration of the Olympic games, to be engraved in letters of gold on a pillar of marble. At present, if we add 1 to the number of the year, and divide it by 19, the remainder will be the golden number; thus, for 1806, the golden number is 2.

If we subtract 1 from the golden number, then multiply by 11, and divide by 30, the remainder will be the epact, which is the moon's age on the first of January, without any material error; thus, for 1806, the epact is 11, and the moon is actually 11 days old on the first of January.

From a combination of chronological periods of various kinds, Scaliger imagined the Julian period, as an epoch to which all past events might with convenience be referred, beginning 4713 years before the birth of Christ. Laplace proposes, as a universal epoch, the time when the earth's apogee was at right angles with its nodes, in the year 1250, calling the vernal equinox of that year the first day of the first year. But the fewer changes of this kind that we make, the less confusion we introduce into our chronology. The astronomical year begins at noon on the 31st of December, and the date of an observation expresses the days and hours actually elapsed from that time. Thus, the first of January 1806, at 10 in the morning, would be called, in astronomical language, 1805 December 31 days 22 hours, or more properly 1806 January 0 day 22 hours.

For ascertaining, by immediate measurement, the position of any of the heavenly bodies, it is usual to determine its meridian altitude by quadrants, and the time of its passing the meridian by transit instruments. The large quadrants, generally used for this purpose in observatories, are fixed to vertical walls, in order to give them greater stability, and are thence called mural quadrants; sometimes a smaller portion of an arc only is adapted for observations near the zenith, under the name of a zenith sector. A transit instrument is a telescope so fixed on an axis as to remain always in the plane of the meridian; the axis being perpendicular to this plane, and consequently in a horizontal position, and directed east and west. Those who are in the constant habit of observing with attention, can estimate, in this manner, the precise time of the passage of a celestial object over the meridian; without an error of the tenth of a second, so that its angular right ascension may be thus determined within about a second of the truth. A very convenient mode of adjusting a transit instrument is to direct it to the north polar star, at the same time that the last of the three horses in the wain is perpendicularly above or below it: this process, in 1751, gave precisely the true meridian; but since that time, the precession of the equinoxes, which produces a slight change in the places of the stars, has made it necessary to wait 1 minute $13\frac{1}{2}$ seconds for every ten years that have elapsed. Thus, in 1806, if we wait $6\frac{1}{2}$ minutes, the pole star will then be precisely in the meridian, and will serve for the correct adjustment of the instrument. (Plate XXXV. Fig. 507 . . 510.)

The quadrant in most common use, especially for nautical observations, was first proposed by Newton, but improved, or perhaps reinvented, by Hadley. Its operation depends on the effect of two mirrors which bring both the objects, of which the angular distance is to be measured, at once into the field of view; and the inclination of the speculums by which this is performed serves to determine the angle. The ray proceeding from one of the objects is made to coincide, after two reflections, with the ray coming immediately from the other, and since the inclination of the reflecting surfaces is then half the angular distance of the objects, this inclinatoin is read off on a scale in which every actual degree represents two degrees of angular distance, and is marked accordingly. There is also a second fixed speculum, placed a right angles to the moveable one, when in its remotest situation, which then produces a deviation of two right angles in the apparent place of one of the objects, and which enables us, by moving the index, to measure any angle between 180° and 90°. This operation is called the back observation; it is however seldom employed, on account of the difficulty of adjusting the speculum for it with accuracy. The reflecting instrument originally invented by Hooke was arranged in a manner somewhat different. (Plate XXXV. Fig. 511.)

From the meridian altitude of any point, it is easy, when the elevation of the pole is known, to deduce its declination; and its right ascension may be found from the time of its passage over the meridian after that of the equinoctial point, allowing 15 degrees for each sidereal hour. (Plate XXXV. Fig. 512.)

In all astronomical observations it is necessary to make proper corrections, according to the rules of optics, for the effects of atmospherical refraction; and also, in observations on the moon more especially, for those of parallax, or the difference of the apparent place of the luminary with respect to the earth's centre, and to the place of the spectator, which is equal to the angle subtended at the centre of the luminary by the semidiameter of the earth passing through the place of observation; since all calculations of the geocentric places of the heavenly bodies are referred to the earth's centre. This angle, which is to be added to the apparent altitude, amounts sometimes, in the case of the moon, when near the horizon, to more than a degree; the refraction, which is in a contrary direction, and is to be subtracted from the altitude, being at the horizon about 33 minutes. (Plate XXXV. Fig. 513.)

The most important applications of practical astronomy are in the determination of the latitudes and longitudes of places on the earth's surface. The latitude, which is the angular distance of the place from the equator, or the angle formed by the plane of its horizon with the earth's axis, is easily ascertained by finding the meridian altitude of a body, of which the declination is known; since, by deducting or adding the declination, we have at once the elevation of the equinoctial, or of the plane of the equator, above the horizon, and subtracting this from a right angle, we find the elevation of the pole, or the latitude. (Plate XXXV. Fig. 512.)

It is also common to determine the latitude of a place by means of two altitudes observed at different times in the same day, noticing accurately the interval of time that elapses between the observations. This method has a great advantage in cloudy weather, when it is not possible to insure an observation of a meridian altitude.

The longitude of a place, or the relative position of its meridian, is by no means so readily determined. For this purpose, it becomes necessary to ascertain the time that elapses between the passages of a given point in the heavens over its meridian and some other meridian which serves as a standard of comparison. Thus, if the sun arrives three hours later at the meridian of any place than at the meridian of London, that place must necessarily be 45 degrees west of London, or in 45° west longitude: and if we know, when it is noon at the given place, that it is three o'clock in the afternoon at Greenwich, we may be certain that we are in some part of a meridian 45° west of that of Greenwich. Had we perfect timekeepers, we might easily adjust them to the time of our first meridian, and then, by comparison with the usual determinations of time in any other place, to which they might be carried, the longitude of this place might be found with perfect accuracy. Such timekeepers as we have are indeed sufficiently correct, to be of considerable utility, but it is necessary to compare them frequently with astronomical observations of phenomena, which occur at times capable of a correct calculation. Sometimes the transits of Mercury and Venus, or the eclipses of the moon, are employed for this purpose, but more usually the eclipses of the satellites of Jupiter; these, however, cannot be well observed without a more powerful telescope than can be employed at sea; and the theory of the moon's motion, has of late years been so much improved, that her distance from the sun or from a fixed star can be calculated, with sufficient accuracy, for determining the time in London or at Paris without an error of one third of a minute; so that supposing the observation could be rendered perfectly correct, the longitude might be thus ascertained within about one twelfth of a degree, or at most five nautical miles.

The observed parallax of the sun and moon may be employed for the determination of their distances from the earth; but in the case of the sun, the simple comparison of his calculated with his apparent altitude is insufficient for ascertaining the magnitude of the parallax with accuracy. Sometimes the parallax of Mars, which is considerably greater than the sun's, has been directly measured; but the most correct mode of ascertaining the actual dimensions of the solar system is, to observe a transit of Venus over the sun's disc, at two places situated in opposite parts of the earth's surface. For, since the diurnal motion of some parts of the earth is directed the same way with the motion of Venus in her orbit, and that of others the contrary way, the different effects of these motions must furnish a mode of comparing the rotatory velocity of the earth, with the progressive velocity of Venus, and consequently of inferring, from the known velocity with which the earth's surface revolves, the actual velocity of Venus, and her distance from the sun; whence the distances of all the other planets may be readily deduced. (Plate XXXV. Fig. 514.)

Our countryman Horrox was the first that particularly attended to the phenomena of a transit of Venus over the sun's disc: Dr. Halley, when he observed a transit of Mercury at St. Helena, thought that he could ascertain the times of immersion and emersion without an error of a single second; and hence he concluded, that by means of a transit of Venus, the sun's distance might be determined within a five hundredth part. The most advantageous places for the experiment being such as differ most in longitude, and are most remote from each other, Captain Cook was sent by the British government to the South Seas, in the years 1761 and 1769, in order to observe the transits of Venus in the island of Otaheite. These observations were compared with those which were made at Wardhuys, in Danish Lapland; the difference of the times occupied by the transit at these places was found to be 23 minutes 10 seconds, and from this comparison, corrected by a number of collateral observations, the sun's mean parallax was found to be 8 seconds and two thirds, or perhaps $8\frac{3}{4}$; for it does not appear that we are sure of having avoided even an error of one fortieth part of the whole; although Mr. Laplace's determination of the sun's distance, from the lunar motions, agrees very well with that which is usually considered as the result of the observations of the transit of Venus.

The comparative densities of the sun, and of such planets as have satellites, may be calculated from the periods and distances of the bodies revolving round them; the densities of the other planets have sometimes been assigned from conjecture only, but of late years the mathematical theory of the planetary perturbations has been rendered so perfect, that some dependence may perhaps be placed on the density assigned to them from calculations of this kind. It was formerly supposed that the densities of the planets were regularly greater as they were nearer to the sun; but it is now certain that the Georgian planet is more dense than Saturn, and it is probable that Venus is somewhat less dense than the earth. The mass of the moon is deduced from a comparison of the effects of her attraction on the earth and sea with those of the sun's attraction.

The artificial globe serves as a useful instrument for determining, in a rough manner, without calculation, the affections of the heavenly bodies at particular times; their places being first ascertained from tables, or, in the case of the sun, merely from a scale on the globe's horizon, or on its surface. We have only to adjust the elevation of the pole of the globe in such a manner, that its axis may form the same angle with its horizon as the axis of the earth does with the real horizon of the place; then finding a point on its surface corresponding to the place of the sun or planet, we may represent its apparent motion by the motion of this point, and the time occupied by that motion will be shown by the index of the globe; thus we may find the length of the day and night, and the time and place of rising and setting; and by means of a graduated circle, perpendicular to the horizon, we may measure the altitude of the sun or planet at any other time, and also its azimuth, or the distance of this circle from the north or south point of the horizon. If we have a ring of any kind parallel to the horizon, and 33 minutes below it, we may consider this ring as the apparent horizon, allowing for the effects of refraction; if it be still 15 or 16 minutes lower, it will represent the rising or setting of the extreme margin of the sun or moon: we might also have a circle about a degree above either of these, which might represent the sensible or apparent horizon with regard to the moon, including the correction for her parallax; and a similar ring, placed still lower, would show the duration of twilight, on any supposition that might be formed respecting the depression of the sun required for producing total darkness. By means of the celestial globe, the apparent motions of the fixed stars may be represented in a manner nearly similar, proper attention being paid to the situation of the sun in the ecliptic, as determining the time corresponding.

Many of these operations may also be performed with equal convenience with a planisphere, which is a stereographical projection of the globe on a plane surface. Professor Bode's planisphere comprehends in one view all the stars that are ever visible at Berlin: he has added to it a moveable circle, representing the horizon of that place, carrying with it the circles of altitude and azimuth, delineated on a transparent paper, which is adjusted, by graduations at the margin of the chart, to the day and hour for which we wish to ascertain the apparent places of the heavenly bodies. Any other chart of the stars, having the pole in its centre, may be applied to a similar use, by cutting out a circle, or a part of a circle, to represent the horizon of a place of which the latitude is given; and if the stars are projected, as is usual, on two equal charts, they must have two equal arcs to represent the respective parts of the horizon belonging to them. A simple construction may also often be made to serve for solving many problems of a similar nature. (Plate

XXXV. Fig. 515, 516. Plate XXXVI. Fig. 517. Plate XXXVII. Fig. 518.)

For representing the real as well as the apparent motions of the different parts of the solar system, planetariums or orreries have sometimes been employed, in which the comparative periods of the revolutions have been expressed by various combinations of wheelwork. Of these instruments Archimedes was the original inventor, and Huygens revived them, with many improvements, in modern times. The construction of the large planetarium, which has been made in the house of the Royal Institution, was principally directed by Mr. Pearson. I suggested to him, that the instrument might be placed in a vertical position, and that the eccentricities of the planetary orbits might be shown by the revolution of short arms, retained in their situation by weights, and their deviation from the plane of the ecliptic by inclining the axes of these arms, in a proper angle, to the plane of the instrument. The other parts of the arrangement, which have any claim to novelty, were entirely of Mr. Pearson's invention, and he appears to have rendered the instrument in many respects more accurate than any other planetarium that has ever been constructed.

LECTURE XLVI.

ON GEOGRAPHY.

FROM the consideration of the stars, the sun, and the planets in general, we are now to descend to that of the earth, the particular planet which we inhabit, and which we can examine more minutely than the other parts of the solar system. Its external form, its divisions, whether astronomical or natural, its most remarkable features, and its internal structure, will require to be separately investigated.

The general curvature of the earth's surface is easily observable in the disappearance of distant objects, and in particular, when the view is limited by the sea, the surface of which, from the common property of a fluid, becomes naturally smooth and horizontal: for it is well known that the sails and rigging of a ship come into view long before her hull, and that each part is the sooner seen as the eye is more elevated. On shore, the frequent inequalities of the solid parts of the earth usually cause the prospect to be bounded by some irregular prominence, as a hill, a tree, or a building, so that the general curvature is the less observable.

The surface of a lake or sea must be always perpendicular to the direction of a plumb line, which may be considered as the direction of the force of gravity; and by means either of a plumb line or of a spirit level, we may ascertain the angular situation of any part of the earth's surface with respect to a fixed star passing the meridian; by going a little further north or south, and repeating the observation on the star, we may find the difference of the inclination of the surfaces at both points; of course, supposing the earth a sphere, this difference in latitude will be the angle subtended at its centre by the given portion of the surface, whence the whole circumference may be determined; and on these principles the earliest measurements of the earth were conducted. The first of these, which can be considered as accurate, was executed by Picart in France, towards the end of the seventeenth century.

But the spherical form is only an approximation to the truth; it was calculated by Newton, and ascertained experimentally by the French Academicians, sent to the equator and to the polar circle, that, in order to represent the earth, the sphere must be flattened at the poles, and prominent at the equator. We may therefore consider the earth as an oblate elliptic spheroid; the curvature being greater, and consequently every degree shorter, at the equator, than nearer the poles. If the density of the earth were uniform throughout, its ellipticity, or the difference of the length of its diameters, would be $\frac{1}{230}$ of the whole; on the other hand, if it consisted of matter of inconsiderable density, attracted by an infinite force in the centre, the ellipticity would be only $\frac{1}{577}$;. and whatever may be the internal structure of the earth, its form must be between these limits, since its internal parts must necessarily be denser than those parts which are nearer the surface. If indeed the earth consisted of water or ice, equally compressible with common water or ice, and following the same laws of compression with elastic fluids, its density would be several thousand times greater at the centre than at the surface; and even steel would be compressed into one fourth of its bulk, and stone into one eighth, if it were continued to the earth's centre; so that there can be no doubt but that the central parts of the earth must be much more dense than the superficial. Whatever this difference may be, it has been demonstrated by Clairaut, that the fractions expressing the ellipticity and the apparent diminution of gravity at the equator must always make together $\frac{5}{578}$, and it has been found, by the most accurate observations on the lengths of pendulums in different latitudes, that the force of gravity is less powerful by $\frac{1}{180}$ at the equator than at the pole, whence the ellipticity is found to be $\frac{1}{320}$ of the equatorial diameter, the form being the same as would be produced, if about three eighths of the whole force of gravity were directed towards a central particle, the density of the rest of the earth being uniform.

This method of determining the general form of the earth is much less liable to error and irregularity, than the measurement of the lengths of degrees in various parts, since the accidental variations of curvature produced by local differences of density, and even by superficial elevations, may often produce considerable errors in the inferences which might be deduced from these measurements. For example, a degree measured at the Cape of Good Hope, in latitude 33° south, was found to be longer than a degree in France, in latitude 46° north, and the measurements in Austria, in North America, and in England, have all exhibited signs of similar irregularities. There appears also to be some difference in the length of degrees under the same latitude, and in different longitudes. We may, however, imagine a regular elliptic spheroid to coincide very nearly with any small portion of the earth's surface, although its form must be somewhat different for different parts: thus, for the greater part of Europe, that is, for England, France, Italy, and Austria, if the measurements have been correct, this osculating spheroid must have an ellipticity of $\frac{1}{130}$.

The earth is astronomically divided into zones, and into climates. The torrid zone is limited by the tropics, at the distance of 23° 28′ on each side of the equator, containing all such places as have the sun sometimes vertical, or immediately over them; the frigid zones are within the polar circles, at the same distance from the poles, including all places which remain annually within the limit of light and darkness, for a whole diurnal rotation of the earth, or longer; the temperate zones, between these, have an uninterrupted alternation of day and night, but are never subjected to the sun's vertical rays. At the equator, therefore, the sun is vertical at the equinoxes, his least meridian altitude is at the solstices, when it is 66° 32′, that is, more than with us at midsummer, and this happens once on the north and once on the south side of the hemisphere. Between the equator and the tropics, he is vertical twice in the year, when his declination is equal to the latitude of the place, and his least meridian altitudes, which are unequal between themselves, are at the solstices. At the tropics, the meridian sun is vertical once only in the year, and at the opposite solstice, or the time of midwinter, his meridian altitude is 43° 4′, as with us in April, and the beginning of September. At the polar circles, the sun describes on midsummer day a complete circle, touching the north or south point of the horizon; and in midwinter he shows only half his disc above it for a few minutes in the opposite point; that is, neglecting the elevation produced by refraction, which, in these climates especially, is by no means inconsiderable. At either pole, the corresponding pole of the heaven being vertical, the sun must annually describe a spiral, of which each coil is nearly horizontal, half of the spiral being above the horizon, and half below; the coils being much opener in the middle than near either end.

The climates, in the astronomical sense of the word, are determined by the duration of the longest day in different parts of the earth's surface; but this division is of no practical utility, nor does it furnish any criterion for judging of the climate in a meteorological sense.

The natural division of the surface of the globe is into sea and land: about three fourths of the whole being occupied by water, although probably no where to a depth comparatively very considerable, at most of a few miles on an average. The remaining fourth consists of land, elevated more or less above the level of the sea, interspersed, in some parts, with smaller collections of water, at various heights, and, in a few instances, somewhat lower than the general surface of the main ocean. Thus the Caspian sea is said to be about 300 feet lower than the ocean, and in the interior part of Africa there is probably a lake equally depressed.

We cannot observe any general symmetry in this distribution of the earth's surface, excepting that the two large continents, of Africa and South America, have some slight resemblance in their forms, and that each of them is terminated to the eastward by a collection of numerous islands. The large capes projecting to the southward have also a similarity with respect to their form, and the islands near them: to the west the continents are excavated into large bays, and the islands are to the east: thus Cape Horn has the Falkland Islands, the Cape of Good Hope Madagascar, and Cape Comorin Ceylon, to the east. (Plate XLII, XLIII.)

The great continent, composed of Europe, Asia, and Africa, constitutes about a seventh of the whole surface of the earth, America about a sixteenth, and Australasia or New South Wales about a fiftieth; or, in hundredth parts of the whole, Europe contains 2, Asia 7, Africa 6, America 6, and Australasia 2, the remaining 77 being sea; although some authors assign 72 parts only out of 100 to the sea, and 28 to the land. These proportions may be ascertained with tolerable accuracy by weighing the paper made for covering a globe, first entire, and then cut out according to the terminations of the different countries: or, if still greater precision were required, the greater part of the continents might be divided into known portions of the whole spherical surface, and the remaining irregular portions only weighed.

The general inclinations and levels of the continents are discovered by the course of their rivers. Of these the principal are, the River of Amazons, the Senegal, the Nile, the River St. Laurence, the Hoangho, the River Laplata, the Jenisei, the Mississippi, the Volga, the Oby, the Amur, the Oronooko, the Ganges, the Euphrates, the Danube, the Don, the Indus, the Dnieper, and the Dwina; and this is said to be nearly the order of their magnitudes. But if we class them according to the length of country through which they run, the order will, according to Major Rennel's calculation, be somewhat different: taking the length of the Thames for unity, he estimates that of the River of Amazons at $15\frac{3}{4}$, the Kian Kew, in China, $15\frac{1}{2}$, the Hoangho $13\frac{1}{2}$, the Nile $12\frac{1}{2}$, the Lena $11\frac{1}{2}$, the Amur 11, the Oby $10\frac{1}{2}$, the Jenisei 10, the Ganges, its companion the Burrampooter, the river of Ava, and the Volga, each $9\frac{1}{2}$, the Euphrates $8\frac{1}{2}$, the Mississippi 8, the Danube 7, the Indus $5\frac{1}{2}$, and the Rhine $5\frac{1}{4}$.

We may form a tolerably accurate idea of the levels of the ancient continent, by tracing a line across it in such a direction as to pass no river, which will obviously indicate a tract of country higher than most of the neighbouring parts. Beginning at Cape Finisterre, we soon arrive at the Pyrenees, keeping to the south of the Garonne and the Loire. After taking a long turn northwards, to avoid the Rhine, we come to Swisserland, and we may approach very near to the Mediterranean in the state of Genoa, taking care not to cross the branches of the Po. We make a circuit in Swisserland, and pass between the sources of the Danube and of the branches of the Rhine in Swabia. Crossing Franconia, we leave Bohemia to the north, in order to avoid the Elbe, and coming near to the borders of Austria, follow those of Hungary, to the south of the Vistla. The Dnieper then obliges us to go northwards through Lithuania, leaving the Don wholly to the right; and the Volga, to pass still further north, between Petersburg and Moscow, a little above Bjelesero. We may then go eastwards to the boundary of Asia, and thence northwards to Nova Zembla. Hence we descend to the west of the Oby, and then to the east of the branches of the Volga, and the other inland rivers flowing into the lake Aral and the Caspian sea. Here we are situated on the widely extended elevation of India, in the neighbourhood of the sources of the Indus: and, lastly, in our way from hence towards Kamschatka, we leave the Jenisei and Lena on the left, and the Ganges, the Kiang Kew, the Hoangho, and the Amur to the right.

The direction of the most conspicuous mountains is, however, a little different from this, the principal chain first constitutes the Pyrenees, and divides Spain from France, then passes through Vivarais and Auvergne, to join the Alps, and through the south of Germany to Dalmatia, Albania, and Macedonia; it is found again beyond the Euxine, under the names of Taurus, Caucasus, and Imaus, and goes on to Tartary and to Kamschatka. The peninsula of India is divided from north to south by the mountains of Gate, extending from the extremity of Caucasus to Cape Comorin. In Africa, Mount Atlas stretches from Fez to Egypt, and the mountains of the moon run nearly in the same direction: there is also a considerable elevation between the Nile and the Red Sea. In the new world, the neighbourhood of the western coast is in general the most elevated; in North America the Blue mountains, or Stony mountains, are the most considerable; and the mountains of Mexico join the Andes or Cordeliers, which are continued along the whole of the west coast of South America.

There are several points in both hemispheres from which we may observe rivers separating to run to different seas; such are Swisserland, Bjelosero, Tartary, Little Tibet, Nigritia or Guinea, and Quito. The highest mountains are Chimboraçao and some others of the Cordeliers in Peru, or perhaps Descabesado in Chili, Mont Blanc, and the Peak of Teneriffe. Chimboraçao is about 7000 yards, or nearly 4 miles, above the level of the sea; Mont Blanc 5000, or nearly 3 miles; the Peak of Teneriffe about 4000, or 2 miles and a quarter: Ophir, in Sumatra, is said to be 5 or 6 hundred feet higher. It has, however, been, asserted that some of the snowy mountains, to the north of Bengal, are higher than any of those of South America. The plains of Quito, in Peru, are so much elevated, that the barometer stands at the height of 15 inches only, and the air is reduced to half its usual density. But none of these heights is equal to a thousandth part of the earth's semidiameter, and the greatest of them might be represented on a six inch globe by a single additional thickness of the paper with which it is covered. Mount Sinai in Japan, Mount Caucasus, Etna, the Southern Pyrenees, St. George among the Azores, Mount Adam in Ceylon, Atlas, Olympus, and Taurus are also high mountains: and there are some very considerable elevations in the island Owhyhee. Ben Nevis, in Scotland, is the loftiest of the British hills, but its height is considerably less than a mile. (Plate XXXVIII. Fig. 519.)

The most elevated mountains, excepting the summits of volcanos, consist of rocks, more or less mixed, without regular order, and commonly of granite or porphyry. These are called primary mountains; they run generally from east to west in the old world, and from north to south in the new; and many of them are observed to be of easier ascent on the east than on the west side. The secondary mountains accompany them in the same direction, they consist of strata, mostly calcarious and argillaceous, that is, of the nature of limestone and clay, with a few animal and vegetable remains, in an obscure form, together with salt, coals, and sulphur. The tertiary mountains are still smaller; and in these, animal and vegetable remains are very abundant; they consist chiefly of limestone, marble, alabaster, building stone, mill stone, and chalk, with beds of flint. Where the secondary and tertiary mountains are intersected by vallies, the opposite strata often correspond at equal heights, as if the vallies had been cut or washed from between them, but sometimes the mountains have their strata disposed as if they had been elevated by an internal force, and their summits had afterwards crumbled away, the strata which are lowest in the plains being highest in the mountains. The strata of these mountains are often intermixed with veins of metal, running in all possible directions, and occupying vacuities which appear to be of somewhat later date than the original formation of the mountains. The volcanic mountains interrupt those of every other description without any regularity, as if their origin were totally independent of that of all the rest.

The internal constitution of the earth is little known from actual observation, for the depths to which we have penetrated are comparatively very inconsiderable, the deepest mine scarcely descending half a mile perpendicularly.

It appears that the strata are more commonly in a direction nearly horizontal than in any other; and their thickness is usually almost equable for some little distance; but they are not disposed in the order of their specific gravity, and the opinion of their following each other in a similar series, throughout the greater part of the globe, appears to rest on very slight foundations.

From observations on the attraction of the mountain Shehallion, Dr. Maskelyne inferred the actual mean density of the earth to be to that of water as $4\frac{1}{2}$ to 1, judging from the probable density of the internal substance of the mountain, which he supposed to be a solid rock. Mr. Cavendish has concluded more directly, from experiments on a mass of lead, that the mean density of the earth is to that of water as $5\frac{1}{2}$ to 1. Mr. Cavendish's experiments, which were performed with the apparatus invented and procured by the late Mr. Michell, appear to have been conducted with all possible accuracy, and must undoubtedly be preferred to conclusions drawn from the attraction of a mountain, of which the internal parts are perfectly unknown to us, except by conjectures founded on its external appearance. Supposing both series of experiments and calculations free from error, it will only follow that the internal parts of Shehallion are denser, and perhaps more metallic, than was before imagined. The density assigned by Mr. Cavendish is not at all greater than might be conjectured from observations on the vibrations of pendulums; Newton had long ago advanced it as a probable supposition that the mean density of the earth might be about 5 or 6 times as great as that of water, and the perfect agreement of the result of many modern experiments with this conjecture affords us a new proof, in addition to many others, of the accuracy and penetration of that illustrious philosopher.

LECTURE XLVII.

ON THE TIDES.

THE form and structure of the solid parts of the globe have afforded but few remarkable features capable of arresting our attention, except the general distribution of land and water, and the permanent differences of elevation of different parts of the earth. But the sea exhibits a series of phenomena far more interesting to the mathematical philosopher, because they admit of a methodical investigation, and of a deduction from general causes, the action of which may be traced in detail. For the height of the surface of the sea at any given place is observed to be liable to periodical variations, which are found to depend on the relative position of the moon, combined in some measure with that of the sun. These variations are called tides; they were too obvious to escape the observation even of the ancients, who inhabited countries where they are least conspicuous: for Aristotle mentions the tides of the northern seas, and remarks that they vary with the moon, and are less conspicuous in small seas than in the ocean: Caesar, Strabo, Pliny, Seneca, and Macrobius give also tolerably accurate accounts of them.

There are in the tides three orders of phenomena which are separately distinguishable: the first kind occurs twice a day, the second twice a month, and the third twice a year. Every day, about the time of the moon's passing over the meridian, or a certain number of hours later, the sea becomes elevated above its mean height, and at this time it is said to be high water. The elevation subsides by degrees, and in about six hours it is low water, the sea having attained its greatest depression; after this it rises again when the moon passes the meridian below the horizon, so that the ebb and flood occur twice a day, but become daily later and later by about $50\frac{1}{2}$ minutes, which is the excess of a lunar day above a solar one, since $28\frac{1}{2}$ lunar days are nearly equal to $29\frac{1}{2}$ solar ones.

4

The second phenomenon is, that the tides are sensibly increased at the time of the new and full moon; this increase and diminution constitute the spring and neap tides; the augmentation becomes also still more observable when the moon is in its perigee, or nearest the earth. The lowest as well as the highest water is at the time of the spring tides; the neap tides neither rise so high nor fall so low.

The third phenomenon of the tides is the augmentation which occurs at the time of the equinoxes: so that the greatest tides are when a new or full moon happens near the equinox, while the moon is in its perigee. The effects of these tides are often still more increased by the equinoctial winds, which are sometimes so powerful as to produce a greater tide before or after the equinox, than that which happens in the usual course, at the time of the equinox itself.

These simple facts are amply sufficient to establish the dependence of the tides on the moon; they were first correctly explained by Newton as the necessary consequences of the laws of gravitation, but the theory has been still further improved by the labours of later mathematicians. The whole of the investigations has been considered as the most difficult of all astronomical problems; some of the circumstances depend on causes which must probably remain for ever unknown to us; and unless we could every where measure the depth of the sea, it would be impossible to apply a theory, even if absolutely perfect, to the solution of every difficulty that might occur. A very injudicious attempt has been made to refer the phenomena of the tides to causes totally different from these, and depending on the annual melting of the polar ice: the respectability of its author is the only claim which it possesses even to be mentioned; and a serious confutation of so groundless an opinion would be perfectly superfluous.

A detached portion of a fluid would naturally assume, by its mutual gravitation, a spherical form, but if it gravitate towards another body at a distance, it will become an oblong spheroid of which the axis will point to the attracting body: for the difference of the attraction of this body on its different parts will tend to separate them from each other in the greatest part of the sphere, that is, at all places within the angular distance of $79\frac{1}{2}°$ from the line passing through the attracting body, either in the nearer, or in the remoter hemisphere; but to urge them towards the centre, although with a smaller force, in the remaining part. Hence, in order that there may be an equilibrium, the depth of the fluid must be greatest where its gravitation, thus composed, is least; that is, in the line directed towards the attracting body, and it may be shown that it must assume the form of an oblong elliptic spheroid.

If the earth were wholly fluid, and the same part of its surface were always turned towards the moon, the pole of the spheroid being immediately under the moon, the lunar tide would remain stationary, the greatest elevation being at the points nearest to the moon and furthest from her, and the greatest depression in the circle equally distant from these points; the elevation being, however, on account of the smaller surface to which it is confined twice as great as the depression. The actual height of this elevation would probably be about 40 inches, and the depression 20, making together a tide of 5 feet. If also the waters were capable of assuming instantly such a form as the equilibrium would require, the summit of a spheroid equally elevated would still be directed towards the moon, notwithstanding the earth's rotation. This may be called the primitive tide of the ocean: but on account of the perpetual change of place which is required for the accommodation of the surface to a similar position with respect to the moon, as the earth revolves, the form must be materially different from that of such a spheroid of equilibrium. The force employed in producing this accommodation may be estimated by considering the actual surface of the sea as that of a wave moving on the spheroid of equilibrium, and producing in the water a sufficient velocity to preserve the actual form. We may deduce, from this mode of considering the subject, a theory of the tides which appears to be more simple and satisfactory than any which has yet been published: and by comparing the tides of narrower seas and lakes with the motions of pendulums suspended on vibrating centres, we may extend the theory to all possible cases.

If the centre of a pendulum be made to vibrate, the vibrations of the pendulum itself, when they have arrived at a state of permanence, will be performed in the same time with those of the centre; but the motion of the pendulum will be either in the same direction with that of the centre, or in a contrary direction, accordingly as the time of this forced vibration is longer or shorter than that of the natural vibration of the pendulum; and in the same manner it may be shown that the tides either of an open ocean or of a confined lake may be either direct or inverted with respect to the primitive tide, which would be produced if the waters always assumed the form of the spheroid of equilibrium, according to the depth of the ocean, and to the breadth as well as the depth of the lake. In the case of a direct tide, the time of the passage of the luminary over the meridian must coincide with that of high water, and in the case of an inverted tide with that of low water.

In order that the lunar tides of an open ocean may be direct, or synchronous, its depth must be greater than 13 miles, and for the solar tides than 14. The less the depth exceeded these limits, the greater the tides would be, and in all cases they would be greater than the primitive tides. But in fact the height of the tides in the open ocean is always far short of that which would be produced in this manner; it is therefore improbable that the tides are ever direct in the open ocean, and that the depth of the sea is so great as 13 miles.

In order that the height of the inverted or remote lunar tides may be five feet, or equal to that of the primitive tides, the depth of the open sea must be $6\frac{1}{4}$ miles; and if the height is only two feet, which is perhaps not far from the truth, the depth must be 3 miles and five sevenths.

The tides of a lake or narrow sea differ materially from those of the open ocean, since the height of the water scarcely undergoes any variation in the middle of the lake; it must always be high water at the eastern extremity when it is low water at the western: and this must happen at the time when the places of high and low water, with respect to the primitive tides, are equally distant from the middle of the lake. (Plate XXXVIII. Fig. 520.)

The tides may be direct in a lake 100 fathoms deep and less than 8 degrees wide; but if it be much wider, they must be inverted. Supposing the depth a mile, they will be direct when the breadth is less than 25°; but if a sea, like the Atlantic, were 50 or 60 degrees wide, it must be at least four miles deep, in order that the time of high water might coincide with that of the moon's southing.

Hitherto we have considered the motion of the water as free from all resistance; but where the tides are direct, they must be retarded by the effect of a resistance of any kind; and where they are inverted, they must be accelerated; a small resistance producing, in both cases, a considerable difference in the time of high water.

Where a considerable tide is observed in the middle of a limited portion of the sea, it must be derived from the effect of the elevation or depression of the ocean in its neighbourhood; and such derivative tides are probably combined in almost all cases with the oscillations belonging to each particular branch of the sea. Mr. Laplace supposes that the tides, which are observed in the most exposed European harbours, are produced almost entirely by the transmission of the effect of the main ocean, in about a day and a half; but this opinion does not appear to be justified by observation; for the interval between the times of the high water belonging to the same tide, in any two places between Brest and the Cape of Good Hope, has not been observed to exceed about twelve hours at most; nor can we trace a greater difference by comparing the state of the tides at the more exposed situations of St. Helena, the Cape Verd Islands, the Canaries, the Madeiras, and the Azores, which constitute such a succession as might be expected to have indicated the progress of the principal tide, if it had been such as Mr. Laplace supposes. The only part of the ocean, which we can consider as completely open, lies to the south of the two great continents, chiefly between the latitudes 30° and 70° south, and the original tide, which happens in this widely extended ocean, where its depth is sufficiently uniform, must take place, according to the theory which has been advanced, at some time before the sixth lunar hour. It sends a wave into the Atlantic, which is perhaps 12 or 13 hours in its passage to the coast of France, but certainly not more. This tide, which would happen at the sixth lunar hour after the moon's transit, if there were no resistance, is probably so checked by the resistance, that the water begins to subside about the fourth, and in some seas even somewhat earlier, although in others it may follow more nearly its natural course. There is scarcely a single instance which favours the supposition of the time of high water in the open sea being within an hour of the moon's southing, as it must be if the depth were very great: so that neither the height of the tides nor the time of high water will allow us to suppose the sea any where quite so deep as 4 miles.

The tide entering the Atlantic appears to advance northwards at the rate of about 500 miles an hour, corresponding to a depth of about 3 miles, so as to reach Sierra Leone at the 8th hour after the moon's southing; this part of Africa being not very remote from the meridian of the middle of the south Atlantic ocean, and having little share in the primitive tides of that ocean. The southern tide seems then to pass by Cape Blanco and Cape Bojador, to arrive at Gibraltar at the 13th hour, and to unite its effects with those of other tides at various parts of the coast of Europe.

We may therefore consider the Atlantic as a detached sea about 3500 miles broad and 3 miles deep; and a sea of these dimensions is susceptible of tides considerably larger than those of the ocean, but how much larger we cannot determine without more accurate measures. These tides would happen on the European coasts, if there were no resistance, a little less than 5 hours after the moon's southing, and on the coast of America, a little more than seven hours after; but the resistance opposed to the motion of the sea may easily accelerate the time of high water in both cases about two hours, so that it may be a little before the third hour on the western coasts of Europe and of Africa, and before the fifth on the most exposed parts of the eastern coast of America; and in the whole of the Atlantic, this tide may be combined more or less both with the general southern tide, and with the partial effects of local elevations or depressions of the bottom of the sea, which may cause irregularities of various kinds. The southern tide is, however, probably less considerable than has sometimes been supposed, for, in the latitudes in which it must originate, the extent of the elevation can only be half as great as at the equator; and the Islands of Kergulen's Land and South Georgia, in the latitudes of about 50° and 55°, have their tides delayed till the 10th and 11th hours, apparently because they receive them principally from distant parts of the ocean, which are nearer to the equator.

4 C

On the western coasts of Europe, from Ireland to Cadiz, on those of Africa, from Cape Coast to the Cape of Good Hope, and on the Coast of America, from California to the streights of Magellan, as well as in the neighbouring islands, it is usually high water at some time between two and four hours after the moon's southing; on the eastern coast of South America between four and six, on that of North America between seven and eleven; and on the eastern coasts of Asia and New Holland between four and eight. The Society islands are perhaps too near the middle of the Pacific ocean to partake of the effects of its primitive tide, and their tide, being secondary, is probably for this reason a few hours later. At the Almirantes, near the eastern coast of Africa, the tide is at the sixth hour; but there seem to be some irregularities in the tides of the neighbouring islands.

The progress of a tide may be very distinctly traced from its source in the ocean into the narrow and shallow branches of the sea which constitute our channels. Thus the tide is an hour or two later at the Scilly Islands than in the Atlantic, at Plymouth three, at Cork, Bristol, and Weymouth four, at Caen and Havre six, at Dublin and Brighthelmstone seven, at Boulogne and Liverpool eight, at Dover near nine, at the Nore eleven, and at London bridge twelve and a half. Another portion appears to proceed round Ireland and Scotland into the North Sea; it arrives from the Atlantic at Londonderry in about three hours, at the Orkneys in six, at Aberdeen in eleven, at Leith in fourteen, at Leostoffe in twenty, and at the Nore in about twenty four, so as to meet there the subsequent tide coming from the south. From the time occupied by the tide in travelling from the mouth of the English channel to Boulogne, at the rate of about 50 miles an hour, we may calculate that the mean depth of the channel is about 28 fathoms, independently of the magnitude of the resistances of various kinds to be overcome, which require us to suppose the depth from 30 to 40 fathoms. In the great river of Amazons, the effects of the tides are still sensible at the streights of Pauxis, 500 miles from the sea, after an interval of several days spent in their passage up: for the slower progressive motion of the water no more impedes the progress of a wave against the stream, than the velocity of the wind prevents the transmission of sound in a contrary direction. (Plate XXXVIII. Fig. 521.)

Such are the general outlines of the lunar tides; they are, however, liable to a great variety of modifications, besides their combination with the tides produced by the sun. When the moon is exactly over the equator, the highest part of the remoter, or inferior, as well as of the nearer or superior tides, passes also over the equator, and the effect of the tide in various latitudes decreases gradually from the equator to the pole, where it vanishes; but when the moon has north or south declination, the two opposite summits of the spheroid describe parallels of latitude, remaining always diametrically opposite to each other. Hence the two successive tides must be unequal at every place except the equator, the greater tide happening when the nearer elevation passes its meridian: and the mean between both is somewhat smaller than the equal tides which happen when the moon passes the equator. This inequality is, however, much less considerable than it would be if the sea assumed at once the form of the spheroid of equilibrium; and the most probable reasons for this circumstance, are, first, that our tides are partly derived from the equatorial seas; secondly, that the effects of a preceding tide are in some measure continued so as to influence the height of a succeeding one; and, thirdly, that the tides of a narrow sea are less affected by its latitude than those of a wide ocean. The height of the sea at low water is the same whatever the moon's declination may be. There is also a slight difference in the tides, according to the place of the moon's nodes, which allows her declination to be greater or less, and this difference is most observable in high latitudes, for instance, in Iceland; since, in the neighbourhood of the poles, the tides depend almost entirely on the declination.

In all these cases, the law of the elevation and depression of each tide may be derived, like that of the vibrations of a pendulum and of a balance, from the uniform motion of a point in a circle. Thus, if we conceive a circle to be placed in a vertical plane, having its diameter equal to the whole magnitude of the tide, and touching the surface of the sea at low water, the point, in which the surface meets the circumference of the circle, will advance with a uniform motion, so that if the circle be divided into 12 parts, the point will pass over each of these parts in a lunar hour. It sometimes happens, however, in confined situations, that the rise and fall of the water deviates considerably from this law, and the tide rises somewhat more rapidly than it falls; and in rivers, for example in the Severn, the tide frequently advances suddenly with a head of several feet in height. These deviations probably depend on the magnitude of the actual displacement of the water, which in such cases bears a considerable proportion to the velocity of the tide, while in the open ocean a very minute progressive motion is sufficient to produce the whole elevation. The actual progress of the tides may be most conveniently observed, by means of a pipe descending to some distance below the surface, so as to be beyond the reach of superficial agitations, and having within it a float, carrying a wire, and indicating the height of the water on a scale properly divided.

We have hitherto considered the tides so far only as they are occasioned by the moon; but in fact the tides, as they actually exist, depend also on the action of the sun, which produces a series of effects precisely similar to those of the moon, although much less conspicuous, on account of the greater distance of the sun, the solar tide being only about two fifths of the lunar. These tides take place independently of each other, nearly in the same degree as if both were single; and the combination resulting from them is alternately increased and diminished, accordingly as they agree, or disagree, with respect to the time of high water at a given place; in the same manner as if two series of waves, equal among themselves, of which the breadths are as 29 to 30, be supposed to pass in the same direction over the surface of a fluid, or if two sounds similarly related be heard at the same time, a periodical increase and diminution of the joint effect will in either case be produced. Hence are derived the spring and neap tides, the effects of the sun and moon being united at the times of conjunction and opposition, or of the new and full moon, and opposed at the quadratures, or first and last quarters. The high tides at the times of the equinoxes are produced by the joint operation of the sun and moon, when both of them are so situated as to act more powerfully than elsewhere.

The lunar tide being much larger than the solar tide, it must always determine the time of high and low water, which, in the spring and neap tides, remains unaltered by the effect of the sun; so that in the neap tides, the actual time of low water is that of the solar high water; but at the intermediate times, the lunar high water is more or less accelerated or retarded. The progress of this alteration may easily be traced by means of a simple construction. If we make a triangle of which two of the sides are two feet and five feet in length, the external angle which they form being equal to twice the distance of the luminaries, the third side will show precisely the magnitude of the compound tide, and the halves of the two angles opposite to the first two sides the acceleration, or retardation, of the times of high water belonging to the separate tides respectively. Hence it appears that the greatest deviation of the joint tide from the lunar tide amounts to 11° 48′ in longitude, and the time corresponding, to 47 minutes, supposing the proportion of the forces to remain always the same; but in fact the forces increase in proportion as the cubes of the distances of their respective luminaries diminish, as well as from other causes; and in order to determine their joint effects, the lengths of the sides of the triangle must be varied accordingly. In some ports, from a combination of circumstances in the channel, by which the tides reach them, or in the seas, in which they originate, the influence of the sun and moon may acquire a proportion somewhat different from that which naturally belongs to them: thus at Brest, the influence of the moon appears to be three times as great as that of the sun; when it is usually only twice and a half as great. (Plate XXXVIII. Fig. 522.)

The greatest and least tides do not happen immediately at the times of the new and full moon, but at least two, and commonly three tides after, even at those places which are most immediately exposed to the effects of the general tide of the ocean. The theory which has been advanced will afford us a very satisfactory reason for this circumstance; the resistance of fluids in general is as the square of the velocity, consequently it must be much greater for the lunar than for the solar tide, in proportion to the magnitude of the force, and the acceleration of the lunar tide produced by this cause must be greater than that of the solar; hence it may happen that when the lunar tide occurs two or three hours after the transit of the moon, the solar tide may be three or four hours after that of the sun, so as to be about an hour later, at the times of conjunction and opposition, and the tides will be highest when the moon passes the meridian about an hour after the sun; while at the precise time of the new and full moon, the lunar tide will be retarded about a quarter of an hour by the effect of the solar tide.

The particular forms of the channels, through which the tides arrive at different places, produce in them a great variety of local modifications; of which the most usual is, that from the convergence of the shores of the channels, the tides rise to a much greater height than in the open sea. Thus at Brest the height of the tides is about 20 feet, at Bristol 30, at Chepstow 40, at St. Maloes 50; and at Annapolis Royal, in the Bay of Fundy, as much sometimes as 100 feet; although perhaps in some of these cases a partial oscillation of a limited portion of the sea may be an immediate effect of the attraction of the luminary. In the Mediterranean the tides are generally inconsiderable, but they are still perceptible; at Naples they sometimes amount to a foot, at Venice to more than two feet, and in the Euripus, for a certain number of days in each lunation, they are very distinctly observable, from the currents which they occasion. In the West Indies, also, and in the gulf of Mexico, the tides are less marked than in the neighbouring seas, perhaps on account of some combinations derived from the variations of the depth of the ocean, and from the different channels by which they are propagated.

In order to understand the more readily the effects of such combinations, we may imagine a canal, as large as the river of Amazons, to communicate at both its extremities with the ocean, so as to receive at each an equal series of tides, passing towards the opposite extremity. If we suppose the tides to enter at the same instant at both ends, they will meet in the middle, and continue their progress without interruption: precisely in the middle the times of high and low water belonging to each series will always coincide, and the effects will be doubled; and the same will happen at the points, where a tide arrives from one extremity at the same instant that an earlier or a later tide comes from the other; but at the intermediate points the effects will be diminished, and at some of them completely destroyed, where the high water of one tide coincides with the low water of another. The tides at the port of Batsha in Tonkin have been explained by Newton from considerations of this nature. In this port there is only one tide in a day; it is high water at the sixth lunar hour, or at the moon's setting, when the moon has north declination, and at her rising, when she has south declination; and when the moon has no declination there is no tide. In order to explain this circumstance, we may represent the two unequal tides which happen in succession every day, by combining with two equal tides another tide, independent of them, and happening only once a day; then, if a point be so situated in the canal which we have been considering, that the effects of the two equal semidiurnal tides may be destroyed, those of the daily tides only will remain to be combined with each other; and their joint result will be a tide as much greater than either, as the diagonal of a square is greater than its side; the times of high and low water being intermediate between those which belong to the diurnal tides considered separately. Thus, in the port of Batsha, the greater tide probably arrives at the third lunar hour directly from the Pacific ocean, and at the ninth from the gulf of Siam, having passed between Sumatra and Borneo; so shat the actual time of high water is at the sixth lunar hour. The magnitude of this compound tide is by no means inconsiderable; it sometimes amounts to as much as 13 feet. (Plate XXXVIII. Fig. 523, 524.)

Besides the variations in the height of the sea, which constitute the tides, the attractions of the sun and moon are also supposed to occasion a retardation in its rotatory motion, in consequence of which it is left a little behind the solid parts of the earth; and a current is produced, of which the general direction is from east to west. This current comes from the Pacific and Indian oceans, round the Cape of Good Hope, along the coast of Africa, then crosses to America, and is there divided and reflected southwards towards the Brazils, and northwards into the Gulf stream, which travels round the gulf of Mexico, and proceeds north eastwards into the neighbourhood of Newfoundland, and then probably eastwards and south eastwards once more across the Atlantic. It is perhaps on account of these currents that the Red Sea is found to be about 25 feet higher than the Mediterranean: their direction may possibly have been somewhat changed in the course of many ages, and with it the level of the Mediterranean also; since the floor of the cathedral at Ravenna is now several feet lower with respect to the sea than it is supposed to have been formerly, and some steps have been found in the rock of Malta, apparently intended for ascending it, which are at present under water.

The atmosphere is also liable to elevations and depressions analogous to those of the sea, and perhaps these changes may have some little effect on the winds and on the weather; but their influence must be very inconsiderable, since the addition of two or three feet to the height of the atmosphere at any part can scarcely be expected to be perceptible. The height of an aerial tide must be very nearly the same with the observed height of the principal tides of the sea; and the variation of atmospherical pressure, which is measured by the difference between the actual form and the spheroid of equilibrium, must be equivalent to the weight of a column of about 10 feet of air, or only $\frac{1}{100}$ of an inch of mercury. A periodical variation five times as great as this has indeed been observed near the equator, where the state of the atmosphere is the least liable to accidental disturbances; but this change cannot in any degree be referred to the effect of the moon's action, since it happens always about the same hour of the day or night. The atmosphere is also affected by a general current from east to west, like that of the sea, and there is reason, from astronomical observations, to suppose that a similar circumstance happens in the atmosphere of Jupiter, on account of the actions of his satellites, which must be considerably more powerful than that of the moon.

LECTURE XLVIII.

ON THE HISTORY OF ASTRONOMY.

WE have now taken a general view of the most striking phenomena of the universe at large, of the great features of the solar system, and of the peculiarities of the planet which we inhabit, with respect both to its solid and to its fluid parts. All these are departments of astronomy, and we shall conclude our examination of the subject with a summary of the history of the science, principally extracted and abridged from Laplace's Exposition du systême du monde.

In all probability the astronomy of the earliest ages was confined to observations of the obvious motions and eclipses of the sun and moon, the rising, setting, and occultations of the principal stars, and the apparent motions of the planets. The progress of the sun was followed, by remarking the stars as they were lost in the twilight, and perhaps also by the variation of the length of the shadow of a detached object, observed at the time of the day when it was shortest. In order to recognise the fixed stars, and their different motions, the heavens were divided into constellations; and twelve of these occupied the zone denominated the zodiac, within the limits of which the sun and planets were always found.

The entrance of the sun into the constellation aries, or the ram, denoted, in the time of Hipparchus, the beginning of the spring; and as the season advanced, the sun continued his progress through the bull, the twins, and the other signs in order; some of which appear to have been denominated from their relation to the agriculture and to the climates of the countries in which they were imagined, and others from the celestial phenomena attending the sun's passage through them; the crab, for example, denoting his retrograde motion after the time of the solstice, and the balance the equality of day and night at the autumnal equinox. But the motion of the equinoctial points having changed in some degree the course of the seasons with regard to the stars, the signs of the ecliptic, by which the places of the sun and planets are described, no longer coincide precisely with the constellations of the zodiac from which they derive their names.

The most ancient observations of which we are in possession, that are sufficiently accurate to be employed in astronomical calculations, are those made at Babylon in the years 719 and 720 before the Christian era, of three eclipses of the moon. Ptolemy, who has transmitted them to us, employed them for determining the period of the moon's mean motion, and, therefore, had probably none more ancient on which he could depend. The Chaldeans, however, must have made a long series of observations before they could discover their Saros or lunar period of $6585\frac{1}{3}$ days, or about 18 years, in which, as they had learnt at a very early time, the place of the moon, her node, and apogee, return nearly to the same situation with respect to the earth and sun, and of course a series of nearly similar eclipses recurs. The observations attributed to Hermes indicate a date seven hundred years earlier than those of the Babylonians, but their authenticity appears to be extremely doubtful.

The Egyptians were very early acquainted with the length of the year, as consisting nearly of 365 days and a quarter, and they derived from it their Sothic period of 1460 years, containing 365 days each. The accurate correspondence of the faces of their pyramids with the points of the compass is considered as a proof of the precision of their observations: but their greatest merit was the discovery that Mercury and Venus revolve round the sun, and not round the earth, as it had probably been before believed: they did not, however, suppose the same of the superior planets. (Plate XXXVIII. Fig. 525, 526.)

In Persia and in India, the origin of astronomy is lost in the darkness which envelopes the early history of those countries. We find the annals of no country so ancient and so well authenticated as those of China, which are confirmed by an incontestable series of historical monuments. The regulation of the calendar, and the prediction of eclipses, were regarded in this country as important objects, for which a mathematical tribunal was established at a very early period. But the scrupulous attachment of the Chinese to their ancient customs, extending itself even to their astronomy, has impeded its progress, and retained it in a state of infancy. The Indian tables indicate a much higher degree of perfection in the early state of the science, than it had attained in China; but we have every reason to believe that they are not of very remote antiquity. "Here", says Mr. Laplace, who must be allowed to be free from prejudices in favour of established opinions, "I am sorry to be obliged to differ from an illustrious philosopher, Mr. Bailly, who, after having distinguished his career by a variety of labours useful to the sciences, and to mankind at large, fell a victim to the most sanguinary tyranny that ever disgraced a civilised nation. The Indian tables are referred to two principal epochs, which are placed the one 3102 years before Christ, the other 1491. These are connected by the mean motions, and not the true motions, of the sun, the moon, and the planets; so that one of the epochs must necessarily be fabulous. The celebrated author, who has been mentioned, has sought to establish, in his treatise on Indian astronomy, that the former of these epochs is founded on observation. But if we calculate from our own improved tables, we shall find that the general conjunction of the sun, moon, and planets, which the Indian tables suppose, in reality never happened, although it may be deduced, according to those tables, by ascending from the later series. The equation of the sun's centre, depending on the eccentricity of the earth's orbit, appears indeed to indicate a still higher antiquity; but its magnitude, as deduced from eclipses, must have been affected by a contrary error with respect to the moon's place: and the determination of the mean motion of the moon seems to make it probable that these tables are even of a later date than Ptolemy."

In astronomy, as well as in other sciences, the Greeks were the disciples of the Egyptians; they appear to have divided the stars into constellations 13 or 1400 years before Christ. Newton attributes this arrangement to Chiron, and he supposes that he made the middle of the constellations correspond to the beginning of the respective signs. But until the time of the foundation of the school of Alexandria, the Greeks treated astronomy as a science purely speculative, and indulged themselves in the most frivolous conjectures respecting it. It is singular that amidst the confusion of systems heaped up on each other, without affording the least information to the mind, it should never have occurred to men of so great talents, that the only way to become accurately acquainted with nature, is to institute experimental inquiries throughout her works.

Thales of Miletus, who was born in the year 640 before Christ, having travelled and studied in Egypt, founded, on his return, the Ionian school of philosophy, in which he taught the sphericity of the earth, and the obliquity of the ecliptic with respect to the equator. He also explained the true causes of eclipses, which he was even able to foretel, unquestionably by means of the information that he had obtained from the Egyptian priests.

Pythagoras of Samos was born 590 years before Christ; he probably profited by the information which Thales had acquired, and travelled also into Egypt for his further improvement. It is conjectured that he was acquainted with the diurnal and annual motions of the earth, but he did not publicly profess the true system of the world. It was taught after his death, by his disciple Philolaus, about the year 450, as well as by Nicetas, and by others of the school. They considered all the planets as revolving round the sun, and as inhabited globes; and they understood that the comets were only eccentric planets. Some time after this, the lunar period of Meto was publicly made known at the Olympic games, and was universally adopted as the basis of the calendar. (Plate XXXVIII. Fig. 527.)

The next occurrence which deserves to be noticed, with respect to astronomy is the foundation of the school of Alexandria, which was the first source of accurate and continued observations. Upon the death of Alexander, and the subsequent division of his empire, the province of Egypt fell to the lot of Ptolemy Soter; a prince whose love of science, and whose munificence towards its professors, attracted to his capital a great number of learned men from various parts of Greece. His son, Ptolemy Philadelphus, continued and increased the benefits conferred on them by his father, and built the magnificent edifice which contained, together with the celebrated library, collected by Demetrius Phalereus, an observatory, furnished with the necessary books and instruments. The first astronomers, who were appointed to occupy this building, were Aristyllus and Timocharis; they flourished about 300 years before Christ, and observed with accuracy the places of the principal stars of the zodiac. Aristarchus of Samos was the next: he imagined a method of finding the sun's distance, by observing the portion of the moon's disc that is enlightened, when she is precisely in the quadrature, or 90° distant from the sun; and although he failed in his attempt to determine the sun's distance with accuracy, yet he showed that it was much greater than could at that time have been otherwise imagined; and he asserted that the earth was but as a point in comparison with the magnitude of the universe. His estimation of the distance of the sun is made by Archimedes the basis of a calculation of the number of grains of sand that would be contained in the whole heavenly sphere, intended as an illustration of the powers of numerical reckoning, and of the utility of a decimal system of notation, which was the foundation of the modern arithmetic.

Eratosthenes, the successor of Aristarchus, is known by his observation of the obliquity of the ecliptic, and his measurement of a certain portion of the earth's circumference; the whole of which he determined to be 250 000 stadia; but the length of his stadium is uncertain. Ptolemy, calculating perhaps from the same measures, or from some others still more ancient, calls it 180 000; which, if the stadium is determined from the Nilometer at Cairo, and from the base of the pyramid, is within one thousandth part of the truth, the length of the base of the pyramid being equal to 400 Egyptian cubits, or to 729 feet 10 inches English.

Hipparchsu of Bithynia flourished at Alexandria about the year 140 before Christ. Employing the observations of Timocharis, and comparing them with his own, he discovered the precession of the equinoxes. He also observed that the summer was 9 days longer than the winter, and that the solstices divided each of these seasons a little unequally. In order to explain this, Hipparchus supposed the sun to move uniformly in an eccentric circle, the distance of its centre from that of the earth being $\frac{1}{24}$ of the radius, and placed the apogee in the sixth degree of gemini. Probably the annual equation of the moon, which has some influence on the time of eclipses, was the cause of his making the eccentricity too great; had he assumed it but one fifth part less, the supposition would have represented the sun's place with tolerable accuracy. Hipparchus appears to have been the first that employed astronomical observations for determining the latitudes and longitudes of places.

The interval of three centuries, which elapsed between Hipparchus and Ptolemy, offers us little that is remarkable in the progress of astronomy, except the reformation of the calendar, by Julius Caesar, who was assisted in making the arrangement by Sosigenes, an astronomer of the same school that gave birth to all the preceding discoveries, as well as to the improvements of Ptolemy. This great astronomer was born at Ptolemais in Egypt, and flourished about the year 140 of our era. He continued the vast project, begun by Hipparchus, of reforming the whole science which he studied. He discovered the evection of the moon, or the change of her velocity, occasioned by the position of the apogee with respect to the sun; he determined the quantity of this equation with great precision; and in order to represent it, he supposed the moon to perform a subordinate revolution in an epicycle, or a smaller circle, of which the centre was carried round in the line of the general orbit, which he considered as an eccentric circle. This mode of approximation is exceedingly ingenious; it is said to have been the invention of Apollonius of Perga, the mathematician, and although it sometimes becomes complicated, yet it is very convenient for calculation; and it may be employed with advantage in the representation of the planetary motions by machinery. Ptolemy adopted the most ancient opinion with respect to the solar system, supposing all the heavenly bodies to revolve round the earth; the moon being nearest, then Mercury, Venus, the Sun, Mars, Jupiter, and Saturn. This opinion had long been the most general, although some astronomers had placed Mercury and Venus at greater distances than the sun, and some attributed to the earth a diurnal motion only; but the doctrine of the Pythagoreans appears to have been wholly exploded or forgotten. Ptolemy determined the quantity of the precession of the equinoxes from a comparison of his own observations with those of Hipparchus; but he made it smaller than the truth; and he probably formed his table of the places of the stars by applying this erroneous correction to the tables of Hipparchus, in order to accommodate them to his own time. Both these errors may, however, be otherwise explained, by supposing him to have followed Hipparchus in the length of the tropical year, which being somewhat too great, caused an error in the calculation of the sun's place, to which that of the stars was referred; but upon this supposition, he must also have been mistaken in three observations of the place of the equinoctial points. Ptolemy's principal work is his mathematical system of astronomy, which was afterwards called the great syntax or body of astronomy, and is at present frequently quoted by the Arabic name Almagest. He also wrote a treatise on optics, in which the phenomena of atmospherical refraction are described, and which is extant in manuscript in the National library at Paris. (Plate XXXVIII. Fig. 528.)

Ptolemy was the last as well as the greatest of the Alexandrian astronomers, and the science made no further progress till the time of the Arabians. The first of these was Almamoun, was the son of the celebrated Aaron Reschid; he reigned at Bagdad in 814, and having conquered the Greek emperor, Michael the Third, he made it a condition of peace, that a copy of the works of each of the best Greek authors should be delivered to him; and among them were the works of Ptolemy, of which he procured an Arabic translation. Almamoun also observed the obliquity of the ecliptic, and measured the length of a degree in the plains of Mesopotamia.

Among the astronomers protected by this prince and his successors, Albategni was the most eminent. He ascertained with great accuracy, in 880, the eccentricity of the solar motion, and discovered the change of the place of the sun's apogee, or of the earth's aphelion.

Ibn Junis made his observations at Cairo, about the year 1000; he was a very assiduous astronomer, and determined the length of the year within 2 seconds of the truth. At this time the Arabians were in the habit of employing, in their observations, the vibrations of a pendulum.

The Persians soon after applied themselves to astronomy; and in the eleventh century they invented the approximation of reckoning 8 bissextiles in 33 years, which was afterwards proposed by Dominic Cassini as an improvement of the Gregorian calendar. The most illustrious of this nation was Ulugh Beigh, who observed in his capital Samarcand, about the year 1437, with very elaborate instruments. In the mean time Cocheouking had made in China, some very accurate observations, which are valuable for the precision with which they ascertain the obliquity of the ecliptic: their date is about 1278.

It was not long after the time of Ulugh Beigh, that Copernicus laid the foundation of the more accurate theories which modern improvements have introduced into astronomy. Dissatisfied with the complicated hypotheses of the Ptolemaean system, he examined the works of the ancients, in quest of more probable opinions. He found from Cicero that Nicetas and other Pythagoreans had maintained, that the sun is placed in the centre of the system, and that the earth moves round him in common with the other planets. He applied this idea to the numerous observations which the diligence of astronomers had accumulated, and he had the satisfaction to find them all in perfect conformity with this theory. He quickly discarded the Ptolemaean epicycles, imagined in order to explain the alternations of the direct and retrograde motions of the planets; in these remarkable phenomena, Copernicus saw nothing but the consequences necessarily produced by the combination of the motions of the earth and planets round the sun; and from a minute examination of these circumstances he calculated the relative distances of the planets from the sun, which till then had remained unknown. In this system, every thing had the marks of that beautiful simplicity which pervades all the works of nature, and which, when once understood, carries with itself sufficient evidence of its truth. Copernicus was born at Thorn, in Polish Prussia, in the year 1475; he studied in Italy; he taught mathematics at Rome, and afterwards settled on a canonicate at Frauenberg, where, in 36 years of retirement and meditation, he completed his work on the celestial revolutions, which was scarcely published when he died.

About this time, William the Fourth, Landgrave of Hesse Cassel, not only enriched astronomy by his own observations, but also exerted his influence with Frederic, King of Denmark, to obtain his patronage for the celebrated Tycho Brahe. Frederic agreed to give him the little island Huen, at the entrance of the Baltic, where Tycho built his observatory of Uraniburg, and, in a period of 21 years, made a prodigious collection of accurate observations. After the death of his patron, his progress was impeded, and he sought an establishment at Prague, under the emperor Rudolph. Here he died soon after, at the age of 55. Struck with the objections made to the system of Copernicus, principally such as were deduced from a misinterpretation of the scriptures, he imagined a new theory, which, although mechanically absurd, is still astronomically correct; for he supposed the earth to remain at rest in the centre, the stars to revolve round it, together with the sun and all the planets, in a sidereal day, and the sun to have, besides, an annual motion, carrying with him the planets in their orbits. Here the apparent or relative motions are precisely the same as in the Copernican system; the argument that Tycho Brahe drew from the scriptures in favour of his theory was, therefore, every way injudicious; for it is not to be imagined that any thing but relative motion or rest could be intended in the scriptures, when the sun is said to move, or to stand still. But in the Copernican system, there was an evident regularity in the periods of all the planets, that of the earth being longer than that of Venus, and shorter than that of Mars, which were the neighbouring planets on each side; and when Tycho imagined the sun to move round the earth, this analogy was entirely lost. Tycho Brahe was the discoverer of the variation and of the annual equation of the moon, the one being an irregularity in its velocity, dependent on its position with respect to the sun, the other a change in the magnitude of all the perturbations produced by the sun, dependent on his distance from the earth. (Plate XXXVIII. Fig. 529.)

Kepler was the pupil and assistant of Tycho, whose observations were the basis of his important discoveries: he succeeded him in his appointments at Prague, and enjoyed the title of Imperial Mathematician. Adopting the Copernican system, which was then becoming popular, he proceeded to examine the distances of the celestial bodies from each other at various times; and after many fruitless attempts to reconcile the places of the planets with the supposition of revolutions in eccentric circles, at last discovered that their orbits are ellipses, and demonstrated, chiefly from his observations on the planet Mars, that the revolving radius, or the line drawn from the sun to the planet, always describes equal areas in equal times. By comparing the periods and the mean distances of the different planets with each other, he found, after 17 years calculation, that the squares of the times of revolution are always proportional to the cubes of the mean distances from the sun.

Kepler died in 1630: before his death he had the satisfaction of applying his theory to the motions of the satellites of Jupiter, which, as well as the phases of Venus, and the spots of the sun, had lately been discovered in Italy by the telescopic observations of Galileo. This great man, celebrated as well for his theory of projectiles, as for his zealous defence of the Copernican system, was born at Pisa in 1564, and lived to the age of 78, full of that enthusiasm which made him despise the threats of the Inquisition, and submit patiently to its persecutions. He died in 1642, the year in which Newton was born.

The invention of logarithms, by Baron Napier, requires to be noticed for its importance to practical astronomy, and the laborious observations of Hevelius deserve also to be mentioned with commendation. The discoveries of the form of the ring of Saturn, and of one of his satellites, by Huygens, and of four more, together with the belts and rotation of Jupiter, by Dominic Cassini, were among the early improvements derived from the introduction of the telescope. But, without dwelling on any of these subjects, we hasten to the establishment of the system of gravitation, which has immortalised the name of Newton, and done unrivalled honour to the country that gave him birth.

The mutual attraction of all matter seems to have been suspected by the Epicureans, but Lucretius never speaks of it in such terms as are sufficient to convey by any means a distinct idea of a reciprocal force. Gregory, in the preface of his Astronomy, has endeavoured to prove that Pythagoras must have been acquainted even with the law of the decrease of gravitation; and Lalande appears to assent to his arguments; but they rest only on the bare possibility that Pythagoras might have deduced an analogy from the tension of chords, which we have no reason to suppose that he even completely understood: and this merely because he fancifully imagined, that there was a correspondence between the planets and the strings of a lyre. But the nature of gravitation had long been in some measure suspected; Plutarch had asserted that the moon is retained by it in her orbit, like a stone in a sling; and Bacon, Copernicus, Kepler, Fermat, and Roberval were aware of its efficacy. Bacon, in his Novum organum, calls the descent of heavy bodies the motion of "general congregation", and attributes the tides to the attraction of the moon. Kepler mentions also the perfect reciprocality of the action of gravitation, and considers the lunar irregularities as produced by the attraction of the sun. But our most ingenious countryman, Dr. Hooke, was still more decided in attributing the revolutions of the planets to the combination of a projectile motion with a centripetal force; he expresses his sentiments on the subject very clearly in his Attempt to prove the motion of the earth, published in 1674, and had his skill in mathematics been equal to his practical sagacity, he would probably have completed, or at least have published, the discovery before his great cotemporary.

It must be confessed that Newton's good fortune was equal to his talents and his application; for had he lived earlier, he might probably have confined his genius to speculations purely mathematical; had he been later, his discoveries in natural philosophy might have been anticipated by others; and yet Newton would perhaps have improved still more on their labours than they have done on his. It was in 1676, when he was 34 years old, that he first demonstrated the necessary connexion of the planetary revolutions in elliptic orbits, with an attractive force varying inversely as the square of the distance. But he had collected the law of the force, from the discoveries of Kepler respecting the periods of the different planets, some time before 1671, as he asserts to Dr. Halley, and, to the best of his recollection, about 1668, although in his Principia he allows, with the most laudable candour, to Wren, Hooke, and Halley, the merit of having made the same discovery, without any connexion with each other's investigations, or with his own. The manner, in which Newton was led to attend particularly to the subject, is thus related by Pemberton, in the preface to his View of Sir Isaac Newton's philosophy.

"The first thoughts," says Pemberton, "which gave rise to his Principia, he had, when he retired from Cambridge in 1666, on account of the plague. As he sat alone in a garden, he fell into a speculation on the power of gravity: that as this power is not found sensibly diminished at the remotest distance from the centre of the earth, to which we can rise, neither at the tops of the loftiest buildings, nor even on the summits of the highest mountains; it appeared to him reasonable to conclude, that this power must extend much further than was usually thought; why not as high as the moon?

said he to himself; and if so, her motion must be influenced by it; perhaps she is retained in her orbit thereby. However, though the power of gravity is not sensibly weakened in the little change of distance, at which we can place ourselves from the centre of the earth; yet it is very possible that so high as the moon this power may differ much in strength from what it is here. To make an estimate, what might be the degree of this diminution, he considered with himself, that if the moon be retained in her orbit by the force of gravity, no doubt the primary planets are carried round the sun by the like power. And by comparing the periods of the several planets with their distances from the sun, he found, that if any power like gravity held them in their courses, its strength must decrease in the duplicate proportion of the increase of distance. This he concluded by supposing them to move in perfect circles concentrical to the sun, from which the orbits of the greatest part of them do not much differ. Supposing, therefore, the power of gravity, when extended to the moon, to decrease in the same manner, he computed whether that force would be sufficient to keep the moon in her orbit. In this computation, being absent from books, he took the common estimate in use among geographers and our seamen, before Norwood had measured the earth, that 60 English miles were contained in one degree of latitude on the surface of the earth. But as this is a very faulty supposition, each degree containing about 69½ of our miles, his computation did not answer expectation; whence he concluded that some other cause must at least join with the action of the power of gravity on the moon. On this account he laid aside for that time any further thoughts upon this matter. But some years after, a letter, which he received from Dr. Hooke, put him on inquiring what was the real figure, in which a body let fall from any high place descends, taking the motion of the earth round its axis into consideration. Such a body, having the same motion, which by the revolution of the earth the place has from whence it falls, is to be considered as projected forwards, and at the same time drawn down to the centre of the earth. This gave occasion to his resuming his former thoughts concerning the moon; and Picart, in France, having lately measured the earth, by using his measures, the moon appeared to be kept in her orbit purely by the power of gravity; and consequently, that this power decreases as you recede from the centre of the earth, in the manner our author had formerly conjectured. Upon this principle he found the line described by a falling body to be an ellipsis, the centre of the earth being one focus. And the primary planets moving in such orbits round the sun, he had the satisfaction to see, that this inquiry, which he had undertaken merely out of curiosity, could be applied to the greatest purposes. Hereupon he composed near a dozen propositions relating to the motion of the primary planets about the sun. Several years after this, some discourse he had with Dr. Halley, who at Cambridge made him a visit, engaged Sir Isaac Newton to resume again the consideration of this subject; and gave occasion to his writing the treatise which he published under the title of Mathematical principles of natural philosophy. This treatise, full of such variety of profound inventions, was composed by him, from scarce any other materials than the few propositions before mentioned, in the space of one year and a half."

The astronomers of Great Britain have not been less diligent in the practical, than successful in the theoretical part of the science. The foundation of the observatory at Greenwich was laid in 1675, some years before the completion and publication of the discoveries of Newton. It is with the erection of this edifice that the modern refinements in practical astronomy may be said to have commenced; its immediate object was to assist in the perfection of the science of navigation, and the series of observations, which have been made in it, has afforded an invaluable fund of materials to astronomers of every country. A reward had been proposed, more than half a century before, by Philip the Third, of Spain, for the discovery of a mode of determining the longitude of a ship at sea; and the states of Holland had followed his example: a large reward was also offered by the French government in the minority of Louis the Fifteenth. In 1674, Mr. St. Pierre, a Frenchman, had undertaken to determine the longitude of a place from observations of the moon's altitude, and King Charles the Second had been induced to appoint a commission to examine his proposals. Mr. Flamsteed was consulted by the commissioners, and was added to their number: he showed the disadvantages of the method proposed by Mr. St. Pierre, and the inaccuracy of the existing tables of the lunar motions, as well as of the catalogues of the places of the stars, but expressed his opinion, that, if the tables were improved, it would be possible to determine the longitudes of places with sufficient accuracy by lunar observations. The king, being informed of Flamsteed's representations, is said to have replied with earnestness, that he "must have the places of the stars anew observed, examined, and corrected, for the use of his seamen"; upon this Flamsteed was appointed Astronomer Royal, with a salary of £100 a year, and it was proposed to have an observatory built either in Hyde Park, or at Chelsea college; but, upon Sir Christopher Wren's recommendation, the situation of Greenwich Park was preferred.

In the year 1714, the British Parliament offered £20 000 for a determination of the longitude of a ship at sea, without an error of 30 miles, and a smaller sum for a less accurate method, appointing at the same time a Board of Longitude for the examination of the methods which might be proposed. Under this act several rewards were assigned, and in 1774, it was superseded by another, which offers £5000 for the invention of any timekeeper, or other method, capable of determining the longitude of a place within 1 degree, and £10 000 if within 30 miles; and a reward of £5000 to the author of any lunar tables, which should be found within 15 seconds of the truth; allowing the Board also the power of granting smaller sums at their discretion. Timekeepers are at present very commonly employed in the British navy, and some of them have been capable of determining the longitude within half a degree, after having been two or three months at sea. The lunar tables, which have been employed for the Nautical Almanacs, are those of Professor Mayer, who adopted the methods of calculation invented by Leonard Euler; but the tables of Mr. Burg, of Vienna, are still more accurate, and are said to be always within about ten seconds of the truth.

The progress of astronomy, since the death of Newton, in 1727, has been fully adequate to what its most sanguine votaries could have hoped. The great discoveries of the aberration of the fixed stars, and of the nutation of the earth's axis, were made by our countryman Bradley, with the assistance of the instruments for which he was indebted to the delicate workmanship of our artists. Among these the names of Bird, Short, Sisson, Graham, Dollond, Harrison, and Ramsden have long been celebrated throughout Europe. The geographical operations, which have been performed in every part of the globe, have been chiefly conducted by the liberality of the French and English governments, although other countries have not been deficient in taking their share of the labour. The observations of the transit of Venus were twice made in the south seas by British navigators, whom the munificence of our present sovereign enabled to undertake so arduous a voyage for this express purpose; and we are indebted to the fund which was granted on the occasion, as well as to the zeal of the Astronomer Royal, for the experiments on the attraction of mountains, which were instituted after their return. In this country also, Dr. Herschel, besides many other important additions to our astronomical knowledge, has discovered a primary planet, and eight secondary ones, unknown before. The astronomers of Sicily and Germany have, however, the honour of the first discovery of the three humbler members of the solar system which have been last introduced to our acquaintance, Ceres by Piazzi, Pallas by Olbers, and Juno by Harding: and the mathematicians of France have excelled all their predecessors in the elaborate and refined application of the theory of gravitation, to the investigation of the most minute and intricate details of the celestial motions.

CHRONOLOGY OF ASTRONOMERS.

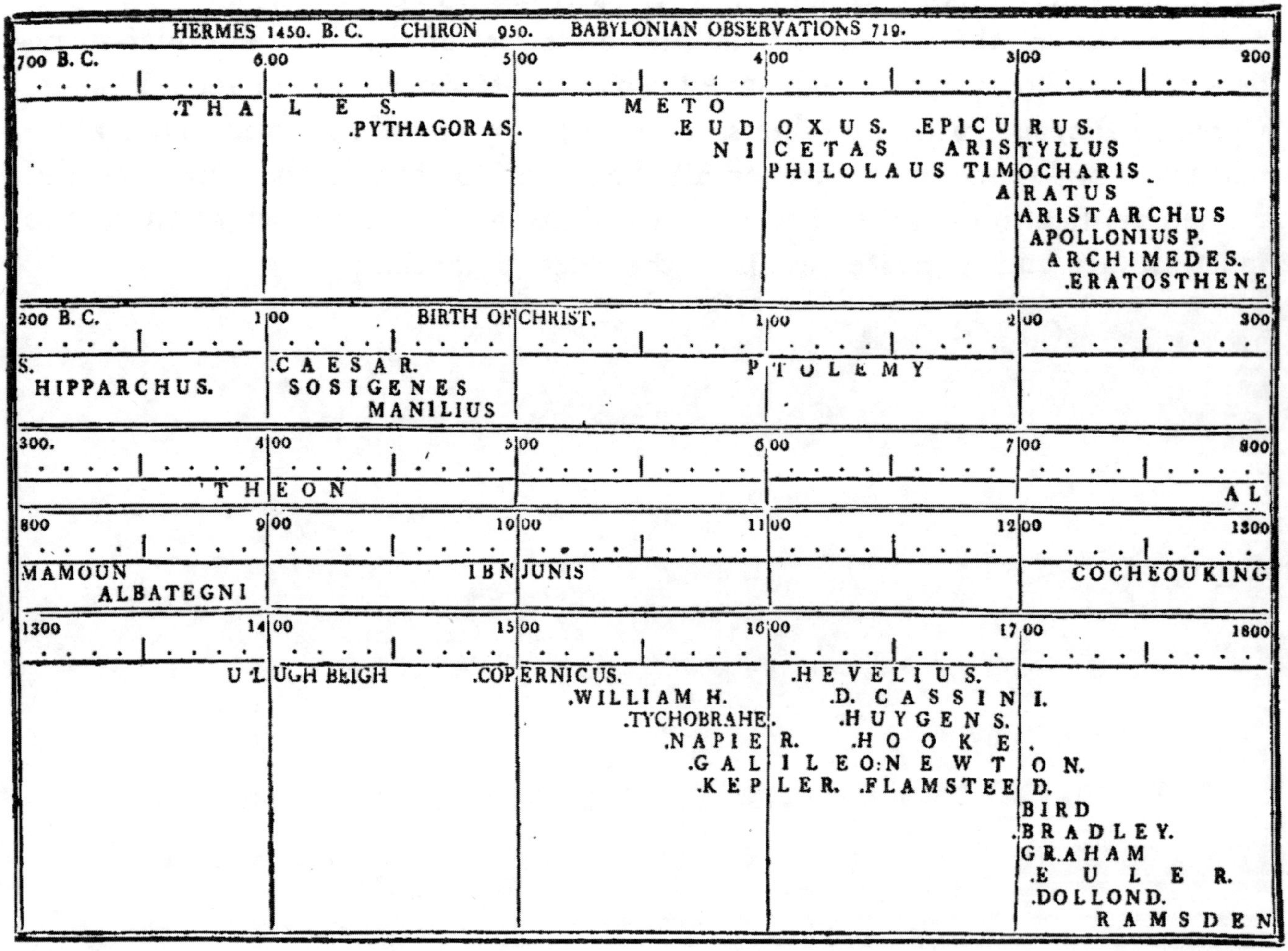

LECTURE XLIX.

ON THE ESSENTIAL PROPERTIES OF MATTER.

THE objects, which have lately occcupied our inquiries, are the most sublime and magnificent that nature any where exhibits to us, and the contemplation of them naturally excites, even in an uncultivated mind, an admiration of their dignity and grandeur. But all magnitude is relative; and if we examine with more calm attention, we shall find still greater scope for our investigation and curiosity, in the microscopic, than in the telescopic world. Pliny has very justly observed, that nature no where displays all her powers with greater activity, than in the minutest objects perceptible to our senses; and we may judge how wide a field of research the corpuscular affecti ons of matter afford, from the comparatively small progress that has hitherto been made in cultivating it. For while the motions of the vast bodies, which roll through the heavens, have been completely subjected to the most rigorous calculations, we know nothing, but from experience only, of the analogies by which the minute actions of the particles of matter are regulated. It is probable, however, that they all depend ultimately on the same mechanical principles. We have seen, for example, that the widely extended elevations and depressions of the ocean, which are raised by the attractive powers of the two great luminaries, and cover at once a half of the globe, are governed and combined according to the same laws, which determine the motions of the smaller waves excited by different causes in a canal, the rapid tremors of a medium transmitting sound; or the inconceivably diminutive undulations which are capable of accounting for the phenomena of light, and which must be exerted in spaces as much smaller than those of sound, as a grain of sand is smaller than a mountain. Thus the annihilation of the effects of the semidiurnal changes of the tide; and the preservation of the diurnal change, in the harbour of Batsha, may be explained precisely in the same manner as the reflection of red light from a transparent substance, of such a thickness, as to be capable of destroying a portion of violet light under the same circumstances.

We are at present to descend from the affections of the large masses of matter, which form the great features of the universe, to the particular properties of the matter which constitutes them, as far as they are common to all matter in general; but those properties which are peculiar to certain kinds of matter only, being the subjects of chemical science, are not to be included in the discussion. If we are asked for a definition of matter, it will be somewhat difficult to avoid all circuitous expressions. We may make gravitation a test of matter, but then we must say, that whatever is attracted by other matter, is also to be denominated matter, and this supposes the subject of our definition already known; besides that the property of attraction may also possibly belong to substances not simply material; for the electrical fluid, if such a fluid exists, is probably attracted by matter, and yet it seems to be different in most respects from any modification of common matter. A similar difficulty would occur if we attempted to define matter by its impenetrability or mutual repulsion, or if we considered every thing as material that is capable of affecting the senses. We must, therefore, take it for granted that matter is known without a definition, and we may describe it as a substance occupying space, or as a gravitating or ponderable substance.

It cannot be positively determined whether matter is originally of one kind, owing its different appearances only to the form and arrangement of its parts; or whether there are various kinds of simple matter, essentially distinct from each other; but the probability appears to be in favour of the former supposition. However this may be, the properties of matter are by no means so simple in their nature, nor so easily reducible to general laws, as the more mathematical doctrines of space and motion; and since our knowledge of them depends more on experience than on abstract principles, they may properly be considered as belonging to particular physics. We have found no inconvenience from the omission of the doctrine of matter as a part of the subject of mechanics; although, in treating of the strength of materials, as subservient to practical mechanics, it was necessary to consider the effects of some of these properties as deduced from experiment; but it will appear that it was impossible to examine their origin and mutual connexion, without supposing a previous knowledge of many other departments of natural philosophy.

We may distinguish the general properties of matter into two principal classes, those which appear to be inseparable from its constitution, and those which are only accidental, or which are not always attached to matter of all kinds. The essential properties are chiefly extension and divisibility, density, repulsion, or impenetrability, inertia, and gravitation; the accidental properties are in great measure dependent on cohesion, as liquidity, solidity, symmetry of arrangement, cohesive elasticity, stiffness, toughness, strength, and resilience.

The extension of matter can scarcely be considered as a property separate from its impenetrability, unless we conceive that it can occupy space, without excluding other bodies from it. This opinion has indeed been maintained by some philosophers, who have imagined that the minute particles which they suppose to constitute light, may penetrate the ultimate atoms of other matter without annihilating or displacing them; and if this hypothesis were admitted, it would be necessary to consider each particle of matter as a sphere of repulsion, extended without being impenetrable.

The divisibility of matter is great beyond the power of imagination, but we have no reason for asserting that it is infinite; for the demonstrations, which have sometimes been adduced in favour of this opinion, are obviously applicable to space only. The infinite divisibility of space seems to be essential to the conception that we have of its nature; and it may be strictly demonstrated, that it is mathematically possible, to draw an infinite number of circles between any given circle and its tangent, none of which shall touch either of them, except at the general point of contact; and that a ship, following always the same oblique course with respect to the meridian, for example, sailing north eastwards, would continue perpetually to approach the pole without ever completely reaching it. But when we inquire into the truth of the old maxim of the schools, that all matter is infinitely divisible, we are by no means able to decide so positively. Newton observes, that it is doubtful whether any human means may be sufficient to separate the particles of matter beyond a certain limit; and it is not impossible that there may be some constitution of atoms, or single corpuscles, on which their properties, as matter, depend, and which would be destroyed if the units were further divided; but it appears to be more probable that there are no such atoms; and even if there are, it is almost certain that matter is never thus annihilated in the common course of nature.

It remains to be examined how far we have any experience of the actual extent of the divisibility of matter; and we shall find no appearance of any thing like a limit to this property. The smallest spherical object, visible to a good eye, is about $\frac{1}{1000}$ of an inch in diameter; by the assistance of a microscope, we may perhaps distinguish a body one hundredth part as large, or $\frac{1}{100000}$ of an inch in diameter. The thickness of gold leaf is less than this, and the gilding of lace is still thinner, probably in some cases not above one ten millionth of an inch; so that $\frac{1}{2000}$ of a grain would cover a square inch, and a portion, barely large enough to be visible by a microscope, might weigh only the 80 million millionth part of a grain. A grain of musk is said to be divisible into 320 quadrillions of parts, each of which is capable of affecting the olfactory nerves. There are even living beings, visible to the microscope, of which a million million would not make up the bulk of a common grain of sand. But it is still more remarkable, that, as far as we can discover, many of these animalcules are as complicated in their structure as an elephant or a whale. It is true that the physiology of the various classes of animals is somewhat more simple as they deviate more from the form of quadrupeds, and from that of the human species; the solid particles of the blood do not by any means vary in their magnitude in the same ratio with the bulk of the animal; and some of the lower classes appear to approximate very much to the nature of the vegetable world. But there are single instances that seem wholly to destroy this gradation: Lyonnet has discovered a far greater variety of parts in the caterpillar of the willow butterfly, than we can observe in many animals of the largest dimensions; and among the microscopic insects in particular, we see a prodigality of machinery, subservient to the various purposes of the contracted life of the little animal, in the structure of which nature appears to be ostentatious of her power of giving perfection to her minutest works.

If Newton's opinion, respecting the origin of the colours of natural bodies in general, were sufficiently established, it would afford us a limit to the divisibility of matter with respect to coloured substances; for the colours of thin transparent substances, which he considers as resembling those of most other substances, are no longer observable, in any known medium, when the thickness is less than about $\frac{1}{500000}$ of an inch. But we have positive evidence that coloured substances may be reduced to dimensions far below this limit; besides the instance of the gilt wire, which has already been mentioned, a particle of carmine may still retain its colour, when its thickness is no more than one thirty millionth of an inch, or one sixtieth part of the limit deduced from the supposition of Newton; and it is therefore scarcely possible that the colours of such substances can precisely resemble those of thin plates, although they may perhaps still be in some measure analogous to them.

Impenetrability is usually attributed to matter, from the common observation that two bodies cannot occupy the same place at once. And it is thus that we distinguish matter from space; for example, when we dip an inverted jar into mercury, the air contained in the jar depresses the surface of the mercury, and prevents its occupying the space within the jar: but if the jar had been void of matter, like the space above the mercury of a barometer, nothing would have prevented its being filled by the mercury, as soon as either its weight, or the pressure of the atmosphere, urged it to enter the jar.

But it does not appear that our senses are fully competent to extend this proposition to all substances, whether material or not. We cannot prove experimentally that the influence of gravitation is incapable of pervading even the ultimate particles of solid matter, for this power appears to suffer no diminution nor modification, when a third body is interposed between the two gravitating masses. In the same manner, a magnet operates as rapidly on a needle, through a plate of glass or of gold, whatever its thickness may be, as if a vacuum only intervened. It may, however, be inquired if the gold or the glass has not certain passages or pores, through which the influence may be transmitted: and it may be shown, in many instances, that substances, apparently solid, have abundant orifices into which other substances may enter; thus mercury may easily be made to pass through leather, or through wood, by the pressure of the atmosphere, or by any other equal force: and, however great we may suppose the proportion of the pores to the solid matter, it may be observed, that it requires only a more or less minute division of the matter, to reduce the magnitude of the interstices between the neighbouring particles within any given dimensions. Thus platina contains, in a cubic inch, above 200 thousand times as many gravitating atoms as pure hydrogen gas, yet both of these mediums are free from sensible interstices, and appear to be equally continuous; and there may possibly be other substances in nature that contain in a given space 200 thousand times as many atoms as platina; although this supposition is not positively probable in all its extent; for the earth is the densest of any of the celestial bodies with which we are fully acquainted, and the earth is only one fourth as dense as if it were composed entirely of platina; so that we have no reason to believe that there exists in the solar system any considerable quantity of a substance even so dense as platina.

Besides this porosity, there is still room for the supposition, that even the ultimate particles of matter may be permeable to the causes of attractions of various kinds, especially if those causes are immaterial: nor is there any thing in the unprejudiced study of physical philosophy that can induce us to doubt the existence of immaterial substances; on the contrary we see analogies that lead us almost directly to such an opinion. The electrical fluid is supposed to be essentially different from common matter; the general medium of light and heat, according to some, or the principle of caloric, according to others, is equally distinct from it. We see forms of matter differing in subtility and mobility, under the names of solids, liquids, and gases; above these are the semimaterial existences which produce the phenomena of electricity and magnetism, and either caloric or a universal ether; higher still perhaps are the causes of gravitation, and the immediate agents in attractions of all kinds, which exhibit some phenomena apparently still more remote from all that is compatible with material bodies; and of these different orders of beings, the more refined and immaterial appear to pervade freely the grosser. It seems therefore natural to believe that the analogy may be continued still further, until it rises into existences absolutely immaterial and spiritual. We know not but that thousands of spiritual worlds may exist unseen for ever by human eyes; nor have we any reason to suppose that even the presence of matter, in a given spot, necessarily excludes these existences from it. Those who maintain that nature always teems with life, wherever living beings can be placed, may therefore speculate with freedom on the possibility of independent worlds; some existing in different parts of space, others pervading each other, unseen and unknown, in the same space, and others again to which space may not be a necessary mode of existence.

Whatever opinion we may entertain with respect to the ultimate impenetrability of matter in this sense, it is probable that the particles of matter are absolutely impenetrable to each other. This impenetrability is not however commonly called into effect in cases of apparent contact. If the particles of matter constituting water, and steam, or any other gas, are of the same nature, those of the gas cannot be in perfect contact; and when water is contracted by the effect of cold, or when two fluids have their joint bulk diminished by mixture, as in the case of alcohol, or sulfuric acid, and water, the particles cannot have been in absolute contact before, although they would have resisted with great force any attempt to compress them. Metals too, of all kinds, which have been melted, become permanently more dense when they are hammered and laminated. A still more striking and elegant illustration of the nature of repulsive force is exhibited in the contact of two pieces of polished glass. The colours of thin plates afford us, by comparison with the observations of Newton, the most delicate micrometer that can be desired, for measuring any distances less than the ten thousandth of an inch: it was remarked by Newton himself, that when two plates of glass are within about this distance of each other, or somewhat nearer, they support each other's weight in the same manner as if they were in actual contact, and that some additional force is required, in order to make them approach still nearer; nor does it appear probable that the contact is ever perfect, otherwise they might be expected to cohere in such a manner as to become one mass. Professor Robison has ascertained by experiment the force necessary to produce the greatest possible degree of contact, and finds it equivalent to a pressure of about a thousand pounds for every square inch of glass. It is therefore obvious that in all common cases of the contact of two distinct bodies, it must be this repulsive force that retains them in their situation. I have found that glass, placed on a surface of metal, exhibits this force nearly in the same degree as if placed on another piece of glass; it is also independent of the presence of air; but under water it disappears.

The existence of a repulsive force, extending beyond the actual surface of a material substance, being proved, it has been conjectured by some that such a force, unconnected with any central atom, may be sufficient for producing all the phenomena of matter. This representation may be admitted without much difficulty, provided that it be allowed that the force becomes infinite at or near the centre; but it has been sometimes supposed that it is every where less than infinite, and consequently that matter is not absolutely impenetrable; such a supposition appears however to lead to the necessity of believing that the particles of matter must sometimes be annihilated, which is not a very probable opinion.

The magnitude of the repulsive force, by which the particles of any single body are enabled to resist compression, increases nearly in proportion to the degree of compression, or to the decrease of the distances between the particles. This is almost a necessary consequence of any primary law that can be imagined, for the immediate actions of the particles: for instance, if the repulsion increased either as the square or as the cube of the distance diminished, the effect of a double change of dimensions would always be nearly a double change of the repulsive force; that is, if an elastic substance were compressed one thousandth part of its bulk, it would in either case resist twice as much as if it were only compressed one two thousandth.

It is obvious that if the particles of matter are possessed of a repulsive force decreasing in any regular proportion with the increase of distance, they can never remain at rest without the operation of some external pressure, but will always retain a tendency to expand. This is the case of all elastic fluids, the density of which is found to vary exactly as the compressing force, whence it may be demonstrated, that the primary repulsive force of the particles must increase in the same proportion as the distance decreases. It follows also that this force can only be exerted between such particles as are either actually or very nearly in contact with each other; since it requires no greater pressure, acting on a given surface, to retain a gallon of air in the space of half a gallon, than to retain a pint in the space of half a pint; which could not be, if the particles exercised a mutual repulsion at all possible distances.

Mr. Dalton has proposed a singular theory respecting the constitution and mutual repulsion of elastic fluids; he imagines that when any two gases of different kinds are mixed, the particles of each gas repel only the similar particles of the same gas, without exerting any action on those of the other gas, except when the ultimate solid atoms chance to interfere. The idea is ingenious and original, and may perhaps be of use in connecting some facts together, or in leading to some other less improbable suppositions; but it may easily be shown, that Mr. Dalton's hypothesis cannot possibly be true in all its extent, since it would follow from it, that two portions of gases, of different kinds, could not exist, for a sensible time, in the same vessel, without being uniformly diffused throughout it, while the fact is clearly otherwise; for hydrogen gas remains, when left completely at rest, a very considerable time above, and carbonic acid gas below, a portion of common air with which it is in contact; nor is there any circumstance, attending the mixture of gases, which may not be explained without adopting so paradoxical an opinion! Mr. Dalton thinks that, from the laws of hydrostatics, no two gases, not chemically united, could remain mixed, if their particles acted mutually on each other: but the laws of hydrostatics do not apply to the mixture of single particles of fluids of different kinds; since they are only derived from the supposition of a collection of particles of the same kind.

In liquids and in solids, this repulsive force appears at first sight to be wanting; but when we consider that the particles both of liquids and of solids are actuated by the attractive force of cohesion, we shall see the necessity of the presence of a repulsive force, in order to balance it; it is, therefore, probable that the particles of aeriform fluids still retain their original repulsive powers, when they are reduced to a state of liquidity or of solidity, by being subjected to the action of a second force, which causes them to cohere.

The mutual repulsion of the particles of matter is a reciprocal force, acting equally, in opposite directions, on each of the bodies concerned. It scarcely requires either experiment or argument to show, that if two bodies repel each other, neither of them will remain at rest, but both of them will move, with equal quantities of motion. Thus, if a portion of condensed air be made to act upon the bullet of an air gun, it will force the gun backwards with as much momentum as it impels the bullet forwards.

Inertia is that property of matter, by which it retains its state of rest or of uniform motion, with regard to a quiescent space, as long as no foreign cause occurs to change that state. This property depends on the intimate constitution of matter; it is generally exhibited by means of the force of repulsion, which enables a body in motion to displace another, in order to continue its motion, or by means of some attractive force, which causes two bodies to approach their common centre of inertia with equal momenta.

Another universal property of matter is reciprocal gravitation, of which the force is directly in the joint proportion of the quantities of matter attracting each other, and inversely as the square of their distance. In order to prove that the gravitation towards a given substance, for instance, the weight of a body, or its gravitation towards the earth, is precisely in proportion to the mass or inertia of the moveable matter of which it consists, Sir Isaac Newton made two equal pendulums, with hollow balls of equal size: in order that the resistance of the air might be the same with respect to both, he placed successively within the balls a variety of different substances, and found that the time of vibration remained always the same; whence he inferred that the attraction was proportional in all cases to the quantity of matter possessing inertia. For if any of these substances had contained particles, capable of receiving and communicating motion, yet without being liable to gravitation, they would have retarded the vibrations of the pendulum, by adding to the quantity of matter to be moved, without increasing the moving force. The law of gravitation, which indicates the ratio of its increase with the diminution of the distance, is principally deduced from astronomical observations and computations: it is the simplest that can be conceived for any influence, that either spreads from a centre, or converges towards a centre; for it supposes the force acting on the same substance to be always proportional to the angular space that it occupies.

Newton appears to have considered these laws of gravitation, which he first discovered, rather as derivative than as original properties of matter; and although it has often been asserted that we gain nothing by referring them to pressure or to impulse, yet it is undoubtedly advancing a step in the explanation of natural phenomena, to lessen the number of general principles; and if it were possible to refer either all attraction to a modification of repulsion, or all repulsion to a modification of attraction, we should make an improvement of the same kind, as Newton made, when he reduced all the diversified motions of the heavenly bodies to the universal laws of gravitation only. We have, however, at present, little prospect of such a simplification.

It has been of late very customary to consider all the phenomena of nature as derived from the motions of the corpuscles of matter, agitated by forces varying according to certain intricate laws, which are supposed to be primary qualities, and for which it is a kind of sacrilege to attempt to assign any ulterior cause. This theory was chiefly introduced by Boscovich, and it has prevailed very widely among algebraical philosophers, who have been in the habit of deducing all their quantities from each other by mathematical relations, making, for example, the force a certain function or power of the distance, and then imagining that its origin is sufficiently explained; and when a geometrician has translated this language into his own, and converted the formula into a curve, with as many flexures and reflections as the labyrinth of Daedalus, he imagines that he has depicted to the senses the whole procedure of nature. Such methods may often be of temporary advantage, as long as we are contented to consider them as approximations, or as classifications of phenomena only; but the grand scheme of the universe must surely, amidst all the stupendous diversity of parts, preserve a more dignified simplicity of plan and of principles, than is compatible with these complicated suppositions.

"To show", says Newton, in the preface to the second edition of his Optics, "that I do not take gravity for an essential property of bodies, I have added one question concerning its cause, choosing to propose it by way of a question, because I am not yet satisfied about it, for want of experiments." In the query here mentioned, he proceeds from the supposition of an elastic medium, pervading all space; a supposition, which he advances with considerable confidence, and which he supports by very strong arguments, deduced as well from the phenomena of light and heat, as from the analogy of the electrical and magnetic influences. This medium he supposes to be much rarer within the dense bodies of the sun, the stars, the planets, and the comets, than in the empty celestial spaces between them, and to grow more and more dense at greater distances from them, so that all these bodies are naturally forced towards each other by the excess of pressure.

The effects of gravitation might be produced by a medium thus constituted, if its particles were repelled by all material substances with a force decreasing, like other repulsive forces, simply as the distances increase; its density would then be every where such as to produce the appearance of an attraction varying like that of gravitation. Such an ethereal medium would therefore have the advantage of simplicity, in the original law of its action, since the repulsive force which is known to belong to all matter, would be sufficient, when thus modified, to account for the principal phenomena of attraction.

It may be questioned whether a medium, capable of producing the effects of gravitation in this manner, would also be equally susceptible of those modifications which we have supposed to be necessary for the transmission of light. In either case it must be supposed to pass through the apparent substance of all material bodies with the most perfect freedom, and there would, therefore, be no occasion to apprehend any difficulty from a retardation of the celestial motions; the ultimate impenetrable particles of matter being perhaps scattered as thinly through its external form, as the stars are scattered in a nebula, which has still the distant appearance of a uniform light and of a continuous surface: and there seems no reason to doubt the possibility of the propagation of an undulation through the Newtonian medium with the actual velocity of light. It must be remembered that the difference of its pressure is not to be estimated from the actual bulk of the earth or any other planet alone, but from the effect of the sphere of repulsion of which that planet is the centre; and we may then deduce the force of gravitation from a medium of no very enormous elasticity.

We shall hereafter find that a similar combination of a simple pressure with a variable repulsion is also observable in the force of cohesion; and supposing two particles of matter, floating in such an elastic medium, capable of producing gravitation, to approach each other, their mutual attraction would at once be changed from gravitation to cohesion, upon the exclusion of the portion of the medium intervening between them. This supposition is, however, directly opposite to that which assigns to the elastic medium the power of passing freely through all the interstices of the ultimate atoms of matter, since it could never pass between two atoms cohering in this manner; we cannot therefore, at present, attempt to assert the identity of the forces of gravitation and cohesion so strongly, as this theory would allow us to do, if it could be established. In short, the whole of our inquiries, respecting the intimate nature of forces of any kind, must be considered merely as speculative amusements, which are of no further utility than as they make our views more general, and assist our experimental investigations.

LECTURE L.

ON COHESION.

THOSE properties of matter, which we have lately examined, if they are not absolutely inseparable from its constitution, are, at least, always found attached to such matter as we are able to submit to our experiments. There are, however, many other general affections, to which all matter appears to be liable, although none is perpetually subjected to them, and these are principally, if not entirely, dependent on the force of cohesion.

In order that any two particles of matter may cohere, it is necessary that they be within a very small distance of each other, and the density of any substance, composed of cohesive particles, must probably always be more than half as great as that of water. There are indeed some solids apparently a little lighter than this, but they appear to be extremely porous; and perhaps the solid substances of some of the celestial bodies may also be a little more rare. It frequently happens, that the compression of an elastic fluid alone is sufficient to cause the force of cohesion to take place between its particles; thus, if common steam be exposed, in a close vessel, to a pressure greater than that of the atmosphere, it will be wholly condensed into water, provided that no elevation of temperature be allowed; and the same has been experimentally shown of many other aeriform fluids, which may be reduced to liquids by pressure; but others of these fluids retain their elasticity, notwithstanding any force which human art can apply to them.

It is probable that as soon as the particles of any elastic fluid are brought within the reach of the force of cohesion, it commences at once in its full extent, so as to cause them to rush together, until it is balanced by that of repulsion, which continually increases as the particles approach nearer to each other; they must then remain, perhaps after some vibrations, in a state of equilibrium; and if any cause should tend to separate them, or to bring them nearer together, they would resist it, in either case, with a force proportional to the degree of extension or compression. The distance, at which the force of cohesion commences, is not the same for all kinds of matter, nor even for the same substance at different temperatures; it is smaller for vapours of all kinds, in proportion as their temperature is higher, the cohesion itself being also smaller. If the experiments on the density of steam have been correct, it follows that the force of repulsion must increase more rapidly than the distances diminish, for the elasticity of water is nearly ten times as great as that which would be inferred from the compression of steam into a substance of equal density: this supposition agrees also with the experiments on the mean density of the earth, which is probably not so great as it would be if the force of repulsion increased in the simple ratio of the density. The law of repulsion appears also to be in some degree modified by the effect of heat, which increases its force at greater distances more considerably than at smaller. It appears indeed, from the diminution of the elasticity of a spring by heating it, that the repulsive force of the particles of bodies at very small distances is even diminished by heat, unless the force be again supposed to decrease much more rapidly than the distance diminishes: thus the diminution of the elasticity of iron by heat is about thirty times as great as the increase of the distance of its particles; so that the original repulsive force must probably be somewhat diminished, although less than the cohesive force. At greater distances, however, the force of repulsion is certainly increased; for the elasticity of vapours and gases of all kinds is evidently greater as the temperature is higher. (Plate XXXIX. Fig. 530.)

The cohesion of two or more particles of matter to each other does not interfere with their power of repelling other particles situated in a different direction: thus, two pieces of glass require to be brought together with considerable force, and generally with some friction, before they can begin to cohere; and a small drop of water, falling lightly on the surface of a pond, may remain for some instants without coming into perfect contact with it: the same circumstance is also still more observable in spirit of wine a little warmed.

The first and simplest effect of cohesion is to produce liquidity. That all liquids possess some cohesion, is very obvious, from their tendency to assume a spherical form when they are sufficiently detached from other substances, and from the suspension of a drop from any solid, to which its upper surface adheres with sufficient force. Without cohesion, indeed, a liquid would be only a very fine powder, except that the particles of powders have not the power of moving with perfect freedom on each other, which constitutes fluidity. The apparent weakness of the cohesion of liquids is entirely owing to this mobility, since their form may be changed in any degree without considerably increasing the distances of their particles, and it is only under particular circumstances that the effects of their cohesion can become sensible.

When a liquid is considered as unlimited in its extent, the repulsion of its particles, situated in all possible directions with regard to each other, may be supposed in all cases precisely to balance the cohesion, which is derived from the actions of particles similarly situated; and this must also be the state of the internal parts of every detached portion of a liquid, where they are so remote from the surface as to be beyond the minute distance which is the limit of the action of these forces. But the external parts of the drop will not remain in the same kind of equilibrium: they may be considered as a thin coating of a liquid surrounding a substance which resists only in a direction perpendicular to its surface, and does not interfere with the mutual actions of the particles of the liquid. Now since the repulsive force increases as the distance diminishes, it must be exerted more powerfully by the nearest particles, while the cohesion is directed equally towards all the particles within a certain distance, and wherever the surface is curved, the joint cohesive force will be directed to a remoter part of the curve than the repulsive force opposed to it, so that each particle will be urged, by the combination of these forces, towards the concave side of the curve, and the more as the curvature is greater; hence the coating of the liquid, thus constituted, must exert a force on the parts in contact with it, precisely similar to that of a flexible surface, which is every where stretched by an equal force; and from this simple principle we may derive all the effects produced by a cohesion of this kind, which, from its being most commonly observed in the ascent of water in capillary tubes, has been denominated capillary attraction. (Plate XXXIX. Fig. 531.)

It is, therefore, a general law, that the surface of every detached portion of a fluid must every where have such a curvature, as to be able to withstand the hydrostatical pressure which acts against it; and hence we may calculate in many cases the properties of the curve which it must form; but in other cases the exact calculation becomes extremely intricate, and perhaps impracticable. A drop descending in a vacuum would be perfectly spherical; and if its magnitude were inconsiderable, it would be of the same form when descending through the air; a small bubble rising in a liquid must also be spherical; but where the drop or the bubble is larger, its curvature will be greatest where the internal pressure is greatest, or where the external pressure is least, and in different cases this pressure may be differently distributed. Where a drop is suspended from a solid, its length may be such that the pressure at its upper part may become negative, and its surface will then be concave instead of convex: and when a bubble rises to the surface of a liquid, it often carries with it a film of the liquid, of which the weight is probably smaller than the contractile force with which the surface resists the escape of the air, so that, from the magnitude of the contractile force, we may determine the greatest possible weight of a bubble of given dimensions. A slight imperfection of fluidity probably favours the formation of detached bubbles, by retarding the ascent of the air, but it has a still greater effect in prolonging their duration when formed. (Plate XXXIX. Fig. 532.)

In order to determine the forms of the surfaces of liquids in the cases which most commonly occur, it is necessary to examine how they are affected by the action of other liquids, and of solids of different descriptions. Supposing the horizontal surface of a liquid to be in contact with a vertical plane surface of a solid of half the attractive power, it will remain at rest in consequence of the equilibrium of attractions; for the particles situated exactly at the junction of the surfaces may be considered as actuated by three forces; one deduced from the effect of the liquid, the other two from that of the two equal portions of the solid above and below the surface of the fluid; and it may be shown that the combination of these three forces will produce a joint result in the direction of gravity; consequently the direction of the surface must remain the same as when it is subjected to the force of gravity alone, since the surface of every fluid at rest must be perpendicular to the joint direction of all the forces acting on it. But if the attractive power of the solid be more than half as great as that of the liquid, the result of the forces will be inclined towards the solid, and the surface of the liquid, in order to be perpendicular to it, must be more elevated at the side of the vessel than elsewhere, and therefore concave; consequently the fluid must ascend until it arrives at a position capable of affording an equilibrium in this manner: if, on the contrary, the attractive power of the solid be weaker, the liquid will descend, and its surface will be convex. (Plate XXXIX. Fig. 533.)

It may also be shown, that if the attractive power of the solid be equal to that of the liquid, or still greater, it will be wetted by the liquid, which will rise until its surface acquires the same direction with that of the solid; and in other cases the angle of contact will be greater in proportion as the solid is less attractive. These conclusions are obtained by comparing the common surface of the liquid and solid with the surface of a single liquid, of which the attractive power is equal only to the difference of the respective powers of the substances concerned; and the comparison is equally applicable to the contact of two liquids of different densities.

The magnitude of the superficial cohesion or contractility of a liquid may be expressed, for a certain extent, by a certain weight; thus every inch of the surface of water is stretched each way by a force equal to the weight of the hundredth part of a cubic inch of water, or to two grains and a half: and for each inch of the surface of mercury, the force is equivalent to 17 grains, which is the weight of $\frac{1}{300}$ of a cubic inch of mercury. Thus if a solid of any form, of which the surfaces are vertical, and which is capable of being wetted by either of these fluids, be immersed into a reservoir containing it, the fluid will be elevated around it to such a height, that $2\frac{1}{2}$ or 17 grains, for each inch of the circumference of the solid, will remain above the general level of the reservoir, the surface assuming nearly the same form as a very long and slender elastic rod, fixed horizontally at one end, and bearing a large weight at the other. (Plate XXXIX. Fig. 534.)

The elevation of the summit of an extended surface of water, in contact with the flat and upright surface of a solid which is wetted by it, is one seventh of an inch: but when two such surfaces, for instance, two plates of glass, are brought near to each other, the elevation of the water between them must be greater than this, in order that each inch of the line of contact may support its proper weight: thus, if the distance were one fiftieth of an inch, the elevation would be a whole inch; and if the distance were smaller than this, the elevation would be greater in the same proportion; so that when two plates are placed in such a manner as to touch each other at one of their upright edges, the outline of the water raised between them assumes the form of a hyperbola. (Plate XXXIX. Fig. 535.)

The weight supported by the cohesion of the water in a tube may be determined, in a similar manner, from the extent of the circumference; the height being an inch in a tube one twenty fifth of an inch in diameter, or as much greater as the diameter of the tube is smaller: and in a tube wetted with mercury the height would be half as great. It is obvious that if the lower part of the tube be either contracted or dilated, the height of the fluid will remain unaltered, while its weight may be varied without limit; for the hydrostatical pressure on the surface is the same, in both these cases, as if the diameter of the tube were equal throughout its length. (Plate XXXIX. Fig. 536.)

The attractive force of glass to mercury is less than half as great as the mutual attraction of the particles of mercury, and the surface of mercury in a dense glass vessel becomes, therefore, convex and depressed; the angle of contact being about 140°, and the depression one 17th of an inch. Between two plates of glass, the depression of mercury is an inch when their distance is $\frac{1}{133}$, and in a tube, when its diameter is $\frac{1}{67}$ of an inch. (Plate XXXIX. Fig. 537, 538.)

A liquid may also adhere to a horizontal surface which is gradually raised from it, until the hydrostatical pressure becomes sufficient to overpower the cohesion of its superficial parts; the internal part of the fluid being usually raised, not immediately by the force of cohesion, but by the pressure of the atmosphere. The solid bears the whole weight of the liquid which is elevated above the surface; and when the surface is perfectly wetted, this weight is equal, at the moment of separation, to the hydrostatical pressure, or rather suction, corresponding to the height; but in other cases the weight may be somewhat greater than the hydrostatical pressure on the surface of the solid, on account of the elevation which surrounds the body, and which is not compensated by the excavation immediately under it. A surface thus raised from water will elevate it to the height of one fifth of an inch, and will require a force of $50\frac{1}{4}$ grains for each square inch, in order to overcome the apparent attraction of the water; and for mercury the utmost height is about one seventh of an inch. (Plate XXXIX. Fig. 539, 540.)

A detached portion of a liquid may stand on any surface, which it is not capable of wetting, at a height which is different according to its magnitude, and to the attraction of the surface. If the drop is very small, its form may be nearly spherical; but when its extent becomes considerable, its height must always be less than that at which the liquid would separate from a horizontal surface; and it will approach the nearer to this limit, as its attraction to the surface on which it stands is weaker. Thus a wide portion of mercury stands on glass at the height of $\frac{2}{15}$ of an inch, and on paper nearly at $\frac{1}{7}$; and a portion of water will stand on a cabbage leaf, or on a table strewed with the seeds of lycopodium, nearly at the height of one fifth of an inch. (Plate XXXIX. Fig. 541.)

For the operation of a powder like lycopodium, it appears to be only necessary, that it should possess a weaker power of attraction than water, and should, therefore, be incapable of being readily wetted by it: each particle of the powder, being then but partially in contact with the water, will project beyond its surface, and prevent its coming into contact with any of the surrounding bodies, while the surface assumes such a curvature as is sufficient to withstand the pressure of the internal parts. (Plate XXXIX. Fig. 542.)

When a dry and light substance of any kind is placed on the surface of water, its weight is not sufficient to bring it within the distance at which cohesion commences, and it floats, surrounded by a slight depression. Any substance of this kind, or any other substance surrounded by a depression, as a ball of glass or iron floating on mercury, appears to be attracted by another similar substance in its neighbourhood; for the depression between the two substances is increased, and the pressure of the fluid on that side is consequently lessened, so that they are urged together, by a force which varies inversely as the square of the distance. And in the same manner, when two bodies, surrounded by an elevation, approach each other, they exhibit an attractive force of a similar nature, the pressure of the atmosphere being diminished by the weight of the water, which is raised between them to a greater height than on the opposite sides. But when a body, surrounded by a depression, approaches another, which is surrounded by an elevation, they seem to repel each other, the pressure of the water urging the one, and that of the atmosphere the other, in opposite directions. (Plate XXXIX. Fig. 543.)

If two smooth plates of any kind are perfectly wetted by a fluid, and brought into contact, they exhibit an appearance of cohesion, which is so much the greater as the quantity of fluid is smaller: if we attempt to separate them, the fluid is drawn inwards, so as to have its surface made concave, and it resists the separation of the plates with a certain force, which acts with a hydrostatic advantage so much the greater, as their distance is smaller, and hence produces the appearance of a cohesion varying in proportion to the distance. (Plate XXXIX. Fig. 544.)

Supposing the two plates to be separated at one end, and the fluid between them to assume the form of a drop, one of the marginal surfaces of the drop, being narrower than the other, will act with a greater advantage, like a tube of smaller diameter, and will tend to draw the drop towards it; and the apparent attraction towards the line of contact of the glasses will increase in proportion as the square of the distance decreases. This result was experimentally observed almost a century ago, but it has been usually explained on mistaken grounds. (Plate XXXIX. Fig. 545.)

The attractive power of water being greater than that of oils, a small portion of oil thrown on water is caused to spread on it with great rapidity by means of the force of cohesion; for it does not appear that any want of chemical affinity, between the substances concerned, diminishes their cohesive power; water readily adheres to tallow when solid, and probably essential oils would adhere still more readily to ice. There is, however, some difficulty in understanding how these oils can so suddenly come within the limit of the cohesive force of water, while the drops of water themselves sometimes remain for a few seconds beyond it.

A sponge affords us a familiar instance of the application of capillary attraction to useful purposes: it is well known, that in order to its speedy operation, it requires to be previously moistened, by the assistance of a little pressure, otherwise it exhibits the same appearance of repulsion that is observable in many other cases where the contact is imperfect. The absorption of moisture by sugar depends on the same principle, and here the tubes are so minute, that the height of ascent appears to be almost unlimited.

The magnitude of the cohesion between fluids and solids, as well as of the particles of fluids with each other, is more directly shown by an experiment on the continuance of a column of mercury, in the tube of a barometer, at a height considerably greater than that at which it usually stands, on account of the pressure of the atmosphere. If the mercury has been well boiled in the tube, it may be made to remain in contact with the closed end, at the height of 70 inches or more; and by agitation only it may be made to cohere so strongly, as to occupy the whole length of the tube of a common barometer, which is several inches more than the height at which the pressure of the atmosphere sustains it. A small siphon may also convey mercury from one vessel into another in the vacuum of an air pump: and in both these cases it is obvious that no other force than cohesion can retain the upper surface of the mercury in contact with the glass, or its internal parts in contact with each other.

The force of cohesion may also be exerted by solid substances on other solids, either of the same kind, or of different kinds. Thus two masses of lead, when once united by pressure, assisted by a little friction, require a very considerable force to separate them, and it may be shown either by measuring this force, or by suspending the lead in the vacuum of the air pump, that the pressure of the atmosphere is not materially concerned in producing this appearance of cohesion, since its magnitude much exceeds that of the atmospherical pressure. A cohesion of this kind is sometimes of practical utility in the arts; little ornaments of laminated silver remaining attached to iron or steel, with which they have been made to cohere by the powerful pressure of a blow, so as to form one mass with it.

The contact of two pieces of lead, although intimate enough to produce a considerable cohesion, is by no means so complete as to unite the parts into one mass; the union, however, appears to be nearly of the same kind as the common cohesion of aggregation; and if the lead were softened into an amalgam by the addition of mercury, the cohesion of the two masses would become precisely the same as the internal cohesion of each mass. Harder substances, such as marble or glass, cohere but weakly, perhaps because their surfaces are never so perfectly adjusted to each other as to touch throughout. The interposition of a fluid usually increases the apparent attraction of such substances, but this circumstance has already been explained from the effect of the capillary contraction of its surface; and when the substances are wholly immersed in a fluid, the cohesion is little if at all increased.

The immediate cause of solidity, as distinguished from liquidity, is the lateral adhesion of the particles to each other, to which the degree of hardness or solidity is always proportional. This adhesion prevents any change of the relative situation of the particles, so that they cannot be withdrawn from their places, without experiencing a considerable resistance from the force of cohesion, while those of liquids may remain equally in contact with the neighbouring particles, notwithstanding their change of form. When a perfect solid is extended or compressed, the particles, being retained in their situations by the force of lateral adhesion, can only approach directly to each other, or be withdrawn further from each other, and the resistance is nearly the same, as if the same substance, in a fluid state, were inclosed in an unalterable vessel, and forcibly compressed or dilated. Thus the resistance of ice to extension or compression is found by experiment to differ very little from that of water contained in a vessel; and the same effect may be produced even when the solidity is not the most perfect which the substance admits; for the immediate resistance of iron or steel to flexure is the same whether it may be harder or softer. It often happens, however, that the magnitude of the lateral adhesion is so much limited as to allow a greater facility of extension or compression, and it may yet retain a power of restoring the bodies to their original form by its reaction. This force may even be the principal or perhaps the only source of the body's elasticity: thus when a piece of elastic gum is extended, the mean distance of its particles is not materially increased, for it is said to become rather more than less dense during its extension; consequently the change of form is rather to be attributed to a displacement of the particles, than to their separation to a greater distance from each other, and the resistance must be derived from the lateral adhesion only: some other substances also, approaching more nearly to the nature of liquids, may be extended to many times their original length, with a resistance continually increasing; and in such cases there can scarcely be any material change of the specific gravity of these substances. Professor Robison has mentioned the juice of bryony as affording a remarkable instance of such a viscidity.

It is probable that the immediate cause of the lateral adhesion of solids is a symmetrical arrangement of their constituent parts: it is certain that almost all bodies are disposed, in becoming solid, to assume the form of crystals, which evidently indicates the existence of such an arrangement; and all the hardest bodies in nature are of a crystalline form. It appears, therefore, consistent both with reason and with experience to suppose that a crystallization more or less perfect is the universal cause of solidity. We may imagine that when the particles of matter are disposed without any order, they can afford no strong resistance to a motion in any direction, but when they are regularly placed in certain situations with respect to each other, any change of form must displace them in such a manner, as to increase the distance of a whole rank at once; and hence they may be enabled to cooperate in resisting such a change. Any inequality of tension in a particular part of a solid is also probably so far the cause of hardness, as it tends to increase the strength of union of any part of a series of particles which must be displaced by a change of form.

The immediate resistance of a solid to extension or compression is most properly called its elasticity; although this term has sometimes been used to denote a facility of extension or compression, arising from the weakness of this resistance. A practical mode of estimating the force of elasticity has already been explained, and according to the simplest statement of the nature of cohesion and repulsion, the weight of the modulus of elasticity is the measure of the actual magnitude of each of these forces; and it follows that an additional pressure, equal to that of the modulus, would double the force of cohesion, and require the particles to be reduced to half their distance in order that the repulsion might balance it; and in the same manner an extending force equal to the weight of half the modulus would reduce the force of cohesion to one half, and extend the substance to twice its dimensions. But, if, as there is some reason to suppose, the mutual repulsion of the particles of solids varies a little more rapidly than their distance, the modulus of elasticity will be a little greater than the true measure of the whole cohesive and repulsive force: this difference will not, however, affect the truth of our calculations respecting the properties of elastic bodies, founded on the magnitude of the modulus as already determined.

The stiffness of a solid is measured by its immediate resistance to any force tending to change its form; in this sense, if the force be applied so as to extend or to compress it, or to overcome its lateral adhesion by the effect which we have formerly called detrusion, the primitive elasticity and rigidity of the substance, together with its magnitude, will determine its stiffness: but if the force be otherwise applied, so as to produce flexure or torsion, the form of the body must also be taken into the calculation, in the manner which has already been explained in the lecture on passive strength. The stiffness of a body with respect to any longitudinal force is directly as its transverse section, and inversely as its length; for the same force will compress or extend a rod 100 yards long so as to change its length an inch, that will produce a change of only half an inch in a rod 50 yards long. We have seen that the space through which a body may be extended or compressed, without any permanent alteration of form, constitutes its toughness: that its strength, or the ultimate resistance which it affords, depends on the joint magnitude of its toughness and elasticity or stiffness, and that its resilience, or the power of overcoming the energy or impetus of a body in motion, is proportional to the strength and toughness conjointly.

Softness, or want of solidity, is in general accompanied by a proportional susceptibility of permanent alteration of form without fracture; sometimes, however, from a want of cohesion, a soft body is at the same time brittle. Soft substances, which are capable of direct extension to a considerable degree are called viscous or tenacious; of these, birdlime, sealing wax, and glass sufficiently heated, are some of the most remarkable. Harder substances which have the same property are called ductile, and when the alteration is made by percussion and compression, they are termed malleable. Of all substances gold is perhaps the most ductile; the thinness of leaf gold and of the gilding of silver wire has already been mentioned; and it is said that a single grain of gold has been drawn into a wire 500 yards in length, and consequently little more than $\frac{1}{5000}$ of an inch in diameter. The ductility or tenacity of a spider's web is of a different kind, it is particularly shown by its capability of being twisted, almost without limit, and of accommodating itself to its new position without any effort to untwist.

With respect to the ultimate agent by which the effects of cohesion are produced, if it is allowable to seek for any other agent than a fundamental property of matter, it has already been observed, that appearances extremely similar might be derived from the pressure of a universal medium of great elasticity; and we see some effects, so nearly resembling them, which are unquestionably produced by the pressure of the atmosphere, that one can scarcely avoid suspecting that there must be some analogy in the causes. Two plates of metal, which cohere enough to support each other in the open air, will often separate in a vacuum: when a boy draws along a stone by a piece of wet leather, the pressure of the atmosphere appears to be materially concerned. The well known experiment, of the two exhausted hemispheres of Magdeburg, affords a still more striking instance of apparent cohesion derived from atmospherical pressure; and if we place between them a thick ring of elastic gum, we may represent the natural equilibrium between the forces of cohesion and of repulsion; for the ring would resist any small additional pressure with the same force as would be required for separating the hemispheres so far, as to allow it to expand in an equal degree: and at a certain point the ring would expand no more; the air would be admitted, and the cohesion destroyed, in the same manner as when a solid of any kind is torn asunder. But all suppositions founded on these analogies must be considered as merely conjectural; and our knowledge of every thing which relates to the intimate constitution of matter, partly from the intricacy of the subject, and partly for want of sufficient experiments, is at present in a state of great uncertainty and imperfection. One of the most powerful agents, in changing and modifying the forms of matter, is the operation of heat, by which the states of solidity, liquidity, and elastic fluidity are often produced in succession; and the investigation of the nature and effects of heat will constitute the subject of the two next lectures.

LECTURE LI.

ON THE SOURCES AND EFFECTS OF HEAT.

IT may appear doubtful to some whether the subject of heat belongs most properly to mechanical or to chemical philosophy. Its influence in chemistry is unquestionable and indispensable; but its mechanical effects are no less remarkable: it could not therefore with propriety be omitted either in a course of chemical or of physical lectures, especially by those who are persuaded that what we call heat is, in its intimate nature, rather a mechanical affection of matter than a peculiar substance. We shall first inquire into the nature of the principal sources of heat, and next into the mode of its communication, and its most common effects, whether temporary or permanent: the measures of heat, and the most probable opinions respecting its nature, will afterwards be separately considered.

Heat is an influence capable of affecting our nerves in general with the peculiar sensation which bears its name, and of which the diminution produces the sensation denominated cold. Any considerable increase of heat gives us the idea of positive warmth or hotness, and its diminution excites the idea of positive cold. Both these ideas are simple, and each of them might be derived either from an increase or from a diminution of a positive quality: but there are many reasons for supposing heat to be the positive quality, and cold the diminution or absence of that quality; although we have no more experience of the total absence of heat, than of its greatest possible accumulation, which might be called the total absence of cold. Our organs furnish us, in some cases, with very delicate tests of any increase or diminution of heat; but it is more usually recognised by the enlargement of bulk, generally produced in those bodies to which heat is attached in an increased quantity, and the contraction of those from which it is subtracted.

The simplest modes of exciting heat appear to be the compression of elastic fluids, and the collision or friction of solid bodies; although a more usual and a more powerful source of heat is found in various chemical combinations and decompositions, which are produced by the peculiar elective attractions of different substances for each other, or from the influence of the solar rays, which are probably emitted in consequence of the chemical processes that continually take place at the surface of the sun.

The effects of the condensation and rarefaction of elastic fluids are shown by the condenser and the air pump; when an exhaustion is made with rapidity, the thermometer, placed in the receiver of the air pump, usually sinks a degree or two; and when the air is readmitted abruptly into a partial vacuum, the sudden condensation of the rarefied air raises the mercury: and a similar elevation of temperature is produced by the operation of the condenser. Much of this heat is soon dissipated, but by observing the velocity with which the thermometer rises, Mr. Dalton has estimated that air, compressed to half its dimensions, has its temperature elevated about 50 degrees of Fahrenheit; and some of his experiments indicate, when accurately examined, a still greater change. For the present we may define the sense of the term degree, in Fahrenheit's scale, as corresponding to an expansion of a portion of mercury amounting to one ten thousandth part of its bulk; and a degree of Réaumur originally corresponded to an expansion of a weak spirit of wine, amounting to one thousandth part of its bulk. It may be inferred from the velocity of sound, supposing that the excess of its velocity, above the common calculation, is wholly derived from the heat and cold produced by condensation and expansion, that a condensation amounting to $\frac{1}{180}$ of the bulk of any portion of air will raise its temperature one degree of Fahrenheit. When air is very rapidly compressed in the condenser of an air gun, it is sometimes so much heated as actually to set on fire a small portion of tow, placed near the end of the barrel.

The production of heat by friction is too well known to require an experimental proof; but Count Rumford has taken particular pains to ascertain every circumstance which can be supposed to be concerned in the operation of this cause; and the results of his experiments are so striking, that they deserve to be briefly related. He took a cannon, not yet bored, having a projection of two feet beyond its muzzle, a part which is usually cast with the piece, in order to insure the solidity of the metal throughout, by the pressure which its weight occasions. This piece was reduced to the form of a cylinder, joined to the cannon by a smaller neck, and a large hole was bored in it: the whole cannon was then made to revolve on its axis by means of the force of horses, while a blunt steel borer was pressed against the bottom of the hollow cylinder, by a force equal to about 10 000 pounds avoirdupois; the surface of contact of the borer with the bottom of the cylinder being about 2 square inches. This apparatus was wrapped up in flannel, when its temperature was about 60°. In half an hour, when the cylinder had made 960 turns, the horses being stopped, a mercurial thermometer was introduced into a perforation in the bottom of the cylinder, extending from the side to the axis, and it stood at 130°, which Count Rumford considers as expressing very nearly the mean temperature of the cylinder. The dust or scales, abraded by the borer, weighed only 837 grains, or about $\frac{1}{950}$ of the whole weight of the cylinder. In another experiment, the cylinder was surrounded by a tight deal box, fitted with collars of leather, so as to allow it to revolve freely, and the interval between the cylinder and the box was filled with 19 pounds of cold water, which was excluded from the bore of the cylinder by oiled leathers fixed on the borer; and after two hours and a half, the water was made to boil. Hence Count Rumford calculates that the heat produced in this manner, by the operation of friction, was equal to that of 9 wax candles, each three quarters of an inch in diameter, continuing to burn for the same time.

A still more rapid increase of temperature may be obtained, where the relative velocity of the bodies is more considerable, or where they strike each other with violence. Thus a soft nail may be so heated, by three or four blows of a hammer, that we may light a match with it; and by continuing the operation, it may be made red hot: two pieces of wood may also be set on fire by means of a lathe. When a waggon takes fire, for want of having its wheels properly greased, the friction is probably increased by the tenacity of the hardened tar, which perhaps becomes the more combustible as it dries.

One of the most remarkable circumstances, attending the production of heat by friction, is the discovery of Professor Pictet, that it is often much more powerfully excited by soft substances than by harder ones. In making some experiments in a vacuum, in order to examine how far the presence of air might be concerned in the effects of friction, he accidentally interposed some cotton between the bulb of his thermometer and the cup, which was subjected to the friction of various substances as it revolved; and he found that the soft filaments of the cotton excited much more heat, than any other of the substances employed.

The chemical production of heat is of greater practical importance than its mechanical excitation; but by what means chemical changes operate in exciting heat, we cannot attempt to determine. There is certainly no general law of composition or decomposition that can be applied to all such cases: most commonly heat is produced when oxygen exchanges an aeriform for a solid state, or enters into a new combination, and still remains elastic; but in the case of gunpowder, heat is disengaged while an elastic fluid is produced from a solid; and in some other cases the oxygenous principle is wholly unconcerned. It appears on the whole, that however heat may be excited, the corpuscular powers of cohesion and repulsion are always disturbed and called into action, their equilibrium being destroyed and again restored, whether by mechanical or by chemical means. A wax candle, $\frac{3}{4}$ of an inch in diameter, loses a grain of its weight in 37 seconds, and consumes about three grains, or 9 cubic inches, of oxygen gas, producing heat enough to raise the temperature of about 15 000 grains of water a single degree. According to the experiments of Mr. Lavoisier and Mr. Laplace, the combustion of ten grains of phosphorus requires the consumption of 15 grains of oxygen, the combustion of ten grains of charcoal 26, and of hydrogen gas 56; and by the heat produced during the combustion of a pound of phosphorus, 100 pounds of ice may be melted, during that of a pound of charcoal 96$\frac{1}{2}$, of hydrogen gas 295$\frac{1}{2}$, of wax 133, and of olive oil 149; and during the deflagration of a pound of nitre with about one sixth part of its weight of charcoal, twelve pounds of ice may be melted.

The manner in which heat, when excited or extricated by any of these means, passes from one body to another, requires to be very particularly examined. We shall find that this communication happens in one or both of two ways, by contact, or by radiation; and that it may also differ both with respect to the quantity of heat concerned, and to the time occupied by the process. Whatever heat may be, we may safely conclude that in substances of the same kind, at the same temperature or apparent degree of warmth or coldness, its quantity must be proportional to the mass or weight; for instance, that a quart of the water of a given cistern contains twice as much heat as a pint; and where this is true of the different parts of any substance, they must remain in equilibrium with respect to heat. But if two equal portions of the same substance, containing different quantities of heat, be in contact, they will affect each other in such a manner as to have their temperatures equalised, and the more rapidly as the contact is more perfect. Thus, if two portions of a fluid at different temperatures be mixed together, they will acquire immediately an intermediate temperature; and when two solids are in contact, the quantity of heat, communicated by the hotter to the colder in a given time, is nearly proportional to the difference of the temperatures. Hence it would follow, that they could never become precisely of the same temperature in any finite time; but in fact the difference of temperature is rendered, in a moderate time, too small to be perceptible. The nature of the substances concerned has also a material effect on the velocity with which heat is communicated through their internal parts; metallic bodies in general conduct it the most readily, earthy and vitreous bodies, the least; but the various metals possess this power in different degrees; silver and copper conduct heat more rapidly than iron, and platina transmits it but very slowly. Professor Pictet supposes that heat ascends within solid bodies more readily than it descends; but the effect of the air remaining in the imperfect vacuum of the air pump may be sufficient to explain his experiments; the difference of temperature producing an ascending current in the neighbourhood of the heated body, by means of which the cold air continually approaches its lower parts, and carries the heat upwards: and it has been found that the rarefaction of air does not by any means diminish its power of conducting heat, in proportion to the diminution of its density.

Count Rumford's experiments have shown that all fluids are very imperfect conductors of heat by immediate contact, although it is scarcely credible that they can be absolutely nonconductors; but heat is usually communicated between different portions of the same fluid, almost entirely by the mixture of their particles: hence a fluid heated on its surface transmits the heat very slowly downwards, since the parts which are first heated, being rendered specifically lighter, retain their situation above the colder and heavier parts; while, on the contrary, any cause of heat, applied at the bottom of a vessel, very soon reduces all its contents to a uniform temperature. It appears also, from some late experiments, that the immediate transmission of heat within the internal parts of solids is much slower than has commonly been supposed; and it has been found almost impossible to keep a thermometer, at the centre of a large and solid globe of metal, at the same temperature with that of its superficial parts.

Besides the communication of heat by contact, it is usually, if not always, emitted from the surfaces of bodies in the form of radiant heat, which is thrown off in all directions, wherever it meets no obstacle from a substance impervious to it, and is propagated nearly in the same manner as light, and probably with the same velocity, without producing any permanent effect on the temperature of the medium transmitting it. Thus, a thermometer, suspended by a fine thread under the receiver of an air pump, or in the Torricellian vacuum, will continue to vary its temperature with that of the surrounding bodies: and in this case the whole of the heat must be communicated by radiation. Mr. Leslie has discovered that the quantity of heat thus emitted depends not only on the temperature, but also on the nature of the surface concerned, a polished surface of tin emitting only $\frac{12}{100}$, or less than one eighth part as much, as the same surface blackened. A surface of tin scraped with a file in one direction has its powers of radiation more than doubled; but by crossing the scratches, they are reduced nearly to their original state; and a coating of isinglass, resin, or writing paper, or a glassy surface of any kind, produces an effect nearly approaching to that of black paint. This radiation from a heated surface, like that of light, takes place in almost equal degrees in every direction; and its magnitude is nearly independent of the nature of the fluid in contact with the surface, provided however that it be an elastic fluid; for water does not seem to transmit every kind of radiant heat with freedom. It appears that the radiant heat emitted by a surface of glass, or of black paint, is about one third greater than that which is at the same time carried off by the atmospheric air; but that the radiation from a metallic surface is only one sixth of that which the air receives. Mr. Leslie has also found that the same surfaces which emit heat the most freely, are also the readiest to receive it from the radiation of other bodies.

The solar heat radiates freely through air, glass, water, ice, and many other transparent mediums, without producing any sensible effect on their temperatures, and even when it is concentrated by the effect of a burning mirror, it scarcely affects the air through which it passes, and other transparent mediums but little. But the heat of a fire warms a piece of common glass very rapidly, and its further progress is almost entirely interrupted by the glass, although probably a certain portion still continues to accompany the light in all cases. Hence a screen of glass is sometimes practically convenient for allowing us the sight of a fire, and protecting us at the same time from its too great heat. Mr. Lambert showed that culinary heat was much more strongly reflected by mirrors of metal than of glass, although little difference was observable in the quantity of light, and he very justly attributed this difference to the interception of a part of the heat by the glass, which operated with respect to it like an opaque substance, although it transmitted the light with freedom. Opaque substances in general appear to be wholly impervious to radiating heat of all kinds; but Dr. Herschel has found that dark red glass, which transmits a very small portion of light only, suffers some kinds of radiant heat to pass through it with very little interruption.

In other respects, radiating heat is subject, in all cases, to the optical laws which govern the reflection and refraction of light. Dr. Hoffmann appears to have been the first that collected the invisible heat of a stove into a focus by the reflection of one or more concave mirrors. Buffon, Saussure, Pictet, and Mr. King, made afterwards similar experiments on the heat of a plate of iron and of a vessel of boiling water. Mr. Pictet, as well as Hoffmann, employed two mirrors facing each other; and by means of this arrangement the experiment may be performed when the thermometer is placed at a considerable distance from the heated body.

The temperature of the air, not being affected by the radiation of heat, is probably in all respects indifferent to its emission in this manner; and as the rays of light cross each other freely in all possible directions, so it appears that heat may flow in different directions through the same medium without being interrupted; nor does there seem to be any more reason that a hot body should cease to emit heat while it is receiving heat from another body, than that a luminous body should cease to afford light when another body shines on it. This continual interchange of heat, constituting in common cases a kind of equilibrium of motion, appears to have been first suggested by Mr. Prévost, as an explanation of an experiment on the reflection of cold, revived by Mr. Pictet, but originally made some centuries before, by Plempius, and by the Academicians del Cimento. A thermometer, for example, must be supposed to retain its temperature by means of the continual accession of radiant heat from the surrounding bodies, supplying the place of that which is continually thrown off in all directions towards those bodies. Supposing the thermometer to be placed near the focus of a metallic speculum, not much less than a hemisphere, about one half of the heat, which the thermometer would otherwise have received from the surrounding bodies, must be intercepted by the mirror, which, being metallic, emits itself but little radiant heat, but reflects, notwithstanding, an equal quantity of heat from the objects on the opposite side, so that the temperature of the thermometer remains unaltered. But all the heat, which falls on the thermometer from the mirror, must have passed through the conjugate or corresponding focus; and if a body at the same temperature be placed in that focus, the radiation will still be the same: but if a substance absolutely cold were placed there, the whole of the heat before reflected by the mirror would be intercepted, that is, almost half of that which was received by the thermometer from the surrounding bodies; and if a piece of ice be put in the conjugate focus, a delicate thermometer will instantly show its effect in depressing the temperature; as if the cold were absolutely reflected in the same manner as heat or light.

Dr. Herschel's experiments have shown that radiant heat consists of various parts, which are differently refrangible, and that in general, invisible heat is less refrangible than light. This discovery must be allowed to be one of the greatest that has been made since the days of Newton, although the theories of some speculative philosophers might have led to it a few years earlier. Dr. Herschel was occupied in determining the properties of various kinds of coloured glass, which rendered them more or less fit for enabling the eye to view the sun through a telescope; and for this purpose it was necessary to inquire which of the rays would furnish the greatest quantity of light, without subjecting the eye to the inconvenience of unnecessary heat. He first observed that the heat became more and more considerable as the thermometer approached the extreme red rays in the prismatic spectrum; and pursuing the experiment, he found not only that the heat continued beyond the visible spectrum, but that it was even more intense when the thermometer was at a little distance without the limits of the spectrum, than in any point within it. (Plate XXXIX. Fig. 546, 547.)

Sir Henry Englefield has repeated these experiments with many additional precautions, and Mr. Davy was a witness of their perfect accuracy: the excess of heat beyond the spectrum was even considerable enough to be ascertained by the sense of warmth occasioned by throwing it on the hand. The skin appears, when compared with a thermometer, to have its sensibility more adapted to the perception of radiant heat than to that of heat imparted by contact, perhaps because a much smaller quantity of heat is sufficient to raise the temperature of the thin cuticle very considerably, than would be required in order to affect any thermometer in the same degree.

It was first observed in Germany by Ritter, and soon afterwards in England by Dr. Wollaston, that the muriate of silver is blackened by invisible rays, which extend beyond the prismatic spectrum, on the violet side. It is therefore probable that these black or invisible rays, the violet, blue, green, perhaps the yellow, and the red rays of light, and the rays of invisible heat, constitute seven different degrees of the same scale, distinguished from each other into this limited number, not by natural divisions, but by their effects on our senses: and we may also conclude that there is some similar relation between heated and luminous bodies of different kinds.

The effects of heat, thus originating, and thus communicated, may be divided into those which are temporary only, and those which are permanent. The permanent effects are principally confined to solids, but the temporary effects are different with respect to substances in different states of aggregation, and they also frequently comprehend a change from one of these states to another. The effect of heat on an elastic fluid is the simplest of all these, being merely an expansion of about one five hundredth of its bulk for each degree of Fahrenheit that the temperature is raised; or an equivalent augmentation of the elasticity when the fluid is confined to a certain space. This expansion is very nearly the same for all gases and vapours, amounting to $\frac{1}{500}$ for each degree, at the common temperature of 56° of Fahrenheit, but at higher temperatures it is less than $\frac{1}{500}$ of the bulk of the gas, and at lower temperatures somewhat more, being nearly the same in quantity for the same portion of the fluid at all temperatures.

When an elastic fluid is contracted by cold within certain limits, determined by the degree of pressure to which it is exposed, as well as by the nature of the fluid, its particles become subjected to the force of cohesion; they rush still nearer together, and form a liquid. Thus, when steam, under the common atmospheric pressure, is cooled below the heat of boiling water, it is instantly condensed, and becomes water: but with a pressure of two atmospheres, it would be condensed at a temperature 36° higher, and with the pressure of half our atmosphere only, it might be cooled without condensation 33° lower than the common temperature of boiling water. And similar effects take place in vapours of other kinds at higher or lower temperatures, a double pressure producing, in all cases an equal disposition to condensation, with a depression of temperature of between 20 and 40 degrees, and most commonly of about 35°, of Fahrenheit. Thus, the vapour of spirit of wine is usually condensed at 175° of Fahrenheit; but with a double pressure it is condensed at a temperature 39° higher; and with the pressure of half an atmosphere, at a temperature 35° lower; and the vapour of ether, which is commonly condensed at 102°, requires a temperature 38° higher, with a double pressure, or as much lower, with half the usual pressure. If the temperature be below the freezing point of the liquid, the pressure being sufficiently lessened, the vapour may still retain its elasticity, but a further reduction of temperature or increase of pressure will convert it immediately into a solid.

The expansion of liquids by the effect of heat is much less uniform and regular than that of elastic fluids, since it varies considerably, not only in different liquids, but also in the same liquid at different temperatures, being in general greater as the temperature is more elevated, and sometimes almost in proportion to the excess of the temperature above a certain point, at which it begins. This variation appears to be the least considerable in mercury, although even this fluid expands a little more rapidly as it becomes more heated; but the expansion is always very nearly one ten thousandth for each degree: that of water is equal to this at the temperature 64°, and is greater or less nearly in proportion to the distance from 39°, where it begins, but in high temperatures it varies less, since it is not quite four times as great at the heat of boiling water. The expansion of spirit of wine at 70° is six times as great as that of mercury: its utmost variation is much less than that of water, although it is at least twice as great in some parts of the scale as in others.

It has already been observed that an elevation of temperature considerably diminishes the powers of cohesion and of repulsion in solid bodies: the same is also true of liquids; for the height to which a liquid rises in a capillary tube is diminished somewhat less than $\frac{1}{1000}$ for each degree of Fahrenheit that the temperature is raised, the force of superficial cohesion being diminished both by the diminution of the immediate actions of the particles, and by that of the distances to which they extend.

When the temperature of a liquid is so much elevated as to become equal to that of its vapour in a state capable of sustaining the atmospherical pressure, or any other pressure which may be substituted for it, a certain portion of the liquid is converted into vapour, and the heat being generally applied at the bottom of the vessel, the vapour rises up in bubbles, and the effect is called boiling. The whole liquid is not converted at once into vapour, because a certain quantity of heat appears to be consumed by the change, and a constant supply of heat is necessary, in order that the operation may be continued.

It is not, however, only at the boiling point that a fluid begins to be converted into vapour: the pressure of the atmosphere is not sufficient wholly to prevent the detachment of a certain quantity of vapour from its surface, at a temperature which is incapable of supporting it separately in the form of steam in the open air, and it may be thus suspended, when mixed either with common air, or with any other elastic fluid, at the ordinary temperature of the atmosphere. And it appears that the quantity, which is thus suspended, bears very nearly some constant proportion to the density of which the steam is capable at the given temperature in a separate state, the interposition of the air either not affecting the distance at which the cohesion would take place, or altering it equally in all cases. It seems to be most probable that the density of vapour, suspended in this manner in the atmosphere, is always about twice as great, or at least half as great again, as that of steam existing independently at the same temperature. There is perhaps no liquid absolutely free from a disposition to evaporate: even mercury rises in the vacuum of the barometer, and lines the cavity with small globules; and it is said that the effect of light is favourable to this slow evaporation. At whatever temperature evaporation takes place it is always accompanied by the production of cold; hence it is usual in warm climates, to employ various methods of promoting evaporation, in order to lower the temperature of the air, to cool liquids for drinking, or even to procure ice for domestic uses.

It appears that all aqueous fluids are contracted by cold, until we arrive at a certain point, which is generally about 7 or 8 degrees higher than their freezing point: they then expand again almost in an equal degree as they are still more cooled; and provided that they be free from agitation, they may remain liquid at a temperature considerably below the point, at which they usually freeze, and at which their ice always melts. Water may be cooled in this manner to about 10° of Fahrenheit, and if it be then agitated, or especially if a small particle of ice or snow be thrown into it, a certain part of it will instantly congeal, and its temperature will be raised at once to 32°, in consequence of the heat which is always produced or extricated in the act of freezing. In most cases, although not in all, the solid occupies more space than the fluid; thus, it is probable that ice, when perfectly free from air bubbles, is at least one 16th lighter than water at the same temperature. A saturated solution of Glauber's salts, or sulfate of soda, in hot water, may be cooled slowly to the temperature of the atmosphere, when the pressure of the air is excluded, and may be made to crystallize by admitting it suddenly, the liquor becoming at the same time warm in consequence of the heat which is extricated; and there is no doubt but that the congelation of water, and perhaps of all other substances, is a crystallization of the same kind.

The expansions of solid bodies appear to be more regular than those of liquids or even of elastic fluids; they vary little at any temperature, although it is said that they do not always take place in their full extent at the instant

2

that the substance has become heated, and that a blow, or the agitation produced when they are made to sound by the friction of the bow of a violin, may sometimes be observed to cause them to assume the state of equilibrium with greater rapidity. Brass expands about one hundred thousandth of its length for each degree of Fahrenheit, copper and gold a little less; silver somewhat more; glass and platina less than half as much; iron and steel about two thirds as much; tin one third more, and lead and zinc about half as much more. Wood and earthenware are the least expansible of all known solids. The diminution of the elasticity of iron and steel by the elevation of their temperature amounts to about $\frac{1}{5000}$ of the whole for each degree: but probably various substances are variously affected in this respect.

The liquefaction of solids, and their conversion into fluids by the operation of heat, is liable to fewer irregularities than any other of its effects; the change depending only on the temperature, and not being accelerated or retarded by any accidental circumstances. When the temperature is too low, or the pressure too small, for the existence of the substance in a liquid form, it may still be converted into vapour, either mixed with air, or in a separate state: thus ice loses weight when it is exposed to a dry frosty wind; and camphor, benzoin, and ammonia are sublimed by heat without being melted, although it is probable, that a pressure sufficiently strong might enable them to exist as liquids in elevated temperatures. In all changes from solidity to liquidity or to elastic fluidity, a certain quantity of heat disappears, except some cases in which a chemical decomposition has accompanied the change; thus, in the detonation of gunpowder, a large quantity of gas acquires the state of elasticity, but at the same time a great degree of heat is produced.

The effects of the expansion of bodies by heat, and of their contraction by cold, are observed in the frequent accidents which happen to glass and to porcelain from a sudden change of temperature. Glass conducts heat so slowly, that one side of a vessel may become much heated, and consequently expanded, while the other remains much colder, and if the vessel cannot readily accommodate its form to this change of proportions, it will most commonly crack, the colder parts dividing, in consequence of their being too much stretched by the adjoining hotter parts. Hence the thinner the glass is, the less liable it is to crack from any sudden expansion; and if it be very thick, however simple its form may be, it will still crack; for no flexure, which it can assume, can be sufficient for the equilibrium of the external parts without being too great for that of the parts near the middle.

When glass in fusion is very suddenly cooled, its external parts become solid first, and determine the magnitude of the whole piece; while it still remains fluid within. The internal part, as it cools, is disposed to contract still further, but its contraction is prevented by the resistance of the external parts, which form an arch or vault round it, so that the whole is left in a state of constraint; and as soon as the equilibrium is disturbed in any one part, the whole aggregate is destroyed. Hence it becomes necessary to anneal all glass, by placing it in an oven, where it is left to cool slowly; for, without this precaution, a very slight cause would destroy it. The Bologna jars, sometimes called proofs, are small thick vessels, made for the purpose of exhibiting this effect; they are usually destroyed by the impulse of a small and sharp body, for instance a single grain of sand, dropped into them; and a small body appears to be often more effectual than a larger one; perhaps because the larger one is more liable to strike the glass with an obtuse part of its surface. In the same manner the glass drops, sometimes called Prince Rupert's drops, which are formed by suffering a portion of green glass in fusion to fall into water, remain in equilibrium while they are entire; but when the small projecting part is broken off, the whole rushes together with great force, and rebounding by its elasticity, exhibits the effect of an explosion. The ends of these drops may, sometimes, but not always, be gradually ground off without destroying them, so that the concussion produced by breaking the drop seems to be concerned in the destruction of the equilibrium.

The tempering of metals appears to bear a considerable analogy to the annealing of glass; when they are made red hot, and suddenly cooled, they acquire a great degree of hardness, which renders them proper for some purposes, while the brittleness which accompanies it would be inconvenient for others. By heating them again to a more moderate temperature, and suffering them to cool more gradually, they are rendered softer and more flexible, and the more as the heat which is thus applied is the more considerable.

3

which forms itself on the surface of polished iron or steel, serves as a test of the degree of heat which is applied to it, the yellowish colour which it assumes indicating the first stage of tempering, the violet the second, and the blue the last; and if the heat be raised till the surface becomes grey, the steel will be rendered perfectly soft. The density of metals is also a little increased by tempering them, probably for the same reason as water is more dense than ice. In what manner the unequal distribution of the mutual actions of the particles of bodies contributes to increase their hardness, cannot be very positively ascertained, although some conjectures might be formed which would, perhaps, be in some measure explanatory of the facts; but it is safer, in the present state of our knowledge, to be contented with tracing the analogy between these effects in substances of different kinds, and under different circumstances, without attempting to understand completely the immediate operation of the forces which are concerned.

LECTURE LII.

ON THE MEASURES AND THE NATURE OF HEAT.

THE principal particulars concerning the origin, the progress, and the effects of heat, having been noticed in the last lecture, we now proceed to examine the most usual modes of measuring its degrees and its quantity, and to inquire into the most probable opinions respecting its intimate nature and its immediate operation.

The expansion of solids is measured by a pyrometer, which is calculated for rendering the smallest change of dimensions perceptible either by mechanical or by optical means. The first of these methods was adopted by those who first investigated these effects; a bar of metal being placed in a vessel of water, or of oil, which was heated by lamps, while the extremities of the bar were in contact with a fixed point on one side, and on the other with a series of levers, which multiplied the expansions, so as to render them easily observable by means of the end of the last lever, serving as an index. But it is obvious that the expansion of the fixed part of the instrument, and the irregular changes of temperature of the levers themselves, must very much interfere with the accuracy of such an instrument. A much more correct mode of determination is to employ two microscopes, fixed to an apparatus, which is always kept, by means of ice, at a constant temperature, and to observe with a micrometer the change of place of either end of the heated bar.

For such purposes, the degrees of heat may be ascertained by the natural measures of the freezing and boiling points of certain liquids, and of water in particular; but for subdividing the intervals between these points, other means must be employed. The most natural mode of determining the intermediate degrees of heat, which must be considered as the standard for the comparison of all others, is too laborious and complicated for common use. If we mix together equal quantities of the same liquid at two different temperatures, they will obviously acquire an intermediate temperature, which is the natural mean between the separate temperatures, provided that no heat be lost or gained during the process; and provided that no irregularity be produced from the approach of the liquid to a state of congelation, the existence of which might be detected by a comparison of experiments on various liquids at the same temperatures. By repeating the operation, we may subdivide the intervals as often as we please, or we may mix the liquids in any other proportion, so as to obtain at once any other point of the scale, which may afterwards be identified by a thermometer of any description.

There is also another method of comparing the divisions of a thermometer with those of the natural scale, but it is not wholly free from objections; the instrument being placed in a cone of the sun's rays, made to converge by means of a lens or mirror, the quantity of heat falling on it must be nearly in the inverse proportion of the square of its distance from the focus; and the elevation of a common thermometer appears to be nearly proportional to the heat which is thrown on it in this manner.

The expansion of solids probably approaches the nearest to the steps of the natural scale, although even in this there seems to be some inequality; but that of mercury is scarcely less regular, and a portion of mercury inclosed in a bulb of glass, having a fine tube connected with it, forms a thermometer the most convenient, and most probably the most accurate, of any, for common use; the degrees corresponding very nearly with those of the natural scale, although, according to the most accurate experiments, they appear to indicate, towards the middle of the common scale of Fahrenheit, a temperature 2 or 3 degrees too low. There is an inequality of the same kind, but still greater, in the degrees of the spirit thermometer; and this instrument has also the disadvantage of being liable to burst in a heat below that of boiling water; although it is well calculated for the measurement of very low temperatures, since pure alcohol has never yet been frozen, while mercury has been reduced to a solid by the cold of Siberia and of Hudson's Bay: but both mercury and linseed oil support a heat of between 5 and 600° without ebullition. For higher temperatures than this, a thermometer has been made of

I

semitransparent porcelain, containing a fusible metal, which may be compared with the upper part of the mercurial scale, and then continued further; and the expansion of such of the metals, as are difficult of fusion, affords another mode of determining the highest degrees of heat. Mr. Wedgwood's thermometer derives its properties from the contraction of a small brick of prepared clay, which contracts the more, as the heat to which it is exposed is higher: it may be extremely useful for identifying the degree of heat which is required for a particular purpose: but for the comparison of temperatures by an extension of the numerical scale, we have not sufficient evidence of its accuracy to allow us to depend on its indications; and it is scarcely credible that the operation of furnaces, of any kind, can produce a heat of so many thousand degrees of a natural scale, as Mr. Wedgwood's experiments have led him to suppose; nor is the supposition consistent with the observations of other philosophers.

Mercurial thermometers are in general hermetically sealed, the tube being perfectly closed at the end, in order to exclude dust, and to prevent the dissipation of the mercury. When a standard thermometer is to be adjusted, its freezing point is readily fixed by immersing it wholly in melting snow or pounded ice; but for the boiling point, some further precautions are required; the easiest method appears to be, to immerse its bulb in an open vessel of boiling water, to cover it with several folds of cloth, and to pour hot water continually over it; for if it were immersed to a considerable depth, the pressure would raise the temperature of the boiling point, and if it were not covered, the mercury in the tube would be too cold. Attention must also be paid to the state of the barometer; it must either stand at 30 inches, or the place of the boiling point must be raised, when the barometer is lower than 30, and lowered when it is higher; the difference of nine tenths of an inch either way requiring an alteration amounting to $\frac{1}{100}$ of the interval between freezing and boiling. This interval is subdivided, in Fahrenheit's thermometer, into 180 degrees; in Réaumur's, into 80, and in the centigrade thermometer of Celsius and of the French, into 100; and in making the subdivision, care must be taken to examine the equality of the bore throughout, by observing the length occupied by a detached portion of mercury, and to allow for any irregularities which may have been thus detected. The scales of Réaumur and of Celsius begin at the freezing point of water; but in that of Fahrenheit the freezing point stands at 32°, the scale beginning from the cold produced by a freezing mixture, which was supposed by Fahrenheit to be the greatest that would ever occur in nature.

The expansion, which is observed in a mercurial thermometer, is in reality only the difference of the expansions of mercury and of glass; but this circumstance produces no difference in the accuracy of the results. The separate effects of the expansion of glass are, however, sometimes perceptible; thus, when a thermometer is plunged suddenly into hot water, the glass, being first heated, expands more rapidly than the mercury, and, for a moment, the thermometer falls. This circumstance would perhaps be still more observable in a thermometer of spirit or of water; for an equal bulk of these liquids would be much longer in acquiring the temperature of the surrounding medium than a mercurial thermometer.

The expansion of elastic fluids affords in some cases a test of heat, which is very convenient from its great delicacy, and because a very small quantity of heat is sufficient to raise their temperature very considerably. The thermometer first invented by Drebel was an air thermometer; but instruments of this kind, when they are subject to the variations of the pressure of the atmosphere as well as to those of its temperature, are properly called manometers, and require, for enabling us to employ them as thermometers, a comparison with the barometer; while on the other hand, they may be used as barometers, if the temperature be otherwise ascertained. They are however, very useful even without this comparison, in delicate experiments of short duration, since the changes of the barometer are seldom very rapid; and they may also be wholly freed from the effects of the pressure of the atmosphere, in various ways. Bernoulli's method consists in closing the bulb of a common barometer, so as to leave the column of mercury in equilibrium with the air contained in the bulb at its actual temperature, and capable of indicating, by the changes of its height and of its pressure, any subsequent changes in the temperature of the air, which must affect both its bulk and its elasticity. Mr. Leslie's photometer, or differential thermometer, has some advantages over this instrument, but it can only be employed where the changes of temperature can be confined to a part only of the instrument. The elasticity of the air contained in the bulb is here counteracted, not by the pressure of a column of mercury, but by the elasticity of another portion of air in a second bulb, which is not to be exposed to the heat or cold that is to be examined: and the difference between the temperatures of the two bulbs is indicated by the place of a drop of a liquid, moving freely in the tube which joins them. (Plate XXXIX. Fig. 548 . . 550.)

The degree of heat, as ascertained by a thermometer, is only to be considered as a relation to the surrounding bodies, in virtue of which a body supports the equilibrium of temperature when it is in the neighbourhood of bodies equally heated: thus, if a thermometer stands at 60°, both in a vessel of water, and in another of mercury, we may infer that the water and the mercury may be mixed without any change of their temperature: but the absolute quantity of heat, contained in equal weights, or in equal bulks, of any two bodies at the same temperature, is by no means the same. Thus, in order to raise the temperature of a pound of water from 50° to 60°, we need only to add to it another pound of water at 70°, which, while it loses 10° of its own heat, will communicate 10° to the first pound; but the temperature of a pound of mercury at 50° may be raised 10°, by means of the heat imparted to it, by mixing with it one thirtieth part of a pound of water, at the same temperature of 70°. Hence we derive the idea of the capacities of different bodies for heat, which was first suggested by Dr. Irvine, the capacity of mercury being only about one thirtieth part as great as that of water. And by similar experiments it has been ascertained, that the capacity of iron is one eighth of that of water, the capacity of silver one twelfth, and that of lead one twenty fourth. But for equal bulks of these different substances, the disproportion is not quite so great; thus, copper contains nearly the same quantity of heat in a given bulk as water; iron, brass, and gold, a little less, silver $\frac{5}{6}$ as much, but lead and glass each about one half only.

It is obvious that if the capacity of a body for heat, in this sense of the word, were suddenly changed, it would immediately become hotter or colder, according to the nature of the change, a diminution of the capacity producing heat, and an augmentation cold. Such a change of capacity is often a convenient mode of representation for some of the sources of heat and cold: thus, when heat is produced by the condensation of a vapour, or by the congelation of a liquid, we may imagine that the capacity of the substance is diminished; and that it overflows, as a vessel would do if its dimensions were contracted. It appears also from direct experiments, in some such cases, that the capacity of the same substance is actually greater in a liquid than in a solid state, and in a state of vapour, than in either; and both Dr. Irvine and Dr. Crawford have attempted to deduce, from a comparison of the proportional capacities of water and ice, with the quantity of heat extricated during congelation, a measure of the whole heat which is contained in these substances, and an estimation of the place which the absolute privation of heat, or the natural zero, ought to occupy in the scale of the thermometer. Thus, when a pound of ice, at 32°, is mixed with a pound of water at 172° of Fahrenheit, the whole excess of 140° is absorbed in the conversion of the ice into water, and the mixture is reduced to the temperature of 32°; and, on the other hand, when a pound of ice freezes, a certain quantity of heat is evolved which is probably capable of raising the temperature of a pound of water 140°, or that of 140 pounds a single degree. Dr. Crawford found, by means of other experiments, that a quantity of heat capable of raising the temperature of water 9° would raise that of ice as much as 10°; hence he inferred that the capacity of ice was $\frac{9}{10}$ as great as that of water, and that if this capacity, instead of being reduced to $\frac{9}{10}$, had been wholly destroyed, the quantity of heat extricated would have been 10 times as great, or about 1400°, which has, therefore, been considered as the whole quantity of heat contained in a pound of water at 32°, and the beginning of the natural scale has been placed about 1368° below the zero of Fahrenheit. Dr. Irvine makes the capacity of ice still less considerable, and places the natural zero about 900 degrees below that of Fahrenheit.

If direct experiments on the quantities of heat, required for producing certain elevations of temperature, in different states of the same substance, compared in this manner with the emission or absorption of heat which takes place while those changes are performed, agreed with similar experiments made on different substances, there could be no objection to the mode of representation. But if it should appear that such comparisons frequently present us with contradictory results, we could no longer consider the theory of capacities for heat as sufficient to explain the phenomena. With respect to the simple changes constituting congelation and liquefaction, condensation and evaporation, and compression and rarefaction, there appears to be at present no evidence of the insufficiency of this theory; it has not perhaps yet been shown that the heat absorbed in any one change is always precisely equal to that which is emitted in the return of the substance to its former state, but nothing has yet been advanced which renders this opinion improbable; and the estimation of the natural zero, which is deduced from this doctrine, may at least be considered as a tolerable approximation.

If, however, we attempt to deduce the heat produced by friction and by combustion from changes of the capacities of bodies, thus estimated, we shall find that the comparison of a very few facts is sufficient to demonstrate the imperfection of such a theory. Count Rumford found no sensible difference between the capacities of solid iron and of its chips; but if we even suppose, for the sake of the argument, that the pressure and friction of the borer had lessened the capacity of the iron one twelfth, so as to make it no greater than that of copper; we shall then find that one twelfth of the absolute heat of the chips, thus abraded, must have amounted to above 60 000 degrees of Fahrenheit, and consequently that the natural zero ought to be placed above 700 000 degrees below the freezing point, instead of 14 or 1500 only. It is, therefore, impossible to suppose that any alteration of capacities can account for the production of heat by friction: nor is it at all easier to apply this theory correctly to the phenomena of combustion. A pound of nitre contains about half its weight of dry acid, and the capacity of the acid, when diluted, is little more than half as great as that of water; the acid of a pound of nitre must therefore contain less heat than a quarter of a pound of water: but Lavoisier and Laplace have found, that the deflagration of a pound of nitre produces a quantity of heat sufficient to melt twelve pounds of ice, consequently the heat extricated by the decomposition of a pound of dry nitrous acid must be sufficient to melt 24 pounds of ice; and even supposing the gases, extricated during the deflagration, to absorb no more heat than the charcoal contained, which is for several reasons highly improbable, it follows that a pound of water ought to contain at least as much heat, as would be sufficient to melt 48 pounds of ice, that is, about 6720 degrees of Fahrenheit.

In short, the further we pursue such calculations, the more we shall be convinced of the impossibility of applying them to the phenomena. In such a case as that of the nitrous acid, Dr. Black's term of latent heat might be thought applicable, the heat being supposed to be contained in the substance, without being comprehended in the quantity required for maintaining its actual temperature. But even this hypothesis is wholly inapplicable to the extrication of heat by friction, where all the qualities of the substances concerned remain precisely the same after the operation, as before it. If any further argument were required in confutation of the opinion, that the heat excited by friction is derived from a change of capacity, it might be obtained from Mr. Davy's experiment on the mutual friction of two pieces of ice, which converted them into water, in a room at the temperature of the freezing point: for in this case it is undeniable that the capacity of the water must have been increased during the operation ; and the heat produced could not, therefore, have been occasioned by the diminution of the capacity of the ice.

This discussion naturally leads us to an examination of the various theories which have been formed respecting the intimate nature of heat; a subject upon which the popular opinion seems to have been lately led away by very superficial considerations. The facility with which the mind conceives the existence of an independent substance, liable to no material variations, except those of its quantity and distribution, especially when an appropriate name, and a place in the order of the simplest elements has been bestowed on it, appears to have caused the most eminent chemical philosophers to overlook some insuperable difficulties attending the hypothesis of caloric. Caloric has been considered as a peculiar elastic or ethereal fluid, pervading the substance or the pores of all bodies, in different quantities, according to their different capacities for heat, and according to their actual temperatures; and being transferred from one body to another upon any change of capacity, or upon any other disturbance of the equilibrium of temperature: it has also been commonly supposed to be the general principle or cause of repulsion ; and in its passage from one body to another, by radiation, it has been imagined by some to flow in a continued stream, and by others in the form of separate particles, moving, with inconceivable velocity, at great distances from each other.

The circumstances which have been already stated, respecting the production of heat by friction, appear to afford an unanswerable confutation of the whole of this doctrine. If the heat is neither received from the surrounding bodies, which it cannot be without a depression of their temperature, nor derived from the quantity already accumulated in the bodies themselves, which it could not be, even if their capacities were diminished in any imaginable degree, there is no alternative but to allow that heat must be actually generated by friction; and if it is generated out of nothing, it cannot be matter, nor even an immaterial or semimaterial substance. The collateral parts of the theory have also their separate difficulties: thus, if heat were the general principle of repulsion, its augmentation could not diminish the elasticity of solids and of fluids; if it constituted a continued fluid, it could not radiate freely through the same space in different directions; and if its repulsive particles followed each other at a distance, they would still approach near enough to each other, in the focus of a burning glass, to have their motions deflected from a rectilinear direction.

If heat is not a substance, it must be a quality; and this quality can only be motion. It was Newton's opinion, that heat consists in a minute vibratory motion of the particles of bodies, and that this motion is communicated through an apparent vacuum, by the undulations of an elastic medium, which is also concerned in the phenomena of light. If the arguments which have been lately advanced, in favour of the undulatory nature of light, be deemed valid, there will be still stronger reasons for admitting this doctrine respecting heat, and it will only be necessary to suppose the vibrations and undulations, principally constituting it, to be larger and stronger than those of light, while at the same time the smaller vibrations of light, and even the blackening rays, derived from still more minute vibrations, may, perhaps, when sufficiently condensed, concur in producing the effects of heat. These effects, beginning from the blackening rays, which are invisible, are a little more perceptible in the violet, which still possess but a faint power of illumination; the yellow green afford the most light; the red give less light, but much more heat, while the still larger and less frequent vibrations, which have no effect on the sense of sight, may be supposed to give rise to the least refrangible rays, and to constitute invisible heat.

It is easy to imagine that such vibrations may be excited in the component parts of bodies, by percussion, by friction, or by the destruction of the equilibrium of cohesion and repulsion, and by a change of the conditions on which it may be restored, in consequence of combustion, or of any other chemical change. It is remarkable that the particles of fluids, which are incapable of any material change of temperature from mutual friction, have also very little power of communicating heat to each other by their immediate action, so that there may be some analogy, in this respect, between the communication of heat and its mechanical excitation.

The effects of heat on the cohesive and repulsive powers of bodies have sometimes been referred to the centrifugal forces and mutual collisions of the revolving and vibrating particles; and the increase of the elasticity of aeriform fluids has been very minutely compared with the force which would be derived from an acceleration of these internal motions. In solids and in liquids, however, this increase of elasticity is not observable, and the immediate effect of heat diminishes not only the force of cohesion, but also in some degree, that of repulsion, so that these vibrations, if they exist, must derive their effect on the corpuscular forces from the alterations which they produce on the situation of the particles, with respect to the causes of these forces.

The different chemical effects of heat and light are far from furnishing any objection to this system; it is extremely easy to imagine the attraction between two or three bodies to be modified by the agitations, into which their particles are thrown. If certain undulations be capable of affecting one of the three bodies only, its cohesion with both the others may be weakened, and hence their mutual attraction may be comparatively increased; and from various combinations of such differences, in the operation of different kinds of heat and light, a great diversity of effects of a similar kind may be derived.

If heat, when attached to any substance, be supposed to consist in minute vibrations, and when propagated from one body to another, to depend on the undulations of a medium highly elastic, its effects must strongly resemble those of sound, since every sounding body is in a state of vibration, and the air, or any other medium, which transmits sound, conveys its undulation to distant parts by means of its elasticity. And we shall find that the principal phenomena of heat may actually be illustrated by a comparison with those of sound. The excitation of heat and sound are not only similar, but often identical; as in the operations of friction and percussion: they are both communicated sometimes by contact and sometimes by radiation; for besides the common radiation of sound through the air, its effects are communicated by contact, when the end of a tuning fork is placed on a table, or on the sounding board of an instrument, which receives from the fork an impression that is afterwards propagated as a distinct sound. And the effect of radiant heat, in raising the temperature of a body upon which it falls, resembles the sympathetic agitation of a string, when the sound of another string, which is in unison with it, is transmitted to it through the air. The water, which is dashed about by the vibrating extremities of a tuning fork dipped into it, may represent the manner in which the particles at the surface of a liquid are thrown out of the reach of the force of cohesion, and converted into vapour; and the extrication of heat, in consequence of condensation, may be compared with the increase of sound produced by lightly touching a long chord which is slowly vibrating, or revolving in such a manner as to emit little or no audible sound; while the diminution of heat by expansion, and the increase of the capacity of a substance for heat, may be attributed to the greater space afforded to each particle, allowing it to be equally agitated with a less perceptible effect on the neighbouring particles. In some cases, indeed, heat and sound not only resemble each other in their operations, but produce precisely the same effects; thus, an artificial magnet, the force of which is quickly destroyed by heat, is affected more slowly in a similar manner, when made to ring for a considerable time; and an electrical jar may be discharged, either by heating it, or by causing it to sound by the friction of the finger.

All these analogies are certainly favourable to the opinion of the vibratory nature of heat, which has been sufficiently sanctioned by the authority of the greatest philosophers of past times, and of the most sober reasoners of the present. Those, however, who look up with unqualified reverence to the dogmas of the modern schools of chemistry, will probably long retain a partiality for the convenient, but superficial and inaccurate, modes of reasoning, which have been founded on the favourite hypothesis of the existence of caloric as a separate substance; but it may be presumed that in the end a careful and repeated examination of the facts, which have been adduced in confutation of that system, will make a sufficient impression on the minds of the cultivators of chemistry, to induce them to listen to a less objectionable theory.

LECTURE LIII.

ON ELECTRICITY IN EQUILIBRIUM.

THE phenomena of electricity are as amusing and popular in their external form as they are intricate and abstruse in their intimate nature. In examining these phenomena, a philosophical observer will not be content with such exhibitions as dazzle the eye for a moment, without leaving any impression that can be instructive to the mind, but he will be anxious to trace the connexion of the facts with their general causes, and to compare them with the theories which have been proposed concerning them: and although the doctrine of electricity is in many respects yet in its infancy, we shall find that some hypotheses may be assumed, which are capable of explaining the principal circumstances in a simple and satisfactory manner, and which are extremely useful in connecting a multitude of detached facts into an intelligible system. These hypotheses, founded on the discoveries of Franklin, have been gradually formed into a theory, by the investigations of Aepinus and Mr. Cavendish, combined with the experiments and inferences of Lord Stanhope, Coulomb, and Robison.

We shall first consider the fundamental hypotheses on which this system depends, and secondly the conditions of equilibrium of the substances concerned in it; determining the mode of distribution of the electric fluid, and the forces or pressures derived from its action when at rest; all which will be found to be deduced from the theory precisely as they are experimentally observable. The motions of the electric fluid will next be noticed, as far as we can form any general conclusions respecting them; and the manner in which the equilibrium of electricity is disturbed, or the excitation of electricity, will also be considered; and, in the last place, it will be necessary to take a view of the mechanism or the practical part of electricity, and to examine the natural and artificial apparatus concerned in electrical phenomena, as well as in those effects, which have been denominated galvanic.

It is supposed that a peculiar ethereal fluid pervades the pores, if not the actual substance, of the earth and of all other material bodies, passing through them with more or less facility, according to their different powers of conducting it: that the particles of this fluid repel each other, and are attracted by the particles of common matter: that the particles of common matter also repel each other: and that these attractions and repulsions are equal among themselves, and vary inversely as the squares of the distances of the particles.

The effects of this fluid are distinguished from those of all other substances by an attractive or repulsive quality, which it appears to communicate to different bodies, and which differs in general from other attractions and repulsions, by its immediate diminution or cessation, when the bodies, acting on each other, come into contact, or when they are touched by other bodies. The name electricity is derived from electrum, amber; for it was long ago observed that amber, when rubbed, continues for some time to attract small bodies; but at present electricity is usually excited by other means. In general a body is said to be electrified, when it contains, either as a whole, or in any of its parts, more or less of the electric fluid than is natural to it; and it is supposed that what is called positive electricity depends on a redundancy, and negative electricity on a deficiency of the fluid.

These repulsions and attractions are supposed to act, not only between two particles which are either perfectly or very nearly in contact with each other, but also between all other particles at all distances, whatever obstacles may be interposed between them. Thus, if two electrified balls repel each other, the effect is not impeded by the interposition of a plate of glass: and if any other substance interposed appears to interfere with their mutual action, it is in consequence of its own electrical affections. In these respects, as well as in the law of their variation, the electrical forces differ from the common repulsion which operates between the particles of elastic fluids, and resemble more nearly that of gravitation. Their intensity, when separately considered, is much greater than that of gravitation, and they might be supposed to be materially concerned in the great phenomena of the universe; but in the common neutral state of all bodies, the electrical fluid, which is every where present, is so distributed, that the various forces hold each other exactly in equilibrium, and the separate results are destroyed; unless we choose to consider gravitation itself as arising from a comparatively slight inequality between the electrical attractions and repulsions.

The attraction of the electric fluid to common matter is shown by its communication, from one body to, another, which is less copiously supplied with it, as well as by many other phenomena; and this attraction of the fluid of the first body, to the matter of the second, is precisely equal to its repulsion for the quantity of the fluid, which naturally belongs to the second, so as to saturate the matter. For the excess or deficiency of the fluid in the first body does not immediately produce either attraction or repulsion, so long as the natural distribution of the fluid in the second body remains unaltered.

Since also two neutral bodies, the matter which they contain being saturated by the electric fluid, exhibit no attraction for each other, the matter in the first must be repelled by the matter in the second; for its attraction for the fluid of the second would otherwise remain uncompensated. We are, however, scarcely justified in classing this mutual repulsion among the fundamental properties of matter; for useful as these laws are in explaining electrical appearances, they seem to deviate too far from the magnificent simplicity of nature's works, to be admitted as primary consequences of the constitution of matter: they may, however, be considered as modifications of some other more general laws, which are yet wholly unknown to us.

When the equilibrium of these forces is destroyed, the electric fluid is put in motion; those bodies, which allow the fluid a free passage, are called perfect conductors; but those which impede its motion, more or less, are nonconductors, or imperfect conductors. For example, while tbe electric fluid is received into the metallic cylinder of an electrical machine, its accumulation may be prevented by the application of the hand to the cylinder which receives it, and it will pass off through the person of the operator to the ground; hence the human body is called a conductor. But when the metallic cylinder, or conductor, of the machine is surrounded only by dry air, and supported by glass, the electric fluid is retained, and its density increased, until it becomes capable of procuring itself a passage, some inches in length, through the air, which is a very imperfect conductor. If a person, connected with the conductor, be placed on a stool with glass legs, the electricity will no longer pass through him to the earth, but may be so accumulated, as to make its way to any neighbouring substance, which is capable of receiving it, exhibiting a luminous appearance, called a spark; and a person or a substance, so placed as to be in contact with nonconductors only, is said to be insulated. When electricity is subtracted from the substance thus insulated, it is said to be negatively electrified, but the sensible effects are nearly the same, except that in some cases the form of the spark is a little different.

Perfect conductors, when electrified, are in general either overcharged or undercharged with electricity in their most distant parts at the same time; but nonconductors, although they have an equal attraction for the electric fluid, are often differently affected in different parts of their substance, even when those parts are similarly situated in every respect, except that some of them have had their electricity increased or diminished by a foreign cause. This property of nonconductors may be illustrated by means of a cake of resin, or a plate of glass, to which a local electricity may be communicated in any part of its surface, by the contact of an electrified body; and the parts thus electrified may afterwards be distinguished from the rest, by the attraction which they exert on any small particles of dust or powder projected near them; the manner, in which the particles arrange themselves on the surface, indicating also in some cases the species of electricity, whether positive or negative, that has been employed; positive electricity producing an appearance somewhat resembling feathers; and negative electricity an arrangement more like spots. The inequality in the distribution of the electric fluid in a nonconductor may remain for some hours, or even some days, continually diminishing till it becomes imperceptible.

These are the fundamental properties of the electric fluid, and of the different kinds of matter as connected with that fluid. We are next to examine its distribution, and the attractive and repulsive effects exhibited by it, under different forms. Supposing a quantity of redundant fluid to exist in a spherical conducting body, it will be almost wholly collected into a minute space contiguous to the surface, while the internal parts remain but little overcharged. For we may neglect the actions of the portion of fluid which is only occupied in saturating the matter, and also the effect of the matter thus neutralised, since the redundant fluid is repelled as much by the one as it is attracted by the other; and we need only to consider the mutual actions of the particles of this superfluous fluid on each other. It may then be shown, in the same manner as it is demonstrated of the force of gravitation, that all the spherical strata which are remoter from the centre than any given particle, will have the whole of their action on it annihilated by the balance of their forces, and that the effective repulsion of the interior strata will be the same, as if they were all collected in the centre. This repulsion will, therefore, impel the particles of the fluid towards the surface, as long as it exists, and nothing will impede the condensation of the redundant fluid there, until it is exhausted from the neighbourhood of the centre. In the same manner it may be shown, that if there be a deficiency of fluid, it will be only in the external parts, the central parts remaining always in a state of neutrality: and since the quantity of electric fluid taken away from a body, in any common experiment, bears but a very small proportion to the whole that it contains, the deficiency will also be found in a very small portion of the sphere, next to its surface. And if, instead of being spherical, the body be of any other form, the effects of electricity will still be principally confined to its surface. This proposition was very satisfactorily investigated by Mr. Cavendish; and it was afterwards more fully shown, by Dr. Gray's experiments, that the capacities of different bodies, for receiving electricity, depend much more on the quantity of their surfaces, than on their solid contents: thus, the conductor of an electrical machine will contain very nearly or quite as much electricity if hollow as if solid.

If two spheres be united by a cylindrical conducting substance of small dimensions, there will be an equilibrium, when the actions of the redundant fluid in the spheres, on the whole fluid in the cylinder, are equal; that is, when both the spheres have their surfaces electrified in an equal degree: but if the length of the cylinder is considerable, the fluid within it can only remain at rest when the quantities of redundant fluid are nearly equal in both spheres, and consequently when the density is greater in the smaller. And for a similar reason, in bodies of irregular forms, the fluid is always most accumulated in the smallest parts; and when a conducting substance is pointed, the fluid becomes so dense at its extremity, as easily to overcome the forces which tend to retain it in its situation. (Plate XXXIX. Fig. 551.)

In this distribution we find a very characteristic difference between the pressure of the electric fluid and the common hydrostatic pressure of liquids or of simple elastic fluids; for these exert on every surface similarly situated a pressure proportionate to its magnitude; but the electric fluid exerts a pressure on small and angular surfaces greater, in proportion to their magnitudes, than the pressure on larger parts: so that if the electric fluid were in general confined to its situation by the pressure of the atmosphere, that pressure might easily be too weak to oppose its escape from any prominent points. It does not appear, however, that this pressure is the only cause which prevents the escape of the electric fluid; nor is it certain that this fluid can pass through a perfect vacuum, although it has not yet been proved, that a body placed in a vacuum is perfectly insulated. Whatever the resistance may be, which prevents the dissipation of electricity, it is always the more easily overcome, as the electrified substance is more pointed, and as the point is more prominent; and even the presence of dust is often unfavourable to the success of electrical experiments, on account of the great number of pointed terminations which it affords.

The general effect of electrified bodies on each other, if their bulk is small in comparison with their distance, is, that they are mutually repelled when in similar states of electricity, and attracted when in dissimilar states. This is a consequence immediately deducible from the mutual attraction of redundant matter and redundant fluid, and from the repulsion supposed to exist between any two portions either of matter or of fluid, and it may also easily be confirmed by experimental proof. A neutral body, if it were a perfect nonconductor, would not be affected either way by the neighbourhood of an electrified body: for while the whole matter contained in it remains barely saturated with the electric fluid, the attractions and repulsions balance each other. But in general, a neutral body appears to be attracted by an electrified body, on account of a change of the disposition of the fluid which it contains, upon the approach of a body either positively or negatively electrified. The electrical affection produced in this manner, without any actual transfer of the fluid, is called induced electricity.

When a body positively electrified approaches to a neutral body, the redundancy of the fluid expels a portion of the natural quantity from the nearest parts of the neutral body, so that it is accumulated at the opposite extremity; while the matter, which is left deficient, attracts the redundant fluid of the first body, in such a manner as to cause it to be more condensed in the neighbourhood of the second than elsewhere; and hence the fluid of this body is driven still further off, and all the effects are redoubled. The attraction of the redundant fluid of the electrified body, for the redundant matter of the neutral body, is stronger than its repulsion for the fluid which has been expelled from it, in proportion as the square of the mean distance of the matter is smaller than that of the mean distance of the fluid: so that in all such cases of induced electricity, an attraction is produced between the bodies concerned. And a similar attraction will happen, under contrary circumstances, when a neutral body and a body negatively electrified, approach each other.

The state of induced electricity may be illustrated by placing a long conductor at a little distance from an electrified substance, and directed towards it; and by suspending pith balls or other light bodies from it, in pairs, at different parts of its length: these will repel each other, from being similarly electrified, at the two ends, which are in contrary states of electricity, while at a certain point towards the middle, they will remain at rest, the conductor being here perfectly neutral. It was from the situation of this point that Lord Stanhope first inferred the true law of the electric attractions and repulsions, although Mr. Cavendish had before suggested the same law as the most probable supposition.

The attraction, thus exerted by an electrified body upon neutral substances, is strong enough, if they are sufficiently light, to overcome their gravitation, and to draw them up from a table at some little distance: upon touching the electrified body, if it is a conductor, they receive a quantity of electricity from it, and are again repelled, until they are deprived of their electricity by contact with some other substance, which, if sufficiently near to the first, is

4

usually in a contrary state, and therefore renders them still more capable of returning, when they have touched it, to the first substance, in consequence of an increased attraction, assisted also by a new repulsion. This alternation has been applied to the construction of severalelectrical toys; a little hammer, for example, has been made to play between two bells; and this instrument has been employed for giving notice of any change of the electrical state of the atmosphere. The repulsion, which takes place between two bodies, in a similar state of electricity, is the cause of the currents of air which always accompany the discharge of electricity, whether negative or positive, from pointed substances; each particle of air, as soon as it has received its electricity from the point, being immediately repelled by it; and this current has also been supposed to facilitate the escape of the electricity, by bringing a continual succession of particles not already overcharged.

If two bodies approach each other, electrified either positively or negatively in different degrees, they will either repel or attract each other, according to their distance: when they are very remote, they exhibit a repulsive force, but when they are within a certain distance, the effects of induced electricity overcome the repulsion, which would necessarily take place, if the distribution of the fluid remained unaltered by their mutual influence.

When a quantity of the electric fluid is accumulated on one side of a non-conducting substance, it tends to drive off the fluid from the other side; and if this fluid is suffered to escape, the remaining matter exerts its attraction on the fluid which has been imparted to the first side, and allows it to be accumulated in a much greater quantity than could have existed in an equal surface of a conducting substance. In this state, the body is said to be charged; and for producing it the more readily, each surface is usually coated with a conducting substance, which serves to convey the fluid to and from its different parts with convenience. The thinner any substance is, the greater quantity of the fluid is required for charging it in this manner, so as to produce a given tension, or tendency to escape: but if it be made too thin, it will be liable to break, the attractive force of the fluid, for the matter on the opposite side overcoming the cohesion of the substance, and perhaps forcing its way through the temporary vacuum which is formed.

When a communication is made in any manner by a conducting substance between the two coatings of a charged plate or vessel, the equilibrium is restored, and the effect is called a shock. If the coatings be removed, the plate will still remain charged, and it may be gradually discharged by making a communication between its several parts in succession, but it cannot be discharged at once, for want of a common connexion: so that the presence of the coating is not absolutely essential to the charge and discharge of the opposite surfaces. Such a coated substance is most usually employed in the form of a jar. Jars were formerly filled with water, or with iron filings; the instrument having been principally made known from the experiments of Musschenbroek and others at Leyden, it was called the Leyden phial; but at present a coating of tin foil is commonly applied on both sides of the jar, leaving a sufficient space at its upper part, to avoid the spontaneous discharge, which would often take place between the coatings, if they approached too near to each other; and a ball is fixed to the cover, which has a communication with the internal coating, and by means of which the jar is charged, while the external coating is allowed to communicate with the ground. A collection of such jars is called a battery, and an apparatus of this kind may be made so powerful, by increasing the number of jars, as to exhibit many striking effects by the motion of the electric fluid, in its passage from one to the other of the surfaces.

The conducting powers of different substances are concerned, not only in the facility with which the motions of the electric fluid are directed into a particular channel, but also in many cases of its equilibrium, and particularly in the properties of charged substances, which depend on the resistance opposed by nonconductors to the ready transmission of the fluid. These powers may be compared, by ascertaining the greatest length of each of the substances to be examined, through which a spark or a shock will take it course, in preference to a given length of air, or of any other standard of comparison. The substances, which conduct electricity the most readily, are metals, well burnt charcoal, animal bodies, acids, saline liquors, water, and very rare air. The principal nonconductors are glass, ice, gems, dry salts, sulfur, amber, resins, silk, dry wood, oils, dry air of the usual density, and the barometrical vacuum. Heat commonly increases the conducting powers of bodies; a jar of glass may be discharged by a moderate heat, and liquid resins are capable of transmitting shocks, although they are by no means good conductors: it is remarkable also that a jar may be discharged by minute agitation, when it is caused to ring by the friction of the finger. It has been observed that, in a great variety of cases, those substances, which are the best conductors of heat, afford also the readiest passage to electricity; thus, copper conducts heat more rapidly, and electricity more readily, than iron, and platina less than almost any other metal; glass also presents a considerable resistance to the transmission of both these influences. The analogy is, however, in many respects imperfect, and it affords us but little light, with regard either to the nature of heat, or to that of the electric fluid.

LECTURE LIV.

ON ELECTRICITY IN MOTION.

THE manner in which the electric fluid is transferred from one body to another, the immediate effects of such a transfer, the causes which originally disturb the equilibrium of electricity, and the practical methods, by which all these circumstances are regulated and measured, require to be considered as belonging to the subject of electricity in motion. Among the modes of excitation by which the equilibrium is originally disturbed, one of the most interesting is the galvanic apparatus, which has been of late years a very favourite subject of popular curiosity, and of which the theory and operation will be briefly examined, although the subject appears rather to belong to the chemical than to the mechanical doctrine of electricity.

The progressive motion of the electric fluid through conducting substances is so rapid, as to be performed in all cases without a sensible interval of time. It has indeed been said, that when very weakly excited, and obliged to pass to a very great distance, a perceptible portion of time is actually occupied in its passage; but this fact is somewhat doubtful, and attempts have been made in vain, to estimate the interval, employed in the transmission of a shock through several miles of wire. We are not to imagine that the same particles of the fluid, which enter at one part, pass through the whole conducting substance, any more than that the same portion of blood, which is thrown out of the heart, in each pulsation, arrives at the wrist, at the instant that the pulse is felt there. The velocity of the transmission of a spark or shock far exceeds the actual velocity of each particle, in the same manner as the velocity of a wave exceeds that of the particles of water concerned in its propagation; and this velocity must depend both on the elasticity of the electric fluid, and on the force with which it is confined to the conducting substance. If this force were merely derived from the pressure of the atmosphere, we might infer the density of the fluid from the velocity of a spark or shock, compared with that of sound; or we might deduce its velocity from a determination of its density. It has been supposed, although perhaps somewhat hastily, that the actual velocity is nearly equal to that of light.

When a conducting substance approaches another, which is electrified, the distribution of the electric fluid within it is necessarily altered by induction, before it receives a spark, so that its remoter extremity is brought into a state similar to that of the first body: hence it happens that when the spark passes, it produces less effect at the remoter end of the substance, while the part presented to the electrified body is most affected, on account of its sudden change to an opposite state. But if both ends approach bodies in opposite states of electricity, they will both be strongly affected when the shock takes place, while the middle of the circuit undergoes but little change.

The manner in which the electric fluid makes its way, through a more or less perfect nonconductor, is not completely understood: it is doubtful whether the substance is forced away on each side, so as to leave a vacuum for the passage of the fluid, or whether the newly formed surface helps to guide it in its way; and in some cases it has been supposed that the gradual communication of electricity has rendered the substance more capable of conducting it, either immediately, or, in the case of the air, by first rarefying it. However this may be, the perforation of a jar of glass by an overcharge, and that of a plate of air by a spark, appear to be effects of the same kind, although the charge of the jar is principally contained in the glass, while the plate of air is perhaps little concerned in the distribution of the electricity.

The actual direction of the electric current has not in any instance been fully ascertained, although there are some appearances which seem to justify the common denominations of positive and negative. Thus, the fracture of a charged jar of glass, by spontaneous explosion, is well defined on the positive, and splintered on the negative side, as might be expected from the passage of a foreign substance from the former side to the latter; and a candle, held between a positive and a negative ball, although it apparently vibrates between them, is found to heat the negative ball much more than the positive. We cannot, however, place much dependence on any circumstance of this kind, for it is doubtful whether any current of the fluid, which we can produce, possesses sufficient momentum to carry with it a body of sensible magnitude. It is in fact of little consequence to the theory, whether the terms positive and negative be correctly applied, provided that their sense remain determined; and that, like positive and negative quantities in mathematics, they be always understood of states which neutralise each other. The original opinion of Dufay, of the existence of two distinct fluids, a vitreous and a resinous electricity, has at present few advocates, although some have thought such a supposition favoured by the phenomena of the galvanic decomposition of water.

When electricity is simply accumulated without motion, it does not appear to have any effect, either mechanical, chemical, or physiological, by which its presence can be discovered; the acceleration of the pulse, and the advancement of the growth of plants, which have been sometimes attributed to it, have not been confirmed by the most accurate experiments. An uninterrupted current of electricity, through a perfect conductor, would perhaps be also in every respect imperceptible, since the best conductors appear to be the least affected by it. Thus, if we place our hand on the conductor of an electrical machine, the electricity will pass off continually through the body, without exciting any sensation. A constant stream of galvanic electricity, passing through an iron wire is, however, capable of exciting a considerable degree of heat, and if it be transmitted through the hands of the operator, it will produce a slight numbness, although in general some interruption of the current is necessary in order to furnish an accumulation sufficient to produce sensible effects; and such an interruption may even increase the effect of a single spark or shock; thus, gunpowder is more readily fired by the discharge of a battery passing through an interrupted circuit, than through a series of perfect conductors.

The most common effect of the motion of the electric fluid is the production of light. Light is probably never occasioned by the passage of the fluid through a perfect conductor; for when the discharge of a large battery renders a small wire luminous, the fluid is not wholly confined to the wire, but overflows a little into the neighbouring space. There is always an appearance of light whenever the path of the fluid is interrupted by an imperfect conductor; nor is the apparent contact of conducting substances sufficient to prevent it, unless they are held together by a considerable force; thus, a chain, conveying a spark or shock, appears luminous at each link, and the rapidity of the motion is so great, that we can never observe any difference in the times of the appearance of the light in its different parts; so that a series of luminous points, formed by the passage of the electric fluid, between a string of conducting bodies, represents at once a brilliant delineation of the whole figure in which they are arranged. A lump of sugar, a piece of wood, or an egg, may easily be made luminous in this manner; and many substances, by means of their properties as solar phosphori, retain for some seconds the luminous appearance thus acquired. Even water is so imperfect a conductor, that a strong shock may be seen in its passage through it; and when the air is sufficiently moistened or rarefied to become a conductor, the track of the fluid through it is indicated by streams of light, which are perhaps derived from a series of minute sparks passing between the particles of water or of rarefied air. When the air is extremely rare, the light is greenish; as it becomes more dense, the light becomes blue, and then violet, until it no longer conducts. The appearance of the electrical light of a point enables us to distinguish the nature of the electricity with which it is charged; a pencil of light, streaming from the point, indicating that its electricity is positive, while a luminous star, with few diverging rays, shows that it is negative. The sparks, exhibited by small balls, differently electrified, have also similar varieties in their forms, according to the nature of their charges. (Plate XL. Fig. 552.)

The production of heat by electricity frequently accompanies that of light, and appears to depend in some measure on the same circumstances. A fine wire may be fused and dissipated by the discharge of a battery; and without being perfectly melted, it may sometimes be shortened or lengthened, accordingly as it is loose or stretched during the experiment. The more readily a metal conducts, the shorter is the portion of it which the same shock can destroy; and it has sometimes been found that a double charge of a battery has been capable of melting a quadruple length of wire of the same kind.

The mechanical effects of electricity are probably in many cases the consequences of the rarefaction produced by the heat which is excited; thus, the explosion, attending the transmission of a shock or spark through the air, may easily be supposed to be derived from the expansion caused by heat; and the destruction of a glass tube, which contains a fluid in a capillary bore, when a spark is caused to pass through it, is the natural consequence of the conversion of some particles of the fluid into vapour. But when a glass jar is perforated, this rarefaction cannot be supposed to be adequate to the effect. It is remarkable that such a perforation may be made by a very moderate discharge, when the glass is in contact with oil or with sealing wax; and no sufficient explanation of this circumstance has yet been given.

A strong current of electricity, or a succession of shocks or sparks, transmitted through a substance, by means of fine wires, is capable of producing many chemical combinations and decompositions, some of which may be attributed merely to the heat which it occasions, but others are wholly different. Of these the most remarkable is the production of oxygen and hydrogen gas from common water, which are usually extricated at once, in such quantities, as, when again combined, will reproduce the water which has disappeared; but in some cases the oxygen appears to be disengaged most copiously at the positive wire, and the hydrogen at the negative.

When the spark is received by the tongue, it has generally a subacid taste; and an explosion of any kind is usually accompanied by a smell somewhat like that of sulfur, or rather of fired gunpowder. The peculiar sensation, which the electric fluid occasions in the human frame, appears in general to be derived from the spasmodic contractions of the muscles through which it passes; although in some cases it produces pain of a different kind; thus, the spark of a conductor occasions a disagreeable sensation in the skin, and when an excoriated surface is placed in the galvanic current, a sense of smarting, mixed with burning, is experienced. Sometimes the effect of a shock is felt most powerfully at the joints, on account of the difficulty which the fluid finds in passing the articulating surfaces which form the cavity of the joints. The sudden death of an animal, in consequence of a violent shock, is probably owing to the immediate exhaustion of the whole energy of the nervous system. It is remarkable that a very minute tremor, communicated to the most elastic parts of the body, in particular to the chest, produces an agitation of the nerves, which is not wholly unlike the effect of a weak electricity.

The principal modes, in which the electric equilibrium is primarily destroyed, are simple contact, friction, a change of the form of aggregation, and chemical combinations and decompositions. The electricity produced by the simple contact of any two substances is extremely weak, and can only be detected by very delicate experiments: in general it appears that the substance, which conducts the more readily, acquires a slight degree of negative electricity, while the other substance is positively electrified in an equal degree. The same disposition of the fluid is also usually produced by friction, the one substance always losing as much as the other gains; and commonly although not always, the worst conductor becomes positive. At the instant in which the friction is applied, the capacities or attractions of the bodies for electricity appear to be altered, and a greater or less quantity is required for saturating them; and upon the cessation of the temporary change, this redundancy or deficiency is rendered sensible. When two substances of the same kind are rubbed together, the smaller or the rougher becomes negatively electrified; perhaps because the smaller surface is more heated, in consequence of its undergoing more friction than an equal portion of the larger, and hence becomes a better conductor; and because the rougher is in itself a better conductor than the smoother, and may possibly have its conducting powers increased by the greater agitation of its parts which the friction produces. The back of a live cat becomes positively electrified, with whatever substance it is rubbed; glass is positive in most cases, but not when rubbed with mercury in a vacuum, although sealing wax, which is generally negative, is rendered positive by immersion in a trough of mercury. When a white and a black silk stocking are rubbed together, the white stocking acquires positive electricity, and the black negative, perhaps because the black dye renders the silk both rougher and a better conductor.

Those substances, which have very little conducting power, are sometimes called electrics, since they are capable of exhibiting readily the electricity which friction excites on their surfaces, where it remains accumulated, so that it may be collected into a conductor; while the surfaces of such substances, as have greater conducting powers, do not so readily imbibe the fluid from others with which they are rubbed, since they may be supplied from the internal parts of the substances themselves, when their altered capacity requires it; thus, glass, when heated to 110° of Fahrenheit, can with difficulty be excited, becoming an imperfect conductor:

but a thin plate of a conducting substance, when insulated, may be excited almost as easily as an electric, commonly so called.

Vapours are generally in a negative state, but if they rise from metallic substances, or even from some kinds of heated glass, the effect is uncertain, probably on account of some chemical actions which interfere with it. Sulfur becomes electrical in cooling, and wax candles are said to be sometimes found in a state so electrical, when they are taken out of their moulds, as to attract the particles of dust which are floating near them. The tourmalin, and several other crystallized stones, become electrical when heated or cooled, and it is found that the disposition, assumed by the fluid, bears a certain relation to the direction in which the stone transmits the light most readily; some parts of the crystal being rendered always positively and others negatively electrical, by an increase of temperature.

The most remarkable of the phenomena, attending the excitation of electricity by chemical changes, are those which have lately received the appellation of galvanic. Some of the effects which have been considered as belonging to galvanism are probably derived from the electrical powers of the animal body, and the rest have been referred by Mr. Volta, and many other philosophers on the continent, to the mere mechanical actions of bodies possessed of different properties with regard to electricity. Thus, they have supposed that when a circulation of the electric fluid is produced through a long series of substances in a certain direction, the differences of their attractions and of their conducting powers, which must remain the same throughout the process, keep up this perpetual motion, in defiance of the general laws of mechanical forces. In this country it has been generally maintained, that no explanation founded on such principles could be admissible, even if it were in all other respects sufficient and satisfactory, which the mechanical theory of galvanism certainly is not.

The phenomena of galvanism appear to be principally derived from an inequality in the distribution of the electric fluid, originating from chemical changes, and maintained by means of the resistance opposed to its motion, by a continued alternation of substances of different kinds, which furnishes a much stronger obstacle to its transmission than any of those substances alone would have done. The substances employed must neither consist wholly of solids nor of fluids, and they must be of three different kinds, possessed of different powers of conducting electricity; but whether the difference of their conducting powers is of any other consequence than as it accompanies different chemical properties, is hitherto undetermined. Of these three substances, two must possess a power of acting mutually on each other, while the other appears to serve principally for making a separate connexion between them: and this action may be of two kinds, or perhaps of more; the one is oxidation, or the combination of a metal or an inflammable substance with a portion of oxygen derived from water or from an acid, the other sulfuration, or a combination with the sulfur contained in a solution of an alkaline sulfuret.

We may represent the effects of all galvanic combinations, by considering the oxidation as producing positive electricity in the acting liquid, and the sulfuration as producing negative electricity, and by imagining that this electricity is always communicated to the best conductor of the other substances concerned, so as to produce a circulation in the direction thus determined. For example, when two wires of zinc and silver, touching each other, are separately immersed in an acid, the acid, becoming positively electrical, imparts its electricity to the silver, and hence it flows back into the zinc: when the ends of a piece of charcoal are dipped into water and into an acid, connected together by a small tube, the acid, becoming positive, sends its superfluous fluid through the charcoal into the water; and if a wire of copper be dipped into water and a solution of alkaline sulfuret, connected with each other, the sulfuret, becoming negative, will draw the fluid from the copper on which it acts; and in all these cases the direction of the current is truly determined, as it may be shown by composing a battery of a number of alternations of this kind, and either examining the state of its different parts by electrical tests, or connecting wires with its extremities, which, when immersed into a portion of water, will exhibit the production of oxygen gas where they emit the electric fluid, and of hydrogen where they receive it. These processes of oxidation and of sulfuration may be opposed to each other, or they may be combined in various ways, the sum or difference of the separate actions being obtained by their union; thus it usually happens that both the metals employed are oxidable in some degree, and the oxidation, which takes place at the surface of the better conductor, tends to impede the whole effect, perhaps by impeding the passage of the fluid through the surface. The most oxidable of the metals, and probably the worst conductor, is zinc; the next is iron; then come tin, lead, copper, silver, gold, and platina. (Plate XL. Fig. 553..555.)

In the same manner as a wire charged with positive electricity causes an extrication of oxygen gas, so the supply of electricity through the more conducting metal promotes the oxidation of the zinc of a galvanic battery; and the effect of this circulation may be readily exhibited, by fixing a wire of zinc, and another of silver or platina, in an acid, while one end of each is loose, and may be brought together or separated at pleasure: for at the moment that the contact takes place, a stream of bubbles rising from the platina, and a white cloud of oxid falling from the zinc, indicate both the circulation of the fluid and the increase of the chemical action. But when, on the other hand, a plate of zinc is made negative by the action of an acid on the greater part of its surface, a detached drop of water has less effect on it, than in the natural state: while a plate of iron, which touches the zinc, and forms a part of the circle with it, is very readily oxidated at a distant point: such a plate must therefore be considered, with regard to this effect, as being made positive by the electricity which it receives from the acid or the water; unless something like a compensation be supposed to take place, from the effects of induced electricity. Instead of the extrication of hydrogen, the same causes will sometimes occasion a deposition of a metal which has been dissolved, will prevent the solution of a metal which would otherwise have been corroded, or produce some effects which appear to indicate the presence of an alkali, either volatile or fixed. All these operations may, however, be very much impeded by the interposition of any considerable length of water, or of any other imperfect conductor. (Plate XL. Fig. 556.)

It is obvious, that since the current of electricity, produced by a galvanic circle, facilitates those actions from which its powers are derived, the effect of a double series must be more than twice as great as that of a single one: and hence arises the activity of the pile of Volta, the discovery of which forms the most important era in the history of this department of natural knowledge. The intensity of the electrical charge, and the chemical and physiological effects of a pile or battery, seem to depend principally on the number of alternations of substances; the light and heat more on the joint magnitude of the surfaces employed. In common electricity, the greatest heat appears to be occasioned by a long continuation of a slow motion of the fluid; and this is perhaps best furnished in galvanism by a surface of large extent, while some other effects may very naturally be expected to depend on the intensity of the charge, independently of the quantity of charged surface. It may easily be imagined, that the tension of the fluid must be nearly proportional to the number of surfaces, imperfectly conducting, which are interposed between the ends of a pile or battery, the density of the fluid becoming greater and greater by a limited quantity at each step; and it is easily understood, that any point of the pile may be rendered neutral, by a connexion with the earth, while those parts, which are above it or below it, will still preserve their relations unaltered with respect to each other: the opposite extremities being, like the opposite surface of a charged jar, in contrary states, and a partial discharge being produced, as often as they are connected by a conducting substance. The various forms, in which the piles or troughs are constructed, are of little consequence to the theory of their operation: the most convenient are the varnished troughs, in which plates of silvered zinc are arranged side by side, with intervening spaces for the reception of water, or of an acid. (Plate XL. Fig. 557.)

It is unquestionable that the torpedo, the gymnotus electricus, and some other fishes, have organs appropriated to the excitation of electricity, and that they have a power of communicating this electricity at pleasure to conducting substances in their neighbourhood. These organs somewhat resemble in their appearance the plates of the galvanic pile, although we know nothing of the immediate arrangement, from which their electrical properties are derived; but the effect of the shock, which they produce, resembles in all respects that of the weak charge of a very large battery. It has also been shown by the experiments of Galvani, Volta, and Aldini, that the nerves and muscles of the human body possess some electrical powers, although they are so much less concerned in the phenomena which were at first attributed to them by Galvani, than he originally supposed, that many philosophers have been inclined to consider the excitation of electricity as always occasioned by the inanimate substances employed, and the spasmodic contractions of the muscles as merely very delicate tests of the influence of foreign electricity on the nerves.

Such is the general outline of the principal experiments and conclusions which the subject of galvanism afforded before Mr. Davy's late ingenious and interesting researches, which have thrown much light, not only on the foundation of the whole of this class of phenomena, but also on the nature of chemical actions and affinities in general. Mr. Davy is inclined to infer from his experiments, that all the attractions, which are the causes of chemical combinations, depend on the opposite natural electricities of the bodies concerned; since such bodies are always found, by delicate tests, to exhibit, when in contact, marks of different species of electricity; and their mutual actions may be either augmented or destroyed, by increasing their natural charges of electricity, or by electrifying them in a contrary way. Thus, an acid and a metal are found to be negatively and positively electrical with respect to each other; and by further electrifying the acid negatively, and the metal positively, their combination is accelerated; but when the acid is positively electrified, or the metal negatively, they have no effect whatever on each other. The acid is also attracted, as a negative body, by another positively electrified, and the metal by a body negatively electrified, so that a metallic salt may be decomposed in the circuit of Volta, the positive point attracting the acid, and the negative point the metal: and these attractions are so strong, as to carry the particles of the respective bodies through any intervening medium, which is in a fluid state, or even through a moist solid; nor are they intercepted in their passage, by substances which, in other cases, have the strongest elective attractions for them. Alkali, sulfur, and alkaline sulfurets, are positive with respect to the metals, and much more with respect to the acids: hence they have a very strong natural tendency to combine with the acids and with oxygen: and hydrogen must also be considered as belonging to the same class with the alkalis.

Supposing now a plate of zinc to decompose a portion of water: the oxygen, which has a negative property, unites with the zinc, and probably tends to neutralise it, and to weaken its attractive force; the hydrogen is repelled by the zinc, and carries to the opposite plate of silver its natural positive electricity; and if the two plates be made to touch, the energy of the plate of zinc is restored, by the electricity which it receives from the silver: and it receives it the more readily, as the two metals, in any case of their contact, have a tendency to become electrical, the zinc positively, and the silver negatively. Mr. Davy therefore considers this chemical action as destroying, or at least counteracting, the natural tendency of the electric fluid to pass from the water to the zinc, and from modifications of this counteraction he explains the effects of galvanic combinations in all cases. Thus, in a circle composed of copper, sulfuret, and iron, the fluid tends to pass from the iron towards the sulfuret, and from the copper to the iron, in one direction, and in the opposite direction from the copper to the sulfuret, with a force which must be equal to both the others, since there would otherwise be a continual motion without any mechanical cause, and without any chemical change; but the action of the sulfuret on the copper tends to destroy its electromotive, or rather electrophoric, power, of directing the current towards the sulfuret, and its combination with the sulfur makes it either positively electrical, or negatively electrical in a less considerable degree; consequently the fluid passes, according to its natural tendency, from the copper to the iron, and from the iron to the sulfuret. In a third case, when copper, an acid, and water, form a circle, the natural tendency is from the acid to the copper on one side, and from the acid to the water, and from the water to the copper on the other; here we must suppose the first force to be only a little weakened by the chemical action, while the third is destroyed, so that the first overcomes the second, and the circulation is determined, although very feebly, in such a direction, that the fluid passes from the acid to the copper. When, in the fourth place, the combination consists of copper, sulfuret, and water, the tendencies are, first, from the copper to the sulfuret, and from the water to the copper; and secondly, from the water to the sulfuret: in this instance a chemical action must be supposed between the oxygen of the water and the sulfuret, which lessens the electromotive tendency, more than the action that takes place between the sulfuret and the copper, so that the fluid passes from the copper to the sulfuret; and the current has even force enough to prevent any chemical action between the water and the copper, which would tend to counteract that force, if it took place.

Mr. Davy has observed that the decomposition of the substances, employed in the battery of Volta, is of much more consequence to their activity than either their conducting power, or their simple action on the other elements of the series: thus, the sulfuric acid, which conducts electricity better, and dissolves the metals more readily, than a neutral solution, is, notwithstanding, less active in the battery, because it is not easily decomposed. Mr. Davy has also extended his researches, and the application of his discoveries, to a variety of natural as well as artificial phenomena, and there can be no doubt but that he will still make such additions to his experiments, as will be of the greatest importance to this branch of science.

The operation of the most usual electrical machines depends first on the excitation of electricity by the friction of glass on a cushion of leather, covered with a metallic amalgam, usually made of mercury, zinc, and tin, which probably, besides being of use in supplying electricity readily to different parts of the glass, undergoes in general a chemical change, by means of which some electricity is extricated. The fluid, thus excited, is received into an insulated conductor by means of points, placed at a small distance from the surface which has lately undergone the effects of friction, and from this conductor it is conveyed by wires or chains to any other parts at pleasure. Sometimes also the cushion, instead of being connected with the earth, is itself fixed to a second conductor, which becomes negatively electrified; and either conductor may contain within it a jar, which may be charged at once by the operation of the machine, when its internal surface is connected either with the earth, or with that of the jar contained in the opposite conductor. The glass may be either in the form of a circular plate or of a cylinder, and it is uncertain which of the arrangements affords the greatest quantity of electricity from the same surface; but the cylinder is cheaper than the plate, and less liable to accidents, and appears to be at least equally powerful. (Plate XL. Fig. 558, 559.)

The plate machine in the Teylerian museum, employed by Van Marum, when worked by two men, excited an electricity, of which the attraction was sensible at the distance of 38 feet, and which made a point luminous at 27 feet, and afforded sparks nearly 24 inches long. A battery charged by it, melted at once twenty five feet of fine iron wire. Mr. Wilson had also a few years ago, in the Pantheon in London, an apparatus of singular extent; the principal conductor was 150 feet long, and 16 inches in diameter, and he employed a circuit of 4800 feet of wire.

The electrophorus derives its operation from the properties of induced electricity. A cake of a nonconducting substance, commonly of resin or of sulfur, is first excited by friction, and becomes negatively electric: an insulated plate of a conducting substance, being placed on it, does not come sufficiently into contact with it to receive its electricity, but acquires by induction an opposite state at its lower surface, and a similar state at its upper; so that when this upper and negative surface is touched by a substance communicating with the earth, it receives enough of the electric fluid to restore the equilibrium. The plate then being raised, the action of the cake no longer continues, and the electricity, which the plate has received from the earth, is imparted to a conductor or to a jar; and the operation may be continually repeated, until the jar has received a charge, of an intensity equal to that of the plate when raised. Although the quantity of electricity, received by the plate, is exactly equal to that which is emitted from it at each alternation, yet the spark is far less sensible; since the effect of the neighbourhood of the cake is to increase the capacity of the plate, while the tension or force impelling the fluid is but weak; and at the same time the quantity received is sufficient, when the capacity of the plate is again diminished, to produce a much greater tension, at a distance from the cake. (Plate XL. Fig. 560.)

The condenser acts in some measure on the same principles with the electrophorus, both instruments deriving their properties from the effects of induction. The use of the condenser is to collect a weak electricity from a large substance into a smaller one, so as to make its density or tension sufficient to be examined. A small plate, connected with the substance, is brought nearly into contact with another plate communicating with the earth; in general a thin stratum of air only is interposed; but sometimes a nonconducting varnish is employed; this method is, however, liable to some uncertainty, from the permanent electricity which the varnish sometimes contracts by friction. The electricity is accumulated by the attraction of the plate communicating with the earth, into the plate of the condenser; and when this plate is first separated from the substance to be examined, and then removed from the opposite plate, its electricity is always of the same kind with that which originally existed in the substance, but its tension is so much increased as to render it more easily discoverable. This principle has been variously applied by different electricians, and the employment of the instrument has been facilitated by several subordinate arrangements. (Plate XL. Fig. 561.)

Mr. Cavallo's multiplier is a combination of two condensers; the second or auxiliary plate of the first, like the plate of the electrophorus, is moveable, and carries a charge of electricity, contrary to that of the substance to be examined, to the first or insulated plate of the second condenser, which receives it repeatedly, until it has acquired an equal degree of tension; and when the two plates of this condenser are separated, they both exhibit an electricity much more powerful than that of the first condenser. The force is, however, still more rapidly augmented by the instruments of Mr. Bennet and Mr. Nicholson, although it has been supposed that these instruments are more liable to inconvenience from the attachment of a greater portion of electricity to the first plate of the instrument, which leaves, for a very considerable time, a certain quantity of the charge, not easily separable from it. Mr. Bennet employs three varnished plates laid on each other, but Mr. Nicholson has substituted simple metallic plates, approaching only very near together, so that there can be no error from any accidental friction. In both of these instruments, the second plate of a condenser acquires an electricity contrary and nearly equal to that of the first, by means of which it brings a third plate very nearly into the same state with the first; and when the first and third plates are connected and insulated, they produce a charge nearly twice as great in the second plate, while the first plate becomes at the same time doubly charged; so that by each repetition of this process, the intensity of the electricity is nearly doubled: it is therefore scarcely possible that any quantity should be so small as to escape detection by its operation. (Plate XL. Fig. 562, 563.)

The immediate intensity of the electricity may be measured, and its character distinguished, by electrical balances, and by electrometers of different constructions. The electrical balance measures the attraction or repulsion exerted by two balls at a given distance, by the magnitude of the force required to counteract it; and the most convenient manner of applying this force is by the torsion of a wire, which has been employed for the purpose by Mr. Coulomb. The quadrant electrometer of Henley expresses the mutual repulsion of a moveable ball and a fixed column, by the divisions of the arch to which the ball rises. These divisions do not exactly denote the proportional strength of the action, but they are still of utility in ascertaining the identity of any two charges, and in informing us how far we may venture to

4

proceed in our experiments with safety; and the same purpose is answered, in a manner somewhat less accurate, by the electrometer, consisting of two pith balls, or of two straws, which are made to diverge by a smaller degree of electricity. Mr. Bennet's electrometer is still more delicate; it consists of two small portions of gold leaf, suspended from a plate, to which the electricity of any substance is communicated by contact: a very weak electricity is sufficient to make them diverge, and it may easily be ascertained whether it is positive or negative, by bringing an excited stick of sealing wax near the plate, since its approach tends to produce by induction a state of negative electricity in the remoter extremities of the leaves, so that their divergence is either increased or diminished, accordingly as it was derived from negative or from positive electricity: a strip of gold leaf or tin foil, fixed within the glass which covers the electrometer, opposite to the extremities of the leaves, prevents the communication of any electricity to the glass, which might interfere with the action of the instrument. When the balls of an electrometer stand at the distance of 4 degrees, they appear to indicate a charge nearly 8 times as great as when they stand at one degree: a charge 8 times as great in each ball producing a mutual action 64 times as great at any given distance, and at a quadruple distance a quadruple force; in the same manner a separation of 9 degrees is probably derived from an intensity 27 times as great as at 1. In Lane's electrometer the magnitude of a shock is determined by the quantity of air through which it is obliged to pass, between two balls, of which the distance may be varied at pleasure; and the power of the machine may be estimated by the frequency of the sparks which pass at any given distance. It appears from Mr. Lane's experiments, that the quantity of electricity required for a discharge is simply as the distance of the surfaces of the balls, the shocks being twice as frequent when this distance is only $\frac{1}{24}$ of an inch as when it is $\frac{1}{12}$. Mr. Volta says, that the indications of Lane's and Henley's electrometer agree immediately with each other; but it seems difficult to reconcile this result with the general theory. Sometimes the force of repulsion between two balls in contact is opposed by a counterpoise of given magnitude, and as soon as this is overcome, they separate and form a circuit which discharges a battery; whence the instrument is called a discharger. (Plate XL. Fig. 564..568.)

It must be confessed that the whole science of electricity is yet in a very imperfect state: we know little or nothing of the intimate nature of the substances and actions concerned in it: and we can never foresee, without previous experiment, where or how it will be excited. We are wholly ignorant of the constitution of bodies, by which they become possessed of different conducting powers; and we have only been able to draw some general conclusions respecting the distribution and equilibrium of the supposed electric fluid, from the laws of the attractions and repulsions that it appears to exert. There seems to be some reason to suspect, from the phenomena of cohesion and repulsion, that the pressure of an elastic medium is concerned in the origin of these forces; and if such a medium really exists, it is perhaps nearly related to the electric fluid. The identity of the general causes of electrical and of galvanic effects is now doubted by few; and in this country the principal phenomena of galvanism are universally considered as depending on chemical changes; perhaps, also, time may show, that electricity is very materially concerned in the essential properties, which distinguish the different kinds of natural bodies, as well as in those minute mechanical actions and affections which are probably the foundation of all chemical operations; but at present it is scarcely safe to hazard a conjecture on a subject so obscure, although Mr. Davy's experiments have already in some measure justified the boldness of the suggestion.

LECTURE LV.

ON MAGNETISM.

THE theory of magnetism bears a very strong resemblance to that of electricity, and it must therefore be placed near it in a system of natural philosophy. We have seen the electric fluid not only exerting attractions and repulsions, and causing a peculiar distribution of neighbouring portions of a fluid similar to itself, but also excited in one body, and transferred to another, in such a manner as to be perceptible to the senses, or at least to cause sensible effects, in its passage. The attraction and repulsion, and the peculiar distribution of the neighbouring fluid, are found in the phenomena of magnetism; but we do not perceive that there is ever any actual excitation, or any perceptible transfer of the magnetic fluid from one body to another distinct body; and it has also this striking peculiarity, that metallic iron is very nearly, if not absolutely, the only substance capable of exhibiting any indications of its presence or activity.

For explaining the phenomena of magnetism, we suppose the particles of a peculiar fluid to repel each other, and to attract the particles of metallic iron with equal forces, diminishing as the square of the distance increases; and the particles of such iron must also be imagined to repel each other, in a similar manner. Iron and steel, when soft, are conductors of the magnetic fluid, and become less and less pervious to it as their hardness increases. The ground work of this theory is due to Mr. Aepinus, but the forces have been more particularly investigated by Coulomb and others. There are the same objections to these hypotheses as to those which constitute the theory of electricity, if considered as original and fundamental properties of matter: and it is additionally difficult to imagine, why iron, and iron only, whether apparently magnetic or not, should repel similar particles of iron with a peculiar force, which happens to be precisely a balance to the attraction of the magnetic fluid for iron. This is obviously improbable; but the hypotheses are still of great utility in assisting us to generalise, and to retain in memory, a number of particular facts which would otherwise be insulated. The doctrine of the circulation of streams of the magnetic fluid has been justly and universally abandoned, and some other theories, much more ingenious and more probable, for instance that of Mr. Prévost, appear to be too complicated, and too little supported by facts, to require much of our attention.

The distinction between conductors and nonconductors is, with respect to the electric fluid, irregular and intricate: but in magnetism, the softness or hardness of the iron or steel constitutes the only difference. Heat, as softening iron, must consequently render it a conductor; even the heat of boiling water affects it in a certain degree, although it can scarcely be supposed to alter its temper; but the effect of a moderate heat is not so considerable in magnetism as in electricity. A strong degree of heat appears, from the experiments of Gilbert, and of Mr. Cavallo, to destroy completely all magnetic action.

It is perfectly certain that magnetic effects are produced by quantities of iron incapable of being detected either by their weight or by any chemical tests. Mr. Cavallo found that a few particles of steel, adhering to a hone, on which the point of a needle was slightly rubbed, imparted to it magnetic properties; and Mr. Coulomb has observed that there are scarcely any bodies in nature which do not exhibit some marks of being subjected to the influence of magnetism, although its force is always proportional to the quantity of iron which they contain, as far as that quantity can be ascertained; a single grain being sufficient to make 20 pounds of another metal sensibly magnetic. A combination with a large proportion of oxygen deprives iron of the whole or the greater part of its magnetic properties; finery cinder is still considerably magnetic, but the more perfect oxids and the salts of iron only in a slight degree; it is also said that antimony renders iron incapable of being attracted by the magnet. Nickel, when freed from arsenic and from cobalt, is decidedly magnetic, and the more so as it contains less iron. Some of the older chemists supposed nickel to be a compound metal containing iron, and we may still venture to assume this opinion as a magnetical hypothesis. There is indeed no way of demonstrating that it is impossible for two substances to be so united as to be incapable of separation by the art of the chemist; had nickel

4

been as dense as platina, or as light as cork, we could not have supposed that it contained any considerable quantity of iron, but in fact the specific gravity of these metals is very nearly the same, and nickel is never found in nature but in the neighbourhood of iron; we may therefore suspect, with some reason, that the hypothesis of the existence of iron in nickel may be even chemically true. The aurora borealis is certainly in some measure a magnetical phenomenon, and if iron were the only substance capaple of exhibiting magnetic effects, it would follow that some ferruginous particles must exist in the upper regions of the atmosphere. The light usually attending this magnetical meteor may possibly be derived from electricity, which may be the immediate cause of a change of the distribution of the magnetic fluid, contained in the ferruginous vapours, that are imagined to float in the air.

We are still less capable of distinguishing with certainty in magnetism, than in electricity, a positive from a negative state, or a real redundancy of the fluid from a deficiency. The north pole of a magnet may be considered as the part in which the magnetic fluid is either redundant or deficient, provided that the south pole be understood in a contrary sense: thus, if the north pole of a magnet be supposed to be positively charged, the south pole must be imagined to be negative; and in hard iron or steel these poles may be considered as unchangeable.

A north pole, therefore, always repels a north pole, and attracts a south pole. And in a neutral piece of soft iron, near to the north pole of a magnet, the fluid becomes so distributed by induction, as to form a temporary south pole next to the magnet, and the whole piece is of course attracted, from the greater proximity of the attracting pole. If the bar is sufficiently soft, and not too long, the remoter end becomes a north pole, and the whole bar a perfect temporary magnet. But when the bar is of hard steel, the state of induction is imperfect, from the resistance opposed to the motion of the fluid; hence the attraction is less powerful, and an opposite pole is formed, at a certain distance, within the bar; and beyond this another pole, similar to the first; the alternation being sometimes repeated more than once. The distribution of the fluid within the magnet is also affected by the neighbourhood of a piece of soft iron, the north pole becoming more powerful by the vicinity of the new south pole, and the south pole being consequently strengthened in a certain degree; so that the attractive power of the whole magnet is increased by the proximity of the iron. A weak magnet is capable of receiving a temporary induction of a contrary magnetism from the action of a more powerful one, its north pole becoming a south pole on the approach of a stronger north pole; but the original south pole still retains its situation at the opposite end, and restores the magnet nearly to its original condition, after the removal of the disturbing cause.

The polarity of magnets, or their disposition to assume a certain direction, is of still greater importance than their attractive power. If a small magnet, or simply a soft wire, be poised on a centre, it will arrange itself in such a direction, as will produce an equilibrium of the attractions and repulsions of the poles of a larger magnet; being a tangent to a certain oval figure, passing through those poles, of which the properties have been calculated by various mathematicians. This polarity may easily be imitated by electricity; a suspended wire being brought near to the ends of a positive and negative conductor, which are placed parallel to each other, as in Nairne's electrical machine, its position is perfectly similar to that of a needle attracted by a magnet, of which those conductors represent the poles. (Plate XLI. Fig. 569.)

The same effect is observable in iron filings placed near a magnet, and they adhere to each other in curved lines, by virtue of their induced magnetism, the north pole of each particle being attached to the south pole of the particle next it. This arrangement may be seen by placing the filings either on clean mercury, or on any surface that can be agitated; and it may be imitated by strewing powder on a plate of glass, supported by two balls, which are contrarily electrified. (Plate XLI. Fig. 570.)

The polarity of a needle may often be observed when it exhibits no sensible attraction or repulsion as a whole; and this may easily be understood by considering that when one end of a needle is repelled from a given point, and the other is attracted towards it, the two forces, if equal, will tend to turn it round its centre, but will wholly destroy each other's effects with respect to any progressive motion of the whole needle. Thus, when the end of a magnet is placed under a surface on which iron filings are spread, and the surface is shaken, so as to leave the particles for a moment in the air, they are not drawn sensibly towards the magnet, but their ends, which are nearest to the point over the magnet, are turned a little downwards, so that they strike the paper further and further from the magnet, and then fall outwards, as if they were repelled by it. (Plate XLI. Fig. 571.)

The magnets, which we have hitherto considered, are such as have a simple and well determined form; but the great compound magnet, which directs the mariner's compass, and which appears to consist principally of the metallic and slightly oxidated iron, contained in the internal parts of the earth, is probably of a far more intricate structure, and we can only judge of its nature from the various phenomena derived from its influence.

The accumulation and the deficiency of the magnetic fluid, which determine the place of the poles of this magnet, are probably in fact considerably diffused, but they may generally be imagined, without much error in the result, to centre in two points, one of them nearer to the north pole of the earth, the other to the south pole. In consequence of their attractions and repulsions, a needle, whether previously magnetic or not, assumes always, if freely poised, the direction necessary for its equilibrium; which, in various parts of the globe, is variously inclined to the meridian and to the horizon. Hence arises the use of the compass in navigation and in surveying: a needle, which is poised with a liberty of horizontal motion, assuming the direction of the magnetic meridian, which for a certain time remains almost invariable for the same place; and a similar property is also observable in the dipping needle, which is moveable only in a vertical plane; for when this plane is placed in the magnetic meridian, the needle acquires an inclination to the horizon, which varies according to the situation of the place with respect to the magnetic poles. (Plate XLI. Fig. 572, 573.)

The natural polarity of the needle may be in some measure illustrated by inclosing an artificial magnet in a globe; the direction of a small needle, suspended over any part of its surface, being determined by the position of the poles of the magnet, in the same manner as the direction of the compass is determined by the magnetical poles of the earth, although with much more regularity. In either case the whole needle is scarcely more or less attracted towards the globe than if the influence of magnetism were removed; except when the small needle is placed very near to one of the poles of the artificial magnet, or, on the other hand, when the dipping needle is employed in the neighbourhood of some strata of ferruginous substances, which, in particular parts of the earth, interfere materially with the more general effects, and alter the direction of the magnetic meridian.

A bar of soft iron, placed in the situation of the dipping needle, acquires from the earth, by induction, a temporary state of magnetism, which may be reversed at pleasure by reversing its direction; but bars of iron, which have remained long in or near this direction, assume a permanent polarity; for iron, even when it has been at first quite soft, becomes in time a little harder. A natural magnet is no more than a heavy iron ore, which, in the course of ages, has acquired a strong polarity from the great primitive magnet. It must have lain in some degree detached, and must possess but little conducting power, in order to have received and to retain its magnetism.

We cannot, from any assumed situation of two or more magnetic poles, calculate the true position of the needle for all places; and even in the same place, its direction is observed to change in the course of years, according to a law which has never yet been generally determined, although the variation which has been observed, at any one place, since the discovery of the compass, may perhaps be comprehended in some very intricate expressions; but the less dependence can be placed on any calculations of this kind, as there is reason to think that the change depends rather on chemical than on physical causes. Dr. Halley indeed conjectured that the earth contained a nucleus, or separate sphere, revolving freely within it, or rather floating in a fluid contained in the intermediate space, and causing the variation of the magnetic meridian; and others have attributed the effect to the motions of the celestial bodies: but in either case the changes produced would have been much more regular and universal than those which have been actually observed. Temporary changes of the terrestrial magnetism have certainly been sometimes occasioned by other causes; such causes are, therefore, most likely to be concerned in the more permanent effects. Thus, the eruption of Mount Hecla was found to derange the position of the needle considerably; the aurora borealis has been observed to cause its north pole to move

6 or 7 degrees to the westward of its usual position; and a still more remarkable change occurs continually in the diurnal variation. In these climates the north pole of the needle moves slowly westwards from about 8 in the morning till 2, and in the evening returns again; a change which has with great probability been attributed to the temporary elevation of the temperature of the earth, eastwards of the place of observation, where the sun's action takes place at an earlier hour in the morning, and to the diminution of the magnetic attraction in consequence of the heat thus communicated. In winter this variation amounts to about 7 minutes, in summer to 13 or 14.

Important as the use of the compass is at present to navigation, it would be still more valuable if its declination from the true meridian were constant for the same place, or even if it varied according to any discoverable law; since it would afford a ready mode of determining the longitude of a place by a comparison of an astronomical observation of its latitude with another of the magnitude of the declination. And in some cases it may even now be applied to this purpose, where we have a collection of late and numerous observations. Such observations have from time to time been arranged in charts, furnished with lines indicating the magnitude of the declination or variation at the places through which they pass, beginning from the line of no variation, and proceeding on the opposite sides of this line to show the magnitude of the variation east or west. It is obvious that the intersection of a given parallel of latitude, with the line showing the magnitude of the variation, will indicate the precise situation of the place at which the observations have been made.

The line of no variation passed in 1657 through London, and in 1666 through Paris: its northern extremity appears to have moved continually eastwards, and its southern parts westwards; and it now passes through the middle of Asia. The opposite portion seems to have moved more uniformly westwards; it now runs from North America to the middle of the South Atlantic. On the European side of these lines, the declination is westerly; on the South American side, it is easterly. The variation in London has been for several years a little more than 24°. In the West Indies it changes but slowly; for instance it was 5° near the island of Barbadoes, from 1700 to 1756. (Plate XLI. Fig. 574 . . 576. Plate XLII. XLIII.)

The dip of the north pole of the needle in the neighbourhood of London is 72°. Hence the lower end of a bar standing upright, as a poker, or a lamp iron, becomes always a north pole, and the temporary south pole of a piece of soft iron being uppermost, it is somewhat more strongly attracted by the north pole of a magnet placed over it, than by its south pole; the distribution of the fluid in the magnet itself being also a little more favourable to the attraction, while its north pole is downwards. It is obvious that the magnetism of the northern magnetic pole of the earth must resemble that of the south pole of a magnet, since it attracts the north pole; so that if we considered the nature of the distribution of the fluid, rather than its situation in the earth, we should call it a south pole. Although it is impossible to find any places for two, or even for a greater number of magnetic poles, which will correctly explain the direction of the needle in every part of the earth's surface, yet the dip may be determined with tolerable accuracy, from the supposition of a small magnet placed at the centre of the earth, and directed towards a point in Baffin's Bay, about 75° north latitude, and 70° longitude west of London; and the variation of the dip is so inconsiderable, that a very slow change of the position of this supposed magnet would probably be sufficient to produce it; but the operation of such a magnet, according to the general laws of the forces concerned, could not possibly account for the very irregular disposition of the curves indicating the degree of variation or declination; a general idea of these might perhaps be obtained from the supposition of two magnetic poles situated in a line considerably distant from the centre of the earth; but this hypothesis is by no means sufficiently accurate to allow us to place any dependence on it. (Plate XLI. Fig. 577, 578.)

The art of making magnets consists in a proper application of the attractions and repulsions of the magnetic fluid, by means of the different conducting powers of different kinds of iron and steel, to the production and preservation of such a distribution of the fluid in a magnet, as is the best fitted to the exhibition of its peculiar properties.

We may begin with any bar of iron that has long stood in a vertical position; but it is more common to employ an artificial magnet of greater strength. When one pole of such a magnet touches the end of a bar of hard iron or steel; that end assumes in some degree the opposite character, and the opposite end the same character: but in drawing the pole along the bar, the first end becomes neutral, and afterwards has the opposite polarity; while the second end has its force at first a little increased, then becomes neutral, and afterwards is opposite to what it first was. When the operation is repeated, the effect is at first in some measure destroyed, and it is difficult to understand why the repetition adds materially to the inequality of the distribution of the fluid; but the fact is certain, and the strength of the new magnet is for some time increased at each stroke, until it has acquired all that it is capable of receiving. Several magnets, made in this manner, may be placed side by side, and each of them being nearly equal in strength to the first, the whole collection will produce together a much stronger effect; and in this manner we may obtain from a weak magnet others continually stronger, until we arrive at the greatest degree of polarity of which the metal is capable. It is, however, more usual to employ the process called the double touch: placing two magnets, with their opposite poles near to each other, or the opposite poles of a single magnet, bent into the form of a horseshoe, in contact with the middle of the bar: the opposite actions of these two poles then conspire in their effort to displace the magnetic fluid, and the magnets having been drawn backwards and forwards repeatedly, an equal number of times to and from each end of the bar, with a considerable pressure, they are at last withdrawn in the middle, in order to keep the poles at equal distances.

Iron filings, or the scoriae from a smith's forge, when finely levigated, and formed into a paste with linseed oil, are also capable of being made collectively magnetic. A bar of steel, placed red hot between two magnets, and suddenly quenched by cold water, becomes in some degree magnetic, but not so powerfully as it may be rendered by other means. For preserving magnets, it is usual to place their poles in contact with the opposite poles of other magnets, or with pieces of soft iron, which, in consequence of their own induced magnetism, tend to favour the accumulation of the magnetic power in a greater quantity than the metal can retain after they are removed. Hence the ancients imagined that the magnet fed on iron.

A single magnet may be made of two bars of steel, with their ends pressed into close contact; and it might be expected that when these bars are separated, or when a common magnet has been divided in the middle, the portions should possess the properties of the respective poles only. But in fact the ends which have been in contact are found to acquire the properties of the poles opposite to those of their respective pieces, and a certain point in each piece is neutral, which is at first nearer to the newly formed pole than to the other end, but is removed by degrees to a more central situation. In this case we must suppose, contrarily to the general principles of the theory, that the magnetic fluid has actually escaped by degrees from one of the pieces, and has been received from the atmosphere by the other.

There is no reason to imagine any immediate connexion between magnetism and electricity, except that electricity affects the conducting powers of iron or steel for magnetism, in the same manner as heat or agitation. In some cases a blow, an increase of temperature, or a shock of electricity, may expedite a little the acquisition of polarity; but more commonly any one of these causes impairs the magnetic power. Professor Robison found, that when a good magnet was struck for three quarters of an hour, and allowed in the mean time to ring, its efficacy was destroyed; although the same operation had little effect when the ringing was impeded; so that the continued exertion of the cohesive and repulsive powers appears to favour the transmission of the magnetic as well as of the electric fluid. The internal agitation, produced in bending a magnetic wire round a cylinder, also destroys its polarity, and the operation of a file has the same effect. Mr. Cavallo has found that brass becomes in general much more capable of being attracted when it has been hammered, even between two flints; and that this property is again diminished by fire: in this case it may be conjectured that hammering increases the conducting power of the iron contained in the brass, and thus renders it more susceptible of magnetic action. Mr. Cavallo also observed that a magnetic needle was more powerfully attracted by iron filings during their solution in acids, especially in the sulfuric acid, than either before or after the operation: others have not always succeeded in the experiment; but there is nothing improbable in the circumstance, and there may have been some actual difference in the results, dependent on causes too minute for observation. In subjects so little understood as the theory of magnetism, we are obliged to admit some paradoxical propositions, which are only surprising on account of the imperfect state of our knowledge. Yet, little as we can understand the intimate nature of magnetical actions, they exhibit to us a number of extremely amusing as well as interesting phenomena; and the principles of crystallization, and even of vital growth and reproduction, are no where so closely imitated, as in the arrangement of the small particles of iron in the neighbourhood of a magnet, and in the production of a multitude of complete magnets, from the influence of a parent of the same kind.

LECTURE LVI.

ON CLIMATES AND WINDS.

THE science of meteorology relates principally to the natural history of the air, and to such temporary changes in the earth and sea as are produced by causes not mechanical only. The subject is of a very complicated and intricate nature; it comprehends many effects derived from such causes, as belong separately to every department of physics which we have hitherto examined; and although it has occupied the attention of several philosophers of considerable eminence, we cannot yet boast of having made any great advancement in it. Whether we shall ever be able to carry our theories to so high a degree of perfection, as to furnish us with much information applicable to the purposes of common life, to agriculture, or to medicine, is at present uncertain; although some advantage has already been derived from the indications of meteorological instruments; and the philosophy of the science is in many respects much more advanced than has commonly been supposed. We shall divide this extensive subject into two parts, the first relating principally to the effects of heat on the atmosphere, including the phenomena of winds; the second to the nature and consequences of evaporation, comprehending atmospherical electricity, and to the effects of subterraneous fires and igneous meteors.

The variations of temperature, in different parts of the earth's surface, require to be examined in the first place; since they are not only of considerable importance in themselves, but are also among the principal causes of other changes in the state of the winds and weather. These changes are measured by thermometers, of various kinds, which have already been described; but, for meteorological purposes, some additions are frequently made to the simple thermometer. In Six's thermometer, the tube is twice bent, so as to return in a parallel direction: the bulb is in the form of a long cylinder, and is usually filled with spirit of wine, which is in contact with a portion of mercury occupying the lower part of the tube; and this is succeeded by a second portion of spirit. The mercury carries on each of its surfaces an index, which is retained in its remotest situation by means of a weak spring; and consequently shows the greatest degree of heat or of cold that has happened since the last observation. The indexes are of iron or steel, and may be brought back to the surface at pleasure by means of a magnet; they are carried up by the mercury, more by its capillary action, than by the difference of the specific gravities. A similar effect is obtained in Rutherford's arrangement of a pair of thermometers, one with mercury, the other with spirit of wine, placed in a horizontal position; one index being without the surface of the mercury, the other within that of the spirit: the thermometers being in contrary directions, both indexes may be brought back to their places, by merely raising the end of the instrument. Self registering thermometers have also sometimes been constructed, for keeping a still more accurate account of all the variations of temperature that have occurred, by describing a line on a revolving barrel, which shows the height for every instant during the whole time of their operation. (Plate XLI. Fig. 579, 580.)

The climates of different parts of the earth's surface are unquestionably owing in great measure to their position with respect to the sun. At the equator, where the sun is always nearly vertical, any given part of the surface receives a much greater quantity of light and heat, than an equal portion near the poles; and it is also still more affected by the sun's vertical rays, because their passage through the atmosphere is shorter than that of the oblique rays. As far as the sun's mean altitude only is concerned, it appears from Simpson's calculations, that the heat received at the equator in the whole year, is nearly twice and a half as great as at the poles; this proportion being nearly the same as that of the meridian heat of a vertical sun, to the heat derived, at the altitude $23\frac{1}{2}°$, in the middle of the long annual day at the poles. But the difference is rendered still greater, by the effect of the atmosphere, which interrupts a greater portion of the heat at the poles than elsewhere. Bouguer has calculated, upon the supposition of the similarity of the affections of heat and light, that in latitude 45°, 80 parts out of 100 are transmitted at noon in July, and 55 only in December. The heat intercepted by the atmosphere is perhaps not wholly, but very nearly, lost with respect to the climate of the neighbouring places. It is obvious that, at any individual place, the climate in summer must approach in some degree to the equatorial climate, the sun's altitude being greater, and in winter to the climate of the polar regions.

While the earth is becoming warmer at any particular spot, the heat thrown off by radiation into the atmosphere, and thence into the empty space beyond it, together with that which is transmitted to the internal parts of the earth, must be less than the heat received from the sun; and when the earth is growing colder, more heat must pass off than is received: but whenever the heat of the surface is stationary, neither increasing nor diminishing, as at the times of the greatest and least heat, it is obvious that the heat received from the sun must be precisely equal to the heat which is thrown off. Now this quantity may be estimated by the degree of refrigeration in the night; and hence Mr. Prévost has very ingeniously deduced the proportion of the sun's heat arriving at the surface of the earth in the latitude of Geneva, in July, and in December; which he finds to be as 7 or 8 to 1; and this result agrees very well with a calculation deduced from the length of the day, the sun's altitude, and the interception of his rays by the atmosphere.

In London the temperature generally varies, in the course of the day and night, somewhat more than 5°, and less than 20°. In January, the mean diurnal variation of temperature is 6°, in March 20°, in July 10°, and in September, 18°. Hence, says Mr. Kirwan, we may understand the reason of the great frequency of colds in spring and in autumn.

Some philosophers have supposed the earth to become progressively warmer in the course of ages, while others have imagined that its heat is exhausted. Both these opinions appear in general improbable. The greater heat the earth receives by day, the more it throws off, both by day and by night; so that in the course of a few ages the heat must probably have attained its maximum. Local changes may indeed arise from local circumstances; thus, the climate of America is said to have become considerably warmer, since a large part of its surface has been cleared from its dense forests by human labour: and to judge from the descriptions of the ancients, it appears that even in Europe the winters were formerly much colder than they are at present. If, however, Dr. Herschel's opinion of the variation of the heat of the sun be onfirmed, it will introduce a great uncertainty into all theories upon the subject: since in these calculations the original heat of the sun has always been supposed unalterable.

The sea is less heated than the land, partly because a greater quantity of water evaporates from it, and partly because the sun's rays penetrate to a considerable depth, and have less effect on the surface, while the water is also mixed, by the agitation of its waves and currents, with the colder water below. It is also more slowly cooled than the land, since, when the temperature of the superficial particles is depressed, they become heavier, and sink to the bottom. For similar reasons, the sea is colder than the land in hot climates, and by day, and warmer in cold climates, and by night. These circumstances, however, nearly balance each other, so that the mean temperatures of both are equal, that of the sea being only less variable. Although the process of evaporation must cool the sea, yet when the vapours are condensed without reaching the land, their condensation must compensate for this effect by an equal extrication of heat.

There is another cause which perhaps contributes in some degree, in temperate climates, to the production of cold; that is, the alternation of freezing and thawing. Mr. Prévost observes that congelation takes place much more suddenly than the opposite process of liquefaction; and that of course the same quantity of heat must be more rapidly extricated in freezing than it is absorbed in thawing; that the heat, thus extricated, being disposed to fly off in all directions, and little of it being retained by the neighbouring bodies, more heat is lost than is gained by the alternation: so that where ice has once been formed, its production is in this manner redoubled. This circumstance must occur wherever it freezes, that is, on shore, in latitudes above 35°; and it appears that from about 30° to the pole, the land is somewhat colder than the sea, and the more as it is further distant from it; and nearer the equator the land is warmer than the sea: but the process of congelation cannot by any means be the principal cause of the difference, and it is probable that the different capacity of earth and water for heat is materially concerned in it.

Since the atmosphere is very little heated by the passage of the sun's rays through it, it is naturally colder than the earth's surface;

and for this reason, the most elevated tracts of land, which are the most prominent, and the most exposed to the effects of the atmosphere, are always colder than situations nearer the level of the sea. The northern hemisphere is somewhat warmer than the southern, perhaps because of the greater proportion of land that it contains, and also in some measure on account of the greater length of its summer than that of the southern; for although, as it was long ago observed by Simpson, the different distance of the sun compensates precisely for the different velocity of the earth in its orbit, with respect to the whole quantity of heat received on either side of the equinoctial points, yet Mr. Prévost has shown, that in all probability the same quantity of heat must produce a greater effect when it is more slowly applied; because the portion lost by radiation from the heated body is greater, as the temperature is higher. Since, therefore, on account of the eccentricity of the earth's orbit, the north pole is turned towards the sun 7 or 8 days longer than the south pole, the northern winters must be milder than the southern: yet the southern summers, though shorter, ought to be somewhat warmer than the northern: but in fact they are colder, partly perhaps from the much greater proportion of sea, which in some degree equalises the temperature, and partly for other reasons. The comparative intensity of the southern summer and winter is not exactly known; but in the island of New Georgia the summer is said to be extremely cold.

The northern ice extends about 9° from the pole: the southern 18° or 20°; in some parts even 30°; and floating ice has occasionally been found in both hemispheres as far as 40° from the poles, and sometimes, as it has been said, even in latitude 41° or 42°. Between 54° and 60° south latitude, the snow lies on the ground, at the sea side, throughout the summer. The line of perpetual congelation is three miles above the surface at the equator, where the mean heat is 84°; at Teneriffe, in latitude 28°, two miles; in the latitude of London, a little more than a mile; and in latitude 80° north, only 1200 feet. At the pole, according to the analogy deduced by Mr. Kirwan, from a comparison of various observations, the mean temperature should be 31°. In London the mean temperature is 50°; at Rome and at Montpelier, a little more than 60°; in the island of Madeira, 70°; and in Jamaica, 80°.

There are frequently some local causes of heat and cold which are independent of the sun's immediate action. Thus, it has been observed, that when the weather has been clear, and a cloud passes over the place of observation, the thermometer frequently rises a degree or two almost instantaneously. This has been partly explained by considering the cloud as a vesture, preventing the escape of the heat which is always radiating from the earth, and reflecting it back to the surface: the cloud may also have been lately condensed, and may itself be of a higher temperature than the earth. Mr. Six has observed that in clear weather, the air is usually some degrees colder at night, and warmer by day, close to the ground, than a few feet above it; but that in cloudy weather there is less difference: and it is possible that this circumstance may be derived from the difference of the quantity of evaporation from the earth's surface, which occasions a different degree of cold in different states of the atmosphere.

The motions of the air, which constitute winds, are in general dependent, in the first instance, on variations of temperature. They are so accidental and uncertain, as to be subjected to no universal laws; as far however as any regularity can be observed in their recurrence, it may in most cases be sufficiently explained.

The principal phenomena of the periodical winds may be reduced to six distinct heads: first the general tendency from north east and south east towards the equator, in latitudes below 30°; secondly, the deviation of this tendency from the precise situation of the equator; thirdly, the prevalence of westerly winds between 30° and 40° or more, especially in the southern hemisphere; fourthly, the local modifications to which these general effects are subjected; fifthly the monsoons, which vary every half year; and lastly the diurnal changes of land and sea breezes.

With respect to the general tendency of the trade winds to the west, it may be sufficiently explained by Hadley's theory of the difference of the rotatory motion of different parts of the atmosphere, combined with the currents occasioned by the greater heat at the equator. For the sun's rays, expanding the air in the neighbourhood of the equator, and causing it to ascend, produce a current in the lower parts of the atmosphere, which rush southwards and northwards towards the equator, in order to occupy the place of the heated air as it rises: and since the rotatory motion of the earth is greatest at the equator, and is directed eastwards, the air coming from the poles has of course a relative motion westwards; and hence the joint motion of the current is directed, in the northern hemisphere, from north east to south west, and in the southern, from south east to north west. Dr. Halley supposed that the air was made in some measure to follow the sun round the earth, simply by means of the expansion of the atmosphere, which takes place immediately under him, and accompanies him round the globe; but it does not seem evident that the air could have any greater tendency to follow the sun than to meet him. Astronomers have, however, deduced an additional cause for an easterly wind from the attractions of the moon and of the sun, which appear, from the laws of gravitation, to have a slight tendency to retard the rotatory motion of the atmosphere: and a similar instance has been observed in the motions of the atmosphere of the planet Jupiter, by means of the appearances of spots of different kinds on his disc, some of which seem to revolve less rapidly than the body of the planet. At so great a distance, the influence of the sun's heat must be comparatively inconsiderable, and the want of a tendency in the spots towards the equator appears to show, that the atmosphere, in which they float, is not put in motion by the same causes, which we have supposed to be most concerned in the production of our own trade winds. It has been remarked that the friction of the atmosphere, thus retarded by the attraction of the sun and moon, must in the course of ages have impaired the uniformity of the earth's diurnal motion; and it has been observed, on the other hand, that even this effect would be partially counteracted by the gradual filling up of valleys, by means of the descent of the superficial parts of mountains, which, at a greater distance from the centre, were revolving with a rapidity somewhat greater than the valleys in which they are deposited; but probably neither of these changes would become sensible in millions of years.

The second circumstance is easily explained by the greater heat of the northern than of the southern hemisphere; so that instead of coinciding with the equator, the neutral portion of the atmosphere lies between 3° and 5° of north latitude; the north east wind not reaching the equator, and the south east continuing about 3° beyond it. But the situation of the neutral portion varies with the sun's declination, accordingly as different parallels of latitude become in succession somewhat hotter than the neighbouring parts. Where the northern and southern currents meet, their joint effect must naturally be to produce a due east wind; but in some parts of the ocean, temporary calms and irregular squalls have been observed to take place of this easterly wind, which generally prevails in the neutral parts near the equator.

The third fact, that is, the frequency of westerly winds between the latitudes 30° and 40°, has not yet been sufficiently explained. The most probable cause of this circumstance is, that the current of heated air, which we have hitherto neglected, and which passes, in the upper parts of the atmosphere, from the equator each way towards the poles, and which, being the converse of the trade wind, must be a south west and north west wind, in the different hemispheres, becomes here sufficiently cool to descend and mix with the lower parts of the atmosphere, or to carry them along by its lateral friction; and while it descends to complete the circle, necessary for supplying the current to the equator, its motion with respect to the horizon must become at a certain time due west, since the cause which stops its progress northwards, has no tendency to impede its motion eastwards. The outward bound East India ships generally make their easting in about 36° south latitude. It is probably also on account of the rotatory motion of the earth, that south west winds are more common in our latitudes than south east, and north east than north west.

Among the local modifications to be considered in the fourth place, we may reckon the greater indistinctness of the third effect in the northern than in the southern hemisphere, a circumstance which is explained from the more irregular distribution of sea and land: for between 30° and 40° south latitude the ocean is scarcely any where interrupted. In lower latitudes also, near the west coast of Africa, the winds are so much deflected towards the land, as to become in general westerly instead of easterly.

The monsoons, which constitute the fifth remarkable circumstance, are so called from a Malay word, denoting season. They are occasioned by the peculiar situation of the continent of Asia, on the north side of the equator. From April to September, the sun having north declination, the heat on this continent, a little north of the tropic, is very intense, and the general current is consequently towards the north. The air, therefore, coming from south latitudes towards the equator, becomes, on account of the deficiency of rotatory motion, a south east wind, as usual, which is found to prevail between Madagascar and New Holland, as far as the equator. In consequence perhaps of friction in its passage, it gradually loses its impetus towards the west, and at the equator is nearly a south wind: but in proceeding north from the equator, it becomes, from an excess of rotatory motion, a south west wind, which blows into the Arabian gulf, and the bay of Bengal. Both these winds are however variously modified by the particular situations of the islands and continents. From October to March, on the contrary, the sun having south declination, the south east trade wind stops at 10° south latitude; the trade winds on the north side of the equator are as usual north east; and beyond the equator they become for some degrees north west, the circumstances being the reverse of those which happen in the summer months, at greater distances, on the other side of the equator. (Plate XLII. XLIII.)

The last fact is the simplest of all. The land and sea breezes are produced by the ascent of the air over the land in the day time, while the land is hotter than the sea; and its descent at night, when the land is become colder: hence the breeze comes from the sea by day, and from the land by night.

The violent agitations of the air, which constitute hurricanes and whirlwinds, occur more commonly in tropical climates than in others. The causes of these storms are little understood: their course is said to be generally opposite to that of the trade winds; but tornados, which are less regular hurricanes, originate indifferently from every quarter.

The variations of the weight of the air, which occasion the winds, and other changes in its density, which are the effects of the winds themselves, are indicated by the height of the barometer, which is in general the more variable as the winds are more liable to sudden changes. Hence in the neighbourhood of the equator the height of the barometer is scarcely ever a quarter of an inch more or less than 30 inches, which is very nearly its mean height on the level of the sea in every part of the globe: in Great Britain it is sometimes as low as 28 inches, but never higher than 31. We have already seen that the elevation of any place above the sea reduces the height of the barometer according to a law which is determined by the general properties of elastic fluids: thus, at an elevation of 1 mile above the sea, the mean height of the barometer is $24\frac{1}{4}$ inches, and at 2 miles, 20 inches only. The use of the barometer, in foretelling variations of weather, is perhaps more limited than has sometimes been supposed; but by a careful observation, conclusions may be drawn from it, which may in many cases be of considerable utility: and it has even been applied with success, by some late navigators, to the prediction of changes of wind, at times when they could not have been suspected from any other circumstances.

LECTURE LVII.

ON AQUEOUS AND IGNEOUS METEORS.

THE phenomena originating from the evaporation of water constitute a large proportion of the subjects of meteorology: they are materially influenced by the diversities of climates and winds, which we have lately considered; and they appear to contribute to the electrical changes, which form a principal part of luminous or igneous meteors: nor is the action of water wholly unconcerned in many of the effects of subterraneous fires, which have also a slight connexion with atmospherical electricity; and it has been conjectured that the only igneous meteors, which appear wholly independent of any of these phenomena, may originate from volcanic commotions in other worlds.

The action of heat appears to detach continually from the surface of water, and perhaps of every other liquid, and even solid, a certain quantity of vapour, in the form of an invisible gas; but when the space above the liquid is already charged with as much vapour as can exist in it at the actual temperature, the vapour, thus continually thrown off, either remains suspended in the form of visible particles, or falls back immediately into the liquid. This is the simplest mode of explaining the continuance of evaporation, under the pressure of any dry gas, however dense, and its apparent suppression in the presence of moist air, however rare. Sometimes also, when the temperature of the liquid is elevated, so that minute globules either of steam or of air rise through it, some visible particles are projected upwards by each globule, and continue to float in the air; this appears, however, to be an irregularity unconnected with the principal process of slow evaporation.

The quantity of vapour, which can exist in the space above any portion of water, has been supposed by Deluc, Volta, and Dalton, to be wholly independent of the nature, the density, or even the presence of the air or gas which that space contains: and we may easily imagine that the smallest distance, at which the particles of water, constituting vapour, can exist, without coming within the reach of their mutual cohesion, is the same, whatever other particles may be scattered through the intervening space. It appears, however, more consistent with some experiments, to suppose, that the presence of air of the usual density allows the particles of water to approach a little nearer together without cohering, so that the utmost quantity of moisture, that can be contained in a cubic foot of air at a given temperature, is not exactly the same as would make a cubic foot of pure vapour, but always in a certain proportion to it; and it seems to follow, from the experiments of Saussure, compared with those of Pictet, that the weight of the vapour contained in a cubic foot of air is about one half greater than that of a cubic foot of pure vapour at the same temperature.

When the air, in the neighbourhood of the surface of the water, has become thus saturated with moisture, the evaporation proceeds very slowly, the vapour being precipitated as soon as it rises: but if the air be continually changed, so that the moistened portion may be removed, and dry air substituted for it, the process will be greatly expedited; and such a change may be effected either by wind, or by the natural circulation, occasioned by any elevation of temperature communicated by the water to the neighbouring air; but when this circulation is prevented, the evaporation is much diminished, although the temperature may be considerably elevated. In moderate exposures, the depth of the quantity of water, evaporating in 24 hours from any surface, is expressed, according to Mr. Dalton's experiments, by the height of the column of mercury equivalent to the force of steam at the given temperature, deducting, however, the effect of the elasticity of the moisture already existing in the air.

Since the quantity of moisture, which the air is capable of receiving, is greater as its temperature is greater, we may obtain a natural measure of the quantity which it contains, by reducing it to the temperature at which the moisture begins to be deposited. Thus, if we take a glass of cold water, and add to it some common salt, or some muriate of lime, we may cool the air near it so much, as to cause it to deposit a part of its moisture on the glass: and by measuring the temperature of the water when the precipitation begins,

Mr. Dalton estimates the true state of the air with respect to moisture. Thus, if the glass begins to be moistened when the water is at 40°, he infers from the known elasticity of steam at that temperature, that the quantity of moisture contained in the air is equivalent to the pressure of a column of mercury about a quarter of an inch in height; and if the actual temperature of the air be 50°, the corresponding elasticity of steam being a little more than one third of an inch, the daily evaporation in such air will amount to about one ninth of an inch, making 40 inches in the whole year. In fact, however, the air is usually moister than this, and the mean evaporation of all England is, according to Mr. Dalton, about 23 inches only.

In hotter climates, and in particular situations, the evaporation may be considerably greater. The Mediterranean Sea, being surrounded by land, is more heated than the ocean, and the winds which blow over it are drier; consequently its evaporation is greater than that of the Atlantic, and its specific gravity is increased by the increased proportion of salt; so that at the straights of Gibraltar, a current runs inwards at the surface and outwards near the bottom, for the same reason as the air, when it is denser in a passage than in the adjoining room, blows a candle towards the room at the lower part of the door, and draws it towards the passage at the upper. Had there been a continual current inwards through the Straights, at all parts, the Mediterranean must in the course of ages have become a rock of salt. It is indeed remarkable that all lakes, into which rivers run without any further discharge, are more or less salt, as well as lakes in general near the sea: but where a river runs through a lake into the sea, it must necessarily, in the course of time, have carried the salt of the lake with it, if it had ever existed.

Experiments on the deposition of moisture, like those of Mr. Dalton, are liable to a slight inaccuracy, on account of the effects of an apparent elective attraction, by means of which, some substances seem to attract humidity at a temperature a little higher than others. Thus, a surface of metal often remains dry, in the neighbourhood of a piece of glass which is covered with moisture. It is certain that some substances attract moisture from the air, even when the quantity which it contains is incomparably less than that which would saturate it, since it is on this circumstance that the construction of hygrometers depends; and it is probably by a property somewhat similar, that even surfaces of different kinds possess different attractive powers for moisture nearly ready to be deposited. It is, however, only necessary to employ, for Mr. Dalton's experiment, a substance which has a very weak attraction for moisture; and any kind of metal will perhaps be found sufficiently correct in its indications.

It has been observed, that a piece of metal, placed on glass, usually protects also the opposite side of the glass from the deposition of dew; and Mr. Benedict Prévost has shown, that in general, whenever the metal is placed on the warmer side of the glass, the humidity is deposited more copiously either on itself, or on the glass near it; that when it is on the colder side, it neither receives the humidity, nor permits its deposition on the glass; but that the addition of a second piece of glass, over the metal, destroys the effect, and a second piece of metal restores it. It appears that, from its properties with respect to radiant heat, the metallic surface produces these effects, by preventing the ready communication either of heat or of cold to the glass.

The quantity of invisible moisture, contained in air, may be, in some degree, estimated from the indications of hygrometers, although these instruments have hitherto remained in a state of great imperfection. A sponge, a quantity of caustic potash, or of sulfuric acid, or a stone of a peculiar nature, has sometimes been employed for determining the degree of moisture of the air, from which it acquires a certain augmentation of its weight. A cord dipped in brine, or the beard of an oat, is also often used for the same purpose: the degree in which it untwists, from the effect of moisture, being shown by an index. But the extension of a hair, or of a slip of whalebone, which have been employed by Saussure and Deluc, appear to be more certain and accurate in their indications. The hair hygrometer acquires more speedily the degree corresponding to any given state of the air, but it seems to reach the utmost extent of its scale before it arrives at perfect humidity; while the whalebone hygrometer appears to express a greater change upon immersion in water than from the effect of the moistest transparent air, which has also been considered by some as an imperfection. Both these instruments are impaired by time, and acquire contrary errors, so that a mean between both is more likely to be correct than either separately. Their indications are at all times widely different from each other, and the mean appears to approach much nearer to a natural scale than either of them. Mr. Leslie employs a very delicate thermometer, of which the bulb is moistened, for measuring the dryness of the air, by the cold produced during evaporation, when the thermometer is exposed to it; but this mode of estimating the quantity of moisture appears to be liable to considerable uncertainty. (Plate XLI. Fig. 581.)

In order that the scale of a hygrometer should be perfectly natural, it ought to express, at all temperatures, the proportion of the quantity of moisture in the air to that which is required for its saturation; thus, at 100 degrees, it should imply that the slightest depression of temperature would produce a deposition; at 50 degrees, that the air contains only half as much water as would saturate it, or, supposing the thermometer at 52°, that a deposition would be produced in it by a depression of 17°. And if we know the actual temperature, and the temperature at which the deposition takes place, we may find the height of the natural hygrometer, by the proportion of the corresponding elasticities of steam. The mean height of the natural hygrometer in London is probably about 80°; that of Deluc's hygrometer, with proper corrections, being nearly 70°: so that a depression of 6° must usually be sufficient to cause a deposition of moisture.

The quantity of water actually contained in a cubic foot of air, saturated with moisture, appears to be about 2 grains at the freezing point, 4 grains at 48°, 6 at 60°, and 8 at 68°; and the density of the vapour, thus mixed with air, is, according to Saussure's experiments, about three fourths as great as that of the air itself; so that moist air is always a little lighter than dry air: and the more so as the air is warmer, provided that it be saturated with moisture by means of the presence of water. It follows from the properties of moisture thus determined, that if any two portions of perfectly humid air, at different temperatures, be mixed together, there must be a precipitation: thus, a cubic foot of air at 32° being mixed with another at 60°, their common temperature must be 46°; if they are saturated with moisture, they must contain 8 grains of water when separate; but when mixed they will be too cold by 2° to contain the same quantity; since air at 48° can only contain 4 grains for each foot; and it has been supposed that such mixtures frequently occasion a precipitation in nature. Thus, it often happens that the breath of an animal, which is in itself transparent, becomes visible when mixed with a cold atmosphere; and in such cases the deposition may perhaps be facilitated by the cooling of the warmer air to a certain degree, even before a perfect mixture has taken place

When visible vapour has been thus deposited from transparent air, by means either of cold or of mixture, it generally remains for some time suspended, in the form of a mist or of a cloud: sometimes, however, it appears to be at once deposited on the surface of a solid, in the form of dew or of hoar frost; for it is not probable that the crystallized form, in which hoar frost is arranged, can be derived from the union of the particles already existing in the air as distinct aggregates.

The dew, which is commonly deposited on vegetables, is partly derived, in the evening, from the vapours ascending from the heated earth, since it is then found on the internal surface of a bell glass; and towards the morning, from the moisture descending from the air above, as it begins to cool. Sometimes, however, in warmer weather, the dew begins to descend in the evening; this the French call serein: the humidity deposited by mists on trees, and by moist air on windows, generally within, but sometimes without, they call givre.

Mists are said to consist sometimes of other particles than pure water: these are called dry mists, and they have been supposed to blight vegetables. Such mists are sometimes attended by a smell, resembling that which is occasioned by an electric spark. Rain falling after a dry season deposits, when it has been suffered to stand, some particles of foreign matter which it has brought down from the atmosphere. There must indeed frequently be a multiplicity of substances of various kinds floating in the air; the wind has been found to carry the farina of plants as far as 30 or 40 miles, and the ashes of a volcano more than 200. It only requires that the magnitude of the particles of any substance be sufficiently reduced in size, in order to render them incapable of falling with any given velocity; and when this velocity is very small, it may easily be overpowered by any accidental motions of the air. The diameter of a sphere of water, falling at the rate of one inch only in a second, ought to be one six hundred thousandth of an inch, which is about the thickness of the upper part of a soap bubble at the instant when it bursts; but the particles of mists are incomparably larger than this, since they would otherwise be perfectly invisible as separate drops: the least particle, that could be discovered by the naked eye, being such as would fall with a velocity of about a foot in a second, if the air were perfectly at rest. But it is very probable that the resistance, opposed to the motion of particles so small, may be considerably greater, than would be expected from a calculation, derived from experiments made on a much larger scale, and their descent consequently much slower.

When the particles of a mist are united into drops capable of descending with a considerable velocity, they constitute rain; if they are frozen during their deposition, they exhibit the appearance of a perfect crystallization, and become snow: but if the drops already formed are frozen, either by means of external cold, or on account of the great evaporation produced by a rapid descent through very dry air, they acquire the character of hail, which is often observed in weather much too hot for the formation of snow.

It cannot be doubted but that there is a connexion between the descent of the barometer and the fall of rain; but no satisfactory reason has yet been assigned for the circumstance; nor is it possible to foretel, with certainty, that rain will follow any changes in the height of the barometer that have been observed. The immediate dependence of rain, or of any other atmospherical phenomena, on the influence of the moon, appears to be rendered highly improbable, not only by mathematical calculations of the effects of the moon's attraction, but also by the irregularity of the very observations, which have been adduced in favour of such a connexion. But however uncertain the ultimate causes of rain may be in general, their effects in some places are sufficiently constant, to be attributed to permanent local circumstances, and in particular to the periodical recurrence of similar winds.

In low and level countries, clouds may often begin to descend from the upper regions of the atmosphere, and may be redissolved by the warmer air below; but when they descend in an equal degree among mountains, they fall on the earth; and besides the quantity of water which they furnish for vegetation, and that which is carried off by evaporation, they afford, by means of springs and rivers, a constant supply for the use of man and of other animals in distant parts. The upper regions of the atmosphere are however by no means the principal sources of rain in ordinary climates, since a gage placed on a very high building seldom collects more than two thirds as much rain as another standing on the ground below: and the effects of mountains in collecting rain are perhaps chiefly derived from the ascending currents which they occasion, and by which the air saturated with moisture is carried to a higher and a colder region.

The Abyssinian rains are the causes of the inundation of the Nile; they last from April to September; but for the first three months the rain is only in the night. The inundation, in Egypt, begins at present about the 17th of June; it increases for 40 days, and subsides in the same time; but the ancient accounts, as well as some modern ones, assign a longer duration to it. The river Laplata rises and falls at the same times as the Nile. The Ganges, the Indus, the Euphrates, the river of Ava or Pegu, and many other large rivers, have also considerable inundations at regular periods. In many other countries there are seasons at which the rains seldom fail to recur; and sometimes the periodical rains are different in different parts of the same country. Thus the coast of Malabar, which is to the west of the Gate mountains, or Gauts, enjoys summer weather, without rain, from September to April, while that of Coromandel, which is on the eastern side, experiences all the rigours of its winter; being at this time exposed to the influence of the north east trade wind. Vicissitudes of a similar nature are also observed on the north and south sides of the island of Jamaica. The mean fall of rain in London is about 23 inches; at Exeter, which is nearer to the Atlantic, 33; the average of England and Wales is 31.

The evaporations and precipitations, and probably also the condensations and expansions, which take place on a large scale in the atmosphere and in the clouds, cannot fail of producing changes in their electrical qualities; and these changes appear to be the principal sources of the phenomena of thunder and lightning. The clouds, when electrified, being more or less insulated by the interposition of the air, exhibit attractive and repulsive effects, and are discharged by explosions, either among themselves, or communicating with the earth, in the same manner as bodies which have been electrified by artificial means; they also sometimes produce, in the neighbouring parts of the earth, and in the animals on its surface, a state of induced electricity; and in this case the returning stroke, or the sudden restoration of the equilibrium, when the electricity of the nearest clouds is imparted to the more remote, may be fatal, without any appearance of an immediate discharge, at the place where the animal stands.

We can, however, by no means precisely ascertain in what manner all the electrical phenomena of the atmosphere are produced. It appears from the experiments of Beccaria and Cavallo that the air is in general positively electrical, and most so in cold and clear weather; in cloudy weather more slightly: and that during rain, the air is generally in a negative state. Mr. Read has found that air charged with putrid vapours of any kind, and in particular the air of close rooms, is almost always negatively electrified. The electricity is more readily communicated to an electrometer in an elevated situation, and in damp weather, than in other circumstances; a candle is also very useful in collecting it. When a wire is connected with a kite, being continued along the string, we may frequently obtain from it sparks a quarter of an inch long.

We find a complete and interesting description of the effects of a violent thunder storm in a paper by Mr. Brereton, inserted in the Philosophical Transactions. The circumstance happened in September 1780, at East Bourn, in a house occupied by Mr. Adair: it was built of stone, and stood facing the sea. About nine o'clock, in a very stormy morning, a black cloud approached the house; several balls of fire were seen to drop from it successively into the sea, and one in particular, appearing like an immense sky rocket, broke against the front of the house in different directions. Mr. Adair was standing at a window on the first floor, with his hands clasped together, and extended against the middle of the frame: his hands were forced asunder, he was thrown several yards off on the floor, and remained for some time speechless and motionless, although not insensible: his clothes were much torn; several articles of metal about his person were partially melted, while others, apparently in similar circumstances, and in particular a silver buckle, escaped; and his skin was in many parts much scorched and lacerated. The whole of the glass in the window, and a pier glass near it, were completely destroyed, and scattered about the room; most of the furniture was broken to pieces, and all the bell wires were melted. In the room above this, a lady and her maid were driven to a distant part, and rendered insensible for some time, but not hurt; in the room below, two servants, who were near the windows, were struck dead: both the bodies were turned black; one of them had a wound near the heart; and neither of them became stiff after death; a third servant, who was a little behind one of them, escaped with the loss of a telescope, which he held in his hand, and with the sensation of a violent pressure on his head and on his back. A large stone was forced out of the wall near them, and thrown into the room, and some other similar effects were observed, which marked the progress of the explosion.

For guarding against accidents so dreadful, Dr. Franklin's great invention of metallic conductors may be very advantageously employed: for, when properly fixed, they afford a degree of security which leaves very little room for apprehension. A conductor ought to be continued deep into the earth, or connected with some well or drain; it should be of ample dimensions, and where smallest, of copper, since copper conducts electricity more readily than iron. In one instance a conductor of iron, four inches wide, and half an inch thick, appears to have been made red hot by a stroke of lightning. It seems to be of some advantage that a conductor should be pointed, but the circumstance is of less consequence than has often been supposed. Mr. Wilson exhibited some experiments in which a point was struck at a greater distance than a ball, and therefore argued against the employment of pointed conductors. Mr. Nairne, on the contrary, showed that a ball is often struck in preference to a point. But it has been observed, that if a point attracts the lightning from a greater distance, it must protect a greater extent of building. It is easy to show, by hanging cotton or wool on a conductor, that a point repels light electrical bodies, and that a pointed conductor may, therefore, drive away some fleecy clouds; but this effect is principally derived from a current of air repelled by the point; and such a current could scarcely be supposed to have any perceptible effect on clouds so distant as those which are concerned in thunder storms. In order to escape personal danger in a thunder storm, the best precautions are, to avoid eminences, and all exposed situations, as well as a near approach to conductors. The neighbourhood of windows, looking glasses, fire places, and trees, must always be considered as hazardous.

It has been supposed that a sudden condensation of the air, arising from cold, accompanied by a deposition of moisture, and propagated by a continuation of the cause, by means of the cold occasioned by expansion, produces frequently the noise of thunder, without any lightning, and without any electrical agitation; but it does not appear that the opinion is well established.

The phenomena of waterspouts, if not of electrical origin, appear to have some connexion with electrical causes. A waterspout generally consists of large drops, like a dense rain, much agitated, and descending or ascending with a spiral motion, at the same time that the whole spout is carried along horizontally, accompanied in general by a sound like that of the dashing of waves. Spouts are sometimes, although rarely, observed on shore, but generally in the neighbourhood of water. They are commonly largest above; sometimes two cones project, the one from a cloud, the other from the sea below it, to meet each other, the junction being accompanied by a flash of lightning: and when the whole spout has exhibited a luminous appearance, it has perhaps served to conduct electricity slowly from the clouds to the earth. Some of these circumstances may be explained by considering the spout as a whirlwind, carrying up drops of water, which it has separated from the surface of the waves; and the remainder may perhaps be deduced from the cooperation of electricity, already existing in a neighbouring cloud.

It is doubtful whether the light of the aurora borealis may not be of an electrical nature: the phenomenon is certainly connected with the general cause of magnetism; the primitive beams of light are supposed to be at an elevation of at least 50 or 100 miles above the earth, and every where in a direction parallel to that of the dipping needle; but perhaps, although the substance is magnetical, the illumination, which renders it visible, may still be derived from the passage of electricity, at too great a distance to be discovered by any other test.

Earthquakes and volcanos appear to originate in chemical changes, which take place within the substance of the earth: they have probably little further connexion with electricity, than as causes which occasionally destroy the electrical equilibrium; for although some authors have inferred, from the great velocity with which the shock of an earthquake is transmitted from place to place, that its nature must be electrical, yet others have, with greater probability, attributed the rapid succession of the effects to the operation of a single cause, acting at a great distance below the earth's surface. There are however some circumstances, which indicate such a connexion between the state of the atmosphere and the approach of an earthquake, as cannot easily be explained by any hypothesis.

The shocks of earthquakes, and the eruptions of volcanos, are in all probability modifications of the effects of one common cause; the same countries are liable to both of them; and where the agitation produced by an earthquake extends further than there is any reason to suspect a subterraneous commotion, it is probably propagated through the earth nearly in the same manner as a noise is conveyed through the air. Volcanos are found in almost all parts of the world, but most commonly in the neighbourhood of the sea; and especially in small islands; for instance, in Italy, Sicily, Iceland, Japan, the Caribbees, the Cape Verd islands, the Canaries, and the Azores: there are also numerous volcanos in Mexico and Peru, especially Pichincha and Cotopaxi. The subterraneous fires, which are continually kept up in an open volcano, depend perhaps in general on sulfureous combinations and decompositions, like the heating of a heap of wet pyrites, or the union of sulfur and iron filings: but in other cases they may perhaps approach more nearly to the nature of common fires. A mountain of coal has been burning in Siberia for almost a century, and must probably have undermined in some degree the neighbouring country. The immediate cause of an eruption appears to be very frequently an admission of water from the sea, or from subterraneous reservoirs; it has often happened that boiling water has been discharged in great quantities from a volcano; and the force of steam is perhaps more adequate to the production of violent explosions, than any other power in nature. The consequence of such an admission of water, into an immense collection of ignited materials, may in some measure be understood, from the accidents which occasionally happen in founderies: thus a whole furnace of melted iron was lately dissipated into the air in Colebrook Dale, by the effect of a flood, which suddenly overflowed it.

The phenomena of earthquakes and volcanos are amply illustrated by the particular accounts, transmitted to the Royal Society by Sir William Hamilton, of those which have happened at different times in Italy. The earthquake, which desolated Calabria, in 1783, was fatal to about 40000 persons, continuing its ravages for more than three months; it destroyed the towns and villages occupying a circle of nearly 50 miles in diameter, lying between 38 and 39 degrees latitude, and extending almost from the western to the eastern coast of the southernmost point of Italy, besides doing considerable damage to places at much greater distances from its origin, which is supposed to have been either immediately under the town of Oppido, in the centre of this circle; or under some part of the sea, between the west of Italy, and the volcanic island of Stromboli. This island, as well as Mount Etna, had smoked less than usual before the earthquake, but they both exhibited appearances of an eruption during its continuance; Etna towards the beginning, and Stromboli at the end. Before each shock the clouds were usually motionless for a certain time, and it rained violently; frequently also lightning and sudden gusts of wind accompanied the rain. The principal shocks appeared to consist in a sudden elevation of the ground to a considerable height, which was propagated somewhat like a wave, from west to east: besides this, the ground had also a horizontal motion backwards and forwards, and in some measure in a circular direction. This motion was accompanied by a loud noise; it continued in one instance for ten seconds without intermission: and it shook the trees so violently that their heads nearly reached the ground. It affected the plains more strongly than the hills. In some places luminous exhalations, which Sir William Hamilton thinks rather electrical than igneous, were emitted by the earth: the sea boiled up near Messina, and was agitated as if by a copious discharge of vapours from its bottom; and in several places water, mixed with sand, was thrown up to a considerable height. The most general effect of these violent commotions was the destruction of buildings of all kinds, except the light barracks of wood or of reeds, into which the inhabitants retreated as soon as they were aware of their danger: the beds of rivers were often left dry, while the shock lasted, and the water on its return overflowed their banks: springs were sometimes dried up, and new ones broke out in other places. The hills which formed the sides of steep vallies were often divided by deep chasms parallel to the vallies; and in many cases large portions of them were separated, and removed by the temporary deluge to places half a mile or a mile off; with the buildings and trees still standing on them; and in this manner hills were levelled, and vallies were filled up. But the most fatal accident of this kind happened at Scilla, where so large a portion of a cliff was thrown into the sea, that it raised an immense wave, which carried off more than 2000 inhabitants who were collected on the beach, and even extended its formidable effects to the opposite coast of Sicily, where several persons perished by it in a similar manner.

The eruptions of volcanos are usually attended by some shocks like those of earthquakes, although commonly less violent. Open volcanos continually

4

throw out, in more or less abundance, smoke, ashes, and pumice stones, or light cinders; but their most formidable effects are produced by a torrent of ignited lava, which, like a vast deluge of liquid or semiliquid fire, lays waste the country over which it runs, and buries all the works of human art. In March, 1767, Vesuvius began to throw out a considerable quantity of ashes and stones, which raised its summit in the course of the year no less than 200 feet, forming first a little mountain of pumice stones within the crater, which by degrees became visible above its margin. The smoke, which was continually emitted, was rendered luminous at night, by the light derived from the fire burning below it. In August some lava had broken through this mountain, and in September it had filled the space left between it and the former crater. On the 13th and 14th of October there were heavy rains, which perhaps supplied the water concerned in the eruption that shortly followed. On the morning of the 19th, clouds of smoke were forced, in continual succession, out of the mouth of the volcano, forming a mass like a large pine tree, which was lengthened into an arch, and extended to the island of Caprea, 28 miles off: it was accompanied by much lightning, and by an appearance of meteors like shooting stars. A mouth then opened below the crater, and discharged a stream of lava, which Sir William Hamilton ventured to approach within a short distance, imagining that the violence of the confined materials must have been exhausted; but on a sudden the mountain opened with a great noise at a much lower point, about a quarter of a mile from the place where he stood, and threw out a torrent of lava, which advanced straight towards him, while he was involved in a shower of small pumice stones and ashes, and in a cloud of smoke. The force of the explosions was so great, that doors and windows were thrown open by them at the distance of several miles: the stream of lava was in some places two miles broad, and 60 or 70 feet deep; it extended about six miles from the summit of the mountain, and remained hot for several weeks. In 1794 a still more violent eruption occurred: it was expected by the inhabitants of the neighbourhood, the crater being nearly filled, and the water in the wells having subsided. Showers of immense stones were projected to a great height; and ashes were thrown out so copiously, that they were very thick at Taranto, 250 miles off; some of them also were wet with salt water. A heavy noxious vapour, supposed to be carbonic acid, issued in many places from the earth, and destroyed the vineyards in which it was suffered to remain stagnant. A part of the town of Torre del Greco was overwhelmed by a stream of lava, which ran through it into the sea; yet notwithstanding the frequency of such accidents, the inhabitants had so strong a predilection for their native spot, that they refused the offer of a safer situation for rebuilding their houses.

Convulsions of these kinds must have very materially influenced the disposition of the strata of the earth, as well as the form of its surface; but it is by no means fully determined how far such causes have been concerned, or how far the effects are to be attributed to the intermediation of water only. Mineralogists and geologists have been principally divided into two classes with respect to their theories of the earth, some maintaining the Vulcanian, and some the Neptunian hypothesis. It appears to be impossible to decide with any certainty between these opposite opinions; nor is it perhaps of much consequence for any purpose of practice, or even of science. The Neptunians are certainly able to establish their own theory positively, and to prove that the fluid parts of the earth and sea must have been very materially concerned in producing the changes which have happened to the solid parts; but it may be difficult for them to confute the assertion, that heat, whether caused by volcanos or otherwise, has also been a very powerful agent in these operations, and in some cases the joint effects of heat and of increased pressure appear to have been concerned, in giving to minerals of different kinds their actual form; although on the whole it seems probable that the operation of heat has been much more limited than that of aqueous solutions and precipitations. Mr. Davy has also very justly inferred, from his experiments with the battery of Volta, that the effects of the electricity excited by means of chemical changes within the earth, have probably been very materially concerned in the gradual formation of a variety of mineral productions.

The arguments for establishing the general fact, that great convulsions have actually happened to the earth, are too well known to require minute examination: the variety of fossil substances, many of them marine productions, and some almost preserving a recent appearance, that are found in mountains remote from the sea, are undeniable proofs that the levels of the earth's surface must have undergone considerable changes; although some philosophers are of opinion, that such of the primary mountains as are above 6 or 700 feet high, have never been wholly covered by the sea. It is not at all easy to explain the change of climate, which some of these circumstances appear to indicate; the remains of animals inhabiting hot countries, and the marine productions of hot climates, which are frequently found in high northern latitudes, would induce us to suspect, that the position of the earth's axis was at a former time very different from its present position; and we can scarcely assign any other probable cause for this change, than the casual interference, and perhaps incorporation, of a comet with the earth. The probabilities of such an event, in the whole course of time, are however so small, that we have no reason to be apprehensive of the chance of its occurring in future, for it is not enough that a comet should approach so near to the earth as to be very powerfully attracted by it, its motion must also be directed almost in a straight line towards the earth; otherwise it might only be inflected into a new orbit, and go off again, without having caused any other disturbance than a partial overflow of the sea.

The face of the globe has also been very materially changed in the course of ages, by the gradual operation of the sea and of rivers. The sea has incroached in particular parts, and retired from others; and the mouths of large rivers, running through low countries, have often been variously modified, by a deposition and transfer of the matter washed down from the land. At Havre the sea undermines the steep coast, and recedes at Dunkirk, where the shore is flat: in Holland the Zuyder Zee was probably formed in the middle ages by continual irruptions of the sea, where only the small lake Flevo had before existed; and the mouths of the Rhine have been considerably altered, both in their dimensions and in their directions. The mud, deposited by large rivers, generally causes a Delta, or triangular piece of land, to grow out into the sea; thus the mouth of the Mississippi is said to have advanced above 50 miles since the discovery of America; and the sea has retired from Rosetta above a mile in 40 years. The mouths of the Arno and of the Rhone consist also in great measure of new land.

The meteors denominated shooting stars are observed to move in all directions, as well upwards as downwards, although they frequently seem to have a tendency towards a particular quarter in the course of the same evening. Their height is seldom less than 20 miles, and sometimes as much as 100 or 200, but usually about 50; their velocity is commonly about 20 miles in a second, which differs very little from that of the earth in its orbit. The rapidity of their motion, as well as its occasional deviation from a right line, has generally been considered as a reason for supposing that they depend on electricity; but the opinion is by no means fully established.

Other igneous meteors, which nearly resemble in their appearance the largest of these, are sometimes observed to fall on the earth, either entire or divided; and after their fall, certain stones have been found, which have been supposed to have descended in an ignited state. Mr. Howard has ascertained that almost all these stones agree in their general characters, and in their chemical analysis, especially in the circumstance of containing nickel. It has been conjectured, both in this country and on the continent, that they have been emitted by lunar volcanos, and it has been observed, that since they would find little or no resistance from the very rare atmosphere of the moon, they would require a velocity of projection only four times as great as that which a cannon ball sometimes receives, in order to rise into the sphere of the earth's attraction. Their heat and combustion may not improbably be derived from the great condensation which they must occasion in the air immediately before them, and even their friction might easily produce enough of electric light, to render them visible in the dark. Among many such substances projected from the moon, it is probable that a few only would be directed towards the earth, and many more would be made to revolve in ellipses round it, and become little satellites, too small for human observation, except when they enter far enough into the atmosphere to produce an appearance of light, resembling that of a shooting star; but it is scarcely probable that their velocity could ever be at all comparable with that which has been attributed to these meteors. There is, however, no difficulty in supposing, on the other hand, that the wandering substances, which may be moving through empty space, with a velocity equal to that of the shooting stars, may be so much retarded, when they penetrate deep into our atmosphere, as to make but a moderate impression by their fall on the ground; and if we suppose the meteors to be of one kind only, they must be referred rather to the description of shooting stars than to that of the productions of lunar volcanos; although the undulatory motion, sometimes observed in these meteors seems to be in some measure inconsistent with the progress of a heavy body, moving by means of its natural inertia in a straight line.

LECTURE LVIII.

ON VEGETATION.

It may appear idle to some persons, to attempt to reduce the outlines of natural history into so small a compass, as is required for their becoming a part of this course of lectures; and it would indeed be a fruitless undertaking to endeavour to communicate a knowledge of the particular subjects of this science, even in a much longer time than we shall bestow on it. But many naturalists have spent a great portion of their lives in learning the names of plants and animals, and have known at last less of the philosophy of the science, than might have been told them in a few hours, by persons who had observed with more enlarged views, and who had reasoned on general principles. And we shall perhaps find it possible to collect into a small compass the most useful information, that has hitherto been obtained, respecting the laws of animal and vegetable life, as well as the foundations of the methods, by which the most received systematical classifications have been regulated.

The surface of the earth, as well sea as land, is occupied by innumerable individuals, constituting an immense variety of distinct species of animated and inanimate beings, comprehended in the three grand divisions of natural bodies. The mineral kingdom consists of such substances, as are composed of particles either united without any regular form, or collected together by accretion or external growth only. When mineral substances crystallize, they often imitate the form, and almost assume the external appearance of vegetables: but their particles are never extended to admit others between them, and to be thus enlarged in all their dimensions; their growth is only performed by the addition of similar particles, upon the surface of those that have been already deposited.

Vegetables derive their existence, by seeds, or otherwise, from a parent stock, their parts are extended and evolved from within, and they imbibe their nutriment by superficial absorption only. There is indeed in the crystallization of minerals a slight resemblance to a reproduction or generation, when a small portion of the substance serves as a basis for the formation of subsequent crystals: but this portion becomes a constituent part of the crystal, while it preserves its original form; a seed, on the contrary, is a substance naturally and completely detached from the plant, and containing within itself the simplest rudiments of a new individual, which is afterwards evolved and enlarged. Sometimes, however, vegetables are propagated by means of bulbs, or by spreading roots, by slips, or by ingrafted scions, without a seed detached in the regular manner; but in these cases the new plant is much more identical with the old one, than when it is raised from a seed, being as it were a continuation of the same existence. Plants are nourished in great measure by means of their roots; and sometimes, where they are without roots, their nutriment is probably absorbed by all parts of their surface.

Animals are distinguished from vegetables by the reception of their food, for digestion and assimilation, into an internal cavity constituting a stomach. The existence of a stomach, calculated for the digestion of food, appears to be the best, if not the only criterion of an animal. Some vegetables, indeed, have a power of catching and detaining animals, by curling up their leaves so as to cover them, as the drosera or sundew, and the dionaea muscipula, or catchfly; but this mechanism can scarcely be intended for their immediate nutriment, at least the leaf can scarcely be supposed to assume the character of a stomach. It is true that we imagine all animals to have sensation, and all plants to be without it; and if it were possible to discriminate decisively between sensation and irritation, the distinction would supersede every other: but in many cases it is extremely difficult to say where sensation is present, and where irritation only produces the same apparent effects. We cannot be sure that the hydra, or fresh water polypus, or the trichurus sol, an animalcule described by Dr. Shaw, suffers any sensation of pain when it is divided into two parts; at least the pain seems to agree remarkably well with its constitution, for it lives and thrives with increased vigour, as two distinct animals. On the other hand, many plants are easily stimulated to perform motions, which have the appearance of muscular actions, influenced by sensation: the sensitive plants close or depress their leaves, in consequence of agitation or of electricity; the stamina of the barberry and of the pellitory are thrown into motion, when touched with a needle, and those of rue, and of the grass of parnassus, have at times alternate motions without any apparent cause. A zoophyte is an animal absolutely fixed to one place; and the vallisneria is a vegetable possessed of a certain limited power of locomotion. A plant chooses in preference to turn towards the light; and it has been known that an ash tree on a wall, when incapable of being any longer supported by the wall only, has concentrated all its force in the production of one large root, descending to the ground. Some of these circumstances may be explained without recurring to any thing like volition; but, as far as we know, the same explanations might be applied to some animal motions: and although it is very possible that there may be a certain limit, where the influence of mind and sensation terminates, and the laws of vegetable life only prevail; yet the place of the division is not strongly enough marked, to allow it to form a characteristic in an artificial system. It has been asserted that some worms are nourished by absorption only, without the assistance of a stomach; thus hydatids, which are supposed to be of an animal nature, appear to be simply bags of a fluid without any visible opening; but a few doubtful cases of this kind can scarcely be sufficient, to invalidate the general position, that all bodies decidedly animal have a cavity for the reception of food. There are usually also some chemical distinctions in the component parts of animals and vegetables; animal substances commonly containing greater proportions of azote or nitrogen, and of phosphoric acid; but there are some exceptions to this observation; thus the carica papaya, or papaw, contains nearly the same principles as are usually found in substances of animal origin. In general we may readily distinguish a small portion of an animal from a vegetable substance, by the smell produced in burning it. According to common language, we say, that minerals have growth only, but not always; that vegetables grow and live also; and that animals have sensation, as well as life and increase of magnitude.

Mineralogy is a branch of natural history so nearly allied to chemistry, that it cannot be completely understood without a previous knowledge of that science. It may therefore be more properly considered as belonging to a course of chemical than of physical lectures.

The vegetable kingdom presents to us a spectacle highly interesting by its variety and by its elegance; but the economy of vegetation appears to be little diversified, although little understood. With respect to the apparent perfection of their functions, and the complication of their structure, we may consider all vegetables as belonging to two principal divisions, in one of which the seed is prepared with the assistance of a flower, having its stamina and its pistils, with petals or a calyx; while in the other, the preparation of the seed is less regular and conspicuous, and hence such plants are called cryptogamous. In some of these there is a slight resemblance to the flowers of other vegetables, but on the whole, the class appears to form one of the connecting links between the three kingdoms of nature; its physiology is probably simple, but it has been little examined. The herbs, palms, shrubs, and trees, which constitute the numerous genera of flowering vegetables, exhibit the greatest diversity in the forms and dispositions of the organs of fructification, while they have all a general resemblance in their internal economy.

Every vegetable may be considered as a congeries of vessels, in which, by some unknown means, the aqueous fluids, imbibed by its roots, are subjected to peculiar chemical and vital actions, and exposed in the leaves to the influence of the light and air; so as to be rendered fit for becoming constituent parts of the plant, or of the peculiar substances contained within it.

The first process in the germination of a seed is its imbibing moisture, and undergoing a chemical fermentation, in which oxygen is absorbed, and a part of the mucilage contained in the seed is converted into sugar; a substance probably more nutritive to the young plant. The radicle shoots downwards, and the seed leaves, or cotyledons, which are generally two, although sometimes more or less numerous, raise themselves above the ground, till in a short time they die and drop off, being succeeded by the regular and more adult leaves.

In every transverse section of a vegetable, we commonly discover at least four different substances. The parts next to the axis of the tree or branch consist of medulla or pith, which is supposed by some to be the residence of the vegetable life of the plant; but a tree may live for many years after being in great measure deprived of its medulla. The pith is of a loose and light spongy texture; it sends a ramification into each branch and each leaf, where it appears to serve also as a reservoir of moisture. The pith is surrounded by the woody part, composed of fibres more or less strongly compacted together, but not actually ramifying into each other in any great degree, although there is reason to suspect some lateral communications between them. They are interrupted, at certain intervals, in many trees, by fibres, in a radiating direction, forming what is called the silver grain. Like the bones in animals, the wood constitutes the strongest part of the vegetable; and like them too it is in a certain degree furnished with vessels. It has even been supposed by some, that the fibres themselves are distinct tubes, and by others, that the interstices between them serve the purpose of vessels, but neither of these opinions is at present generally received. The wood consists of a number of concentric layers or strata, formed in successive years; the external part, which is last formed, is called the alburnum, or white wood, and this part is the most vascular. The bark encompasses the wood; and this also consists, in trees, of several layers, which are produced in as many different years; the external parts usually cracking, and allowing us at their divisions to observe their number, the inner layer only being of immediate use. This layer is called the liber, and since this material was once used instead of paper, the Romans called a book also liber. The bark consists of fibres of the same kind as the wood, but more loosely connected. It is covered by the cuticle, which extends itself in a very great degree, as the growth of the vegetable advances, but at last cracks, and has its office supplied by the outer layers of bark. Between the bark and the cuticle a green pulpy substance, or parenchyma, is found, which seems to be analogous to the rete mucosum, interposed between the true skin and the cuticle in animals. Mr. Desfontaines has observed, that in palms, and in several other natural orders of plants, the annual deposition of new matter is not confined to the external surface, but that it takes place in various parts of the plant, as if it were composed of a number of ordinary stems united together.

There are three principal kinds of vessels in the different parts of vegetables: the sap vessels, which are found both in the wood and in the bark, although their nature appears to require further examination: secondly, the air vessels, or tracheae, which are composed of single threads wound into a spiral tube, like the spring of a bell, and capable of being easily uncoiled; these, though they have been called air vessels, and supposed by some to serve the purposes of respiration, are described by others as containing, during the life of the plant, an aqueous fluid: and they are probably little more than sap vessels, with an additional spiral coat: they are not found in the bark, nor in all species of plants; and it has thence been inferred that they are not immediately necessary to the growth of the plant. The third kind are the proper vessels of the plant, which are generally disposed in concentric circles, and appear to be unconnected with the sap vessels, and to contain the milky, resinous, and other peculiar juices, which are found in different kinds of plants; for the sap is nearly the same in all, at least it is independent of the gums and resin, which often distinguish particular plants; it contains a certain portion of mucilage, and probably in some plants, as the sugar maple, a considerable quantity of sugar. Mr. Mirbel has also made a number of still more accurate distinctions respecting the structure of the different kinds of vessels. The circulation of the sap is not completely understood; when an orifice is made near the root of a tree, it flows most copiously from above: when near the summit, from below. Dr. Hope actually reverted the natural course of the juices of a tree, without changing its position; by inoculating a willow with two others, he completely united its existence with theirs, and then, removing its roots, he found that its vegetation was supported by the juices of the two others. A tree may also be actually inverted, and the upper part will strike root, the lower putting out branches and leaves.

Plants perspire very considerably, and also emit a quantity of gases of different kinds; they generate a slight degree of heat, which may be observed by means of the thermometer, and by the melting of snow in contact with them. The growth of every tree takes place at the internal surface of the bark, not only the bark itself being formed there, but the wood also being deposited by the bark; for Dr. Hope separated the whole of the bark of a branch of willow from the wood, leaving it connected only at the ends, so as to constitute a hollow cylinder, parallel to the wood; and he found that new layers were formed within the bark; and in another experiment a part of the wood, deprived of the bark, although protected from the air, was only covered with new bark as it grew over from the old bark above and below. The layers of wood, which are added in successive seasons, and keep a register of the age of the tree, are very easily observed when it is cut across; sometimes as many as 400 have been found in firs, and oaks are said to have lived 1000 years.

Mr. Knight has inferred, from a great variety of experiments, that the sap, either usually or universally, ascends through the wood into the leaves, and then descends through the bark to nourish the plant. The leaves seem to be somewhat analogous to lungs, or rather to the gills of fishes: for plants have need of air, and it has been found, that even seeds will not germinate in a vacuum. As the lungs of animals appear to be concerned in forming the blood, so it may be inferred from Mr. Knight's experiments, that the sap first ascends to the leaves through the external fresh wood or alburnum, and through the central vessels of the young leaves and branches, derived from the alburnum, and accompanied by the spiral tubes; and after being perfected by exposure to light and air in the leaves, it descends in the bark, and serves for the secretion of the alburnum, and of the internal layers of the bark, being conveyed probably by two distinct sets of vessels. The sap, thus prepared by the leaves in the summer and autumn, is supposed to leave its extractive matter in the tree throughout the winter, in such a state as to be ready to unite with the aqueous juices, which ascend from the root, in the succeeding spring. The internal parts of the wood, having served the purposes of vegetation, are hardened, and perhaps dried up, so as to be afterwards principally subservient to strength alone. By subsequent experiments, Mr. Knight has also found, that when a branch hangs downwards, the sap still appears to proceed from the part of the bark which is uppermost; so that the direction of the force of gravity seems to be concerned in determining that of the motion of the sap. There appears also to be some reason to suppose that mechanical means assist in the protrusion of the sap, and the consequent growth of the tree; for if a tree be more agitated by the wind in one direction than in another, its diameter will be greatest in that direction.

The process of grafting depends on a remarkable property of the growth of vegetables; if the cut surface of the inner bark of a small branch, or cutting, be placed in contact with that of the branch of another tree, they will unite sufficiently for the nourishment of the cutting; provided, however, that the nature of the plants be not too different. Something of the same kind occurs in animal life, where a tooth has been transplanted into the socket of another, or where the spur of a cock has been inserted into his comb.

Plants have their natural periods of life, either of a few days, as in the case of some of the fungi, of a year, of a few years, or of many centuries. They have also their diseases; they are often infested by insects, as in the gall of the oak, and the woodruff of the rose, or by animalcules of a still lower order, which are either the causes of the smut of corn, or constant attendants on it. From unnatural and too luxuriant culture, they become sterile, and produce double flowers instead of fruits and seeds. When deprived of sufficient moisture, or nipped by frost, their leaves and branches often die; and if the plants recover their vigour, a separation is affected by a natural process, resembling the sloughing of decayed parts of animals: but when the whole plant sinks, the dead leaves continue to adhere to it. The annual fall of leaves in autumn appears to be a natural separation nearly of the same kind, which takes place when the leaves are no longer wanted; the growth of the plant being discontinued, and their functions being no longer required.

Succulent plants generally die when the cuticle is removed, but not all other plants. The air appears to be injurious to vegetables where it is not natural; hence arises the benefit of Mr. Forsyth's method of completely excluding the air from the wounded parts of trees, by means of which their losses are often in great measure repaired, and they acquire new strength and vigour. Sometimes a diminution of the magnitude of a tree immediately increases its fertility; its force being more concentrated, by lopping off its useless branches and leaves, it produces a larger quantity of fruit, with the juices which would have been expended in their nourishment.

The Linnean system of vegetables is confessedly rather an artificial than a natural one; but it is extremely well adapted for practice, and its universal adoption has been productive of the most important improvements in the science of botany. Of the 24 classes into which Linné has divided the vegetable kingdom, 23 are distinguished by the forms of the flowers and fruit, and the 24th by the want of a regular florescence. The first 10 are named from monandria, in order, to decandria; then follow dodecandria; icosandria, and polyandria; the names expressing the number of the stamina, or filaments, surrounding the seed vessel; and the orders are deduced in a similar manner from the number of pistils, or little columns immediately connected with the seed vessel; and denominated monogynia, digynia, and so forth, as far as polygynia. These classes differ little in general with respect to their natural habits, except the twelfth, icosandria, which is characterized by the attachment of the filaments to the green cup, surrounding the flower, and which comprehends the most common fruit trees: this class has, however, been incorporated by some later botanists with the next. In the third class we find most of the natural order of grasses; the fifth, pentandria, is by far the most numerous of any: the sixth contains the lilies, and many other bulbous plants. The 14th class, didynamia, is known by two longer and two shorter filaments; it is perfectly natural, and comprehends flowers similar in their structure to the foxglove and the deadnettle. The 15th also, tetradynamia, is a class of plants strongly characterized even by chemical properties; two of the filaments are here shorter than the other four: cresses, radishes, and many other acrid and ammoniacal vegetables belong to this class, as well as the turnip and cabbage, which, when cultivated, become mild and nutritious. The class monadelphia contains a few plants similar to the mallow; they are known by the union of the filaments at their bases into a cylinder: those of the next class have generally nine united, and one separate, whence the class is named diadelphia; it contains the papilionaceous flowers, somewhat resembling a butterfly in their form, like the pea, and other leguminous plants, the broom, the furze, and the acacia. The 18th class, polyadelphia, has the filaments of its flowers united into several masses or bundles, as the hypericum or tutsan. The next class is perfectly natural, and contains the composite flowers, which have a peculiar union of the summits of the filaments; it is named syngenesia: sunflowers, daisies, and artichokes, are familiar examples of the plants of this class. The 20th class, gynandria, though it contains the natural family of the orchides, has been omitted by some late botanists; here the filaments are fixed on the pistil; or more properly, in the arums, within the pistils. The three following classes, monoecia, dioecia, and polygamia, differ from the rest in having some flowers with filaments or chives, and some with pistils only, either on the same plant, or on different plants, or mixed with flowers of the more common construction. Most of the forest trees belong to these classes, but the distinctions which separate them from other classes are not always very uniformly preserved, and, for this reason, many later botanists have disused them. The plants of the last class, cryptogamia, are exceedingly numerous; the families of ferns, mosses, algae, or membranous weeds, and fungi or mushrooms, fill up its extensive departments; some have also separated a part of the algae under the name of hepaticae, or gelatinous weeds. In this class the fructifications are extremely various; some of the fuci and confervae approach so much in their general appearance and mode of growth to corallines and zoophytes, that they seem to form an obvious connexion between the lowest ranks of the vegetable and animal kingdoms; while other plants of the class are scarcely distinguishable by their appearance from some of the productions of the mineral kingdom.

The French have introduced into very general use the botanical system of Jussieu. The most prominent feature in this system is the division of all the genera into a hundred natural orders, which are also arranged in fifteen classes. Jussieu begins, like Linné, with the separation of cryptogamic from phanerogamic plants; the seeds of the cryptogamic plants, which form the first class, being without cotyledons or seed leaves, and all other plants being distinguished into such as have seeds with one and with two cotyledons. Accordingly as the stamina or filaments are inserted below the pistil, on the calyx, or on the seed vessel, the first description of seeds affords three distinct classes. The plants which have two cotyledons follow, and are divided into apetalous, monopetalous, and polypetalous, from distinctions respecting the corolla or flower leaves, which are somewhat arbitrarily understood; and lastly diclinous, from the separation of the stamina and pistils. The three first of these divisions are subdivided according to the insertion of the stamina, and the union or separation of the antherae, which they support, into ten classes, making, with the four already mentioned, fourteen, to which the diclinous plants add a fifteenth. The orders are determined without any particular limitation of the parts from which the characters are taken. This system is of acknowledged merit as a philosophical classification of the natural orders of plants; such vegetables as nearly agree in their habits and appearances being brought more uniformly together than in the system of Linné. Hence, in the arrangement of a botanical garden, or in a treatise on the chemical or medical properties of plants, it might be employed with advantage: but for the practical purposes of botanical investigation it appears to be utterly unfit, since its author has sacrificed all logical and systematical laws to the attempt to follow nature, in analogies, which are often discoverable only with great difficulty, and which are seldom reducible to methodical definitions.

LECTURE LIX.

ON ANIMAL LIFE.

THE functions of animal life are not only more complicated in the same individual than those of vegetation, but also more diversified in the different classes into which animals are divided; so that the physiology of each class has its peculiar laws. We are indebted to Linné for the first enlargement of our views of the different classes of animals, and perhaps for the most convenient arrangement, of the animal kingdom; although his method has never been universally adopted by our neighbours on the continent.

A considerable portion of the bulk of all animals is composed of tubular vessels, which originate in a heart; the heart propels through the arteries, with the assistance of their own muscular powers, either a colourless transparent fluid, or a red blood, into the extremities of the veins; through which it again returns to the origin of its motion. Both insects, and vermes, or worms, have their circulating fluids a little warmer than the surrounding medium, and generally colourless; but insects have legs furnished with joints, and worms have nothing but simple tentacula at most in the place of legs. Fishes have cold red blood, which is exposed to the influence of the air contained in water, by means of their gills. The amphibia receive the air into their lungs, but their blood is cold, like that of fishes, and in both these classes the heart has only two regular cavities, while that of animals with warm blood has four; the whole contents of one pair being obliged to pass through the lungs, in order to arrive at the other pair. Of animals with warm blood, the oviparous are birds, and are generally covered with feathers, the viviparous are either quadrupeds or cetaceous animals, and are furnished with organs for suckling their young.

Each of these classes of animals is subdivided by Linné into different orders, of which we shall only be able to take a very cursory view. The first class, denominated mammalia, from the female's suckling its young, comprehends all viviparous animals with warm blood. These, with very few exceptions, have teeth fixed in their jaw bones; and from the form and number of these teeth, the orders are distinguished, except that of cetaceous fishes, which is known by the fins that are found in the place of feet. The distinctions of the teeth are somewhat minute, but they appear to be connected with the mode of life of the animal, and they are tolerably natural. The first order, primates, contains man, monkeys, and bats; the second, bruta, among others, the elephant, the rhinoceros, the ant eater, and the ornithorhynchus, an extraordinary quadruped, lately discovered in New Holland, with a bill like a duck, and sometimes teeth inserted behind it; but there are some suspicions that the animal is oviparous. The order ferae contains the seal, the dog, the cat, the lion, the tiger, the weasel, and the mole, most of them beasts of prey; the opossum and the kangaroo also belong to this order, and the kangaroo feeds on vegetables, although its teeth are like those of carnivorous animals. The fourth order, glires, comprehends beavers, mice, squirrels, and hares, the fifth, pecora, camels, goats, sheep, and horned cattle. The sixth order, belluae, contains the horse, the hippopotamus, and the hog. The cetaceous fishes, or whales, form the seventh and last order: they reside in the water, enveloped in a thick clothing of fat, that is, of oily matter, deposited in cells, which enables their blood to retain its temperature, notwithstanding the external contact of a dense medium considerably colder.

Birds are distinguished from quadrupeds, by their laying eggs; they are also generally feathered, although some few are rather hairy; and instead of hands or fore legs they have wings. Their eggs are covered by a calcarious shell; and they consist of a white, or albumen, which nourishes the chick during incubation, and a yolk, which is so suspended within it, as to preserve the side on which the little rudiment of a chicken is situated, continually uppermost, and next to the mother that is sitting on it. The yolk is in great measure received into the abdomen of the chicken a little before the time of its being hatched, and serves for its support, like the milk of a quadruped, and like the cotyledons of young plants, until the system is become sufficiently strong for extracting its own food out of the ordinary nutriment of the species.

Birds are divided, according to the form of their bills, into six orders: accipitres, as eagles, vultures, and hawks; picae, as crows, jackdaws, humming birds, and parrots; anseres, as ducks, swans, and gulls; grallae, as herons, woodcocks, and ostriches; gallinae, as peacocks, pheasants, turkies, and common fowls; and, lastly, passeres, comprehending sparrows, larks, swallows, thrushes, and doves.

The amphibia are in some respects very nearly allied to birds: but their blood is little warmer than the surrounding medium. Their respiration is not necessarily performed in a continual succession of alternations, since the whole of their blood does not pass through the lungs, and the circulation may continue without interruption in other parts, although it may be impeded in these organs, for want of the motion of respiration. They are very tenacious of life; it has been asserted on good authority that some of them have lived many years without food, inclosed in hollow trees, and even in the middle of stones: and they often retain vestiges of life some days after the loss of their hearts. Their eggs are generally covered with a membrane only. They have sometimes an intermediate stage of existence, in which all their parts are not yet developed, as we observe in the tadpole; and in this respect they resemble the class of insects. They are now universally considered as divided into two orders only; reptilia, as the tortoise, the dragon, or flying lizard, the frog and the toad; all these have four feet: but the animals which belong to the order serpentes are without feet. Most of the serpentes are perfectly innocent, but others have fangs, by which they instil a poisonous fluid into the wounds that they make. In England the viper is the only venomous serpent; it is known by its dark brown colour, and by a stripe of whitish spots running along its back; but to mankind its bite is seldom, if ever, fatal.

The first three classes of animals have lungs, as we have already seen, for respiration, and receive air by the mouth; those which have gills, and red blood, are fishes, residing either in fresh or in salt water, or indifferently in both: their eggs are involved in a membrane, and have no albumen. Of the six orders of fishes, four have regular gills, supported by little bones; and they are distinguished, according to the place of their ventral fins, into apodes, as the eel and lamprey; jugularis, as the cod; thoracici, as the sole and perch, and abdominales, as the salmon and pike: distinctions which appear to be perfectly artificial, although useful in a systematic arrangement. The two remaining orders are without bones in the gills, those of the one being soft, and of the other cartilaginous or gristly. These are the branchiostegi and chondropterygii of Artedi, which Linné, from a mistake, classed among the amphibia. The sun fish, the lump fish, the fishing frog, and the sea horse, are of the former, and the sturgeon, the skate, and the shark, of the latter order.

Insects derive their name from being almost always divided, into a head, thorax, and abdomen, with very slender intervening portions: although these divisions do not exist in all insects. They are usually oviparous: they respire, but not by the mouth; they have a number of little orifices on each side of the abdomen, by which the air is received into their ramified tracheae; and if these are stopped with oil, they are suffocated. Instead of bones, they have a hard integument or shell. Their mouths are formed on constructions extremely various, but generally very complicated: Fabricius has made these parts the basis of his classification; but from their minuteness in most species, the method is, in practice, insuperably inconvenient; and the only way, in which such characters can be rendered really useful, is when they are employed in the subdivision of the genera, as determined from more conspicuous distinctions. Insects have most frequently jaws, and often several pairs, but they are always so placed as to open laterally or horizontally. Sometimes, instead of jaws, they have a trunk, or proboscis. In general, they pass through four stages of existence, the egg, the larva, or stage of growth, the pupa, or chrysalis, which is usually in a state of torpor or complete inactivity, and the imago, or perfect insect, in its nuptial capacity. After the last change, the insect most frequently takes no food till its death.

The Linnean orders of insects are the coleoptera, with hard sheaths to their wings, generally called beetles; the hemiptera, of which the sheaths are of a softer nature, and cross each other, as grasshoppers, bugs, and plant lice; the lepidoptera, with dusty scales on their wings, as butterflies and moths; the neuroptera, as the libellula, or dragon fly, the may fly, and other insects with four transparent wings, but without stings; the hymenoptera, which have stings, either poisonous or not, as bees, wasps, and ichneumons; the diptera, with two wings, as common flies and gnats, which have halteres, or balancing rods, instead of the second pair of wings; and lastly the aptera, without any wings, which form the seventh order, comprehending crabs, lobsters, shrimps and prawns, for these are properly insects; spiders, scorpions, millepeds, centipeds, mites, and monoculi. The monoculus is a genus including the little active insects found in pond water, which are scarcely visible to the naked eye, as well as the Molucca crab, which is the largest of all insects, being sometimes six feet long. Besides these there are several genera of apterous insects which are parasitical, and infest the human race as well as other animals.

The vermes are the last and lowest of animated beings, yet some of them are not deficient either in magnitude or in beauty. The most natural division of vermes is into five orders; the intestina, as earthworms and ascarides, which are distinguished by the want of moveable appendages, or tentacula, from the mollusca; such as the dew snail, the cuttle fish, the sea anemone, and the hydra, or fresh water polype. The testacea have shells of one or more pieces, and most of them inhabit the sea, and are called shell fish, as the limpet, the periwinkle, the snail, the muscle, the oyster, and the barnacle. The order zoophyta contains corallines, sponges, and other compound animals, united by a common habitation, which has the general appearance of a vegetable, although of animal origin; each of the little inhabitants, resembling a hydra, or polype, imitating by its extended arms the appearance of an imperfect flower. The last order, infusoria, is scarcely distinguished from the intestina and mollusca by any other character than the minuteness of the individuals belonging to it, and their spontaneous appearance in animal and vegetable infusions, where we can discover no traces of the manner in which they are produced. The process, by which their numbers are sometimes increased, is no less astonishing than their first production; for several of the genera often appear to divide spontaneously, into two or more parts, which become new and distinct animals, so that in such a case the question respecting the identity of an individual would be very difficult to determine. The volvox, and some of the vorticellae are remarkable for their continual rotatory motion, probably intended for the purpose of straining their food out of the water: while some other species of the vorticella resemble fungi or corallines in miniature.

Among the animals of these different classes, the more perfect are informed of the qualities of external objects by the senses of touch, taste, smell, hearing, and vision. A few quadrupeds are incapable of seeing: the mole has an eye so small as to be with difficulty distinguishable; and the mus typhlus, supposed to be the aspalax of Aristotle, has its eye completely covered by the skin and integuments, without any perforation. Birds have hearing, but no external ears, or auriculae. Insects appear to want the organs of smell; but it is not impossible that their antennae may answer the purpose of hearing. Many of the vermes are totally destitute of sight, and some of all the organs of sense: none of them have either ears or nostrils. The external senses of animals with warm blood are usually liable to a periodical state of inactivity in the night time, denominated sleep. It is said that fishes never sleep; and it is well known that some animals pass the whole of the severest part of the winter in a state nearly resembling their usual sleep.

In animals which approach, in their economy, to that of the human system, the process for supporting life by nutrition begins with the mastication of the food, which has been received by the mouth. The food thus prepared is conveyed into the stomach by the operation of swallowing; but in ruminating cattle, it is first lodged in a temporary receptacle, and more completely masticated at leisure. In the stomach, it undergoes digestion, and being afterwards mixed with the bile and other fluids, poured in by the liver and the neighbouring glands, it becomes fit for affording the chyle, or nutritive juice, which is separated from it by the absorbents of the intestines, in its passage through the convolutions of a canal nearly forty feet in length. Together with the chyle, all the aqueous fluids, which are swallowed, must also be absorbed, and pass through the thoracic duct into the large veins entering the heart, and thence into the general circulation, before they can arrive at the kidneys, by which the superfluous parts are rejected. The chyle passes unaltered, with the blood, through the right auricle and ventricle of the heart, and enters the lungs, to be there more intimately mixed with it, and perhaps to be rendered animal and vital; while the blood receives from the air, in the same place, a supply of oxygen, with a small portion of nitrogen, and emits some superfluous carbonic matter, in the form of carbonic acid. The blood, thus rendered arterial, returning to the left side of the heart, is distributed thence to every part of the system, supplying nutriment throughout, while the glands and arteries secrete from it such fluids, as are become redundant, and such as are required for particular purposes subservient to the animal functions. It is probably in these processes that heat is evolved; for by experiments on living animals, it has been found, that the blood, returning from the lungs, is not warmer than before its entrance into them: we must therefore suppose, that when the florid arterial blood is, by some unknown means, converted, in the extreme ramifications of the arteries, into the purple venous blood, to return to the heart by the converging branches of the veins, there is a much more considerable extrication of heat, than in the conversion of venous into arterial blood, by the absorption of oxygen and nitrogen in the lungs. If the chyle is actually converted into blood in the lungs, it is here that we must look for the formation of the red globules, those singular corpuscles, to which the blood owes its colour, as it does its power of coagulation to a glutinous lymph, mixed with a less coagulable serum. The red particles in the human blood are about $\frac{1}{5000}$ of an inch in diameter, somewhat oblong, and flattened; they have usually the appearance of a dark point in the centre; but there is tome reason to suspect that this is merely an optical deception. In a few animals they are a little smaller, but in most of the amphibia, much larger and flatter than in man. While the lymph remains fluid, after the blood has been withdrawn from the vessels, these globules tend to subside, and to leave it semitransparent: hence arises the appearance of a buff coat on blood left to coagulate, which is thinner or thicker, accordingly as the globules are sooner or later arrested in their descent.

The muscles are probably furnished by the blood with a store of that unknown principle, by which they are rendered capable of contracting, for producing locomotion, or for other purposes, in obedience to the influence transmitted by the nerves from the sensorium; the brain and nervous system in general are also sustained, by means of the vascular circulation, in a fit state for transmitting the impressions, made by external objects on the senses, to the immediate seat of thought and memory, in the sensorium; and for conveying the dictates of the will, and the habitual impulses almost independent of volition, to the muscular parts of the whole frame.

In what manner these reciprocal impressions are transmitted by the nerves, has never yet been fully determined: but it has long been conjectured, that the medium of communication may bear a considerable analogy to the electrical fluid; and the extreme sensibility of the nerves to the slightest portion of electrical influence, as well as the real and apparently spontaneous excitation of that influence in animal bodies, which have been of late years evinced by galvanic experiments, have added very materially to the probability of the opinion. An extremely slender fibre, of a substance capable of conducting electricity with perfect freedom, enveloped in a sheath of a perfect nonconductor, would perhaps serve to communicate an impulse, very nearly in the same manner, as the nerves appear to do. Indeed nothing can be more fit to constitute a connecting link between material and immaterial beings, than some modification of a fluid, which appears to differ very considerably, in its essential properties, from the common gross matter of the universe, and to possess a subtility and an activity, which entitle it to a superior rank in the order of created substances.

When all the functions of animal life are carried on in their perfect and natural manner, the animal is said to be in health: when they are disturbed, a state of disease ensues. The diseases to which the human frame is liable are so various and irregular, that they cannot easily be reduced to any systematical order. Dr. Cullen has divided them into four classes. Febrile diseases, which constitute the first class, consist principally in an increase of the frequency of the pulsations of the heart and arteries, together with an elevation of the temperature, the whole animal economy being at the same time in some measure impaired: they are often accompanied by unnatural or irregular actions of the vessels of particular parts, constituting local inflammations, which were formerly considered as a derivation of diseased humours, falling on those parts: thus, a pleurisy is a fever, with an inflammation of the membrane lining the chest. The incapacity of a part to perform its functions, upon the application of a natural stimulus, or perhaps more frequently the incapacity of the nerves to transmit to it the dictates of the mind, constitutes a palsy: such derangements, and others, by which the actions of the nervous system are peculiarly impaired, form the class of neuroses, including spasmodic affections, madness, melancholy, and epilepsy. A general derangement of the system, without fever, or any peculiar debility of the nerves, constitutes the class of cachectic diseases, such as atrophy, consumption, scrofula, and dropsy. Besides these diseases, we have a fourth class, consisting of local affections only, such as blindness, deafness, tumors, and luxations.

Notwithstanding the labours of men of the greatest learning and genius, continued for many centuries, it must be confessed that the art of healing diseases is still in a state of great imperfection. Happily, however, for mankind, we may observe in almost all cases, where the offending cause is discoverable, and where the system is not at once overwhelmed by its magnitude, a wise and wonderful provision for removing it, by a mechanism admirably simple and efficacious; and it is reasonable to conclude, where the cause is more obscure, that the same benevolent Providence has employed agents equally well adapted for counteracting it, although their operations are utterly beyond the reach of human penetration.

LECTURE LX.

ON THE HISTORY OF TERRESTRIAL PHYSICS.

THROUGHOUT the whole of nature, we discover a tendency to the multiplication of life, of activity, and of enjoyment: man is placed at the head of terrestrial beings, the only one that comprehends, and that can trace, in a faint outline, the whole plan of the universe. We have seen the innumerable luminaries which enliven the widely expanded regions of immeasurable space, with their brilliant, but distant emanations of light and heat. Revolving round them at lesser intervals, and cherished by their fostering influences, are their planets and their comets; those preserving their distances nearly equal, and these, ranging more widely from the upper to the lower regions, without limits to their numbers or to their motions. Having conjectured what might possibly exist on other planetary globes, we descended to our own, and examined its structure and the proportions of its parts. Next we studied the general properties of the matter within our reach, and then the particular substances or qualities that are either not material, or are distinguished by very remarkable properties from other matter, as we found them concerned in the phenomena of heat, of electricity, and of magnetism; and we afterwards examined the combinations of all these, in the great atmospherical apparatus of nature, which serves for the exhibition of meteorological phenomena. The forms and the laws of animal and vegetable life have been the last objects of our inquiries; but the magnitude of some departments of natural history, and the obscurity of others, have prevented our entering more than superficially upon any of them.

Of the gradual advancement of astronomy we have already taken a historical view. With respect to the other sciences comprehended under the denomination of proper physics, the progress of discovery has generally been slow, and frequently casual. The ancients had little or no substantial knowledge of any part of physics, except astronomy and natural history: their opinions were in general mere speculations, derived from fancy, and inapplicable to the real phenomena of nature. Opinions such as these will only require to be so far examined, as to enable us to trace the imperfect rudiments of discoveries, which were only completed after intervals of many ages.

The Chinese are said to have been acquainted with the use of the compass above 3000 years ago; but in such accounts, it is impossible to ascertain how far the spirit of national vanity may have induced a historian to falsify his dates. It has been conjectured that the death of Numa, like that of Professor Richmann, was occasioned by some unguarded experiments on the electricity of the atmosphere, which drew on him the effects of a thunderstorm that was passing by. If, however, the fact was such, the experiments must probably have been suggested rather by an accidental discovery of the light on the point of a spear, than by any rational opinions respecting the nature of the ethereal fire.

Thales is the most ancient of the Grecian philosophers, who appear to have seriously studied the phenomena of nature. He supposed water to be the general principle from which all material things are formed, and into which they are resolved; an opinion which was without doubt suggested to him by the obvious effects of water in the nutrition of plants and of animals. He particularly noticed the properties of the magnet, which had been before observed to attract iron, as well as the effect of friction in exciting the electricity of amber; and he attributed to both of these substances a certain degree of animation, which he considered as the only original source of motion of any kind.

Anaximander appears to have paid some attention to meteorology; he derived the winds from the rarefaction of the air, produced by the operation of heat: thunder and lightning he attributed to the violent explosion or bursting of the clouds, which he seems to have considered as bags, filled with a mixture of wind and water. The same mistaken notion was entertained by Anaximenes, who compared the light attending the explosion, to that which is frequently exhibited by the sea, when struck with an oar.

Pythagoras, great as he was in some other departments of science, reasoned respecting physical effects in a manner too mathematical and visionary, to allow him much claim to be ranked among those, who have studied to investigate the minute operations of nature.

Anaxagoras was so far from confining himself to the supposition of four elements, which was most generally received by the philosophers of antiquity, that he imagined the number of elements nearly if not absolutely infinite. He conceived that the ultimate atoms, composing every substance, were of the same kind with that substance, and his system was thence called the homoeomeria; it erred perhaps less from the truth than many of the more prevalent opinions. Democritus, adopting the sentiments of Leucippus, proposed a still more correct theory of the constitution of matter, supposing it to be ultimately so far homogeneous, that the weight of its atoms was proportional to their bulk. He asserted that the forms of these atoms were different and unalterable; that they were always in motion, and that besides their primitive difference of form, they were also susceptible of a variety in the mode of their arrangement. The space not occupied by the atoms of matter, he considered as a perfect vacuum.

As Thales had supposed water to be the first principle of all things, and Anaximenes air, so Heraclitus fixed on fire as the foundation of his system, attributing to it the property of constant motion, and deriving all kinds of grosser matter from its condensation in different degrees. This doctrine was wholly unsupported by any thing like reason or observation.

Plato introduced into philosophy a variety of imaginations, which resembled the fictions of poetry much more than the truths of science. He maintained, for example, that ideas existed independently of the human mind, and of the external world, and that they composed beings of different kinds, by their union with an imperfect matter. It is observed by Bacon, in his essay on the opinions of Parmenides, that the most ancient philosophers, Empedocles, Anaxagoras, Anaximenes, Heraclitus, and Democritus, submitted their minds to things as they found them; but that Plato made the world subject to ideas, and Aristotle made even ideas, as well as all other things, subservient to words; the minds of men beginning to be occupied, in those times, with idle discussions and verbal disputations, and the correct investigation of nature being wholly neglected. Plato entertained, however, some correct notions respecting the distinction of denser from rarer matter by its greater inertia; and it would be extremely unjust to deny a very high degree of merit to Aristotle's experimental researches, in various parts of natural philosophy, and in particular to the vast collection of real information contained in his works on natural history. Aristotle attributed absolute levity to fire, and gravity to the earth, considering air and water as of an intermediate nature. By gravity the ancients appear in general to have understood a tendency towards the centre of the earth, which they considered as identical with that of the universe; and as long as they entertained this opinion, it was almost impossible that they should suspect the operation of a mutual attraction in all matter, as a cause of gravitation. The first traces of this more correct opinion respecting it are found in the works of Plutarch.

Epicurus appears to have reasoned as justly respecting many particular subjects of natural philosophy, as he did absurdly respecting the origin of the world, and of the animals which inhabit it. He adopted in great measure the principles of Democritus respecting atoms, but attributed to them an innate power of affecting each other's motions, and of declining, in such a manner, as to constitute, by the diversity of their spontaneous arrangements, all the varieties of natural bodies. He considered both heat and cold as material; the heat emitted by the sun he thought not absolutely identical with light, and even went so far as to conjecture that some of the sun's rays might possibly possess the power of heating bodies, and yet not affect the sense of vision. In order to explain the phenomena of magnetism, he supposed a current of atoms, passing, in certain directions, through the magnet and through iron, which produced all the effects by their interference with each other. Earthquakes and volcanos he derived from the violent explosions of imprisoned air.

Among all these opinions and conjectures, there is scarcely any one which was scientifically established upon sure foundations. Some insulated observations had a certain degree of merit; and we find many interesting facts relating to different departments of natural knowledge, not only in Aristotle, but also in Theophrastus, Dioscorides, and Pliny, as well as in some of the historical writers of antiquity. Protagorides of Cyzicum, who is quoted by Athenaeus, relates that in the time of king Antiochus, it was usual, as a luxury, to cool water by evaporation; and it is not impossible that the custom may have been introduced from the east, where even ice is frequently made at present by a similar process; others of the ancients had remarked, according to Dr. Falconer, that water usually froze the more readily for having been boiled; and it is possible that some other detached observations of a similar nature may occur to those who have the curiosity to make them objects of research.

The thirteenth century may be considered as the date of the revival, if not of the commencement, of physical discoveries. Our countryman, Roger Bacon, was one of its principal ornaments: he appears to have anticipated in his knowledge of chemistry, as well as of many other parts of natural philosophy, the labours of later times. The polarity of the magnetic needle is described in some lines which are attributed to Guyot, a French poet, who lived about 1180; but some persons are of opinion that this description was actually written by Hugo Bertius, in the middle of the succeeding century; and it is generally believed that the compass was first employed in navigation by Gioja of Amalfi, about the year 1260; he is said to have marked the north with a fleur de lis, in compliment to a branch of the royal family of France, then reigning at Naples. The declination of the needle from the true meridian is mentioned by Peter Adsiger, the author of a manuscript which bears the date 1269. The poet Dante, who flourished at the close of this century, distinguished himself not only by his literary, but also by his philosophical pursuits; and we find among his numerous works an essay on the nature of the elements.

The learned and voluminous labours, by which Gesner and Aldrovandus enriched the various departments of natural history, may be considered as comprehending the greatest part of what had been done by the ancients in the investigation of the economy of the animal world; but their works have too much the appearance of collections of what others had asserted, rather than of original observations of their own.

The first of the moderns, whose discoveries respecting the properties of natural bodies excite our attention, by their novelty and importance, is Dr. Gilbert, of Colchester: his work on magnetism, published in 1590, contains a copious collection of valuable facts, and ingenious reasonings. He also extended his researches to many other branches of science, and in particular to the subject of electricity. It had been found, in the preceding century, that sulfur, as well as amber, was capable of electric excitation, and Gilbert made many further experiments on the nature of electric phenomena. The change or variation of the declination of the needle is commonly said to have been discovered by Gellibrand, a professor at Gresham college, in the year 1625; but it must have been inferred from Gunter's observations, made in 1622, if not from those of Mair, or of some other person, as early as 1612; for at this time the declination was considerably less than Burrows had found it in 1580.

In the beginning of the seventeenth century, Lord Bacon acquired, by his laudable efforts to explode the incorrect modes of reasoning, which had occupied the schools, the just character of a reformer of philosophy: but his immediate discoveries were neither striking nor numerous. In 1620, he proposed, with respect to heat, an opinion which appears to have been at that time new, inferring, from a variety of considerations, which he has very minutely detailed in his Novum organum, that it consisted in "an expansive motion, confined and reflected within a body, so as to become alternate and tremulous; having also a certain tendency to ascend". A similar opinion, respecting the vibratory nature of heat, was also suggested, about the same time, by David Gorlaeus, and it was afterwards adopted by Descartes, as a part of his hypothesis respecting the constitution of matter; which he imagined to consist of atoms of different forms, possessing no property besides extension, and to derive all its other qualities from the operation of an ethereal and infinitely elastic fluid, continually revolving in different orders of vortices.

A much more important step, than the proposal of any hypothesis concerning the nature of heat, was also made about the year 1620, by Cornelius Drebel, who appears to have been the original inventor of the method of measuring the degrees of heat by a thermometer. The utility of the instrument remained, however, much limited, for want of an accurate method of adjusting its scale, and it was not till the close of the century, that Dr. Hooke's discovery, of the permanency of the temperature of boiling water, afforded a correct and convenient limit to the scale on one side, while the melting of snow served for fixing a similar point on the other; although there would have been no great difficulty in forming a scale sufficiently natural, from the proportion of the expansion of the fluid contained in the thrmomeeter to its whole bulk.

It was about the year 1628, that Dr. Harvey succeeded in demonstrating, by a judicious and conclusive train of experiments, the true course of the circulation of the blood, through the veins and arteries, both in the perfect state of the animal, and during its existence as an embryo. Servetus had explicitly asserted, in his work on the Trinity, as early as the year 1553, that the blood performed, in its passage through the lungs, a complete revolution, beginning and ending in the heart; and Cisalpinus had even expressed, in 1569, some suspicions that the circulation of the whole body was of a similar nature; but neither of these authors had advanced any satisfactory proofs in confirmation of his opinions.

In the middle of the seventeenth century, the barometer was invented by Torricelli; the variation of the atmospheric pressure was discovered by Descartes; and Pascal made several experiments on the difference of its magnitude at different places, which tended to illustrate the principles, on which the method of determining heights by barometrical observations is founded.

What Gesner and Aldrovandus had before done with regard to the animal kingdom, was performed, a century later, for the vegetable world by John and Caspar Bauhin, whose works, as collections of all that was to be found on record respecting the distinctions and properties of plants, have not yet been superseded by the latest publications. Our countrymen, Ray and Willughby, contributed also to add much new matter to the stores of natural history, in all its departments; and their labours, as well as those of Tournefort and Réaumur, are of the more value, as they were far more studious than their predecessors to discriminate truth from fiction.

The foundation of the most celebrated of the philosophical societies of Europe renders the latter half of the seventeenth century a very interesting period in the history of natural knowledge. The Royal Society of London, and the Academy of Sciences of Paris, have always been the most distinguished of these: and the Florentine Academy del Cimento, although its labours were not of long duration, produced at first in a short time a very copious and interesting collection of experiments, relating to various subjects of physical research. In the Royal Society, Boyle, Hooke, and Newton were the most industrious, as well as the most successful investigators of natural phenomena: the elementary doctrines of chemistry, the nature of combustion, the effects of heat and cold, and the laws of attraction, repulsion, and cohesion were attentively examined and discussed. The expansion of water, by a reduction of its temperature, near the freezing point, was first observed by Dr. Croune; although his experiments were considered by Dr. Hooke as inconclusive. The attention of the society was directed by Newton to the phenomena of electricity, some of which had been a short time before particularly noticed by Guericke; the mode of making electrical experiments was greatly improved by Hauksbee; this accurate observer investigated also the nature of capillary attraction with considerable success. Early in the succeeding century, many of the members of the Academy of Petersburg followed the example of other societies with great industry; and the experiments of Richmann on heat were among the first and best fruits of their researches.

Dr. Halley employed himself, with the most laudable zeal, in procuring information respecting the variation of the compass; he undertook a voyage round the world, for the express purpose of making magnetical observations; and he published a chart of variation, adapted to the year 1700. He also collected many particulars respecting the trade winds and monsoons, and he endeavoured to explain them by a theory which has been adopted by some of the latest authors, but which is in reality much less satisfactory than the hypothesis proposed some time afterwards by Hadley. His magnetical investigations were continued with great diligence by Mountaine and Dodson, who published, at different periods, two charts representing the successive states of the variation. Euler, Mayer, and others have attempted, in later times, to discover such general laws as might be sufficient to determine the magnitude of the variation for every part of the globe; but their success has been very much limited.

The science of electricity was diligently cultivated in the middle of the last century by Stephen Gray, Dufay, Winkler, Nollet, Musschenbroek, and Franklin. As early as 1735 it was remarked by Gray, that "the electric fire seemed to be of the same nature as lightning," and their identity was afterwards more strongly asserted by Winkler, and experimentally demonstrated by Franklin. The shock of a charged jar was first discovered by Kleist, in 1745; and the experiment was repeated by Lallamand and Musschenbroek, who described its disagreeable effects on the sensations with an exaggeration not the most philosophical. The theory of the nature of the charge was the second great improvement made by Dr. Franklin in this science.

The introduction of the Linnean system of botany and zoology is to be considered as bringing near to perfection the logic and phraseology of natural history; nor has its celebrated author wholly neglected the philosophy of the science. The number and the diligence of his successors have already furnished to the different departments of natural history a much ampler store of observations than could easily have been expected from the short time which their labours have occupied. Buffon had merit of a different kind, and though his fancy was too little regulated by mathematical accuracy, the elegance of his writings have made their subjects highly interesting to the general reader. Among other modern naturalists of great respectability, Spallanzani, Daubenton, Degeer, Geoffroy, Pennant, the Jussieus, Lacepede and Haüy, have particularly distinguished themselves by the importance of their discoveries, and the accuracy of their descriptions.

The absorption of heat, during the conversion of ice into water, appears to have been separately observed by Deluc, Black, and Wilke, about the year 1755. On this experiment Dr. Black principally founded his doctrine of latent heat, supposed to be retained in chemical combination by the particles of fluids. Dr. Irvine and Dr. Crawford explained the circumstances somewhat differently, by the theory of a change of capacity for heat only. Bergmann, Lavoisier, Laplace, Kirwan, Seguin, and many other philosophers have

2

illustrated, by experiments and calculations, the various opinions which have been entertained on this subject; and few chemists, from the times of Boerhaave, Stahl, and Scheele to those of Priestley and other later authors, have left the properties of heat wholly unnoticed.

The elegant hypothesis of Aepinus, respecting magnetism and electricity, founded in great measure on the theory of Franklin, was advanced in 1759: our venerable countryman, Mr. Cavendish, had invented a similar theory, and had entered in many respects more minutely into the detail of its consequences, without being acquainted with Aepinus's work; although the publication of his paper on the subject was 12 years later. Lambert, Mayer, Coulomb, and Robison have also pursued inquiries of a similar nature, both theoretically and experimentally, with great success. The electrophorus of Wilke, and the condenser of Volta, are among the earliest fruits of the cultivation of a rational system of electricity, and Mr. Cavendish's investigation of the properties of the torpedo may serve as a model of accuracy and precision in the conduct of experimental researches.

The speculations of Boscovich respecting the fundamental properties of matter, and the general laws of the mutual action of bodies on each other, have been considered by some candid judges as deserving the highest commendation; they remain however almost in all cases speculations only; and some of the most intricate of them, being calculated for the explanation of some facts, which have perhaps been much misunderstood, must consequently be both inaccurate and superfluous.

The attention of several experienced philosophers, who are now living, has been devoted, with much perseverance, to the difficult subject of hygrometry. Deluc's experiments have offered us a very useful comparison of the hygrometrical qualities of various substances: Saussure has investigated, with great labour, the indications of the hygrometer and the thermometer, as connected with the presence of a certain portion of vapour, contained in air of various densities; and Pictet has ascertained some similar circumstances respecting vapours of different kinds wholly unmixed with any air. The hypotheses, which have usually accompanied the relation of most of these experiments, have however been in general too little supported by facts to be entitled to universal adoption.

For some years past, the philosophical, as well as the unphilosophical world, has been much occupied and entertained by the discoveries of Galvani, Volta, and others, respecting the operations of the electric fluid. The first circumstance, that attracted Galvani's attention to the subject of animal electricity, was the agitation of a frog, that had a nerve armed, that is, laid bare and covered with a metal, when a spark was taken in its neighbourhood. A person acquainted with the well known laws of induced electricity might easily have foreseen this effect: it proved, however, that a frog so prepared was a very delicate electrometer, and it led Galvani to further experiments. It has been shown by Volta, that an entire frog may be convulsed by a degree of electricity which affects an electrometer but very weakly; but that when prepared in Galvani's manner, it will be agitated by an electricity one fiftieth part as great, which cannot be discovered, by any other means, without the assistance of a condenser. Galvani, however, found that a communication made between the armed nerve and its muscle, by means of any conducting substance, was sufficient to produce a convulsion, without the presence of foreign electricity: hence he concluded that the nerve and muscle, like the opposite surfaces of a charged jar, were in contrary states of electricity, and that the communication produced a discharge between them. He observed, however, a considerable difference in the effects, when different metals were employed for forming the circuit; and this circumstance led to the discovery of the excitation of electricity by means of a combination of different inanimate substances only, which Mr. Davy attributes to Fabroni, Creve, and Dr. Ash. It was, however, still more satisfactorily demonstrated by Volta; and he at first supposed that all the phenomena observed by Galvani were derived from effects of this kind, but on further examination he was obliged to allow the independent existence of animal electricity. This industrious and ingenious philosopher has the sole merit of the invention of the pile or battery, which has rendered every other mode of exciting the galvanic action comparatively insignificant.

No sooner was Volta's essay communicated to the Royal Society, than a

4

pile was constructed by Mr. Carlisle, and its singular effects in the decomposition of water were jointly observed by himself and Mr. Nicholson. The original existence of animal electricity, as asserted by Galvani and Volta, has been in some degree confirmed by the experiments of Aldini, the nephew of Galvani. A number of detached observations, of considerable merit, have also been made by Pfaff, Ritter, Cruikshank, Wollaston, Fourcroy, and many other chemists, both in this country and on the continent. But Mr. Davy's late experiments must be considered as exceeding in importance every thing that has been done upon the subject of electricity, since the discovery of the pile of Volta. The conclusions which they have enabled him to form respecting the electrical properties of such bodies as have the strongest tendencies to act chemically on each other, and the power of modifying and counteracting those tendencies which the electric fluid possesses, have greatly extended our views of the minute operations of nature, and have opened a new field for future investigations. I hope that I shall be pardoned by astronomers for having inserted, on this occasion, in a vacant space among the constellations, in the neighbourhood of Pegasus, the figure of a galvanic battery; which must now be allowed to have as great pretensions to such a distinction as the electrical machine and the chemical furnace.

The late experiments and speculations of Mr. Dalton, on various subjects, belonging to different branches of physics, have tended to place some parts of the science of meteorology in a new light. It is true that many of his hypotheses are very arbitrarily assumed; some of them are manifestly contrary to experiment, and others to analogy and probability; at the same time his remarks appear in some cases to be either perfectly correct, or to lead to determinations which are sufficiently accurate for every practical purpose. I have attempted to borrow from Mr. Dalton's ideas some hints, which I have incorporated with a less exceptionable system; and by a comparison of his experiments with those of many other philosophers, I have deduced some methods of calculation which may perhaps be practically useful; in particular a simple rule for determining the elasticity of steam, and a mode of reducing the indications of hygrometers of different kinds to a natural scale.

Count Rumford's establishment of a prize medal, to be given every three years by the Royal Society to the author of the most valuable discovery respecting heat or light, forms an era less remarkable, than the first adjudication of the medal to himself, and the second to Mr. Leslie. Count Rumford's numerous experiments on the production and communication of heat are highly important, both for the utility which may be derived from their economical application, and for the assistance which they afford us in the investigation of the intimate nature of heat. Mr. Leslie's discovery of the different properties possessed by surfaces of different kinds, with regard to emitting and receiving radiant heat, is in every respect highly interesting; and the multiplicity and diversity of his experiments would have entitled him to still higher commendation than he has obtained, if they had been more simply and circumstantially related. Perhaps, however, none of the modern improvements in speculative science deserves a higher rank than Dr. Herschel's discovery of the separation of heat from light by refraction. Mr. Prévost has made some just remarks on the experiments of other philosophers respecting heat; and his own theory of radiant heat, and his original investigations, on the effect of the solar heat on the earth, have tended materially to illustrate the subject of his researches.

The general laws of the ascent and descent of fluids in capillary tubes, and between plates, of different kinds, had long ago been established by the experiments of Hauksbee, Jurin, and Musschenbroek; many other circumstances, depending on the same principles, had been examined by Taylor, Achard, and Guyton; and some advances towards a theory of the forms assumed by the surfaces of liquids, had been made by Clairaut, Segner, and Monge. In an essay on the cohesion of fluids, read before the Royal Society in the year 1804, I have reduced all effects of this nature to the joint operation of a cohesive and repulsive force, which balance each other; assuming only that the repulsion is more augmented by the approach of the particles to each other than the cohesion; and I have had the satisfaction of discovering in this manner a perfect correspondence between many facts, which had not been supposed to have the slightest connexion with each other. Almost a year after the publication of this paper, Mr. Laplace read to the National Institute a memoir on capillary tubes, in which, as far as he has pursued the subject, he has precisely confirmed the most obvious of my conclusions;

although his mode of calculation appears to be by no means unexceptionable, as it does not include the consideration of the effects of repulsion. Had my paper been so fortunate as to attract Mr. Laplace's attention before his memoir was presented to the Institute, he would perhaps have extended the results of my theory with the same success, which has uniformly distinguished his labours in every other department of natural philosophy.

When we reflect on the state of the sciences in general, at the beginning of the seventeenth century, and compare it with the progress which has been since made in all of them, we shall be convinced that the last two hundred years have done much more for the promotion of knowledge, than the two thousand that preceded them: and we shall be still more encouraged by the consideration, that perhaps the greater part of these acquisitions has been made within fifty or sixty years only. We have therefore the satisfaction of viewing the knowledge of nature not only in a state of advancement, but even advancing with increasing rapidity; and the universal diffusion of a taste for science appears to promise, that, as the number of its cultivators increases, new facts will be continually discovered, and those, which are already known, will be better understood, and more beneficially applied. The Royal Institution, with other societies of a similar nature, will have the merit of assisting in the dissemination of knowledge, and in the cultivation of a taste for its pursuit; and the advantages arising from the general introduction of philosophical studies, and from the adoption of the practical improvements depending on them, will amply repay the labours of those, who have been active in the establishment and support of associations so truly laudable.

CHRONOLOGY OF PHYSICAL AUTHORS.

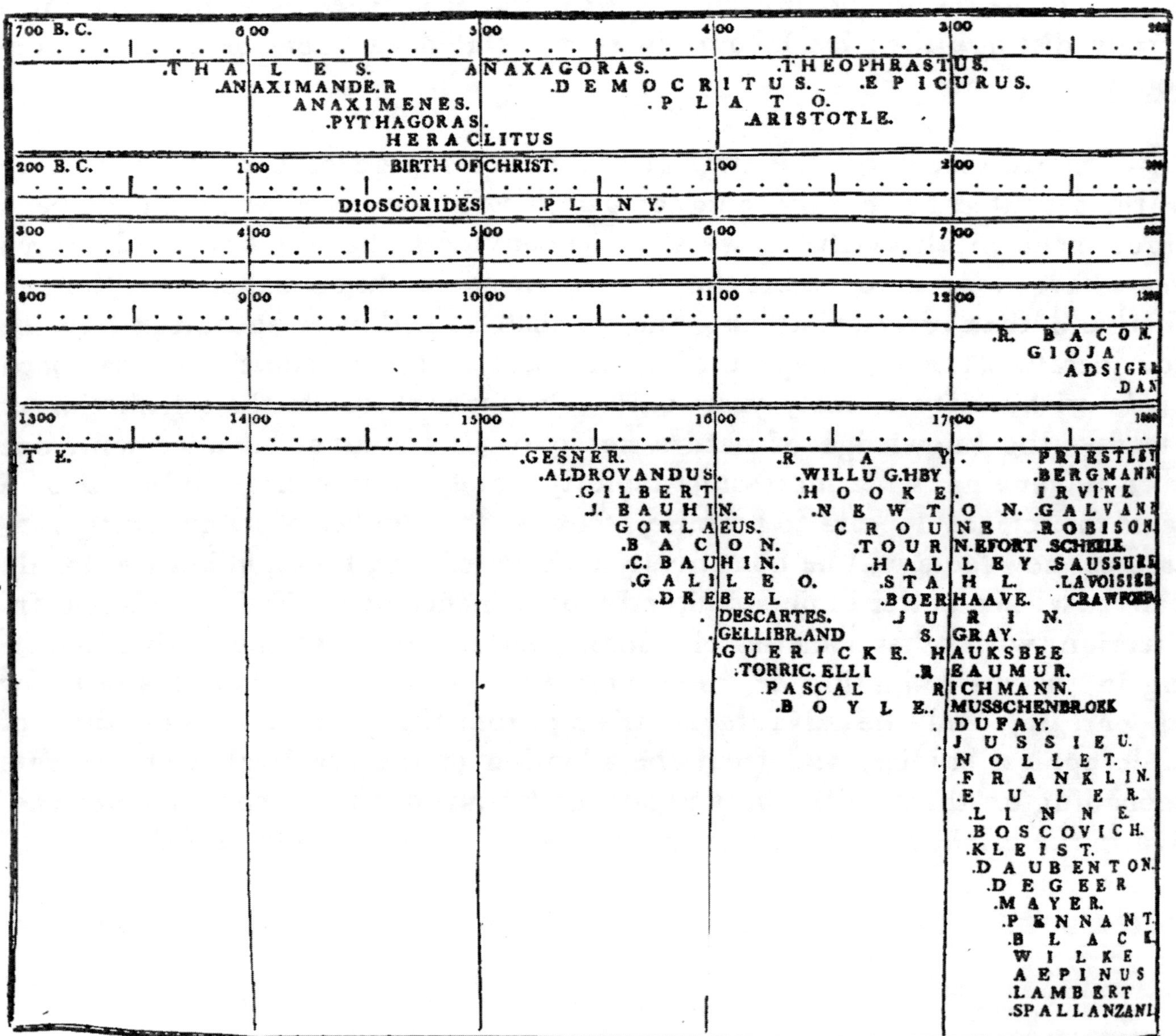

EXPLANATION OF THE PLATES.

5 B

PLATE I.

Fig. 1. The point A being supposed to move in a right line to B, A B is the direction of its motion. P. 21.

Fig. 2. The lines A B, B C, C D, are the successive directions of the point A, moving from A to D in the figure A B C D. P. 21.

Fig 3. The tangent A B is the direction of the motion of the point C, moving in the curve C D, when it arrives at E. P. 21.

Fig. 4. The square A B, moving on the board C D, so that the points E, F, describe the parallel lines E G, E H, with equal velocities, the plane A E F B is in rectilinear motion with respect to the surface C D. P. 24.

Fig. 5. The cycloid A B C, and the trochoid D E F are the results of the rotatory motion of the points B and E round the centre of the wheel, combined with the progressive motion of the wheel along the base A C. P. 24, 44.

Fig. 6. A B is a fixed bar, C D an arm which slides on it, E C F a thread passing round the pulley at C, and either fixed to the pin on the slider F, or passed over the pulley G, and fixed again at H. The arm turns round the same axis that carries the pulley at C, and may be fixed by means of the screw which is cut on the axis, while two other screws keep it steady by pressing on the slider below it. The point I describes, by its compound motion, the oblique line K I. P. 24.

Fig. 7. The diagonal A B of the parallelogram C D is the joint result of the motions, represented by its sides A C, A D. P. 25.

Fig. 8. The line A B may be either simply drawn in the direction A B, or it may be traced by the equal motions A C and A D of the arm and its slider, or by the unequal motions A E and A F. P. 25.

Fig. 9. The body A, moving uniformly along the line A B, first approaches to the point C, and then recedes from it, as if repelled. P. 27.

Fig. 10. When A B and A C approach each other, and coincide, the diagonal A D becomes equal to their sum. P. 30.

Fig. 11. Atwood's machine. The boxes A, B, containing equal weights, are connected by the thread A C B, passing over the pulley C, which is supported either on friction wheels, or by the points of screws, one of which is seen at D. The box A is made to descend either by a flat weight placed on it, or by the bar E, which is intercepted by the ring F, and the box continues to descend till it strikes the stage G; the space being measured on the scale H I, and the time by the pendulum K, which may be kept in motion by a clock scapement with a weight. The machine is levelled by the screws L, M. P. 31.

Fig. 12. The time of the descent of a falling body being represented by any portion A B of the base of a triangle, the velocity will be proportional to B C, which is equal to A B, and the space described during the time D E, supposed infinitely short, will be proportional to the area D E F G, which is expressed by the product of B C and D E; consequently the whole area A E F will represent the space described in the time A E, and A H I the space described in the time A H; but A H I is half of the square H K, and A E F of E L: the space is therefore always as the square of the time, and is equal to half the space which would be described in the same time with the final velocity. P. 32.

Fig. 13. The whirling table. The arms A B, C D, are made to revolve on the axes E F, G H by the string passing over the wheel I, the upper or under pulley of either axis being employed at pleasure: the stages K, L, with their weights, are placed at certain distances from the centre, by means of the racks or teeth below them; they move along the arms by means of friction wheels resting on wires, and they raise the weights M, N, by means of threads passing each over two pullies. P. 35.

Fig. 14. If a body revolving in a curve A B C, by means of a force directed to D, describe the portions A E, B F, C G in equal times, the areas A D E, B D F, C D G, will be equal, and the velocities in A, B, and G, will be inversely as the perpendiculars D H, D I, and D K. P. 36.

PLATE I.

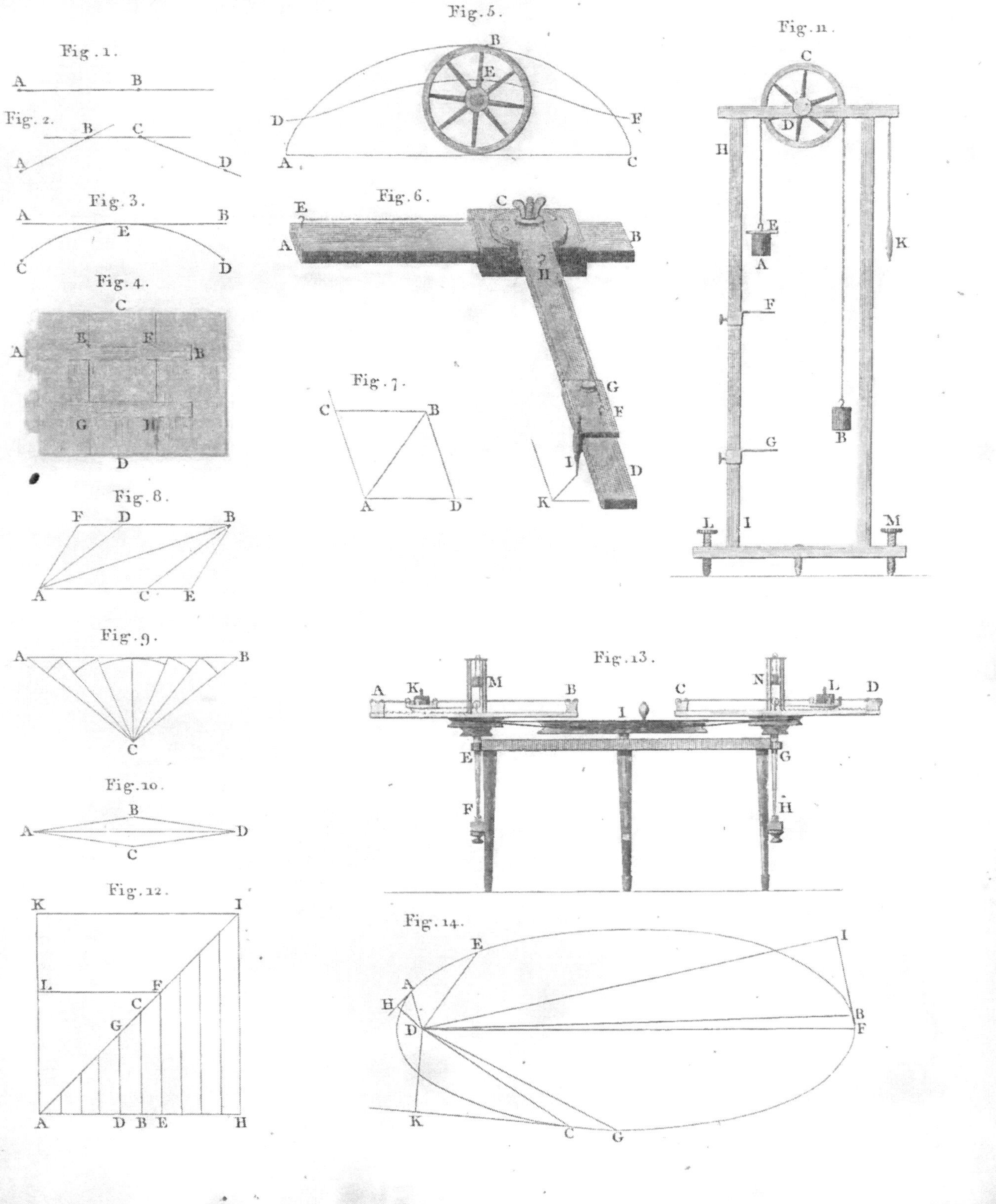

Josh. Skelton sculp.

Pub. by J. Johnson, London 1 July 1806.

PLATE II.

Fig. 15.

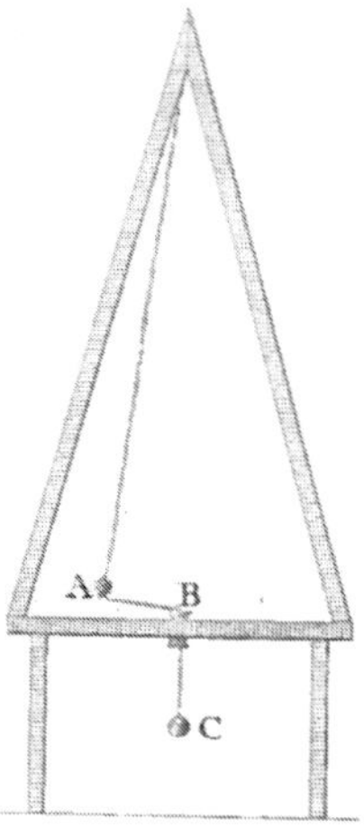

Fig. 17.

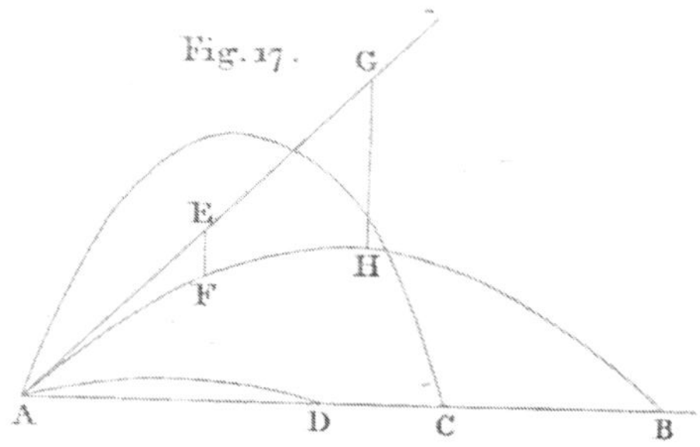

Fig. 24.

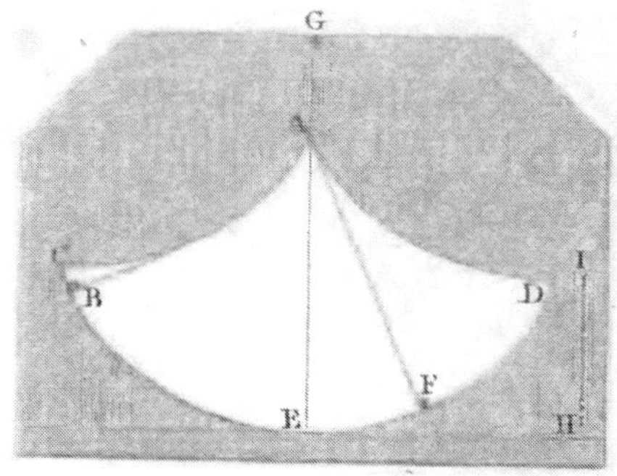

Fig. 18.

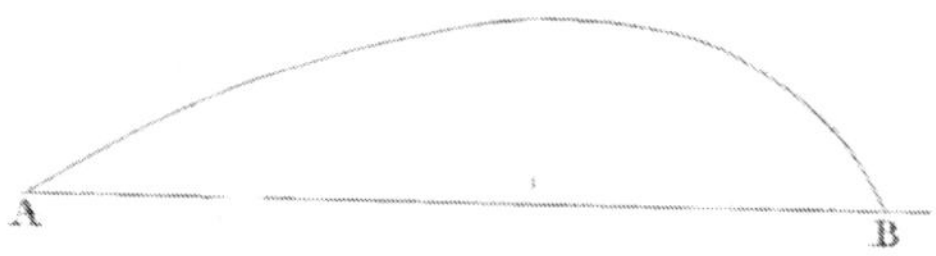

Fig. 26.

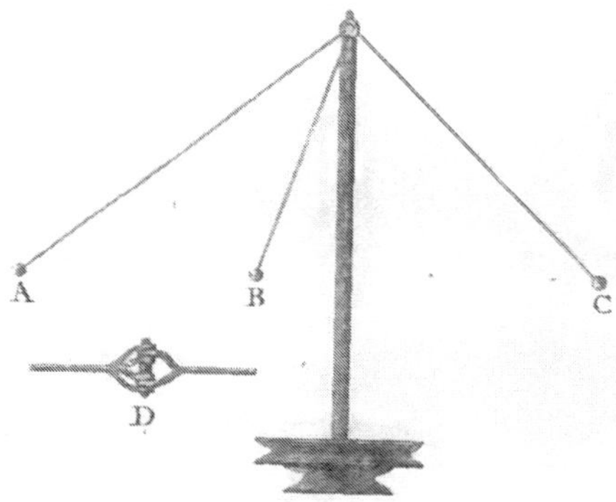

Fig. 16.

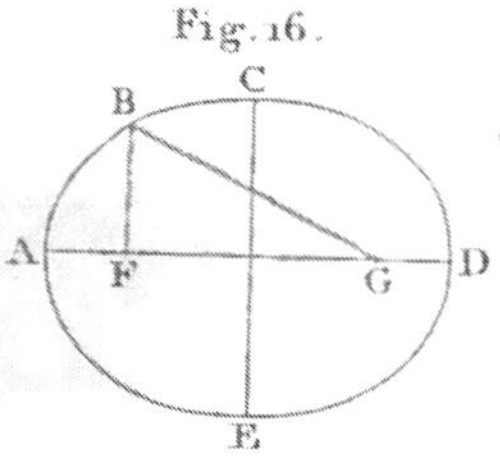

Fig. 20.

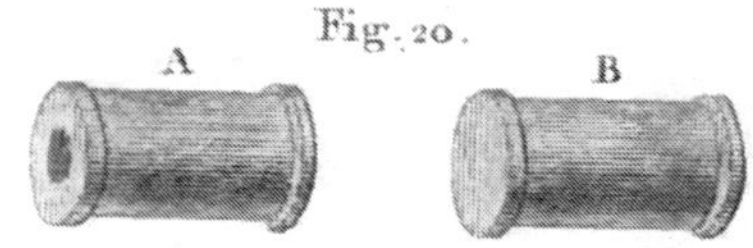

Fig. 21.

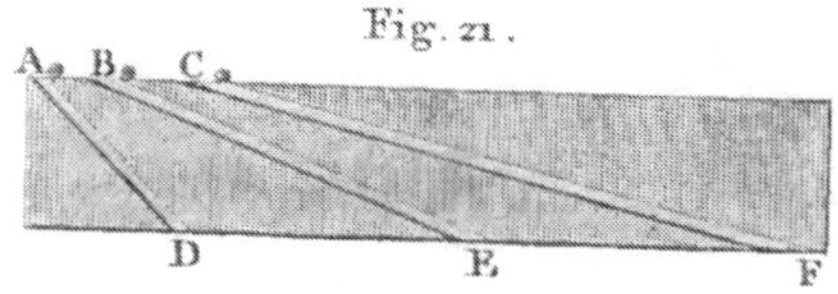

Fig. 19.

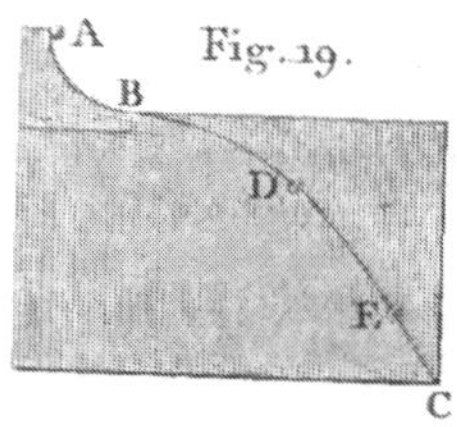

Fig. 23.

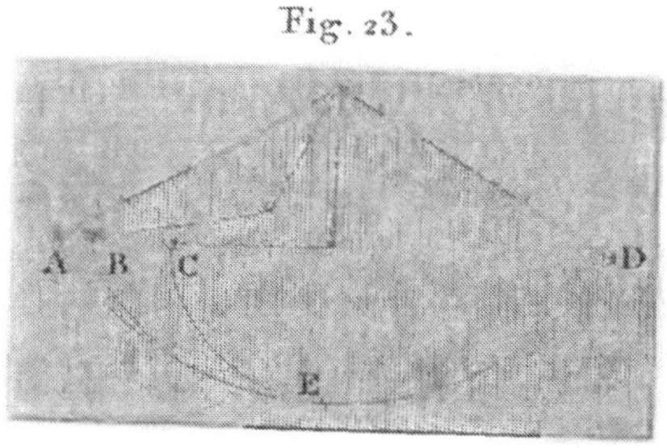

Fig. 28.

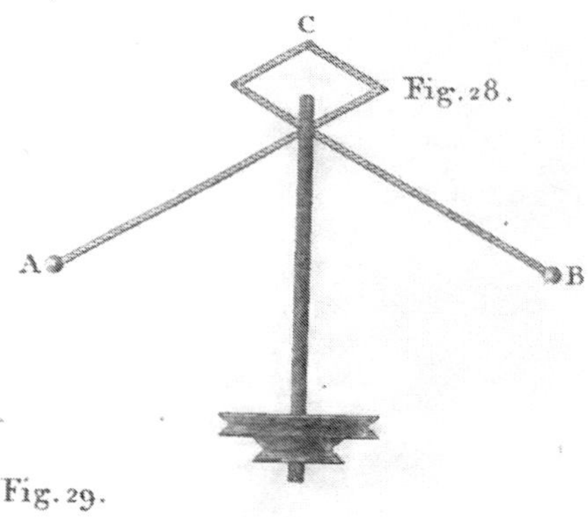

Fig. 22.

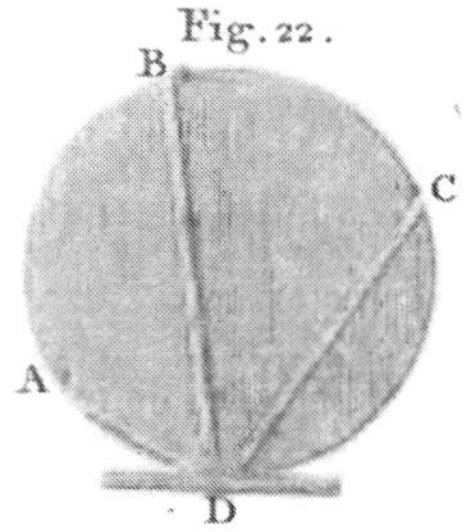

Fig. 29.

Fig. 25.

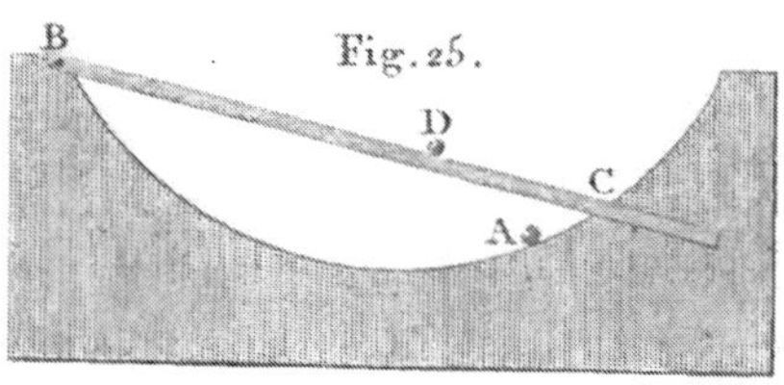

Fig. 27.

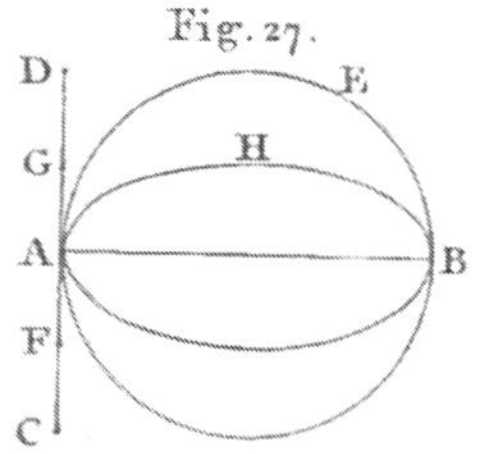

Fig. 30.

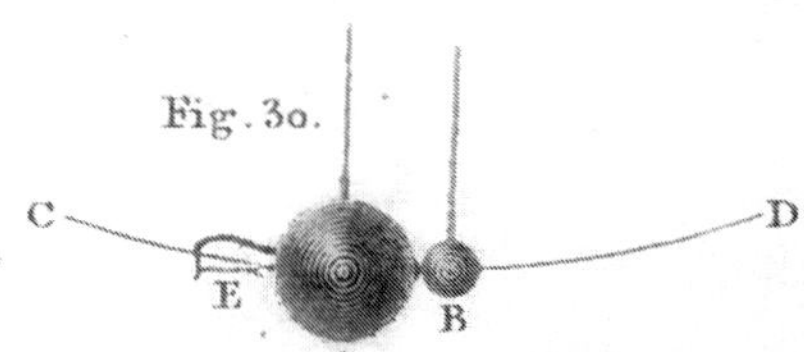

Pub. by J. Johnson, London 1 July 1806.

Joseph Skelton sculp

PLATE II.

Fig. 15. The ball A, revolving round the point B, and being drawn towards it by means of the thread B C, with a force variable at pleasure, its velocity may be observed to vary, according to its distance from the point B. P. 37.

Fig. 16. The curve A B C D E is an ellipsis; E and G are its foci; A D its greater axis, and C E its lesser axis. P. 37.

Fig. 17. The horizontal range, A B, of a body projected at an elevation of 45°, is greater than A C or A D, the ranges of bodies projected with the same velocity at a greater or less elevation. If the parallel lines E F, G H, be always as the squares of A E, A G, the curve A F H will be a parabola; and such is the path of a projectile. P. 39, 40.

Fig. 18. The path of a ball moving swiftly through the atmosphere nearly resembles the curve A B. P. 39.

Fig. 19. The ball A, having descended along the groove A B, describes the parabola B C, passing through the rings D, E. P. 40.

Fig. 20. The cylinder A, loaded at the axis, descends along an inclined plane more rapidly than the cylinder B, loaded with an equal weight at the circumference. P. 42.

Fig. 21. The balls A, B, C descend along the planes A D, B E, CF, of equal height, in times proportional to their lengths. The upper surfaces of the slips A D, B E, C F, are slightly grooved. P. 43.

Fig. 22. The balls A, B, C, descend in equal times along the chords A D, B D, C D. P. 43.

Fig. 23. The same ball, descending from equal heights, at A, B, or C, by different paths, will rise to the same height at D on the opposite side of E. P. 43.

Fig. 24. The thread A B, playing between the cycloidal cheeks A C, A D, describes the cycloid C E D, and the balls B, E, descending from any two points of the curve, will meet at E, in the same time that the ball G falls from a point nearly ¼ of A E above A. The space described by the pendulum in descending is always proportional to the height H I, to which a body setting out from E, and revolving uniformly in a circle, will rise in the same time. The circle E I lies without the cycloid C E D, and is somewhat less inclined to the horizon at equal distances from E. P. 44, 45.

Fig. 25. The ball A, descending from B in the curve B A, arrives at C before the ball D moving in a right line on the plane B C. P. 46.

Fig. 26. The balls A, B, C, being made to revolve by means of the whirling table, they are always found in the same horizontal plane. The joint connecting them with the axis is represented at D, as seen from above. P. 47.

Fig. 27. The equal vibrations, represented by A B, C D, compose, when united, the circular revolution A E B: the unequal vibrations A B, F G, compose the ellipsis A H B; the place of the body being always ascertained by combining the versed sines of two circular arcs increasing uniformly. P. 47.

Fig. 28. The balls A, B, as their revolution becomes more rapid, fly out, and the point C is depressed. P. 48.

Fig. 29. The mass of the body A being 1 and that of B 2, and A C being twice B C, C is the centre of inertia. P. 51.

Fig. 30. The balls A and B are suspended by long threads, which allow them to move in the arcs A C, B D; the ball A is perforated in a horizontal direction, and contains a spiral spring, which is confined by the thread E, and being set at liberty by burning this thread, strikes the ball B, so as to cause each of the balls to move through an arc, of which the chord is proportional to the weight of the other ball. P. 52.

PLATE III.

Fig. 31. The centre of inertia of the bodies A, B, C, D, may be determined either by finding E the centre of inertia of A and B, and supposing a body equal to their sum to be placed in it, then determining F from E and C; and G, the point required, from F and D; or by finding first H and I from A, C, B, D, taken in pairs, and dividing H I in due proportion in the same point G. P. 54.

Fig. 32. The point A being the centre of inertia of the bodies B, C, D, E, the products obtained by multiplying B by B F, C by C G, D by D H, and E by E I, are equal, when added together, to the product of the masses of all the bodies by the distance A K; all the lines drawn to the plane F I being parallel. P. 55.

Fig. 33. The weights A B C will remain at rest when they are in the same proportion to each other as the respective sides of the triangle D E F; D F being parallel to E G. P. 61.

Fig. 34. The bodies A, B, remain in equilibrium when their centre of inertia C is immediately below the point of suspension D. P. 61.

Fig. 35. The system of bodies A, B, C, is at rest when the centre of inertia D is immediately below the point of suspension E. P. 61.

Fig. 36. The bodies A, B, remain at rest when the centre of inertia C is immediately above the point of support D. P. 61.

Fig. 37. The bodies A, B, remain at rest when the centre of inertia C coincides with the fulcrum or point of support. P. 61.

Fig. 38. The irregular body A B remains at rest when the centre of inertia C is immediately below the point of suspension D. P. 61.

Fig. 39. A being the centre of gravity of the board B, C, the point of suspension being D, E, or F, the position of the vertical line will be D A, E A, or F A. P. 62.

Fig. 40. The equilibrium of the vessel A is stable; that of the vessel B tottering, the path of the centre of gravity having its concavity upwards in the first, and downwards in the second. P. 62.

Fig. 41. Paths of the centre of gravity of an oval. P. 62.

Fig. 42. Paths of the centre of gravity of a body resting on a sphere. P. 62.

Fig. 43. A, the path of the centre of gravity of a body standing on a flat basis; B, the tottering equilibrium of the same body inclined. P. 63.

Fig. 44. The effects of a certain inclination of a waggon, loaded with light and heavy materials, are represented at A and B respectively. P. 63.

Fig. 45. The suspension of a weight from a rod projecting over a table. P. 64.

Fig. 46. A shows the path of the centre of gravity of a loaded cylinder on an inclined plane, B that of the centre of gravity of a double cone moving towards the more elevated end of a triangular surface. C is an elevation of the double cone. P. 64.

Fig. 47. A B is a lever of the first kind, the forces acting on different sides of the fulcrum C; D E of the second kind, the forces being applied at D and F, on the same side of E. P. 65.

Fig. 48. A force applied at A may be held in equilibrium by a triple force, applied in the direction B C either at B or at C, or in a direction perpendicular to the arm C D at E, D E and D B being each one third of A D. P. 67.

Fig. 49. A force, acting at A on the lever A B, has a great mechanical advantage in turning the lever C D; but when the levers are in the position B E, D F, the force acts with a similar disadvantage. P. 67.

Fig. 50. The diameter of the cylinder A being three times as great as that of B, the weight C, or an equivalent force applied to the winch D, will support a triple weight at E. P. 67.

PLATE III.

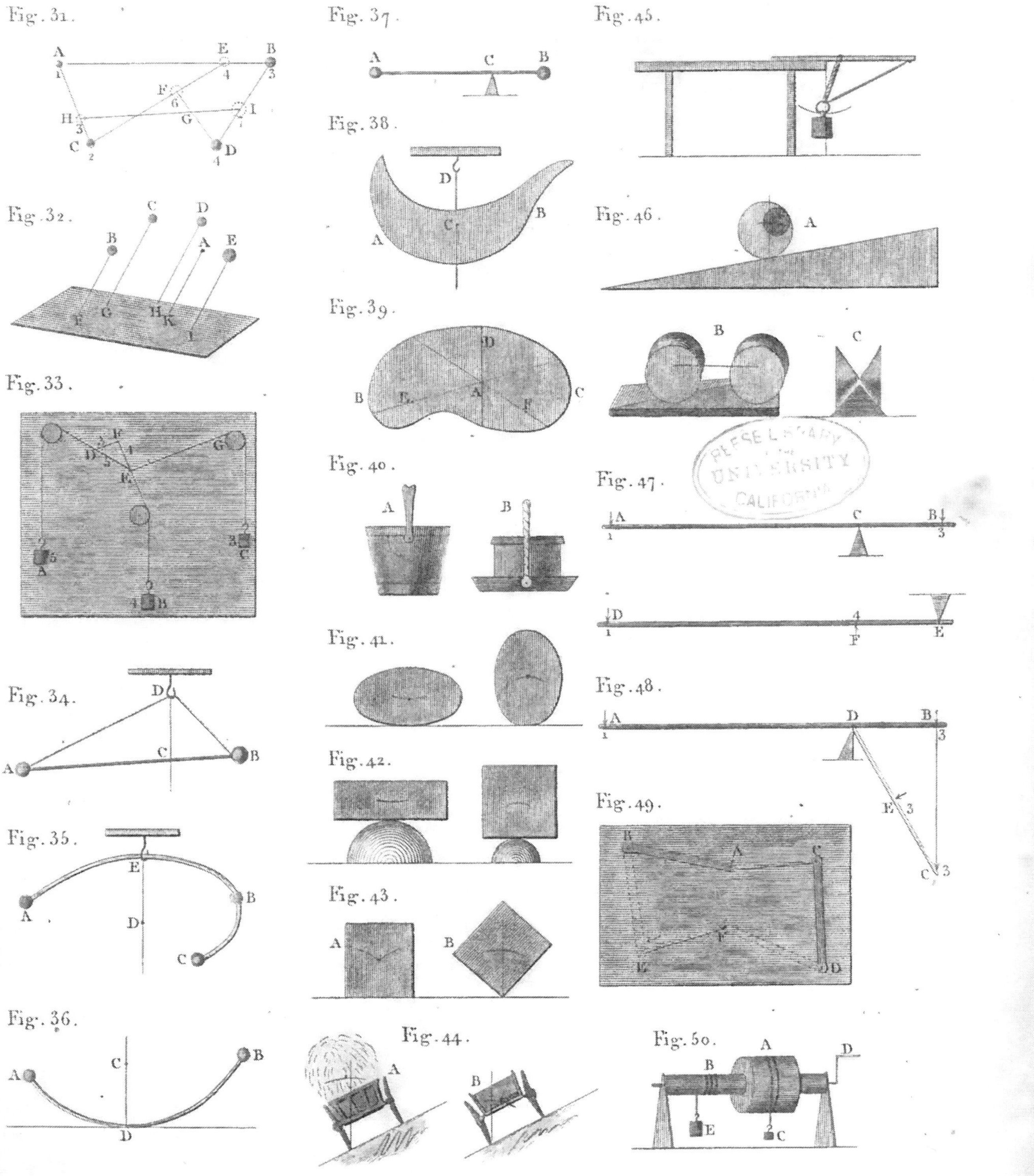

Pub. by J. Johnson, London 1 July 1806.

Joseph Skelton sculp.

PLATE IV.

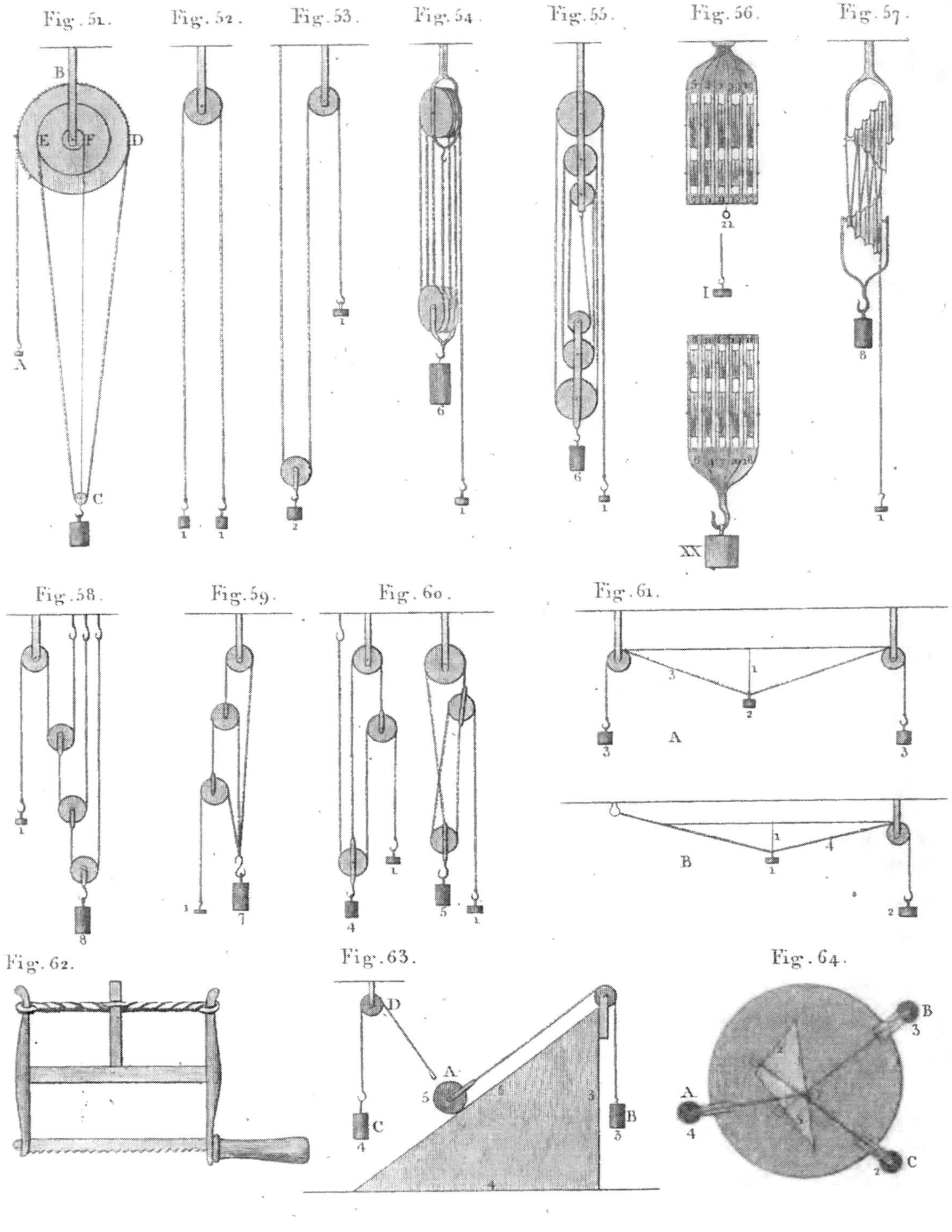

Pub. by J. Johnson, London 1 July 1806.

Joseph Skelton sculp.

PLATE IV.

Fig. 51. The weight A, acting on the double cylinder B, supports the weight C by the pulley running in the angle of the rope D C E, which is wound on the larger cylinder at D, while it is uncoiled from the smaller at E, and the force is the same as if the weight C were attached to the line C F, acting on the axis F, of which the diameter is equal to the difference of the radii of the double cylinder. P. 68, 206.

Fig. 52. A single fixed pulley, supporting two equal weights. P. 68.

Fig. 53. A single moveable pulley, by means of which a weight supports another twice as great. P. 68.

Fig. 54. The arrangement of pullies in ships' tackles, with a force of six to one. P. 69.

Fig. 55. An arrangement of pullies in a vertical line, with a force of six to one. P. 69.

Fig. 56. Mr. Smeaton's blocks, giving a force of twenty to one, the rope being applied in the middle of the outer series, and following the order of the figures from 1 to 21. P. 69, 207.

Fig. 57. A system of pullies fixed on one axis in each block; having a power of 8 to 1. P. 69.

Fig. 58. A system of pullies, each of which doubles the effect; having a power of 8 to 1. P. 69.

Fig. 59. A system of pullies with each rope fixed to the weight, having a force of 7 to 1. P. 69.

Fig. 60. Two systems of pullies, of the kind denominated Spanish bartons, in which two of the pullies are suspended by the same rope: the one has a power of 4, the other of 5. P. 69.

Fig. 61. A. The depression of the middle weight being one third of its distance from the pullies, it sustains two equal weights, which are together three times as great as itself. B. The depression of the smaller weight being one fourth of its distance from the pulley, it supports a weight twice as great as itself. P. 70.

Fig. 62. A joiner's saw, stretched by twisting a double cord, by means of a lever passing through it.

Fig. 63. The weight A, resting on an inclined plane of which the height is to the oblique length as 3 to 5, is sustained by a weight B three fifths as great as itself; and if for the resistance of the plane we substitute the action of the weight C, reduced to the direction A D perpendicular to the plane, this weight must be four fifths of the weight A, the horizontal length of the wedge being four fifths of its oblique length. P. 70.

Fig. 64. The weights A, B, and C, acting, by means of threads passing over pullies, which are fixed to any required part of a horizontal table, on the rollers which press against the sides of a wedge, proportional in length to the respective weights, retain each other in equilibrium, when their directions meet in one point. In order that the threads may pass on each side of the wedge, it may be supported by three or more balls. P. 71.

PLATE V.

Fig. 65. By means of the moveable inclined plane AB, of which the height AC is one third of the horizontal length BC, the weight D, acting horizontally, sustains a triple weight E, acting in a vertical direction. P. 71.

Fig. 66. AB being one fourth of BC, the rope AB must exert a force of tension equal to one fourth of the weight C, in order to support it, supposing the surfaces to be without friction. But if the friction of the end of the beam AC were equal to one fourth of the pressure, it would support the weight C without any other force, whatever might be its magnitude. P. 72.

Fig. 67. AB being half of BC, or one fourth of CD, the force extending the rope CD each way is equal to the weight E. P. 72.

Fig. 68. The thin wedge AB, of which the height is one fifth of the length, being rolled round the cylinder C, makes the screw D, by means of which the weight E is capable of supporting a weight five times as great as F. P. 72.

Fig. 69. A is a screw, and B the nut belonging to it. P. 72.

Fig. 70. The endless screw AB acts on the teeth of the wheel CD. P. 72.

Fig. 71. The distance of the threads of the interior screw is four fifths of that of the exterior or perforated screw, and this distance is one thirtieth of the circumference. Hence the weight A is capable of sustaining a weight B 150 times as great as itself. P. 73.

Fig. 72. The apparatus for experiments on collision. Those balls which are not employed may be left behind the graduated arc, as at A and B; some of the strings have balls of half the weight of the rest, others have a small dish C, on which balls of clay, or of wax softened with one fourth its weight of oil, may be supported. P. 76.

Fig. 73. If the ball A strike the ball B in the oblique direction AC, the ball B will be impelled in the direction CD perpendicular to the surface of contact; and the velocity EC being resolved into EF and FC, the part FC will continue unaltered; and if the balls are equal, the part EF will be destroyed, so that the ball A will move after the stroke in the direction CG, excepting the effect of any accidental disturbance which may be derived from the resistance of the surrounding bodies. If we imagine a ball at C in contact with B, in the direction DB, we may aim a blow at the centre of this ball, in order to drive the ball B to D; and if B happen to be situated any where in the semicircle DCG, the motion of A after the impulse will be in the direction BG or GB, if there be no resistance. When the ball H is reflected by a fixed obstacle, as by the cushion of a billiard table, at I, its velocity KI may be resolved into the parts KL, LI; the part KL continues, and may be represented by LM equal to KL, the part LI is converted into IL in a contrary direction, which when combined with LM makes IM, the angle LIM being equal to LIK. We may find the proper direction for striking any ball by reflection if we suppose a ball N in contact with the nearest point of the cushion, and making NO equal to MN, aim at a ball supposed to be at O. In the same manner if we wish to impel the ball P in the direction PQ by a stroke of the ball R after reflection at S, we first place a ball at T behind P, and determine the direction RS by aiming at a ball U, as if we wished to strike a ball at T with a direct impulse. But in the case of a billiard ball, the rotation of the ball round its axis, which is not destroyed by the collision, will cause the ball to move, on account of the friction of the table, in a direction different from its first direction: thus the ball C will not go on to G, but will strike the cushion between C and D; and the ball H, after reflection at I, will proceed in a direction a little nearer to N than IM; so that the imaginary ball O ought perhaps to be placed as far from the cushion itself as M, in order that the ball may be struck after reflection. P. 82.

Fig. 74. Mr. Smeaton's apparatus for experiments on rotatory motion. P. 84.

Fig. 75. The moveable centre of suspension being fixed at the distance of 5 inches from one of the balls, and 7 from the other, the vibration is performed at the same time as that of a pendulum 37 inches long. P. 85.

Fig. 76. The three weights, supported on wheels, being drawn up the three inclined planes at the same time, by the action of three other equal weights, the middle weight arrives first at the top, the length of its plane being twice the height. P. 88.

PLATE V.

Fig. 65.

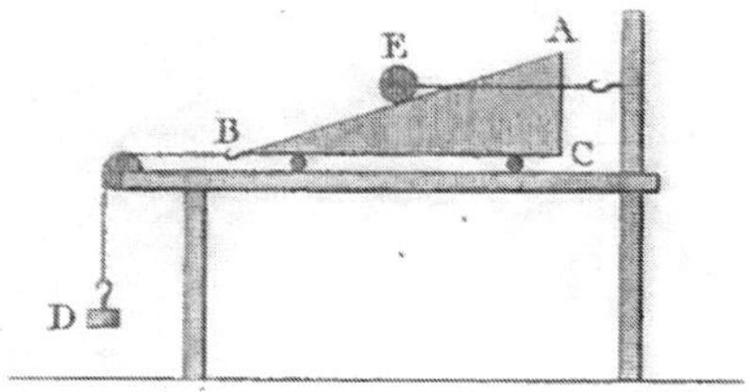

Fig. 66.

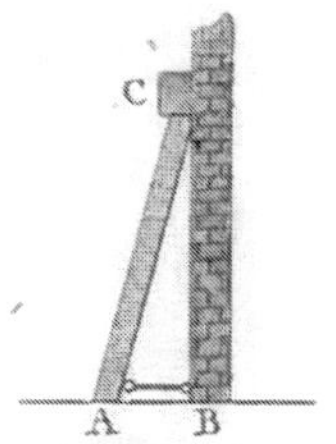

Fig. 67.

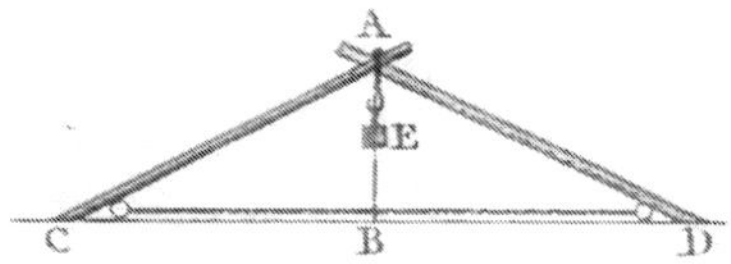

Fig. 69.

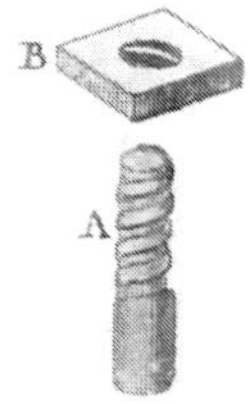

Fig. 68.

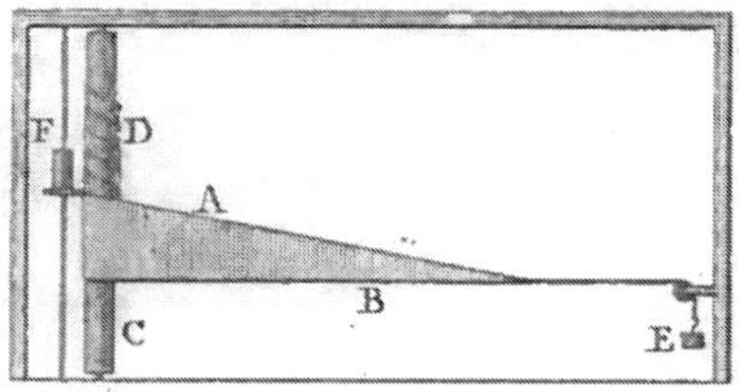

Fig. 70.

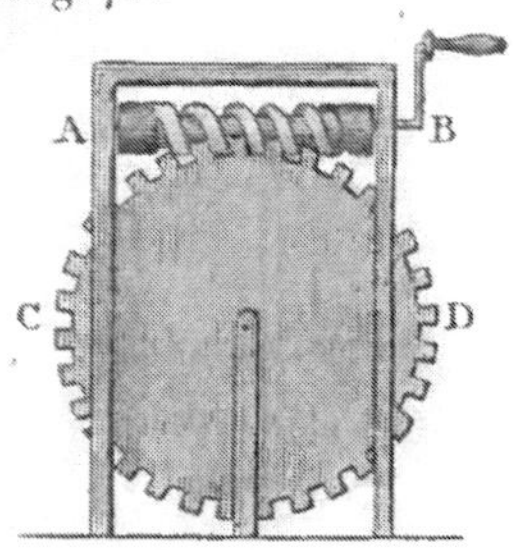

Fig. 71.

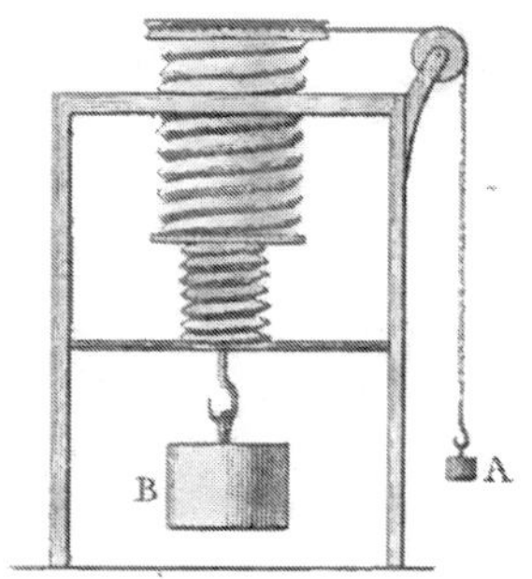

Fig. 75.

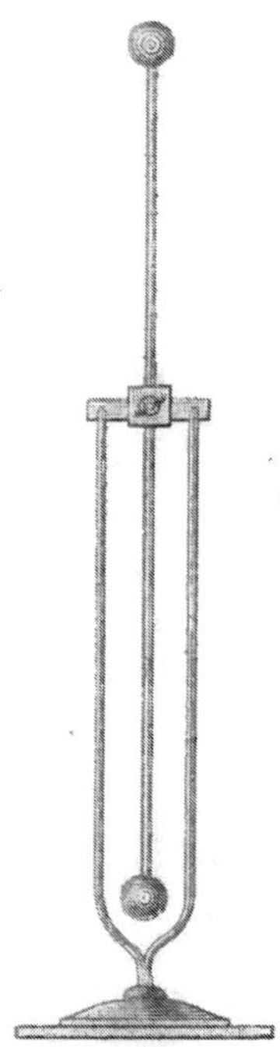

Fig. 72.

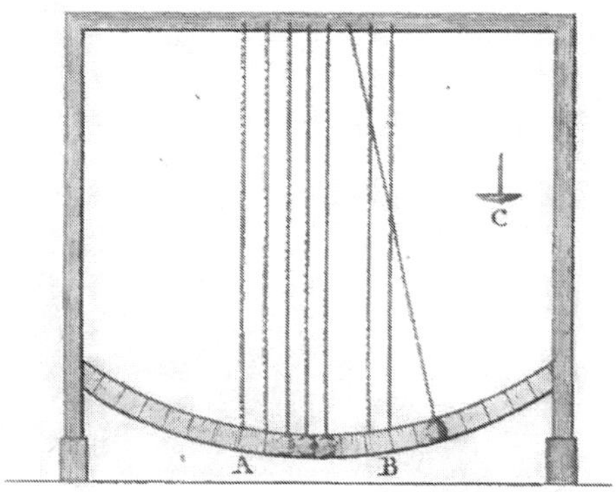

Fig. 73.

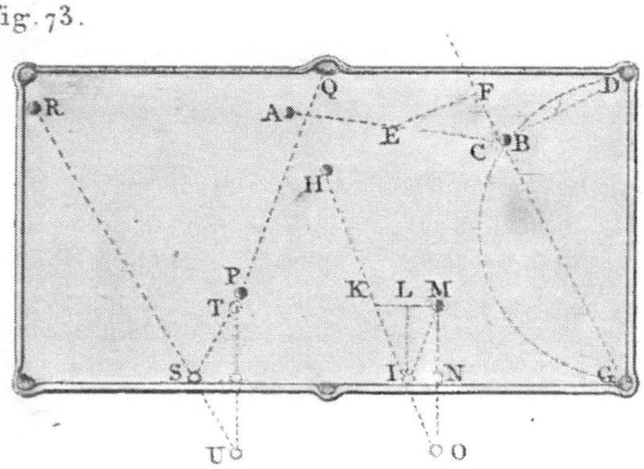

Fig. 74.

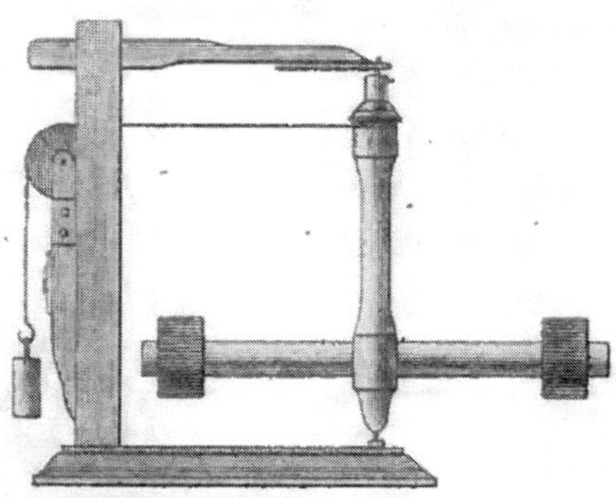

Fig. 76.

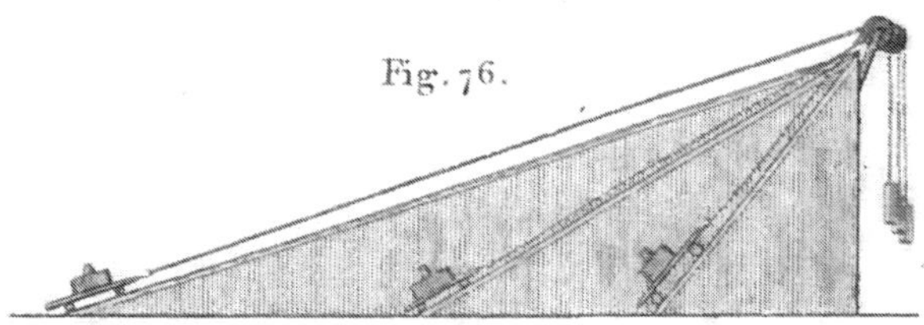

Pub. by J. Johnson, London, 1 July 1806

Joseph Skelton sculp.

PLATE VI.

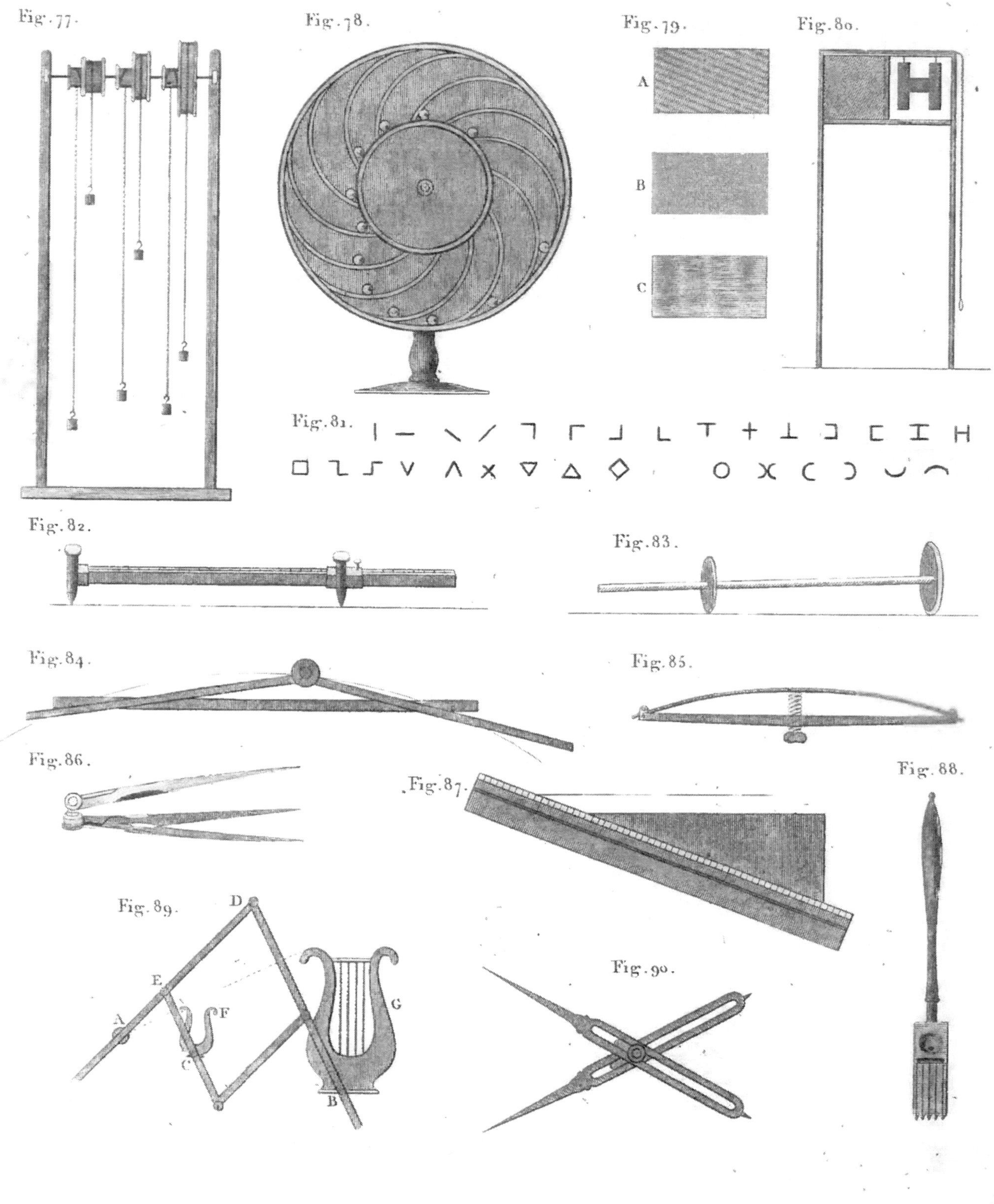

Pub. by J. Johnson, London 1 July 1806.

Joseph Skelton sculp.

PLATE VI.

Fig. 77. The proportions of the diameters of the different parts of the double pullies being 3 to 2, 5 to 2, and 8 to 2, the middle weight may be observed to rise the most rapidly. P. 88.

Fig. 78. A wheel supposed to be capable of producing a perpetual motion; the descending balls, acting at a greater distance from the centre, but being fewer in number, than the ascending. In the model, the balls may be kept in their places by a plate of glass covering the wheel. P. 92.

Fig. 79. A, the inclination of cross lines generally most convenient for producing the effect of a tint, in drawing. B shows the effect of lines crossing each other perpendicularly, and C that of lines crossing too obliquely. Where the surface to be shaded is large, the separate lines or hatches should begin and end with a point, in order that the junction of the different portions may escape observation. P. 95.

Fig. 80. Dr. Hooke's telegraph, in which the characters are arranged behind a screen, and drawn out as they are required. P. 100.

Fig. 81. Dr. Hooke's alphabet, with some other arbitrary characters for his telegraph. P. 100.

Fig. 82. A beam compass, with a scale. P. 102.

Fig. 83 . . 85. Instruments for drawing arcs of large circles. P. 102.

Fig. 86. A pair of triangular compasses. P. 102.

Fig. 87. Marquois's scales, for drawing parallel lines. P. 103.

Fig. 88. A pen for ruling musical lines. P. 103.

Fig. 89. A pantograph. A being the centre of motion, B the tracing point, and C the describing point, A B is always to A C as A D to A E, and the copy F is similar to the original G. P. 103.

Fig. 90. A pair of proportional compasses. P. 104.

PLATE VII.

Fig. 91. A sector. The scale of equal parts is marked L. As A B is to A C, so is B D to C E; and if any line B D be placed with its extremities in the third division of the scale on each leg, the distance C E between the seventh divisions will contain 7 equal parts, of which B D contains 3; and the same is true of any other numbers. P. 104.

Fig. 92. A vernier, indicating $38\frac{1}{67}$ of the divisions of its scale. P. 105.

Fig. 93. A sliding rule. The slider being drawn out, so that the division marked 1 is opposite to 3 on the rule; all the other figures on the rule are triple of those which stand opposite to them. P. 107.

Fig. 94. A circular logarithmic instrument. The inner circle slides within the outer, and as it is represented in the figure, each number stands opposite to another which is twice as great. P. 107.

Fig. 95. A steel chain, made by Ramsden. A, the screw for bringing the mark B precisely to the point required; C a joint between the adjoining links; D, a cross joint at every tenth link; E, a pulley and weight for stretching the chain. P. 112.

Fig. 96. A micrometrical scale made by Troughton. The compound microscopes A and B are fixed nearly at the required distance on the scale C: A is then made to point exactly to a division of the standard scale D by means of the screw E, and B to another division, at the required distance, by means of the screw F, the fractional parts being added by the turns of the screw G. The scale D is then removed, and the object to be compared with it is put in its place. P. 112.

Fig. 97. A diagonal scale. The line A B contains 274 parts, of which the units of the scale contain 100. P. 112.

Fig. 98. The statuary's compass, seen sideways. The pin A B is forced down, till it is stopped by the moveable stud C; the screw D fixes it in its angular position. It is also capable of motion round the axis E F, which is fixed by the screw G. P. 113.

Fig. 99. An instrument for making drawings in perspective; the perforated sight may be drawn out to any required distance. The dotted lines show how a second frame may be applied instead of the sight, so as to answer the same purpose. P. 115.

Fig. 100. Illustration of the principles of perspective. A being the place of the eye, and B C the plane of projection, if A D be parallel to E F, G H, and I K, D will be their vanishing point, and E D, G D, and I D, their whole images: A L being parallel to E M and I N, L will be their vanishing point, and E L, I L, their whole images: and A O being parallel to P Q, O will be its vanishing point. P. 115.

Fig. 101. A being the centre of the picture, A B the horizontal vanishing line, A C the vertical line, and D the point of distance, if a ground plan E F G H of any figure on the horizontal plane be placed in its true position with respect to I K, the bottom of the picture, the vanishing points of all its lines will be found by drawing D L, D M, D N, and D O, parallel to those lines respectively; and the whole images of the lines will be P L, Q M, K N, and I O, determining, by their intersections, the figure R S T U, which will be the projection of E F G H. The plan may also be drawn, in an inverted position, below the line I K, and the point of distance taken above A instead of below it. P. 115.

Fig. 102. A B being the whole image of the line represented by A C as a ground plan, and D the point of distance, we may find E, the image of the point C, by drawing C D; or we may make BF=BD and AG=AC, FG will then also cut A B in the point E. P. 116.

PLATE VII.

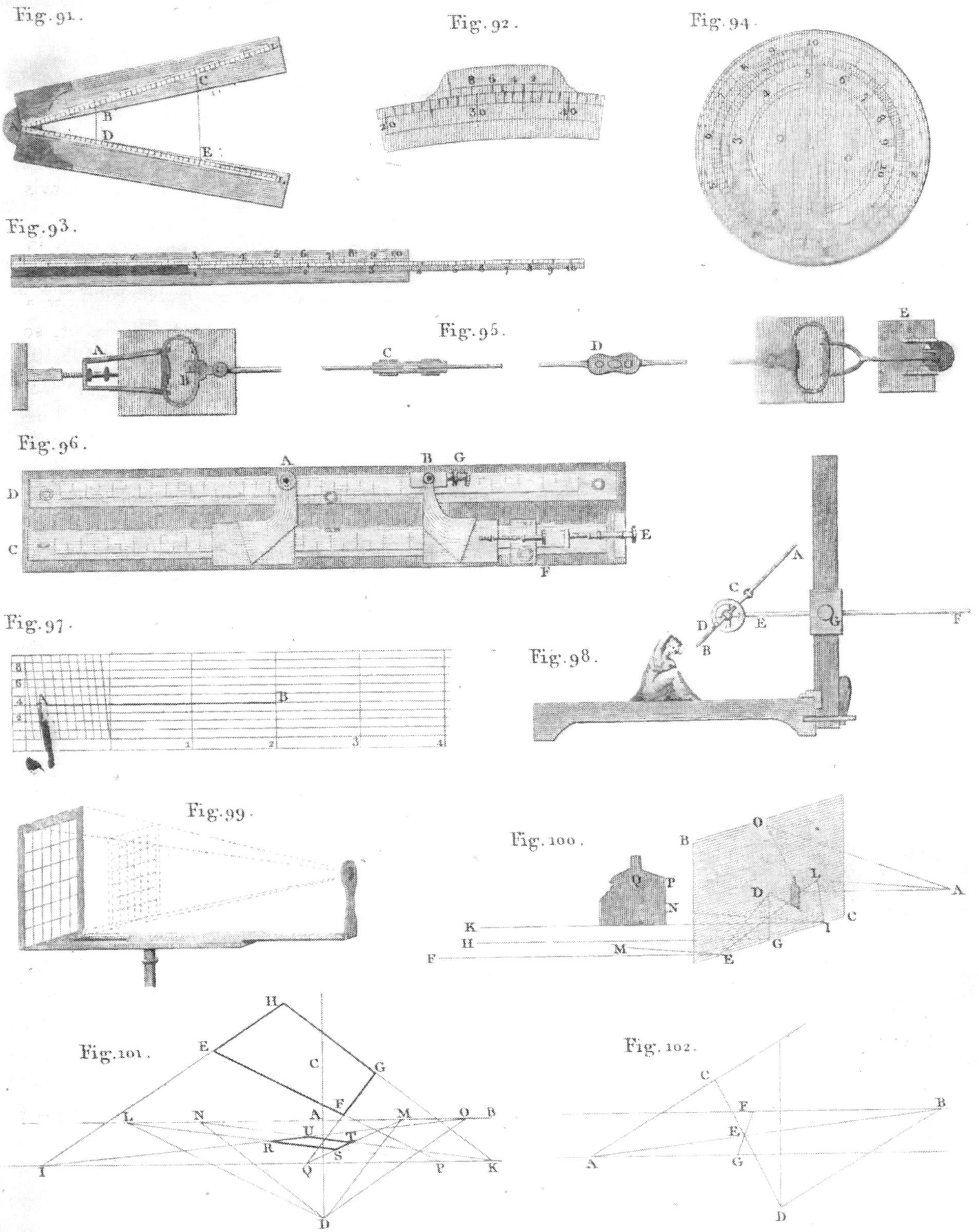

Pub by J. Johnson, London 1 July 1806

Joseph Skelton

PLATE VIII.

Fig. 103.

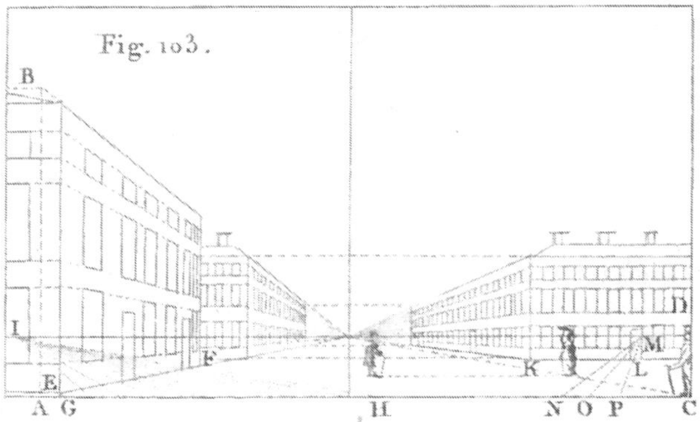

Fig. 104.

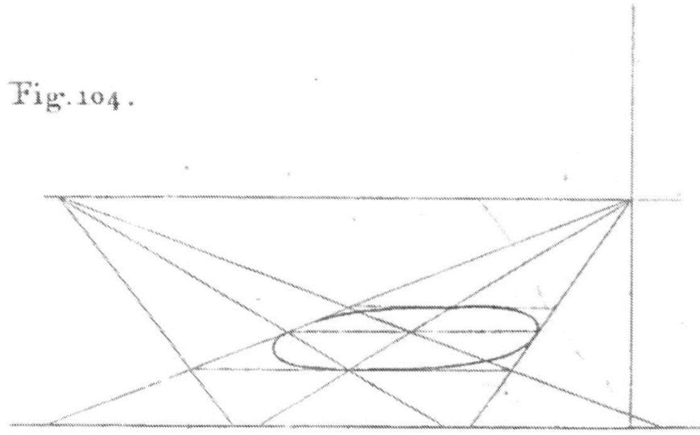

Fig. 105.

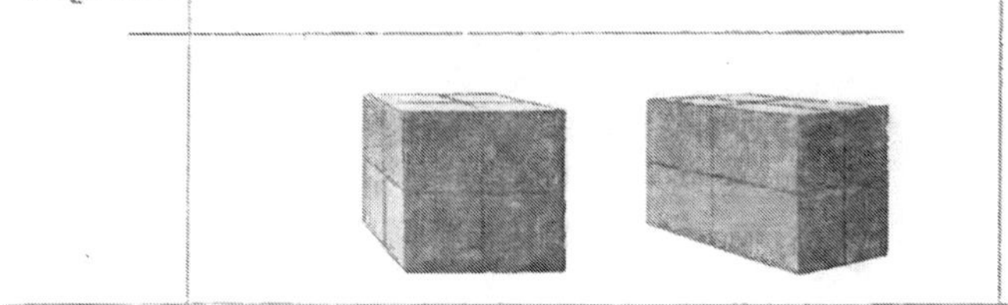

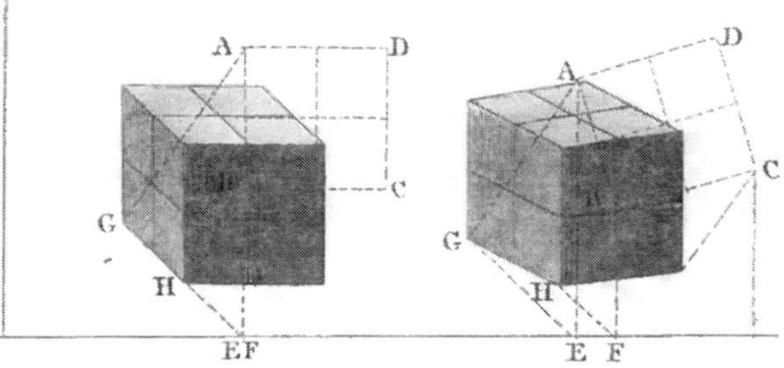

Fig. 106.

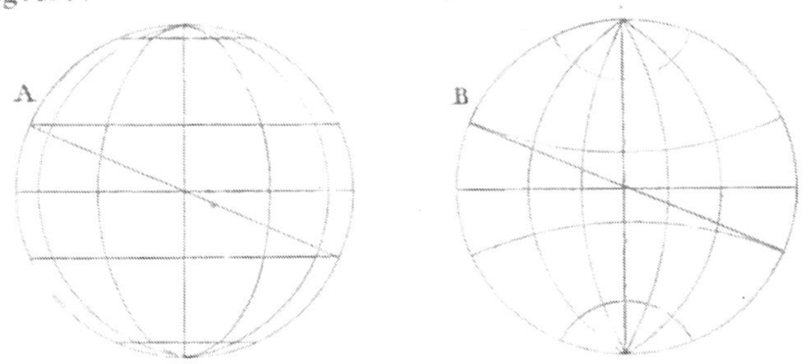

Fig. 107.

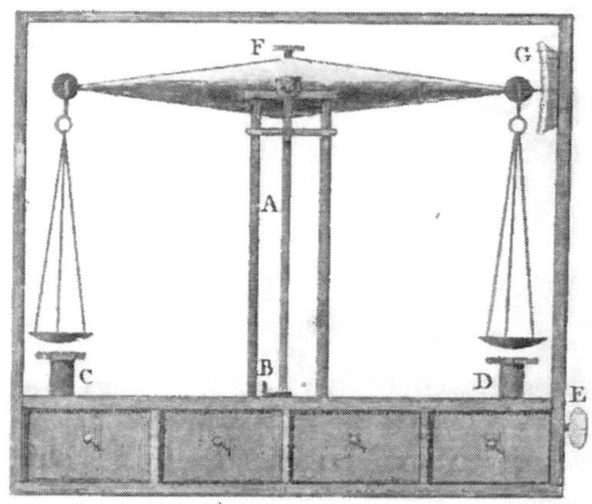

Fig. 108.

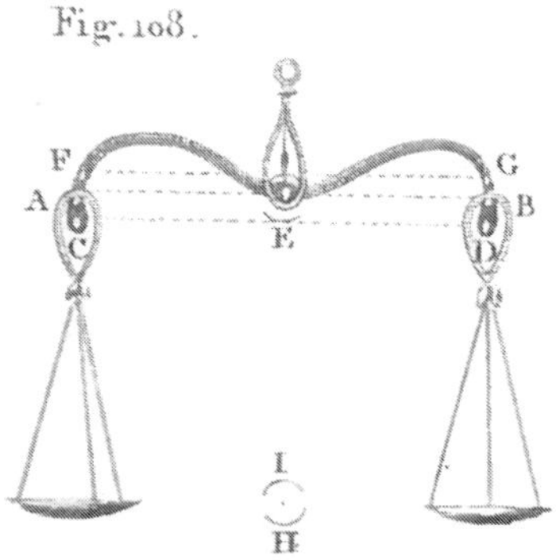

Fig. 109.

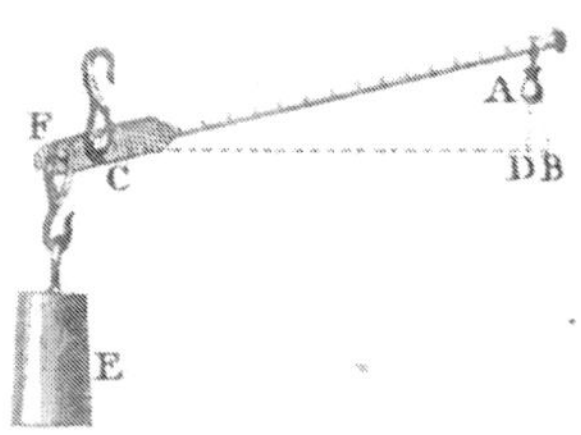

Fig. 110.

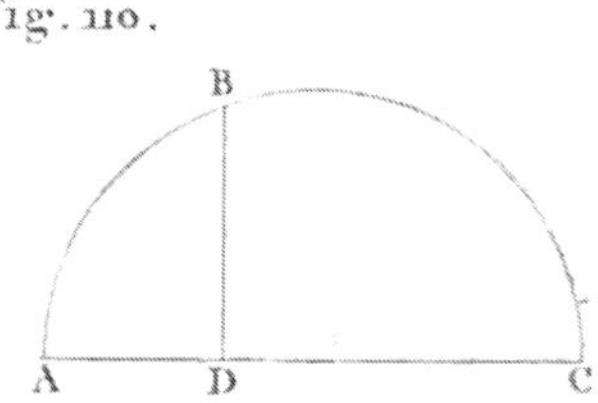

Pub. by J. Johnson, London 1 July 1806.

Joseph Skelton sculp.

PLATE VIII.

Fig. 103. The heights of the houses, windows, doors, and figures are determined by lines directed to the centre of the picture; the true height being measured on the lines A B, C D, where the objects are supposed to touch the plane of projection. The distance E F, and all other parts of lines perpendicular to the picture, are measured by laying off the lengths of the originals, as G H, on the line A C, and drawing I E G I E H, from I, the point of distance; which, in most cases, will be more remote from the centre of the picture than it is here made. The line K L, and others parallel to A C, may be measured by the assistance of any point M in the horizontal line, the distances, N O, O P, being laid off on A C, or simply by reducing the scale in the proportion of M P to M L. P. 116.

Fig. 104. A circle thrown into perspective, by means of the circumscribed square, the points of contact being found by bisecting the sides. P. 116.

Fig. 105. Two perspective delineations, and two orthographical projections of a cube, in different positions. For the orthographical projection, the ground plan being A B C D, the image of any point A, B, may be found by drawing A E, B F, perpendicular to the ground line, E G, F H, parallel to the line assumed for the direction of the centre of the picture, and A G, B H parallel to the line of direction of the point of distance; the intersections G and H will then be the points corresponding to A and B. P. 116.

Fig. 106. A is the orthographical projection of a sphere, with some of its circles; B the stereographical projection of the same circles. P. 117.

Fig 107. A balance made by Fidler for the Royal Institution, nearly resembling those of Ramsden and Troughton. The middle column A is raised at pleasure by the cock B, and carries the round ends of the axis in the forks at its upper part, in order to remove the pressure on the sharp edges of the axis within the forks. The scales are occasionally supported by the pillars C and D, which are elevated or depressed by turning the handle E. The screw F serves for raising or lowering a weight within the conical beam, by means of which the place of the centre of gravity is regulated. The extent of the vibrations is measured on the graduated arc G. P. 125.

Fig. 108. A balance for the illustration of the different kinds of equilibrium. When the scales are hung on the middle pins, A, B, which are in the same horizontal line with the support of the beam, the equilibrium is neutral, the weights acting as if the centre of gravity coincided with the point of suspension. If the scales be hung on the lowest pins C, D, the centre of gravity will be nearly in the line C D, and its path the curve E, which has its concavity upwards; but if the scales are hung on the pins F, G, the path of the centre of gravity will be convex upwards, and the beam will overset. In reality the true paths of the centre of gravity would be nearly in the curves H and I, situated between the weights in the scales: but these are similar to the other curves. P. 125.

Fig. 109. When the equilibrium of a balance is tottering, the lower weight acts with the greatest advantage: thus the effect of the weight A is reduced in the proportion of B C to D C, by the obliquity of the arm C A, while the weight E acts on the whole length of its arm C F. P. 125.

Fig. 110. If A B C be a semicircle, and B D represent a given weight, and A D its counterpoise in one of the scales of an unequal balance D C will be the counterpoise in the other scale. It is obvious that A C is more than twice as great as B D. P. 126.

PLATE IX.

Fig. 111. A weighing machine. The platform supporting the weight rests on the pins A, B, C, D, at equal distances from the fulcra E, F, G, H; so that wherever the weight may be placed, it presses equally on the lever IK, at L, and is counterpoised by a much smaller weight placed in the scale M. P. 126.

Fig. 112. A steelyard resembling that of Mr. Paul, in which different weights may be employed. A, a loop to check the vibrations; B a scale to be suspended by the hook C. If great delicacy be required in the weights, the fractional parts may be expressed by the turns of a micrometer screw D, furnished with an index. P. 126.

Fig. 113. A bent lever balance. P. 127.

Fig. 114. A spring steelyard: half the case being removed, to show the spring. P. 127.

Fig. 115. A B, the path of the centre of gravity of the human body, such as it would be described in walking, if the legs were inflexible. C D, the path described in running, on the same supposition. P. 130.

Fig. 116. The actual path of the centre of gravity, as it is usually described. P. 130.

Fig. 117. An elastic column, compressed by a weight acting at the distance of one third of its depth from the concave surface; the compression being every where as the distance of the lines A B, A C. P. 139.

Fig. 118. An elastic column, extended by a weight acting at the distance of one third of its depth from the convex surface, the extension being every where as the distance of A B, A C. P. 139.

Fig. 119. An elastic column, compressed by a weight acting immediately on the concave surface: the compression extends only to the line A B, the parts beyond this line being extended. P. 139.

Fig. 120. A column bent, by a weight acting longitudinally, into the form of a harmonic curve: the line A B C D is the limit between the parts which are compressed, and those which are extended. P. 139.

Fig. 121. An elastic plate or rod, considerably bent by a weight acting at its extremity. P. 139.

Fig. 122. An elastic rod fixed at one end, and bent by its own weight. P. 139.

Fig. 123. An elastic rod supported at each end, and bent by its own weight. P. 139.

PLATE IX.

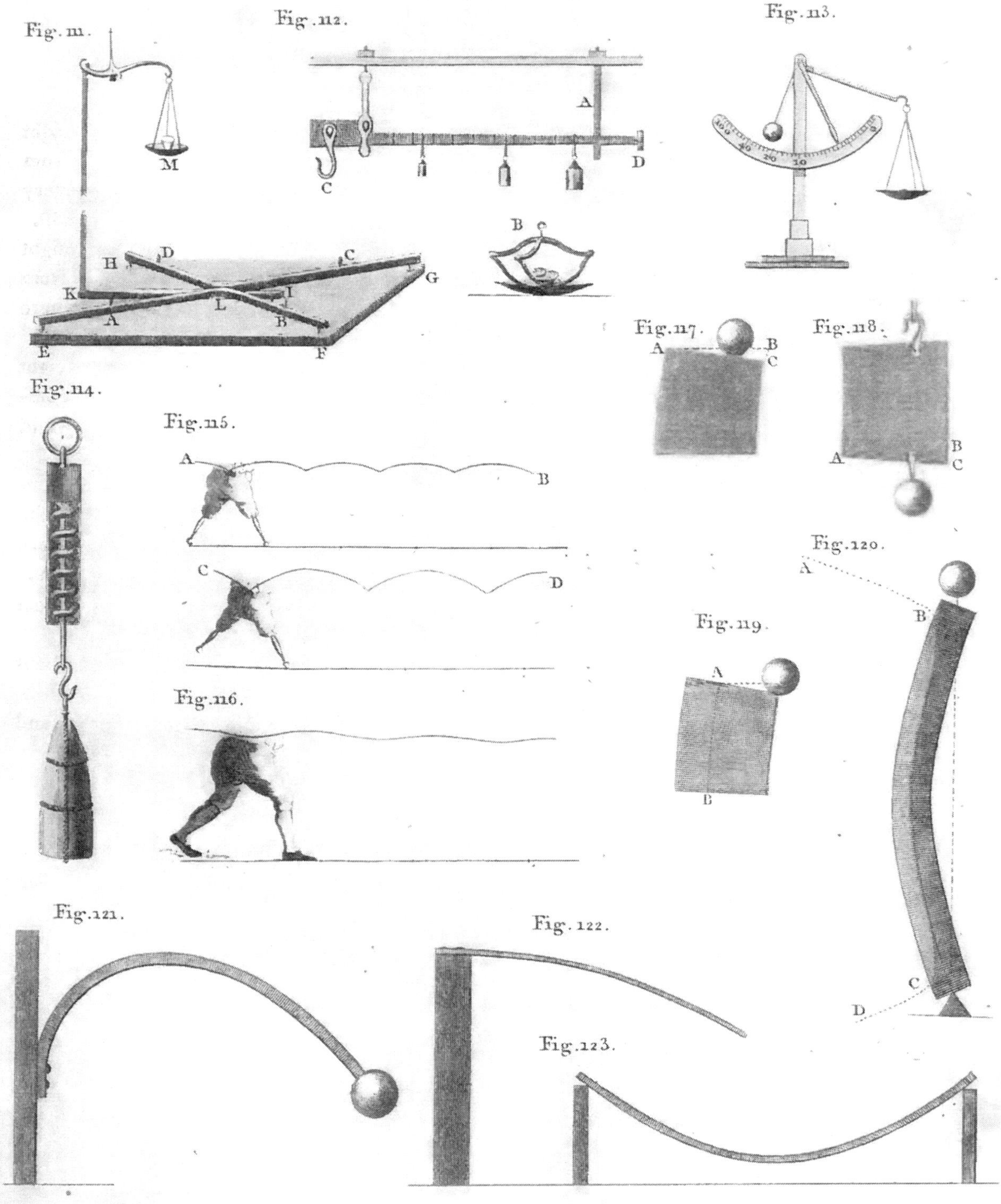

Pub. by J. Johnson, London 2 July 1806. *Joseph Skelton*

PLATE X.

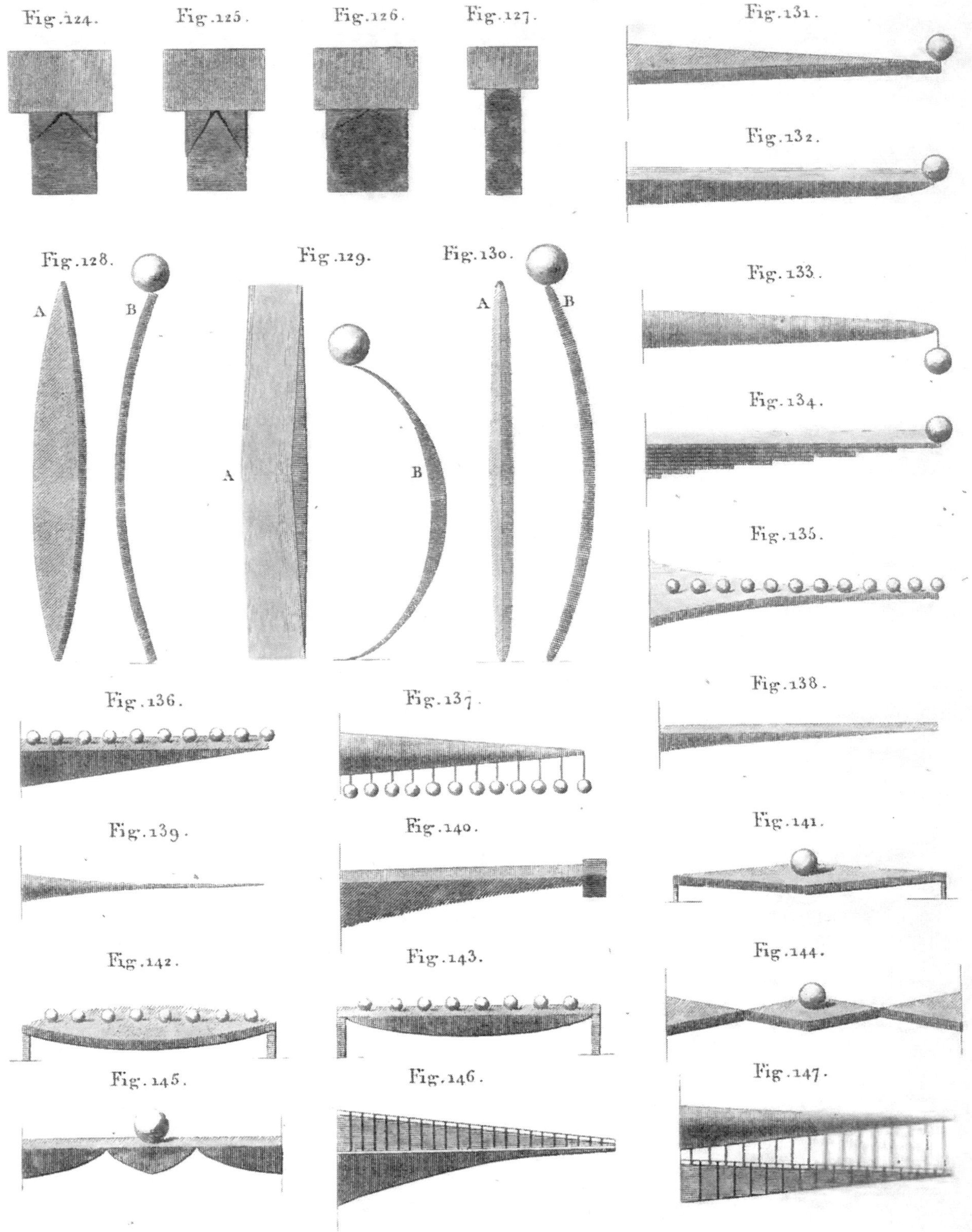

Pub. by J. Johnson, London 1 July 1806.

Joseph Skelton

PLATE X.

Fig. 124. The manner in which a prismatic column is crushed by pressure, supposing the lateral adhesion to be simply proportional to the surface concerned. P. 146.

Fig. 125. The manner in which a column is crushed, supposing the lateral adhesion to be increased by pressure. P. 146.

Fig. 126. The circle is as strong as the circumscribing square, supposing the adhesion proportional to the surface, the relative force of all its chords being equal. P. 146.

Fig. 127. The three circles are as strong as the circumscribing parallelogram. P. 146.

Fig. 128. A, the strongest form for a beam, cut out of a plank of uniform depth, for resisting a longitudinal force; B, the form into which it is bent; both curves being circular. P. 150.

Fig. 129. A, the strongest form for a beam cut out of a plank of equable breadth, for resisting a longitudinal force which bends it into the cycloidal curve seen at B. P. 150.

Fig. 130. A, the strongest form for a square or turned beam or column, slightly bent by a longitudinal force:, B, the form into which it is bent by such a force. P. 150.

Fig. 131. The strongest form of a beam cut out of a horizontal plank, fixed at one end, and supporting a weight at the other. P. 150.

Fig. 132. The strongest form of a beam cut out of a vertical plank, fixed at one end, and supporting a weight at the other; the outline being parabolic. In practice the best method in such a case would be simply to reduce the depth at the end to one half of the whole, keeping the outline straight; in this manner one fourth of the timber would be saved. P. 150.

Fig. 133. The strongest form of a square or turned beam, fixed at one end, and supporting a weight at the other; the outline being a cubic parabola. P. 150.

Fig. 134. The strongest form for the outline of a compound spring, supporting a weight at the end. P. 150.

Fig. 135. The strongest form for a beam cut out of a horizontal plank, fixed at one end, and supporting a weight equally distributed throughout its length; the outline being a parabola. P. 150.

Fig. 136. The strongest form for a beam cut out of a vertical plank, fixed at one end, and supporting a weight equally distributed throughout its length. P. 150.

Fig. 137. The strongest form for a square or turned beam, fixed at one end, and supporting a weight equally distributed throughout its length; the outline being a semicubic parabola, in which the cube of the thickness is as the square of the distance from the end. P. 150.

Fig. 138. The strongest form for a beam cut out of a vertical plank, for supporting its own weight; the outline being a parabola. P. 150.

Fig. 139. The strongest form for a turned beam, for supporting its own weight; the outline being parabolic. P. 150.

Fig. 140. The strongest form of a beam calculated to resist the pressure of its own weight by lateral adhesion only. The outline is a logarithmic curve, which never comes into contact with the axis, and in order that the condition of equal strength may be possible, the beam must be loaded with a weight, at its extremity, equal to that of the portion which is wanting to complete the figure. P. 150.

Fig. 141. The strongest form for a beam cut out of a horizontal plank, supported at both ends, and bearing a weight at the middle. P. 150.

Fig. 142. The strongest form for a beam cut out of a horizontal plank, supported at both ends, and bearing a weight equally distributed throughout its length; the outline being parabolic. P. 150.

Fig. 143. The strongest form for a beam cut out of a vertical plank, supported at both ends, and bearing a weight equally distributed throughout, the outline being elliptic. P. 150.

Fig. 144. The strongest form for a beam cut out of a horizontal plank, firmly fixed at both ends, and supporting a weight at the middle. P. 150.

Fig. 145. The strongest form for a beam cut out of a vertical plank, firmly fixed at both ends, and supporting a weight at the middle, the curves being parabolic. P. 150.

Fig. 146. The strongest form for a beam cut out of a vertical plank, and supporting every where a weight proportional to the distance from the extremity: the outline being a cubic parabola. P. 150.

Fig. 147. The strongest form for a square or turned beam, supporting every where a weight, proportional to the distance from the extremity, and represented by the section of the same figure, which is a pyramid or a cone. P. 150.

PLATE XI.

Fig. 148. A machine for examining the strength of materials. The force is applied by means of the winch A, which winds up the rope BC, passing over the first pulley, and under the second, which is directly under the point D, at which the force acts on the piece EF to be broken; the pullies slide on two parallel bars, fixed in a frame, which is held down by a point projecting at G, from the lever GH, which is graduated like a steelyard, and measures the force. The piece to be broken is held by a double vice, I, K, with four screws, two of them hiding the other two in the figure; if a wire is to be torn, it may be fixed to be the cross bar LM; and a substance to be crushed must be placed under the lever NO, the end N receiving the rope, and the end O being held down by the click, which acts on the double ratchet OP. The lever is double from O to Q, and acts on the substance by a loop, fixed to it by a pin. P. 151.

Fig. 149. The outline of a column diminished one fifth of its diameter, in two different ways: the side A being an arc of an ellipsis, of which the semidiameter AB is the lesser semiaxis, joined at A to a right line AC, of one third of the length of the column, the part AD being cylindrical; the side DE is a cubic parabola, and may be drawn mechanically by fixing a straight ruler EF, in such a position that DF may be twice the diminution at E, and then bending it to D: the diminution being every where as the cube of the distance from D. These two methods are compared in a contracted scale at G: the outer line represents the first method, and the next line the second; the third, which is nearest to, G the conchoid of Nicomedes, recommended by Chambers, said to be found in the columns of the Pantheon; the curve beginning at the base. Palladio fixes the ruler at A, and bends it to H, which makes the curvature abruptly greater at H. P. 158.

Fig. 150. A section of Mr. Smeaton's light house at the Eddystone. P. 159.

Fig. 151. Mr. Smeaton's mode of uniting tiers of stones by wooden pins and wedges. P. 160.

Fig. 152. A string of beads, suspended in equilibrium from two points, and remaining in equilibrium in an inverted position. The ends are supported by two pieces, which slide backwards and forwards, and are fixed by screws: the string is also tightened by turning a pin. P. 161.

Fig. 153. A system of bars, hanging in equilibrium, and supporting each other in the same form when inverted. P. 161.

Fig. 154. A, a chain loaded, at equal distances, with other chains of such a length, as to represent the depth of the materials pressing on an arch of the form shown by the first chain, and holding it in equilibrium. B, an arch of a similar form. P. 161.

Fig. 155. A comparison of the curves which have various advantages for the construction of an arch supporting a horizontal road. The full line is an elliptic arc, somewhat less than half the ellipsis. The outside curve, which is also continued furthest down, is that which is calculated for resisting the pressure of materials acting like a fluid, or in the manner of wedges: the second dotted curve, for supporting the pressure of the materials above each part, supposed to act in a vertical direction only: the third is a circular arc, making one third of a whole circle: the fourth is part of a logarithmic curve, which is nearly of equal strength with respect to the tendency of the materials to give way for want of lateral adhesion, and the fifth is composed of parabolic curves, showing the outline which would be strongest for supporting any additional weight placed on the middle of the arch. If the height were greater in proportion to the span, as usually happens in practice, there would be less difference between the curves. The radius of curvature at the summit being AB, the horizontal thrust is equal to the weight of the portion ABCD of the materials.

PLATE XI.

Fig. 148.

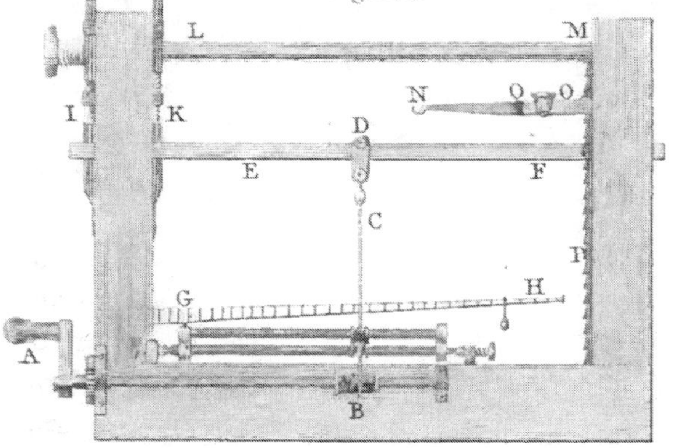

Fig. 149.

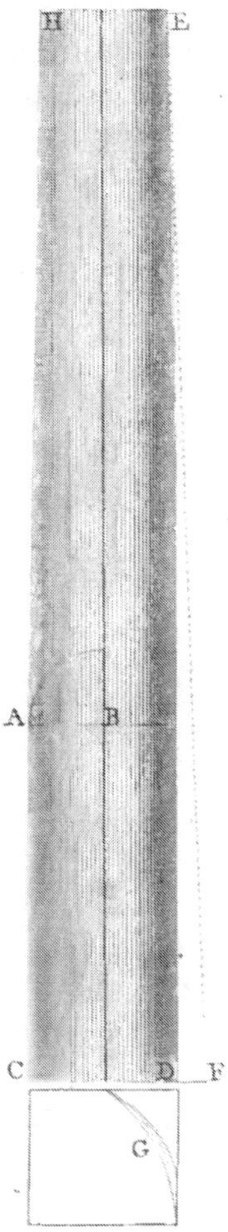

Fig. 150.

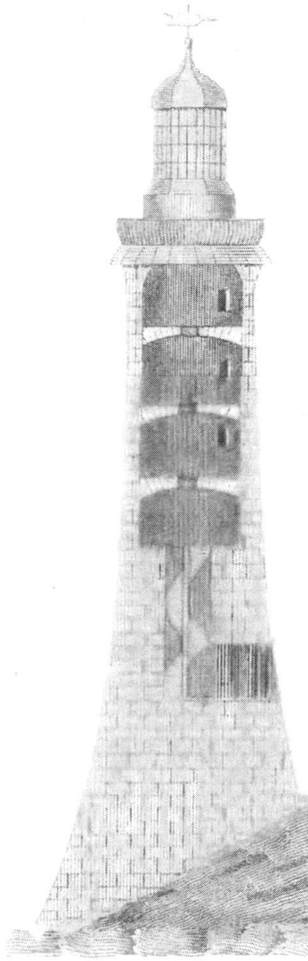

Fig. 151.

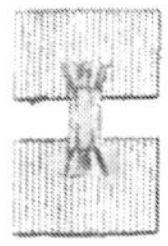

Fig. 152.

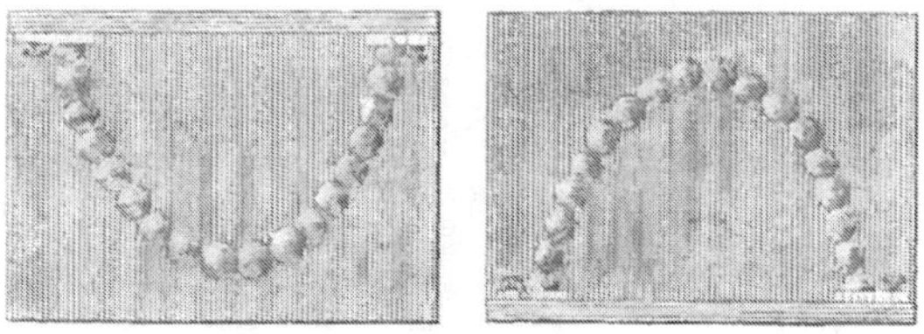

Fig. 153.

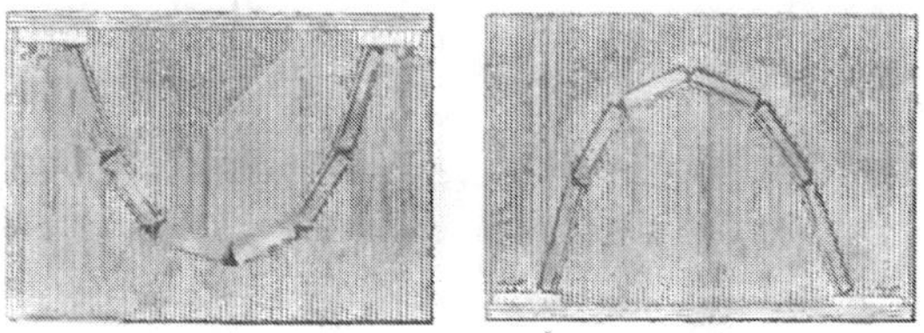

Fig. 154.

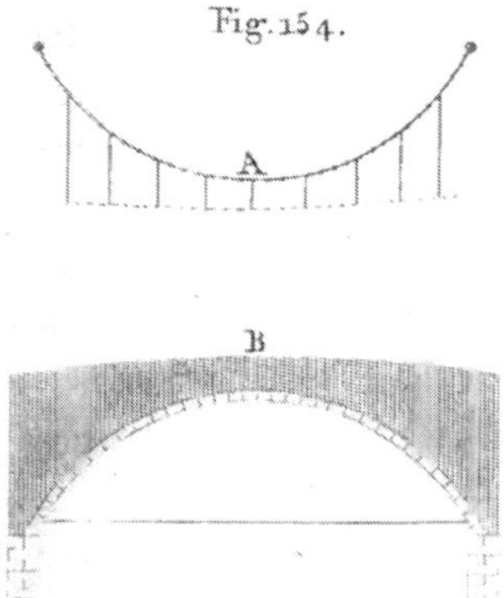

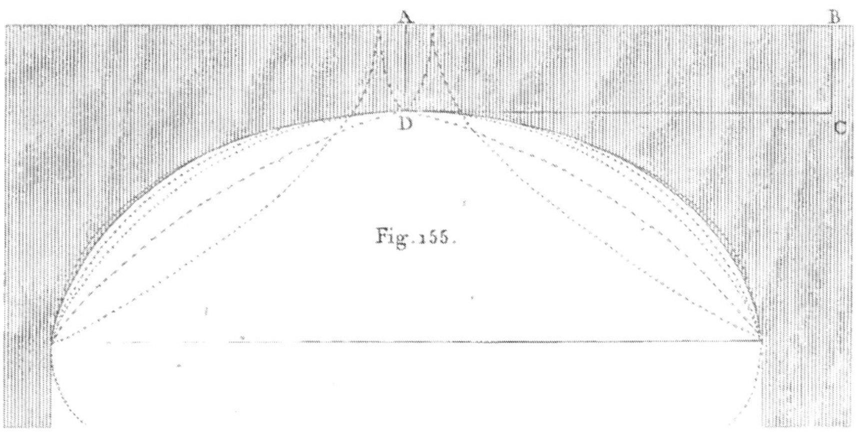

Fig. 155.

Pub. by J. Johnson, London 1 July 1806. *Joseph Skelton sculp.*

PLATE XII.

Fig. 156.

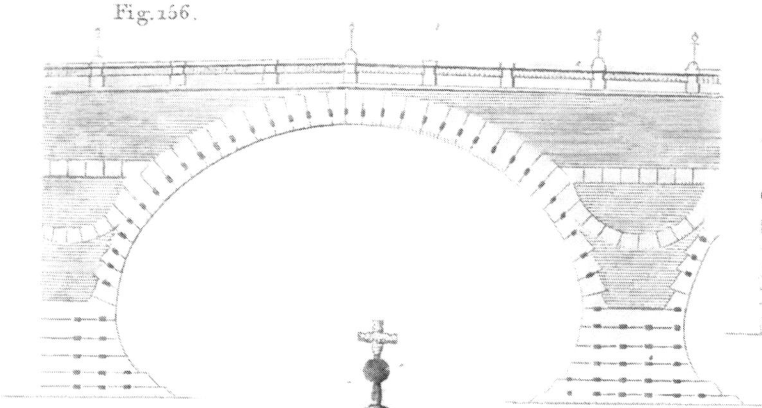

Fig. 157.

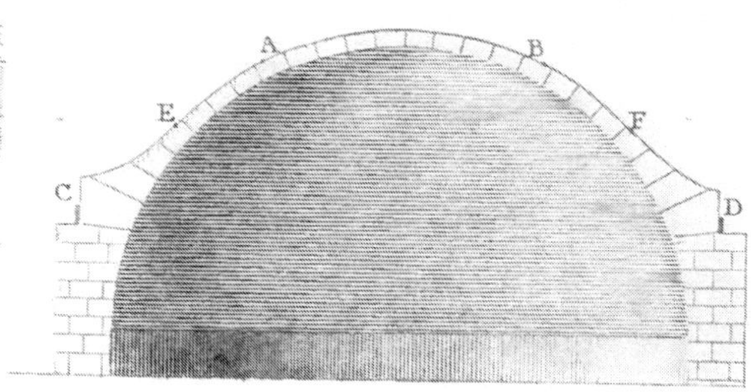

Fig. 158.

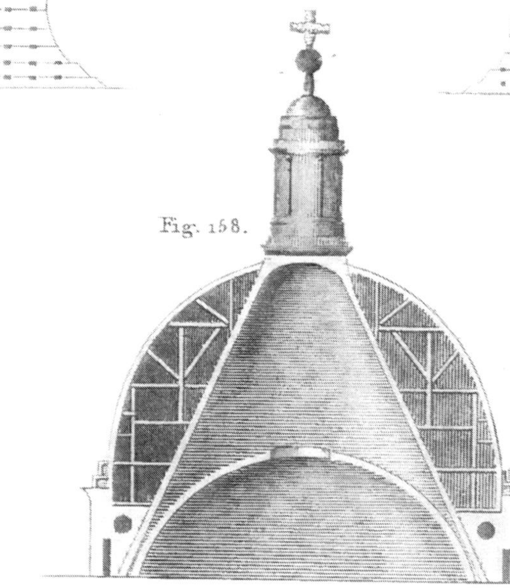

Fig. 159.

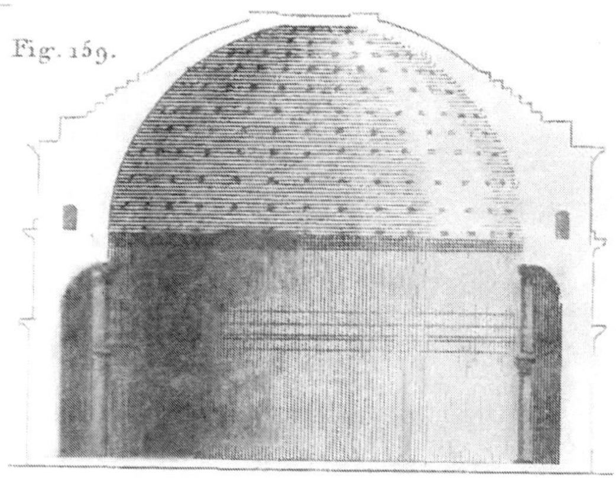

Fig. 160.

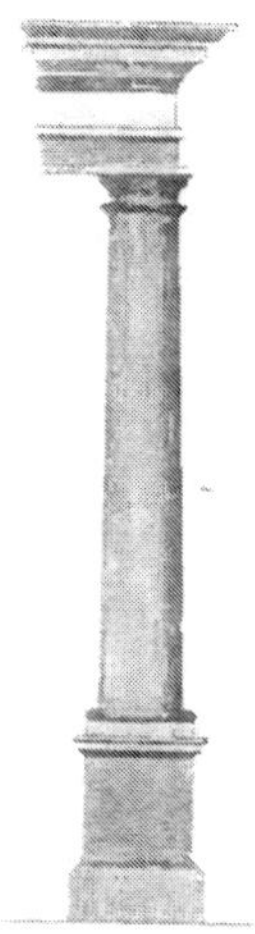

Fig. 161.

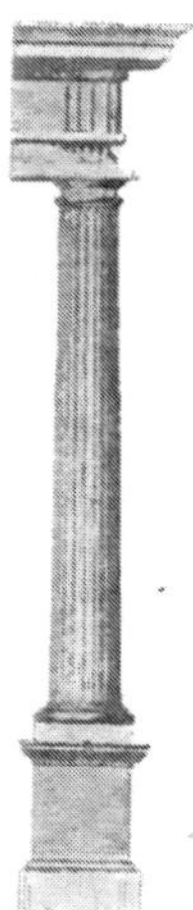

Fig. 162.

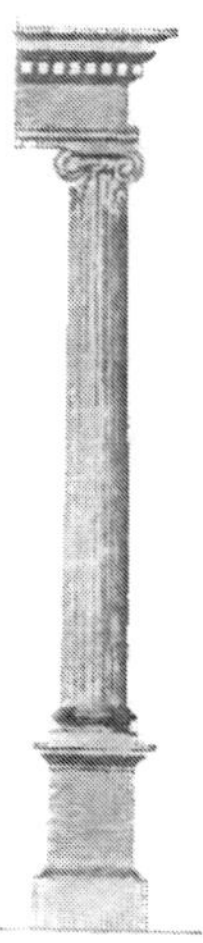

Fig. 163.

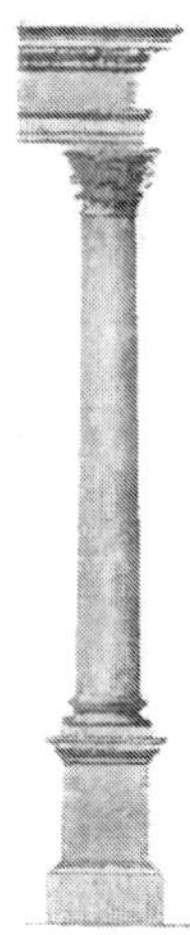

Fig. 164.

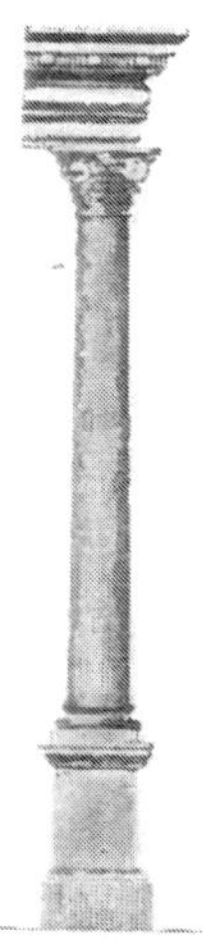

Fig. 165.

Pub. by J. Johnson, London 1 July 1806.

Joseph Skelton sc.

PLATE XII.

Fig. 156. The middle arch of Black Friars Bridge. P. 164.

Fig. 157. A spherical dome, of which the lower parts are made thicker, in order that they may be of equal stability throughout. From A to B the dome is of equable thickness: below C and D the thickness cannot be increased sufficiently to procure an equilibrium, without the application of a chain or hoop, of which the section is represented at C, D. If the thickness were not at all increased, a hoop would be required at E, F, or still higher. P. 165.

Fig. 158. A section of the roof of St. Paul's Cathedral. The section of the dome consists of two circular arcs, of which the centres are a little beyond the axis: it is supported by carpentry, resting on a cone of brickwork. The internal dome is of brickwork only, and is open at the summit. P. 165.

Fig. 159. A section of the dome of the Pantheon at Rome. P. 165.

Fig. 160. A Tuscan column, with its pedestal, capital, and entablature. P. 165.

Fig. 161. A Doric column. P. 165.

Fig. 162. An Ionic column. P. 165.

Fig. 163. A Corinthian column. P. 165.

Fig. 164. A Composite column. P. 165.

Fig. 165. An elevation of the end of King's College Chapel, Cambridge; showing on one side the buttresses, the tower being supposed to be removed, and on the other the tower, which not only supplies the place of a buttress at the end, but assists also in supporting a considerable portion of the throst in the direction or the length of the chapel; the roof, which is of stone, being vaulted in this direction as well as tranversely. There is also a roof of carpentry, covered with lead above the stone roof. P. 166.

PLATE XIII.

Fig. 166. Joints for a tie beam. The joints at A and B cannot be more than half as strong as the entire beam, supposing the adhesion, produced by the pressure of the bolts, as strong as could be required. The joint at C is called a dovetail joint; its strength is a little less than that of A and B, but the adhesion is more easily secured, since a force tending to separate the beams must tighten the joint. P. 167.

Fig. 167. Joints for a tie beam. The joint A, if sufficiently tight, may possess $\frac{5}{7}$ of the strength of the beam. The joint B might be as strong as the beam, if the adhesion were great enough, but it would be difficult to apply sufficient pressure to create such an adhesion, and if the beam were subject to be much shaken, the joint would be a very bad one. P. 167.

Fig. 168. A good joint for a tie beam; the adhesion being secured by a slight diminution of the strength. P. 167.

Fig. 169. A, a simple scarfed joint, which may be tightened by a wedge at the centre; it is not strong. B, a scarfed joint which is much stronger. P. 167.

Fig. 170. A joint for a beam supporting a weight by its transverse strength. The junction might be made, if it were necessary, by means of a third piece, of which the limits are marked by the dotted line. The strength is but little diminished by the joint. P. 168.

Fig. 171. A beam supporting a weight by its transverse strength, joined to another by means of a third piece of half the depth, spliced or fished on, below the beam, and secured by pins, and by blocks or joggles. The strength is a little greater than that of the original beam. The dotted lines show the proportion in which the strata are extended or compressed, the lower part of the original beam remaining in its natural state, without sustaining any pressure, as far as one fourth of the depth, and a little further. P. 168.

Fig. 172. A joint for a beam pressing obliquely against another. The dotted lines show the form of the tenon, which may occupy a considerable part of the breadth of the beam. The upper strap, A, is in the most usual situation, but the lower one, B, appears to afford greater strength, as it presses the beams more closely together, yet without any danger of crippling them; besides the advantage of having a firmer hold of the lower beam. P. 169.

Fig. 173. A joint for a horizontal beam suspended from a vertical one: the end of the tenon being dilated by wedges, and the whole secured by a strong strap. The tenon ought not to be wide, since it diminishes the strength of the horizontal beam. P. 169.

Fig. 174. The straps, bent so as to deviate from the right lines joining their extremities in the degree that is here represented, have their strength reduced to about one seventh of that which they would have if straight. Thus, A B is only one seventh as strong as C D, supposing the substance inflexible. P. 169.

Fig. 175. The simplest form of a roof. A B, A C, are the rafters, and B C the tie beam; the weight of each half being represented by A B, or A C, the thrust in the direction of the rafters will be A D, and the horizontal thrust each way B D or C D. It is obvious that A D will be least when B A C is a right angle. P. 170.

Fig. 176. A common roof, with braces. A B is the king post, and B C, B D the braces. P. 170.

Fig. 177. A kirb or mansard roof, the rafters of which hold each other in equilibrium. A B and C D are queen posts helping to support the tie beam. The piece A C acts as a strut, in supporting the pressure occasioned by the weight of the tie beam. The heads of the queen posts are not much thickened, in order to avoid the change arising from the unequal contraction of the wood. P. 170.

Fig. 166.

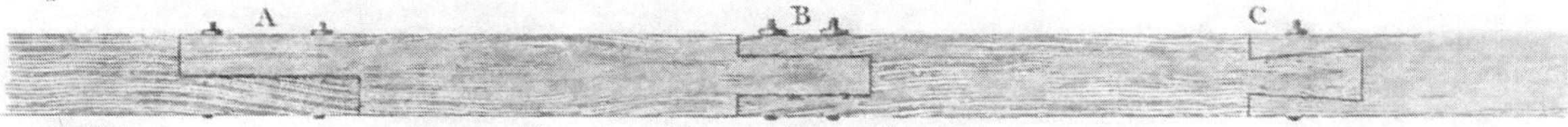

Fig. 167.

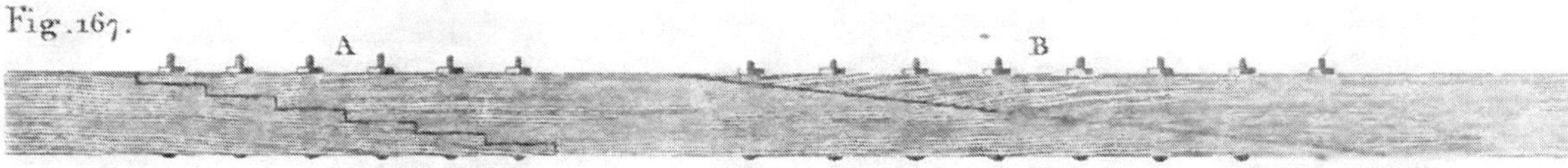

Fig. 168.

Fig. 169.

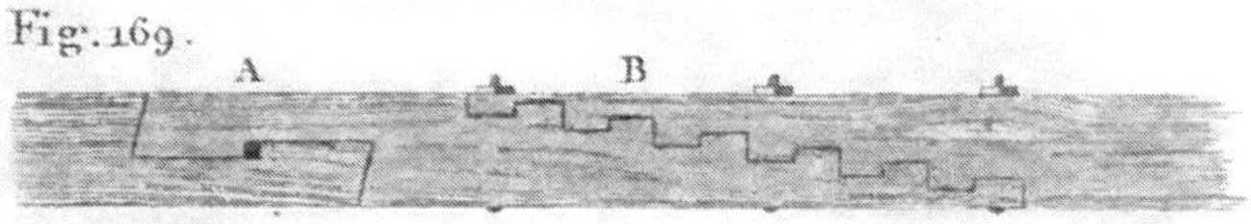

Fig. 170.

Fig. 171.

Fig. 172.

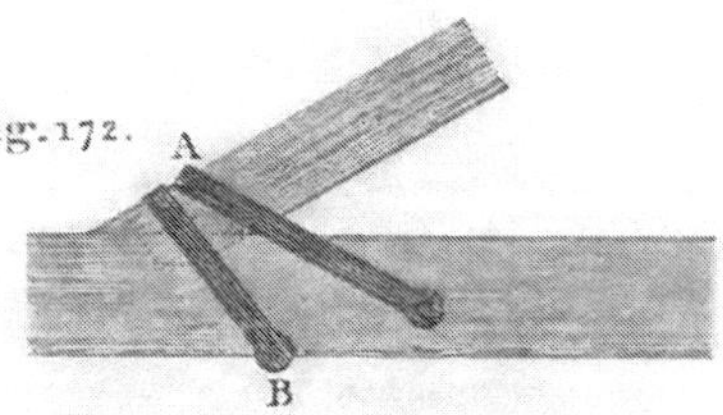

Fig. 173.

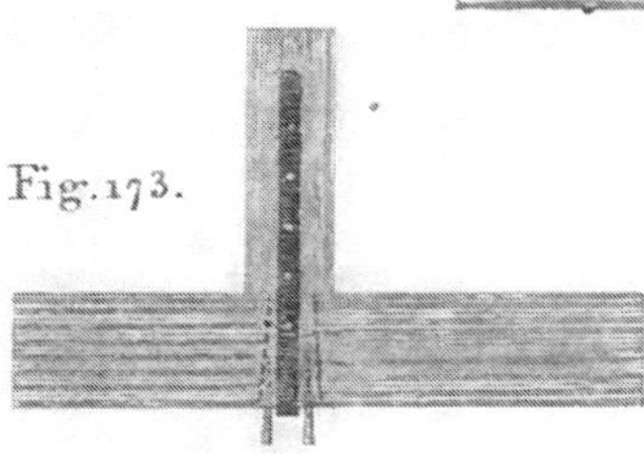

Fig. 174.

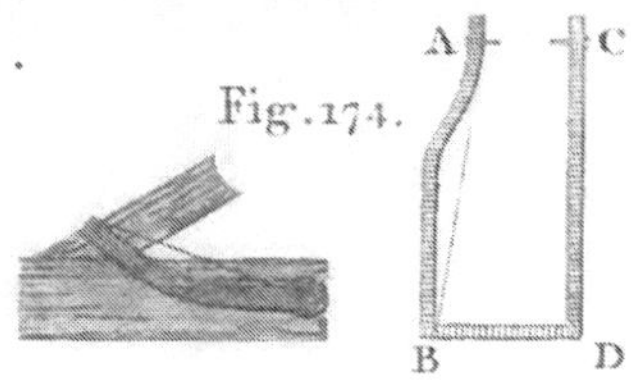

Fig. 175.

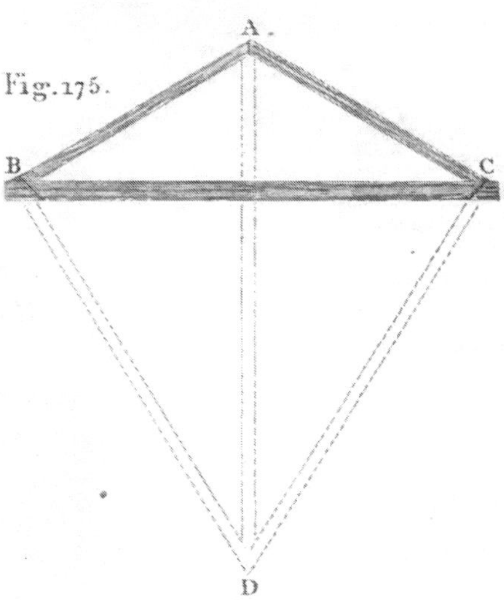

Fig. 176.

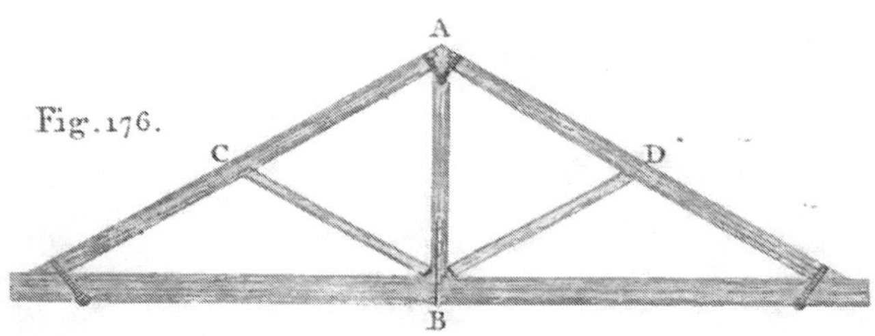

Fig. 177.

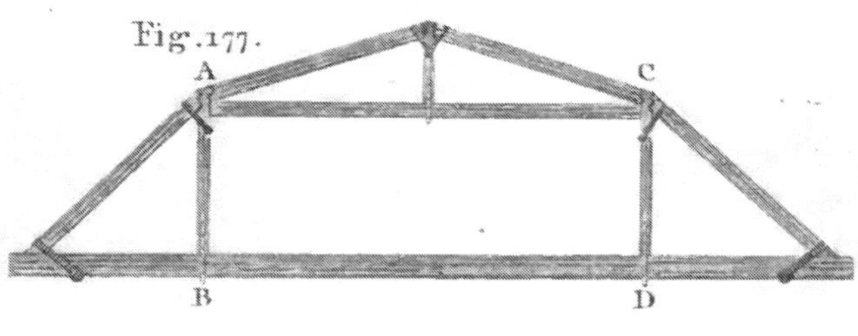

Pub. by J. Johnson, London 1 July 1806.

Joseph Skelton sculp

PLATE XIV.

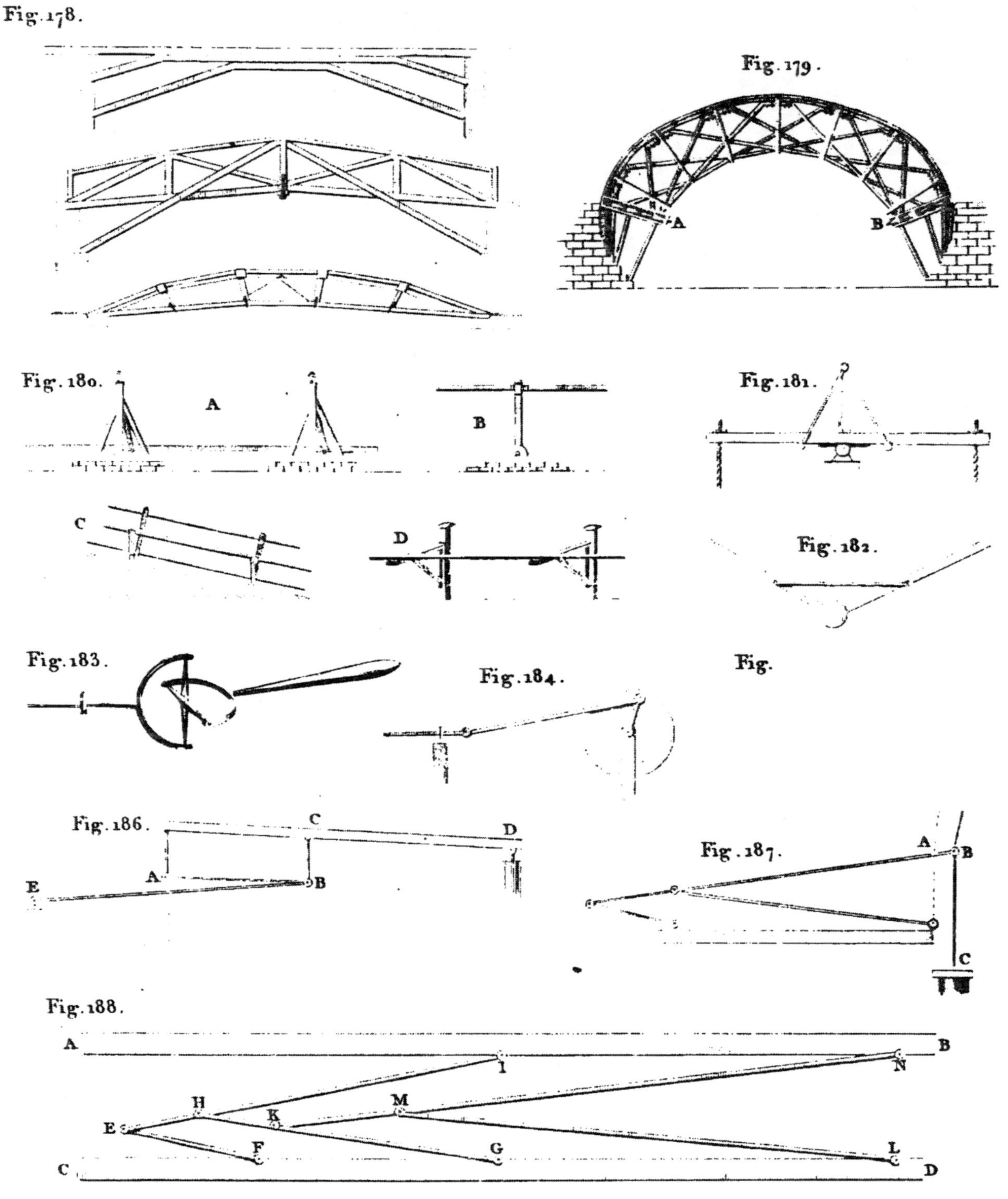

Pub. by J. Johnson, London 1 July 1806. *Joseph Skelton sculp.*

PLATE XIV.

Fig. 178. Three sketches for wooden bridges; the last requires no abutments. P. 171.

Fig. 179. The centring used for building one of the arches of Black Friars Bridge. It was struck, or removed, by forcing back the compound wedges A, B, by the impulse of a battering ram. P. 171.

Fig. 180. Modes of supporting a series of rods, for communicating alternate motion. A is the best and most common method, the rods being suspended from a centre above them: at B the centre of motion is below the rods. Where there is a declivity, the arrangement at C may be useful. The mode shown at D is also recommended in some cases. P. 173.

Fig. 181. A lever strengthened by a projecting frame. P. 173.

Fig. 182. A bent lever strengthened by a cross bar. P. 173.

Fig. 183. Hooke's universal joint. P. 173.

Fig. 184. A wheel with a crank, for producing alternate motion in a rod. P. 174, 336.

Fig. 185. A wheel with an inclined and undulated surface, for producing alternate motion in a rod, with the interposition of a friction wheel. P. 175, 336.

Fig. 186. A frame for guiding the motion of a point A in a direction nearly rectilinear, A B being to C D as C D to B E. The dotted line shows the path of the point A. P. 175, 336.

Fig. 187. A frame for producing a motion nearly rectilinear in the point A. It may be applied to a pump rod BC, worked by a crank, or otherwise. P. 175, 336.

Fig. 188. A compound frame, for keeping two rods A B, C D, in a direction very nearly parallel. E F is 36 parts of the scale, F G, 64, G H and H I each 80, E H and H K 20, G L $\frac{64}{36}$ G K, or $106\frac{2}{3}$, K M $33\frac{1}{3}$, and L M and M N each $133\frac{1}{3}$. P. 175.

PLATE XV.

Fig. 189. The form of a wheel or pulley, on which a broad strap runs, the surface being convex: the wheel which drives it is of a similar form, but its upper part only is shown in the figure. P. 175.

Fig. 190. The teeth of two wheels, formed into epicycloidal curves, acting on planes: the dotted lines show the effective magnitude of the wheels. P. 176.

Fig. 191. The teeth of two wheels, formed into involutes of circles, described by uncoiling a thread from the dotted circles; the point of contact of the teeth being always in the straight line which touches both circles. P. 176.

Fig. 192. Two surfaces formed into involutes of circles, revolving in contact with each other, the equidistant lines, drawn on them, continuing to meet each other throughout the revolution. P. 176.

Fig. 193. The pinion A is of the kind called a spur wheel; B is a crown wheel, or a contrate wheel. P. 177.

Fig. 194. The wheel and pinion are both bevilled: the faces of the teeth being directed to the point A. P. 177.

Fig. 195. Two wheels a little eccentric, acting on each other. P. 178.

Fig. 196. An eccentric contrate wheel, acting on a long pinion. P. 178

Fig. 197. A machine for cutting the teeth of wheels. A is the wheel, of which the teeth are formed by the revolving saw B, turned by the wheel and pinion C, D, by means of the handle E, while the frame, which holds the saw, moving on hinges, and resting on a spring, is depressed by the handle F, its place having been previously adjusted by the screw G. The large plate HI contains a number of concentric circles, variously divided by points, into which the end of the spring I sinks at each step, so as to fix the apparatus in the required position. P. 178.

Fig. 198. A chronometer for measuring minute portions of time. The axis A B being turned, either by the handle A or by the weight C, the balls D, E fly out, and carry the weights F, G further from the axis; in consequence of which the increased effect of friction retards the motion, when it becomes too rapid. The barrel H is turned in the mean time, with the axis, and is allowed to descend as the thread at I is uncoiled, so that the point K, which is pressed against it by a spring, describes on it a spiral, which is interrupted whenever the pin K is touched. P. 191.

Fig. 199. The fusee of a watch or clock, the general outline of which forms part of the hyperbola A B, in which the distance of each point from the axis C D is inversely as its distance from the line D E. P. 192.

1

PLATE XV.

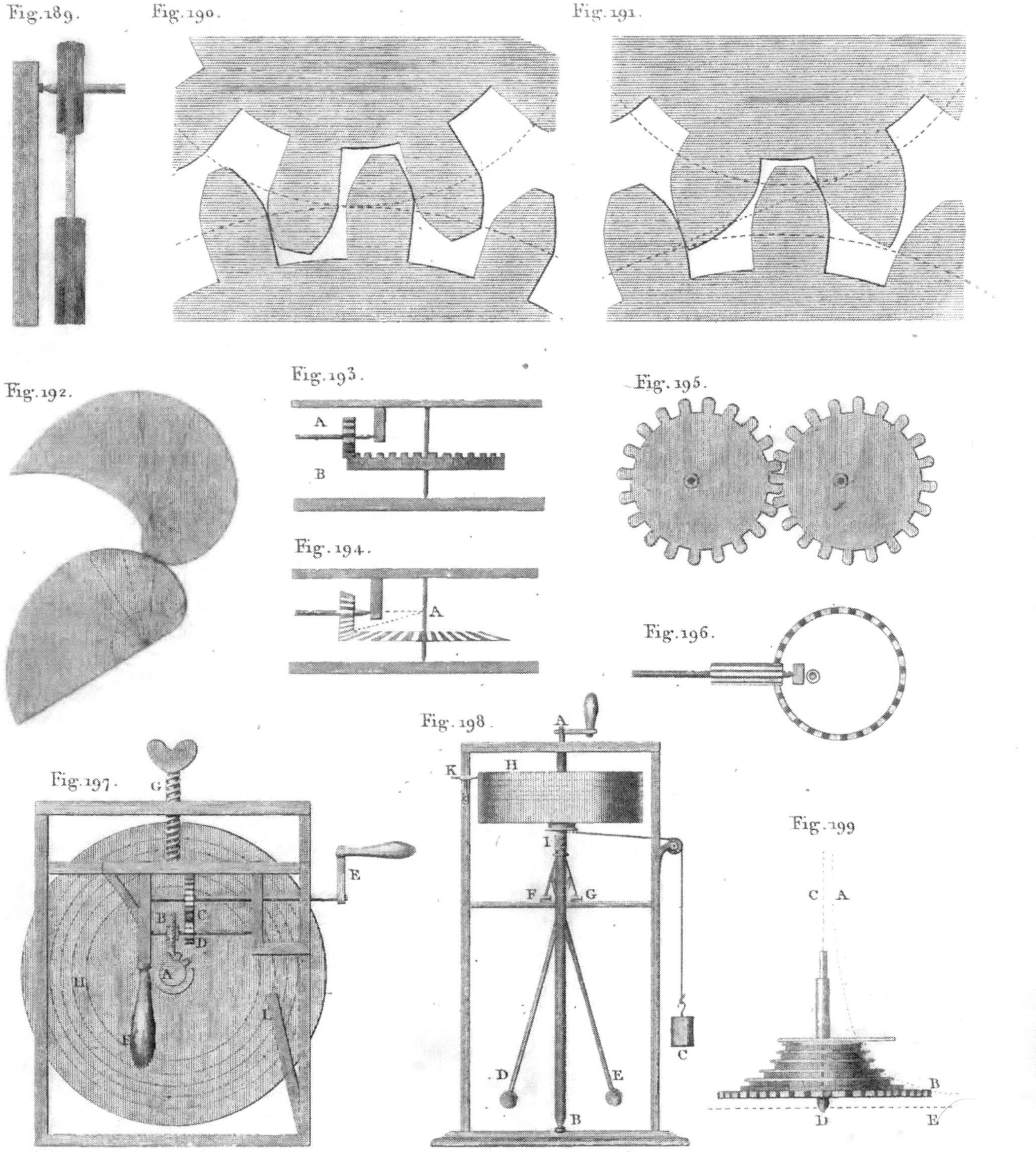

Pub. by J. Johnson, London 2 July 1806.

Joseph Skelton sculp.

PLATE XVI.

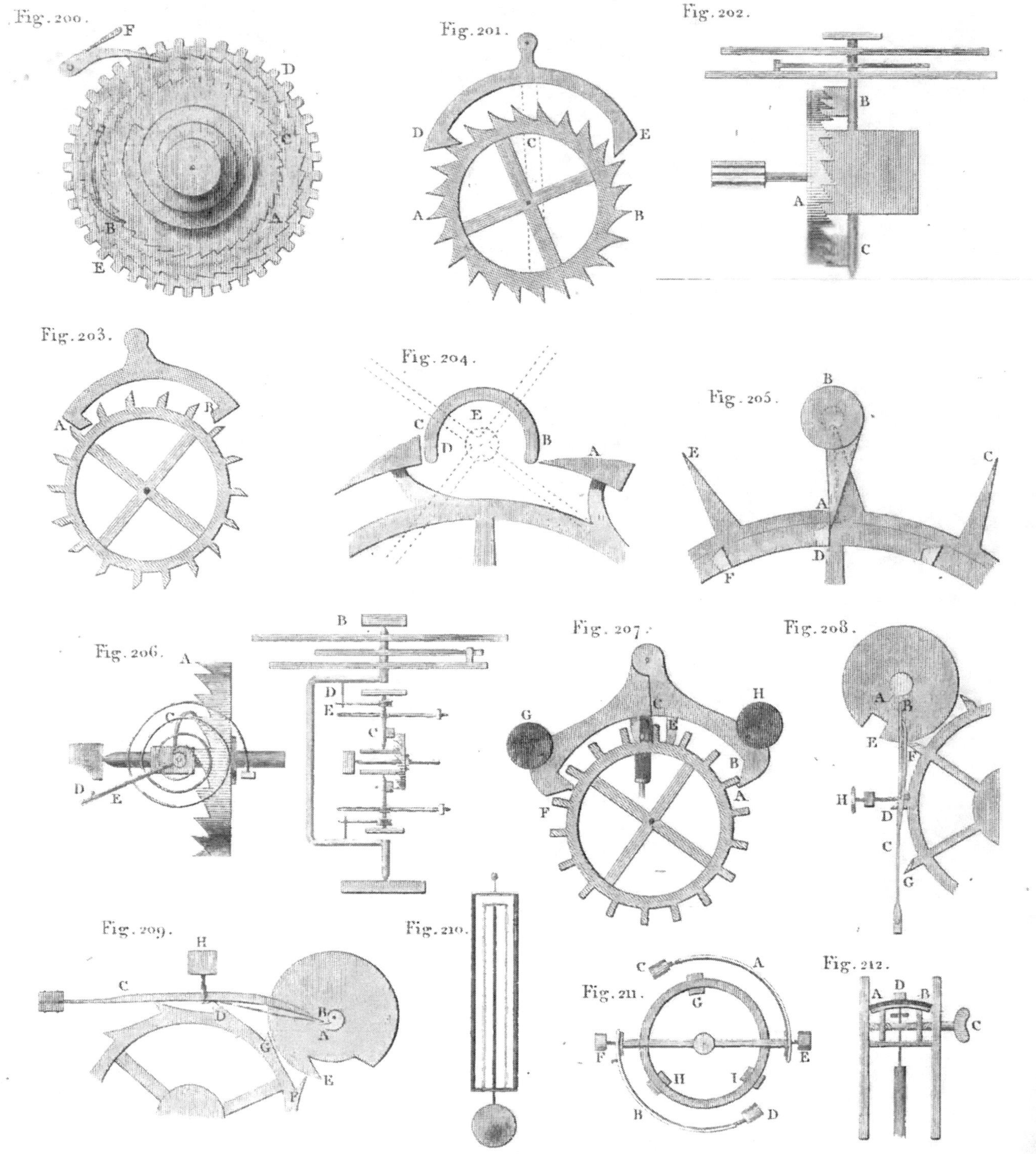

Pub. by J. Johnson, London 1 July 1806.

Joseph Skelton sculp.

PLATE XVI.

Fig. 200. A fusee with an auxiliary spring, for continuing the motion when the watch is wound up. The action of the main spring turns the fusee in the direction A B; the fusee acts on the ratchet wheel A B C by means of the click B, and this wheel impels the toothed wheel D E by the spring C B A, which is supposed to be seen through it. When the watch is wound up, this spring forces back the wheel A B C against the click F, which serves as a fixed point, while the other end continues to act on D E, and to maintain the motion. P. 193.

Fig. 201. The scape wheel A B, moving in the direction A C B, impels the pallets D, E of the crutch or anchor, alternately in contrary directions. P. 194.

Fig. 202. A is the scape wheel, B and C the pallets of the common watch scapement. P. 194.

Fig. 203. The dead beat scapement. The teeth are first received on the flat or rather cylindrical surfaces A, B, on which they rest until the pendulum arrives near the middle of its vibration, when the teeth begin to act on the inclined surfaces terminating the pallets. P. 195.

Fig. 204. The horizontal scapement, for a watch, The tooth A rests first on the external surface of the cylinder, B C, and then impels it by its inclined face, in the direction B C; it afterwards falls on the concave surface D E, and lastly impels the cylinder in the contrary direction. P. 195.

Fig. 205. The duplex scapement. A B is the pallet, through which the cylinder, and the tooth which rests on it, are supposed to be seen, the point of the tooth being about to escape from the notch towards C. The short tooth D next impels the point of the pallet, and the long tooth E falls on the cylinder. It first rests on the convex surface, and then drops into the notch, which causes a slight recoil in the wheel, and passes by, the tooth F being beyond the reach of the pallet; but on its return, the tooth falls again into the notch; and when it escapes, the pallet is impelled as before. P. 196.

Fig. 206. Mr. Mudge's watch scapement. A, the scapewheel, and one of the subsidiary springs, seen from above; B a general view of the balance, with both the subsidiary springs, seen from one side. The point of one of the teeth rests at C on the end of the pallet, which is bent so as to detain it until the pin D, which is attached to the balance, sets it at liberty, by striking against the arm E: this arm is then carried on by the balance, to the end of its vibration, and impels it in its return, until the pallet meets the next tooth. The other spring acts alternately in the same manner, but in a contrary direction. P. 197.

Fig. 207. An improvement on Mr. Cumming's scapement for a clock. The tooth A is seen resting on a flat surface at the end of the pallet B: it is disengaged by the descent of the opposite pallet into the position in which it is represented, the pallet B being impelled by it at C. This pallet continues resting on the flat end of the tooth, until the pin D of the pendulum strikes against the arm E, which is carried before it, and impels the pendulum in its descent, until the pallet B acquires the situation in which the opposite pallet is represented, and sets that pallet at liberty from the tooth E, which has raised it. The situation and magnitude of the weights G, H, may be adjusted at pleasure. P. 197.

Fig. 208. Mr. Arnold's watch scapement. The pin A, projecting from the verge or axis of the balance, moving towards B, carries before it the spring B, and with it the stiffer spring C, so as to set at liberty the tooth D, which rests on a pallet projecting from the spring. The angle E of the principal pallet has then just passed the tooth F, and is impelled by it until the tooth G arrives at the detent. In the return of the balance, the pin A passes easily by the detent, by forcing back the spring B. The screw H serves to adjust the position of the detent, which presses against it. P. 197.

Fig. 209. Mr. Earnshaw's scapement. A is the unlocking pallet, B the spring on which it acts, C the detent, holding the tooth D by a pin; E is the point of the principal pallet first impelled by the tooth F, G is the tooth next locked, and H the adjusting screw. P. 197.

Fig. 210. A gridiron pendulum, consisting of three bars of iron, and two of a mixture of zinc and silver. P. 200.

Fig. 211. A compensation balance, as employed by Arnold. The outside of the hoops A, B is of brass, the inside of steel: the weights C, D are screwed backwards and forwards, in order to obtain the requisite degree of compensation. The weights E, F, are employed to regulate the mean rate of the watch, and G, H, and I, for adjusting it to all positions with respect to the horizon. P. 201.

Fig. 212. The compound plate A B rests on two supports, which are adjusted to a proper distance by turning the double screw C, the flexure of the plate by heat raising the bar D, which supports the pendulum, while its effective length is determined by a fixed clip, which is seen below the plate. P. 211.

PLATE XVII.

Fig. 213. A jack for raising weights by the alternate motions of a lever, the clicks on each side being detained in the teeth of the ratchets by the assistance of the springs in which they terminate, and which are connected together. P. 204.

Fig. 214. The mode of supporting a tackle for raising stones in building; the summit of the triangle, which is composed of three poles, being raised or lowered by means of a rope and pullies. P. 207.

Fig. 215. A method of raising weights obliquely, by means of a rope, passing over a pulley, which is drawn along horizontally. P. 207.

Fig. 216. A B, a section of an inclined plane, belonging to the Duke of Bridgwater's canal: the boats are drawn into the locks at A, which are then filled with water; C is the plan of the windlass, by which the descending and ascending boats are connected together, and which is turned by a winch; D and E are the locks. P. 208.

Fig. 217. A crane, with an oblique walking wheel, for oxen or horses. The wheel is taken from a mill of Leupold. P. 209.

Fig. 218. A crane with a wheel and break like Mr. White's. The man walks at any required distance from the axis of motion, and pushes forwards the lever A, which moves the bar B C, connected to the same axis, and removes the break C D from the circumference of the wheel. P. 210.

Fig. 219. A lewis, for raising stones. P. 210.

Fig. 220. When the centre of gravity A is twice as far from one of the porters B, as from the other C, the first bears one third of the weight, the other two thirds. P. 212.

Fig. 221. When the centre of gravity A is above the line joining the points of support B, C, the load is divided in the ratio of the segments C D, B D, terminated by the vertical line A D; but it may be supported by two equal forces in the directions B E, C F, found by making G H equal to B G, and joining C H; the angle G B E being equal to G H F; the forces and the weight may then be represented by the lines C I, I K, and C K. P. 212.

Fig. 222. A roller with two wheels fixed on its ends, by means of which the slab resting on it may be moved to a considerable distance without leaving the roller behind. P. 213.

Fig. 223. Mr. Garnet's rollers, for diminishing friction: their axes being loosely connected by a ring, in order to keep them in their places. P. 213.

PLATE XVII.

Fig. 213.

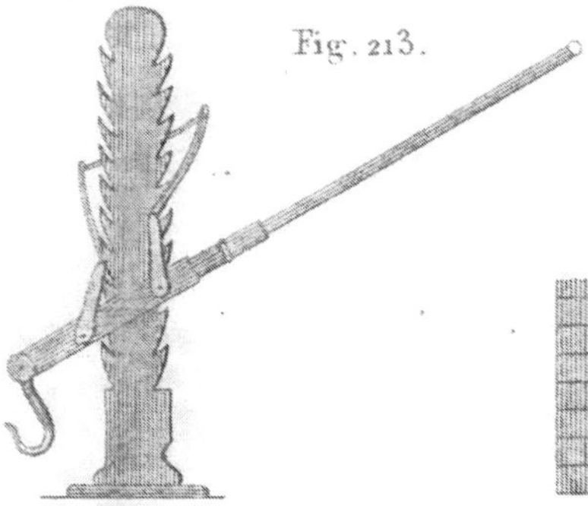

Fig. 214.

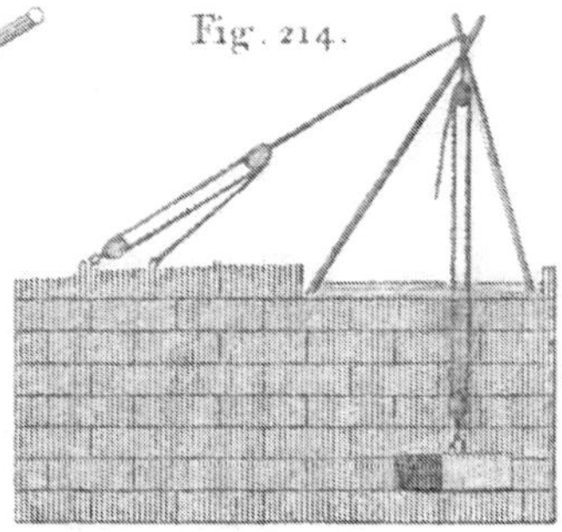

Fig. 215.

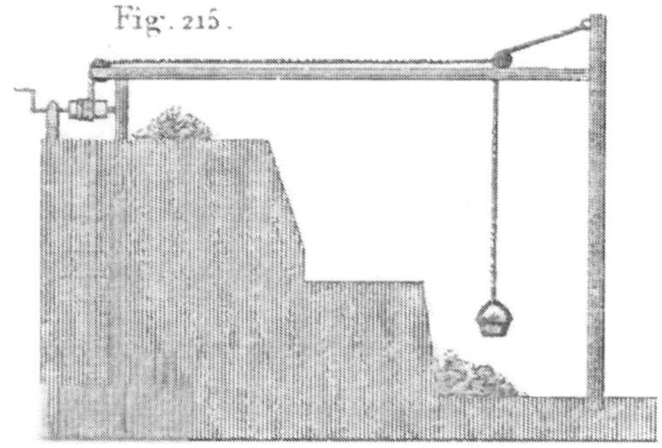

Fig. 216.

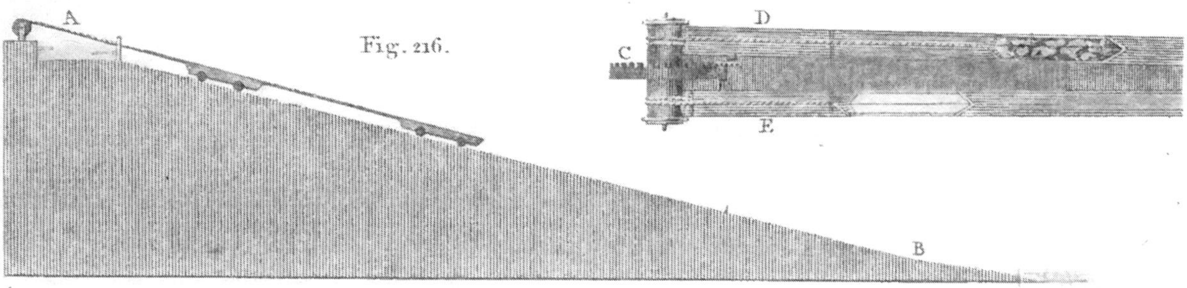

Fig. 217.

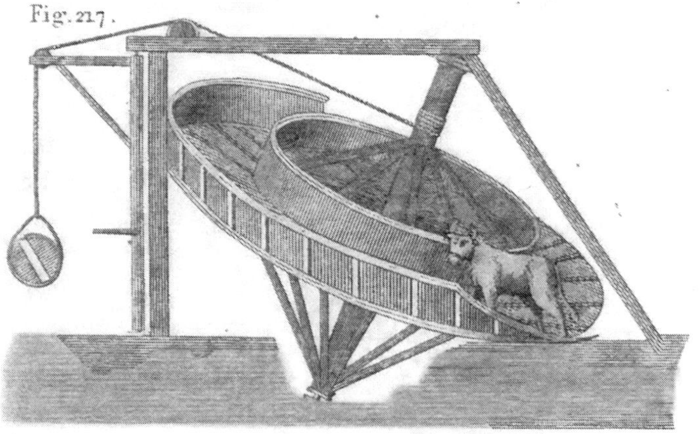

Fig. 218.

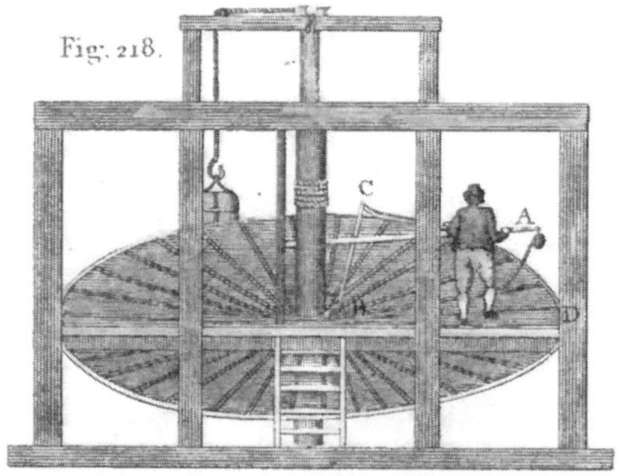

Fig. 219.

Fig. 220.

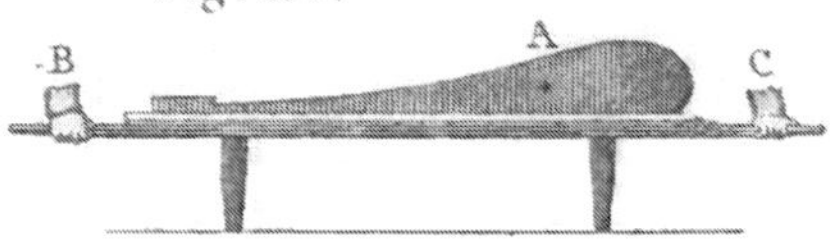

Fig. 221.

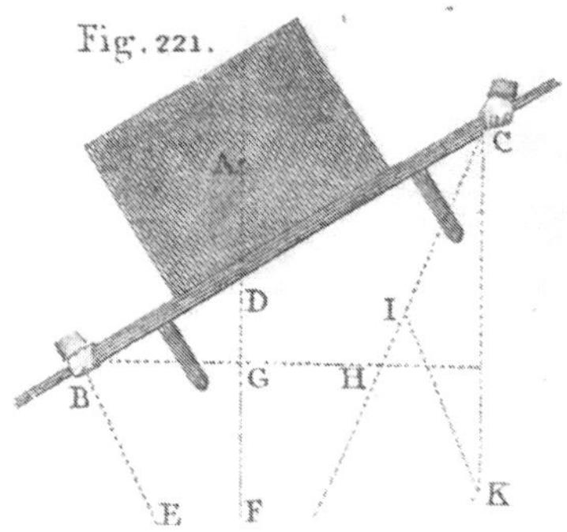

Fig. 222.

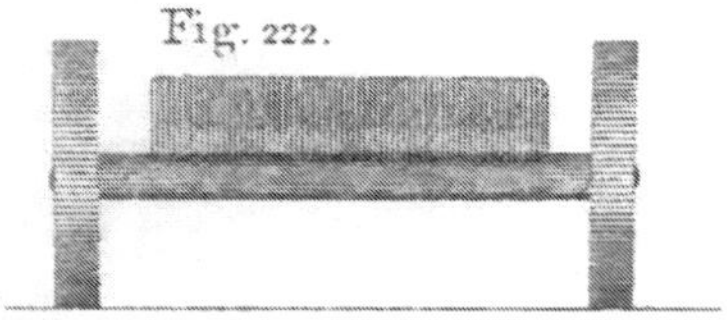

Fig. 223.

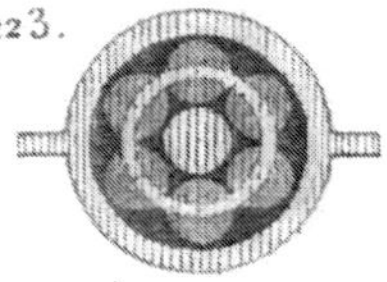

Pub. by J. Johnson, London 1 July 1806.

Joseph Skelton sculp.

PLATE XVIII.

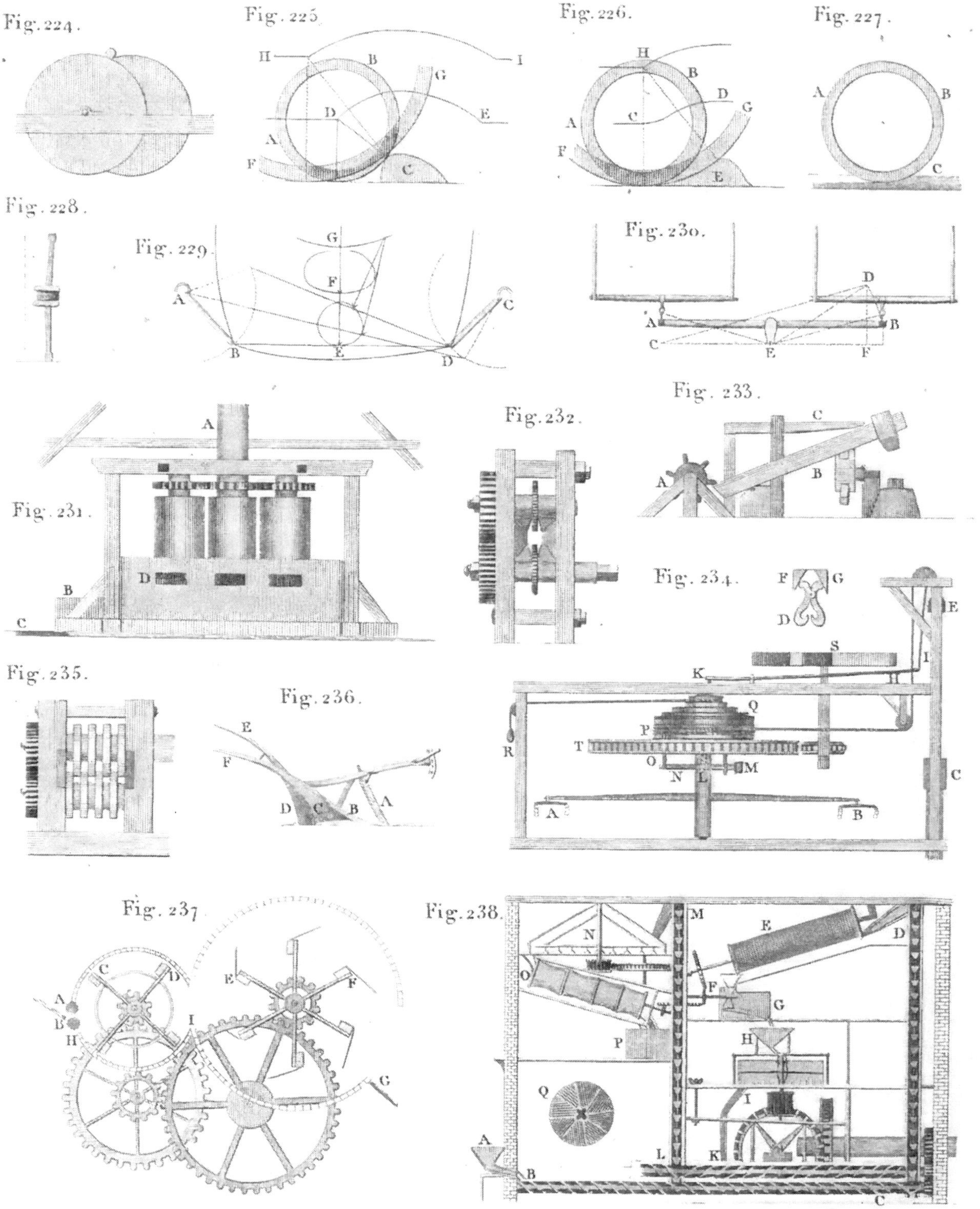

Pub. by J. Johnson, London 1 July 1806.

Joseph Skelton sc.

PLATE XVIII.

Fig. 224. A pair of friction wheels, supporting one end of the axis of a wheel. P. 214.

Fig. 225. The centre of the wheel A B, passing over the obstacle C, describes the path D E; that of the larger wheel F G, the path H I, which is less steep. P. 214.

Fig. 226. The centre of the wheel A B describes the curved path C D, in passing over the obstacle E, while that of the larger wheel F G has an angle at H. P. 214.

Fig. 227. The wheel A B, moving on a soft road towards B, has to overcome the resistance of the earth at C. P. 215.

Fig. 228. A section of the wheel of a carriage, a little dished, or inclined outwards. P. 217.

Fig. 229. A B and C D being the straps or braces by which a coach is suspended, if the centre of gravity be at E, F, or G, it must move, when the carriage swings, in the curve passing through the respective point. P. 218.

Fig. 230. The mode of harnessing two horses, so as to make them draw conveniently together: when either horse advances so far that the bar A B assumes the position C D, the foremost horse has the disadvantage of acting on a lever equivalent only to E F, while the other horse acts on E C. P. 218.

Fig. 231. A sugar mill. The axis A is turned either by animal force or by water: the liquor is collected in the trough B, and runs off in the channel C. The openings D are for the purpose of adjusting the axes of the rollers. The canes are supplied by the hands of the workmen. P. 221.

Fig. 232. A glazier's vice. The vacuity in the middle shows the form of the section of the lead which is drawn through it. P. 223.

Fig. 233. A forge hammer, elevated by the plugs, projecting from an axis, either at A, or, more conveniently, at B, and thrown forcibly against the wooden spring C. P. 224.

Fig. 234. An engine for driving piles, on Vauloué's construction. The horses, drawing at A, B, raise the weight C, held by the tongs D, fixed in the follower E, which are opened, when they reach the summit, by being pressed between the inclined planes F, G, so as to let the weight fall. At the same time the lever H is raised by the rope I, and presses on the pin K L, so as to depress the lever M N, and draw the pin O out of the drum P Q; the follower then descends, and uncoils the rope, its too rapid motion being prevented by the counterpoise R, acting on the spiral barrel Q. The motion is regulated by the fly S, the pinion of which is turned by the great wheel T. P. 226.

Fig. 235. The rollers of the slitting mill. P. 228.

Fig. 236. A simple plough. A is the coulter, for dividing the ground; B the share, fixed on the mould board C, for turning it to the right hand; D is the rest, and E, F, the handles. P. 229.

Fig. 237. Section of a threshing mill. The corn is drawn in by the rollers or feeders A, B: it is beaten by the revolving beaters C, D, and the straw is drawn out by the rakes E F, which discharge it at G; the grain falling through the arched bottoms H I, I G, which are formed like sieves. P. 233.

Fig. 238. A corn mill, with some of the improvements made in America, by Mr. Ellicott and Mr. Evans. The corn, being poured into the funnel A, is conveyed, by the revolutions of a spiral B C, to C, whence it is raised, by the chain of buckets C D, to be cleaned by the revolving sieve E, and the fan F; it is then deposited in the granary G, which supplies the funnel or mill hopper H; this being perpetually agitated by the iron axis of the upper mill stone, shakes it by degrees into the perforation of the stone; it escapes, when ground, at I, and is conveyed, by means of the carrier K L, and the elevator L M, to the cooler N, where it is spread on a large surface; it passes afterwards to the bolter O, and is received in the binn P, from whence it is taken to be packed in sacks or barrels. Q represents the surface of a mill stone, cut into furrows, in order to make it act more readily on the corn. P. 234.

PLATE XIX.

Fig. 239. The surfaces of the fluid in the bent tube A B remain on the same level, in the same manner as if the tube were absent, and the fluid made a part of that which is contained in the reservoir C D, P. 260.

Fig. 240. The bucket A being suspended by the rope B, and made to revolve rapidly round its axis, the surface of the water assumes a parabolic form. P. 261.

Fig. 241. A heavier fluid being contained in the upper part of the bent tube A B, which is immersed in the lighter fluid filling the vessel CD, the fluid in the tube remains in a state of tottering equilibrium, when its surfaces are in the same level. P. 261.

Fig. 242. The fluid A B C presses on the bottom of the vessel BC with the same force as if the vessel were of the form B C D E. P. 261.

Fig. 243. The portion A B C D of the fluid being supposed to be congealed, and then to form a part of the vessel, the pressure on the bottom would remain unaltered. P. 263.

Fig. 244. The weight A may be supported by the pressure of a small quantity of fluid, either by making the surface of the vessel BC very large, and the height of the tube D E moderate, or, while the vessel F remains of a moderate size, by making the height of the tube G H very great. P. 263.

Fig. 245. The pressure on any small part of the side of the vessel A B, at C or D may be represented by the line C E, D F, and the whole pressure on the side by the triangle B G, of which the centre of gravity is at H; and if the side A I be supported by a single prop, it must be placed at the point K, the height of which is equal to that of H. P. 265.

Fig. 246. If the height of the surface A above B be to B C as the specific gravity of the fluid in B C to that of the fluid in A B, the fluids will support each other. P. 265.

Fig. 247. Two square beams floating at the depths shown at A and B, will have a certain degree of stability, but if they sink, as at C, they will overset. But a beam of the breadth shown at D will always float securely. P. 267.

Fig. 248. A jar containing images of fishes, with bubbles of air in them, which sink when the cover of the jar is pressed with the hand. P. 268.

Fig. 249. Dr. Hooke's semicylindrical counterpoise, by means of which a vessel is kept always full. P. 268.

Fig. 250. The form into which the flexible bottom of a cistern would be bent by the pressure of the water: the curve is the same as that into which an elastic rod would be bent by forces acting at A and B. P. 269.

Fig. 251. The bottle A, containing air and mercury, has the tube A B fitted into it: and when the jar C D, in which it is enclosed, is exhausted by means of the air pump, the elasticity of the air in the bottle forces the mercury up the tube. P. 270.

Fig. 252. An instrument for showing the buoyant effect of the air, called by Boyle a statical baroscope; the index A shows, on the scale BC, the degree in which the ball D is obliged to descend, by the diminution of the weight of the air. P. 272.

Fig. 253. The line 0 denoting the natural density of the air, the line 1 A next above it shows the degree in which the air is expanded at the height of a mile, and 1 B the density of the air at the same height: in the same manner 10 C shows the expansion of the air at the height of 10 miles, and 10 D its density; and 5 E, below the line, the density which it would acquire at the depth of 5 miles below the earth's surface. The lines A C, D B E, are of the kind called logarithmic curves. P. 272.

Fig. 254. The box or bason, in which the mercury of the common barometer is contained: A is a float for adjusting the height, by means of the screw B, operating on the leather which forms the bottom of the cavity. P. 276.

PLATE XIX.

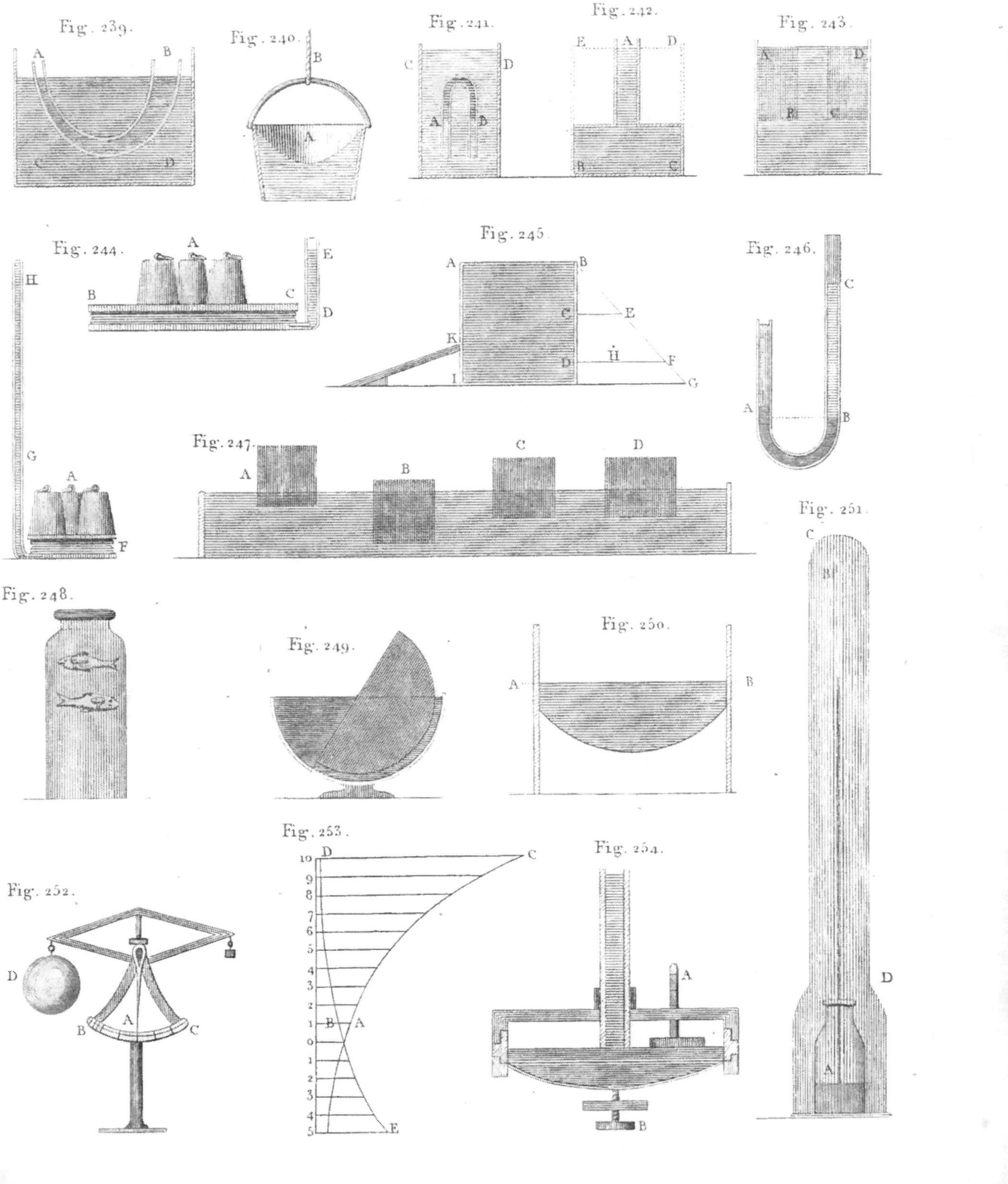

Pub. by J. Johnson, London 1 July 1806.

Joseph Skelton sculp.

Fig. 255.

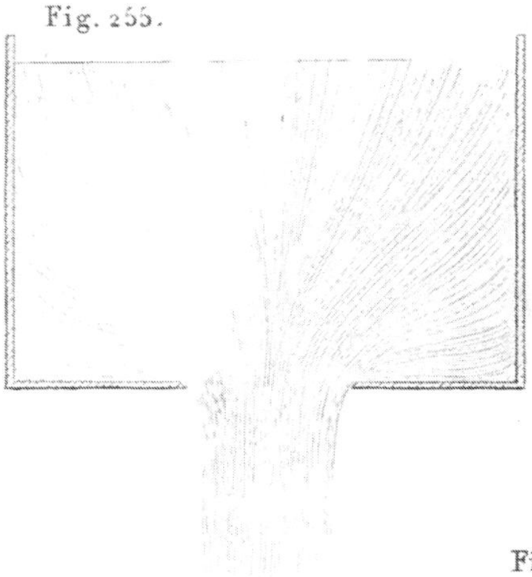

Fig. 256.

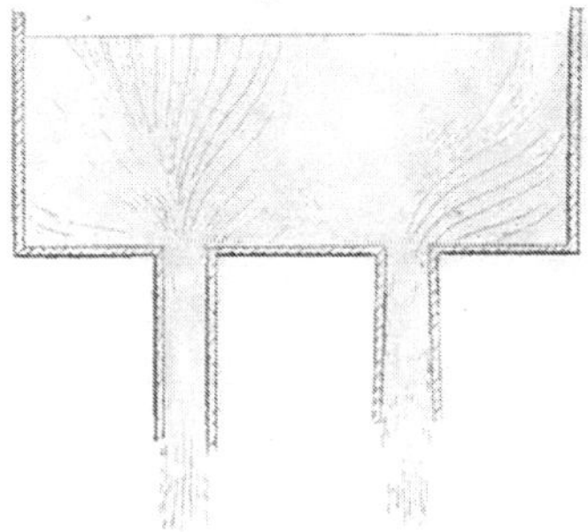

Fig. 257.

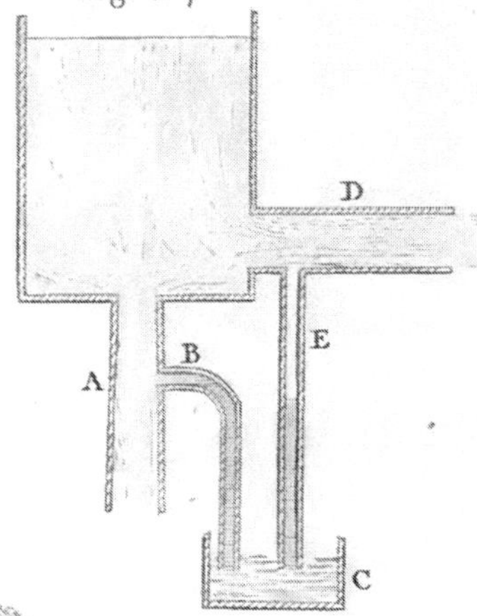

Fig. 260.

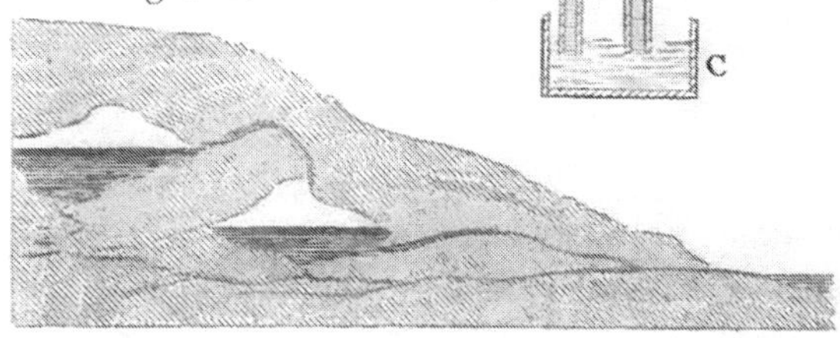

Fig. 259.

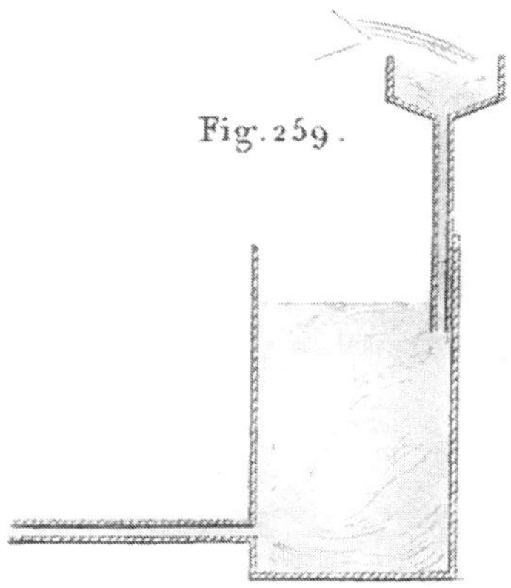

Fig. 258.

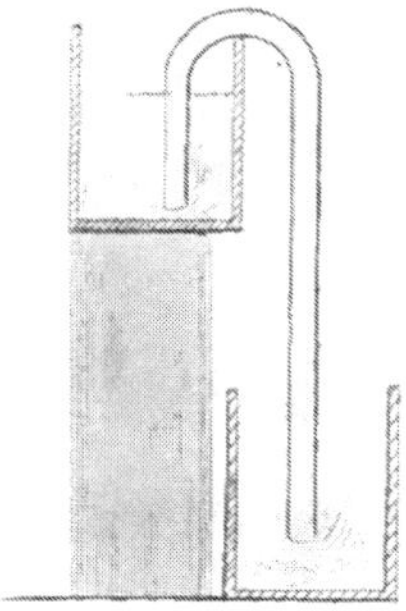

Fig. 263.

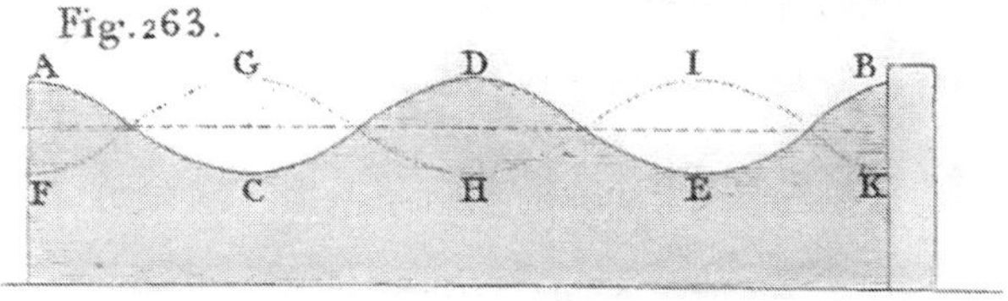

Fig. 261.

Fig. 262.

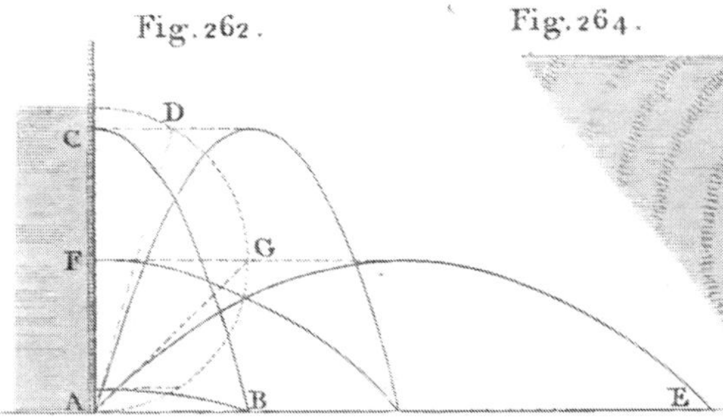

Fig. 264.

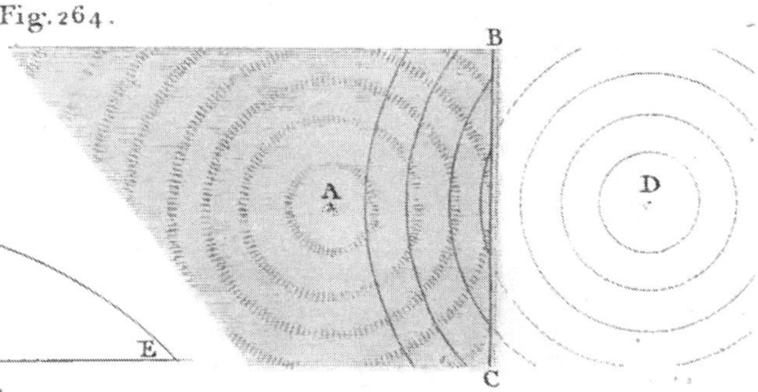

Fig. 265.

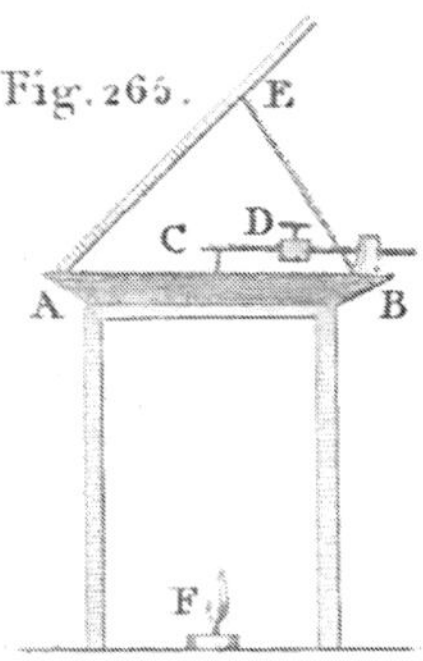

Fig. 266.

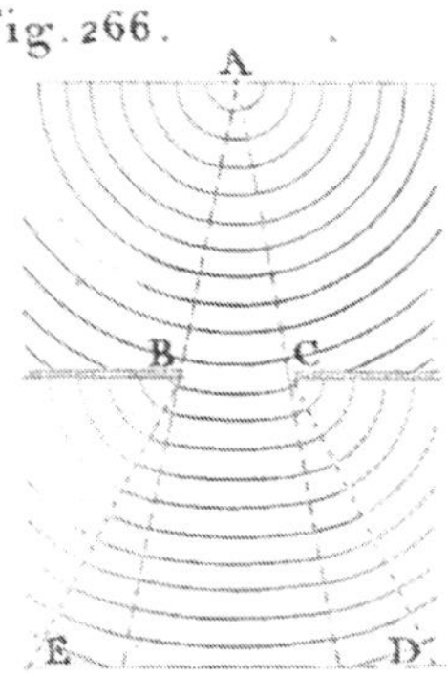

Fig. 267.

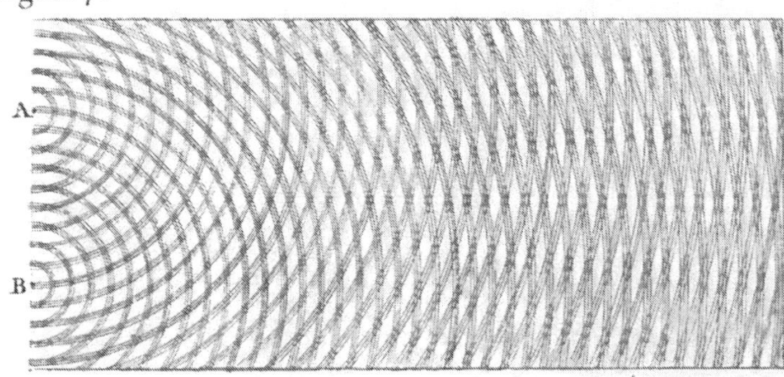

Pub. by J. Johnson London 2 July 1806.

Joseph Skel

PLATE XX.

Fig. 255. A jet or vein of a fluid, passing through an orifice in a thin plate in any direction, and contracted after its escape, in consequence of the lateral motions of the particles which flow towards the stream, nearly in the directions of the lines here drawn. P. 280.

Fig. 256. A stream flowing through a short cylindrical pipe, compared with another flowing through a diverging conical pipe, the directions of the motions of the particles appearing to be nearly similar in both cases. P. 281.

Fig. 257. In an experiment of D. Bernoulli, the water flowing through the conical pipe A drew up water through the tube B from the vessel C; in another of Venturi, the water flowing through the cylindrical pipe D raised water through the tube E. P. 281.

Fig. 258. A siphon, through which a fluid runs from the higher vessel into the lower one. P. 283.

Fig. 259. A fluid flowing through a vertical pipe, and filling a vessel to a height nearly equal to the length of the pipes, while it is discharged through a similar horizontal pipe. P. 284.

Fig. 260. Subterraneous cavities, with outlets in the form of siphons, through which they do not begin to discharge any water till they are nearly full; the lower one will then continue to run till it be empty. In the mean time either of them may keep up a constant stream by other passages. P. 285.

Fig. 261. A tube turned up and contracted, so as to throw out the fluid contained in it, in a jet, which rises very nearly to the height of the fluid in the tube. P. 286.

Fig. 262. The forms of jets issuing from various parts of a reservoir, the amplitude A B being twice C D, and A E four times F G. P. 286.

Fig. 263. A series of waves, moving in the direction A B, and reflected by the obstacle B, loses the appearance of progressive motion, and vibrates up and down within the limits of the curves A C D E B, and F G H I K; the elevation and depression become however twice as great as before reflection. P. 289.

Fig. 264. A series of waves diverging from a centre A, and striking a fixed obstacle B C, are reflected by it into the same form as if they proceeded from the centre D, at an equal distance on the opposite side of the surface B C. P. 289.

Fig. 265. An apparatus for observing the motions of waves excited, in a fluid poured into the trough A B, by the vibrations of the elastic wire C, loaded with a moveable weight D; the shadow of the waves being thrown on a screen E by the lamp F, through the bottom of the trough, which is of glass. P. 290.

Fig. 266. A series of waves, diverging from the centre A, and passing through the aperture B C, extend themselves on each side so as to fill the space B C D E, while they affect the parts without this space much less sensibly. P. 290, 458.

Fig. 267. Two equal series of waves, diverging from the centres A and B, and crossing each other in such a manner, that in the lines tending towards C, D, E, and F, they counteract each other's effects, and the water remains nearly smooth, while in the intermediate spaces it is agitated. P. 290, 461.

PLATE XXI.

Fig. 268. A stream of air being forced through the pipes A and B, the mercury in the barometer C D falls from C to D. P. 297.

Fig. 269. A stream entering the reservoir A, by the pipe B, carries with it all the water C, which stands above the level of its upper surface. P. 297.

Fig. 270. The ball A is permanently supported by the jet B, because, when it falls into the position here represented, the centrifugal force of the water at A carries it back to the middle of the jet. P. 298.

Fig. 271. A plate, bent into the form A B C, turning on the centre B, is impelled by a stream of air D in the direction C D. P. 298.

Fig. 272. A cylinder moveable on an axis, with two curved pipes inserted in its lower part, seen from above. The stream A enters at the top of the cylinder, and is discharged by the orifices B, C, so as to turn the vessel in the direction B D. P. 301.

Fig. 273. A jet of a fluid, striking on an obstacle of equal diameter, and separated by it so as to continue its motion obliquely. P. 302.

Fig. 274. The whole resistance directly opposed to the surface A B being represented by B C, the portion which, according to the principles of the resolution of forces, ought to act on the wedge A B D, is represented by B E; and in the same manner the resistance on A B F is to the whole as B G to B C. P. 303.

Fig. 275. The form of the dead water moving before an obtuse body is nearly like that of A B C; and the form adapted for moving through the water with the least possible resistance like A B D C. P. 304.

Fig. 276. The direction in which the particles of a fluid are supposed to move when they strike against a concave surface. P. 305.

Fig. 277. A hydrostatic balance. P. 309.

Fig. 278. Mr. Nicholson's hydrometer, to be employed with weights, for finding the specific gravity of fluids or solids. P. 309.

Fig. 279. A spirit level. P. 311.

Fig. 280. An overflowing lamp. The hemispherical counterpoise, which is so loaded, that its centre of gravity is at A, raises the surface of the heavy fluid B the higher as it is more exhausted, so that the oil C is always forced up nearly to the level of the wick at D. The oil is poured in by a pipe, in the middle of the cylindrical column. The air holes may be made wherever it is most convenient. P. 311.

Fig. 281. A section of an embankment, of a proper form to be opposed to the sea, with a drain passing through it, and a valve at its opening. P. 312.

Fig. 282. The form recommended for the section of a river or canal. P. 313.

Fig. 283. A B shows the strongest form for a vertical beam, fixed above and below, and calculated to resist the pressure of a fluid; the greatest thickness being at C; and D E is the outline of a series of horizontal planks, of such a thickness as to afford equal strength throughout the sluice or floodgate. P. 314.

Fig. 284. A box, with a valve supported by a hollow ball, for letting out air from pipes, when it is below the level of the reservoir. P. 316.

Fig. 285. Two methods of letting out air from pipes, when it is above the level of the reservoir; A a valve with a stopcock near it; B a vessel of water, screwed on for receiving the air; to be replenished with water as it becomes empty. P. 317.

Fig. 286. A section of a compound stopcock, which receives a fluid from either of the pipes A, B, or C, into a cavity which descends a little in the direction of the axis, and communicates with the pipe D, by means of one of the bores represented by dotted lines, according to the position into which the moveable cylinder is turned. P. 318.

Fig. 287. Valves of different kinds; A the common clack valve; B a double clack valve, consisting of two semicircular valves; C a pyramidical valve, consisting of four triangular pieces; D a circular valve turning on an axis; E, a steam valve of metal, sometimes called a T valve, F, a valve of oiled silk or bladder, supported by a grating, for air. P. 318.

PLATE XXI.

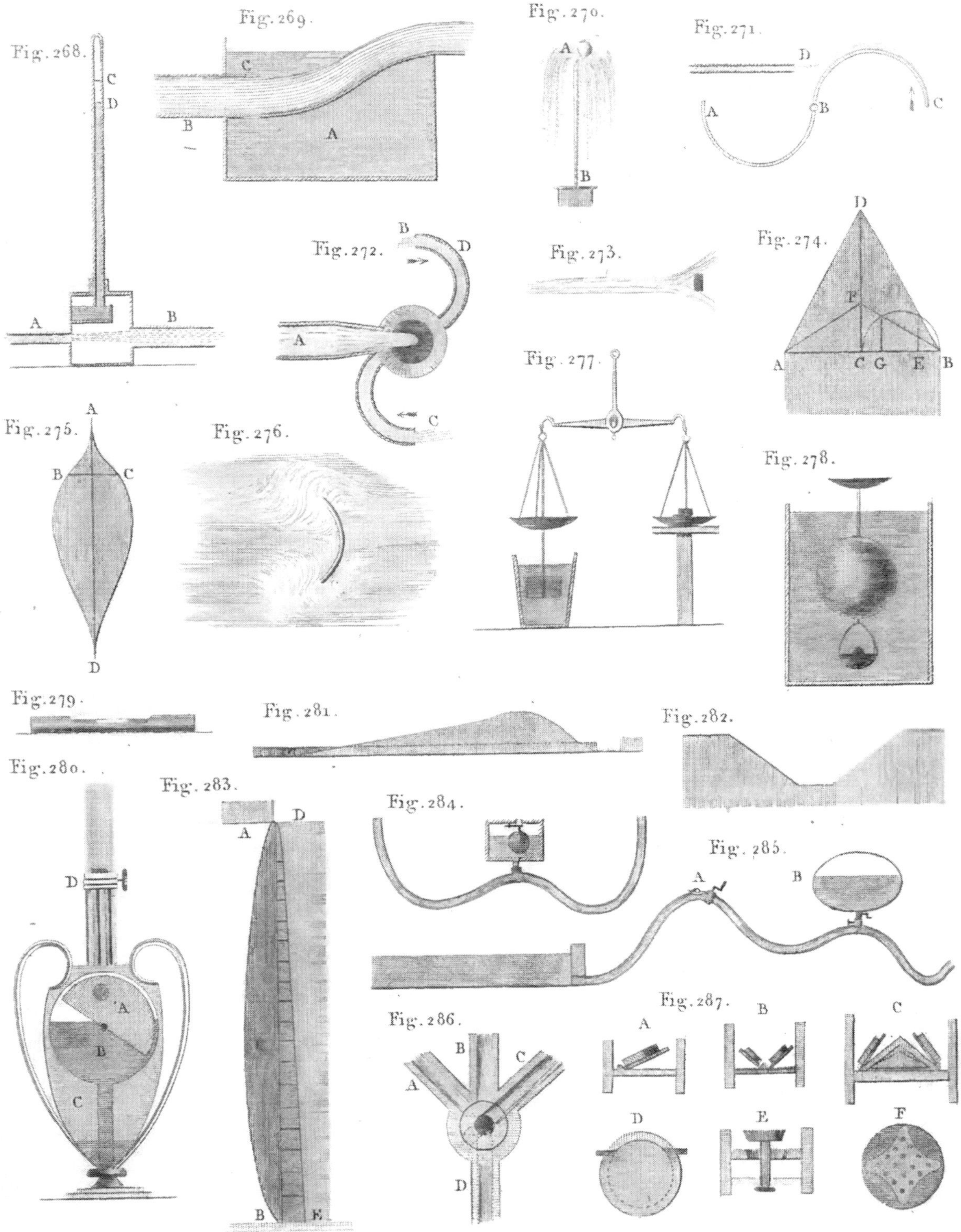

Pub. by J. Johnson, London 1 July 1806. *Joseph Skelton sculp.*

PLATE XXII.

Fig. 288.

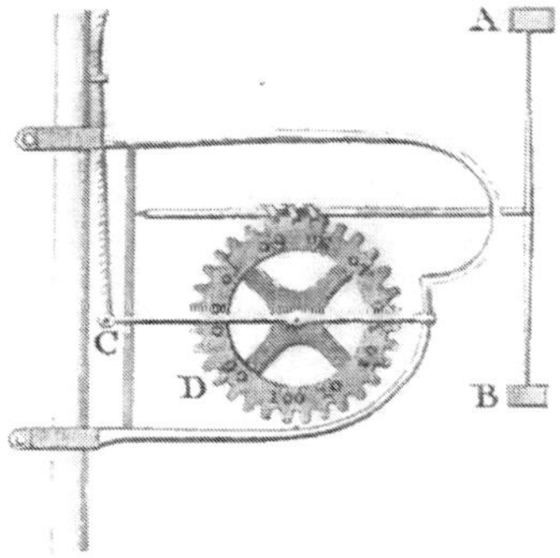

Fig. 289.

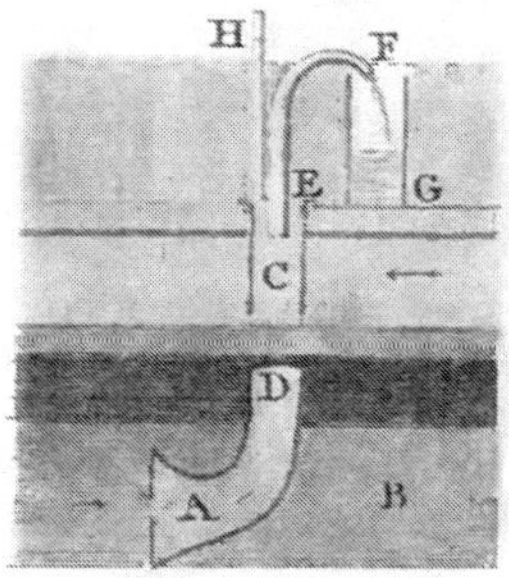

Fig. 290.

Fig. 291.

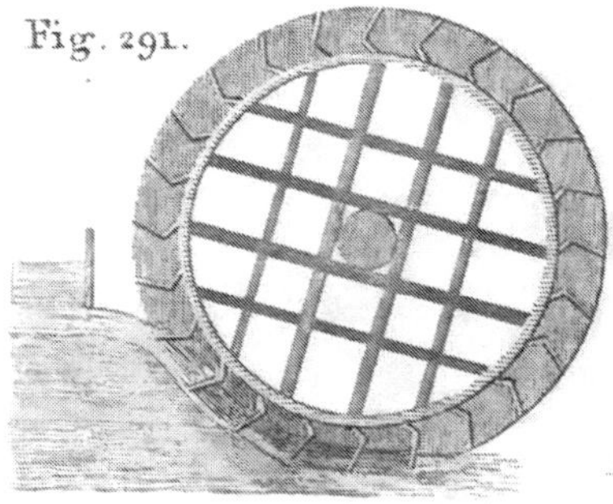

Fig. 292.

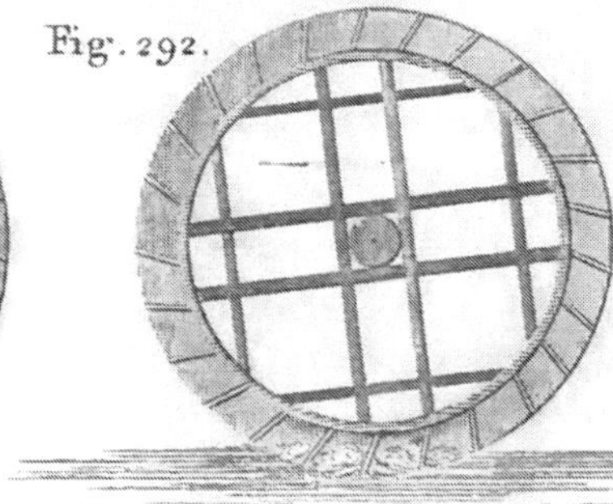

Fig. 293.

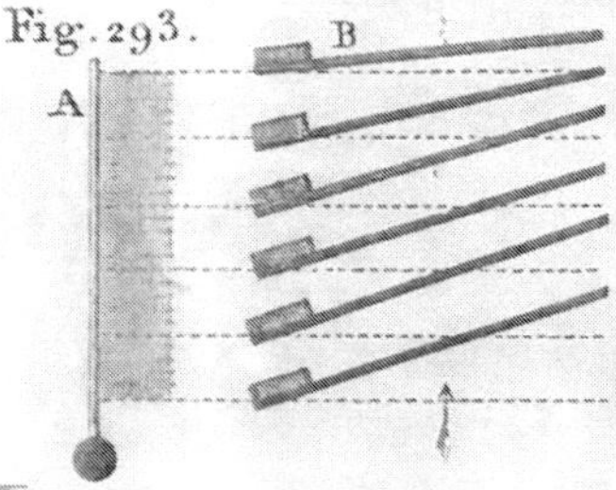

Fig. 294.

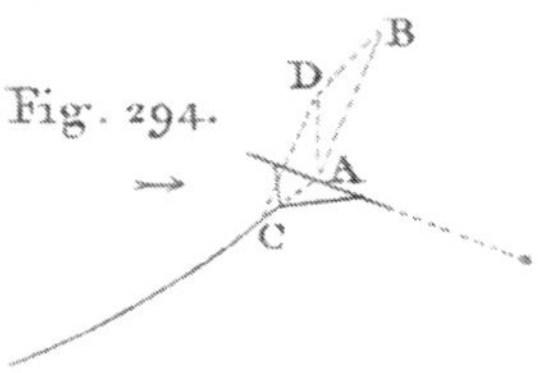

Fig. 295.

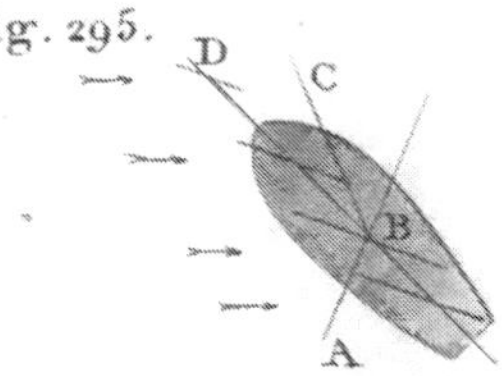

Fig. 296.

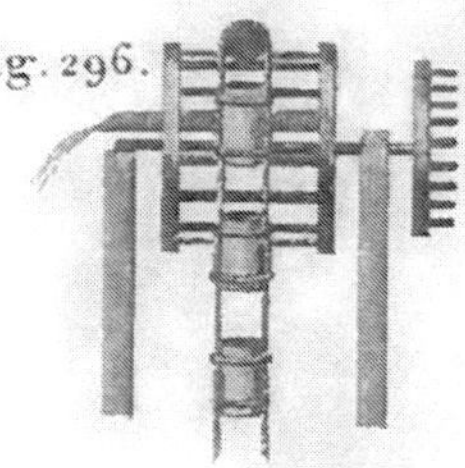

Fig. 299.

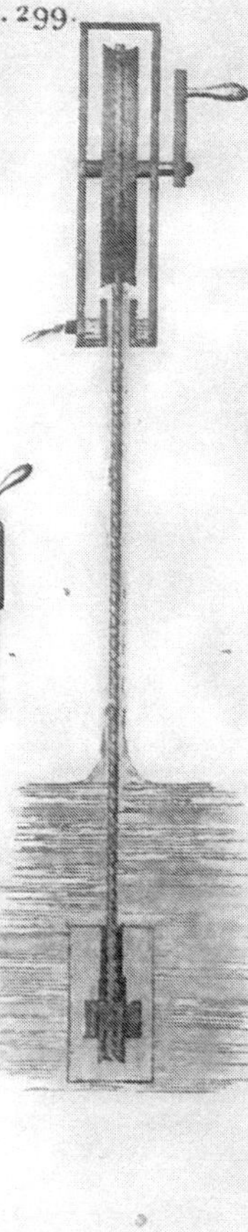

Fig. 297.

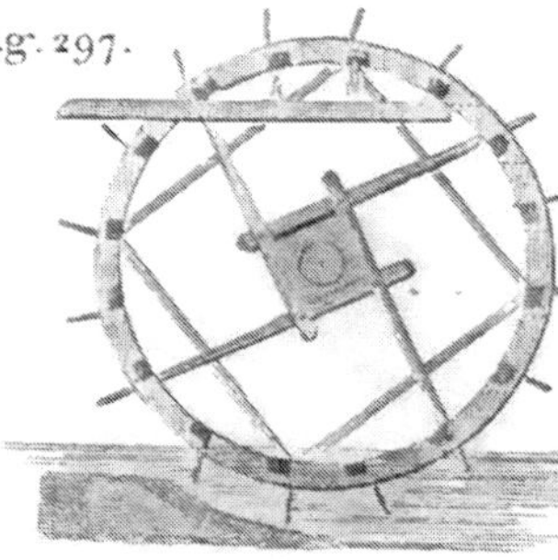

Fig. 298.

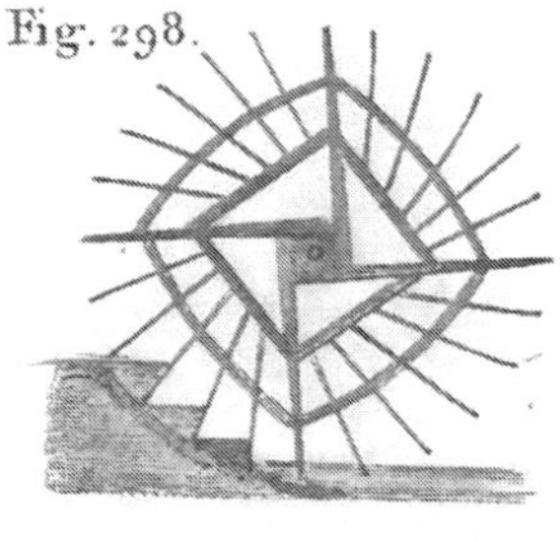

Fig. 300.

Fig. 303.

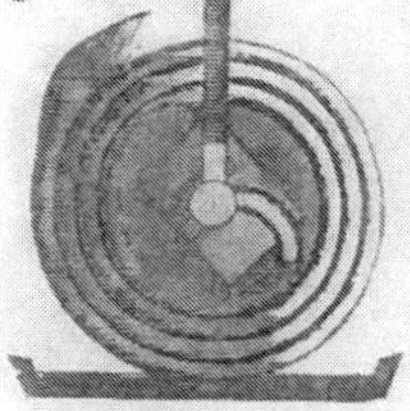

Fig. 301.

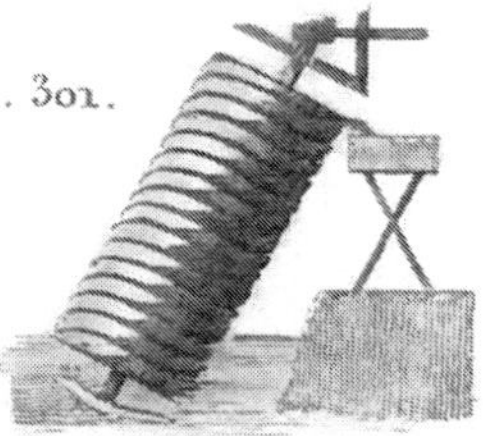

Fig. 302.

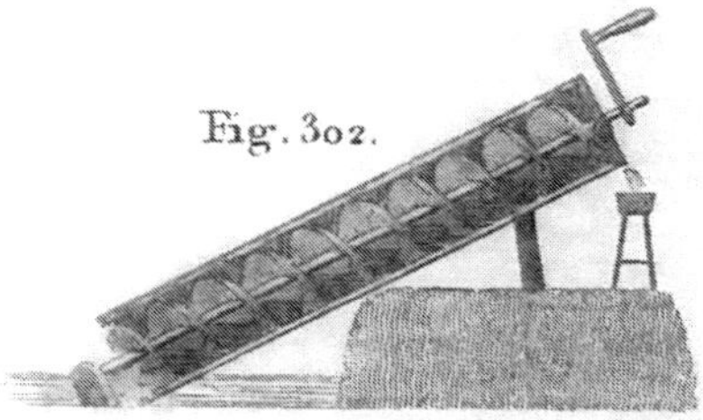

Pub. by J. Johnson, London 1 July 1806.

Joseph Skelt

PLATE XXII.

Fig. 288. Mr. Woltmann's hydrometrical fly. The plates A, B, are so adjusted by experiment, as to move exactly or very nearly with the velocity of the wind, a few degrees being allowed as a compensation for the retardation of friction. The cord C is drawn up, and the wheel D is caused to revolve, at a time observed by a stop watch; and its surface is graduated so as to number the revolutions of the fly. P. 319.

Fig. 289. An apparatus for measuring a ship's way, resembling Captain Hamilton's. A is a funnel partly covered, B a part of the ship's keel, C the upper part of the pipe D, in which the smaller pipe E F slides in a collar of leathers, so as to have the orifice F level with the surface of the water. This pipe has a small aperture at the bottom, which limits the magnitude of the stream discharged into the vessel G, the end F being considerably larger. The tube H serves as a gage, to measure the velocity at any given time. P. 319.

Fig. 290. An overshot wheel, on which the water is admitted in a retrograde direction, so as to run off in a continued stream; at the lower part of the wheel it is retained in the buckets partly by the assistance of a sweep. P. 321.

Fig. 291. A breast wheel, with a sweep. P. 322.

Fig. 292. An undershot wheel. P. 322.

Fig. 293. A the form of the sail of a windmill: B the best inclination for each part of the sail A, according to Smeaton's experiments. P. 324.

Fig. 294. A kite supported by the wind, of which the force acts nearly in the line A B, perpendicular to the surface of the kite; and this, compounded with the force of the cord A C, produces the result A D, which sustains the weight of the kite. P. 324.

Fig. 295. A ship working against a wind; the force of the wind acting nearly in the direction A B, perpendicular to the sails, the ship's real course is B C, the angle C B D being the lee way. P. 326.

Fig. 296. The anoria, or noria, used in Spain, for drawing water, by a series of earthen pitchers, connected by ropes, and passing over a sprocket wheel. P. 327.

Fig. 297. An undershot waterwheel, carrying fixed buckets, which raise a portion of water, and deliver it into a trough, furnished with a projection, which stands under the buckets, at the upper part of the wheel. P. 327.

Fig. 298. A throwing wheel, for draining fens, worked by a windmill or otherwise, and carrying the water upon a sweep from a lower to a higher level. P. 327.

Fig. 299. The rope pump of Vera, for raising water by means of friction: the rope is kept stretched by a pulley under the water, which is loaded with a weight, and slides in a groove. P. 328.

Fig. 300. The screw of Archimedes, nearly as described by Vitruvius. P. 329.

Fig. 301. The screw of Archimedes, as recommended by D. Bernoulli. P. 329.

Fig. 302. A waterscrew, revolving within a fixed cylinder. P. 329.

Fig. 303. The spiral pump of Wirtz. P. 330.

PLATE XXIII.

Fig. 304. A centrifugal pump. The machine is first filled through the funnel A, and when it is made to revolve, the water is discharged into a circular trough, of which a section is seen at B and C. The valve at D remains shut while the pump is filling. P. 331.

Fig. 305. A pump consisting of two plungers, continued nearly to the height at which the water is delivered. P. 332.

Fig. 306. Lahire's double forcing pump. When the piston is depressed, the water enters the barrel at the valve A, and goes out at B; when it is elevated, it enters at C and escapes at D. P. 332.

Fig. 307. The common piston, coated with leather. P. 332.

Fig. 308. Mr. Bramah's press. The pump A forces the water through the pipe B into the barrel C, in which it acts very powerfully on the large piston D, and raises the bottom of the press E. P. 332.

Fig. 309. The common sucking pump. P. 333.

Fig. 310. A bag pump, the bag or puff A being extended and contracted by the motion of the piston. P. 333.

Fig. 311. A lifting pump, the piston rod A B being drawn up by a frame. P. 333.

Fig. 312. A sucking pump, converted, by the addition of a collar of leathers at A, into a forcing pump. P. 333.

Fig. 313. A fire engine, on a construction similar to some machines described by Ramelli. A B is the piston, working within a cylindrical barrel, and moved by the handles C D. When the end C is depressed, the water enters through the valves E and F, and is discharged at G and H; when D is depressed, the water enters at I and K, and is discharged at L and M, into the air vessel N, whence it is expelled by the pipe O. The pipes P and Q may be united, if it be required. P. 334.

Fig. 314. From Ramelli. The wheel A B, revolving in the direction B A, carries a portion of water C between itself and the sweep D E, which is intercepted by the slider F, and forced up the pipe F G. P. 335.

Fig. 315. From Ramelli. The roller A, revolving within the reservoir B C, which is nearly cylindrical, carries with it the slider D E, which is made to sweep the internal surface of the cylinder from C to F, by means of a projecting surface acting on the end D, so that the water G is forced through the pipe F. P. 335.

Fig. 316. From the cabinet of Mr. Servière. The wheels A and B carry, during their revolution, a quantity of water from C to D, or from D to C, according to the direction in which they are turned. P. 335.

Fig. 317. Mr. Gwynn's patent water engine. The valve A is kept, partly by means of the spring B, but still more by the pressure of the water, in contact with the roller or piston C, which revolves within the box D E, and sweeps it from E to F, so that the portion of water G is forced, during each half of a revolution, into the pipe F; or is drawn from F to E, when the roller revolves in a contrary direction. P. 335.

Fig. 318. A chain pump. P. 335.

Fig. 319. The mechanism of Höll's acting pump. In the position of the stopcock A B, here represented, the water flows out of the barrel C, and the piston D is allowed to descend. The rod E then turns the stopcock, and the barrel C communicates only with the pipe F, which fills it, and forces up the piston, until the stopcock is turned back to its former position. P. 336.

Fig. 320. The hydraulic air vessels of Schemnitz. The reservoir A being filled with water, and B with air, and water being poured into the funnel C, the air in B acts by the pipe D on the water in A, and forces it up the pipe E. P. 337.

Fig. 321. A being the high water mark, and B the low water mark, the vessels C and D are filled at high water from below, the air being suffered to escape by a stopcock, which is opened by the fall of the ball F; at low water the air will enter the vessel D at B; and before the next high water, the water C will be forced up the pipe E. P. 337.

Fig. 322. The fountain of Hero. Its operation resembles that of the hydraulic air vessels, fig. 320; but the pipe D here ascends. P. 337.

Fig. 323. The hydraulic ram of Montgolfier. When the water in the pipe A B has acquired a sufficient velocity, it raises the valve B, which stops its passage, so that a part of it is forced through the valve C, into the air vessel D, whence it rises through the pipe E. P. 338.

PLATE XXIII.

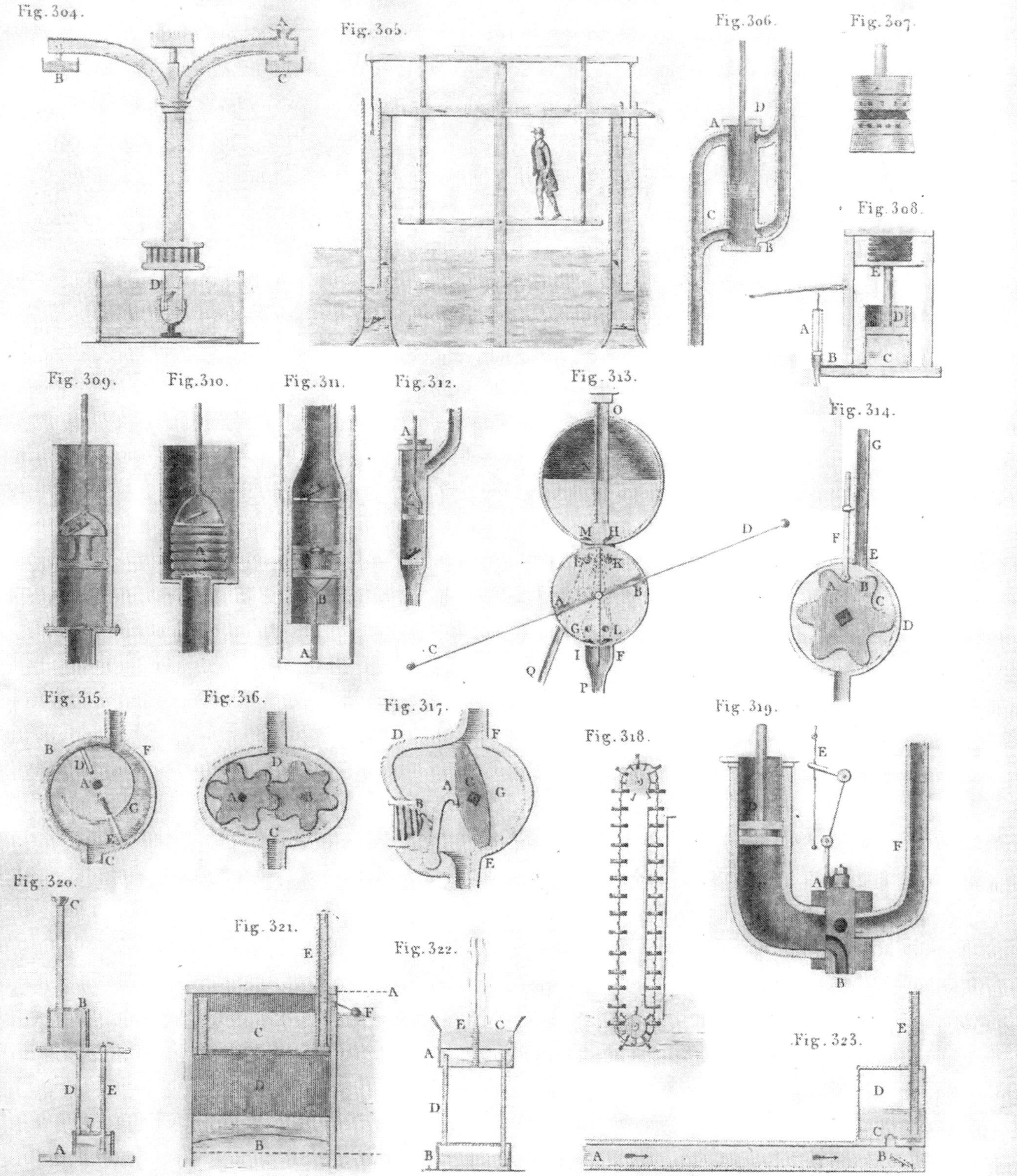

Pub. by J. Johnson, London 1 July 1806. *Joseph Skelton sculp.*

PLATE XXIV.

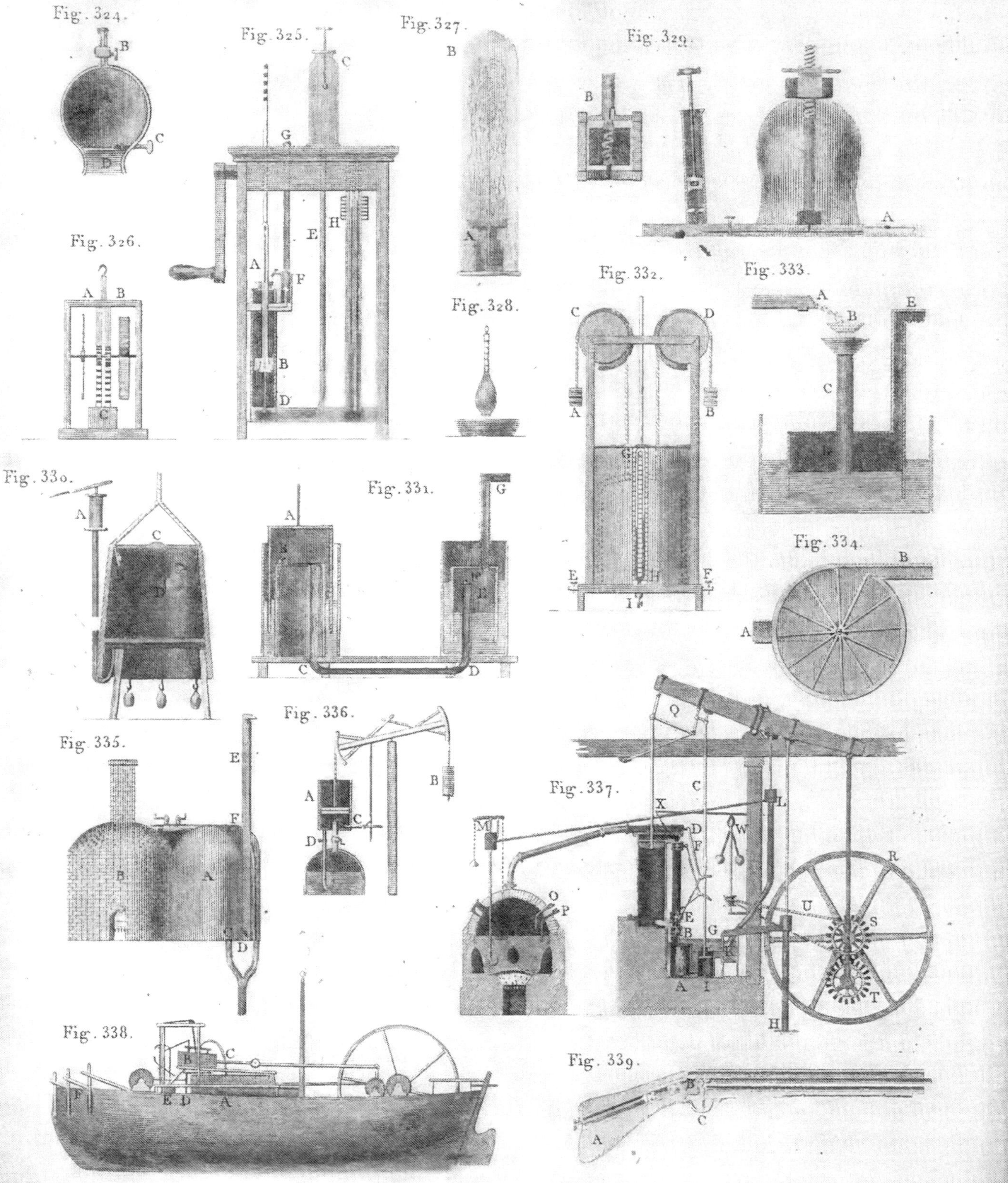

Pub. by J. Johnson, London 1 July 1806. Joseph Skelton.

PLATE XXIV.

Fig. 324, The cupping instrument of Hero. The cavity A was partly exhausted by applying the mouth repeatedly to the pipe B, the stopcock B being turned after each application. When the stopcock C was opened, the air at D in contact with the skin was also rarefied, and the effect of suction was produced. P. 339, 353.

Fig. 325. Mr. Cuthbertson's air pump. When the piston rod A is depressed, it leaves the piston B a little behind it, so as to make an opening between two conical parts which are ground together, and the air escapes from the lower part of the barrel into the upper part; when it is elevated, the whole piston is raised, and a wire, which slides through the axis of the rod, raises a small valve at the bottom of the barrel, which leads to the receiver C, by the tube D E: the air is forced from the upper part of the barrel through a valve in the oil vessel F, whence the oil runs back, when it overflows, by a tube leading to the mouth of the barrel; and if this tube be stopped by turning its cock, the air may be condensed into a receiver fixed at G. At H is a long gage, with a barometer immersed in the same bason of mercury. The piston rod, which is hollow, has a perforation a little above A, to admit the oil, in order that the wire may work freely in it. P. 340.

Fig. 326. The two flies A and B being caused to revolve with equal velocities by the descent of the weight C, they continue to move for an equal length of time in the vacuum of the air pump. P. 341.

Fig. 327. The air in the bottle A expands, when the receiver B is exhausted, and causes the water to rise in a jet. P. 341.

Fig. 328. A pear gage; to be suspended in a receiver by a hook like that which is shown in fig. 325. P. 342.

Fig. 329. A condenser, with screws, for confining the receiver. A is a gage for showing the degree of condensation; B the piston of the syringe, with a valve of the best kind, which is conical, and is confined by a spiral spring. But in common, the valves are made of leather, with a plate of metal to strengthen it. P. 342.

Fig. 330. A diving bell. A is the forcing pump, B a stopcock for letting out the heated air, C a strong glass for giving light, D a float for the security of the diver. P. 343.

Fig. 331. Laurie's hydraulic bellows. When the vessel A is raised, the air enters at the valve B; when it is depressed, the valve B shuts, and the air is forced through the pipe C D, which conducts it to the reservoir E, where it is confined by the valve F, and forced by the pressure of the water through the pipe G. P. 343.

Fig. 332. Mr. Watt's gasometer. The pressure is regulated by the magnitude of the weights A and B, which act by the spiral fusees C, D, so as to sustain a part of the weight of the inverted vessel, represented by the exterior dotted line. The gas is admitted at E or F, and is delivered at G. G H is a gage for showing the height of the water within and without the moveable vessel. I is a cock for letting off the water. P. 344.

Fig. 333. The shower bellows. The stream A, passing through the strainer B, carries with it a quantity of air through the pipe C, which rises to the upper part of the air vessel D, and is discharged by the pipe E. P. 344.

Fig. 334. The centrifugal bellows. By the revolution of the fly, the air is caused to enter at A, and is discharged at B. P. 345.

Fig. 335. The original steam engine of Savery. The vessel A being filled with steam from the boiler B, and the stopcock being turned, the steam cools and is condensed, and water is forced into its place by the pressure of the atmosphere, through the valve C: the steam is then readmitted, and forces the water to ascend through the valve D and the pipe D E. The vessel F acts alternately with A. P. 347.

Fig. 336. The common steam engine of Newcomen and Beighton. The steam being admitted into the cylinder A below the piston, the weight B is allowed to descend: a jet of water is then admitted by the pipe C, which condenses the steam, and the pressure of the atmosphere then depresses the piston: a part of this water is admitted by the pipe D into the boiler, in order to keep it sufficiently full. P. 347.

Fig. 337. Mr. Watt's steam engine. The steam, which is below the piston, is suffered to escape into the condenser A by the cock B, which is opened by the rod C, and at the same time the steam is admitted by the cock D into the upper part of the cylinder; when the piston has descended, the cocks E and F act in a similar manner in letting out the steam from above and admitting it below the piston. The jet is supplied by the water of the cistern G, which is pumped up at H from a reservoir: it is drawn out, together with the air that is extricated from it, by the air pump I, which throws it into the cistern K, whence the pump L raises it to the cistern M; and it enters the boiler through a valve, which opens whenever the float N descends below its proper place. The pipes O and P serve also to ascertain the quantity of water in the boiler. The piston rod is confined to a motion nearly rectilinear by the frame Q; the fly wheel R is turned by the sun and planet wheel S, T; and the strap U turns the centrifugal regulator W, which governs the supply of steam by the valve or stopcock X. P. 349.

Fig 338. Mr. Symington's steam boat. A is the boiler, B the cylinder, C the piston, D the condensation pipe, E the air pump, F stampers for breaking ice. P. 349.

Fig. 339. An air gun. The air is forced by the syringe A into the cavity surrounding the barrel, whence it is discharged by the valve B, which is opened either immediately by the action of the trigger C, or by a spring, which is bent by cocking the gun, and set at liberty by the trigger. P. 351.

PLATE XXV.

Fig. 340. A series of waves or pulses of sound, diverging from one of the foci of an ellipsis, and reflected towards the other. P. 375.

Fig. 341. Waves diverging from a point near the centre of a circle, and converging after reflection to a point at an equal distance on the other side of the centre. P. 375.

Fig. 342. A section of a speaking trumpet and of a hearing trumpet: the lines representing the direction of the sound before and after its reflections. P. 375.

Fig. 343. A string impelled by the bow of a violin, and lightly touched at the same time at a point one third of its length from the end: the small pieces of paper fly off from the middle of the vibrating portions, while the piece situated at the remaining point of division retains its situation. P. 383.

Fig. 344. A vibration compounded with another smaller vibration, three times as frequent, in a transverse direction, the separate vibrations being such that the points may be always opposite to a point moving uniformly in a circle. Thus the vibrations in the lines A B and A C compose the complicated figure D E. P. 384.

Fig 345. A specimen of the manner in which the vibrations of a string are usually performed when it is struck with a bow. P. 384.

Fig. 346. Specimens of the simplest manner in which sand is collected into lines, on a plate of glass or metal, which is made to sound by means of the bow of a violin. P. 385.

Fig. 347. A round plate, performing some of its most complicated vibrations, the lines of division being indicated by the place of the sand. From Chladni. P. 385.

Fig. 348. A square plate divided into a diversity of vibrating portions. From Chladni. P. 385.

Fig. 349. The small bones of the left ear, nearly three times the natural size, supposed to be seen through the membrane of the tympanum, by looking directly into the auditory canal. A B is the membrane of the tympanum, C the hammer, D the anvil, E its attachment to the surrounding bone, F the stirrup, G the round aperture in the bone leading to the cochlea. P. 388.

Fig. 350. A view of the vestibule of the left ear, with the semicircular canals and the cochlea, seen with the eye a little more depressed than by looking straight through the canal, and exactly in the direction of the stirrup. A B C is the vestibule, immediately behind the oval aperture, which is covered by the basis of the stirrup, D are the canals, E the cochlea, the upper spire terminating in the vestibule, the lower in the round aperture at B. The projection of the membrane of the tympanum is marked by an oval line. P. 388.

Fig. 351. The structure of the left ear, seen from above, the upper part of the canal being supposed to be removed. A is the auditory canal, B the membrane of the tympanum, C the hammer, D the anvil, E the stirrup; F the place of the canals, which are higher than the parts represented, G the place of the cochlea, H the round aperture. P. 388.

Fig. 352. A, B, C, a representation of the joint effect of two equal vibrations variously combined, the middle line being always half way between the two outer ones, and showing the compound vibration reduced to half its real extent: D shows the mode of finding the joint effect of vibrations, by cutting a surface into sliders, which are retained in their places by a screw. P. 390.

Fig. 353. The uppermost and lowermost curves represent a series of vibrations, of which 12 occupy any given period of time: the third and sixth lines two series of which 15 and 16 occupy respectively the same time: the joint effect of each pair is shown by the dotted curves which are interposed between them, the middle one representing the effect denominated a beat. P. 391.

Fig. 354. The proportional lengths of a chord or pipe, constituting the different notes of the simple diatonic scale, with their mutual relations, shown by their divisions into aliquot parts. P. 393.

Fig. 355. A good practical mode of temperament; making all the fifths and the third in the first division perfect concords; the three remaining fifths equally imperfect. P. 396.

Fig. 356. The trumpet Marigni, with its bridge, which is supported by the string A B nearly in contact with the sounding board; this string being either stretched by a pin at B, or by a cross string B C. The places, at which the string is to be touched, may be marked by frets fixed under them, as they are here shown by points. At D, the scale of this instrument is exhibited, resembling that of the trumpet and the French horn. P. 399.

PLATE XXV.

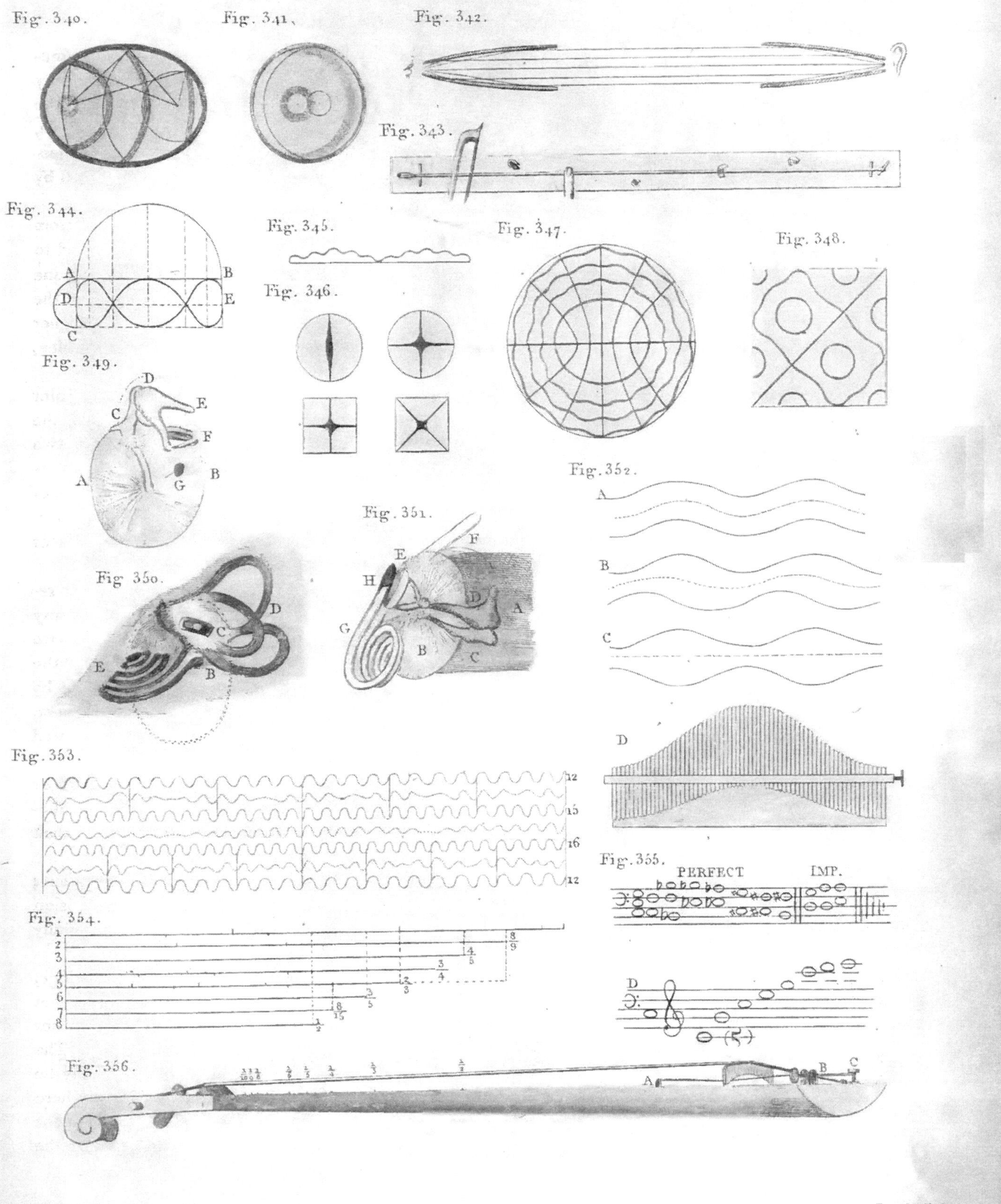

Pub. by J. Johnson London 1 July 1806. Joseph Skelton sculp.

PLATE XXVI.

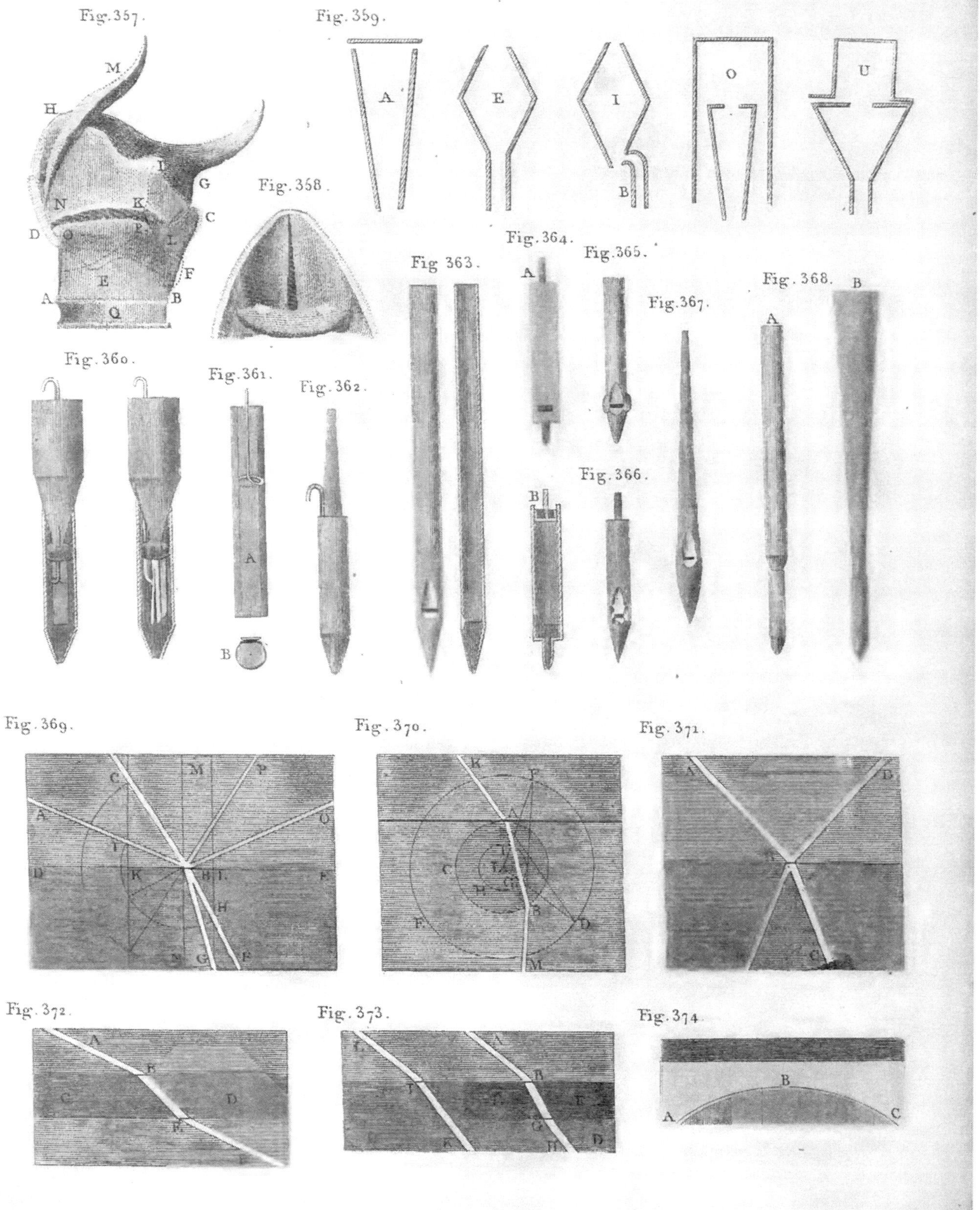

Pub. by J. Johnson London 1 July 1806. *Joseph Skel*

PLATE XXVI.

Fig. 357. The right half of the human larynx. ABC is the outline of the cricoid cartilage, DEFGH of the thyreoid, and I K L of the arytaenoid cartilage; M is the epiglottis, NK the upper ligament of the glottis, OP the lower ligament, and Q the trachea. P. 400.

Fig. 358. A view of the ligaments of the glottis, seen from above, the larynx being divided by a horizontal section a little above them. P. 400.

Fig. 359. Sections of the pipes employed by Kratzenstein for producing the sounds of the different vowels; in general by means of a larynx resembling the mouth piece of a reed organ pipe, but in the case of the vowel I by simple inflation through tho tube B. The pipe for U produces the sound O, except when it is very nearly shut up. P. 401.

Fig. 306. The vox humana organ pipe, with the mouth piece common to reed pipes in general; the lower part in contact with the tongue being nearly semicylindrical: the tongue being adjusted to the proper pitch by means of a sliding wire, which regulates the length of the part that is at liberty to vibrate. P. 401.

Fig. 361. The mouth piece proposed by Kratzenstein, for imitating the human voice, the tongue A passing freely in and out of the tube, which is more than half of a cylinder, as is seen at B. P. 401.

Fig. 362. The form of the regal organ pipe. P. 401.

Fig. 363. A front view and section of the open diapason organ pipe of metal. It is tuned by opening or contracting the upper orifice. P. 402.

Fig. 364. A a front view of the flute organ pipe, of wood, which is tuned by a plug. B a section of the pipe. P. 402.

Fig. 365. A stopped diapason organ pipe, of metal. It is tuned by altering the position of the pieces on each side of the mouth. P. 402.

Fig. 366. A chimney pipe. P. 402.

Fig. 367. A spindle shaped organ pipe, contracted above. P. 402.

Fig. 368. A the form of a cromorn pipe, B, of a trumpet pipe, both having reed mouth pieces. P. 403.

Fig. 369. A ray or pencil of light A B, C B, falling on the surface D E, a portion of the light is reflected, and another portion is transmitted, in the direction BF, BG, so that B G is equal to B C, and B H to B I, C I K and G H L being lines perpendicular to D E at any such distances, that B K may be to B L in a certain proportion, which is that of the sines of the angles of incidence A B M, C B M, to those of the angles of refraction F B N, G B N. B O and B P are the reflected portions of the rays. P. 411.

Fig. 370. A mode of determining the position of a refracted ray, which is particularly convenient in the case of refractions at spherical surfaces. A B C being any circle, either touching the refractive surface at A, or being itself a section of the refracting substance, if another circle D E F be drawn on the same centre, having its diameter to that of the first as the sine of the angle of incidence to that of refraction, and a third circle G H I, which is less than the first in the same proportion as the second is greater; and if the direction of the incident ray K A be continued to D, and L D be drawn from the centre, cutting G H I in G, A G will be the direction of the refracted ray: and if this ray pass again out of the denser medium at B, its direction B M may be found by drawing LIF, and F B M will be thus truly determined. P. 411.

Fig. 371. A ray or pencil A B, refracted at B to C, and there reflected by a perpendicular surface into an opposite direction C B, will return also in the direction B A, a portion of it being reflected in the first place to D, and in the second to E. P. 412.

Fig. 372. A pencil A B passing through a substance C D contained between parallel surfaces, continues its course in the direction E F parallel to A B. P. 413.

Fig. 373. The ray A B, entering the medium C D through the transparent substance E F, contained between parallel surfaces, acquires the direction GH, parallel to I K, into which L I is at once refracted. P. 413.

Fig. 374. The appearance of a prism, of which the lower surface is divided into a bright and a dark portion, separated by a coloured arch A B C. P. 414.

PLATE XXVII.

Fig. 375. A is an actual focus of diverging rays, B an actual focus both of converging and of diverging rays, C a virtual focus of converging rays, and D a virtual focus of diverging rays; A and B, B and C, and C and D are foci conjugate to each other, with respect to the refractions of the three lenses. P. 415.

Fig. 376. The image of the point N, formed by the plane mirror AB, is at an equal distance behind the mirror; and in this manner the whole image of the word is formed in an inverted position. P. 415.

Fig. 377. ABCD represents a pencil of parallel rays falling on the concave mirror CD, and collected into the principal focus at E, which is half way between the surface and its centre. F is the principal focus of the convex mirror G; and H that of the refracting surface I. P. 416.

Fig. 378. A being the centre of the concave mirror B, the image of an object at C will be found at D, and the reverse. P. 416.

Fig. 379. A pencil of light, deflected from its path by a prism of a denser substance, in different positions. P. 416.

Fig. 380. A pencil of light scattered into various directions by a multiplying glass. P. 416.

Fig. 381. A is a section of a double convex lens, B of a double concave. C is a planoconvex, D a planoconcave; and E and F meniscus lenses; but a meniscus of the form represented by F is sometimes called a concavoconvex lens. P. 417.

Fig. 382. The pencils of light A, B are refracted by the convex lens C in the same manner as they would have been by the circumscribed double prism DE; and in the same manner the concave lens F resembles in its operation the prisms G, H. P. 417.

Fig. 383. A, a pencil of parallel rays, made to converge, by a double convex lens of crown glass, to the centre of curvature of one of its surfaces. B a double concave lens, causing the rays to diverge from the centre of curvature. C, D a planoconvex lens, of which the principal focus is at the distance of a diameter. P. 417.

Fig. 384. The lenses represented by the shaded surfaces are equivalent in their effects to those of which the sections are shown by the dotted lines; the figures at A and B being of equal thickness in the middle, and at C at the edges also. P. 417.

Fig. 385. At A, a radiant point and its image are both situated at the distance of twice the focal length from the lens; at B, the one is more remote, the other nearer; and CD is to DE as EF to FG; D and F being the principal foci of the lens. P. 418.

Fig. 386. The oblique pencils of rays A, B, and the direct pencil C, are supposed to be brought to their respective foci in the same plane DE. P. 419.

Fig. 387. The square A intercepts the whole light, proceeding from the point B, which would fall on the surface CD, four times as great, placed at a double distance. P. 421.

Fig. 388. The box of Count Rumford's photometer. The lights, being placed at proper distances on the graduated arms or tables A, B, throw equally dark shadows of the cylinders C, D on a white surface at EF. The wings of the cylinders serve to make the shadows of equal breadth. The shadows are viewed through the aperture at G. P. 421.

Fig. 389. Dr. Wollaston's instrument for the measurement of refractive densities. A is a rectangular prism of flint glass, under which the substance to be examined is attached; BC is a rod, or ruler, 10 inches long, CD and DE are each 15 $\frac{83}{100}$. When the sights at B and C are so placed that the division between the light and dark portion of the lower surface of the prism is seen through them, the rod F, which carries a vernier, shows the index of the refractive density, which, in the situation here represented, would be 1.43. P. 421.

Fig. 390. A is the actual image of the candle B, formed by the convex lens C. P. 422.

Fig. 391. A is the actual image of the candle B, formed by the concave mirror C. P. 422.

Fig. 392. A is the actual image of the candle B, formed by the convex lens C, being as much larger than the object as it is more distant from the lens. P. 422.

Fig. 393. A is the virtual image of the candle B, placed within the focal distance of the concave mirror C, the image remaining erect. P. 422.

Fig. 394. A is the virtual image of the candle B, formed by the concave lens C, and less than the object. P. 422.

Fig. 395. When the object A is placed in the principal focus of the convex lens B, a virtual image is formed at an infinite distance, which subtends, when viewed from C, or from any other point, the same angle as the object subtends at the centre of the lens. P. 422.

Fig. 396. The object A being placed a little within the focus of the lens B, a virtual image C is formed, at such a distance as is most convenient to the eye, which subtends the same angle as the object, from the centre of the lens, and therefore appears somewhat more magnified than when the object is in the principal focus. P. 422.

PLATE XXVII.

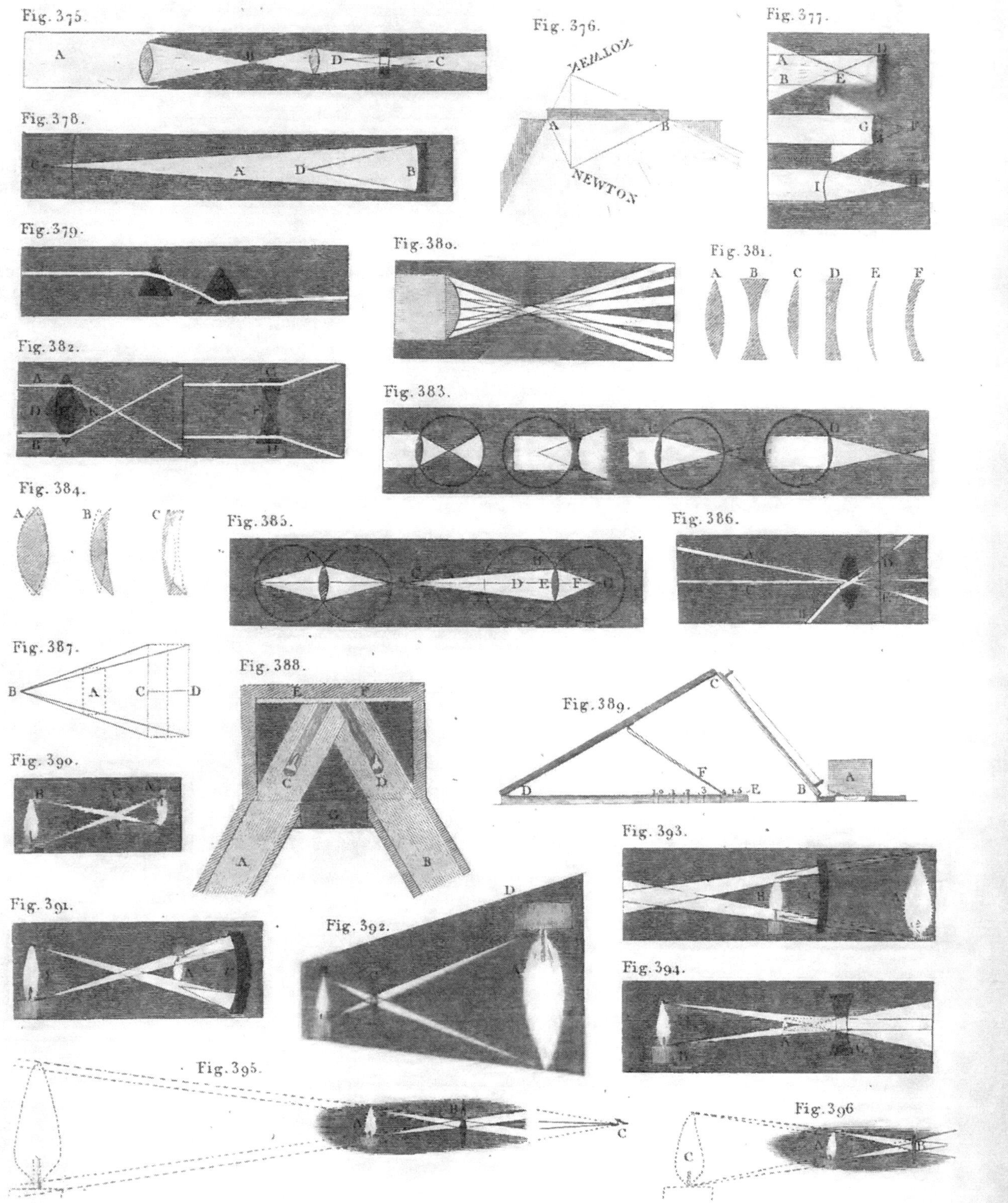

Pub. by J. Johnson, London 1 July 1806.

Joseph Skelton sculp.

PLATE XXVIII.

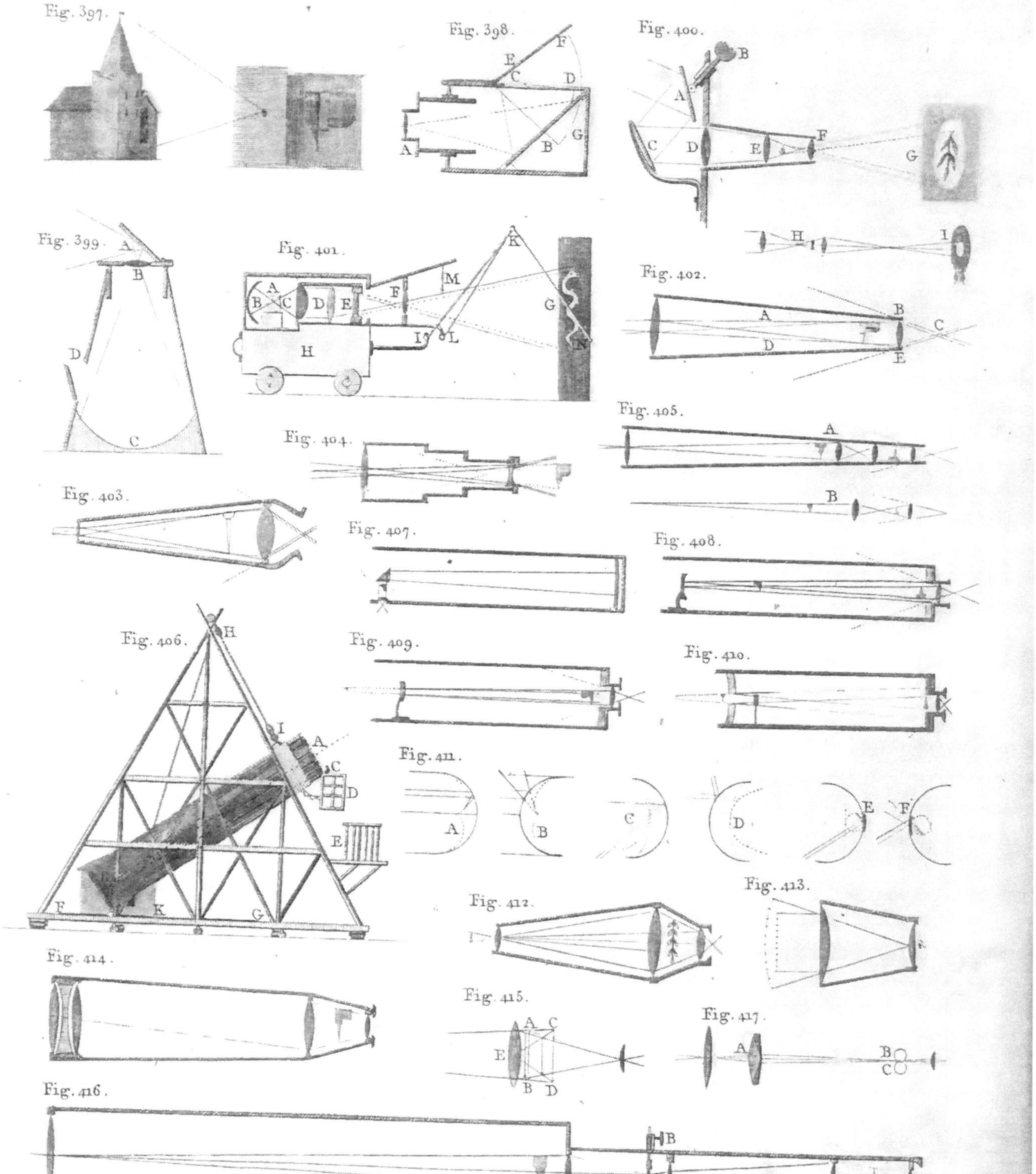

Pub. by J. Johnson London 1 July 1806.

Joseph Skelton

PLATE XXVIII.

Fig. 397. An imperfect image of an external object, painted in a dark room, in an inverted position, by the light coming in right lines through a small aperture. P. 425.

Fig. 398. A portable camera obscura. A is a lens, B a mirror placed obliquely, and throwing the image on a plate of ground glass, C D. E is a moveable cover, and F G a screen attached to it, for excluding foreign light. P. 425.

Fig. 399. A camera obscura, which throws down an image, by means of the mirror A, and the lens B, on the surface C, where it may be seen through the aperture D. The surface C has here the curvature best adapted to receive every where a perfect image of a distant object. P. 425.

Fig. 400. An arrangement proposed for a solar microscope, adapted to a window facing the south. The mirror A is moved by a hinge into the position required for the day, and during the employment of the instrument is turned only round the axis A B, which is parallel to that of the earth. The mirror C is fixed: it receives the beam of light from A, and throws it on the object through the lenses D and E, of which the joint focus is near the magnifying lens F; this lens paints an image of the object in an inverted position on a screen at G. If the focus of the condensing lenses were behind the object, as at H, the light would be liable to be condensed into a spot on the screen at I. P. 426.

Fig. 401. An arrangement proposed for a phantasmagoria. The light of the lamp A is thrown by the mirror B and the lenses C and D on the painted slider at E, and the magnifier F forms the image on the screen at G. This lens is fixed to a slider, which may be drawn out of the general support or box H: and when the box is drawn back on its wheels, the rod I K lowers the point K, and by means of the rod K L adjusts the slider in such a manner, that the image is always distinctly painted on the screen G. When the box advances towards the screen, in order that the images may be diminished and appear to vanish, the support of the lens F suffers the screen M to fall and intercept a part of the light. The rod K N must be equal to I K, and the point I must be twice the focal length of the lens F, before the object, L being immediately under the focus of the lens. The screen M may have a triangular opening, so as to uncover the middle of the lens only, or the light may be intercepted in any other manner. P. 427.

Fig. 402. The construction of the astronomical telescope. A B C and D E C are the central parts of the pencils of rays, coming, from the extremities of the visible field, through the middle of the object glass. P. 427.

Fig. 403. The extreme pencils of rays in the double or compound microscope. P. 428.

Fig. 404. The extreme pencils in the Galilean telescope, or opera glass. P. 428.

Fig. 405. A, the directions of the extreme pencils in the common day telescope of Rheita. If only two eye glasses were employed, as at B, the field would obviously be more contracted. P. 428.

Fig. 406. Dr. Herschel's forty feet telescope. ABC the path of a ray of light, reflected by the mirror at B to the eye glass C. D a chair in which the observer sits. E a moveable gallery, on which several persons may stand. F G a smooth surface, on which the bottom of the telescope is made to roll along, while its opening is raised or depressed by the pullies at H and I. K one of two rooms or huts for the accommodation of the observer's assistants. The wheels, under the frame, serve to turn the whole instrument round its centre. P. 429.

Fig. 407. The Newtonian telescope, with the direction of the central rays. These are not the rays by which the object is actually seen, because they are intercepted by the small speculum, but they afford the simplest determination of the magnitude of the field of view. P. 429.

Fig. 408. The supposed path of the central rays in the Gregorian telescope. P. 429.

Fig. 409. The supposed path of the central rays in Cassegrain's telescope. Here the rays actually represented would not only be intercepted by the small mirror, but they would also fall on the perforation of the great mirror. They, however, serve equally well to determine the magnitude of the field. P. 429.

Fig. 410. The supposed path of the central rays in Dr. Smith's microscope. The rays running directly from the object are intercepted by a screen. P. 429.

Fig. 411. A; the dotted line represents the curve called the caustic of a concave mirror, in which the rays proceeding, in the section represented by the figure, from a distant point, would be collected. B; the dotted line is the caustic of a convex mirror. The eye being supposed to be at a great distance from the hemispherical mirrors C and D, the images of distant objects in all directions will be found between the dotted curves, the distance of those curves showing the degree of confusion. The images of distant objects in all directions formed by the small concave and convex mirrors E and F, are found between the dotted circle and the straight line touching it. P. 430.

Fig. 412. The effect of a field glass in a compound microscope; the inner lines showing what would be the magnitude of the field without it. P. 431.

Fig. 413. The manner in which Mr. Ramsden employed a planoconvex lens in the eye pieces of his telescopes and in his double magnifiers. The curved dotted line shows the image of the straight line divided into equal parts, which is formed by the larger lens, in the focus of the smaller, through which it is viewed. P. 431.

Fig. 414. An achromatic telescope, with a triple object glass, and with Boscovich's achromatic eye piece, consisting of two similar lenses, one of which is every way three times as great as the other, their distance being twice the focal length of the smaller. P. 432.

Fig. 415. The dotted lines A B and C D represent two images of the same object, formed by rays differently refrangible, passing through a simple object glass, which are brought, by the effect of the lens or field glass E, into such places and dimensions as to subtend nearly the same angle from the eye glass F. P. 432.

Fig. 416. A represents Mr. Ramsden's divided eye glass micrometer, the two portions being moved at once in contrary directions by turning the pinion B, until the two extremities of the distance to be measured appear to coincide. P. 433.

Fig. 417. Dr. Maskelyne's micrometer, made by a double achromatic prism A, exhibiting two images B, C, the different parts of which are made to coincide, by moving the prism backwards and forwards in the direction of the axis of the telescope. Mr. Ramsden thinks that any substance thus interposed must interfere greatly with the perfection of the telescope. P. 433.

2

PLATE XXIX.

Fig. 418. If A B and A C represent the comparative velocity of light and of the earth, in their respective directions, a telescope must be placed in the direction B C, in order to see the star D, and the star will appear at E. P. 437.

Fig. 419. The spectrum produced by looking through a prism at a narrow line of light. P. 438.

Fig. 420. The appearance of a portion of the blue light at the bottom of a candle, viewed through a prism. P. 438.

Fig. 421. The appearance of a circular aperture, moderately large, when viewed through a prism. P. 439.

Fig. 422. A B and C D represent the appearance of the two ends of a broad white surface, or a window, when viewed through a prism. The oblique stripes of colour show the degrees by which the lights of different kinds enter into the compound light. It follows from this analysis, that the colours, horizontally opposite each other in A B and C D, would always together make up white light. P. 439.

Fig. 423. The colours on the circle A exhibit, when whirled swiftly round, a whitish light resembling B. P. 440.

Fig. 424 . . 426. The colours of the circle A produce, when made to revolve rapidly, the tints shown at B. P. 440.

Fig. 427. A triangular figure, exhibiting in theory all possible shades of colours. The red, the green, and the violet, are single at their respective angles, and are gradually shaded off towards the opposite sides: a little yellow and blue only are added in their places, in order to supply the want of brilliancy in the colours which ought to compose them. The centre is grey, and the lights of any two colours, which are found at equal distances on opposite sides of it, would always very nearly make up together white light, as yellow and violet, greenish blue and red, or blue and orange. P. 441.

Fig. 428. The appearance of a pin, and of the word POKER, when viewed by looking along the surface of a red hot poker. From Dr. Wollaston. P. 442.

Fig. 429. The appearance of an oblique line, and of the word SPIRIT, viewed simply through rectified spirit of wine, and through a portion of the spirit mixing by degrees with the water on which it floats. From Dr. Wollaston. P. 442.

Fig. 430. The colours of the primary and secondary rainbow, as they usually appear. P. 443.

Fig. 431. The most common form of halos and parhelia. P. 444.

Fig. 432. Magnified figures of the simplest crystals of snow, which are sufficient to account for the production of halos. From Nettis. P. 444.

Fig. 433. A complicated system of halos. From Lowitz. The arcs A, B, and C, were coloured, and, like all the other coloured parts, had the red towards the sun. D and E are two anthelia. P. 444.

Fig. 434. The figures of two complicated flakes of snow. From Nettis. P. 444.

Fig. 435. The ray of light A B, falling perpendicularly on the surface of the piece of calcarious spar C D at B, is divided into the portions B E and B F: the portion B E passing to the point E, where the surface of the spheroid E G H, inscribed in the greater angle of the crystal, becomes parallel to C B. P. 445.

PLATE XXIX.

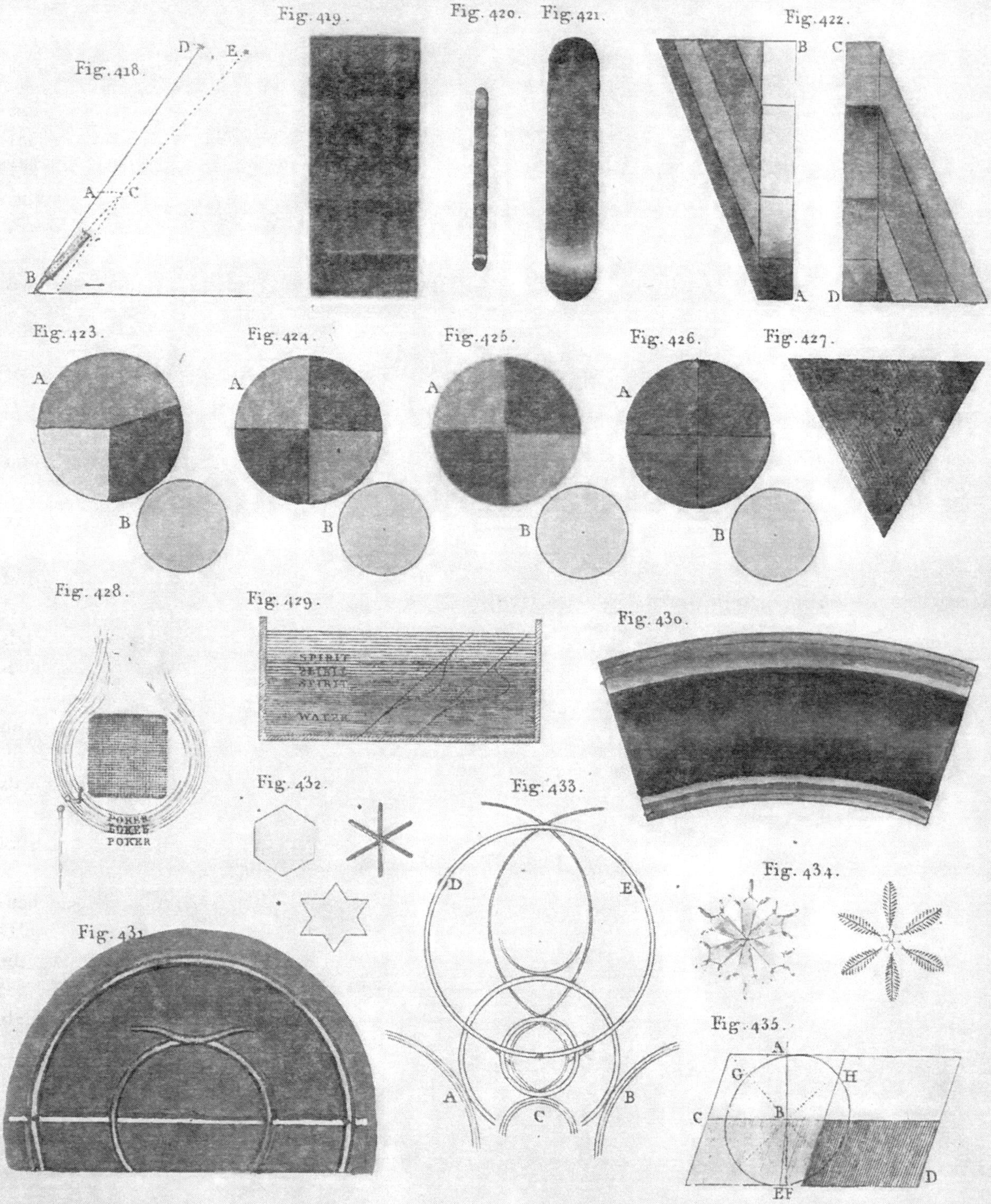

Pub. by J. Johnson London 1 July 1806.

Joseph Skelton sc

PLATE XXX

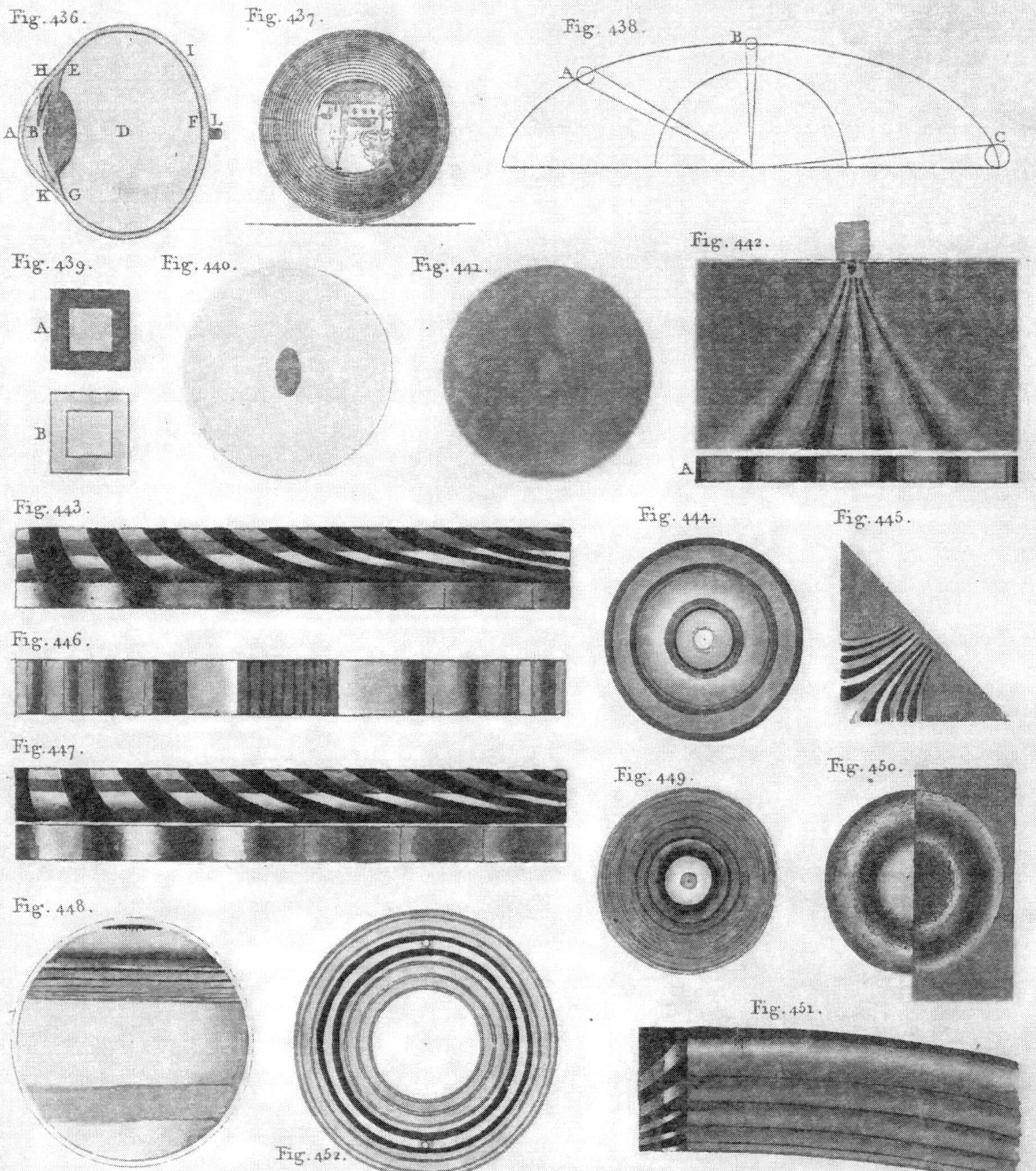

Pub. by J. Johnson. London 1 July 1806.

Joseph Skelton sculp

PLATE XXX.

Fig. 436. A section of the human eye. A is the cornea; B the aqueous humour, in which the uvea hangs; C the crystalline lens; the ciliary processes being between it and the uvea; D the vitreous humour; E F G is the choroid coat, lined by the retina; H I K the sclerotica, and L the optic nerve. P. 447.

Fig. 437. A picture painted on the retina in an inverted position, seen by dissecting off the sclerotica and choroid behind it. P. 448.

Fig. 438. The apparent figure of the heavens being nearly like the curve A B C, the sun or moon at A or C appears to be much larger than at B. P. 454.

Fig. 439. The red square A, inclosing a green square, produces, if viewed attentively, in a strong light, a spectrum resembling B, which is red within and green without, and which appears when we look soon after on any white object. P. 456.

Fig. 440. The spot, which is tinted with black lines only, appears, upon the yellow ground, of a purple hue. P. 456.

Fig. 441. A grey spot on a purple ground appears of a greenish yellow or olive hue. P. 456.

Fig. 442. The manner in which two portions of coloured light, admitted through two small apertures, produce light and dark stripes or fringes by their interference, proceeding in the form of hyperbolas; the middle ones are however usually a little dilated, as at A. P. 465.

Fig. 443. A series of stripes of all colours, of their appropriate breadths, placed side by side in the manner in which they would be separated by refraction, and combined together so as to form the fringes of colours below them, beginning from white. P. 465.

Fig 444. A series of coronae, seen round the sun or moon. P 466.

Fig. 445. The internal hyperbolic fringes of a rectangular shadow. P. 467.

Fig. 446. The external fringes seen on each side of the shadow of a hair or wire, which is also divided by its internal fringes. The dotted lines show the natural magnitude of the shadow, independently of diffraction. P. 468.

Fig. 447. Analysis of the colours of thin plates seen by reflection, beginning from black. A line drawn across the curved fringes would show the portions into which the light of any part is divided when viewed through a prism. P. 469.

Fig. 448. The coloured stripes of a film of soapy water, covering a wine glass. P. 469.

Fig. 449. The colours of a thin plate of air or water, contained between a convex and a plane glass, as seen by reflection. P. 469.

Fig. 450. The colours of a mixed plate; as seen by partially greasing a lens a little convex, and a flat glass, and holding them together between the eye and the edge of a dark object. One half of the series begins from white, the other from black, and each colour is the contrast to that of the opposite half of the ring. P. 470.

Fig. 451. The composition of the colours of the primary rainbow, when attended by supernumerary bows. P. 471.

Fig. 452. The colours of concave mirrors. The small circles in the middle white ring represent the aperture by which the light is admitted, and its image; the coloured rings are formed by the light irregularly dissipated, before and after reflection. P. 471.

PLATE XXXI.

Fig. 453, 454. The appearance of the star Lyra, viewed with telescopes magnifying 460 and 6450 times respectively. From Dr. Herschel. P. 491.

Fig. 455. The appearance of the nebula in Orion, about half a degree in length. From Messier. P. 492.

Fig. 456 . . 463. The appearances of different nebulae. From Dr. Herschel. P. 492.

Fig. 464. A section of the nebula to which the sun is supposed to belong, its projection forming the milky way; taken in a plane perpendicular to its longest diameter. From Dr. Herschel. The large star in the middle represents the sun, and the circle drawn round it is at forty times the distance of the nearest fixed stars, comprehending probably all the stars which are visible to the naked eye. P. 493.

Fig. 465. A large spot, traced through different forms in its path across the sun. From Dr. Wilson. A is its place 23 Nov. 1769; B, 24 Nov. C, 11 Dec. D, 12 Dec. and E, 17 Dec. P. 501.

Fig. 466. A, a large spot on the sun; B, the arrangement of the luminous and opaque strata of clouds by which Dr. Herschel explains the appearance of the spot. P. 501.

Fig. 467. A, a spot with a lighter portion in the middle; B, the arrangement of the strata corresponding to it. P. 501.

Fig. 468. The position assumed by the strata which had formed the spot shown in the last figure, viewed about an hour afterwards. P. 501.

Fig. 469. A and B are the forms of a solar spot, at about two hours distance of time; C, D, and E, are the successive forms of another spot. P. 501.

Fig. 470. The appearance of the zodiacal light, or solar atmosphere, as it is seen in these climates, in the evening, about the beginning of March; A B being the horizon, and C the supposed place of the sun. P. 503.

PLATE XXXI.

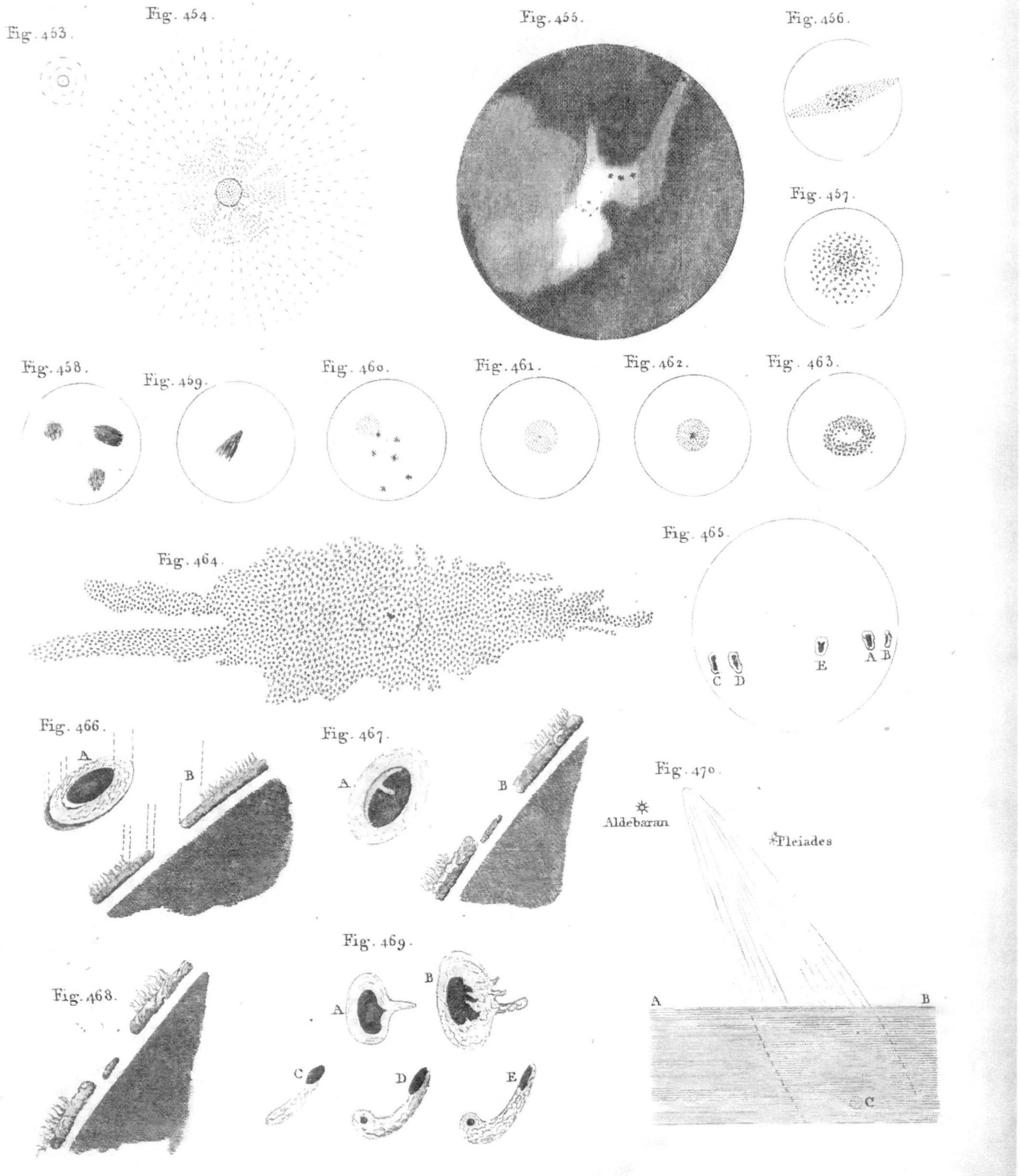

Pub. by J. Johnson, London 1 July 1806.

Joseph Skelton sc

PLATE XXXII.

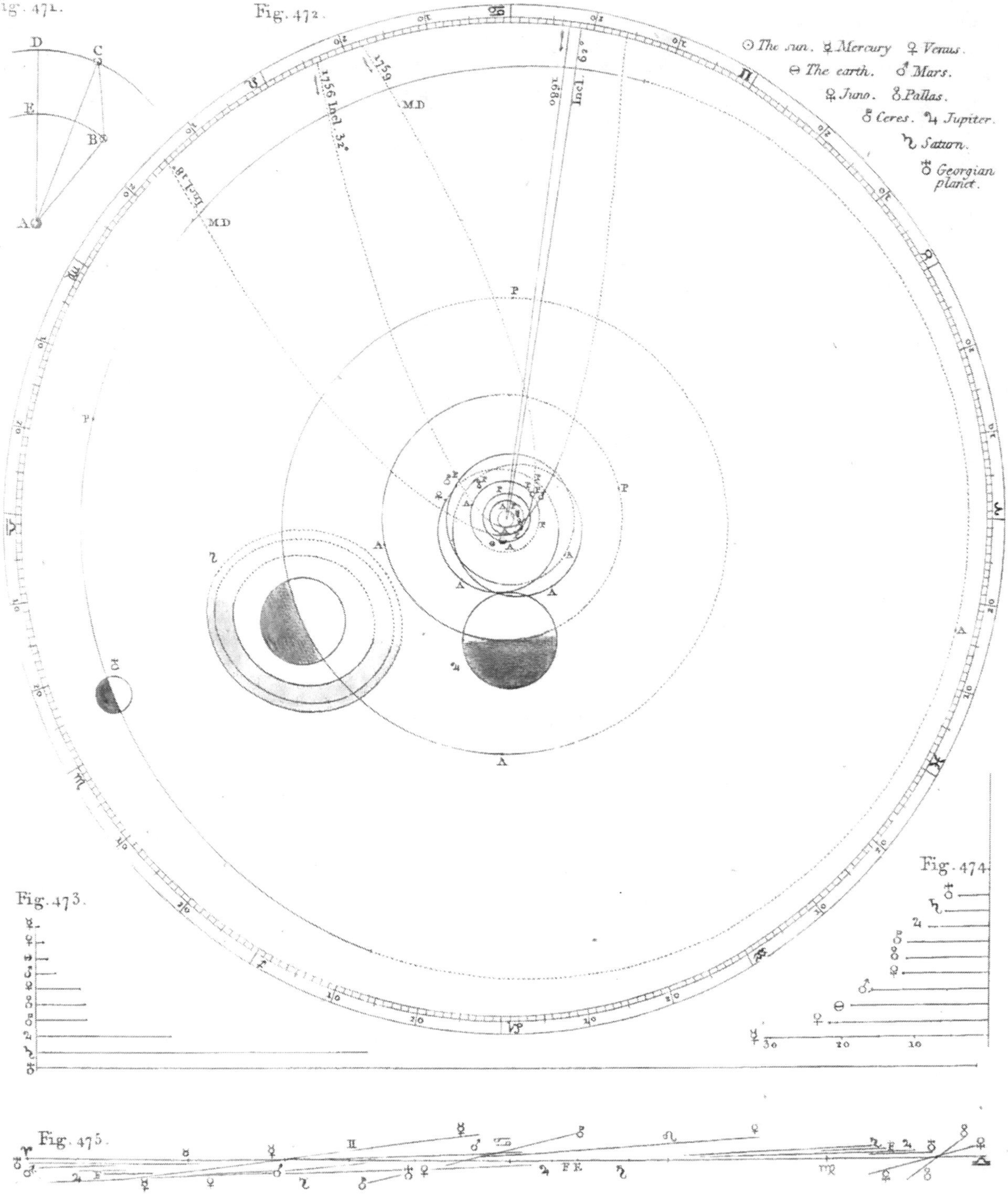

Pub. by J. Johnson, London 1 July 1806.

Joseph Skelton sculp

PLATE XXXII.

Fig. 471. A representing the sun, B the earth, and C the planet Mars; supposing Mars and the earth to set out together from D and E, the angle D A C was determined by Kepler from calculation, and the angles B A D and A B C by observation; whence it was easy to construct the triangle A B C, and to find the proportion of A B to A C. P. 505.

Fig. 472. The solar system, representing the form and proportions of the orbits of all the primary planets, and of three of the comets. The parts of the orbits represented by entire lines are on the north of the ecliptic, the dotted parts on the south: the letters A and P denote the aphelion and perihelion. The point in the centre, which ought to be only $\frac{1}{700}$ of an inch in diameter, represents the sun. The figures of the respective planets show their comparative magnitude, that of the sun being represented by the innermost of the graduated circles which inclose the whole: they are placed according to their actual situations on the 14th June, 1806. The letters M D show the mean distance of the comet of 1759, being placed at the extremity of the lesser axis of the ellipsis in which it must be supposed to revolve. P. 514.

Fig. 473. The periodical times of the different planets, represented by lines of different lengths. P. 514.

Fig. 474. The comparative velocities of the different planets, represented by lines which show the number of English miles described in a second, on the scale marked on the lowest line. P. 514.

Fig. 475. The places of the ascending nodes of all the planets, marked on one half of the ecliptic, supposed to be extended in a straight line; together with the inclinations of their orbits. The line marked F. F. E. E, shows the situation of the fixed ecliptic. P. 514.

PLATE XXXIII.

Fig. 476. A. The appearance of Venus, from Dr. Herschel: B, C, from Mr. Schroeter. P 514.

Fig. 477. A . . D, the appearance of Mars, from Dr. Herschel. The figures are inverted, as they appear in the astronomical telescope. P. 514.

Fig. 478. A, B. The appearance of Jupiter, with his belts, from Dr. Herschel. P. 514.

Fig. 479. The appearance of Saturn, with his ring, from Dr. Herschel. P. 514.

Fig. 480. The appearance of the moon, in an inverted position. The figure is copied from Mr. Nicholson's plate, the references from Cassini and Lalande. Eq. is the place of the moon's equator. P. 514.

Names of the spots, according to Riccioli, and Hevelius.

1	Grimaldus or.	Palus Marcotis
2	Galileus	Mons Andus
3	Aristarchus	Mons Porphyrites
4	Keplerus	Loca paludosa
5	Gassendus	Mons Cataractes
6	Schikardus	Mons Troicus
7	Harpalus	Insula sinus hyperborei
8	Heraclides	Caput mulieris
(*b*)	Vulcanus	
9	Lansbergius	Insula Malta
10	Reinoldus	Mons Neptunus
11	Copernicus	Mons Aetna
12	Helicon	Insula erroris
13	Capuanus	Regio Cassiotis
14	Bullialdus	Insula Creta
15	Eratosthenes	Insula Vulcania
16	Timocharis	Insula Corsica
17	Plato	Locus niger major
18	Archimedes	
(*a*)	Aratus	
19	Insula sinus medii	
20	Pitatus	Mare mortuum
21	Tycho	Mons Sinai
22	Eudoxus	Mons Carpathes
23	Aristoteles	Mons Serrorum
24	Manilius	Insula Berbicus
25	Menelaus	Byzantium
26	Hermes	Mons Bodinus
27	Dionysius	
(*d*)	Albategnius	
29	Plinius	Promontorium Acherusia
30	S. Theophilus	Mons Moschi
31	Fracastorius	Lacus Thospitis
32	Censorinus	Promontorium acutum
33	Messala	
34		Promontorium Somnii
35	Proclus	Mons Corax
36	Cleomedes	Montes Riphaci
37	Snellius	Mons Paropamisus
38	Petavius	Petra Sogdiana
39	Langrenus	Insula major
40	Taruntius	Sinus Phasianus
A	Mare Humorum	
B	Mare Nubium	
C	Mare Imbrium	
D	Mare Nectaris	
E	Mare Tranquilitatis	
F	Mare Serenitatis	
G	Mare Foecunditatis	
H	Mare Crisium	

Fig. 481 . . 483. The satellites of Jupiter, Saturn, and the Georgian planet, at their proper distances, in proportion to the diameters of the planets, shown on the same scale. P. 514.

Fig. 484. The figure of the tail of the comet of 1680, represented in the plane of its orbit, from Newton. A B is the earth's orbit, C and D are the first and last appearances of the tail, and E F is the line of the nodes. P. 514.

Fig. 485. A, B, Two successive appearances of the comet of 1723, from Lord Paisley. P. 514.

PLATE XXXIII.

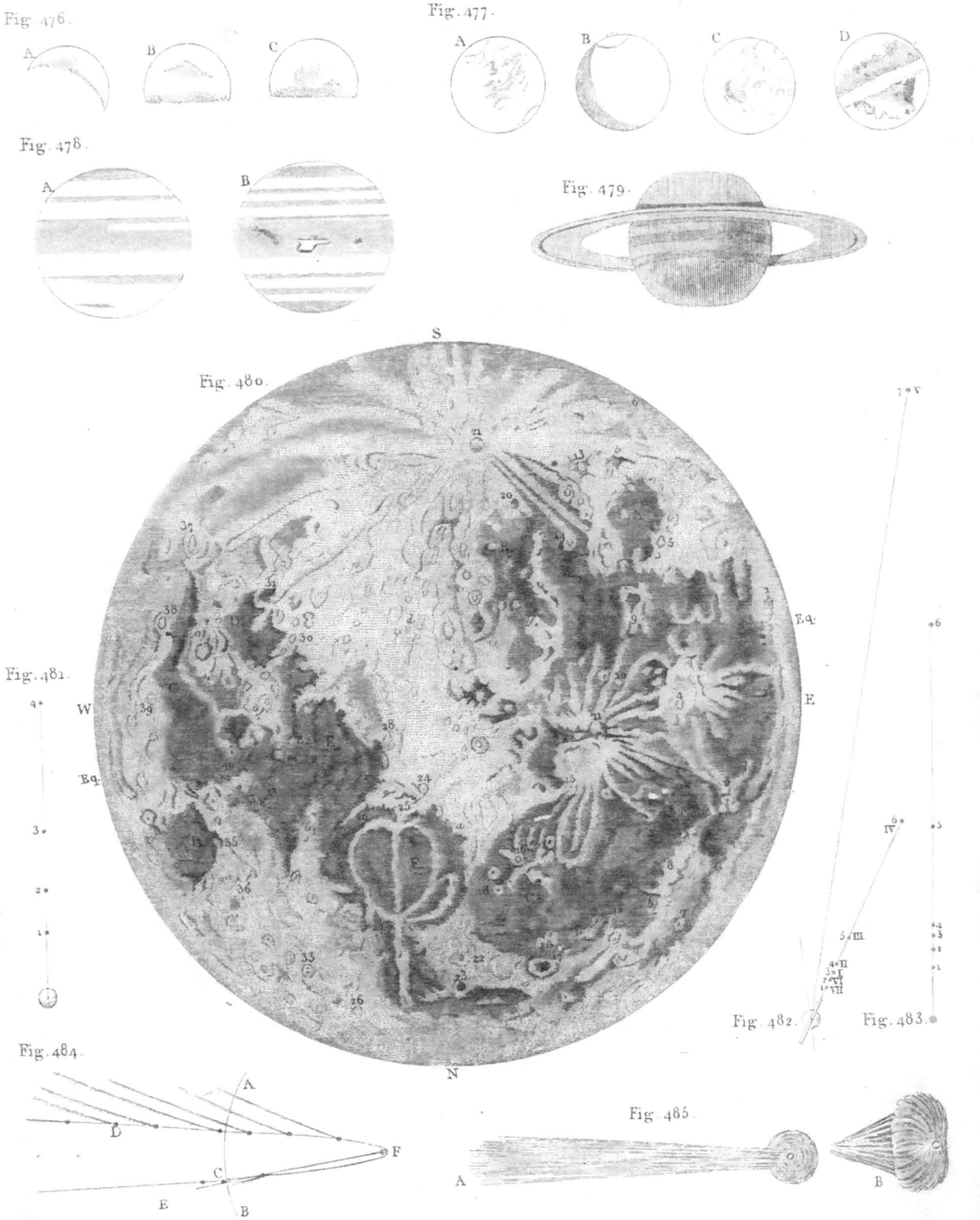

Joseph Skelton sculp.

Pub. by J. Johnson, London 1 July 1806.

PLATE XXXIV.

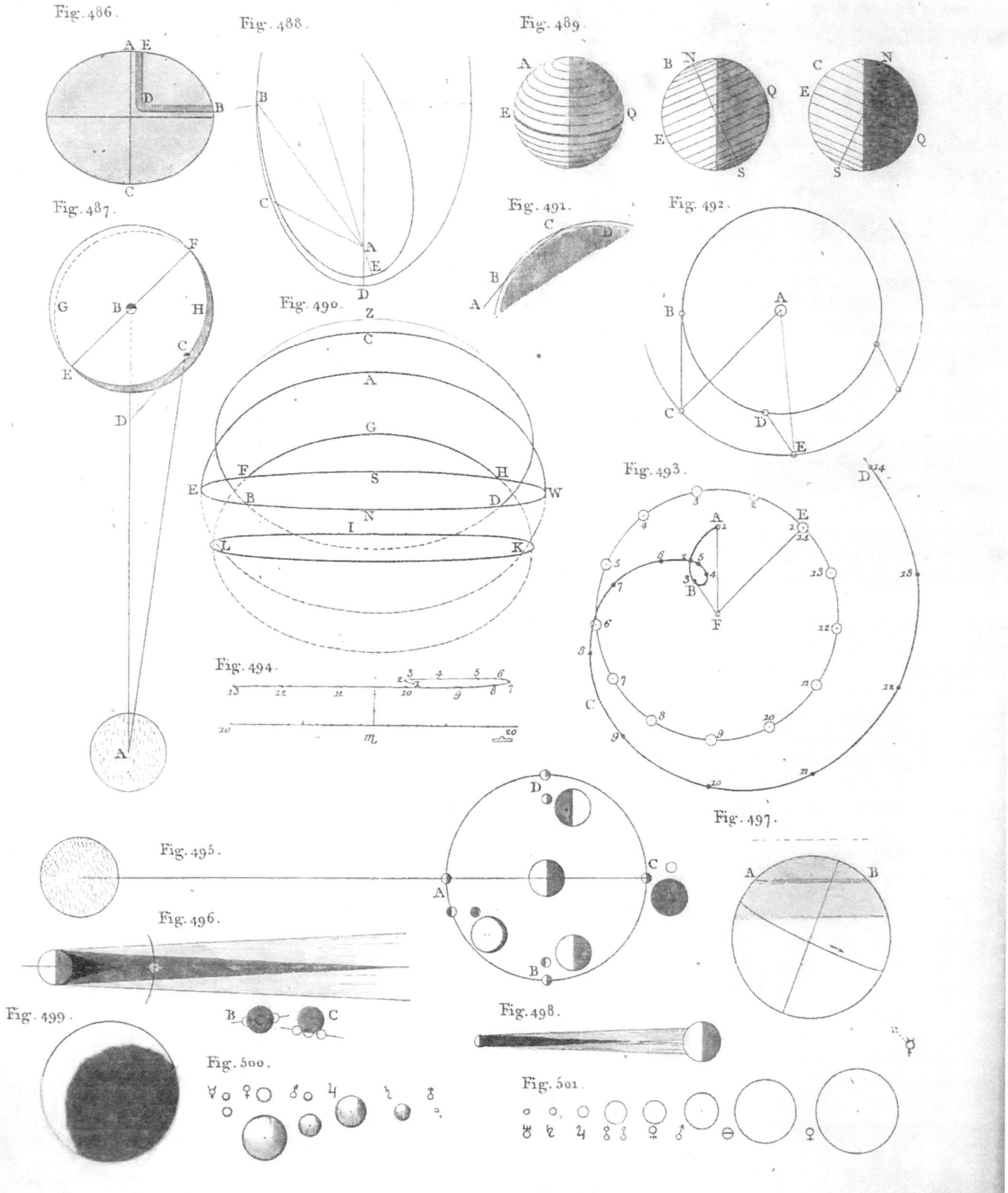

Pub. by J. Johnson, London 1 July 1806.

Joseph Skelton sc.

PLATE XXXIV.

Fig. 486. The gravitating body A B C, being supposed to revolve on the axis A C, the fluid column B D must be longer than E D, in order to support its pressure. P. 519.

Fig. 487. If A represent the place of the sun, B that of the earth, and C that of the moon, taking A D to A C as the square of A C is to the square of A B, A D will represent the sun's attraction acting on the earth, and C D the disturbing force, which, together with A D, makes up A C, the force acting on the moon; and it is obvious that, when the nodes are in any oblique situation, as E F, the force being directed to some point D, between B and A, while the moon moves from G to H, the force C D will tend to lessen the inclination, while the moon is ascending from E towards C, and to cause the node E to move back towards G, and, when it is again descending towards F, the inclination will be increased, and the node F made to recede towards H, until the moon arrives at H, and the force becomes directed to a point on the other side of B; the nodes only advancing while the moon is between H and F, or between G and E. P. 520.

Fig. 488. A body attracted towards the centre A, and descending from B in the ellipsis B C D, has the inclination of its orbit to the revolving radius A B, A C, A D, perpetually changed, until at D it becomes perpendicular to it; but when the force increases more rapidly, the radius does not become perpendicular to the orbit till it arrives at E, and the line of the apsides AD moves forwards to E. P. 521.

Fig. 489. A represents the position of the limit of light and darkness on the earth's surface at the vernal equinox, B at the summer solstice, and C at the winter solstice: E Q denotes the equator, N the north pole, and S the south. P. 525.

Fig. 490. N E S W being the horizon, and Z the zenith, E A W shows the sun's apparent path in London at the time of the equinoxes, B C D at midsummer, and F G H at midwinter, projected orthographically, as if the circles were described on the surface of a globe, and viewed from a great distance. The circle I K L is the boundary of twilight, supposing it 18° below the horizon, and its intersections with the sun's path show the beginning and end of twilight, as at I and K. P. 527.

Fig. 491. The rays of light, coming in the direction A B, are bent by the atmosphere so as to arrive at C, and to illuminate a part of the atmosphere there, which is visible, by means of a second refraction, to a spectator at D, and occasions the first and last twilight. P. 527.

Fig. 492. Venus is at her greatest elongation or angular distance from the sun A, when situated as at B, with respect to the earth at C; and she is stationary at D, when she is moving with the same velocity as the earth, with respect to the direction of the earth's motion, the line E D being then more oblique, with respect to a fixed line, than either before or after. P. 527.

Fig. 493. A B C D is the apparent path of Venus for the year 1806, supposing the sun E to revolve round the earth F. The place of the sun and planet is marked for every four weeks. P. 527.

Fig. 494. The apparent path of Saturn in the heavens for the year 1806, referred to its proper place with respect to the ecliptic. The figures denote the places at the beginning of each month. P. 527.

Fig. 495. The small figures represent the phases of the moon in different parts of her orbit. The smaller detached figures show the appearance of the moon, as seen from the earth; the larger ones, those of the earth at the same times, as seen from the moon, which are always the reverse of the moon's appearance. At A the moon is new; B is the first quarter, C the full moon, and D the last quarter. A and C are sometimes called the syzygies, and B and D the quadratures. P. 528.

Fig. 496. A, the moon passing through the earth's shadow; which is distinguished into three parts, the perfect shadow, the true shadow, and the penumbra. At B and C the moon is shown passing through the section of the shadow. P. 529.

Fig. 497. The path of the moon's shadow passing over the earth, in the solar eclipse of 1764, the earth being supposed at the same time to revolve on its axis. The line A B is the part in which the eclipse appeared annular, C D being the breadth of the whole shadow or penumbra. P. 529.

Fig. 498. The shadow of the moon falling on the earth. The true shadow not extending here to the earth, the cone formed by the continuation of its outlines marks the extent of the parts in which the eclipse appears annular. P. 529.

Fig. 499. The termination of the moon's disc in a solar eclipse. From Dr. Herschel. P. 529.

Fig. 500. The apparent magnitudes of the planets, that of the sun or moon being supposed equal to a circle a foot in diameter: where there are two figures, one of them shows the mean apparent magnitude, and the other the greatest. P. 531.

Fig. 501. The apparent magnitude of the sun, as seen from the different planets; for Mercury, the magnitude is shown by that of the earth in fig. 497. P. 535.

PLATE XXXV.

Fig. 502. A B being the earth's axis, the circle A C B is the meridian of the place C, and C D represents the plane of its horizon. P. 537.

Fig. 503. The effect of the obliquity of the ecliptic in the equation of time is shown by the difference of the angles A B C and D B E, subtended at the pole B by equal portions of the oblique circle A E. P. 538.

Fig. 504. A B being parallel to the earth's axis, the 12 planes passing through it, at equal angular distances, mark, on the circle C D perpendicular to it, the hour lines of an equatorial dial, and on the horizontal surface E F those of a horizontal dial. P. 538.

Fig. 505. A method of constructing a dial on any given plane. A B C is the elevation of the pole, or more generally, the angle which the surface makes with the gnomon A B. The circles are divided into equal parts, and 1, 2, 3, 4, 5, 6 are the hour lines, B being the place of the gnomon. The reason of this construction will appear by comparing the circle in the last figure with the ellipsis which is formed on the horizontal surface. P. 538.

Fig. 506. A dial for a pointed gnomon, or obelisc, drawn on a horizontal surface. P. 538.

Fig. 507. A mural quadrant, with its telescope; A B is the plumb line, for adjusting the instrument, and C the counterpoise for the telescope. P. 542.

Fig. 508. A portable transit instrument. A and B are screws for adjusting the axis by a vertical and a horizontal motion; C D is a spirit level, which may occasionally be hung on the telescope by the pins E and F. G is a small graduated arch, to be viewed through the microscope H, for taking elevations of a few degrees. P. 542.

Fig. 509. A transit circle, resembling Mr. Wollaston's, with a horizontal circle, by means of which both altitudes and azimuths may be measured. A is a microscope for viewing the plumb line, B another for reading off the divisions of the horizontal circle; C and D are spirit levels. P. 542.

Fig. 510. A zenith sector, with its telescope, which has usually a reflecting prism, like that of the Newtonian telescope, for its eyeglass. P. 542.

Fig. 511. The marine octant, introduced by Hadley. The mode of taking the common or front observation, is shown by the lines drawn to the sun and moon: the back observation by the two stars. A is a dark glass to be used in observations of the sun, and which may be fixed at B, when required. P. 542.

Fig. 512. A B being the situation of the earth's axis, if the angle C B D, or the altitude of the body D, be measured, and we subtract from it the elevation of the equinoctial C B E, the remainder will be the declination E B D. P. 542, 543.

Fig. 513. The angle A B C is the moon's horizontal parallax, and D B C the parallax when she is elevated above the horizon D E in the angle B D E. P. 543.

Fig. 514. The situation of the earth at the transit of Venus in June 1769. A spectator at the North Cape was carried during the transit from A to B, and the transit appeared to him to last while Venus moved from C to D: the island of Otaheite, on the contrary, which is situated on the lower part of the illuminated hemisphere, was carried from E to F, and the duration of the transit was there only while Venus moved from G to H. Hence the rotatory motion of the earth was compared with the excess of the motion of Venus in its orbit above that of the earth. P. 544.

Fig. 515. A planisphere nearly resembling that of Professor Bode. The outer circle is fixed to the chart, and is divided either according to the degrees of the ecliptic, or the days of the month; the graduated circle immediately within it is divided into 24 hours, and is fixed to a circle of pasteboard, out of which the circle N E S W, representing the horizon, is cut, the place being filled by thin varnished paper, with circles of azimuth and altitude engraved on it, which is carried round with the hour circle. P. 567.

Fig. 516. A diagram showing the length of the day, and the time of the sun's rising and setting in any part of the globe, within a few minutes; the time of the year being found in the graduated circle representing portions of the ecliptic, and the latitude on the middle line, by following the concentric circles of declination till they meet the horizon passing through the given latitude, the line drawn from the pole through this point will cut the equator in the point showing the length of the day or night. Thus, on the first of March, in latitude 50° north, the length of the day appears to be nearly 10 hours and $\frac{3}{4}$, whence the sun must rise about 37 minutes after six; but in latitude 85° the sun never sets on that day. P. 567.

PLATES XXXVI, XXXVII.

Plate XXXVI. Fig. 517. Projection of the constellations of the northern hemisphere on the plane of the equator. P. 498, 567.

Plate XXXVII. Fig. 518. Projection of the southern hemisphere. P. 498, 567.

PLATE XXXV.

Fig. 502.

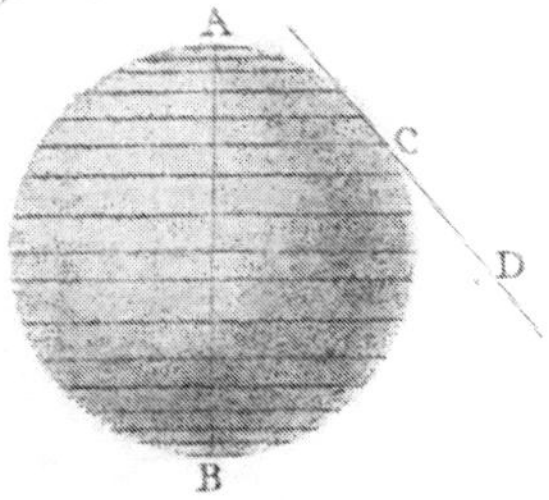

Fig. 503.

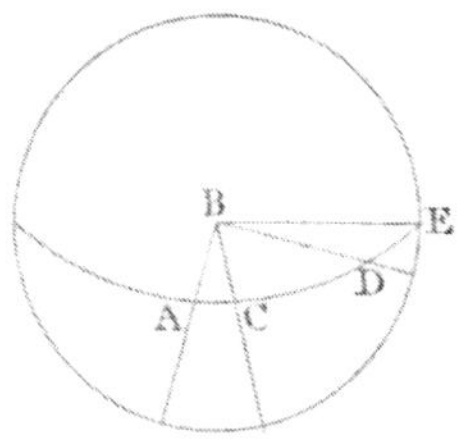

Fig. 504.

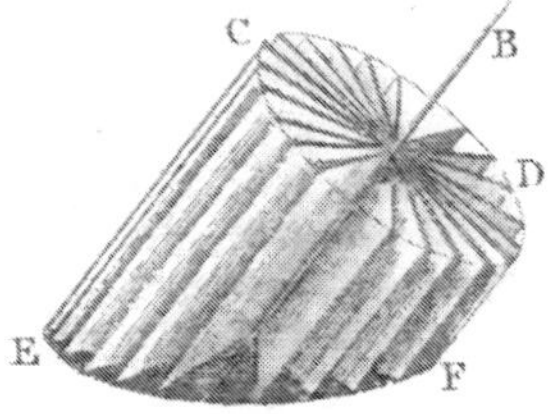

Fig. 505.

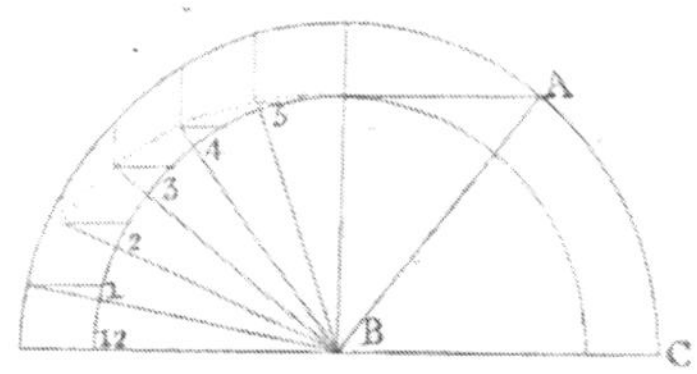

Fig. 506.

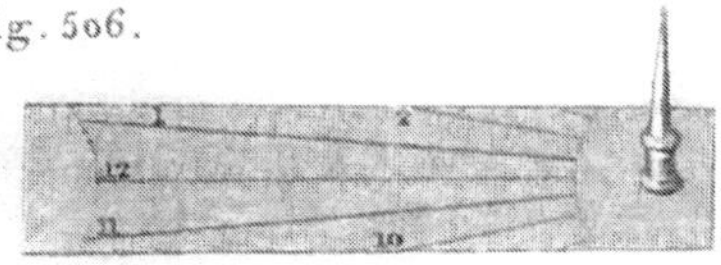

Fig. 507.

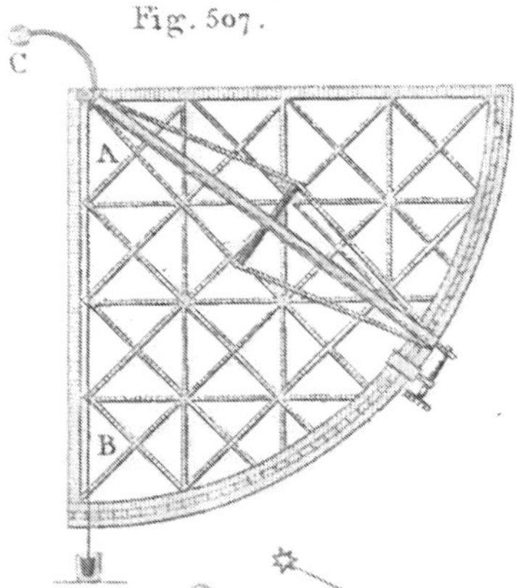

Fig. 508.

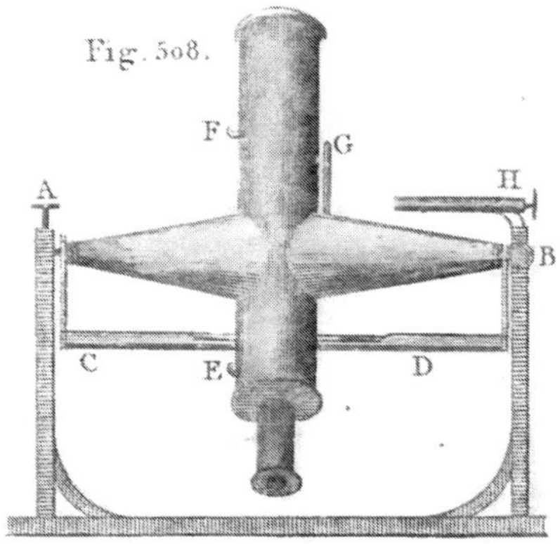

Fig. 509.

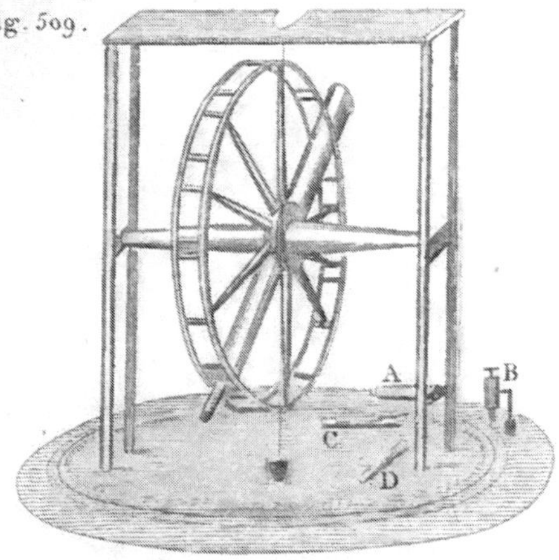

Fig. 511.

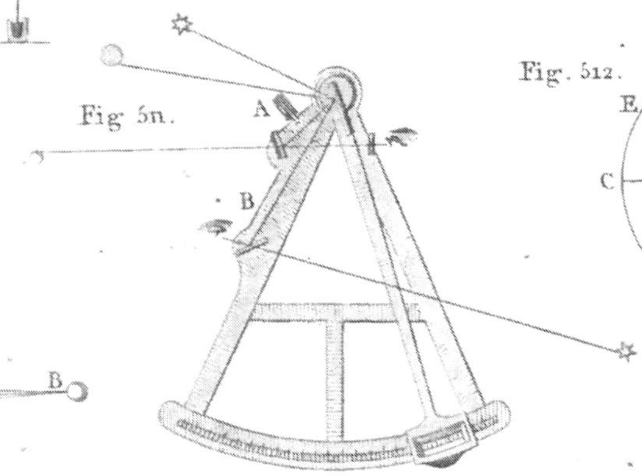

Fig. 512.

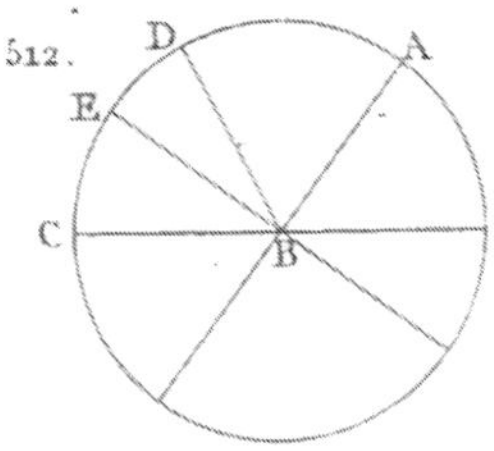

Fig. 513.

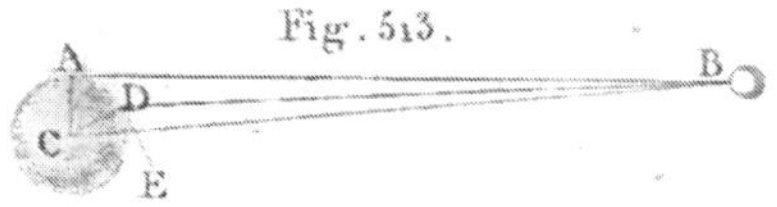

Fig. 510.

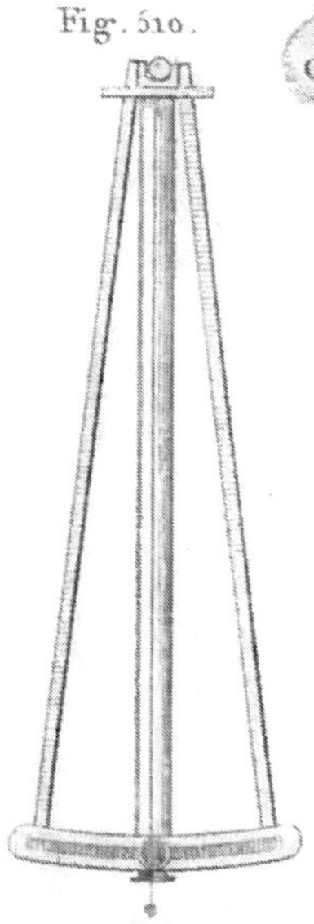

Fig. 514.

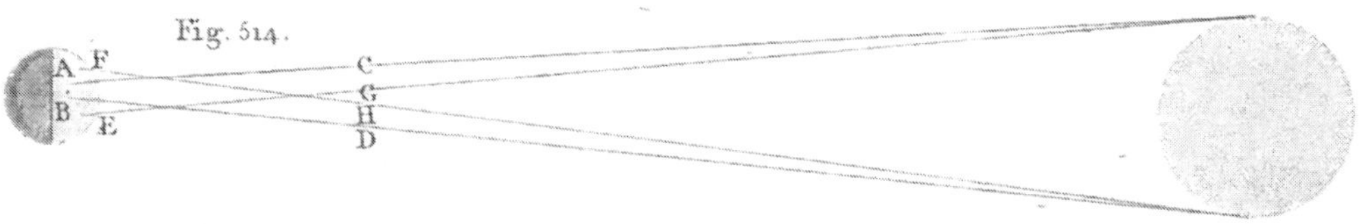

Fig. 515.

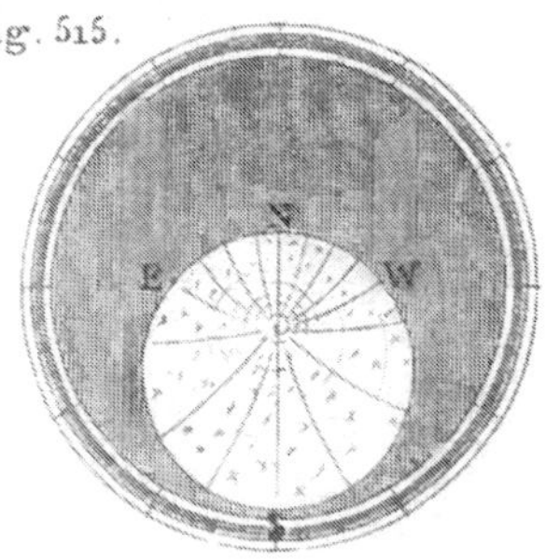

Fig. 516.

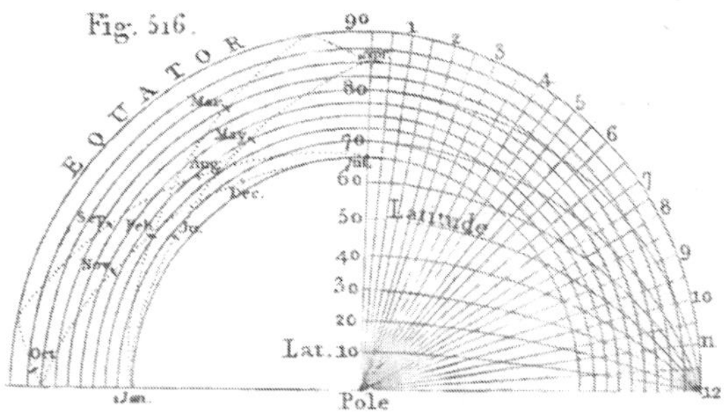

Pub by J. Johnson, London 1 July 1806.

Joseph Skelton sculp

PLATE XXXVI. Fig. 517.

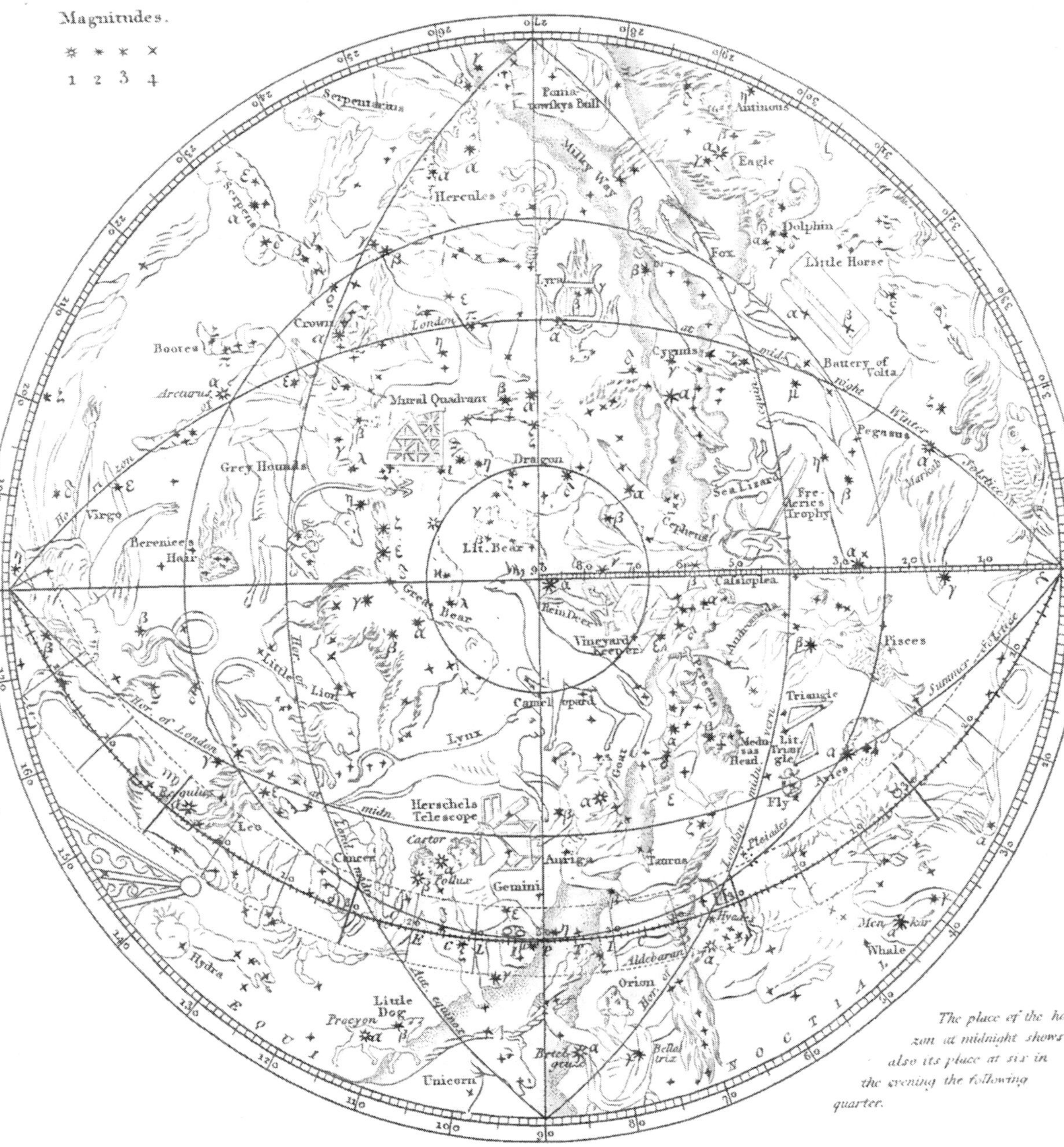

The place of the horizon at midnight shows also its place at six in the evening the following quarter.

Pub. by J. Johnson, London 1 July 1806.

Joseph Skelton sculp.

PLATE XXXVII. Fig. 518.

Magnitudes.

1 2 3 4

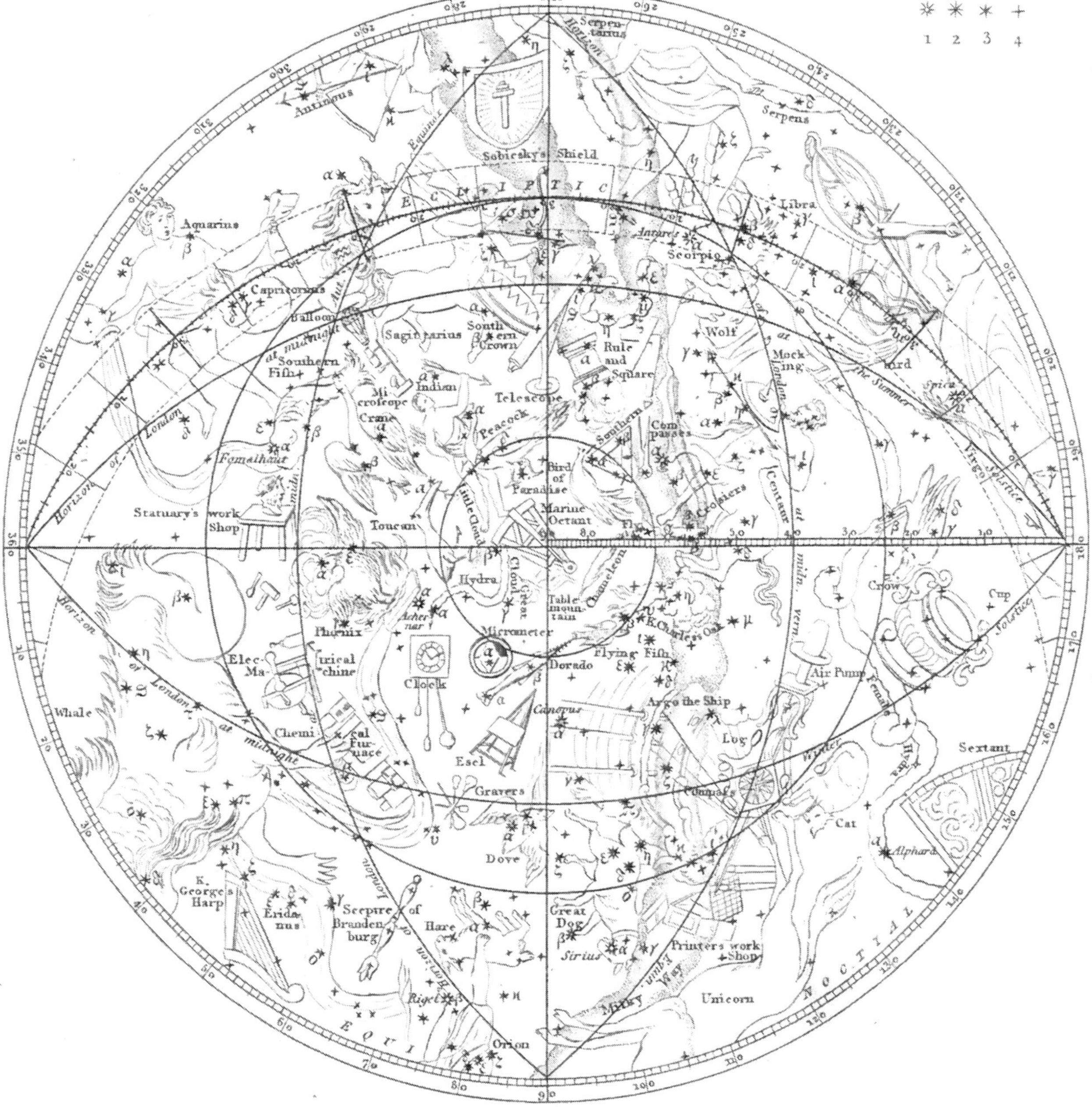

Pub. by J. Johnson London 1 July 1805.

Joseph Skelton sculp.

Fig. 519.

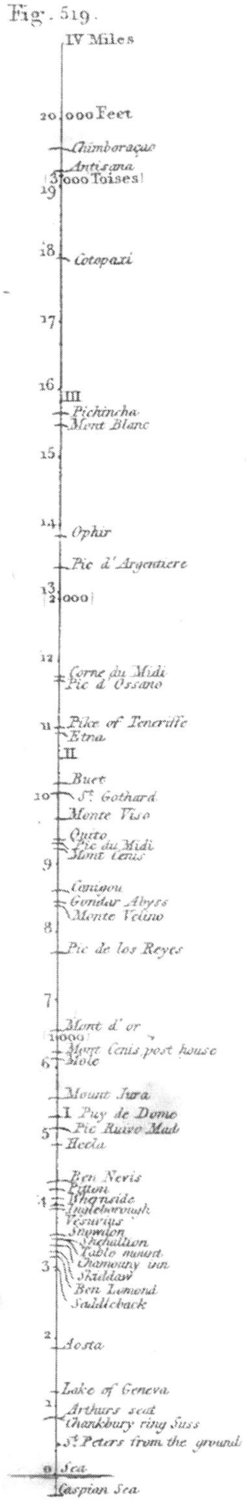

Fig. 520.

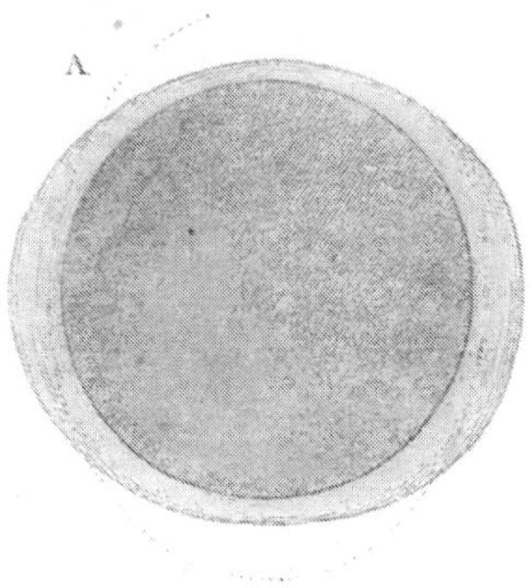

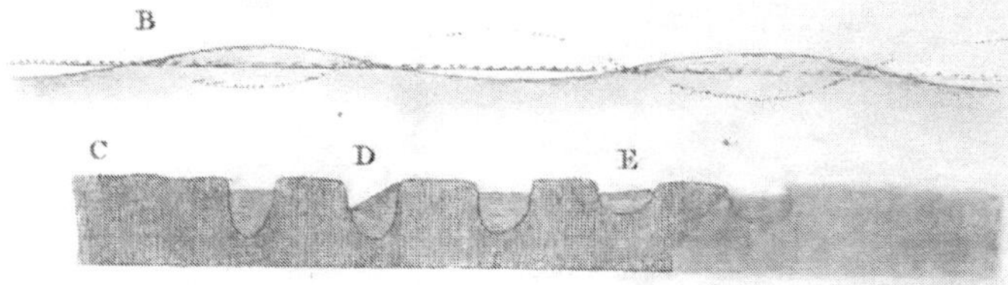

Fig. 521.

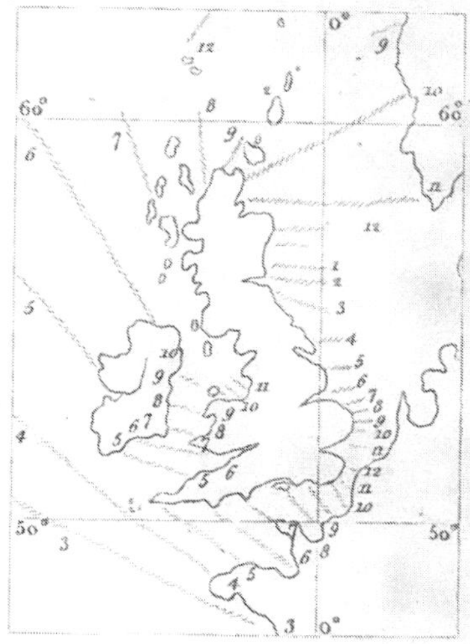

Fig. 522.

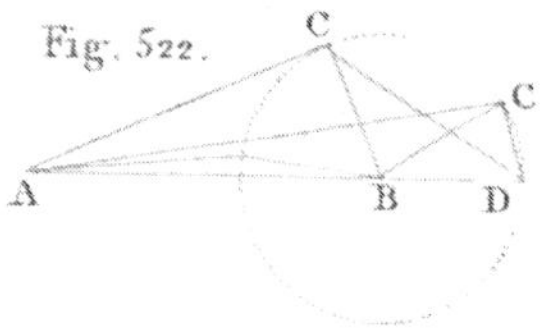

Fig. 523.

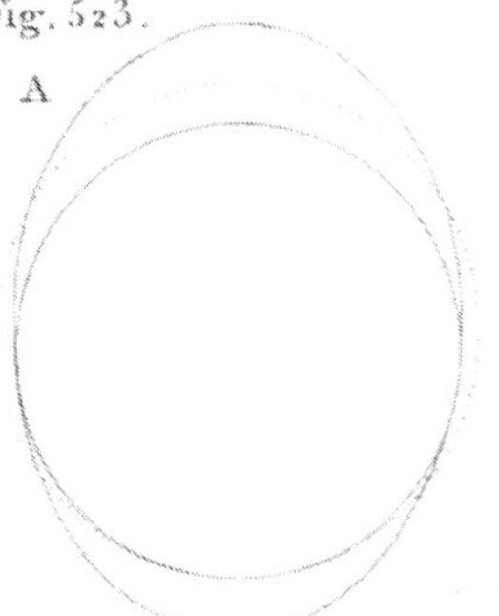

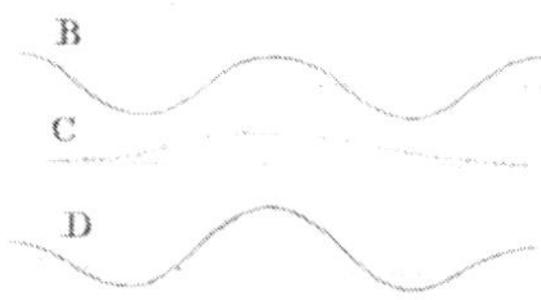

Fig. 524.

Fig. 525.

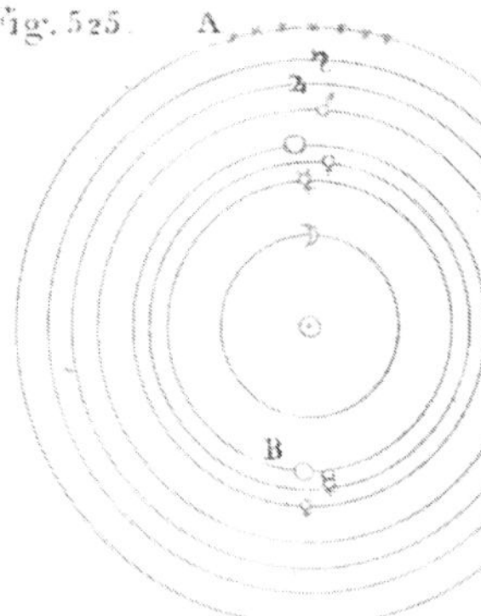

Fig. 526.

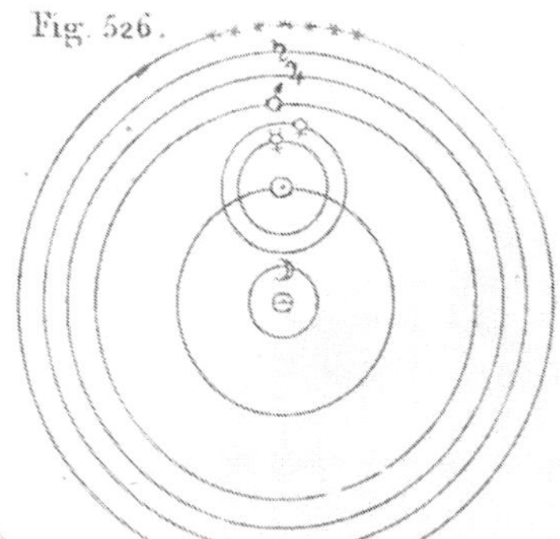

Fig. 528.

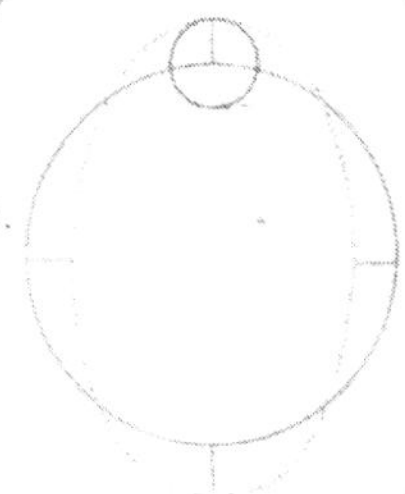

Fig. 527.

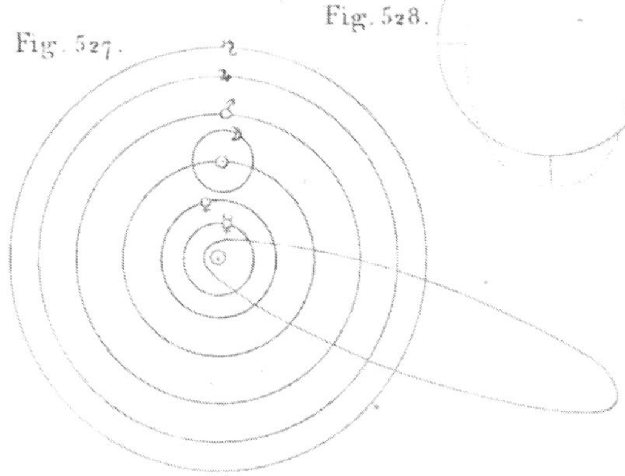

Fig. 529.

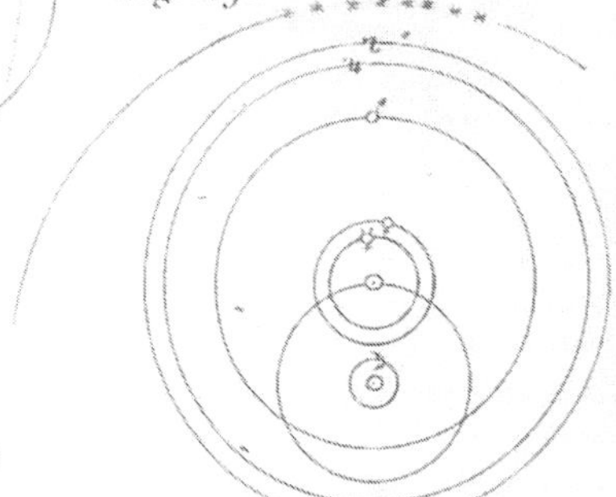

Pub. by J. Johnson, London 1 July 1806. *Joseph Skelton sculp.*

PLATE XXXVIII.

Fig. 519. A scale of the height of different parts of the earth's surface above the level of the sea, in English feet and miles, and in French toises. P. 574.

Fig. 520. A. The dotted ellipsis shows the section of a spheroid, which would be the form of the earth and sea if it were always in a state of equilibrium with the attraction of a distant body, and the shaded ellipsis the actual form assumed in consequence of its rotation round its centre, the depth of the sea being less than 13 miles. B. The surface of the sphere being supposed to be flattened, and the tides spread on it, they would assume the form of the waves here shown. The dotted straight line shows the mean height, which is a little above the surface in the principal sections of the spheroid, although not universally. C. The nature of the tides of lakes, the surface being regulated by that of the dotted line at B, nearly agreeing with it in direction, as at D, when the lake is narrow and deep, but differing from it, as at E, when shallower. P. 579.

Fig. 521. The progress of the tides from the Atlantic through the channels surrounding the British islands, the lunar tides happening in any part of the shaded lines nearly at the hour, after the moon's southing, which is indicated by the figure annexed to it. P. 582.

Fig. 522. The lines A B and B C, representing the heights of the lunar and solar tides, and the angle A B C twice their angular distance, or A D C being simply the angular distance, the line A C shows the height of the compound tide, and the angles B A C and A C B its distance from the lunar and solar tides respectively. P. 585.

Fig. 523. The two unequal tides represented by the elevation of the ellipsis above the smaller circle may be considered as composed of two equal tides cut off by the dotted circle, and the single tide between the two circles; as the tides B and C make the unequal tides at D. P. 587.

Fig. 524. The first and second curves represent two equal semidiurnal and one diurnal tide, which would make together two unequal tides: the third and fourth the same tides six hours more advanced: and when these are combined, the first and third destroy each other, but the second and fourth together compose the fifth, or a large diurnal tide. P. 587.

Fig. 525. A the ancient system of the world, adopted by Ptolemy. B the arrangement supposed by some other astronomers. P. 590.

Fig. 526. The Egyptian system of the world. P. 590.

Fig. 527. The system of the Pythagoreans, and of Copernicus. P. 592.

Fig. 528. The mode of representing the inequalities of the celestial motions employed by Ptolemy, the small circle being carried round the circumference of the larger, while the luminary revolves in it, so as to describe the dotted curve. P. 595.

Fig. 529. The Tychonic system of the world. P. 597.

3

PLATE XXXIX.

Fig. 530. The repulsive force of two particles of matter, situated at the distance A B or A C, is represented by the ordinates or perpendiculars B D, C E, drawn to the curve D E, supposing the force to be inversely as the distance; but the law of the force appears to be more nearly represented by a curve like F E. The line D F G shows the magnitude of the cohesive force, which overcomes the repulsion at the distance A G, and is balanced by it when the particles arrive at the distance A B or A H. The dotted lines represent the nature of the changes made in the lines F E, D F G, and F H, by an elevation of temperature. P. 619.

Fig. 531. The general direction of the cohesive force acting on a particle of a liquid at A being represented by A B or A C, that of the repulsive force will be D A or E A, and in order to maintain the equilibrium, the forces B F and C G, making together H A, must be supplied by the pressure or reaction of the internal parts. P. 620.

Fig. 532. A. The transverse section of a drop, supposed to be of considerable length, and flat at the sides: the curvature of the outline being every where proportional to its distance from the horizontal line A B. B, a round drop, the concavity at the horizontal line being equal to the convexity which would be found by cutting off the drop horizontally; the sum or difference of the curvatures being every where proportional to the distance from this line. P. 621.

Fig. 533. The solid A B possessing half the attractive power of the liquid C D, the surface of the liquid will remain horizontal: for the attractions will be represented by D A, D E, and D C; and of these D A and D E make D B, and D B and D C make D F, which is in a vertical direction. If the solid be more attractive, the forces will be combined nearly as at G, and if less attractive, as at H. P. 622.

Fig. 534. The form of the surface of a liquid in contact with a plane and vertical side of a solid which is wetted by it. The height of the ascent of water is about one fourth of that which is here represented. P. 622.

Fig. 535. The form of the surface of a liquid elevated between two plates which meet at A, and are at a little distance from each other at B; about one third of an inch, supposing the liquid to be water. P. 623.

Fig. 536. The height at which water will stand in tubes of the form and magnitude which are here represented. P. 623.

Fig. 537. The depression of mercury, in contact with a large or flat glass vessel, is one fourth as great as that which is here represented. P. 623.

Fig. 538. The depression of mercury within a small tube of glass. P. 623.

Fig. 539. The actual elevation of a portion of water in contact with a horizontal surface which is wetted by it. P. 624.

Fig. 540. The elevation of mercury in contact with a horizontal surface of glass. P. 624.

Fig. 541. A, a wide drop of water standing on a dry surface, not attracting it. B, a wide drop of mercury, standing on glass. P. 624.

Fig. 542. A magnified representation of the manner in which the seeds of lycopodium prevent a drop of water from wetting the substance on which it stands. P. 624.

Fig. 543. The bodies A and B, and the bodies C and D, appear to attract, and E and F to repel each other. P. 625.

Fig. 544. The apparent cohesion of two plates, between which a fluid is interposed. P. 625.

Fig. 545. The apparent attraction of a drop between two plates, tending to draw it towards the line of their junction, causes the drop to rest in an inclined position of the plates. P. 625.

Fig. 546. Dr. Herschel's figure, representing by the distance of the curve A B C from the line A C the heat thrown on different parts of A C by a prism, while D C is the illuminated part, divided according to Newton's experiments, the quantity of light being expressed by the distance of the line D E C. P. 639.

Fig. 547. Dr. Herschel's figure of the distribution of heat and light corrected according to the division of the coloured spectrum, as ascertained by Dr. Wollaston. P. 639.

Fig. 548. Bernoulli's air thermometer. P. 650.

Fig. 549. A differential air thermometer, or thermoscope, from which the pressure of the atmosphere is excluded. From Kunze. P. 650.

Fig. 550. A differential thermometer on Mr. Leslie's construction. P. 650.

Fig. 551. The distribution of the electric fluid in spheres of different sizes, and at different distances, and in a conical point. The density is represented by the distance of the dotted line from the surface. P. 663.

PLATE XXXIX.

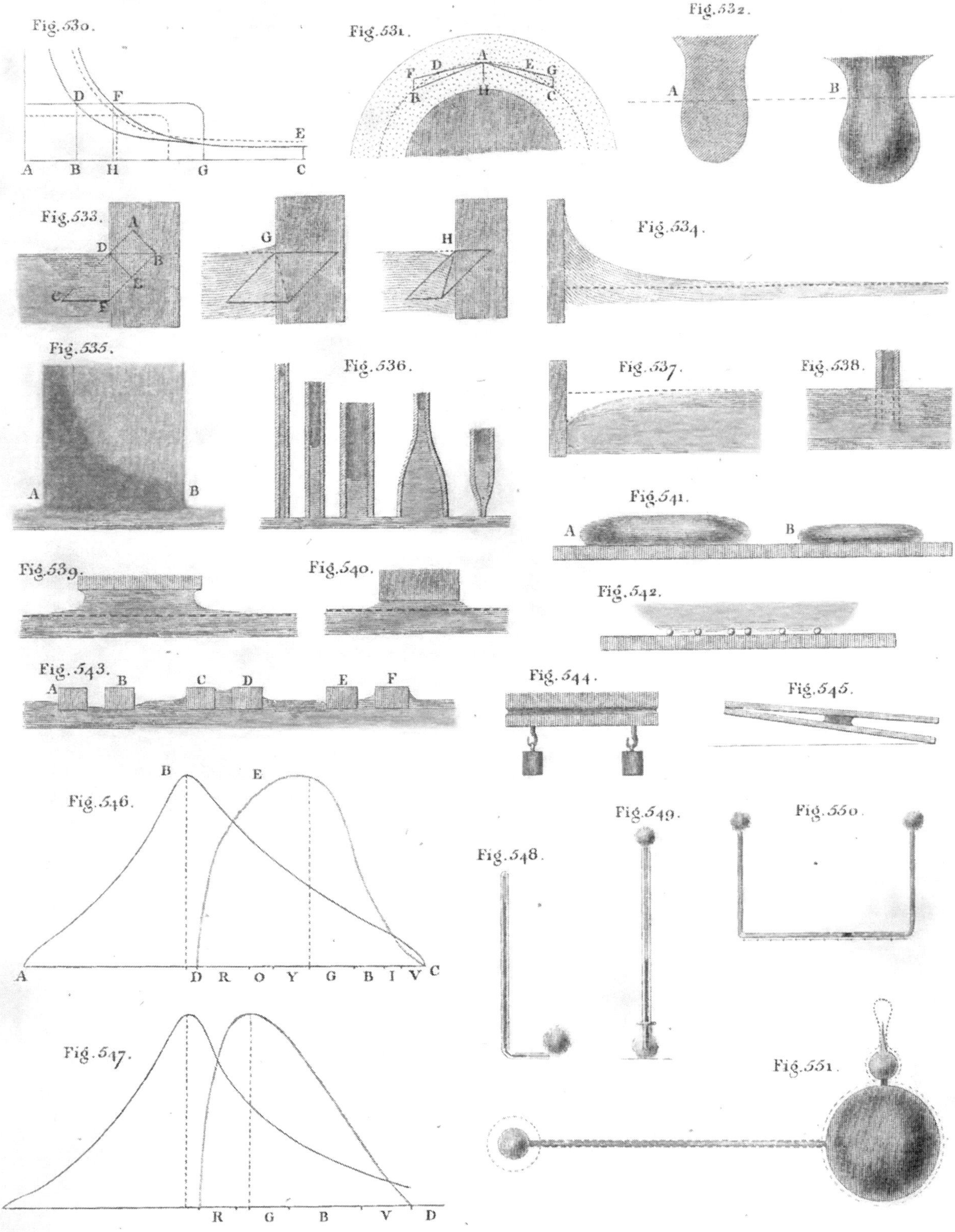

Jos. Skelton sculp.

Pub. by J. Johnson, London, July 1.1806.

PLATE XL.

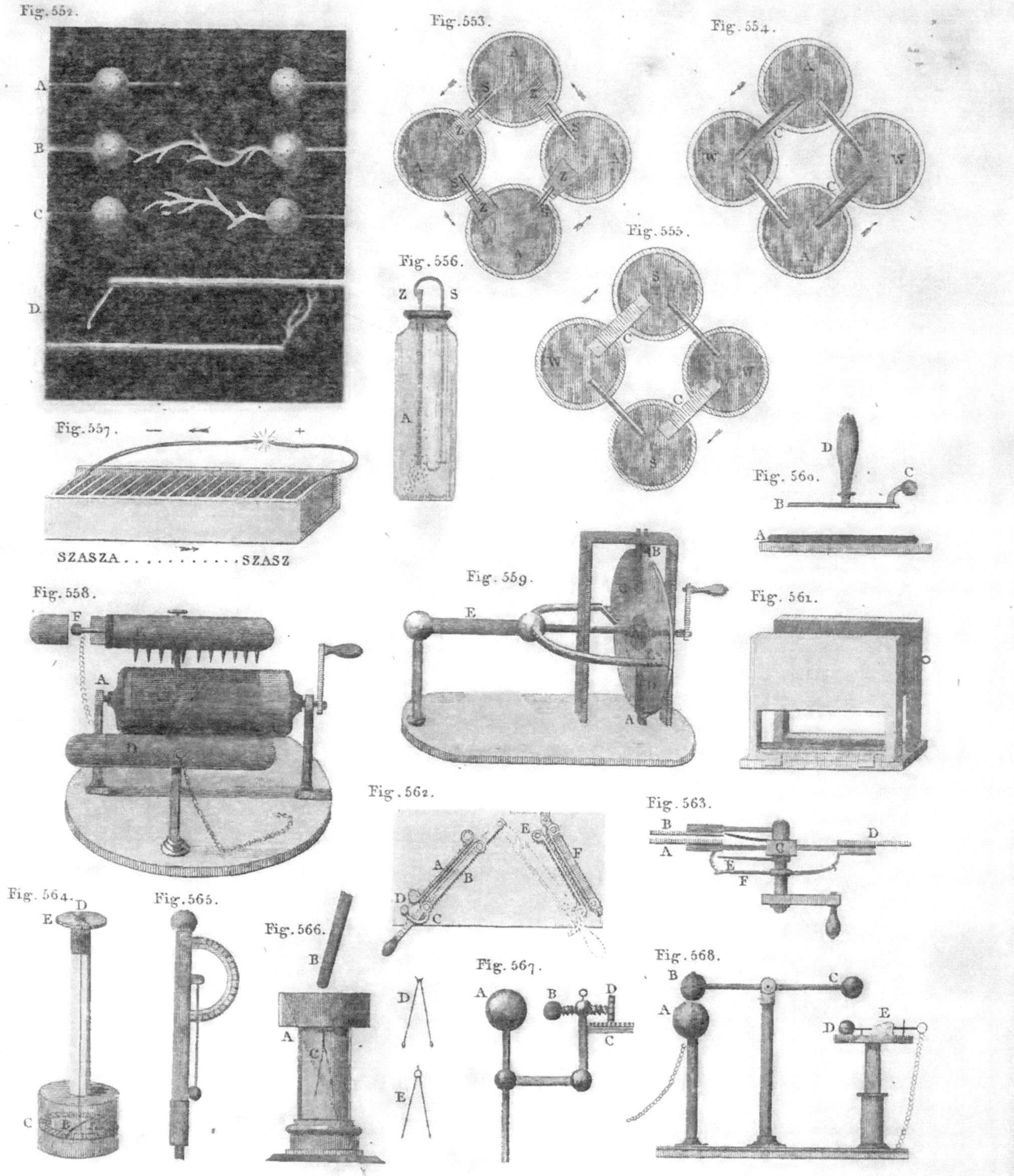

Published by J. Johnson, London, 2 July 1806. *Joseph Skelton sculp.*

PLATE XL.

Fig. 552. A. A spark passing between a negative and a neutral ball; B, between a neutral and a positive ball; C, between a negative and a positive ball. D, two sparks between a negative and a positive cylinder, each of the same form as if it were passing singly from the end of a charged to the side of a neutral cylinder. From Mr. Nicholson. P. 671.

Fig. 553. A compound galvanic circuit, formed by portions of an acid, pieces of zinc, and wires of silver; the arrows show the directions of the electric current. P. 676.

Fig. 554. A compound galvanic circuit, formed by an acid, charcoal and water, the water and the acid communicating by a small siphon. P. 676.

Fig. 555. A compound galvanic circuit, formed by portions of an alkaline sulfuret, and water, and pieces of copper: the liquids being connected by a siphon. P. 676.

Fig. 556. A simple galvanic circuit, formed by wires of zinc and silver, or platina, the lower ends being immersed in an acid, and the upper being brought into contact at pleasure. P. 676.

Fig. 557. A galvanic battery, in the form of a trough, composed of plates of zinc, silvered on one side, with vacant spaces for the reception of an acid: the letters show the order of the elements, and the arrows the direction of the current, from the positive wire + to the negative wire —. P. 677.

Fig. 558. An electrical machine, on Nairne's construction. A, the cylinder of glass; B, the cushion, or rubber; C, the silk flap; D, the negative conductor; E, the positive conductor; F, a ball connected with the internal coating of a glass jar, contained in the conductor. The conductors are insulated by varnished rods of glass. P. 680.

Fig. 559. A plate machine. A and B, the rubbers, which are usually double; C D, double flaps of oiled silk, for confining the electricity; E, the conductor. P. 680.

Fig. 560. An electrophorus. A, the cake of resin; B, the plate of metal; C, the ball for taking the spark; D, the handle of glass. P. 681.

Fig. 561. A condenser, as arranged by Mr. Cavallo, under the name of a collector: the middle plate is insulated: the two outward plates communicate with the earth; they stand near the first plate when the electricity is imparted to it, and are afterwards removed by means of their hinges. P. 681.

Fig. 562. Mr. Cavallo's multiplier. The electricity being first communicated to the insulated plate A, the moveable plate B is brought near it, while the wire C touches the pin D so as to form a communication with the earth; the plate B is then made to communicate with E, which is insulated, and stands near the plate F, which enables it to receive almost the whole of the electricity brought at each alternation by B; and when the plate F is removed from the neighbourhood of E, this plate becomes strongly charged. P. 682.

Fig. 563. A revolving doubler, on the principle of Mr. Bennet's instrument. The fixed and insulated plate A first receives the electricity, and when the moveable plate B stands opposite to it, it receives by a wire from the stand of the instrument C the opposite electricity; when it is brought opposite to D, this plate is made to communicate with the stand by the wire E, and acquires a charge similar and nearly equal to that of A. When B comes again to A, the wire F forming a communication between A, and D, nearly the whole charge of both these plates is brought into A, and B receives a charge almost twice as great as at first. P. 682.

Fig. 564. Mr. Coulomb's electrical balance. The needle A is made of silk, covered with sealing wax; it supports, at the end B, a ball of the pith of elder; another similar ball being fixed at C; the force of attraction or repulsion is ascertained by the torsion of the wire AD, which is measured by a graduated circle E. P. 683.

Fig. 565. Mr. Henley's quadrant electrometer; it is made of box wood, supported by metal; the ball is of cork, the graduated arc of ivory. P. 683.

Fig. 566. A, Mr. Bennet's gold leaf electrometer; B, a piece of excited sealing wax held over it, for distinguishing the electricity. Instead of the pieces of gold leaf C, we may substitute Mr. Cavallo's pith balls D, or the straws E, employed by Volta. P. 683.

Fig. 567. Mr. Lane's discharging electrometer. The distance of the balls A, B is measured by the turns of the screw on the scale C; and the parts of a turn are ascertained by the graduated circle D. P. 683.

Fig. 568. A discharger for a battery. When the repulsion of the balls A, B, becomes greater than the weight of a wire which passes through a perforation in the balls, they separate, and the ball C, descending to D, forms a communication, which completes the circuit, so that the shock passes through any substance placed at E. P. 683.

PLATE XLI.

Fig. 569. The form of the curves which show the direction of the magnetic needle, in consequence of the attraction and repulsion of two poles, situated at A and B. They are found by drawing the lines A C D, B E D, so that the sum or difference of the parts A C, B E, shall be always equal, A C E B being a semicircle: and the direction D F may be found by making A F to B F as the cube of A D to that of B D. P. 688.

Fig. 570. The arrangement of iron filings in the neighbourhood of a magnet. P. 688.

Fig. 571. The particle of iron A B, lying on a card nearly over the magnet C, assumes, when the card is shaken, first the position D, then, falling to E and F, is left a little further from the magnet than at first. P. 689.

Fig. 572. An azimuth compass. The box is turned round, until the shadow of the thread A B or A C falls on the line C D: the position of the needle is then ascertained by that of the card E, which is fixed on it. The compass is kept always in a horizontal position, by means of a double suspension on the gimbals E G. Instead of this suspension, Mr. M'Culloch makes the bottom of the box in the form of a hollow cone, resting on a point, and loaded with a weight, which brings the centre of gravity below the point of support, as at H. P. 689.

Fig 573. A dipping needle. The piece A B is brought into such a situation, that the line drawn on it coincides with the middle of the vibrations of the needle. The position of the needle may be changed, either by turning the stand half round, or by turning the needle within the stand. P. 689.

Fig. 574 . . 576. The situations of the lines of equal declination in 1700, 1744, and 1794, in the hemisphere, which is bisected by the meridian of London. The first two from Mountaine's Tables, the last from Churchman's Chart. P. 691.

Fig. 577. The actual situations of the lines of equal dip. From Churchman's Chart. P. 692.

Fig. 578. The lines of equal dip, calculated from the supposition of a small magnet, situated at the centre of the earth, directed to a point in latitude 75° N. and longitude 70° W. P. 692.

Fig. 579. A, Six's thermometer; B, the wire with a fine spring, which serves as an index. P. 697.

Fig. 580. Rutherford's double thermometer. P. 697.

Fig. 581. Deluc's whalebone hygrometer. A, the slip of whalebone; B, a spiral spring, serving to keep it stretched; C, the index. P. 710.

PLATES XLII, XLIII.

Fig. 582. A chart of the world, on Mercator's projection, from Arrowsmith; with the dip and variation of the compass, principally from Churchman, for the year 1794; and with the trade winds and monsoons. P. 571, 691.

PLATE XLI.

Fig. 569.

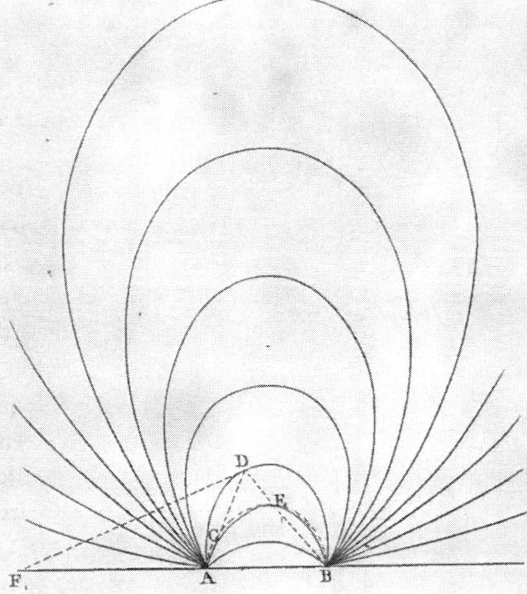

Fig. 570.

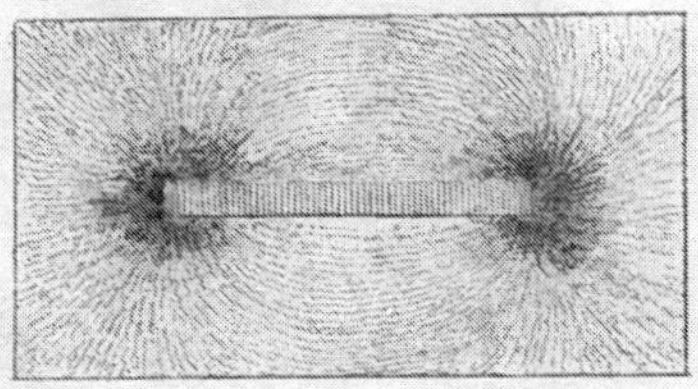

Fig. 571.

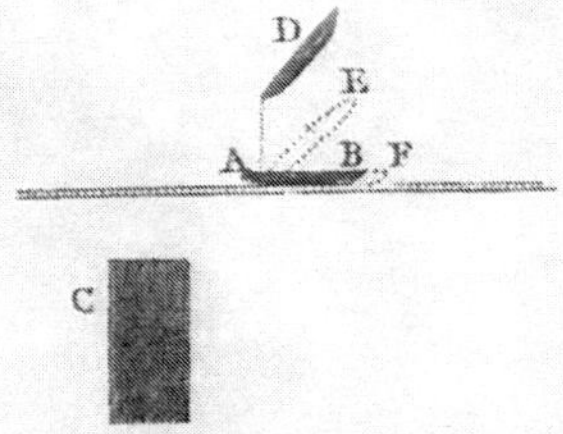

Fig. 572.

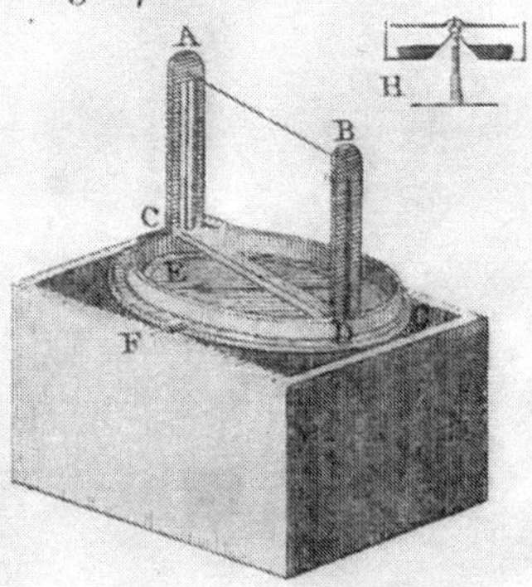

Fig. 573.

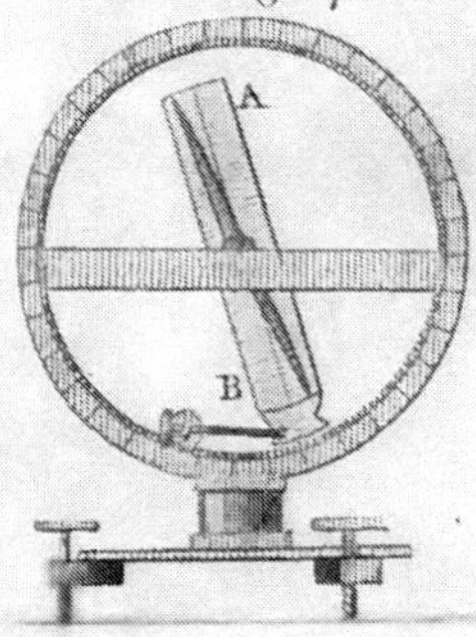

Fig. 574.

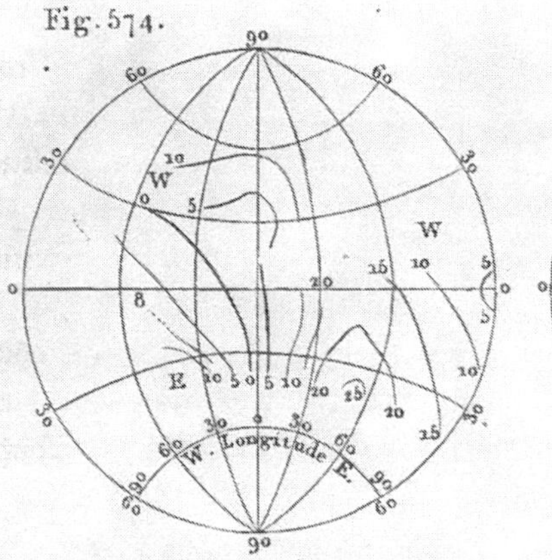

Fig. 575.

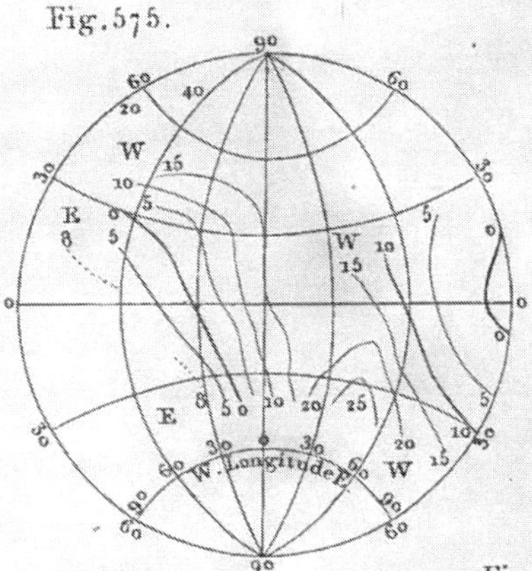

Fig. 576.

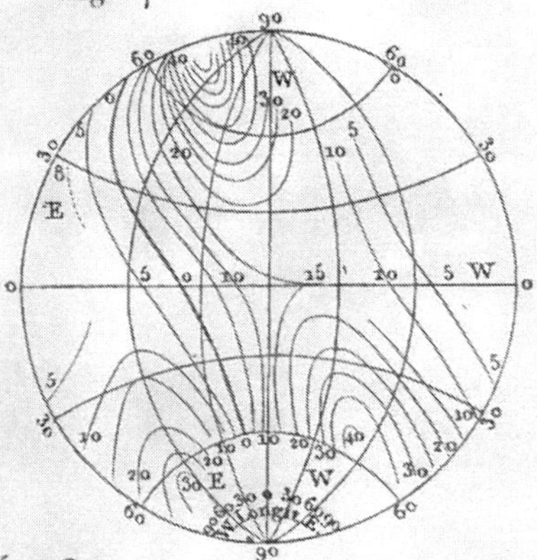

Fig. 577.

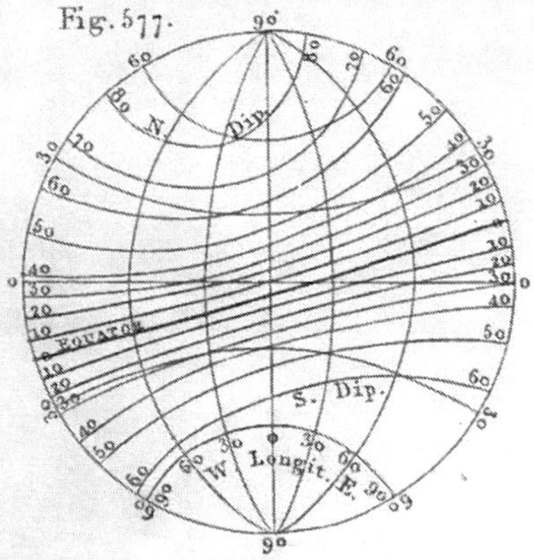

Fig. 578.

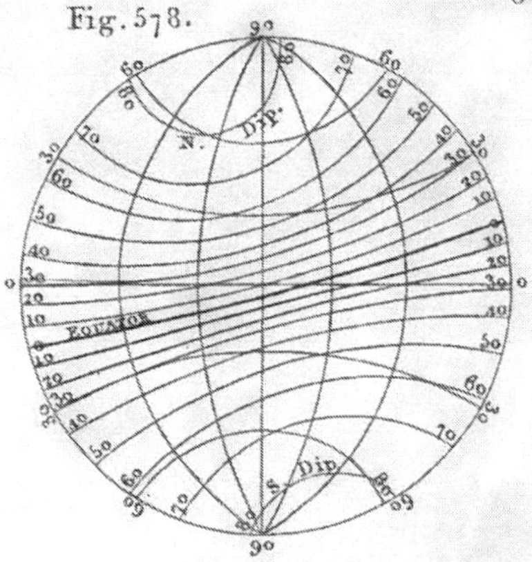

Fig. 579

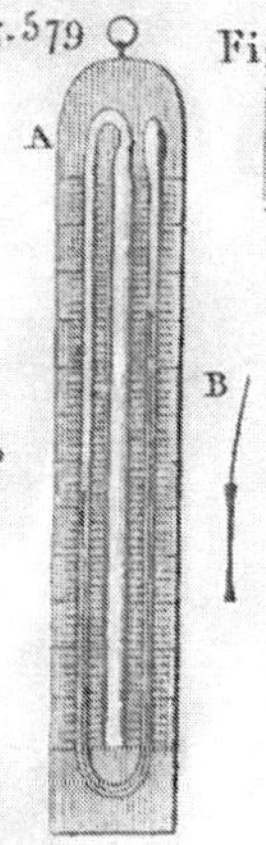

Fig. 580.

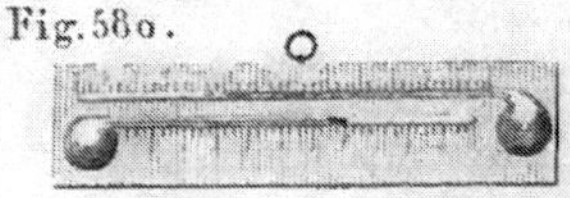

Fig. 581.

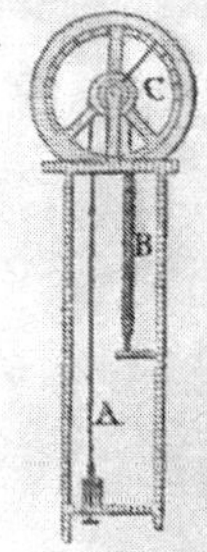

Pub. by J. Johnson, London 1 July 1806. *Joseph Skelton sculp.*

PLATE XLII. Fig. 582.

Pub. by J. Johnson, London 1 July 1806.

To be bound facing Plate XLIII.

PLATE XLIII. Fig. 582.

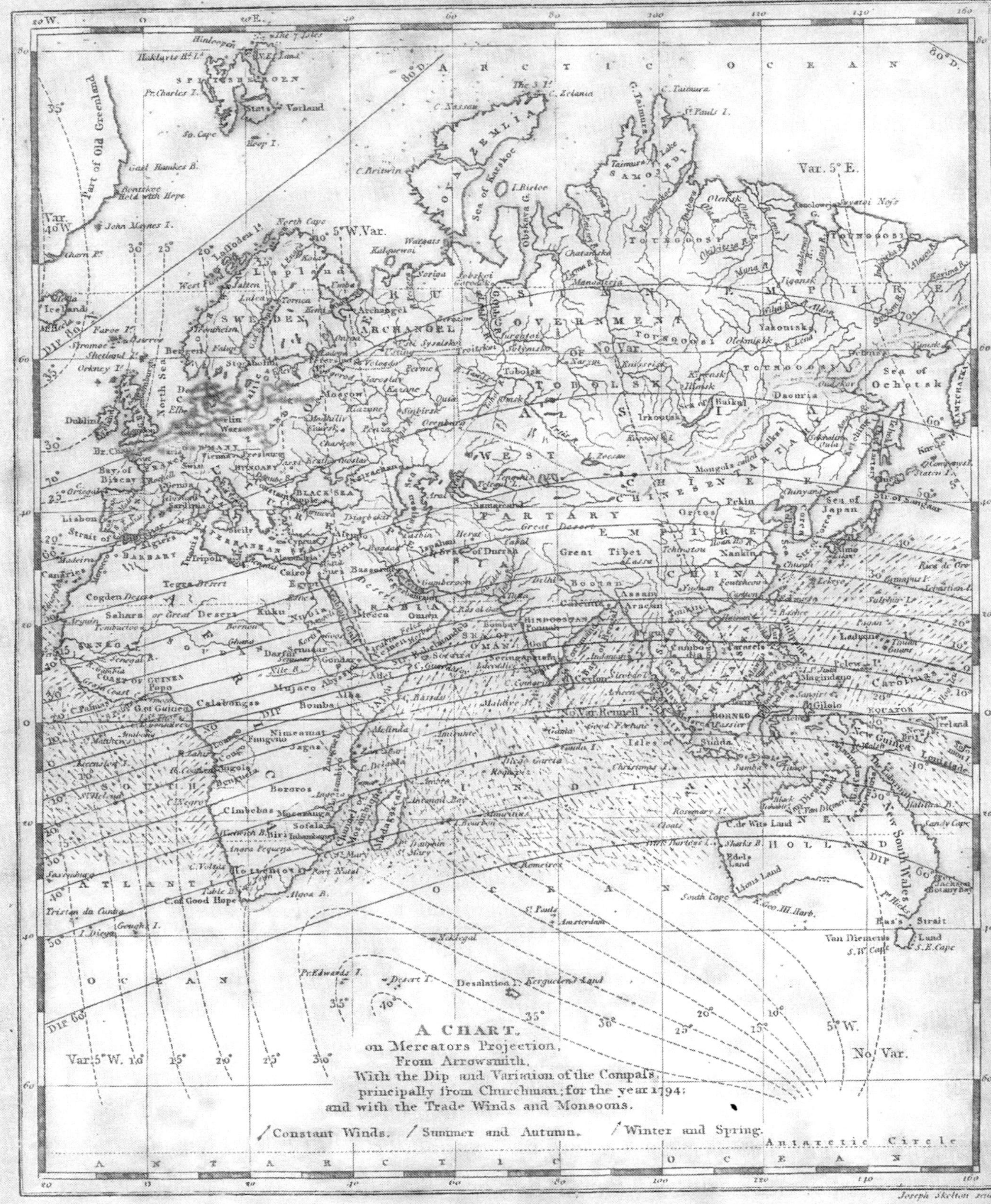

Pub. by J. Johnson London 1 July 1806.

A COURSE OF LECTURES ON NATURAL PHILOSOPHY AND THE MECHANICAL ARTS

CONTENTS.

PART THE FIRST. MECHANICS.

B

actions may perhaps also be considered as tending to illustrate the vibrations of chords. The latest improvement which deserves to be mentioned, with respect to the theory of sound, is Laplace's explanation of the increase of its velocity on account of the effect of heat, which appears to afford a satisfactory explanation of a difficulty so much the more important, as it tended to lessen our confidence in every part of a theory, which differed so widely from the most accurate and best established observations.

CHRONOLOGY OF ACUSTICS.

700 B.C.–600	600–500	500–400	400–300	300–200
TERPANDER	.PYTHAGORAS .SIMO	. NIDES.	.ARISTOTLE.	
200 B.C.–100	**100–BIRTH OF CHRIST.**	**CHRIST.–100**	**100–200**	**200–300**
CTESIBIUS				
300–400	**400–500**	**500–600**	**600–700**	**700–800**
		GREGOR	Y.	
800–900	**900–1000**	**1000–1100**	**1100–1200**	**1200–1300**
		GUIDO		
1300–1400	**1400–1500**	**1500–1600**	**1600–1700**	**1700–1800**
		.F. BAC .GALI .M	ON. .SAUVEU LEO. .TAY ERSENNE. R .KIRCHER. .WALLIS .NEWT	R. LOR. OMIEU .D. BERNOULLI. ..L. EULER. ON.. LAMBERT.

PLATE 1.

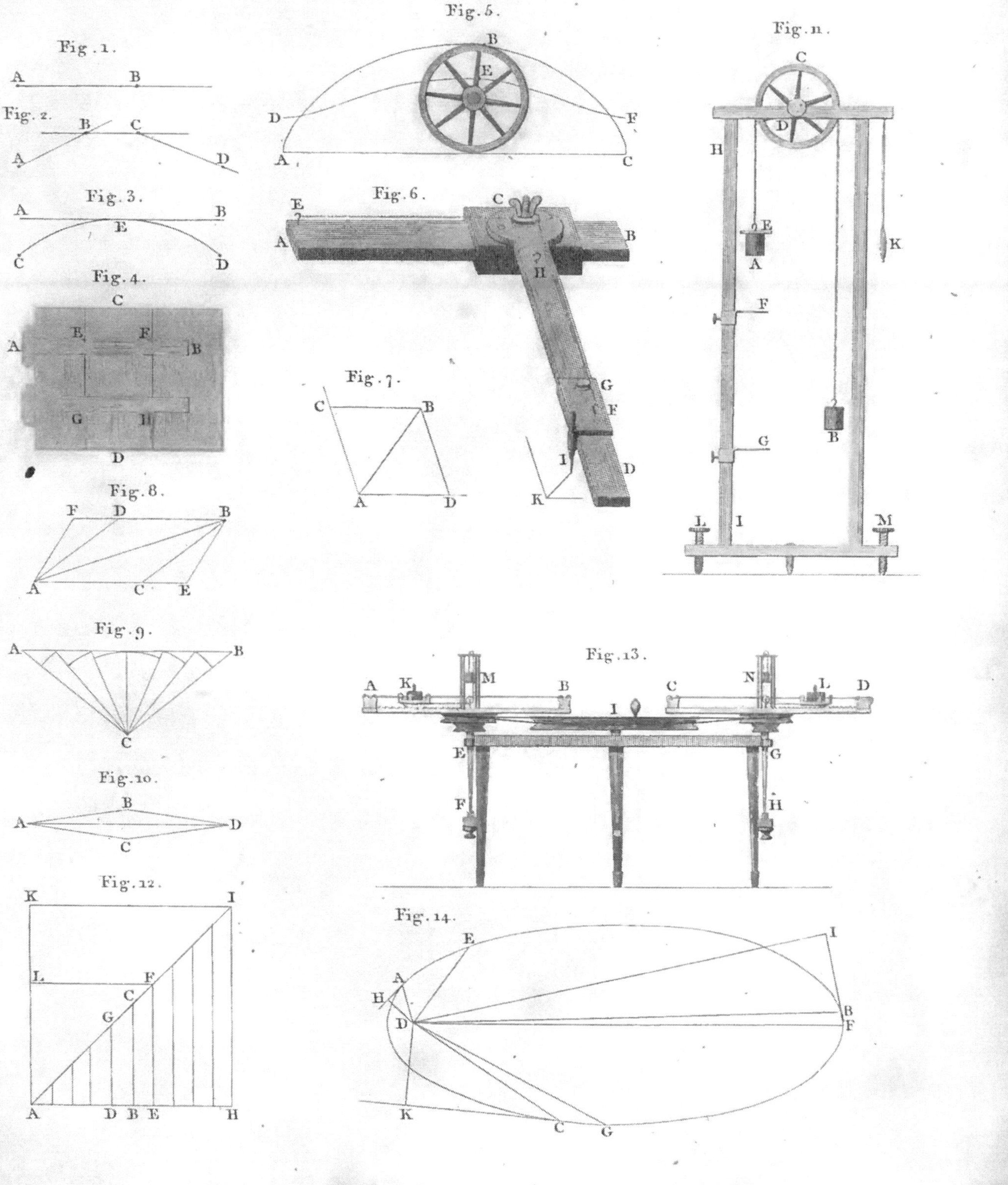

Pub. by J. Johnson, London 1 July 1806.

Jos^h. Skelton sculp.

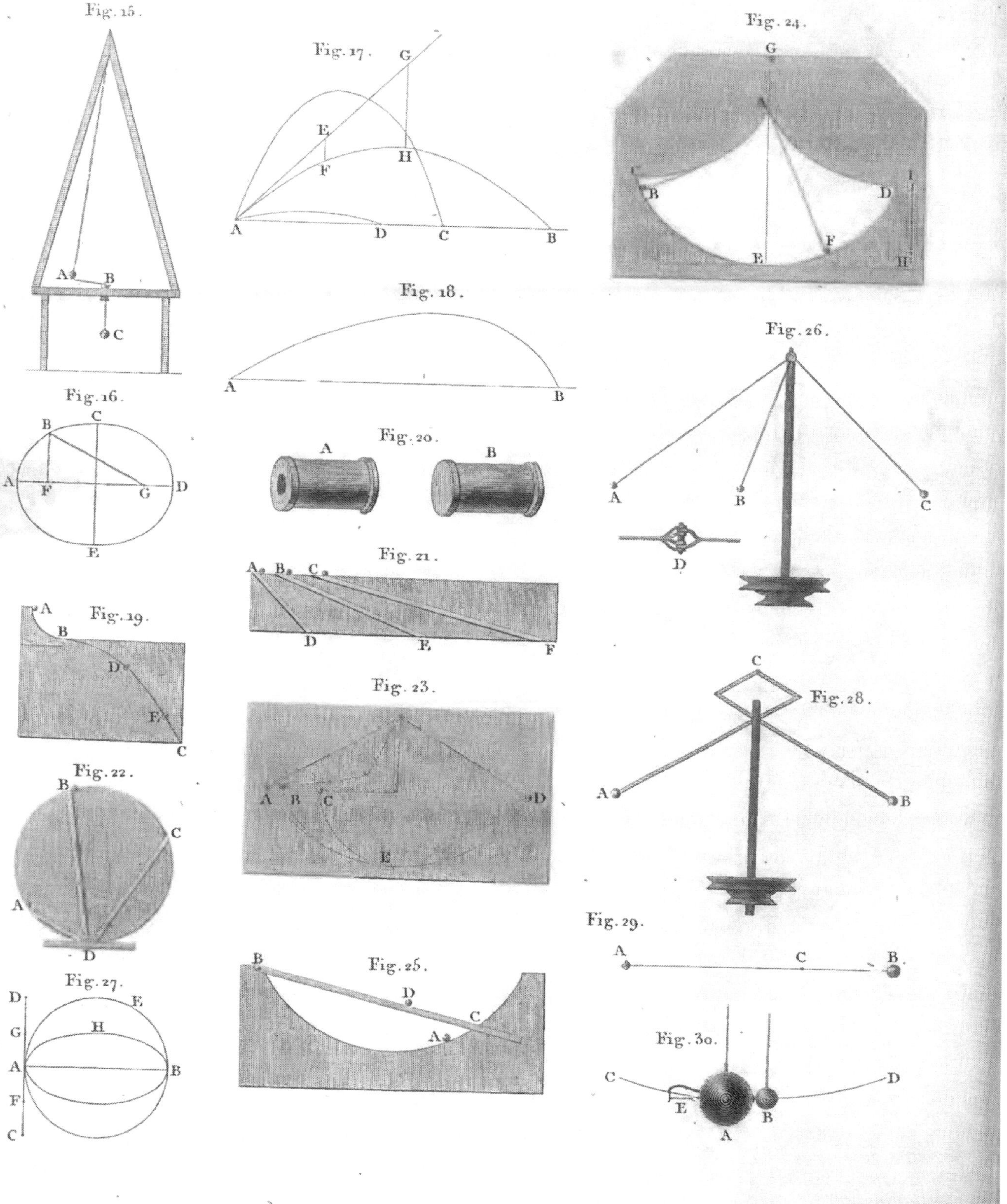

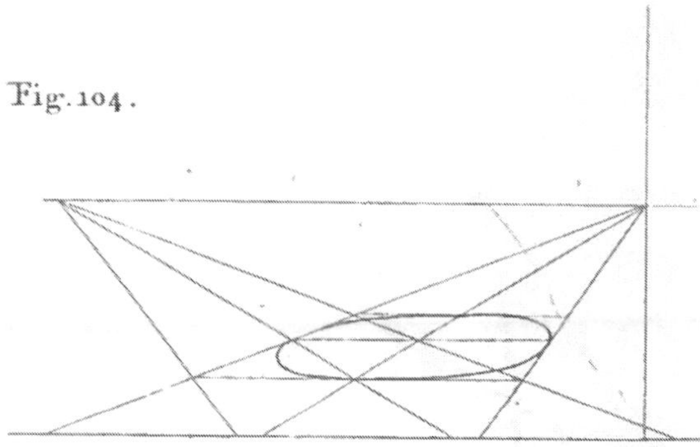

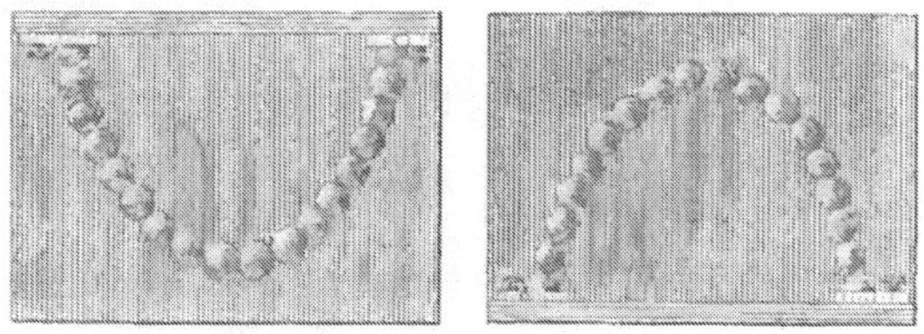

PLATE XII.

Fig. 156.

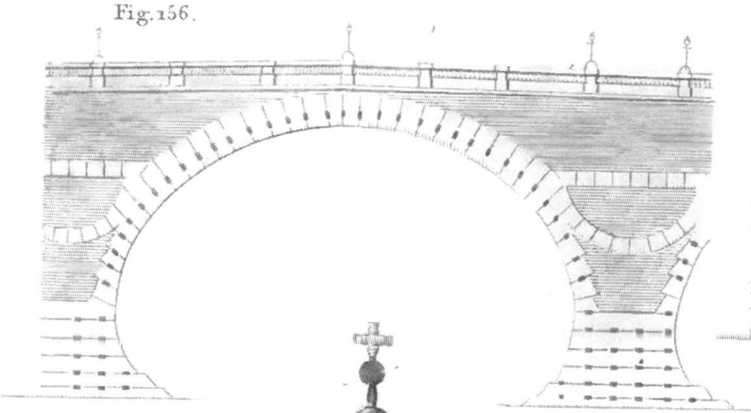

Fig. 157.

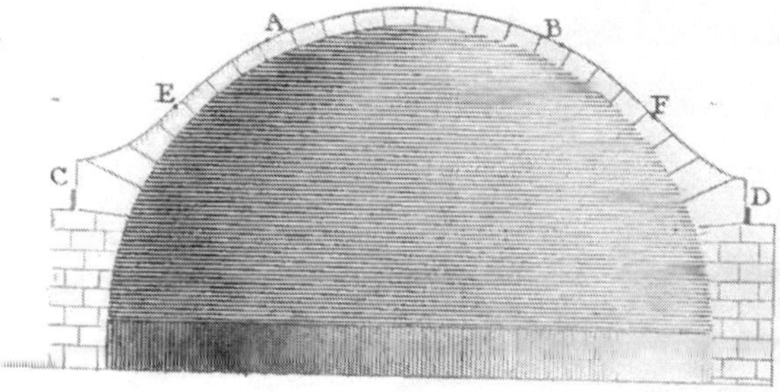

Fig. 158.

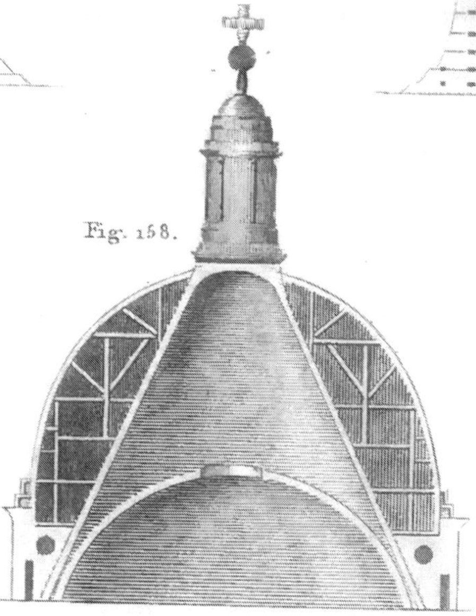

Fig. 159.

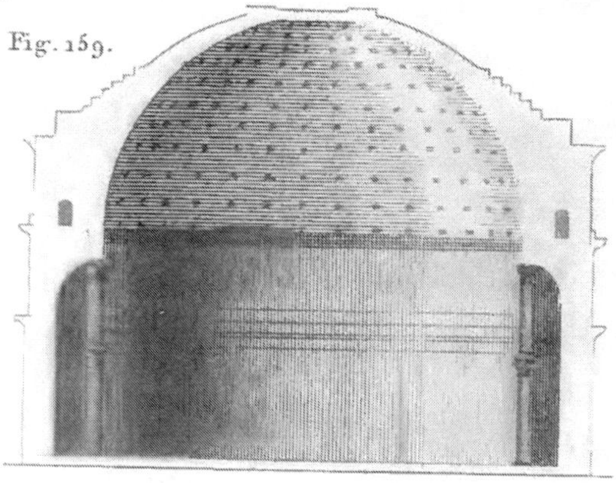

Fig. 165.

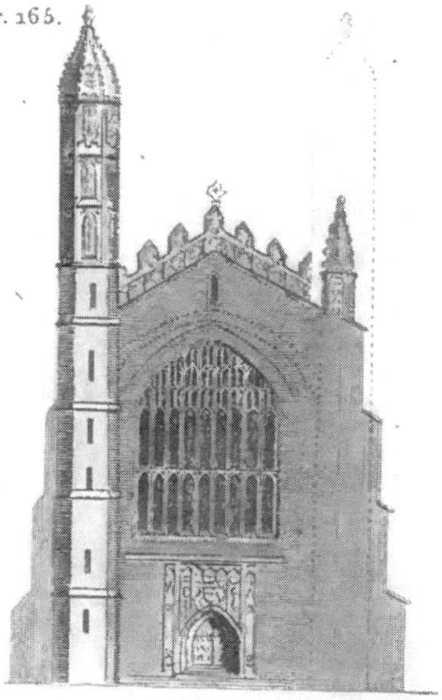

Fig. 160.

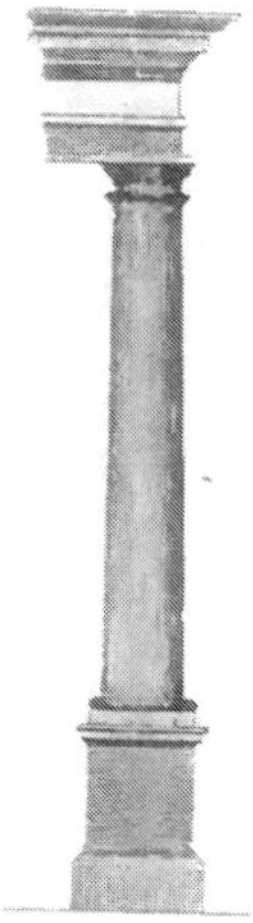

Fig. 161.

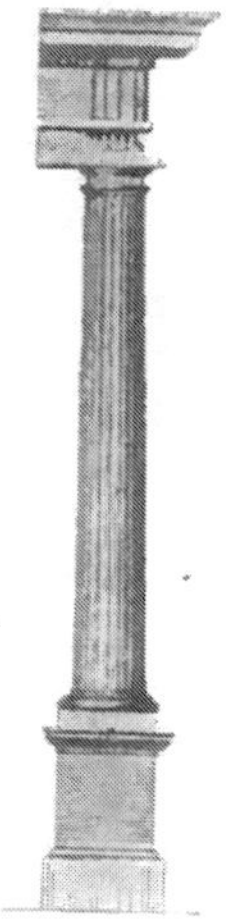

Fig. 162.

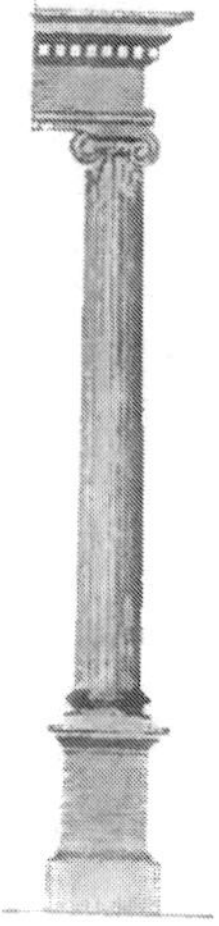

Fig. 163.

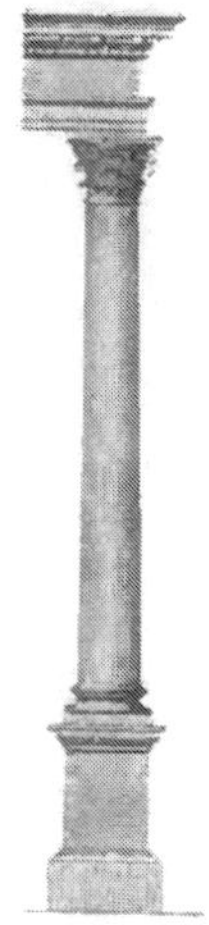

Fig. 164.

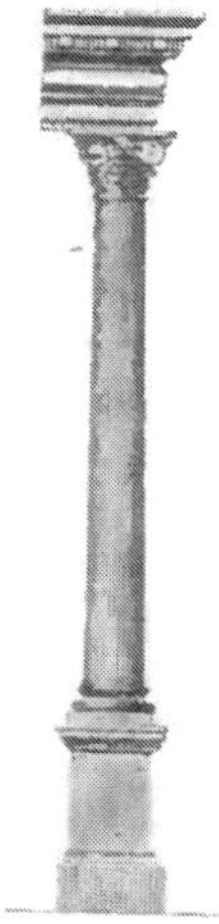

Pub. by J. Johnson, London 1 July 1806.

Joseph Skelton scu

PLATE XIII

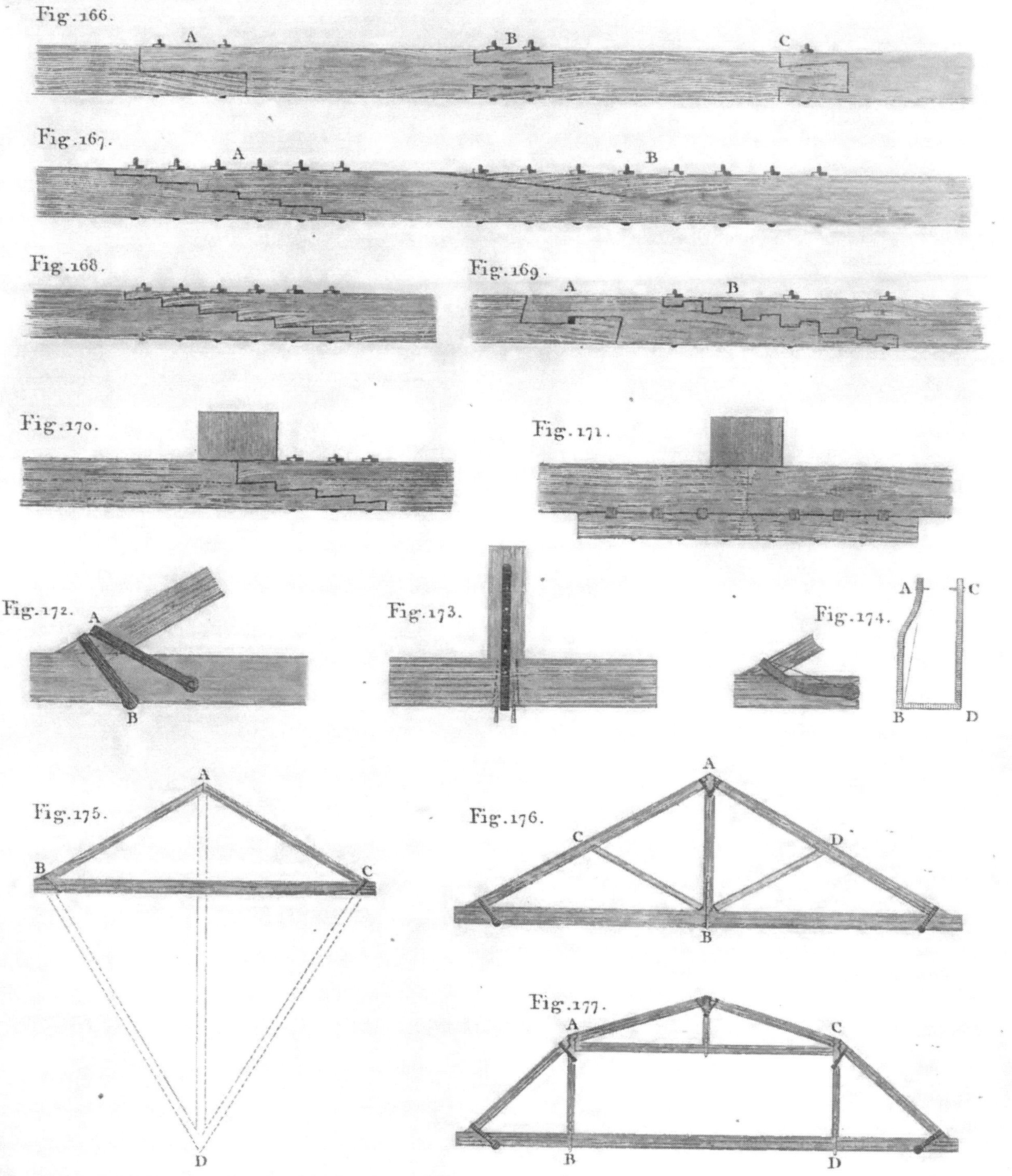

Pub. by J. Johnson, London 1 July 1806. *Joseph Skelton sculp.*

PLATE XIV.

Fig. 178.

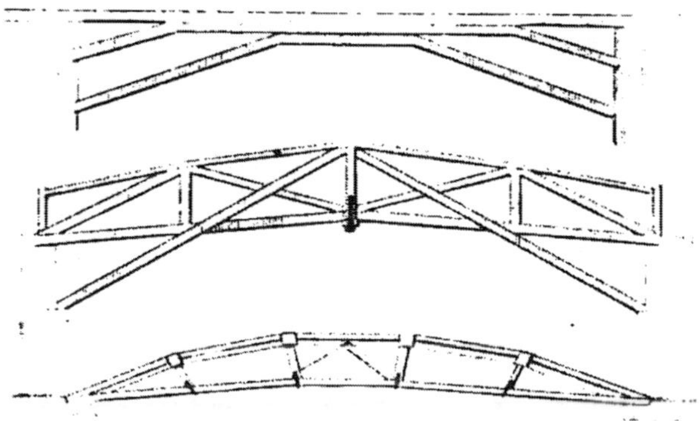

Fig. 179.

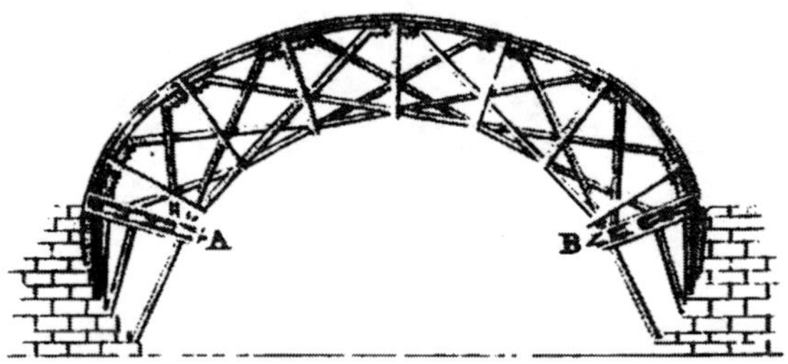

Fig. 180.

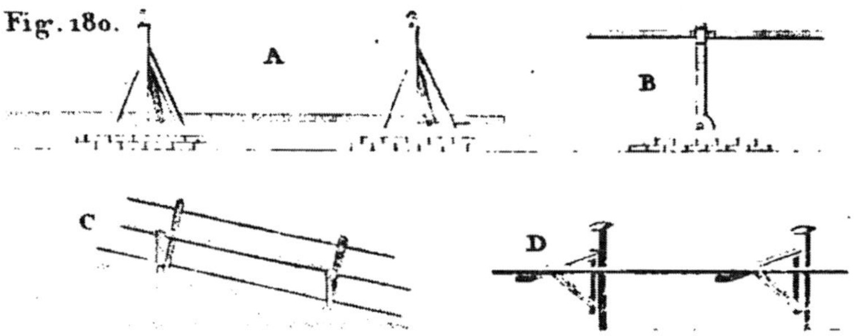

Fig. 181.

Fig. 182.

Fig. 183.

Fig. 184.

Fig.

Fig. 186.

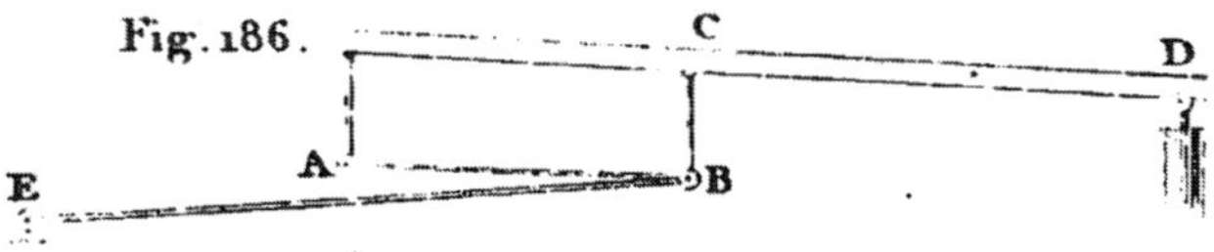

Fig. 187.

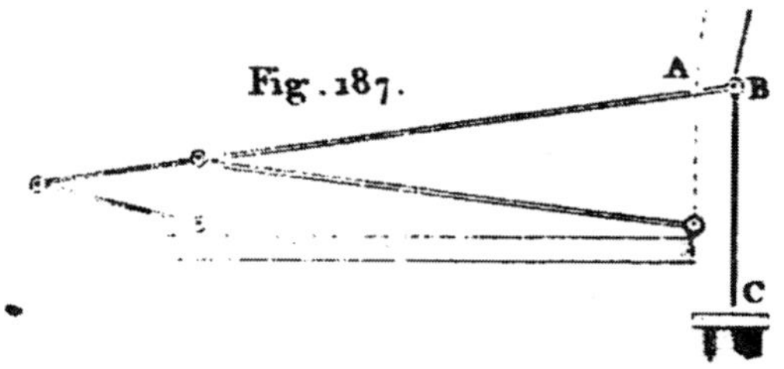

Fig. 188.

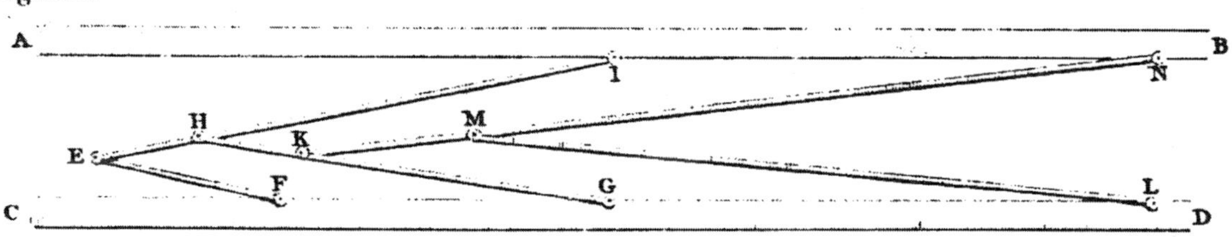

Pub. by J. Johnson, London 1 July 1806. *Joseph Skelton sculp.*

Fig. 189.

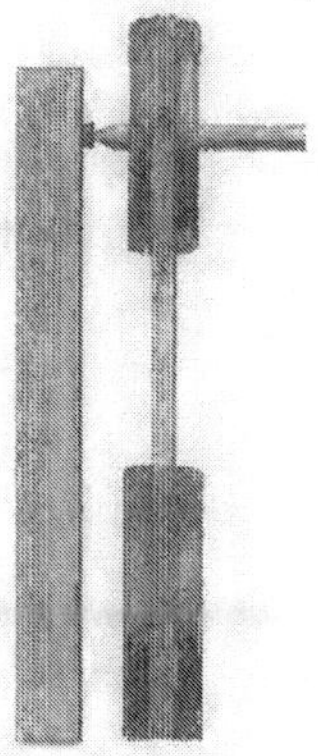

Fig. 190.

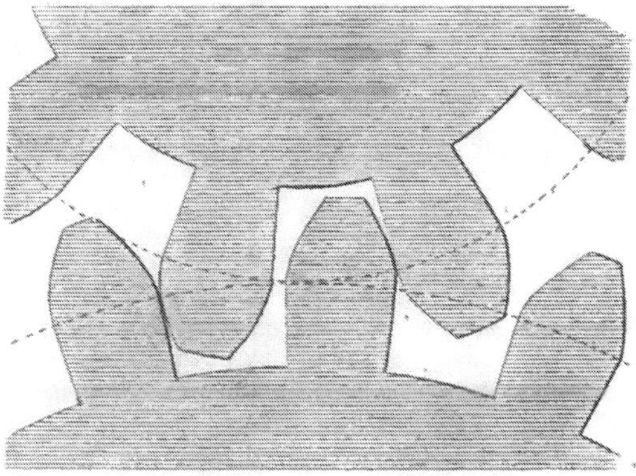

Fig. 191.

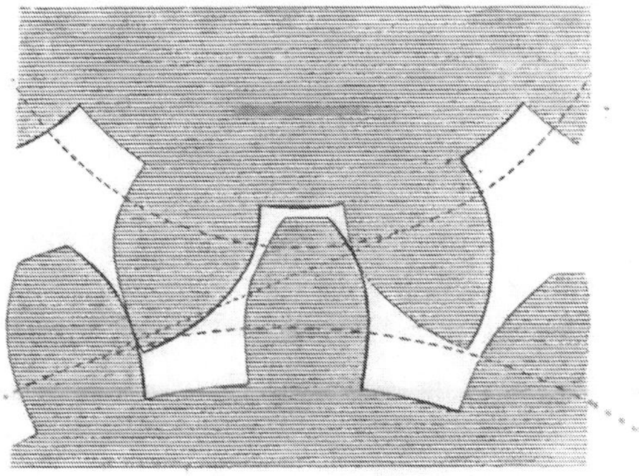

Fig. 192.

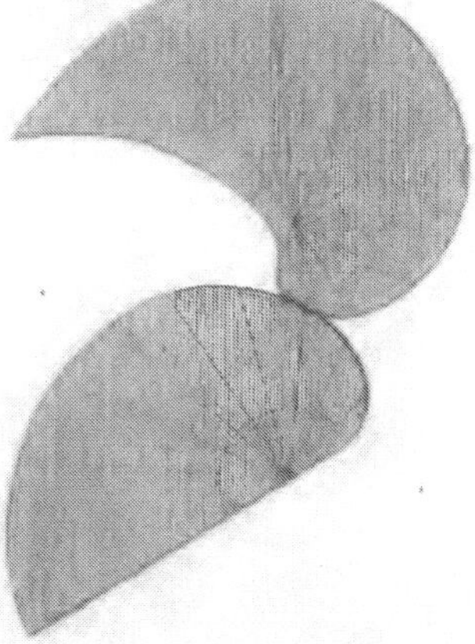

Fig. 193.

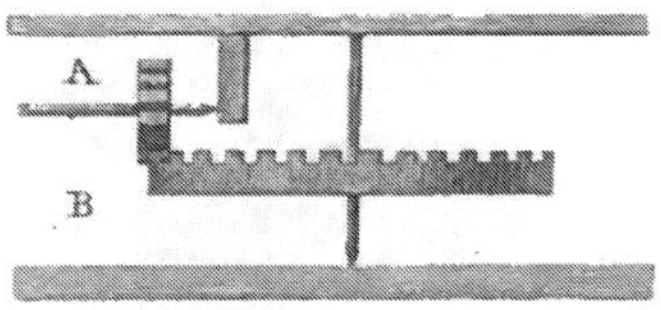

Fig. 194.

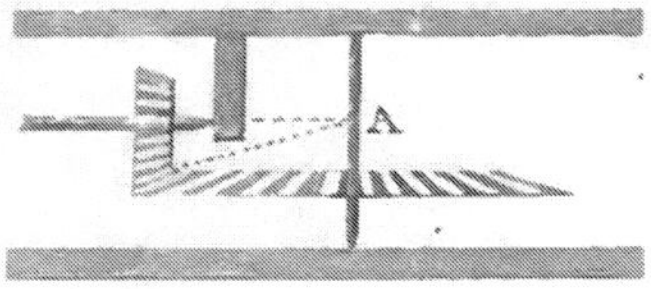

Fig. 195.

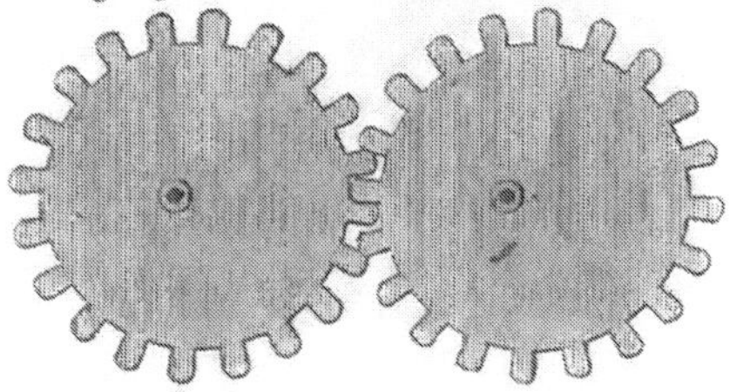

Fig. 196.

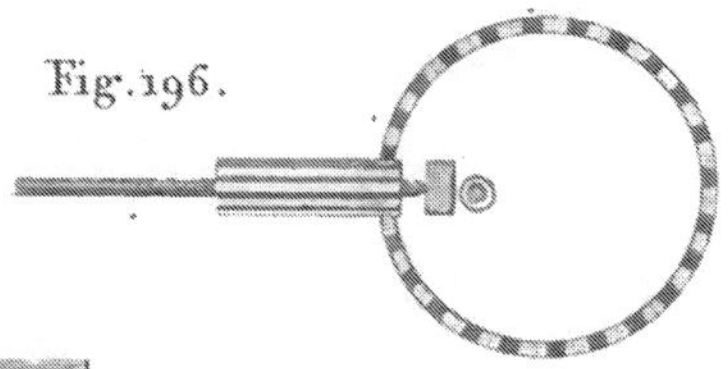

Fig. 197.

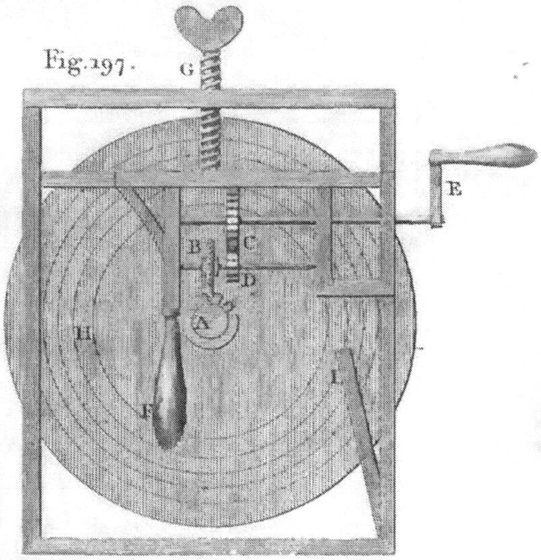

Fig. 198.

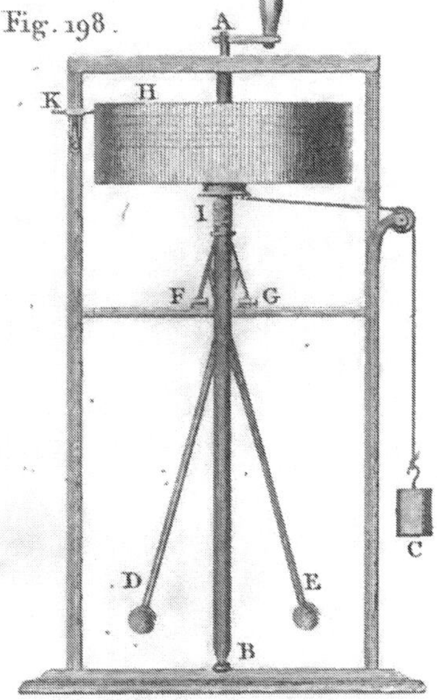

Fig. 199

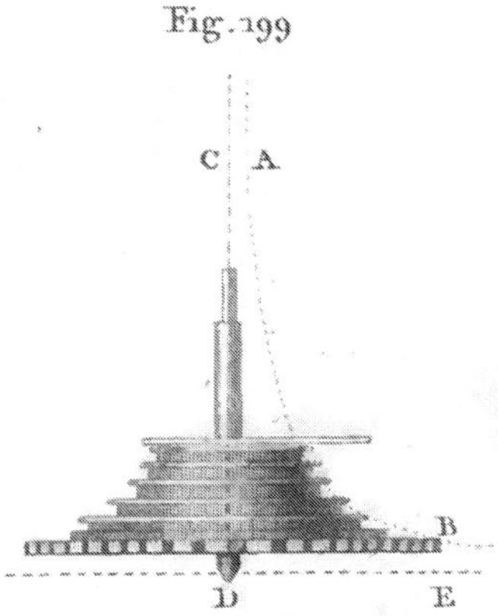

Pub. by J. Johnson, London 1 July 1806. *Joseph Skelton sculp.*

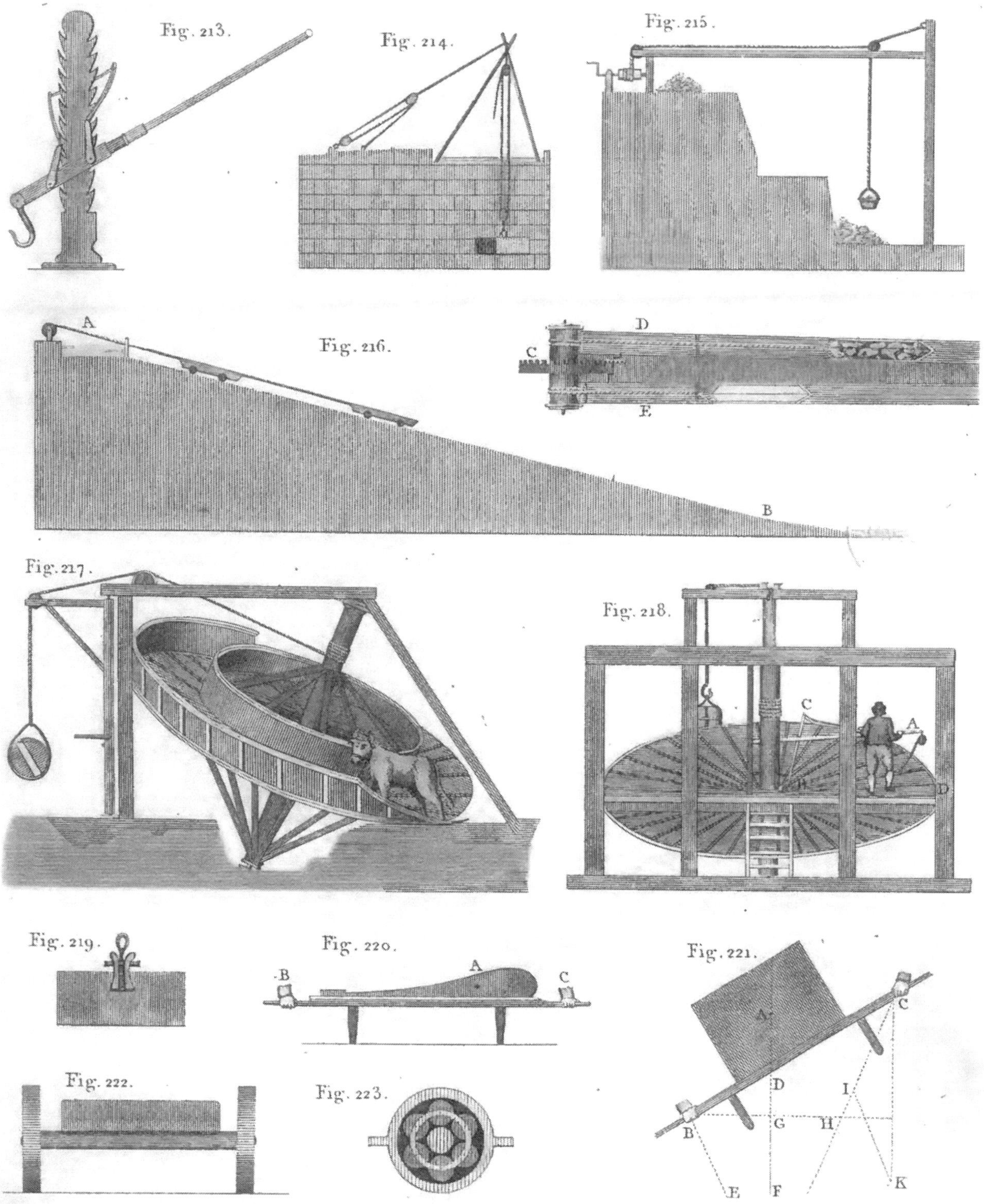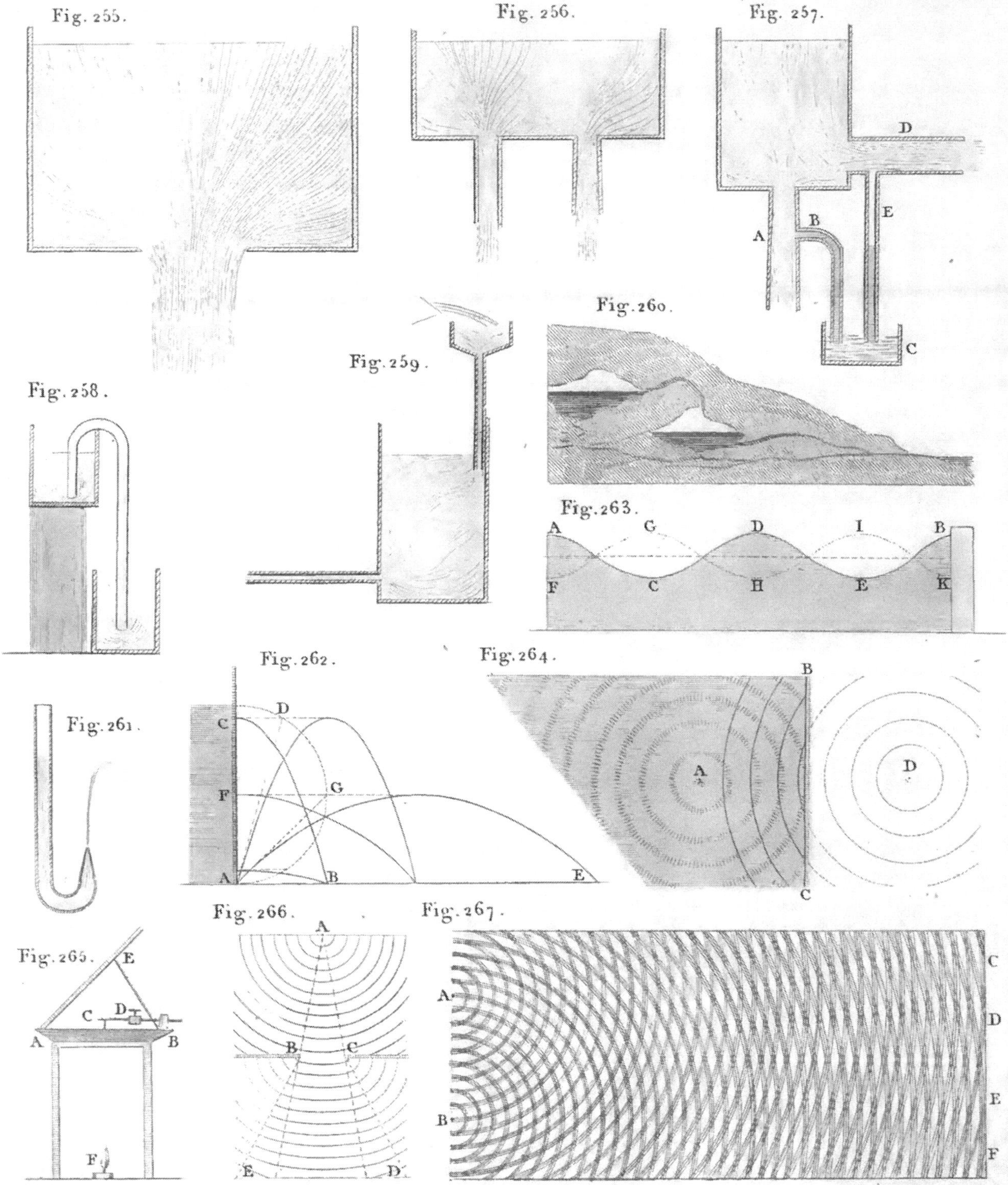

Pub. by J. Johnson London 2 July 1806.

Joseph Skel

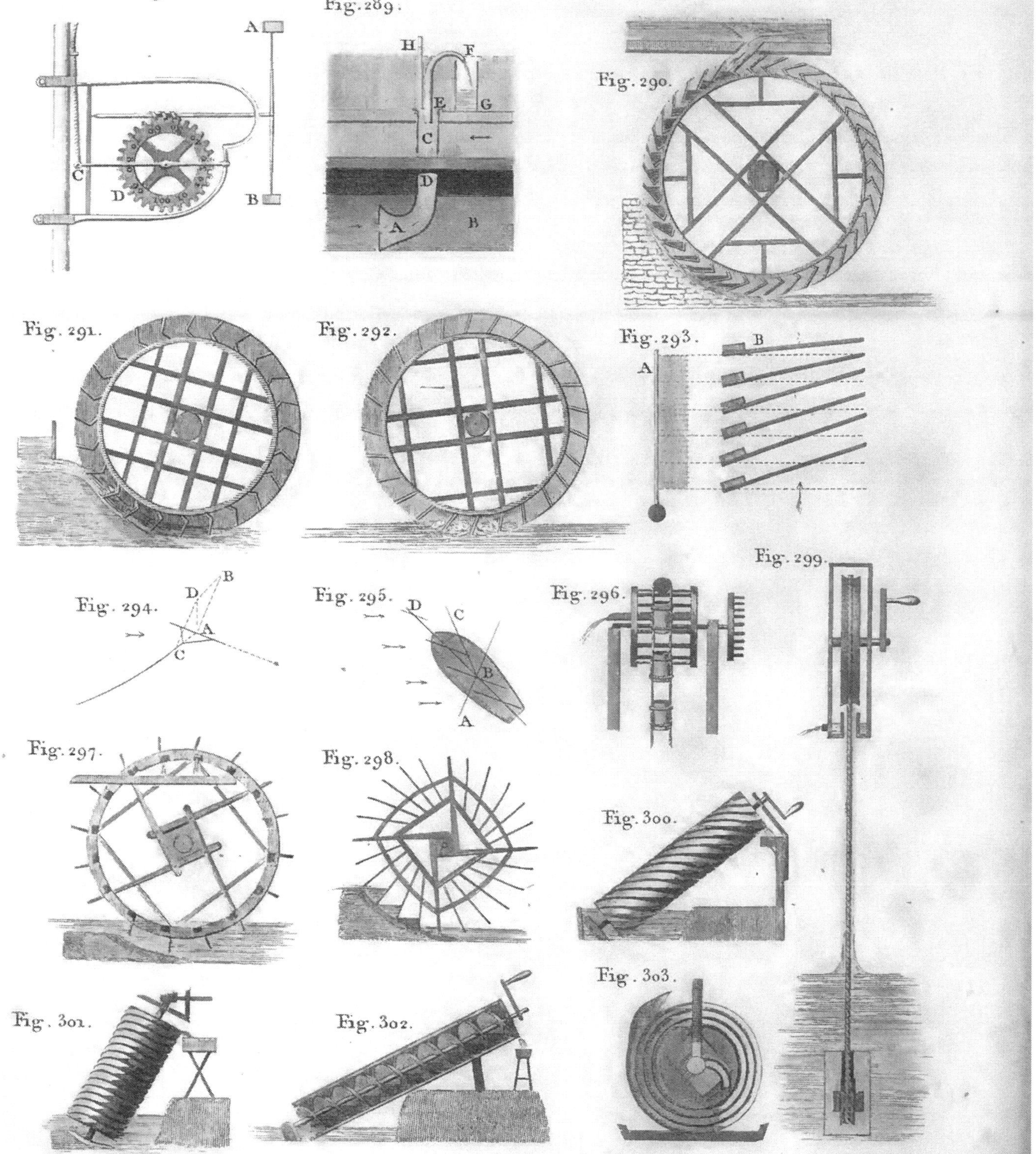

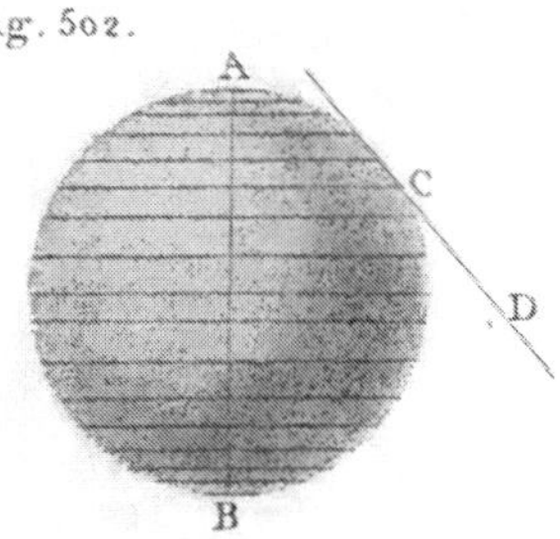

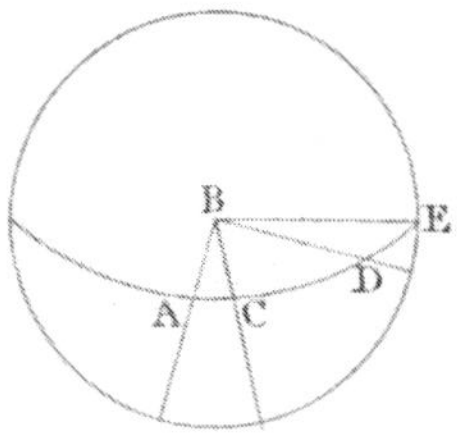

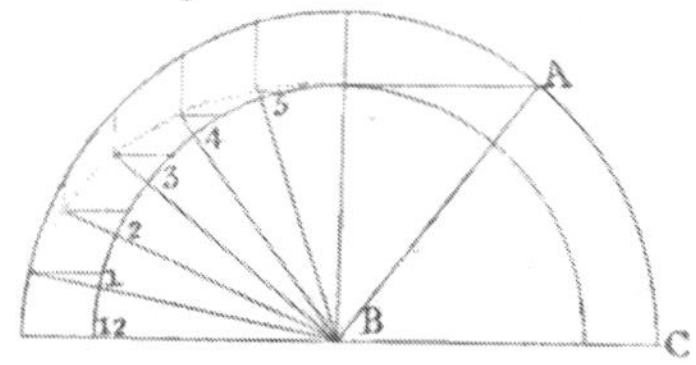

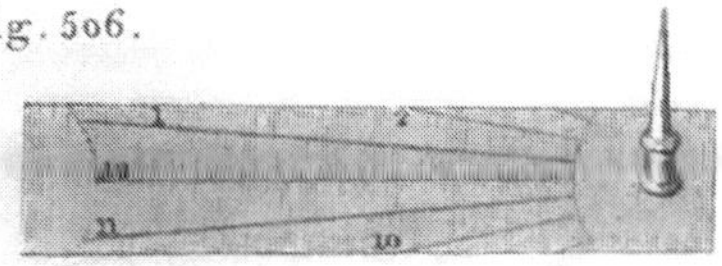

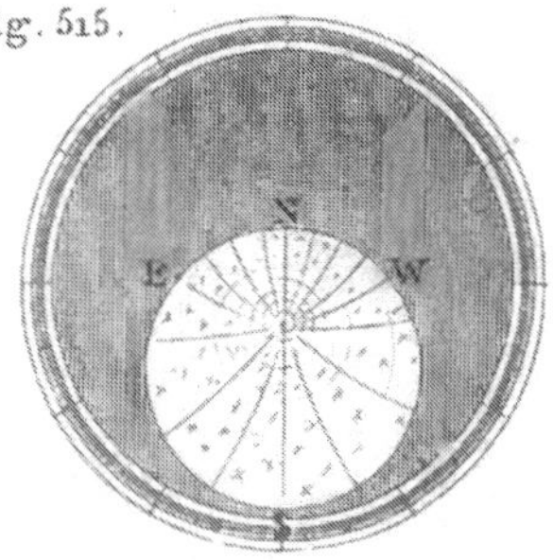

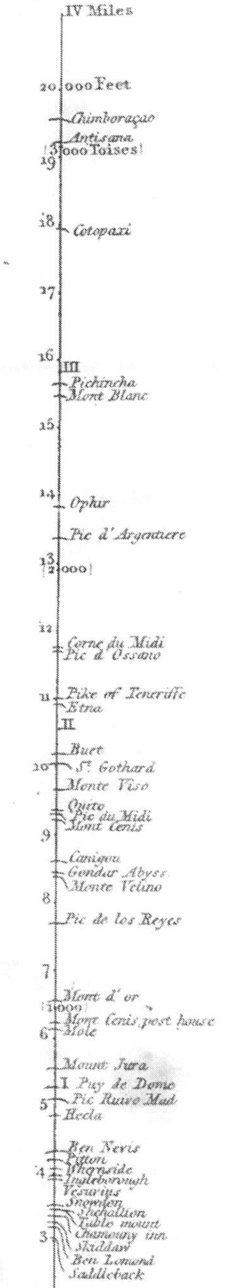

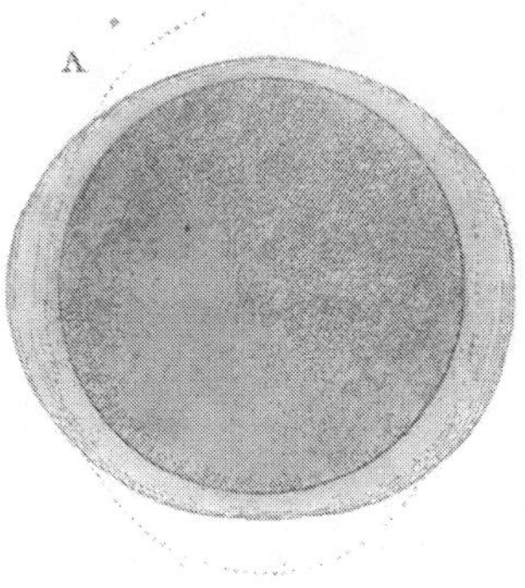

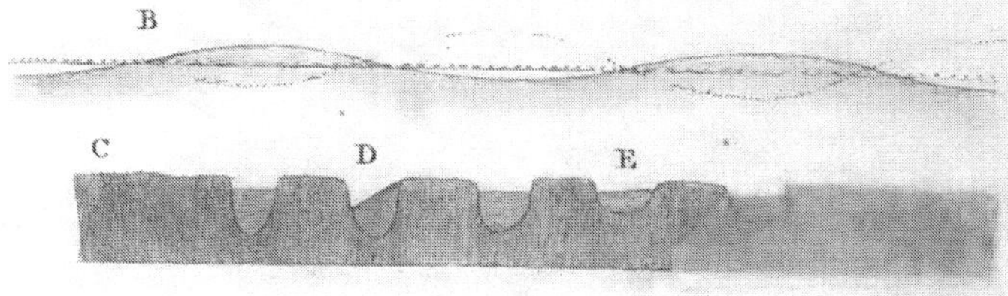

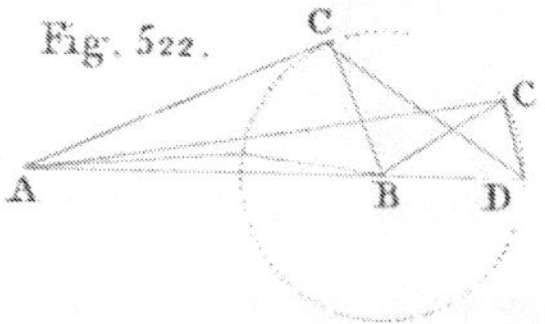

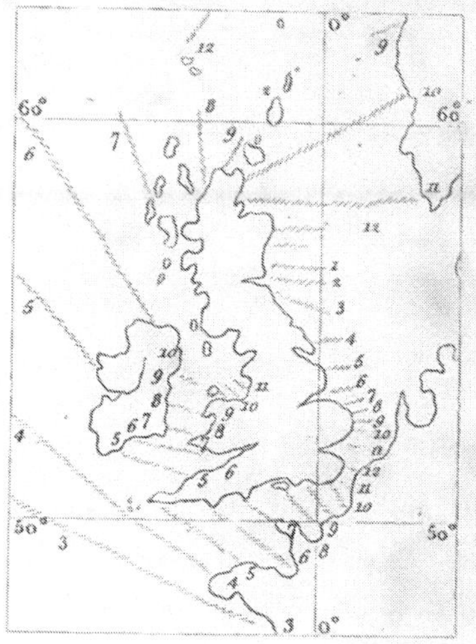

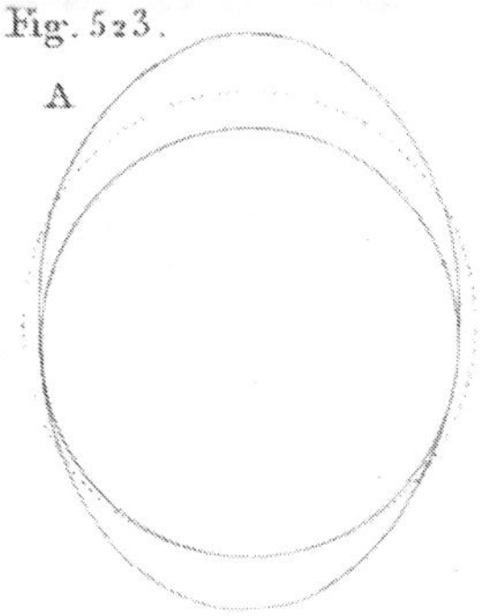

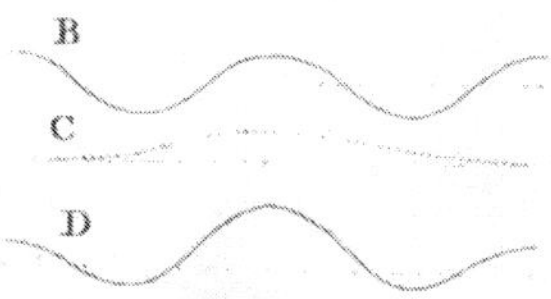

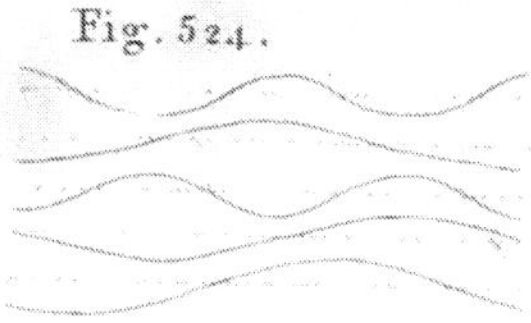

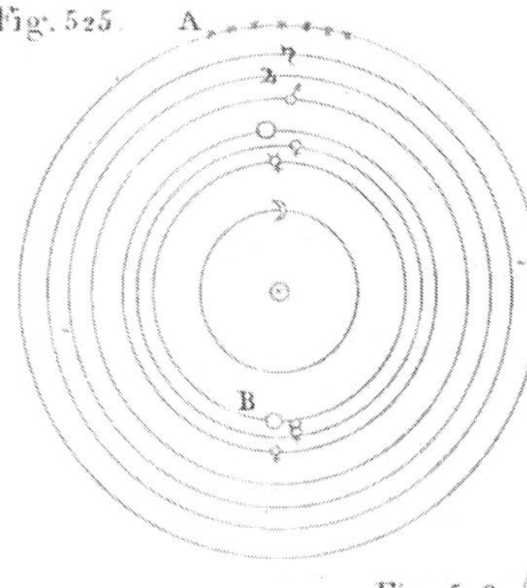

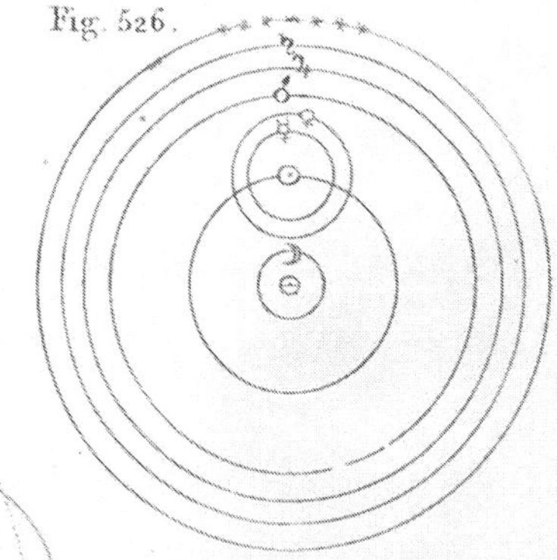

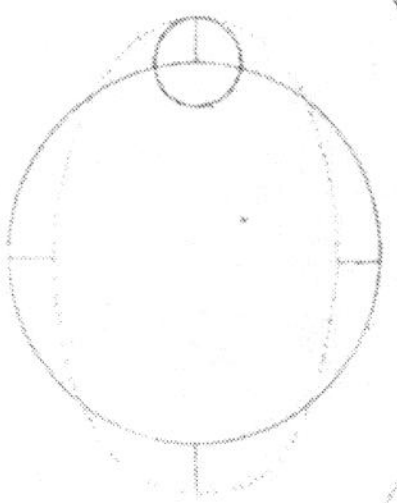

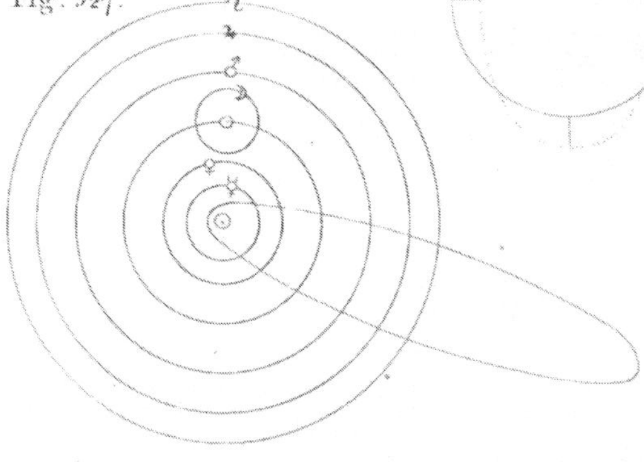

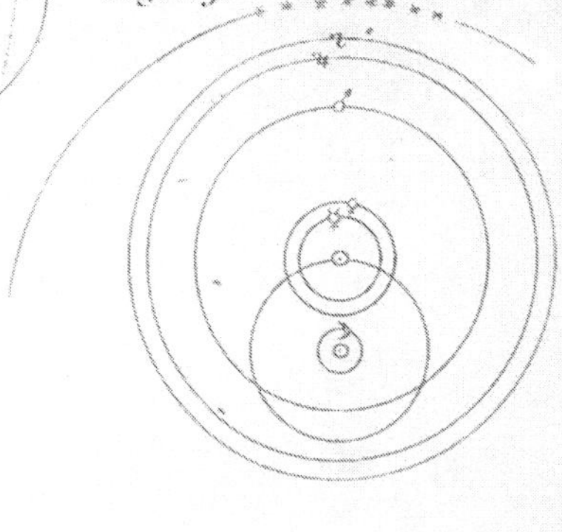

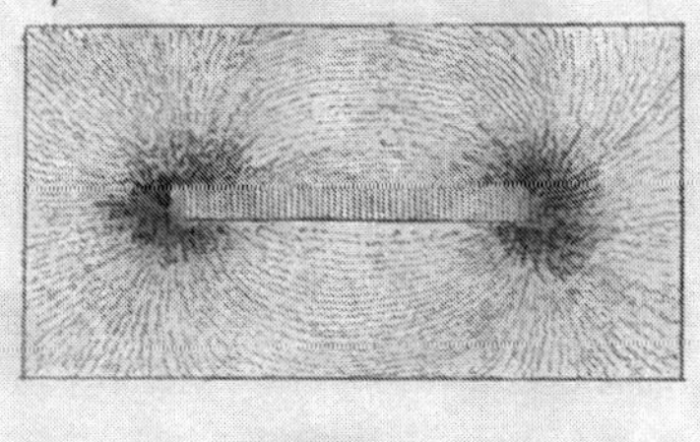

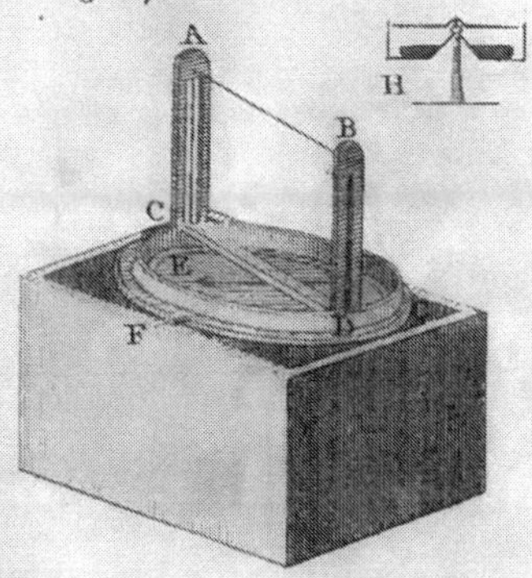

PLATE XLII. Fig. 562.

Joseph Skelton sculp

Pub. by J. Johnson, London 1 July 1806.

To be bound facing Plate XLIII

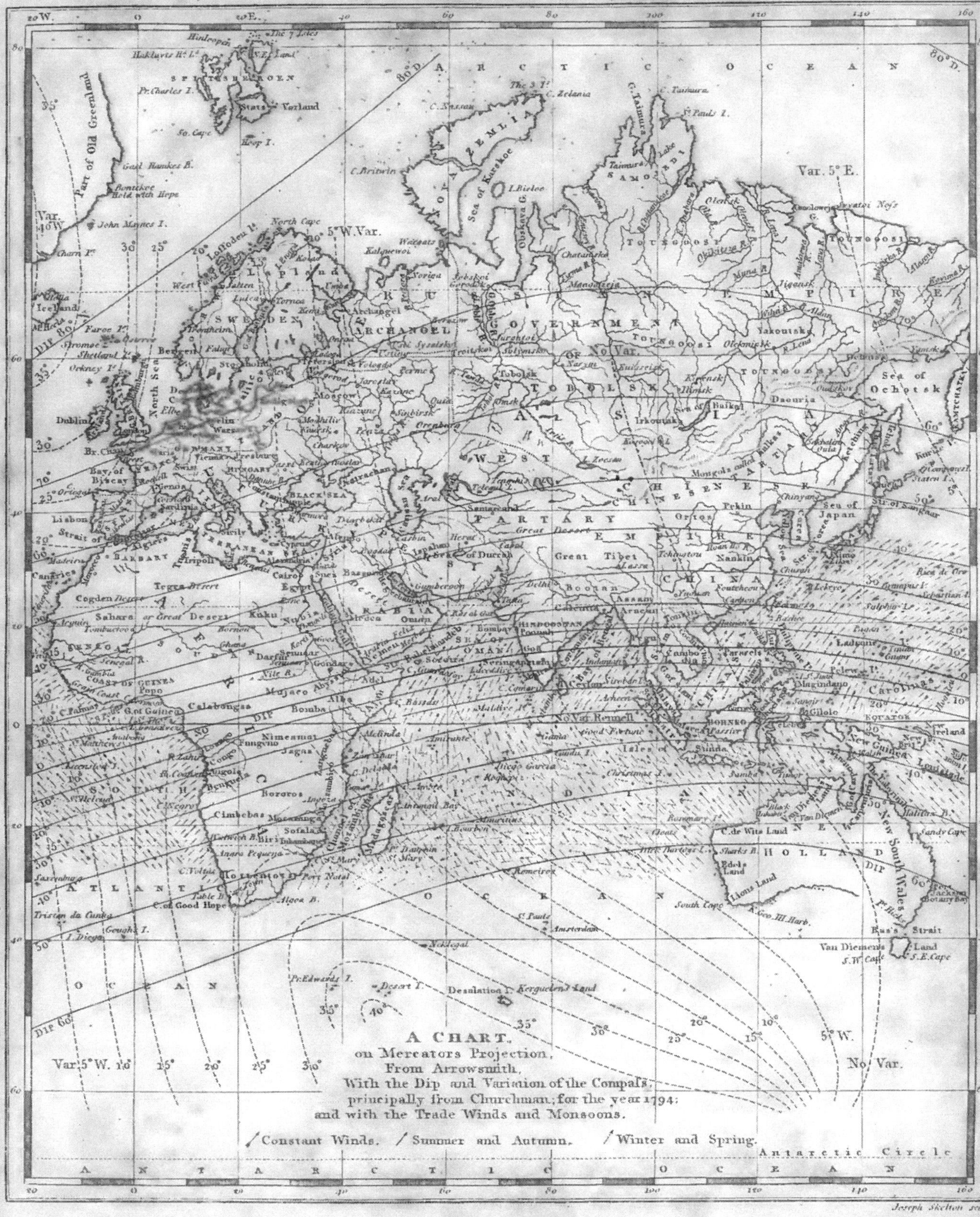

Joseph Skelton sculp.

Pub. by J. Johnson London 1 July 1806.